Pollutant Diseases, Remediation and Recycling

Environmental Chemistry for a Sustainable World

Volume 4

Series Editors

Eric Lichtfouse
INRA, UMR1347 Agroécologie, Dijon, France

Jan Schwarzbauer
Institute of Geology and Geochemistry of Petroleum and Coal, RWTH Aachen University, Aachen, Germany

Didier Robert
Université de Lorraine, ICPEES, Saint-Avold, France

For further volumes:
http://www.springer.com/series/11480

Eric Lichtfouse • Jan Schwarzbauer • Didier Robert
Editors

Pollutant Diseases, Remediation and Recycling

Springer

Editors
Eric Lichtfouse
INRA, UMR1347 Agroécologie
Dijon, France

Didier Robert
ICPEES
Université de Lorraine
Saint-Avold, France

Jan Schwarzbauer
Institute of Geology and Geochemistry of Petroleum and Coal
RWTH Aachen University
Aachen, Germany

ISSN 2213-7114 ISSN 2213-7122 (electronic)
ISBN 978-3-319-02386-1 ISBN 978-3-319-02387-8 (eBook)
DOI 10.1007/978-3-319-02387-8
Springer Cham Heidelberg New York Dordrecht London

Library of Congress Control Number: 2014930164

Printed on acid-free paper

Springer is part of Springer Science+Business Media (www.springer.com)

Preface

It has become appallingly obvious that our technology has exceeded our humanity.
Intellectuals solve problems, geniuses prevent them.

Albert Einstein

Pollution has no borders. This popular 1970s saying from early ecologists is surprisingly still true nowadays despite overwhelming scientific evidence and public awareness of the occurrence of artificial toxic substances in water, food, air, living organisms – including humans – and the environment. On the positive side, considerable efforts have been made by most governments, environmental agencies, industries and individuals to reduce the release of pollutants in the environment. So why are we still breathing, eating and drinking man-made toxicals? What is the real problem? In the wise quotes above, Albert Einstein clearly foresaw the profound grounds of technological side effects. The progressive separation of physical sciences from humanities during the last century has indeed favoured the design of advanced technologies without taking into account possible hazardous consequences. In other words, today we accept paradoxically both the benefits and drawbacks of technology, the later being measured in thousands of deaths caused by pollutants. So what is the solution?

The solution is very simple: since humanities and other social sciences have been historically split up within universities and research bodies, we should gather them again and make them really work together to design both innovative and safe technology. A such reunification of sciences into nexialism – the science of all sciences – has been imagined by the science fiction author Alfred Elton van Vogt in his novel *The Voyage of the Space Beagle*. Trained in integrated science and thought, the novel's hero was able to see the connection between many aspects of a problem that other specialists could not see because of their narrow training. The story provides an uncanny allegory for the situation we face aboard our planetary 'spaceship'.

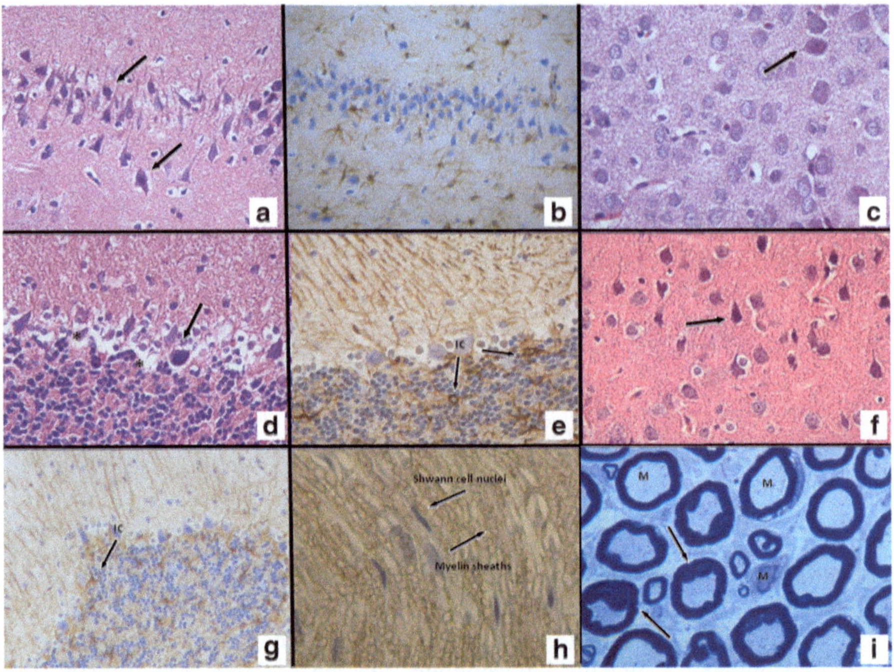

Fig. 1 Effect of lead (Pb) on rat brain showing cell death, oedema, dendritic breakdown and myelin damage (Source: Nava-Ruíz and Méndez-Armenta. This book)

This book presents advanced reviews on pollutant occurrence, transfer, toxicity and remediation. The chapter on school air quality by Dambruoso et al. highlights the overlooked health issue of airborne pollutants in buildings. Children are particularly threatened because they spend 90 % of their time indoors, even in summer. The chapter on industrial wastewater pollutants by Dsikowitzky and Schwarzbauer reviews pollutants from textile, petrochemical, paper, tire, chemical and pharmaceutical plants. The authors describe advanced analytical methods and ecotoxicity tests. Industrial pollutants such as dioxins and furans are also reviewed in the chapter by Mudhoo et al. The chapter on fly ash by Gianoncelli et al. presents many techniques to treat fly ash and, in turn, decrease pollutant concentrations. The authors also explain that fly ash can be recycled in agriculture, buildings and geopolymers. The chapter on antifouling paints used for ship protection, by Sousa et al., highlights the occurrence of toxic organotins in human organs such as heart, liver and breast milk. The chapter on surfactants by Rebello et al. focuses on safety concerns for humans and the ecosystems. Remediation techniques and green surfactants are presented. The chapters on toxic metals by Nava-Ruíz and

Méndez-Armenta, and Abarikwu and Ristić et al. describe sources, monitoring and diseases induced by lead, mercury, cadmium and thallium (Fig. 1). The chapter on carcinogenic nitrosamines by Li et al. presents techniques and materials such as zeolites to remediate liquids and smoke containing nitrosamines.

Thanks for reading

Eric Lichtfouse*, Jan Schwarzbauer and Didier Robert
Founders of the journal Environmental Chemistry Letters and of the European Association of Environmental Chemistry
E-mail: Eric.Lichtfouse@dijon.inra.fr, Jan.Schwarzbauer@emr.rwth-aachen.de, Didier.Robert@univ-lorraine.fr

*Author of the best-selling textbook "Scientific Writing for Impact Factor Journals"

Other Publications by the Editors

Books

Environmental Chemistry
ISBN: 978-3-540-22860-8 (Print) 978-3-540-26531-3 (Online)

Organic contaminants in riverine and ground water systems. Aspects of the anthropogenic contribution
ISBN: 978-3-540-31169-0 (Print) 978-3-540-31172-0 (Online)

Sustainable Agriculture
Vol. 1: ISBN: 978-90-481-2665-1 (Print) 978-90-481-2666-8 (Online)
Vol. 2: ISBN: 978-94-007-0393-3 (Print) 978-94-007-0394-0 (Online)

Rédiger pour être publié!
ISBN: 978-2-8178-0288-6 (Print) 978-2-8178-0289-3 (Online)

Journals

Environmental Chemistry Letters
ISSN: 1610–3653
http://www.springer.com/10311

Agronomy for Sustainable Development
ISSN: 1774–0746
http://www.springer.com/13593

Book series

Environmental Chemistry for a Sustainable World
ISSN: 2213–7114
http://www.springer.com/series/11480

Sustainable Agriculture Reviews
ISSN: 2210–4410
http://www.springer.com/series/8380

Contents

Chapter 1
School Air Quality: Pollutants, Monitoring and Toxicity

Paolo Rosario Dambruoso, Gianluigi de Gennaro, Annamaria Demarinis Loiotile, Alessia Di Gilio, Pasquale Giungato, Annalisa Marzocca, Antonio Mazzone, Jolanda Palmisani, Francesca Porcelli, and Maria Tutino

Contents

P.R. Dambruoso • G. de Gennaro (✉) • A.D. Loiotile • A. Di Gilio • P. Giungato
A. Marzocca • A. Mazzone • J. Palmisani • F. Porcelli • M. Tutino
Laboratory of Environmental Sustainability, Department of Chemistry,
University of Bari 'A. Moro', Via Orabona 4, 70126 Bari, Italy
e-mail: gianluigi.degennaro@uniba.it

E. Lichtfouse et al. (eds.), *Pollutant Diseases, Remediation and Recycling*,
Environmental Chemistry for a Sustainable World 4, DOI 10.1007/978-3-319-02387-8_1,

Abstract In recent years the use of synthetic materials in building and furnishing, the adoption of new lifestyles, the extensive use of products for environmental cleaning and personal hygiene have contributed to the deterioration of the indoor air quality (IAQ) and introduced new sources of risk to humans. Indoor environments include home work places such as offices, public buildings such as hospitals, schools, kindergartens, sports halls, libraries, restaurants and bars, theatres and cinemas and finally cabins of vehicles. Indoor environments in schools have been of particular public concern. According to recent studies, children aged between 3 and 14 spend 90 % of the day indoors both in winter and summer. Adverse environmental effects on the learning and performance of students in schools could have both immediate and lifelong consequences, for the students and for society. In fact, children have greater susceptibility to some environmental pollutants than adults, because they breathe higher volumes of air relative to their body weights and their tissues and organs are actively growing. This review describes methods for the assessment of indoor air quality in schools. To this aim, monitoring strategies for sampling and measurement of indoor air pollutants will be discussed. The paper's goal involves four major points: (1) characteristics of indoor environments, chemical pollutants and their sources within school; (2) monitoring strategies; (3) sampling and analysis techniques; (4) an overview of findings from scientific literature. Finally, we summarize available knowledge about IAQ in schools highlighting key gaps and suggesting priority topics and strategies for research. Moreover, it provides useful tools to support the stakeholder for development of strategies of prevention and mitigation in school environments in order to improve the indoor air quality.

Keywords Carbon Dioxide (CO_2) • Formaldehyde and carbonyl compounds • Guidelines • Health effects • Indoor Air Quality (IAQ) • Inorganic gases • Monitoring strategies • Ozone (O_3) • Particulate Matter (PM) • School environments sources • Volatile Organic Compounds (VOCs)

1.1 Introduction

In recent years numerous scientific studies highlighted that citizens spend most of their time in indoor environments (e.g. home, offices, schools, hospitals, kindergartens, sports halls, libraries, restaurants, bars, theatres, vehicles, etc.) and are more exposed to pollution indoors than outdoors (Blondeau et al. 2005; Bruno et al. 2008; Pegas et al. 2010). Indoor air quality has a considerable impact on public health (Jones 1999) because indoor exposure may pose harmful health effects such as respiratory and cardiopulmonary pathologies and asthma, especially for children

(Yang et al. 2009; Sohn et al. 2012). There is a considerable interest in the assessment of the association between air pollution exposure and health effects in school environments, as shown by more than 70 epidemiological publications currently available (e.g. Guo et al. 1999; Venn et al. 2000).

The phenomena of indoor air pollution are characterized by a large variability: pollutant concentrations can vary widely between different indoor environments and may also vary within a specific environment as a function of location and time. The extent of these variations depends on factors such as the emission characteristics of the sources, the occupants' behavior and the microclimatic and ventilation conditions (Report EUR 16051 EN 1994; UNI EN ISO 16000-1: 2006). Thus indoor air pollution and human exposure are highly dynamic processes rather than staticphenomena.

In this chapter the attention will be focused on Indoor Air Quality (IAQ) in school buildings. Children spend large amount of time in these environments and are more sensitive subjects to indoor pollutants (Faustman et al. 2000; Mendell and Health 2005; WHO 2006a, b; Chithra and Shiva Nagendra 2012). Several studies reported that indoor air pollution can increase the chance of long-term and short-term health problems for students and teachers in terms of comfort, productivity and academic performance (Daisey et al. 2003; Shendell et al. 2004; Dijken et al. 2005; Mendell and Health 2005; Wargocki et al. 2005; Mi et al. 2006; Shaughnessy et al. 2006; Croome et al. 2008). Despite the magnitude of the school population and the sensibility of the children, information on this concern is very limited (Daisey et al. 2003; Godwin and Batterman 2007). The indoor pollution observed inside school buildings can be traced back to a variety of causes, such as the use of high emitting materials for building construction and furnishing, minimal landscaping with poor drainage, heating, ventilation and air conditioning units (HVAC), the lack of preventive maintenance, crowded conditions (Godwin and Batterman 2007) and cleaning products that release chemicals into the air (UBA 2008). Each school environment is uniquely characterized, thus each personal exposure is determined by a combination of the outdoor and indoor pollutants levels (Stranger et al. 2007, 2008). In fact, age and location of school buildings, pollutants transport from outdoor, chemical reactions in indoor air and heterogeneous processes at the air-solid interfaces are the other factors that influence the pollutants concentrations (Poupard et al. 2005). In developed countries many studies were conducted during the past decade in order to assess IAQ in school environments (Seppanen et al. 1999; Daisey et al. 2003; Bartlett et al. 2004; Shendell et al. 2004; Ramachandran et al. 2005; Shaughnessy et al. 2006; Godwin and Batterman 2007) and a large number of indoor air pollutants were measured including Carbon Dioxide (CO_2), Ozone (O_3), Nitrogen Oxides (NO_x), Carbon Oxide (CO), Sulphur Dioxide (SO_2) (Lee and Chang 1999; Scheff et al. 2000a, b; Bartlett et al. 2004; Shendell et al. 2004; Blondeau et al. 2005; Ramachandran et al. 2005; Godwin and Batterman 2007), Volatile Organic Compounds (VOCs) and Particulate Matter (PM). Several studies worldwide focused the attention on the assessment of pupils' exposure to PM in classrooms (Scheff et al. 2000a, b; Liu et al. 2004; Blondeau et al. 2005; Fromme et al. 2005; Lahrz and Piloty 2005; Lahrz 2006; Heudorf et al. 2009), but only few authors reported results about PM chemical composition (e.g. Molnar et al. 2007; Stranger et al. 2008; Almeida et al. 2011; Oeder et al. 2012) and size. The aim of the present chapter is to explore the methodological approaches for the assessment of IAQ in schools so the main characteristics of school environments, the chemical pollutants and their sources, the monitoring strategies and an overview of the main scientific findings are discussed. Moreover, the chapter summarizes available knowledge about IAQ in schools highlighting key gaps and suggesting priority topics and strategies for research. It is important to underline that for a correct assessment of IAQ the bioaerosol contaminants and the concentration of radon should be considered but in this chapter these pollutants will not be discussed. The biological indoor air pollutants (which include pollen from plants, mite, hair from pets, fungi, parasites and some bacteria) and radon emission in fact should be discussed in different chapters because of the different nature of this substances (WHO 2009a) and because of the variable characteristics of their sources (WHO 2009b; Trevisi et al. 2012; Bem et al. 2012; Clouvas and Xanthos 2012).

Finally, this chapter provides useful tools to support the stakeholder for development of corrective measures to improve IAQ in schools.

1.2 Characteristics of the Indoor Environments, Chemical Pollutants and Their Sources Within School

The wide range of school building design leads to large variations in indoor pollutants levels and hence personal exposure (Ashmore and Dimitroulopoulou 2009).Children spend their school hours in different environments: classrooms, laboratories where available, playgrounds and other locations within the school. As a result, individual exposure change related to the variation in pollutants levels inside the several school locations (Mejía et al. 2011).

Pollutants emission can occur in many school settings where different activities take place:

- classrooms (Lee and Chang 2000; Hulin et al. 2011; Bertoni et al. 2002; Blondeau et al. 2005; Mi et al. 2006; Ekmekcioglu and Keskin 2007; Fromme et al. 2007; Godwin and Batterman 2007; Diapouli et al. 2008; Weichenthal et al. 2008; Yang et al. 2009; Sofuoglu et al. 2010; Wu et al. 2010; Goyal and Khare 2011; Gul et al. 2011; Mejía et al. 2011; Mullen et al. 2011; Park et al. 2011; Smedje et al. 2011; Szoboszlai et al. 2011; Sohn et al. 2012; Zhang and Zhu 2012);
- cafeterias (Godwin and Batterman 2007; Hochstetler et al. 2011; Zhang and Zhu 2012);
- gyms (Godwin and Batterman 2007; Branis et al. 2009; Branis and Safránek 2011; Hochstetler et al. 2011; Szoboszlai et al. 2011);
- swimming pools (Godwin and Batterman 2007);
- science labs (often without fume hoods) (Godwin and Batterman 2007; Yang et al. 2009; Jo and Kim 2010; Goyal and Khare 2011; Park et al. 2011; Szoboszlai et al. 2011);

- arts and crafts labs (Blondeau et al. 2005; Godwin and Batterman 2007);
- computer rooms (Yang et al. 2009; Wu et al. 2010; Szoboszlai et al. 2011; Sohn et al. 2012);
- dining halls (Gul et al. 2011);
- kitchen (MacIntosh et al. 2012);
- office/clerical rooms (Godwin and Batterman 2007; Goyal and Khare 2011; Zhang and Zhu 2012);
- miscellaneous use rooms (e.g. music room, library) (Godwin and Batterman 2007; Goyal and Khare 2011; Gul et al. 2011; Zhang and Zhu 2012);
- stairwells (Godwin and Batterman 2007; Sohn et al. 2012).

In order to make a systematic assessment of IAQ in schools several aspects should be taken into account. The following list describes the building characteristics, the micrometeorological parameters and the indoor activities (Stranger et al. 2008; Pegas et al. 2012):

- *School sites*: is it in an industrial (Scheepers et al. 2010; Tran et al. 2012), rural (Blondeau et al. 2005; Fromme et al. 2007; Hulin et al. 2011; Tran et al. 2012; Zhang and Zhu 2012), traffic (Blondeau et al. 2005; Hochstetler et al. 2011; Raysoni et al. 2011; Szoboszlai et al. 2011; Chithra and Shiva Nagendra 2012), suburban (Branis and Safránek 2011), urban (Fromme et al. 2007; Hulin et al. 2011; Mullen et al. 2011; Tran et al. 2012; Zhang and Zhu 2012), or background site? The level of exposure is closely linked to the proximity of the school to outdoor emission sources and then geographic location is an important determinant of children's exposure (Mejía et al. 2011). Several researches documented that a great percentage of schools are located in close proximity to major roadways, resulting in a potentially increased health risk (Janssen et al. 1997, 2001; Green et al. 2004; Wu and Battermann 2006; Van Roosbroeck et al. 2007; Appatova et al. 2008; Branis and Safránek 2011; Hochstetler et al. 2011; Mejía et al. 2011; De Giuli et al. 2012). The Guidelines for indoor air hygiene in school buildings of Federal Environment Agency, according to the US Environmental Protection Agency's recent initiative (USEPA 2006; Raysoni et al. 2011) give indications about the location of the school respect to outdoor pollutants levels, suggesting that school building surroundings should be traffic and industrial emissions free.
- *Age of the buildings*: this characteristic play a key role because many school buildings need to be renovated on account of their great age and many years of intensive use. The purpose of renovation is often to conserve energy by means of thermal insulation walls, replacing windows and modernizing heating systems. Recent studies showed that these improvements can adversely affect the indoor air quality (Godwin and Batterman 2007; Ashmore and Dimitroulopoulou 2009; Yang et al. 2009; Hochstetler et al. 2011; Mullen et al. 2011; Zhang and Zhu 2012) since air tightness cause the accumulation of pollutants both in winter and summer. On the other hand, the old buildings, although allowed greater dispersion, were built and furnished with high emitting materials because of the low knowledge about their emissive processes and the little attention on the choice of the materials.

- *The size of the environments* i.e. the *room design* (floor area and room volume) and the *level of occupancy* (Daisey et al. 2003; Godwin and Batterman 2007; Theodosiou and Ordoumpozanis 2008; Weichenthal et al. 2008; Mumovic et al. 2009; Mejía et al. 2011; Goyal and Khare 2011; Mullen et al. 2011; Chithra and Shiva Nagendra 2012): the Guidelines for indoor air hygiene in school buildings of the Federal Environment Agency recommend that the classrooms should be of sufficient size. The level of occupancy is commonly evaluated by the indoor CO_2 levels used as a surrogate of the rate of outside supply air per occupant (Daisey et al. 2003).
- *The type and quality of ventilation*: number of doors and windows, natural or mechanical ventilation systems (Ashmore and Dimitroulopoulou 2009; Goyal and Khare 2011; Mullen et al. 2011, Mejía et al. 2011, Grimsrud et al. 2006; Lee and Chang 2000; Mumovic et al. 2009; Wåhlinder et al. 1997; Theodosiou and Ordoumpozanis 2008; Blondeau et al. 2005). The importance of ventilation to achieve a healthy indoor environment for occupants was highlighted by several studies (Daisey et al. 2003; Stranger et al. 2008; Smedje et al. 2011; Guo et al. 2008; Zhang and Zhu 2012). One strategy to ensure good IAQ in buildings is the use of mechanical ventilation systems: adequate ventilation systems should not only provide thermal comfort but also distribute fresh air to occupants and remove pollutants (Sohn et al. 2012). In addition, in environments in which hazardous substances are generally used, such as science laboratories or art and crafts labs, a mechanical ventilation is absolutely essential (UBA 2008; Yang et al. 2009).
- *Heating systems*: the thermal condition is crucial to ensure a pleasant indoor environment in school facilities (MacIntosh et al. 2012; Park et al. 2011; De Giuli et al. 2012; Corgnati et al. 2007).
- M*icrometeorological parameters*: mean temperature and relative air humidity (Godwin and Batterman 2007, Park et al. 2011; Smedje et al. 2011; Weichenthal et al. 2008; Zhang and Zhu 2012; Fraga et al. 2008; Yang et al. 2009; De Giuli et al. 2012) play a decisive role in the emissive process of indoor pollutants, e.g. VOCs emission from materials.
- *Materials* of board, desks, chairs, floor, etc.: wood, particle board, medium density fiberboard, plywood, polymeric materials, etc. (Pegas et al. 2010; Yang et al. 2009; Goyal and Khare 2011; Chithra and Shiva Nagendra 2012). Furnishings and equipments in classrooms should be as free as possible from health hazardous pollutants (UBA 2008).
- *Occupant's activities:* collage and painting activities, the use of cleaning products, etc. (Chithra and Shiva Nagendra 2012).

The sources of pollutants in school environments can be classified in two categories: continuous sources, which are active over relatively long time periods (months or years), and intermittent sources which are active only over shorter periods (usually hours). The emission of continuous sources may depend on microclimate parameters while the intermittent sources are generally only slightly affected by these factors. The continuous sources can be further subdivided into two groups: those in which the

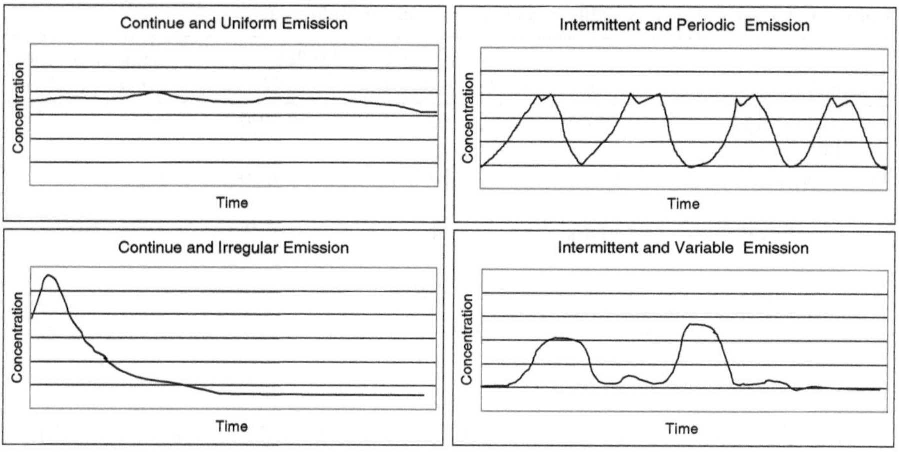

Fig. 1.1 Source temporal profiles (ISS 2012)

pattern is uniform and those in which it is irregular. Continuous sources which generate uniform emissions in time are for example the surface of a desk that emits (above all VOCs and formaldehyde) for long periods of time or other building materials such as linoleum, cork, wood floors, wood finishes, etc. Continuous sources which generate irregular emissions with a pattern that increase or decrease over time in dependence with changes in microclimatic conditions (air speed, humidity and ambient temperature): for example painted walls, wallpapers and wall adhesives can emit organic solvents and degradation products of protection materials (UNI EN ISO 16000-1:2006; UNI EN ISO 16000-2:2006; UNI EN ISO 16000-5:2007). Intermittent sources can be subdivided into two groups: those in which the pattern is periodic and those in which it is variable. A typical example of intermittent source with a periodic time pattern is cooking in dining hall of school. In fact the preparation of meals presents a very regular emission pattern from day to day, since it is often subject to a regular schedule. The occasional use of a particular cleaning product or of a marker is an example of intermittent sources with a variable pattern.

Figure 1.1 gives an overview of temporal emission profiles (ISS 2012).

The most common pollutants found in schools and childcare facilities are listed and deeply described in following paragraphs.

1.2.1 Volatile Organic Compounds (VOCs)

Volatile Organic Compounds (VOCs) are organic chemical compounds whose composition makes it possible for them to evaporate under normal indoor atmospheric conditions of temperature and pressure. Normal indoor atmospheric conditions of temperature and pressure refer to the range of conditions usually found in buildings

Table 1.1 Classification of volatile organic compounds of indoor air

Description	Abbreviation	Boiling point range		Saturation vapour pressures (kPa)
		From °C	To °C	
Very volatile organic compounds	VVOCs	<0	50–100	>15
Volatile organic compounds	VOCs	50–100	240–260	>10^{-2}
Semi-volatile organic compounds	SVOCs	240–260	380–400	10^{-2} to 10^{-8}
Particulate organic matters	POMs	>380		

occupied by people. Thus, depending on the type of building and its geographic location, most used definition is that of the National Institute of Standards and Technology (NIST): temperature of 20 °C (293.15 K) and an absolute pressure of 101.325 kPa (14.696 psi) (WHO 1989). So in accordance with international recommendations the term VOC refers to any organic compound having an initial boiling point less than or equal to 250 °C measured at a standard atmospheric pressure of 101.3 kPa. The World Health Organization, WHO, classified VOCs of indoor air by their boiling points and distinguished between the very volatile, volatile, semi-volatile and non-volatile (particulate organic material) organic compounds (see Table 1.1).

SVOCs are found in significant amounts in both the gas and particle phase and POMs are found in significant amounts in the particle/solid phase in the indoor environment. According to European Collaborative Action (ECA 1997) and Committee for Health-related Evaluation of Building Products (Ausschuss zur gesundheitlichen Bewertung von Bauprodukten, AgBB 2010), organic compounds which can be detected in the elution range between n-hexane and n-hexadecane are designated Total Volatile Organic Compounds (TVOCs) (ISO 16000-6:2011). Moreover AgBB defined VVOC the organic compounds which are found in the retention region below n-hexane while SVOCs those which are found above n-hexadecane up to C_{22}.

It is difficult to establish a comprehensive list of emitted VOCs from each source because of the variation of products and the resulting change in the composition of the mixture of VOCs emitted. These compounds mostly belong to one of the following chemical classes: aliphatic hydrocarbons, such as alkanes and cycloalkanes, aromatic hydrocarbons, terpenes, aldehydes, ketones, alcohols, alkoxyalcohols, esters, ethers and halocarbons. This listing does not include groups of compounds such as carboxylic acids, isocyanates or amines. Although these VOCs may be present in the air of school environments they cannot readily be detected with the analytical methods routinely applied for VOCs (ISO 16000-6:2011).

Among the VOCs, the high priority pollutants that are regulated in indoor environments and that significantly affect children health are Benzene, Naphthalene, Formaldehyde, Toluene, Xylenes, Styrene, Limonene, Alpha-pinene and

Dichloromethane. Benzene, Toluene, Xylenes and Styrene could be emitted from solvent-based paints and consumer products, such as collage and painting materials, used in the art and craft rooms, from Poly Vinyl Chloride (PVC) flooring and adhesive used for gyms covering and from printed materials (Kotzias 2005).

Dichloromethane is found in adhesives, spray paints, while the presence of Limonene and Alpha-pinene is more related to the emission from cleaning products (aerosol and liquid) (Priscilla et al. 2010). Polymeric materials that are used for construction, decorating and furnishing of schools are high VOCs emitters due to their composition and large surface areas. Moreover wood-based products used for construction of writing desks and cabinets are important sources of these pollutants and Formaldehyde in these environments.

Some of the VOCs are associated with variety of serious health effects (Shendell et al. 2004; Sax et al. 2006) and symptoms such as asthma and allergic reactions (Sofuoglu et al. 2011). Moreover, several studies reported a strong association between mucous membrane irritation, central nervous system symptoms and total exposure to VOCs amongst office workers; these symptoms are very similar to those which are frequently attributed as a cause of Sick Building Syndrome (SBS) (Mølhave et al. 1986; Hodgson et al. 1991). In case of extreme concentrations, some VOCs may result in impaired neurobehavioral function (Burton 1997). Exposure to high concentrations of several VOCs commonly found in indoor air is associated with cancers in laboratory animals (Jones 1999). Among this big chemical classes there are different compounds which are characterized by both acute and chronic health effects and it is more correct to study the single compounds with its specific health effect.

Formaldehyde and Carbonyl Compounds

Among the carbonyl compounds the most relevant in indoor environments are Formaldehyde, Acetaldehyde, Acetone, Benzaldehyde, Butyraldehyde, Capronaldehyde, 2,5-Dimethylbenzaldehyde, Formaldehyde, Isovaleraldehyde, Propionaldehyde, m-Tolualdehyde, o-Tolualdehyde, p-Tolualdehyde, Valeraldehyde (ISO 16000-3: 2011).

As a result of their several industrial uses in the manufacture of sheet and insulation materials, paints, cleaning agents and cosmetics, the carbonyl compounds can usually be detected in school environments. Wood-based materials made for indoor use are the following: (1) Particleboard (PB) used as sub-flooring and shelving and in cabinetry and furniture; (2) hardwood plywood paneling used for decorative wall covering and used in cabinets and furniture; (3) Medium Density Fiberboard (MDF) used for drawer fronts, cabinets and furniture tops. Therefore articles produced from wood-based materials such as furniture, doors and paneling are still the most important sources of these compounds in schools. Formaldehyde (HCHO) is the most abundant airborne indoor carbonyl. It is the simplest carbonyl compound and, at a room temperature, is a colorless and reactive gas with a distinct pungent smell.

HCHO is an important constituent of adhesives in the sheet material industry. Amongst those used there are Urea-Formaldehyde resins (UF glues), Phenol-Formaldehyde resins (PF glues), Melamine Formaldehyde resins (MF glues) and Melamine-Urea-Formaldehyde resins (MUF glues). Medium density fiberboard contains a higher resin-to-wood ratio than any other UF pressed wood product and is generally recognized as being the highest formaldehyde-emitting pressed wood product. Several studies showed that indoor HCHO concentrations of schools constructed within 1 year were significantly higher, indicating that school buildings are characterized by several indoor HCHO sources such as furnishings made of PB and MDF.

Several carbonyls, such as formaldehyde, acetaldehyde, and propionaldehyde, are included in the list of air toxics in the Clean Air Act Amendments of 1990 (USEPA 1991). Although formaldehyde is a natural metabolic product of the human body, epidemiological studies demonstrated that high-dose exposure increases the risk of acute poisoning and that long-term exposure can lead to chronic toxicity and cancer. Formaldehyde is defined as a human carcinogen on the basis of a sufficient evidence of carcinogenicity from studies in humans and of supporting data on mechanisms of carcinogenesis. As a result of human exposure to HCHO, various symptoms may occur: irritation of the eyes and of the upper respiratory tract, allergy, asthma, pulmonary damage. California Office of Environmental Health Hazard Assessment (OEHHA) set an 8-h, chronic and acute inhalation reference exposure level (REL) for formaldehyde of 9 $\mu g/m^3$, 9 $\mu g/m^3$ and 55 $\mu g/m^3$ respectively (OEHHA 2008). In recent years, these scientific findings have led an increasing interest in HCHO detection inside school buildings due to the high risk of children exposure (WHO 2002; NIOSH/IPCS 2004; U.S. Department of Health and Human Services Public Health Service 2011; IARC 2012). Acetaldehyde, another abundant carbonyl in ambient air, has been classified as probable human carcinogen by USEPA (2003). Acrolein is a severe lung irritant that, at high acute exposures, can induce oxidative stress and delayed-onset lung injury, including asthma, congestion, and decreased pulmonary function. Because of concerns about adverse human health effects posed by acrolein in ambient air, OEHHA set an 8-h, chronic and acute inhalation reference exposure level (REL) for acrolein of 0.70 $\mu g/m^3$, 0.35 $\mu g/m^3$ and 2.5 $\mu g/m^3$ respectively (OEHHA 2008).

1.2.2 Particulate Matter (PM)

Particulate matter (PM) is the general term used to describe solid particles and liquid droplets found in the air. PM shows differences in size, physical and chemical properties and thus in the toxicological and carcinogenic effects they cause (Sager and Castranova 2009). Conventionally the International Standards Organization classifies PM in two size classes: PM2.5 and PM10 fractions. PM10 includes particles that have aerodynamic diameter less than or equal to 10 μm (or more precisely, the definitions specify the inlet cut-offs for which 50 % efficiency is obtained for these sizes) and PM2.5 is the subset of PM10 particles that have aerodynamic diameter less than or equal to 2.5 μm. Another size classification of PM fractions discriminates course, fine and ultrafine particles which correspond to particles with aerodynamic diameter ranging from 2.5 to 10 μm, smaller than 2.5 μm and smaller than 0.1 μm, respectively.

Particles within different size ranges behave differently in the air. Fine particles, can remain airborne for long periods and travel through hundreds of miles. Coarse particles do not remain airborne for a long time and their spatial impact is typically limited because they tend to deposit on the ground downwind of emissions sources. Larger coarse particles are not readily transported across urban or broader areas because they are generally too large to follow air streams and they tend to be removed easily on contact with surfaces (Branis and Safránek 2011).

Although the school environment normally lacks typical indoor particulate matter (PM) sources such as smoking and cooking, many children are present in a limited space over a period of several hours. Therefore there is growing evidence of comparatively high concentrations of PM in classrooms (Koutrakis et al. 1992; Ozkaynak et al. 1996; Daisey et al. 2003).

The use of cleaning products and floor polish can also temporarily affect the IAQ determining an increase in chemical pollutants in school environments. On the other hand, the floor surface type and level of cleaning are important factors in maintaining low dust levels. The presence of PM can be related to: (1) insufficient ventilation in schools (especially in winter), (2) infrequently and unthoroughly cleaned indoor surfaces, (3) a large number of pupils in relation to room area and volume, (4) low class level related to floor numbers of school buildings and (5) resuspension of particles from room surfaces(Sexton and Ryan 1988). Resuspension of PM coarse fraction plays a major role on PM presence in school classrooms. In fact the daily indoor PM10 levels were always higher than those outdoors, except on weekends, suggesting that the physical activity of the pupils leads to resuspension of coarse particles and greatly contributes to enhance PM10 in classrooms (Fromme et al. 2008; Almeida et al. 2011). Numerous studies showed that gas phase reactions between O_3 and terpenes (for example used in cleaning products) can contribute significantly to the growth of indoor secondary organic aerosols (Weschler and Shields 1999; Long et al. 2000; Wainman et al. 2000; Li et al. 2002; Fan et al. 2003; Sarwar et al 2003).

The chemical composition of PM besides particle size distributions is the most significant factor affecting air quality and human health, although less investigated. In particular, hazardous pollutants such as Polycyclic Aromatic Hydrocarbons (PAHs) and heavy metals are strongly associated with finer particles, increasing PM toxicity and carcinogenicity. In fact, finer particles can convey these pollutants in deep human airways determining negative effects on human health. Among the PM elemental components, calcium (Ca) represents the most abundant element in classrooms and its high concentrations are associated to chalk dusts. The other elements such as aluminium (Al), iron (Fe), strontium (Sr), and titanium (Ti) mainly come from outdoor sources as a consequence of transport of outdoor soil dust (Clayton et al. 1993; Tran et al. 2012).

Concerning the impact of PM on human health, the American Conference of Governmental Industrial Hygienists (ACGIH), the International Organization for Standardization (ISO), and the European Standards Organization (CEN) reached agreement on definitions of the inhalable, thoracic and respirable fractions. The inhalable fraction is constituted by particles that are inhaled through the nose and mouth; thoracic fraction is constitued by particles that can penetrate the head airways and enter the airways of the lung, while respirable fraction is defined as the mass fraction of inhaled particles that can penetrate beyond the terminal bronchioles into the gas-exchange region of the lungs. The effects of PM on human health were deeply investigated in several papers (Englert 2004; Zanobetti and Schwartz 2005; Forbes et al. 2009; Pope et al. 2009): the exposure to high PM10 concentrations has been associated to increased risk of death for cardiovascular or respiratory causes. These effects may be largely caused by finer particles that, as a consequence of their greater surface area, could be an effective media to transport different kinds of pollutants (PAHs, heavy metals, asbestos, etc.) deeply into the lung (Nadadur et al. 2007; Sager and Castranova 2009; Reich et al. 2009). In particular, the exposure to these finer particles can cause short and long-term effects such as increased respiratory symptoms, decreased lung function, alterations in tissue and structure lung, in respiratory tract and premature death (Prieditis and Adamson 2002; Damek-Poprawa and Sawicka-Kapusta 2003; Wahab and Basma 2004; Huang and Ghio 2006; Hong et al. 2007; Wild et al. 2009; Daresta et al. 2010; Liuzzi et al. 2011).

Several epidemiological studies indicated a strong association between high concentrations of inhalable particles and increased mortality and morbidity (Lin and Lee 2004; Arditsoglou and Samara 2005; Namdeo and Bell 2005; Lammel et al. 2010).

1.2.3 Carbon Dioxide CO_2

The carbon dioxide CO_2 concentration is an important indicator in assessing IAQ and has a particular importance in school classrooms that are densely occupied by large number of people in a restricted space. Although CO_2 is not itself a pollutant, its indoor concentration is an indicator of IAQ and is related to pupil's occupancy, productivity and health symptoms (Daisey et al. 2003; Shendell et al. 2004).

Levels of 600–800 ppm are typical in indoor environments. ASHRAE Standard 62-1989 (1989) suggested that CO_2 levels should not exceed 1,000 ppm (700 ppm above outdoor levels) although some individuals are sensitive to CO_2 levels lower than 600 ppm. According to Daisey et al. (2003) and Clements-Croome (2006), CO_2 concentrations beyond these recommendations are common inside school classrooms where values even greater than 4,000 ppm are often detected. Generally, it is difficult to adequately characterize indoor CO_2 concentrations because they are a function of occupancy level, ventilation rate, room design and air exchange rate (Wåhlinder et al. 1997; Grimsrud et al. 2006; Theodosiou and Ordoumpozanis 2008; Mumovic et al. 2009). CO_2 real-time monitoring is a useful tool to evaluate both the indoor air exchange rate and the occupancy level of children inside the classrooms (Ajiboye et al. 2006). Measurements in school buildings suggest that CO_2 concentrations increase when inadequate ventilation conditions occur. As a result, CO_2 concentrations are lowered throughout the day when the windows in classrooms are left open.

Therefore an accurate assessment of risk of disease transmission throughout the classroom should take into account CO_2 concentrations (Rudnick and Milton 2003). In fact Mi et al. (2006) showed that exposure to this pollutant is associated with asthma and Shendell et al. (2004) found that an increase in 1,000 ppm is associated with a 10–20 % increase in student absences.

1.2.4 *Ozone (O_3)*

Ozone (O_3) is a secondary air pollutant. The formation processes of O_3 are not linked directly to primary sources such as traffic, industrial and natural emissions. In fact O_3 concentrations in outdoor air depend on the exchange between upper and lower layers of atmosphere and on photochemical reactions involving nitrogen oxides and VOCs.

In contrast to CO, NO_x and SO_2, the outdoor O_3 concentration has increased over the past five decades. On the contrary O_3 concentrations registered inside the indoor environments are very low and often below the detection limit (Grøntoft and Raychaudhuri 2004). The indoor sources of O_3 are nowadays more common than they were several decades ago. Ozone can be released into the air from some office equipment such as laser printers and copiers and from some types of electrostatic air cleaners. Photocopiers are considered the most important sources of O_3 because they are ubiquitous in offices and schools (Leovic et al. 1996; Destaillats et al. 2008) although, in most buildings, outdoor O_3 appears to be the major source (Weschler 2000). Indoor O_3 concentrations track outdoor concentrations with a slight time lag that depends on the air exchange rate.

O_3 reacts rapidly with indoor surfaces as well as by gas-phase reactions, resulting in a lower concentration indoors than outdoors, varying from 10 to 90 % of the outdoor level (Weschler 2006). Indoor O_3 levels are dependent on the generation rate, leakage, ventilation, degree of mixing, and air filtration (Gold et al. 1996) and its decomposition rate is dependent on the quantity and type of materials in a building. O_3 reacts with many organic chemicals characterized by unsaturated carbon–carbon bonds (e.g., isoprene, styrene, terpenes, sesquiterpenes, squalene, and unsaturated fatty acids and their esters) because such compounds react much faster than do saturated organic compounds. The common sources of O_3-reactive chemicals include soft woods, carpets, linoleum, paints, polishes, cleaning products and air fresheners, soiled fabrics, soiled ventilation filters and the occupants themselves. These ubiquitous sources result in substantial quantities of indoor chemicals that can react with O_3 whenever outdoor concentrations are elevated. The concentrations of O_3-reactive compounds tend to be higher indoors than outdoors (Brown et al. 1994;

Wolkoff 1995; Hodgson and Levin 2003), reflecting more sources and larger emission rates per volumetric flow rate (Weschler 2006).

Many toxicological and field studies of both adults and children (Tager 1999; Lee et al. 2004) established the short-term reversible effects of O_3 on respiratory and pulmonary function, associating O_3 with lung function decrements (Gold et al. 1996). Hubbell et al. (2005) demonstrated the association between O_3 and respiratory-related hospital admissions, school absence, restricted activity days, asthma related emergency department visits and premature mortality (Weschler 2006). Wolkoff and colleagues also demonstrated that ozone/terpene reactions produce strong airway irritants (Wolkoff et al. 1999; Clausen et al. 2001; Wilkins et al. 2001; Rohr et al. 2002, 2003). Some of these products (e.g. formaldehyde, acrolein, peroxyactyl nitrate, hydroperoxides) are known to have adverse health effects (Weschler 2006).

1.2.5 Other Inorganic Gases: Nitrogen Oxides (NO_x), Carbon Oxide (CO), Sulphur Dioxide (SO_2)

Inorganic gases commonly found in school indoor air are CO, SO_2 and NO_2. Sometimes, high H_2S and NH_3 concentrations are determined inside school buildings near industrial plants such as water treatment plants, waste treatment, desulphurization plants, etc.

Nitrogen oxides (NO_x=sum of nitric oxide (NO) and nitrogen dioxide (NO_2)) enter indoor air mainly from outside and the most source of these pollutants is vehicular traffic. Several studies showed that the most important factors in increased exposures to NO_x were both the position of school buildings in the city centre and the use of gas appliances for heating (Oie et al. 1993; Alm 1999; Coward et al. 2001; Dimitroulopoulou et al. 2005; WHO 2006a, b). Long-term exposure to high NO_2 concentrations promotes the onset of diseases of the respiratory tracts; in fact epidemiological studies suggested that NO_2 represents a modest risk factor for respiratory illnesses compared to the use of electric stoves (Basu and Samet 1999).

Carbon monoxide is a vehicular pollutant, therefore vehicle exhaust from roads and parking areas nearby school buildings represents the most important contributor to CO indoor exposure. The carbon monoxide levels are generally very low inside schools since the emissive indoor sources influencing long-term CO levels can be gas cooking, unflued heaters and smoking (Alm et al. 1994; Coward et al. 2001; Ott et al. 2007). Exposure to high CO concentrations can cause acute intoxication since this compound combined with the hemoglobin of human blood produces carboxyhemoglobin, and therefore disrupts the transfer of oxygen to human tissues. Various symptoms of neuropsychological impairment were associated with acute low-level exposure of CO concentration (Raub et al. 2000). Epidemiological studies reported increased relative risks of daily mortality and morbidity of the population by 0.9–4.7 % in prevailing urban air (Touloumi et al. 1994; Burnett et al. 1997, 1998).

Sulphur dioxide is the main oxide of sulphur found in indoor air. Indoor concentrations determined inside school buildings are generally lower than those outdoors

(Weschler 2009). The most sources of SO_2 are outdoors in fact several studies reported that higher indoor SO_2 concentrations were determined inside buildings near open coal fires. Moreover SO_2 is readily absorbed onto indoor material surfaces such as emulsion paints that are considered the most important sink for SO_2 (Ashmore and Dimitroulopoulou 2009). Epidemiological studies on health effects associated with exposure to sulphur dioxide are complicated by a paucity of representative exposure data and by confounding factors such as exposure to other indoor pollutants. However, several studies provided some useful data concerning exposure-effect relationships showing that mortality was observed in populations exposed to 24-h pollution episodes in which sulphur dioxide concentrations exceeded 300–400 μg/m^3 (0.12–0.15 ppm) (Health Canada 1995).

1.3 Monitoring Strategies

1.3.1 Measurement Objective

In order to develop a sampling strategy for assessment of indoor air quality in school environments, it is necessary to clarify the goal of the assessment as a function of the specific characteristics of the investigated environments (UNI EN ISO 16000-1 2006). To this aim, it is important to provide cross-sectional information related to exposure to harmful factors and it could be necessary to monitor trends and provide feedback about the effect of policy on reducing exposure to indoor air pollutants in schools. Furthermore it is important to produce information useful for health risk assessment. Before a sampling strategy is devised for school indoor air monitoring, it is necessary to perform an inspection inside the classrooms, gyms, computer rooms etc. including measurements of room volume, floor area, number of students.

Monitoring activity in school environments can be conducted following complaints by occupants (e.g. resulting from the perception of odors or the occurrence of illnesses such as headaches, nausea, respiratory irritation) in order to identify major emission sources, know the average concentrations of main air pollutants, often grouped according to their chemical-physical and toxicological characteristics, and determine personal exposure. In particular, it can be evaluated the acute effects on the health of occupants by determining the maximum concentration of short-term or chronic effects by monitoring over long periods (average concentration). Sometimes, however, the monitoring can be adopted as a tool for preliminary assessment of air quality, useful to identify any areas of concern or potential sources through investigation of screening or to verify compliance with guide values or to the effectiveness of a remedy adopted. The use of a questionnaire is necessary to obtain a systematic record of the complaint episodes. In the questionnaire all the factors affecting indoor air quality should be registered and reported (indoor and outdoor sources of combustion, cooking, and experimental exercises) as the outdoor sources of contamination as distance to a busy road, industrial sources of emissions such as the start and end time of presence of pupils in classes, the start and duration of school breaks. If the aim of sampling is to study the association between pupils' asthmatic symptoms and indoor and outdoor air pollution in schools, it is important to collect data on their respiratory health, parental asthma or allergy, and selected factors in the home environment. Questions on respiratory health are mainly based on the International Study of Asthma and Allergy in Childhood (ISAAC) (Asher et al. 1995), the European Community Respiratory Health Survey (ECRHS) (Janson et al. 2001) and previous school studies (Smedje et al. 1997a, b; Mi et al. 2006). Questionnaires generally include yes/no questions on cumulative asthma, doctor-diagnosed asthma, current asthma and allergies to furry pets or pollen. Moreover, there are questions on respiratory symptoms including wheeze or whistling in the chest, daytime or nocturnal attacks of breathlessness in the preceding 12 months and recent respiratory infections (cold, upper respiratory infection, or middle ear infection in the preceding 3 months). Questions on parental asthma or allergy and about renovations of home environments (recent home paintings, new floor material and new furniture in the preceding 12 months) are also included. Another important parameter to report is the presence of Environmental Tobacco Smoke (ETS) at home which is classified into four categories: never smoking, smoking one to three times per month, smoking one to four times per week, smoking every day. Subjects with a lack of information on ETS are classified as a separate category. The survey should be performed 1 week before the classroom inspections and measurements, distributed in the school by the class teachers and answered at home in cooperation with the parents (Zhao et al. 2008). The choice of questions and the way in which they are structured should be the result of careful considerations made in cooperation with professors of Psychology, Statistics and Science of Education (De Giuli et al. 2012).

1.3.2 Sampling Procedure

Sampling methods used in outdoor environments can be used also for sampling indoor air, provided that equipment is suitable for the measurement purposes and doesn't have an adverse effect on the normal use of the school environments (because of size, sampling rate and noise). In positioning the monitoring equipment it is important that the sampling rate should not interfere with the normal ventilation rate because the concentration of the indoor air may not be homogeneous (UNI EN ISO 16000-1:2006). Sampling should be conducted during one school week. Ventilation and activities within and outside the school building, affect the value of most indoor pollutants concentration during the schooldays in the evenings, nights and weekends. In defining sampling strategy, selection of the classroom, type of activities that take place and proximity of sea, industrial activities and urban roads were of particular concern. Selection of area (urban – heavy traffic street nearby, residential, rural, rural seaside, urban seaside and urban industrial area nearby), type of room (classroom, former head office, plastic arts and/or video room), ventilation (naturally or mechanical ventilated) and type of windows (PVC, wood, metal, damaged or not) have been a crucial factor in defining sampling strategy to study the influence of outdoor environment to the indoor air quality (Blondeau et al. 2005; Diapouli et al. 2008; Chithra and Shiva Nagendra 2012). Sometimes the choice of optional pollutants to be monitored depends on economic factors.

1.3.3 Sampling Duration and Frequency

The duration of sampling can be determined considering the nature of the indoor pollutants, the potential health effects of the monitored compounds, the emission characteristics of the sources, the limits of quantification of the analytical method employed and the aim of the measurements. According to the potential health effects of the targeted substances, short-term sampling should be used for compounds causing acute effects. Long-term sampling methods, instead, are recommended for compounds causing chronic effects on health. The sampling duration chosen is particularly important because long-term sampling is not able to detect short-term events. The long-term sampling therefore is often used to determine the air quality under normal condition of occupancy. On the other hand short-term sampling is performed in case of extreme situation such as the emission from sources which emit temporarily (for example during cleaning activities). Therefore, the monitoring planning depends on the measurements objective and the emission characteristics of possible sources and should take into account any changes in concentration related to variations of the main factors involved (such as ventilation rate, level of occupancy, intermittent sources and seasonal changes). Finally the sampling duration should also consider the limit of quantification of the analytical method employed. If the monitoring objective is the evaluation of children exposure to indoor pollutants, the sampling should be carried out during class hours (WHO 2011). The final sampling strategy should also take into account the seasonal variability in concentration of some pollutants. During the winter in fact it can be observed variability in concentration due to poorer ventilation and increased emission from materials and products (e.g. paint and furniture) exposed to high temperature originated from heating systems. However summer peaks, related to either higher outdoor temperature or low air exchange rate, can be revealed (UNI EN ISO 16000-1:2006).

1.3.4 Sampling Location

The assessment of indoor air quality in schools should take into account not only the changes in the concentration of the pollutants over time, but also the spatial variation. Therefore the school environments to be monitored and a suitable sampling location should be specified. In each school building at least three sampling locations should be selected (WHO 2011). Classrooms selected for monitoring should be representative of the entire school building and routinely occupied during school hours. Furthermore the choice of the sampling sites is related to the purpose of the measurements (pollutants sources identification, pollutants exposure assessment). If the indoor environments to be monitored are large (e.g. gyms, auditorium) it is recommended to subdivide them in order to define accurately sampling locations. Sampling should be carried out at the centre of the classroom at the approximate height of the average breathing zone (about 1–1.5 m above the floor). Sampling locations nearby heating and ventilation systems should be avoided (WHO 2011; UNI EN ISO 16000-1:2006). When ventilation systems are present in the indoor environments monitored it is advisable to take into account that the sampled volume during 1 h should be less than 10 % of the air volume introduced by forced ventilation. If the ventilation rate cannot be measured and this information is not available, the hourly sampled volume should be less than 10 % of the volume of the investigated room. In order to monitor naturally ventilated environments, it is advisable to perform ventilation of the room for at least 15 min opening doors and windows before starting the sampling activities. After this period, doors and windows shall be closed for about 8–12 h and the sampling conducted.

1.3.5 Parallel Outdoor Air Measurement

Indoor air at schools in urban areas can be contaminated by ambient urban air pollution and traffic pollutants and vicinity to busy roads may affect children's respiratory health (Janssen et al. 2007; Kim et al. 2004; Holguin et al. 2007; Morgenstern et al. 2007). The presence of industrial sources in the vicinity of the school can adversely affect indoor air quality too. So it may be important to supplement indoor air measurements with outdoor simultaneous measurements if possible at the same level or floor of the classroom, or in the vicinity of the school but not closer than 1 m, because it should be contemplated the vertical concentration gradients may occur (UNI EN ISO 16000-1:2006).

1.4 Sampling and Analysis Techniques

The sampling procedure depends on the measurement purpose and the emission characteristics of possible sources. Several procedures, sampling equipments, sample preparation steps and analytical methods for VOCs, PM, CO_2, O_3 and inorganic gases are described in detail in the following paragraphs.

1.4.1 Volatile Organic Compounds (VOCs)

Regarding VOCs, preliminary screening assessment can be conducted performing direct measures with automatic instruments as flame ionization detector (FID) or photo ionization detector (PID) in order to evaluate the sum of VOCs (Fig. 1.2) into

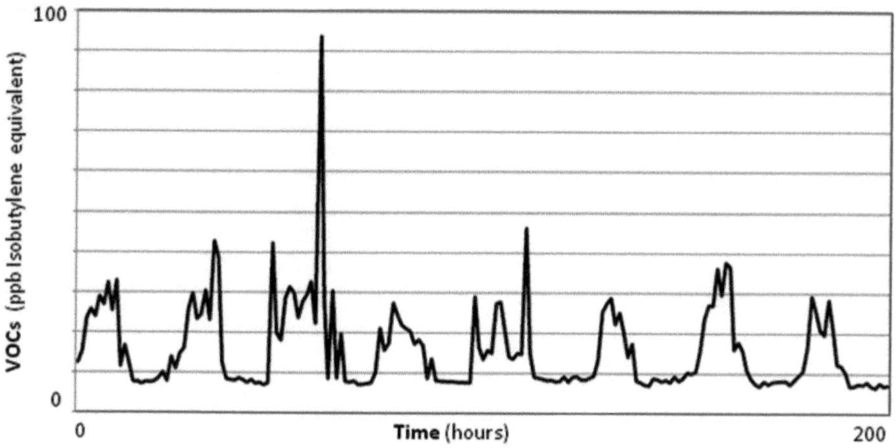

Fig. 1.2 Examples of VOCs behavior by photo ionization detector (PID)

the investigated environments (Hodgson 1995; Pegas et al. 2010). Procedures for monitoring VOCs in indoor environment may be divided in short-term and long-term measurement methods (UNI EN ISO 16000-5: 2007). Short-term measurements generally involve a sampling period from less than 1 h to a few hours depending on the purpose of the measurement. In this case active sampling on adsorbent cartridges is used. VOCs are collected on stainless steel tube packed with different adsorbent beds (for example Tenax TA, Carbograph 4, Carbotrap, Anasorb) using low flow sample pumps (UNI EN ISO 16017-1: 2007; Fraga et al. 2008; Jo and Kim 2010; Pegas et al. 2010; Scheepers et al. 2010; ISO 16000-6: 2011). The sampling flow rate and the final sampling volume shall be selected as a function of the breakthrough volumes of single VOC and considering the Limit Of Detection (LOD) for the analytical method and the expected concentrations in the investigated environments. Instead diffusive sampling is the recommended method to perform long-term measurements (usually from few days to several days or weeks), but it is possible to use also active sampling with low air flow rate (UNI EN ISO 16017-2: 2007; Bruno et al. 2005; Angiuli et al. 2003; Pennequin-Cardinal et al. 2005). Diffusive samplers predominantly work according to the diffusion principle and give an integrated measurement value as a mean over the sampling period. The sampler is composed of an adsorbing cartridge and a diffusive body. Depending on the pollutant to be sampled, several different kinds of adsorbing or chemiadsorbing cartridges are used (Kotzias 2005; Godwin and Batterman 2007; Bruno et al. 2008; Pegas et al. 2010, 2012; Sohn et al. 2012). The sampling rate, preliminarily determined or provided by the manufacturer, is function of diffusive coefficient which is a thermodynamic property of each chemical substance. Diffusive coefficient varies with ambient temperature and pressure, so the sampling rate should be corrected to reflect the actual sampling conditions.

VOCs collected actively or passively onto adsorbent cartridges are thermally or chemically desorbed and analysed by gas chromatography coupled to a flame

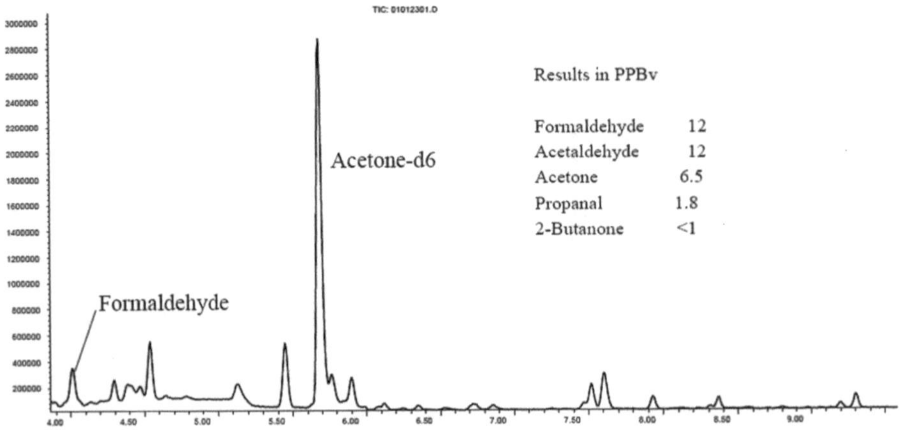

Fig. 1.3 Analysis of indoor air by gas chromatography-mass spectrometry (GC-MS)

ionization detector (GC-FID), or to a mass spectrometer (GC-MS). Another method to collect VOCs in indoor air implies the use of canisters (Fig. 1.3). They are stainless steel containers with internal surfaces passivated suitable for sampling of a wide range of VOCs which have been shown to be stable at low concentrations for at least 30 days (EPA Method TO-15 1999). Canisters are cleaned, certified and evacuated in the laboratory before sampling. Since the vacuum inside the canister drives sampling, no additional equipment is necessary. A fraction of the sampled air is then transferred onto a sorption bed and analysed by thermo desorption and GC-MS (see Fig. 1.3) or GC-FID (Meininghaus et al. 2003; Guo et al. 2004).

Formaldehyde and Carbonyl Compounds

The measurement of HCHO and other carbonyl compounds (Acetaldehyde, Acetone, Benzaldehyde, Butyraldehyde, Capronaldehyde, 2,5-Dimethylbenzaldehyde, Formaldehyde, Isovaleraldehyde, Propionaldehyde, m-Tolualdehyde, o-Tolualdehyde, p-Tolualdehyde, Valeraldehyde) can be performed according to ISO 16000-3: 2011. The method is suitable for determination of these compounds in the approximate concentration range of 1 $\mu g/m^3$ to 1 mg/m^3 and involves drawing air through a cartridge containing silica gel coated with 2,4-DiNitroPhenylHydrazine (DNPH) reagent. The principle of the method is based on the specific reaction of the carbonyl group with DNPH in the presence of an acid to form stable 2,4-dinitrophenylhydrazones. The DNPH derivatives are analysed utilizing High Performance Liquid Chromatography (HPLC) and Ultraviolet (UV) absorption detector operating at 360 nm (Lee and Chang 2000; Daisey et al. 2003; Meininghaus et al. 2003; Yang et al. 2009; Pegas et al. 2010; Park et al. 2011; Sohn et al. 2012). A blank cartridge is likewise desorbed and analysed. HCHO and other carbonyl compounds in the sample are identified and quantified by comparison of their retention times and peak heights or peak areas with those of standard solutions. Interferences in this method are caused by certain isomeric aldehydes or ketones that may be unresolved by the HPLC system and by organic compounds having the same retention times and significant absorbance at 360 nm as the DNPH derivatives. Such interferences can often be overcome by altering the separation conditions (e.g. using alternative HPLC columns or mobile phase composition). Acrolein and Crotonaldehyde could not be accurately quantified by the method because of the formation of multiple derivative peaks and of the instability of the peak ratios. High NO_2 concentrations can negatively affect HCHO sampling and measurement because of NO_2-DNPH reaction. It was observed, in fact, that the retention time of NO_2-DNPH derivative can be similar to that observed for the HCHO-DNPH derivative, depending on the HPLC column and the parameters.

1.4.2 Particulate Matter (PM)

Nowadays there are many different techniques for monitoring PM in indoor environments. The choice of PM monitoring equipment depends by several factors such as cost, dimensions, ease of use and noise. Therefore in dependence on the aim of monitoring, the knowledge of the number and the size particle distributions rather than the mass per unit volume of PM, it is possible to choose among several PM monitoring techniques. When the PM mass concentration is needed, a sampler to accumulate PM on a filter for gravimetric measurement (EN1234-1) or an automated PM analyzer have to be used. The first allows to determine 'off- line' mass concentration of PM, whereas the second provides an indirectly measurement of PM mass measuring β-rays attenuation or frequency of oscillation of a glass tapered tube (TEOM). The filter-based methods are widely used since samplers are relatively not expensive and filters can be used for a possible subsequent chemical or physical characterization of collected PM. On the contrary, when time-resolved or semi-continuous information about particle number concentration and particle size distribution are needed, an optical automated device has to be used.

Mass Concentration

The filter-based gravimetric sampling allows measuring PM mass concentration by using a pumping sampler. The sampler consists of an inlet that is directly connected to the filter substrate and a regulated flow controller. Before and after sampling, the filter is usually conditioned for 48 h period, at 20 °C and 50 % relative humidity and then weighed on an analytical balance. The difference in weight divided by the volume of air pulled through the filter gives the mass concentration of particulate matter. Automated analyzers allow sampling and field mass measurements avoiding sample transportation. The analyzers based on β-rays attenuation principle, were designed to allow filter changing after each 24-h period, so that the instrument does not need to be visited each day (Diapouli et al. 2008; Fromme et al. 2007, 2008; Sohn et al. 2012). Both gravimetric and automated instruments rely on a size-selective inlet to exclude undesired larger particles before the concentration of PM is quantified in order to sample PM_{10} or $PM_{2.5}$ fraction. Moreover multi-stage impactors can be used in indoor air quality monitoring to evaluate the chemical and physical characterization of the airborne particles inside the different PM dimensional fractions. This sampler usually contains six or eight stages (i.e. collection plates) with numerous, successively smaller, evenly distributed holes (orifices) in each stage. A constant flow rate is maintained through the sampler which causes the aerosol stream velocity to increase at each stage, resulting in a deposition of particles in size fractions. Each stage of the sampler consists of a perforated disc located above a collection medium (Fromme et al. 2008). Nowadays in order to perform PM sampling in indoor environments causing the least possible discomfort, portable, user friendly and low noise air samplers have been developed and used (Yang et al. 2009). Moreover, increasingly wide range of measurement techniques for various metrics of PM are being applied to personal exposure analysis. The requirement of exposure analysis is that sampling should be performed from the breathing zone of an individual, which extends approximately 30 cm from their nose and mouth, and moves with them from one microenvironment to the next.

Number and Size Distribution

In order to determine particle number concentration and size distribution, automated analyzers are used. These devices are commonly optical particle monitors that utilise the interaction between airborne particles and laser light. Automated samplers are commonly portable, ease of use and allow to measure particle concentrations over short time intervals. They can therefore be used to assess spatial and temporal variations in particle concentrations. The advantage of this approach is that a single analyzer with a Total Suspended Particles (TSP) inlet can monitor particles of a wide range of size simultaneously (Chithra and Shiva Nagendra 2012; Fromme et al. 2008). Optical particle counters/sizers (OPC/OPS) utilise the interaction between airborne particles and laser light to determine the number and the size of particles (Fig. 1.4). Condensation Particle Counter is based on the light scattering principle such as the previous instruments but it senses ultrafine particles by causing their size growth that is efficiently detected by light scattering. Nowadays, modern and smallest version of CPC and OPC are also used for indoor air quality and personal exposure measurement.

In order to characterize nanoparticles involved in indoor air pollution Scanning Mobility Particle Sizer (SMPS) and Fast Mobility Particle Sizer (FMPS) can be used. These instruments are based on the principle of the mobility of a charged particle in an electric field and are used to measure particle number concentrations (PNC) for a discrete size distribution of aerosols within a range of 10–487 nm for SMPS and 5.6–560 nm for FMPS (Fig. 1.5) (Fromme et al. 2007). Frequently optical particle monitors, OPC, OPS, CPC and other optical devices such as

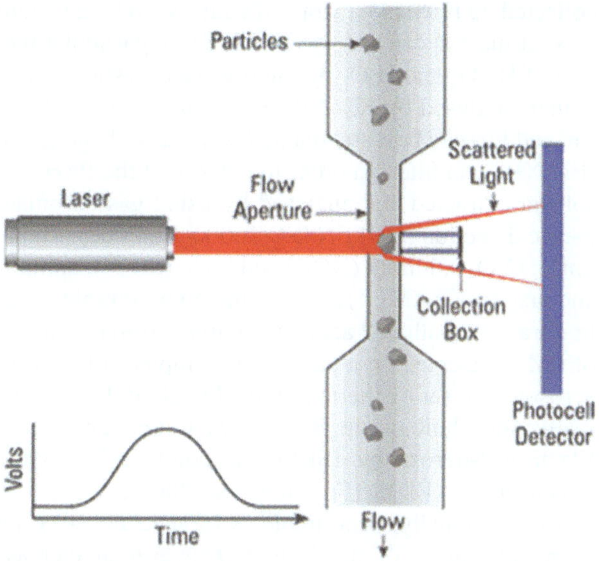

Fig. 1.4 Optical Particle Counter (OPC) scheme

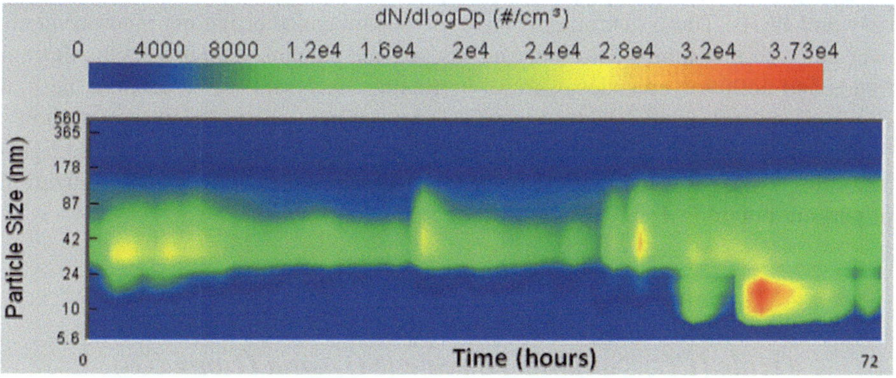

Fig. 1.5 A typical output of a Fast Mobility Particle Sizer (FMPS)

nephelometer, are used for measuring PM mass concentration in school environment: to convert the particle counts to mass concentration approximate assumptions whose validity may vary due to the nature of the particles sampled.

Chemical Characterization

Samples of PM collected using the techniques described above can be subjected to chemical analysis for the subsequent determination of metals, PAHs, major ions and organic (OC) and elemental carbon (EC). Nickel, Arsenic, Cadmium and Lead collected on filters are extracted with a solvent and analyzed by inductively coupled plasma-mass spectrometry (ICP-MS). In particular the filter is dissolved in a solution of HNO_3 and H_2O_2 by microwave digestion in a closed vessel and the extract is then analysed by ICP-MS. Elemental content (S, Cl, K, Ca, Ti, V, Mn, Fe, Ni, Zn, and Pb) can be also obtained by means of energy dispersive X-ray fluorescence (ED-XRF) on filter. In order to determine the PAH concentrations the samples are solvent extracted and analysed by using gas chromatography with mass spectrometric detection (UNI EN ISO 16000-12: 2008). Major ions such as chloride (Cl^-), nitrite (NO_2^-), nitrate (NO_3^-), sulphate (SO_4^{2-}), ammonium (NH_4^+), calcium (Ca^{2+}) and magnesium (Mg^{2+}) are determined on samples as a result of aqueous extraction in ultrasonic bath and analysis by ion chromatography. The carbonaceous fraction of PM is detected on a part of the sampled filter by an Organic/Elemental carbon aerosol analyser by thermal/optical method. In order to characterize the particulate matter morphology, the indoor PM filters can be further examined by Scanning Electron Microscopy (SEM) in combination with Energy Dispersive X-ray Spectrometry (EDX) (Fromme et al. 2008).

Conventionally, measurements of speciated PM are carried out by subsequent chemical analysis of the PM filter. However, an increasing number of direct reading analysers are now available, which permit continuous sampling and measurements for a range of particulate species including elemental/organic carbon, black carbon only and PAHs. These instruments have the advantage of providing measurements over much shorter time resolutions and can provide valuable data in tracking diurnal trends and pollution episodes. Carbon particulate analysers are based on a thermal CO_2 analysis technique. Black carbon (BC) (sometimes termed "elemental carbon", "light-absorbing carbon", or "soot") analyzers, instead, are based on visible light-absorbing principle. Finally PAH analyzer is based on principle of photoionization of particle-bound PAH.

1.4.3 Inorganic Gases: Nitrogen Oxides (NO_x), Carbon Oxide (CO), Sulphur Dioxide (SO_2) and Ozone (O_3)

Suitable methods for measuring nitrogen oxides (NO_x) in indoor environments can be divided into short-term measurement methods and long-term measurement methods UNI EN ISO 16000-15: 2008. Short-term measurements can be performed by continuous analytical monitoring instrument and by manual methods. The continuous monitoring instruments are based on principle of Chemiluminescence and are characterized by high time resolution (10–20 s). In manual methods instead, NO_2 is enriched actively onto a sorbent medium by means of suction pumps and the concentrations obtained by these methods are average concentrations for the duration of sampling. Long-term measurements are generally carried out using diffusive sampler (manual methods) since the noise produced by continuous analytical monitoring instrument could discourage their use inside confined environments (Lee and Chang 1999, 2000; Blondeau et al. 2005; Poupard et al. 2005; Pegas et al. 2010,

2012; Gul et al. 2011; Raysoni et al. 2011; Sohn et al. 2012; Stranger et al. 2012). Both indoor and outdoor measurements of carbon oxides (CO and CO_2) are conducted using a Non-Dispersive Infrared analyzers (NDIR) (Lee and Chang 1999; Chaloulakoua and Mavroidisb 2002; Yang et al. 2009; Pegas et al. 2010, 2012; Park et al. 2011; Smedje et al. 2011; Sohn et al. 2012). Continuous measurements of CO_2 and CO can be performed with specific automatic portable sensors (Pegas et al. 2012). Diffusive portable probes reveal CO_2 concentrations based on the ability of CO_2 to absorb infrared radiation at a certain wavelength (2.3–4.6 μm) such as CO_2 above cited devices, whereas CO concentrations on the basis of electrochemical reactions. Sulphur dioxides SO_2 in indoor environments are continuously measured by Electron Pulsed Fluorescence SO_2 Analyser. The operating principle of this instrument is based on measuring the fluorescence emitted consequently the absorption of ultraviolet light having wavelength in the range of 190–230 nm. The intensity emitted in the range from 300 to 390 nm is directly proportional to the sulphur dioxide concentration (Lee and Chang 1999; Meininghaus et al. 2003). Moreover SO_2 concentration can also be determined using radial passive samplers (Stranger et al. 2012). Indoor ozone concentrations can be monitored using an UV Absorption Ozone Analyzer (Blondeau et al. 2005; Poupard et al. 2005; Sohn et al. 2012) in order to give real-time a synoptic flow diagram. To perform long-term measurements, instead, it can be used specific diffusive adsorbing cartridges and the extract analyzed by UV–VIS spectrophotometry after chemical desorption (Stranger et al. 2007, 2008).

1.5 An Overview of Findings from Scientific Literature

Several studies were conducted in elementary and middle schools in order to evaluate the levels of indoor air pollutants, to identify their emission sources, to evaluate any differences within and between schools, to evaluate seasonal variability and how the location of the school (city centre, suburbs or industrial area) affects the indoor air quality.

1.5.1 Volatile Organic Compounds (VOCs)

Godwin and Batterman (2007), monitoring VOCs concentrations over one work-week in 64 elementary and middle school classrooms in Michigan, found that most VOCs had low concentrations (mean of individual species <4.5 μg/m^3) also if they were higher than outdoor air concentrations (mean of individual species <0.51 μg/m^3). For example Benzene and Toluene concentrations in indoor air were 0.09 μg/m^3 and 2.81 μg/m^3 respectively while their outdoor concentrations were 0.06 μg/m^3 and 0.52 μg/m^3 respectively; the total concentration of chlorinated compounds was 0.24 μg/m^3 in indoor air and <0.07 μg/m^3 in outdoor air. These findings suggested

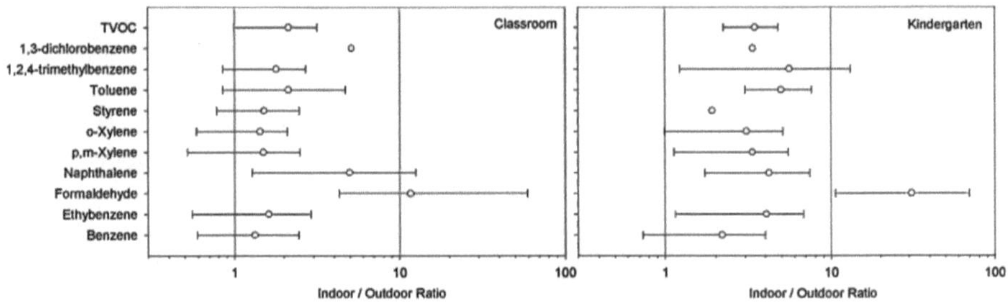

Fig. 1.6 Ratio of indoor/outdoor volatile organic compounds (VOC, Sofuoglu et al. 2011)

that none of the sampled rooms were contaminated and that no building-wide relevant contamination sources were present. Otherwise, higher indoor levels of many VOCs were registered in two studies involving 14 elementary schools in Lisbon, Portugal (Pegas et al. 2010, 2012). Almost all identified VOCs (up to 40 compounds) showed I/O ratios higher than one, suggesting that indoor sources, micrometeorological parameters and building conditions might have negative effects on indoor air. Moreover it was found that increasing ventilation rates and using low-emission materials contribute towards improving indoor air quality (Pegas et al. 2010). The same results were found by Kotzias (2005). In this study measuring campaigns were performed in several cities in Southern and Central Europe (Catania, Athens, Arnhem and Nijmegen, Brussels, Milan, Thessaloniki, Nicosia) to estimate indoor/outdoor relationships and personal exposure concentrations for selected compounds in schools and kindergartens. Results showed that the sum of VOCs monitored were higher and/or similar to the outdoor air concentrations. The sum of concentration inside school buildings in fact varies from a few micrograms (ca. 8 μg/m^3) to 281 μg/m^3. Outdoor concentrations varied instead from 7 to 153 μg/m^3. Moreover in kindergartens and in schools of Arnhem and Nijmegen VOCs concentrations were two to four times higher than the outdoor concentrations, ranging from ca. 19 to 36 μg/m^3.

Among VOCs Benzene, Toluene, Ethylbenzene and Xylenes were most abundant compounds in classrooms of several primary schools and kindergartens with I/O ratios exceeding unity (Norback et al. 1995; Shendell et al. 2004; Sofuoglu et al. 2011; Stranger et al. 2008). Sofuoglu et al. (2011) also measured VOCs concentration in three primary schools in Izmir in Turkey. Indoor/outdoor ratio for TVOC and for the most of them was higher than one (Fig. 1.6): the concentrations of VOCs were utilized to classify the indoor air pollutants with respect to potential health effects. A huge increase in the VOCs concentrations was observed when art work or science activities were undertaken concurrently or just prior to the measurements(Godwin and Batterman 2007; Pegas et al. 2010; Ventilation and Indoor Air Quality in Schools – Guidance Report 202825). In particular Pegas et al. (2010) found that there was an increase in VOCs concentrations around 13 ppm, when pupil's art class was occurring with the use of glue and paints. This makes evident that collage and painting materials increase the VOC levels in indoor air.

Shendell et al. (2004a) found that activities performed in art rooms, science rooms and indoor pools were the main indoor sources of specific VOCs in particular of different aldehydes.

Formaldehyde and Carbonyl Compounds

Many literature papers from the last 10 years investigated formaldehydes and other carbonyls in school buildings (Lee and Chang 2000; Righi et al. 2002; Kotzias 2005; Mentese and Gullu 2006; Vaizoglu et al. 2003; Hanoune et al. 2006; Yang et al. 2009; Sofuoglu et al. 2011; Yamashita et al. 2012; Pegas et al. 2011a, b; Barro et al. 2009). Pegas et al. (2011a) measured indoor and outdoor concentrations of formaldehyde and other carbonyls at 14 elementary schools in Lisbon, Portugal. In all environments, indoor aldehyde levels were higher than those observed outdoors, particularly for formaldehyde. In another study (Pegas et al. 2011b), conducted in the same schools, the indoor and outdoor levels of carbonyls were evaluated during spring, autumn and winter. Most of the assessed carbonyls occurred at I/O ratios above unity in all seasons, thus showing the importance of indoor sources and building conditions in indoor air quality. However, it was observed that carbonyl concentrations were higher in the warm months. Other researchers focused only on formaldehyde because defined a human carcinogen. Yang et al. (2009) characterized HCHO concentrations within 55 school buildings in Korea, selected on the basis of the year of construction, in order to relate indoor levels to the age of school buildings. HCHO levels were measured inside three different school building environments: classrooms, laboratories and computer rooms. Results showed that mean HCHO concentrations inside classrooms and computer rooms exceeded the acute reference exposure levels established by California's Office of Environmental Health Hazards Assessment (OEHHA, USA). Moreover, HCHO concentrations inside schools constructed within 1 year were significantly higher than the Korean Indoor Air Standard, suggesting that renovated schools have important indoor HCHO sources, such as furnishings principally made of particleboard and medium density fiberboard. Measurements were also conducted before and after the operation of ventilation system while air-conditioner system was turned on and the results showed remarkable difference in HCHO levels according to the operation of mechanical ventilation system.

Kotzias (2005) reported the results of measuring campaigns performed in several cities and showed that HCHO concentrations inside the buildings/kindergartens were up to 7–8 times higher than outside. This result means that strong HCHO indoor sources exist. Lee and Chang (2000) also showed the results of a study conducted to characterize HCHO levels inside selected classrooms in Hong Kong in order to compare the measured concentrations with the established standards and to suggest policy interventions to improve indoor air quality. HCHO concentrations reported in this paper were substantially lower than Honk Kong Indoor Air Quality (HKIAQ) standard indicating that there were no apparent HCHO indoor sources and that classroom furnishing did not add an important contribution (Pang 1994).

Sofuoglu et al. (2011) reported HCHO levels measured in primary school classrooms and kindergartens in Turkey. Experimental data revealed that HCHO was one of the most abundant indoor pollutants and that concentrations were related to both spatial and seasonal variability. Similar HCHO levels between urban and suburban schools but different HCHO levels between two urban schools can be explained by the relative strength of the indoor HCHO sources compared to the outdoor sources. The HCHO concentrations measured in classrooms were in the literature range (10–400 μg/m^3) although they resulted high if compared with data related to schools in Sweden (Norback et al. 1990; Smedje et al. 1997a, b) China (Lee et al. 2004) and Australia (Zhang et al. 2006). Furthermore HCHO levels in classrooms were lower than the concentrations measured in homes and offices (Mentese and Gullu 2006; Vaizoglu et al. 2003) but similar to those in libraries (Righi et al. 2002; Hanoune et al. 2006).

Regarding kindergartens, HCHO levels were higher than those measured in classrooms and difference in urban and suburban concentrations was not significant. It can be asserted that there were consistent sources of HCHO inside kindergartens because neither seasonal nor spatial differences were significant. The overall average of the concentrations measured in this study (85 μg/m^3) was clearly higher than the Dutch kindergartens (from ca. 6 to 11 μg/m^3) (Kotzias et al. 2005) and in the range of the Danish and Korean kindergartens (Yang et al. 2009).

1.5.2 Particulate Matter

Chithra and Shiva Nagendra (2012) studied the relationship between outdoor and indoor air quality in eight French schools, founding that indoor PM10, PM2.5and PM1concentrations were higher in winter than in summer. Moreover the indoor – outdoor (I/O) ratios of PM were higher than 2 for coarse fraction and minor than 1 for finer fraction. The high I/O value of PM10 concentration indicated significant contribution from the activities of occupants inside classroom. On the contrary the lower I/O values for PM1and CO confirmed the vehicular emission contribution from the nearby road (Fig. 1.7). This evidence was confirmed by a strong seasonal variability of finer PM fraction. Moreover, investigating the influence of classroom occupancy the authors found that higher particulate matter concentrations were determined for classroom during the periods when the classroom was occupied. Moreover, Fromme et al. (2007) found that PM2.5 indoor concentrations in several schools in Munich, increased by 1.7 μg/m^{3}per 10 % increase in humidity, by 0.5 μg/m^3 per increase in CO_2 indoor concentration by 100 ppm. The higher PM concentrations in winter and their correlation with CO_2 concentrations suggested that inadequate ventilation plays a major role in the establishment of poor indoor air quality. In addition, the PM_{10} concentration measured in low level classes and in rooms with high number of pupils suggested that the physical activity of pupils contributed to a constant process of resuspension of sedimented particles (Lee and Chang 1999, 2000; Blondeau et al. 2005). In fact, Yang et al. (2009) evaluating

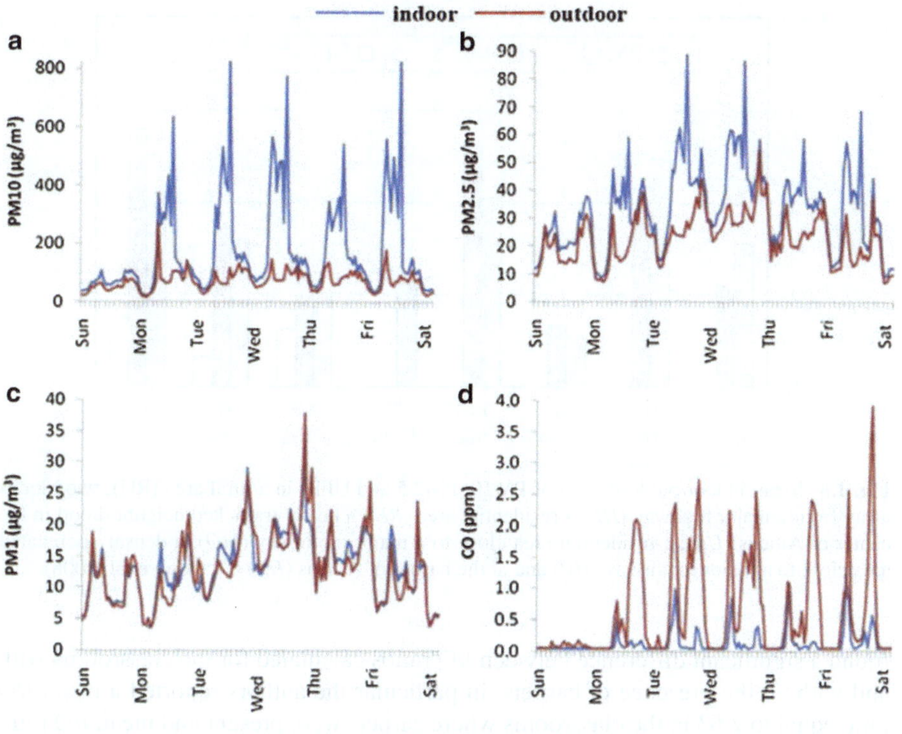

Fig. 1.7 Weekly variations in indoor–outdoor (**a**) PM10, (**b**) PM2.5, (**c**) PM1 and (**d**) CO concentrations inside classroom (Chithra and Shiva Nagendra 2012)

indoor air quality inside three different school environments in Korea found that the mean I/O PM10 ratios were higher in the classrooms than in laboratories and computer rooms, respectively. In addition, Diapouli et al. (2008) showed higher I/O ratio for PM10 and PM2.5 inside gymnasium, where intense activity took place, smoking office and classrooms. On the contrary the corresponding ratio for Ultra Fine Particles (UFP) was smaller than 1 in all investigated indoor environments since vehicular traffic was the main source of UFP (Fig. 1.8). This evidence confirmed that the most important contribution to indoor PM10 is the resuspension of particles. Another parameter influencing the indoor air quality is the presence of mechanical ventilation systems in school environments. Sohn et al. (2012), in fact evaluated the influence of mechanical ventilation systems on indoor air quality in school buildings in Korea. The results showed remarkable difference in indoor air pollutants' level according to the operation of mechanical ventilation system, and in particular showed that the ventilation systems decreased the levels of indoor pollutants in the all selected classrooms. Therefore, increasing the ventilation rate by means of a mechanical ventilation system can play key roles in improving the indoor air quality within schools. The presence of carpets in schools building also contributed to poor indoor air quality. Stranger et al. (2007), in a study regarding Belgium schools,

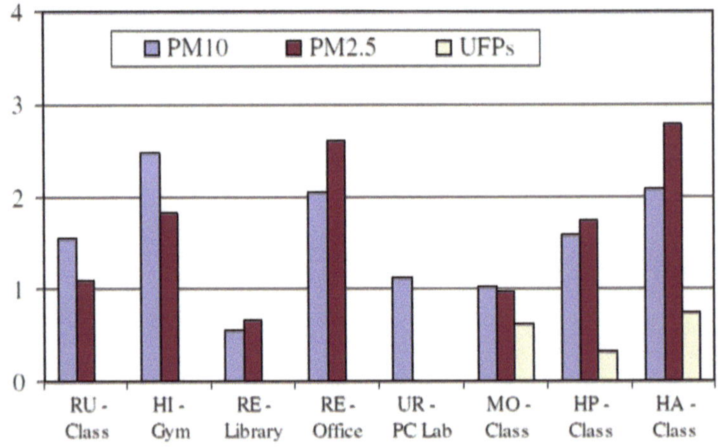

Fig. 1.8 Mean indoor/outdoor ratio of PM10, PM2.5 and UFPs in a rural area (RU), two blocks away from a major highway (*HI*), a residential area (*RE*), a heavy-trafficked neighborhood in the center of Athens (*UR*), a residential area close to a major motorway (*MO*), a densely populated area close to a major motorway (*HP*) and at the harbor of Athens (*HA*) (Diapouli et al. 2008)

found a significant difference between I/O ratios calculated for the classrooms with and without the presence of carpets, in particular the authors reported a mean I/O ratio equal to 2.63 in the classrooms where carpets were present and mean I/O ratio equal to 1.03 in the classroom where tiles or linoleum floor coverage were present. The more recent study conducted by the same authors in 2008 (Stranger et al. 2008), focused the attention on the chemical composition of PM collected in 27 primary schools in the urban and suburban areas of Antwerp (Belgium). The authors showed the elemental composition of indoor particulate matter (PM2.5) determined in classroom were different than that evaluated in outdoor air. In particular the authors found that the elements such S, Si, Fe, Ca, and Al were the highest contributions to local outdoor PM while Ca, S, Si, Cl and Al to indoor PM in schools. In particular the authors suggested that Cl could derive by detergents used for cleaning activities inside the classroom while Ca concentrations could probably be determined by the chalk (mainly $CaSO_4$) used on the blackboards and/or the gypsum walls and plasters used as construction materials. Moreover, the I/O ratios reported for elements typically constituting dust particles were higher than 1 probably due to re-suspension of dust because of room occupation. Fromme et al. (2008) also reported the composition of the classrooms' PM (e.g., high calcium concentrations) and the findings obtained applying scanning electron microscopy (SEM) and energy dispersive microanalysis (EDX) on PM filter and found that the indoor PM consists mainly of earth crustal materials, detrition of the building materials and chalk. Therefore, the authors suggested that increase of PM10 concentrations in classrooms were due to a physical activity of the pupils and to resuspension of mainly indoor coarse particles, concluding that indoor-generated PM could be less toxic compared to PM in outdoor air.

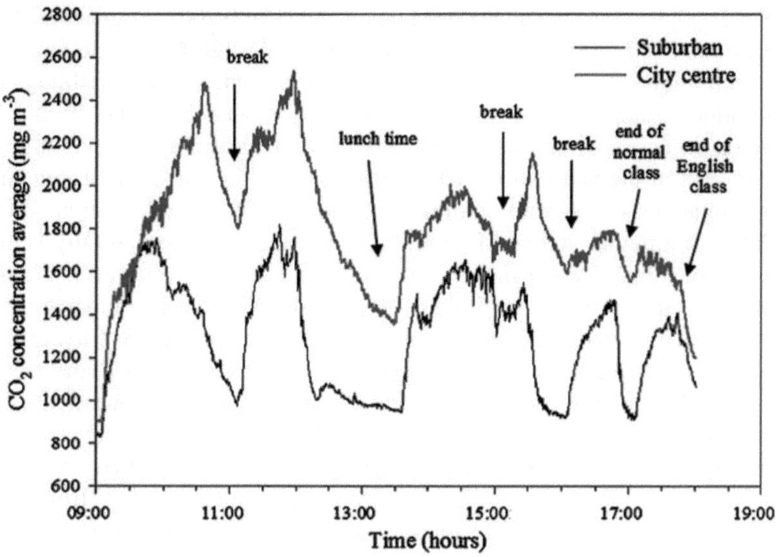

Fig. 1.9 Classroom CO_2 concentrations during a typical occupation period (Pegas et al. 2012)

1.5.3 Carbon Dioxide (CO_2)

Many studies attested that indoor CO_2 concentrations classrooms were primarily influenced by human occupancy (Lee and Chang 1999, 2000; Grimsrud et al. 2006; Van Dijken et al. 2006; Fromme et al. 2007; Theodosiou and Ordoumpozanis 2008; Mumovic et al. 2009). In fact, as shown in Fig. 1.9, lower CO_2 concentrations were determined when the classrooms were un-occupied, during time-breaks and when adequate ventilation was ensured (Fromme et al. 2007; Yang et al. 2009; De Giuli et al. 2012; Pegas et al. 2012). On the contrary higher concentrations were observed when students started physical activities inside the classrooms (Almeida et al. 2011). Moreover, Park et al. (2011) showed higher CO_2 concentrations in winter because the classrooms were not ventilated well in this season respect of summer. In Daisey et al. (2003) the authors studied the influence of ventilation on the IAQ in schools of United States; they found that indoor CO_2 concentrations above 1,000 ppm, which is the ASHRAE Standard 62.1 (1999) were generally regarded as indicative of inadequate ventilation rates, that were unacceptable with respect to body odors (Seppanen et al. 1999; Apte et al. 2000).

1.5.4 Ozone (O_3)

Several studies found that I/O ratios for O_3 at schools were smaller than 1 (in the range 0.13–0.8), suggesting that this pollutant mainly came from outdoor (Gold et al. 1996;

Weschler 2000; Blondeau et al. 2005; Mendell and Health 2005; Poupard et al. 2005; Stranger et al. 2007, 2008; Mejía et al. 2011). Weschler (2000) reported that indoor O_3 concentrations are related to seasonal, temporal and spatial variability. Under normal indoor conditions, the half-life time of O_3 is between 7 and 10 min and is determined primarily by air exchange and removal processes at air/solid interface.

The impact of the ventilation on O_3 concentrations in school environments was also investigated by numerous studies (Gold et al. 1996; Blondeau et al. 2005; Mi et al. 2006; Sohn et al. 2012). Mi et al. (2006) investigated 10 naturally ventilated schools in Shanghai finding that mean weekly indoor O_3 level was 1–9 μg/m^3 in spite of 17–28 μg/m^3 outdoors. Gold et al. (1996) conducted a comparison of outdoor and classroom ozone exposures for school children in Mexico City, evaluating the effect of three distinct classroom conditions: (a) windows/doors open, air cleaner off; (b) windows/doors closed, air cleaner off; (c) windows/doors closed, air cleaner on. They found that, when outdoor O_3 concentrations increased, indoor concentrations increased more rapidly when windows/doors were open. These results suggested that O_3 in indoor environments mostly comes from outdoor sources and so the air exchange rate plays an important role. Moreover, higher indoor O_3 concentrations were found in schools located in areas affected by industrial or urban pollution (Gonzales et al. 2005; Mi et al. 2006; Mejía et al. 2011).

Blondeau's study (2005), performed in eight school buildings in La Rochelle (France) and its suburbs, confirmed that indoor O_3 concentrations were strongly dependent on building air-tightness (the more airtight the building envelope, the lower the I/O ratio). The infiltrated O_3 undergoes fast homogeneous reactions and deposits onto material surfaces (e.g. textiles and brick) in indoor environments (Grøntoft and Raychaudhuri 2004). Given that O_3 is known to be highly reactive (Cano-Ruiz et al. 1993; Weschler 2000; Grontoft 2002), there is evidence that low I/O are determined from the deposition process on solid surfaces or the decomposition in the indoor air.

1.5.5 Other Inorganic Gases: Nitrogen Oxides (NO_x), Carbon Oxide (CO), Sulphur Dioxide (SO_2)

NO_x determination in French schools showed that vehicular exhaust emission from nearby traffic was the most important contribute to indoor concentrations. In fact I/O ratios of NO_2 calculated by Blondeau et al. (2005) and Poupard et al. (2005) varied in a narrow range from 0.88 to 1. In fact indoor and outdoor NO_2 concentrations were positively correlated since indoor concentrations reflected the outdoor ones despite varying of building air-tightness. On the contrary, I/O of NO lied in a wider range ($0.5 < I/O < 1$) and there was no apparent correlation with the air-tightness of the buildings. Therefore, the authors suggested that this evidence was probably related to differences in the contribution of indoor homogeneous and heterogeneous reactions that NO undergoes. Similar considerations were elaborated by

Stranger et al. (2008) in Belgium, by Pegas et al. (2012) in Lisbon and by Lee and Chang (2000) in Hong Kong.

Gul et al. (2011), confirming the results reported in previous study, showed that I/O ratios for NO_2 at high schools located in Eskisehir (Turkey) were >1 in dining hall or teacher's room where cooking and smoking activities took place ($1.8 < I/O < 3$). Moreover, Sohn et al. (2012) studied the relationship between NO_2 concentrations with indoor ventilation rate and showed that a direct correlation existed.

Carbon Oxide (CO) is a pollutant purely outdoor, therefore few studies have been conducted to evaluate the incidence of carbon monoxide on indoor environments. Chaloulakoua and Mavroidisb (2002) measured indoor and outdoor CO concentrations at a school near the center city of Athens. Authors found that the indoor and outdoor diurnal concentration cycles followed similar patterns and indoor concentrations showed a mild and slightly delayed response respect to outdoor concentration changes. In addition, they observed that CO concentrations measured during winter were higher than the respective concentrations measured during summer (3.96 ppm and 1.92 ppm, respectively). These results were linked to the higher traffic volume and to winter meteorological conditions that favor the accumulation of pollutants. Similar results and considerations were obtained by Chithra and Shiva Nagendra (2012) in a study conducted in a school building located close to an urban roadway in India. Finally Yang et al. (2009) showed that renovation works had negative effects on the IAQ in fact, significantly higher concentrations of CO were registered at schools constructed within 1 year (1.03 ppm) respect to those built in previous years (0.59 ppm). These results might be caused by the new electric heating systems.

SO_2 is the less investigated pollutant for the evaluation of the indoor air quality in schools. Several studies reported that mean indoor SO_2 concentrations were about 20–30 % of those outdoors confirming that SO_2 mainly come from outdoors (Lee and Chang 2000; Meininghaus et al. 2003; Stranger et al. 2008). This result was confirmed by Ashmore and Dimitroulopoulou (2009) that found higher concentrations inside school buildings near open coal fires. Finally Spedding (1974) suggested that lower SO_2 indoor concentrations could be linked with the capacity of indoor materials to absorb SO_2. Among the wide variety of materials, the emulsion paints were identified as the most important sink for SO_2.

1.6 Conclusion

The main goal of this chapter was to summarize remarkable findings about IAQ inside school buildings. More specifically chemical pollutants, related sources and monitoring methodologies were reported. The outcomes provide suggestive evidence that certain conditions, commonly found in schools, can have adverse effects on the IAQ and therefore on occupant's health. In particular it was highlighted that the location, the age and air-tightness of school buildings, the room design, the ventilation rate, the building and furnishing materials, the occupant's activities and outdoor pollution play an important role on the indoor pollutants concentrations. Therefore in order to safeguard the health of the occupants and in particular of children that are more sensitive to environmental pollutants some good practices should be followed. These actions include the construction of school buildings equipped with adequate systems for air exchange as well as the use of low emitting building and furniture materials. Indoor concentrations of many pollutants are strongly influenced by outdoors. Therefore it is important that schools are not located in areas affected by high traffic or industrial pollution in order to improve IAQ and reduce the impact on students health. At this regard, several States are nowadays working to define Guidelines for suggesting best practices in order to improve IAQ inside school buildings, for defining reference values and for regulating the control methodologies. This need arises from the lack of available reference values for most of the pollutants monitored in indoor environments (WHO 2010).

Acknowledgement The authors would like to thank the Apulia Region for funding the project "VOC and ODOR: evaluation of volatile organic compounds emissions and odour impacts produced by new plants and new technologies for the productive systems" admitted to the financing under the Announcement for the constitution of research laboratories networks (2009–2012).

References

AgBB (2010) Ausschuss zur gesundheitlichen Bewertung von Bauprodukten. Bewertungsschema fur VOC aus Bauprodukten. Committee for Health-related Evaluation of Building Products. Health-related evaluation procedure for volatile organic compounds emissions (VOC and SVOC) from building products. Available at: http://www.umweltbundesamt.de/produkte-e/bauprodukte/archive/AgBB-Evaluation-Scheme2010.pdf. Accessed 20 Nov 2012

Ajiboye P, White M, Graves H, Ross D (2006) Building research establishment for office of the Deputy Prime Minister: London. Ventilation and indoor air quality in schools – guidance report 202825

Alm S (1999) Personal exposures of pre-school children to carbon monoxide and nitrogen dioxide in urban air. Dissertation, National Public Health Institute A19/99, Kuopio, Finland

Alm S, Reponen A, Mukala K, Pasanen P, Tuomisto J, Jantunen MJ (1994) Personal exposures of preschool children to carbon monoxide: roles of ambient air quality and gas stoves. Atmos Environ 28:3577–3588

Almeida SM, Canha N, Silva A, Do Carmo Freitas M, Pegas P, Alves C, Evtyugina M, Pio CA (2011) Children exposure to atmospheric particles in indoor of Lisbon primary schools. Atmos Environ 45(40):7594–7599

Angiuli L, Bruno P, Caputi M, Caselli M, De Gennaro G, De Rienzo M (2003) Radial passive samplers for air quality monitoring in field comparison with a BTEX automatic analyser preliminary results. Fresenius Environ Bull 12(10):1167–1172

Appatova AS, Ryan PH, LeMasters GK, Grinshpun SA (2008) Proximal exposure of public schools and students to major roadways: a nationwide U.S. survey. J Environ Plan Manag 51:631–646

Apte MG, Fisk WJ, Daisey JM (2000) Associations between indoor (CO_2) concentrations and sick building syndrome symptoms in US Office Buildings: an analysis of the 1994–1996 BASE Study Data (LBNL 44385). Indoor Air 10:246–257

Arditsoglou A, Samara C (2005) Levels of total suspended particulate matter and major and trace elements in Kosovo: a source identification and apportionment study. Chemosphere 59:669–678

Asher MI, Keil U, Anderson HR, Beasley R, Crane J, Martinez F, Mitchell EA, Pearce N, Sibbald B, Stewart AW, Strachan D, Weiland SK, Williams HC (1995) International study of asthma and allergies in childhood (ISAAC): rationale and methods. Eur Respir J 8(3):483–491. doi:10.1183/09031936.95.08030483

Ashmore MR, Dimitroulopoulou C (2009) Personal exposure of children to air pollution. Atmos Environ 43:128–141. doi:10.1016/j.atmosenv.2008.09.024

ASHRAE Standard 62 (1989) American Society of Heating, Refrigerating and Air-conditioning Engineers. Ventilation for acceptable indoor air quality. ASHRAE Standard 62, Atlanta

ASHRAE Standard 62.1 (1999) American Society of Heating, Refrigerating and Air-conditioning Engineers. Ventilation for acceptable indoor air quality. ASHRAE Standard 62, Atlanta

Barro R, Regueiro J, Llompart M, Garcia-Jares C (2009) Analysis of industrial contaminants in indoor air: Part 1. Volatile organic compounds, carbonyl compounds, polycyclic aromatic hydrocarbons and polychlorinated biphenyls. J Chromatogr A 1216:540–566

Bartlett KH, Kennedy SM, Brauer M, Netten CV, Dill B (2004) Evaluation and determinants of airborne bacterial concentrations in school classrooms. J Occup Environ Hyg 1:639–647

Basu R, Samet JM (1999) A review of the epidemiological evidence on health effects of nitrogen dioxide exposure from gas stoves. J Environ Med 1:173–187. doi:10.1002/jem.28

Bem H, Bem EM, Krawczyk J, Płotek M, Janiak S, Mazurek D (2013) Radon concentrations in kindergartens and schools in two cities: Kalisz and Ostrow Wielkopolski in Poland. J Radioanal Nucl Chem 295(3):2229–2232

Bertoni G, Ciuchini C, Pasini A, Tappa R (2002) Monitoring of ambient BTX at Monterotondo (Rome) and indoor-outdoor evaluation in school and domestic sites. J Environ Monit 4:903–909. doi:10.1039/b206959h

Blondeau P, Iordache V, Poupard O, Genin D, Allard F (2005) Relationship between outdoor and indoor air quality in eight French schools. Indoor Air 15:2–12. doi:10.1111/j.1600-0668.2004.00263.x

Branis M, Safránek J (2011) Characterization of coarse particulate matter in school gyms. Environ Res 111:485–491

Branis M, Safránek J, Hytychová A (2009) Exposure of children to airborne particulate matter of different size fractions during indoor physical education at school. Build Environ 44:1246–1252. doi:10.1016/j.buildenv.2008.09.010

Brown S, Sim M, Abramson M, Gray C (1994) Concentrations of volatile organic compounds in indoor air – a review. Indoor Air 4:123–134

Bruno P, Caputi M, Caselli M, de Gennaro G, de Rienzo M (2005) Reliability of a BTEX radial diffusive sampler for thermal desorption: field measurements. Atmos Environ 39:1347–1355

Bruno P, Caselli M, de Gennaro G, Iacobellis S, Tutino M (2008) Monitoring of volatile organic compounds in non-residential indoor environments. Indoor Air 18:250–256

Burnett RT, Dales RE, Brooks JR, Raizenne ME, Krewski D (1997) Association between ambient carbon monoxide levels and hospitalizations for congestive heart failure in the elderly in 10 Canadian cities. Epidemiology 8:162–167

Burnett RT, Cakmak S, Raizenne ME, Stieb D, Vincent R, Krewski D, Brooks JR, Philips O, Ozkayanak H (1998) The association between ambient carbon monoxide levels and daily mortality in Toronto, Canada. J Air Waste Manag Assoc 48:689–700

Burton BT (1997) Volatile organic compounds. In: Bardana EJ, Montanaro A (eds) Indoor air pollution and health. Marcel Dekker, New York, pp 127–153

Cano-Ruiz JA, Kong D, Balas RB, Nazaroff WW (1993) Removal of reactive gases at indoor surfaces: combining mass transport and surface kinetics. Atmos Environ 27A:2039–2050

Chaloulakoua A, Mavroidisb I (2002) Comparison of indoor and outdoor concentrations of CO at a public school. Evaluation of an indoor air quality model. Atmos Environ 36:1769–1781

Chithra VS, Shiva Nagendra SM (2012) Indoor air quality investigations in a naturally ventilated school building located close to an urban roadway in Chennai, India. Build Environ 54:159–167. doi:10.1016/j.buildenv.2012.01.016

Clausen PA, Wilkins CK, Wolkoff P, Nielsen GD (2001) Chemical and biological evaluation of a reaction mixture of R-(+)- limonene/ozone—formation of strong airway irritants. Environ Int 26(7–8):511–522

Clayton CA, Perritt RL, Pellizzari ED, Thomas KW, Whitmore RW, Wallace LA, Özkaynak H, Spengler JD (1993) Particle total exposure assessment methodology (PTEAM) study: distributions of aerosol and elemental concentrations in personal, indoor, and outdoor air samples in a southern California community. J Exp Anal Environ Epidemiol 3:227–250

Clements-Croome D (2006) Ventilation rates in schools. Build Environ 43(3):362–367

Clouvas A, Xanthos S (2012) A comparison study between radon concentration in schools and other workplaces. Radiat Prot Dosimetry 149(2):207–210

Corgnati SP, Filippi M, Viazzo S (2007) Perception of the thermal environment in high school and university classrooms: subjective preferences and thermal comfort. Build Environ 42:951–959

Coward S, Llewellyn J, Raw G, Brown V, Crump D, Ross D (2001) Indoor air quality in homes in England. BRE report BR 433. CRC Ltd, Watford, UK

Croome DJC, Awbi HB, Biro ZB, Kochhar N, Williams M (2008) Ventilation rates in schools. Build Environ 43:362–367

Daisey JM, Angell WJ, Apte MG (2003) Indoor air quality, ventilation and health symptoms in schools: an analysis of existing information. Indoor Air 13:53–64

Damek-Poprawa M, Sawicka-Kapusta K (2003) Damage to the liver, kidney, and testis with reference to burden of heavy metals in yellow-necked mice from areas around steelworks and zinc smelters in Poland. Toxicology 186:1–10

Daresta BE, Liuzzi VC, de Gennaro G, De Giorgi C, De Luca F, Caselli M (2010) Evaluation of the toxicity of PAH mixtures and organic extract from Apulian particulate matter by the model system 'Caenorhabditis elegans'. Fresen Environ Bull 19:2002–2005

De Giuli V, Da Pos O, De Carli M (2012) Indoor environmental quality and pupil perception in Italian primary schools. Build Environ 56:335–345. doi:10.1016/j.buildenv.2012.03.024

Destaillats H, Maddalena RL, Singer BC, Hodgson AT, McKone TE (2008) Indoor pollutants emitted by office equipment. A review of reported data and information needs. Atmos Environ 42:1371–1388

Diapouli E, Chaloulakou A, Mihalopoulos N, Spyrellis N (2008) Indoor and outdoor PM mass and number concentrations at schools in the Athens area. Environ Monit Assess 136:13–20. doi:10.1007/s10661-007-9724-0

Dijken FV, Bronswijk JV, Sundell J (2005) Indoor environment in Dutch primary schools and health of the pupils. In: Proceedings of indoor air, Beijing, China, vol 1, p 623

Dimitroulopoulou C, Crump D, Coward SKD, Brown B, Squire R, Mann H, White M, Pierce B, Ross D (2005) Ventilation, air tightness and indoor air quality in new homes. BR477. Building Research Establishment Press, Garston/Watford/Hertfordshire

ECA (1997) European collaborative action. Indoor air quality and its impact on man. Report N. 19

Ekmekcioglu D, Keskin SS (2007) Characterization of indoor air particulate matter in selected elementary schools in Istanbul, Turkey. Indoor Built Environ 16:169–176. doi:10.1177/14203 26X07076777

Englert N (2004) Fine particles and human health – a review of epidemiological studies. Toxicol Lett 149:235–242

EPA Method TO-15 (1999) Compendium of methods for the determination of toxic organic compounds in ambient air- second edition compendium method TO-15- determination of volatile organic compounds (VOCs) in air collected in specially prepared canisters and analyzed by gas chromatography/mass spectrometry (GC/MS) U.S. Environmental Protection Agency Cincinnati, OH 45268

European Collaborative Action (1994) Indoor air quality & its impact on man – environment and quality of life – report no 14: sampling strategies for volatile organic compounds (VOCs) in indoor air. Available at: http://www.inive.org/medias/ECA/ECA_Report14.pdf. Accessed 20 Nov 2012

Fan Z, Lioy P, Weschler CJ, Fiedler N, Kipen H, Zhang J (2003) Ozone-initiated reactions with mixtures of volatile organic compounds under simulated indoor conditions. Environ Sci Tech 37:1811–1821

Faustman EM, Silbernagel SM, Fenske RA, Burbacher TM, Ponce RA (2000) Mechanisms underlying children's susceptibility to environmental toxicants. Environ Health Perspect 108(Suppl 1):13–21

Forbes L, Patel MD, Rudnicka AR, Cook DG, Bush T, Stedman JR, Strachan DP, Anderson HR (2009) Chronic exposure to outdoor air pollution and diagnosed cardiovascular disease: meta-analysis of three large cross-sectional surveys. Environ Health 8(30):1–9

Fraga S, Ramos E, Martins A, Samúdio MJ, Silva G, Guedes J, Oliveira Fernandes E, Barros H (2008) Indoor air quality and respiratory symptoms in Porto Schools. Rev Port Pneumol XIV(4):487–507

Fromme GH, Lahrz T, Hainisch A, Oddoy A, Piloty M, Rude H (2005) Elemental carbon and respirable particulate matter in the indoor air of apartments and nursery schools and ambient air in Berlin (Germany). Indoor Air 15:335–341

Fromme H, Twardella D, Dietrich S, Heitmann D, Schierl R, Liebl B, Ruden H (2007) Particulate matter in the indoor air of classrooms-exploratory results from Munich and surrounding area. Atmos Environ 41:854–866. doi:10.1016/j.atmosenv.2006.08.053

Fromme H, Diemer J, Dietrich S, Cyrys J, Heinrich J, Lang W, Kiranoglua M, Twardella D (2008) Chemical and morphological properties of particulate matter (PM10, PM2.5) in school classrooms and outdoor air. Atmos Environ 42:6597–6605. doi:10.1016/j.atmosenv.2008.04.047

Godwin C, Batterman S (2007) Indoor air quality in Michigan schools. Indoor Air 17:109–121

Gold DR, Allen G, Damokosh A, Serrano P, Hayes C, Castiilejos M (1996) Comparison of outdoor and classroom ozone exposures for school children in Mexico city. J Air Waste Manag Assoc 46:335–342

González A, Alcalá R, Casillas J, Cordón O, Herrera F (2005) A genetic rule weighting and selection process for fuzzy control of heating, ventilating and air conditioning systems. Eng Appl Artif Intel 18(3):279–296

Goyal R, Khare M (2011) Indoor air quality modeling for PM10, PM2.5, and PM1.0 in naturally ventilated classrooms of an urban Indian school building. Environ Monit Assess 176:501–516. doi:10.1007/s10661-010-1600-7

Green RS, Smorodinsky S, Kim JJ, McLaughlin R, Ostro B (2004) Proximity of California public schools to busy roads. Environ Health Perspect 112:61–66

Grimsrud D, Bridges B, Schulte R (2006) Continuous measurements of air quality parameters in schools. Build Res Inform 34:447–458

Grontoft T (2002) Dry deposition of ozone on building materials. Chamber measurements and modelling of the time-dependent deposition. Atmos Environ 36:5661–5670

Grøntoft T, Raychaudhuri MR (2004) Compilation of tables of surface deposition velocities of O3, NO2 and SO2 to a range of indoor surfaces. Atmos Environ 38(4):533–544

Gul H, Gaga EO, Döğeroğlu T, Özden O, Ayvaz O, Özel S, Güngör G (2011) Respiratory health symptoms among students exposed to different levels of Air pollution in a Turkish city. Int J Environ Res Public Health 8:1110–1125. doi:10.3390/ijerph8041110

Guo YL, Lin YC, Sung FC, Huang SL, Ko YC, Lei JS, Su HJ, Shaw CK, Lin RS, Dockery DW (1999) Climate, traffic-related air pollutants, and asthma prevalence in middle-school children in Taiwan. Environ Health Perspect 107:1001–1006

Guo H, Lee SC, Chan LY, Li WM (2004) Risk assessment of exposure to volatile organic compounds in different indoor environments. Environ Res 94:57–66

Guo H, Morawska L, He C, Gilbert D (2008) Impact of ventilation scenario on air exchange ratesand on indoor particle number concentrations in an air-conditioned classroom. Atmos Environ 42:757–768

Hanoune B, LeBris T, Allou L, Marchand C, Calve LS (2006) Formaldehyde measurements in libraries: comparison between Infrared Diode Laser Spectroscopy and a DNPH-derivatization method. Atmos Environ 40:5768–5775

Health Canada (1995) Exposure guidelines for residential indoor air quality. Report of the Federal-Provincial Advisory Committee on environmental and occupational health. Health Canada Publications – Communications, Ottawa, ON. Available at: http://www.hc-sc.gc.ca/ewh-semt/alt_formats/hecs-sesc/pdf/pubs/air/exposure-exposition/exposure-exposition-eng.pdf. Accessed 20 Nov 2012

Heudorf U, Neitzert V, Spark J (2009) Particulate matter and carbon dioxide in classrooms – the impact of cleaning and ventilation. Int J Hyg Environ Health 212:45–55

Hochstetler HA, Yermakov M, Reponen T, Ryan PH, Grinshpun SA (2011) Aerosol particles generated by diesel-powered school buses at urban schools as a source of children's exposure. Atmos Environ 45:1444–1453. doi:10.1016/j.atmosenv.2010.12.018

Hodgson AT (1995) A review and a limited comparison of methods for measuring total volatile organic compounds in indoor Air. Indoor Air 5(4):247–257

Hodgson AT, Levin H (2003) Volatile organic compounds in indoor air: a review of concentrations measured in North America since 1990. Report LBNL-51715. Berkeley National Laboratory, Lawrence. Available at: http://energy.lbl.gov/ie/pdf/LBNL-51715.pdf. Accessed 20 Nov 2012

Hodgson MJ, Frohlinger J, Permar E, Tidwell C, Traven ND, Olenchock SA, Karpf M (1991) Symptoms and micro-environmental measures in non-problem buildings. J Occup Med 33(4):527–533

Holguin F, Flores S, Ross Z, Cortez M, Molina M, Molina L, Rincon C, Jerrett M, Berhane K, Granados A, Romieu I (2007) Traffic-related exposures, airway function, inflammation and respiratory symptoms in children. Am J Respir Crit Care Med 176:1236–1242. doi:10.1164/rccm.200611-1616OC

Hong H, Cheng LS, Jun-Ji C, Chang-Wei Z, Xin-Geng C, Shao-Jia F (2007) Characteristics of indoor/outdoor PM2.5 and elemental components in generic urban, roadside and industrial plant areas of Guangzhou City China. J Environ Sci 19:35–43

Huang YC, Ghio AJ (2006) Vascular effects of ambient pollutant particles and metals. Curr Vasc Pharmacol 4:199–208

Hubbell BJ, Hallberg A, McCubbin DR, Post E (2005) Health related benefits of attaining the 8-hr ozone standard. Environ Health Perspect 113:73–82

Hulin M, Annesi-Maesano I, Caillaud D (2011) Indoor air quality at school and allergy and asthma among schoolchildren. Differences between rural and urban areas. Rev Fr D'allergol 51:419–424. doi:10.1016/j.reval.2011.01.002

IARC (2011) Monographs on the evaluation of Carcinogenicity to Humans, volume 100 F, Formaldehyde, 2012) U.S. Department of Health and Human Services Public Health Service, National Toxicology Program, Report on Carcinogens, 12th edn

IARC (2012) A review of human carcinogens: chemical agents and related occupations, vol 100F, IARC monographs. IARC, Lyon

Norback D, Wieslander G, Edling C (1995) Occupational exposure to volatile organic compounds (VOCs), and other air pollutants from the indoor application of water-based paints. Ann Occup Hyg 39(6):783–794

ISO 16000-3:2011. Indoor air-part 3: determination of formaldehyde and other carbonyl compounds in indoor air and test chamber air-active sampling method

ISO 16000-6:2011. Indoor air-part 6: determination of volatile organic compounds in indoor and test chamber air by active sampling on Tenax TA sorbent, thermal desorption and gas chromatography using MS/FID

Janson C, Anto J, Burney P, Chinn S, de Marco R, Heinrich J, Jarvis D, Kuenzli N, Leynaert B, Luczynska C, Neukirch F, Svanes C, Sunyer J, Wjst M (2001) The European community respiratory health survey: what are the main results so far ? European Community Respiratory Health Survey II. Eur Respir J 18(3):598–611

Janssen NAH, Hoek G, Harssema H, Brunekreef B (1997) Childhood exposure to PM10: relation between personal, classroom, and outdoor concentrations. Occup Environ Med 54(12):888–894

Janssen NAH, van Vliet PHN, Aarts F, Harssema H, Brunekreef B (2001) Assessment of exposure to traffic related air pollution of children attending schools near motorways. Atmos Environ 35:3875–3884

Janssen NA, Brunekreef B, van Vliet P, Aarts F, Meliefste K, Harssema H, Fischer P (2007) The relationship between air pollution from heavy traffic and allergic sensitization, bronchial hyperresponsiveness, and respiratory symptoms in Dutch schoolchildren. Environ Health Perspect 111:1512–1518. doi:10.1289/ehp.6243

Jo WK, Kim JD (2010) Personal exposure of graduate students attending the college of natural sciences or social sciences to volatile organic compounds on campus. Chemosphere 81:1272–1279

Jones AP (1999) Indoor air quality and health. Atmos Environ 33(28):4535–4564

Kim JJ, Smorodinsky S, Lipsett M, Singer BC, Hodgson AT, Ostro B (2004) Traffic-related air pollution near busy roads: the East Bay Children's Respiratory Health Study. Am J Respir Crit Care Med 170(5):520–526. doi:10.1164/rccm.200403-281OC

Kotzias D (2005) Indoor air and human exposure assessment – needs and approaches. Exp Toxicol Pathol 57:5–7

Kotzias D, Koistinen K, Kephalopoulos S, Schlitt C, Carrer P, Maroni M, Jantunen M, Cochet C, Kirchner S, Lindvall T, McLaughlin J, Mølhave L, de Oliveira Fernandes E, Seifert B (2005) The INDEX project: critical appraisal of the setting and implementation of indoor exposure limits in the EU. Final report, EUR 21590 EN

Koutrakis P, Briggs SLK, Leaderer BP (1992) Source apportionment of indoor aerosols in Suffolk and Onondaga Counties, New York. Environ Sci Tech 23:521–527

Lahrz T (2006) Innenraumluftqualitat in Berliner Schulen – Feinstaub und Adsorbierte Substanzen. In: Bayerisches Landesamt fur Gesundheit und Lebensmittelsicherheit: Aktuelle umweltmedizinische Probleme in Innenraumen, Teil 1. Band 13 der Schriftenreihe: Materialien zur Umweltmedizin. Erlangen, pp 88–113

Lahrz T, Piloty M (2005) Innenraumluftqualitat in Berliner Schulen – Feinstaub und adsorbierte Substanzen. Bericht im Auftrag des Landesamtes fur Arbeitsschutz, Gesund-heitsschutz und technische Sicherheit, Berlin (LAGetSi)

Lammel G, Novák J, Landlová L, Dvorská A, Klánová J, Èupr P, Kohoutek J, Reimer E, Škrdlíková L (2010) Sources and distributions of polycyclic aromatic hydrocarbons and toxicity of polluted atmosphere aerosols. In: Zereini F, Wiseman CLS (eds) Urban airborne particulate matter: origins, chemistry. Fate and health impact. Springer, Berlin, pp 39–62

Lee SC, Chang M (1999) Indoor Air quality investigations at five classrooms. Indoor Air 9:134–138

Lee SC, Chang M (2000) Indoor and outdoor air quality investigation at schools in Hong Kong. Chemosphere 41:109–113

Lee K, Parkhurst WJ, Xue J, Ozkaynak A, Neuberg D, Spengler JD (2004) Outdoor/indoor/personal ozone exposures of children in Nashville, Tennessee. Air Waste Manag Assoc 54:352–359

Leovic KW, Sheldon LS, Whitaker DA, Hetes RG, Calcagni JA, Baskir JN (1996) Measurement of indoor air emissions from dry-process photocopy machines. J Air Waste Manag Assoc 46(9):821–829

Li TH, Turpin BJ, Shields HC, Weschler CJ (2002) Indoor hydrogen peroxide derived from ozone/d-limonene reactions. Environ Sci Tech 36:3295–3302

Lin JJ, Lee LC (2004) Characterization of the concentration and distribution of urban submicron (PM1) aerosol particle. Atmos Environ 38:469–475

Liu Y, Chen R, Shen X, Mao X (2004) Wintertime indoor air levels of PM10, PM2.5 and PM1 at public places and their contributions to TSP. Environ Int 30:189–197

Liuzzi VC, Daresta BE, de Gennaro G, De Giorgi C (2011) Different effects of polycyclic aromatic hydrocarbons in artificial and in environmental mixtures on the free living nematode C. elegans. J Appl Toxicol 32:45–50

Long CM, Suh HH, Koutrakis P (2000) Characterization of indoor particle sources using continuous mass and size monitors. J Air Waste Manag Assoc 50:1236–1250

MacIntosh DL, Minegishi T, Fragala MA, Allen JG, Coghlan KM, Stewart JH, McCarthy JF (2012) Mitigation of building-related polychlorinated biphenyls in indoor air of a school. Environ Health 11:24. doi:10.1186/1476-069X-11-24

Meininghaus R, Kouniali A, Mandin C, Cicolella A (2003) Risk assessment of sensory irritants in indoor air a case study in a French school. Environ Int 28:553–557

Mejía JF, Low Choy S, Mengersen K, Morawska L (2011) Methodology for assessing exposure and impacts of air pollutants in school children: data collection, analysis and health effects – a literature review. Atmos Environ 45:813–823. doi:10.1016/j.atmosenv.2010.11.009

Mendell JM, Health GA (2005) Do indoor pollutants and thermal conditions in schools influence student performance? A critical review of the literature. Indoor Air 15:27–52

Mentese S, Gullu G (2006) Variations and sources of formaldehyde levels in residential indoor air in Ankara, Turkey. Indoor Built Environ 15:273–281

Mi YH, Norback D, Tao J, Mi YL, Ferm M (2006) Current asthma and respiratory symptoms among pupils in Shanghai, China: influence of building ventilation, nitrogen dioxide, ozone, and formaldehyde in classrooms. Indoor Air 16(6):454–464. doi:10.1111/j.1600-0668.2006.00439.x

Mølhave L, Bach B, Pedersen OF (1986) Human reactions to low concentrations of volatile organic compounds. Environ Int 12(1–4):167–175

Molnar P, Bellander T, Sallsten G, Boman J (2007) Indoor and outdoor concentrations of PM2.5 trace elements at homes, preschools and schools in Stockholm, Sweden. J Environ Monit 9:348–357

Morgenstern V, Zutavern A, Cyrys J, Brockow I, Gehring U, Koletzko S, Bauer CP, Reinhardt D, Wichmann HE, Heinrich J (2007) Respiratory health and individual estimated exposure to traffic-related air pollutants in a cohort of young children. Occup Environ Med 64(1):8–16. doi:10.1136/oem.2006.028241

Mullen NA, Bhangar S, Hering SV, Kreisberg NM, Nazaroff WW (2011) Ultrafine particle concentrations and exposures in six elementary school classrooms in northern California. Indoor Air 21:77–87. doi:10.1111/j.1600-0668.2010.00690.x

Mumovic D, Palmer J, Davies M, Orme M, Ridley I, Oreszczyn T, Jud C, Critchlow R, Medina HA, Pilmoor G, Pearson C, Way P (2009) Winter indoor air quality, thermal comfort and acoustic performance of newly built secondary schools in England. Build Environ 44:1466–1477. doi:10.1016/j.buildenv.2008.06.014

Nadadur SS, Miller A, Hopke PK, Gordon T, Vedal S, Vandenberg JJ, Costa DL (2007) The complexities of air pollution regulation: the need for an integrated research and regulatory perspective. Toxicol Sci 100:318–327

Namdeo A, Bell MC (2005) Characteristics and health implications of fine and coarse particulates at roadside, urban background and rural sites in UK. Environ Int 31:565–573

NIOSH/IPCS (2004) International chemical safety cards: formaldehyde

Norback D, Torgen M, Edling C (1990) Volatile organic compounds, respirable dust, and personal factors related to prevalence and incidence of sick building syndrome in primary schools. Br J Ind Med 47:733–741

Oeder S, Dietrich S, Weichenmeier I, Schober W, Pusch G, Jörres RA, Schierl R, Nowak D, Fromme H, Behrendt H, Buters J (2012) Toxicity and elemental composition of particulate matter from outdoor and indoor air of elementary schools in Munich, Germany. Indoor Air 22:148–158

OEHHA (2008) Office of environmental health hazard assessment, Sacramento, CA. Air toxics hot spots, risk assessment guidelines, technical support document for the derivation of noncancer reference exposure levels. Appendix D. Available at: http://oehha.ca.gov/air/hot_spots/2008/NoncancerTSD_final.pdf. Accessed 20 Nov 2012.

Oie L, Nafstad P, Botten G (1993) Infants exposure to NO_2. In: Proceedings of indoor air '93, Helsinki, Finland, vol 3, p 183

Ott WR, Steinemann AC, Wallace LA (2007) Exposure analysis. CRC Press Taylor and Francis Group, New York

Ozkaynak H, Xue J, Spengler JD, Wallace LA, Pellizzari ED, Jenkins P (1996) Personal exposure to airborne particles and metals: results from the particle TEAM study in Riverside, California. J Expo Anal Environ Epidemiol 6:57–78

Park JC, Chung MH, Rhee EK (2011) Field survey on the indoor environment of elementary schools for planning of environment friendly school facilities. J Asian Archit Build Eng 10(2):461–468

Pang SW (1994) Interim indoor air quality guidelines for Hong Kong, EP/TP/1/94. Environmental Protection Department, Hong Kong

Pegas PN, Evtyugina MG, Alves CA, Nunes T, Cerqueira M, Franchi M, Pio C, Almeida SM, Freitas M (2010) Outdoor/indoor air quality in primary schools in Lisbon: a preliminary study. Quim Nova 33(5):1145–1149

Pegas PN, Alves CA, Evtyugina MG, Nunes T, Cerqueira M, Franchi M, Pio CA, Almeida SM, Freitas MC (2011a) Indoor air quality in elementary schools of Lisbon in spring. Environ Geochem Health 33(5):455–468

Pegas PN, Alves CA, Evtyugina MG, Nunes T, Cerqueira M, Franchi M, Pio CA, Almeida SM, Verde SC, Freitas MC (2011b) Seasonal evaluation of outdoor/indoor air quality in primary schools in Lisbon. J Environ Monit 13(3):657–667

Pegas PN, Nunes T, Alves CA, Silva JR, Vieira SLA, Caseiro A, Pio CA (2012) Indoor and outdoor characterisation of organic and inorganic compounds in city centre and suburban elementary schools of Aveiro, Portugal. Atmos Environ 55:80–89

Pennequin-Cardinal A, Plaisance A, Locoge N, Ramalho O, Kirchner S, Galloo JC (2005) Dependence on sampling rates of Radiello diffusion sampler for BTEX measurements with concentration level and exposure time. Talanta 65:1233–1240

Pope CA, Ezzati M, Dockery DW (2009) Fine-particulate air pollution and life expectancy in the United States. N Engl J Med 360:376–386

Poupard O, Blondeau P, Iordache V, Allard F (2005) Statistical analysis of parameters influencing the relationship between outdoor and indoor air quality in schools. Atmos Environ 39:2071–2080

Prieditis H, Adamson IYR (2002) Comparative pulmonary toxicity of various soluble metals found in urban particulate dust. Exp Lung Res 28:563–576

Priscilla N, Pegas M, Evtyugina G, Alves CA, Nunes T, Cerqueira M, Franchi M, Pio C (2010) Outdoor/indoor Air quality in primary schools in Lisbon: a preliminary study. Quim Nova 33(5):1145–1149

Ramachandran G, Adgate JL, Banerjee S, Church TR, Jones D, Fredrickson A et al (2005) Indoor air quality in two urban elementary schools-measurements of airborne Fungi, Carpet allergens, CO_2, temperature, and relative humidity. J Occup Environ Hyg 2:553–66

Raub JA, Mathieu-Nolf M, Hampson NB, Stephen RT (2000) Carbon monoxide poisoning – a public health perspective. Toxicology 145:1–14

Raysoni AU, Sarnat JA, Sarnat SE, Garcia JH, Holguin F, Luèvano SF, Li WW (2011) Binational school-based monitoring of traffic-related air pollutants in El Paso, Texas (USA) and Ciudad Juárez, Chihuahua (México). Environ Pollut 159(10):2476–2486. doi:10.1016/j.envpol.2011.06.024

Reich BJ, Fuentes M, Burke J (2009) Analysis of the effects of ultrafine particulate matter while accounting for human exposure. Environmetrics 20:131–146

Righi E, Aggazzotti G, Fantuzzi G, Ciccarese PG (2002) Air quality and well-being perception in subjects attending university libraries in Modena (Italy). Sci Total Environ 286:41–50

Rohr AC, Wilkins CK, Clausen PA, Hammer M, Nielsen GD, Wolkoff P (2002) Upper airway and pulmonary effects of oxidation products of (+)-alpha-pinene, d-limonene, and isoprene in BALB/c mice. Inhal Toxicol 14(7):663–684

Rohr AC, Shore SA, Spengler JD (2003) Repeated exposure to isoprene oxidation products causes enhanced respiratory tract effects in multiple murine strains. Inhal Toxicol 15(12):1191–1207

Rudnick S, Milton DK (2003) Risk of indoor airborne infection transmission estimated from carbon dioxide concentration. Indoor Air 13:237–245

Sager TM, Castranova V (2009) Surface area of particle administered versus mass in determining the pulmonary toxicity of ultrafine and fine carbon black: comparison to ultrafine titanium dioxide. Part Fibre Toxicol 6:1–12

Sarwar G, Corsi R, Allen D, Weschler CJ (2003) The significance of secondary organic aerosol formation and growth in buildings: experimental and computational evidence. Atmos Environ 37:1365–1381

Sax SN, Bennett DH, Chillrud SN, Ross J, Kinney PL, Spengler JD (2006) A cancer risk assessment of inner-city teenagers living in New York City and Los Angeles. Environ Health Perspect 114(10):1558–1566

Scheepers PTJ, Konings J, Demirel G, Gaga EO, Anzion R, Peer PGM, Dogeroglu T, Ornektekin S, van Doorn W (2010) Determination of exposure to benzene, toluene and xylenes in Turkish primary school children by analysis of breath and by environmental passive sampling. Sci Total Environ 408:4863–4870

Scheff PA, Paulius VK, Curtis L, Conroy LM (2000a) Indoor air quality in a middle school, part II: development of emission factors for particulate matter and bioaerosols. Appl Occup Environ Hyg 15:835–842

Scheff PA, Paulius VK, Huang SW, Conroy LM (2000b) Indoor air quality in a middle school, part I: use of CO_2 as a tracer for effective ventilation. Appl Occup Environ Hyg 15:824–834

Seppanen OA, Fisk WJ, Mendell MJ (1999) Association of ventilation rates and CO_2 concentrations with health and other responses in commercial and institutional buildings. Indoor Air 9(4):226–252

Sexton K, Ryan PB (1988) Assessment of human exposure to air pollution: methods, measurements and models. In: Ann YW, Bates RR, Kennedy D (eds) Air pollution, the automobile and public health, by the health effects institute. National Academy Press, Washington, DC

Shaughnessy RJ, Haverinen-Shaughnessy U, Nevalainen A, Moschandreas D (2006) A preliminary study on the association between ventilation rates in classrooms and student performance. Indoor Air 16:465–468

Shendell DG, Prill R, Fisk WJ, Apte MG, Blake D, Faulkner D (2004) Associations between classroom CO_2 concentrations and student attendance in Washington and Idaho. Indoor Air 14:333–341

Smedje G, Norback D, Edling C (1997a) Asthma among secondary schoolchildren in relation to the school environment. Clin Exp Allergy 27(11):1270–1278. doi:10.1111/j.1365-2222.1997.tb01171.x

Smedje G, Norback D, Edling C (1997b) Subjective indoor air quality in schools in relation to exposure. Indoor Air 7:143–150

Smedje G, Mattsson M, Walinder R (2011) Comparing mixing and displacement ventilation in classrooms: pupils' perception and health. Indoor Air 21:454–461. doi:10.1111/j.1600-0668.2011.00725.x

Sofuoglu A, Kiymet N, Kavcar P, Sofuoglu SC (2010) Polycyclic and nitro musks in indoor air: a primary school classroom and a women's sport center. Indoor Air 20:515–522. doi:10.1111/j.1600-0668.2010.00674.x

Sofuoglu SC, Aslan G, Inal F (2011) An assessment of indoor air concentrations and health risks of volatile organic compounds in three primary schools. Int J Hyg Environ Health 214:38–46. doi:10.1016/j.ijheh.2010.08.008

Sohn JR, Moon HJ, Ryu SH, Hwang T (2012) Indoor air quality in school classrooms with mechanical ventilation systems. In: Proceedings of the 7th international symposium on sustainable healthy buildings, Seoul, Korea, p 349

Spedding JD (1974) Air pollution. Clarendon Press, Oxford

Stranger M, Potgieter-Vermaak SS, Van Grieken R (2007) Comparative overview of indoor air quality in Antwerp, Belgium. Environ Int 33:789–797. doi:10.1016/j.envint.2007.02.014

Stranger M, Potgieter-Vermaak SS, Van Grieken R (2008) Characterization of indoor air quality in primary schools in Antwerp, Belgium. Indoor Air 18:454–463

Stranger M, Madureira J, Paciência I, Ventura G, De Oliveira Fernandes E (2012) Field study on schoolchildren's exposure to indoor air in Porto, Portugal – Preliminary results. In: 10th international conference on healthy buildings 2012, vol 1, pp 851–856

Szoboszlai Z, Furu E, Angyal A, Szikszai Z, Zs K (2011) Investigation of indoor aerosols collected at various educational institutions in Debrecen, Hungary. X-Ray Spectrom 40:176–180. doi:10.1002/xrs.1323

Tager IB (1999) Air pollution and lung function growth—is it ozone? Am J Resp Crit Care Med 160:387–389

Theodosiou TG, Ordoumpozanis KT (2008) Energy, comfort and indoor air quality in nursery and elementary school buildings in the cold climatic zones of Greece. Energy Build 40:2207–2214. doi:10.1016/j.enbuild.2008.06.011

Touloumi G, Pocock SJ, Katsouyanni K, Trichopoulos D (1994) Short term effects of air pollution on daily mortality in Athens: a time-series analysis. Int J Epidemiol 23:957–967

Tran DT, Alleman LY, Coddeville P, Galloo JC (2012) Elemental characterization and source identification of size resolved atmospheric particles in French classrooms. Atmos Environ 54:250–259. doi:10.1016/j.atmosenv.2012.02.021

Trevisi R, Leonardi F, Simeoni C, Tonnarini S, Veschetti M (2012) Indoor radon levels in schools of South-East Italy. J Environ Radioact Pages 112:160–164

UBA (2008) Guidelines for indoor air hygiene in school buildings. In: Dr. Heinz-Jörn Moriske, Dr. Regine Szewzyk (eds) Federal Environment Agency (UBA) Innenraumlufthygiene-Kommission

des Umweltbundesamtes. Available at: http://www.umweltdaten.de/publikationen/fpdf-l/4113.pdf. Accessed 20 Nov 2012
UNI EN ISO 16000-1:2006. Indoor air – Part 1: General aspects of sampling strategy
UNI EN ISO 16000-2: 2006. Indoor air – Part 2: Sampling strategy for formaldehyde
UNI EN ISO 16000-5:2007. Indoor air – Part 5: Sampling strategy for volatile organic compounds (VOCs)
UNI EN ISO 16017-1:2007. Campionamento ed analisi di composti organici volatili mediante tubo di adsorbimento/desorbimento termico/cromatografia gassosa capillare. Parte 1: Campionamento mediante aspirazione con pompa
UNI EN ISO 16017-2:2007 Campionamento ed analisi di composti organici volatili mediante tubo di adsorbimento/desorbimento termico/cromatografia gassosa capillare. Parte 2: Campionamento per diffusione.
UNI EN ISO 16000-12:2008. Indoor air – Part 12: Sampling strategy for polychlorinated biphenyls (PCBs), polychlorinated dibenzo-p-dioxins (PCDDs), polychlorinated dibenzofurans (PCDFs) and polycyclic aromatic hydrocarbons (PAHs).
UNI EN ISO 16000-15:2008. Indoor air – Part 15: Sampling strategy for nitrogen dioxide (NO_2)
USEPA (1991) The Clean Air Act. Office of Research and Development, Washington, DC
USEPA (2003) Integrated risk information system. USEPA, Washington, DC. Available at: http://www.epa.gov/IRIS/subst/2003
USEPA (2006) Integrated report guidance. USEPA, Washington, DC. Available at: http://water.epa.gov/lawsregs/lawsguidance/cwa/tmdl/2006IRG_index.cfm
Vaizoglu SA, Aycan S, Deveci MA, Acer T, Bulut B, Bayraktar UD, Akyollu B, Celik M, Arslan U, Akpinar F, Baris Z, Arslan S, Deniz A, Didem Evci E, Güler C (2003) Determining domestic formaldehyde levels in Ankara, Turkey. Indoor Built Environ 12:329–335. doi:10.1177/142032603035546
Van Dijken F, Van Bronswijk JEMH, Sundell J (2006) Indoor environment and pupils' health in primary schools. Build Res Inf 34(5):437–446
Van Roosbroeck S, Jacobs J, Janssen NAH, Oldenvening M, Hoek G, Brunekreef B (2007) Long-term personal exposure to PM2.5 soot and NOx in children attending schools near busy roads a validation study. Atmos Environ 41:3381–3394. doi:10.1016/j.atmosenv.2006.12.023
Venn A, Lewis S, Cooper M, Hubbard R, Hill I, Boddy R, Bell M, Britton J (2000) Local road traffic activity and the prevalence, severity, and persistence of wheeze in school children: combined cross sectional and longitudinal study. Occup Environ Med 57:152–158
Wahab SAA, Basma Y (2004) Total suspended dust and heavy metal levels emitted from a workplace compared with nearby residential houses. Atmos Environ 38:745–750
Wåhlinder R, Norbäck D, Wieslander G, Smedje G, Erwall C (1997) Nasal mucosal swelling in relation to low air exchange rate in schools. Indoor Air 7:198–205
Wainman T, Zhang J, Weschler CJ, Lioy PJ (2000) Ozone and limonene in indoor air: a source of submicron particle exposure. Environ Health Perspect 108:1139–1145
Wargocki P, Wyon DP, Matysiak B, Irgens S (2005) The effects of classroom air temperature and outdoor air supply rate on the performance of school work by children. In: Proceedings of indoor air 2005, Beijing, vol 1, p 368
Weichenthal S, Dufresne A, Infante-Rivard C, Joseph L (2008) Characterizing and predicting ultrafine particle counts in Canadian classrooms during the winter months: model development and evaluation. Environ Res 106:349–360. doi:10.1016/j.envres.2007.08.013
Weschler CJ (2000) Ozone in indoor environments: concentration and chemistry. Indoor Air 10:269–288
Weschler CJ (2006) Ozone's Impact on public health: contributions from indoor exposures to ozone and products of ozone-initiated chemistry. Environ Health Perspect 114(10):1489–1496. doi:10.1289/ehp.9256
Weschler CJ (2009) Changes in indoor pollutants since the 1950s. Atmos Environ 43:153–169
Weschler GJ, Shields HC (1999) Indoor ozone/terpene reactions as a source of indoor particles. Atmos Environ 33:2301–2312

Wild P, Bourgkard E, Paris C (2009) Lung cancer and exposure to metals: the epidemiological evidence. Method Mol Biol 472:139–167

Wilkins CK, Clausen PA, Wolkoff P, Larsen ST, Hammer M, Larsen K, Hansen V, Nielsen GD (2001) Formation of strong airway irritants in mixtures of isoprene/ozone and isoprene/ozone/nitrogen dioxide. Environ Health Perspect 109:937–941

Wolkoff P (1995) Volatile organic compounds-sources, measurements, emissions, and the impact on indoor air quality. Indoor Air 5(suppl 3):9–73

Wolkoff P, Clausen PA, Wilkins CK, Hougaard KS, Nielsen GD (1999) Formation of strong airway irritants in a model mixture of (+)-alpha-pinene/ozone. Atmos Environ 33(5):693–698

WHO, World Health Organization (1989) Indoor air quality: organic pollutants. Report on a WHO meeting, Berlin, 23–27 Aug 1987. EURO reports and studies 111. World Health Organization Regional Office for Europe, Copenhagen

WHO Formaldehyde (2002) Concise international chemical assessment document 40.

WHO, World Health Organisation (2006a) Air quality guidelines – global update 2005. WHO Regional Office for Europe. Available at: http://www.euro.who.int/__data/assets/pdf_file/0005/78638/E90038.pdf. Accessed 20 Nov 2012

WHO, World Health Organization (2006b) Principles for evaluating health risks in children associated with exposure to chemicals. Available at: http://www.who.int/ipcs/publications/ehc/ehc237.pdf. Accessed 20 Nov 2012

WHO, World Health Organization (2009a) Guidelines for indoor air quality: dampness and mould. ISBN: 7989289041683. Available at: http://www.euro.who.int/__data/assets/pdf_file/0017/43325/E92645.pdf

WHO, World Health Organization (2009b) Handbook on indoor radon a public health perspective. ISBN: 9789241547673. Available at: http://whqlibdoc.who.int/publications/2009/9789241547673_eng.pdf

WHO, World Health Organization (2010) Guidelines for indoor air quality: selected pollutants. ISBN 978 92 890 0213 4. Available at: http://www.euro.who.int/__data/assets/pdf_file/0009/128169/e94535.pdf

WHO, World Health Organisation (2011) Methods for monitoring indoor air quality in schools. Report of a meeting Bonn, Germany, 4–5 April 2011. WHO Regional Office for Europe. Copenhagen, Denmark. Available at: http://www.euro.who.int/__data/assets/pdf_file/0011/147998/e95417.pdf. Accessed 19 Nov 2012

Wu YC, Batterman SA (2006) Proximity of schools in Detroit, Michigan to automobile and truck traffic. J Expo Sci Environ Epidemiol 16:457–470

Wu Q, Baek SY, Fang M, Chang YS (2010) Distribution and fate of polybrominated diphenyl ethers in indoor environments of elementary schools. Indoor Air 20:263–270. doi:10.1111/j.1600-0668.2010.00652.x

Yamashita S, Kume K, Horiike T, Honma N, Masahiro F, Amagai T (2012) Emission sources and their contribution to indoor air pollution by carbonyl compounds in a school and a residential building in Shizuoka, Japan. Indoor Built Environ 21(3):392–402

Yang W, Sohn J, Kim J, Son B, Park J (2009) Indoor air quality investigation according to age of the school buildings in Korea. J Environ Manag 90:348–354

Zanobetti A, Schwartz J (2005) The effect of particulate air pollution on emergency admissions for myocardial infarction: a multicity case-crossover analysis. Environ Health Perspect 113:978–982

Zhang Q, Zhu Y (2012) Characterizing ultrafine particles and other air pollutants at five schools in South Texas. Indoor Air 22:33–42. doi:10.1111/j.1600-0668.2011.00738.x

Zhang G, Spickett J, Rumchev K, Lee AH, Stick S (2006) Indoor environmental quality in a 'low allergen' school and three standard primary schools in Western Australia. Indoor Air 16:74–80

Zhao Z, Zhang Z, Wang Z, Ferm M, Liang Y, Norbäck D (2008) Asthmatic symptoms among pupils in relation to winter indoor and outdoor air pollution in schools in Taiyuan, China. Environ Health Perspect 116(1):90–97. doi:10.1289/ehp.10576

Chapter 2
Organic Contaminants from Industrial Wastewaters: Identification, Toxicity and Fate in the Environment

Larissa Dsikowitzky and Jan Schwarzbauer

Contents

Abstract Industrialization and urbanization in the industrial nations and increasingly also in emerging and developing nations have led to an intensive and still increasing use of freshwater resources. Industries are one of the most important pollution sources and the discharged wastewaters may contain very diverse organic compound groups. Among those, lipophilic contaminants possessing functional groups which are not common in nature rank among the compounds which are most persistent. There have been strong efforts in environmental sciences for a comprehensive

L. Dsikowitzky (✉) • J. Schwarzbauer
Institute of Geology and Geochemistry of Petroleum and Coal,
RWTH Aachen University, Lochnerstraße 4-20, 52056 Aachen, Germany
e-mail: larissa.dsikowitzky@emr.rwth-aachen.de; schwarzbauer@lek.rwth-aachen.de

E. Lichtfouse et al. (eds.), *Pollutant Diseases, Remediation and Recycling*,
Environmental Chemistry for a Sustainable World 4, DOI 10.1007/978-3-319-02387-8_2,

characterization of chemical contamination and the related impacts, in order to provide data as a basis for management measures. Here we review the current state of knowledge about organic contaminants from industrial sources. We present information about (i) the identification of organic contaminants in industrial wastewaters, (ii) how these compounds are traced in aquatic systems, and (iii) on their toxicity for aquatic organisms as observed in laboratory experiments and in the field.

Major advancements concerning this theme include the development of new analytical techniques which allow for the identification of previously unknown, emerging contaminants. Overall, these studies proved the heterogeneity of the chemical composition of industrial wastewaters, even from the same industry branches. The observed differences are related to different industrial production processes, leading to the presence of varying synthesis educts, additives, products and byproducts in the wastewaters, of which many might occur in the environment but have not yet been identified. Attempts which were made to trace industrial discharges in aquatic systems proved accordingly the presence of very heterogenic organic contaminant mixtures in the environment. We therefore conclude that our knowledge about the chemical composition of industrial wastewaters and about the occurrence of industrial organic contaminants in the environment is as yet very limited.

A further advancement was the development of a combination of chemical and toxicological methods to identify causative organic constituents including emerging contaminants which contribute to the toxicity of industrial wastewaters. Only a few studies in this field are available, although the results obtained so far are promising. Several studies demonstrated the toxic activity of field samples and relate the observed effects to the presence of industrial organic contaminants. Exposure experiments in the field proved the bioaccumulation of specific industrial contaminants and assessed the associated toxic effects. The effects of chemical contamination on aquatic invertebrate communities was shown in field studies combining chemical and biological methods, although industrial inputs were not disentangled from other pollution sources. Ecological surveys of sites contaminated by industrial point sources combined with a comprehensive chemical characterization and toxicity evaluation are missing.

Some of the identified organic contaminants are related to characteristic industrial production processes and their presence in water, sediment or biota therefore indicates the input of specific industrial wastewaters. Accordingly, these compounds can be used as industrial markers. As synthesis of the reviewed information we present a list of potential industrial marker compounds. We suggest the proceeding application of the marker concept which helps to verify the input of specific industrial wastewaters to aquatic systems and to investigate the spatial distribution of the emission. Such information is useful to disentangle different emission sources for the subsequent investigation of their potential impacts in the environment. We present further research strategies which are in our opinion promising to address the identified gaps in knowledge.

Keywords Emerging contaminants • Industrial chemicals • Industrial pollution • Organic markers • Toxicity-based fractionation • Aquatic systems • Aquatic organisms

List of Abbreviations

CAS No.	Unique identifier of a chemical, assigned by the Chemical Abstracts Service
CPE	Chlorophenylethanol
DCPE	Chlorophenylethanoldichloroethanol
DBP	Dichlorobenzophenone
DDA	Bis(chlorophenyl)acetic acid
DDABE	Bis(chlorophenyl)acetic acid butylester
DDAMA	Bis(chlorophenyl)acetic acid methylamide
DDCN	Bis(chlorophenyl)acetonitrile
DDD	Bis(chlorophenyl)dichloroethane
DDE	Bis(chlorophenyl)dichloroethene
DDMU	Bis(chlorophenyl)chloroethene
DDT	Bis(chlorophenyl)trichloroethane
DOC	Dissolved organic carbon
DTPA	Diethylenetriaminepentaacetic acid
EC_{50}	Concentration of a chemical which induces a defined response of 50 % of the members of a tested population after a specified test duration
EDTA	Ethylenediaminetetraacetic acid
FIA-MS-MS	Flow injection analysis-tandem mass spectrometry
GC/MS	Gas chromatography–mass spectrometry
HCB	Hexachlorobenzene
HCHs	Hexachlorocyclohexanes
HPLC	High performance liquid chromatography
HPLC-ED	High performance liquid chromatography with electrochemical detection
HT-GC/MS	High temperature-gas chromatography-mass spectrometry
IPC-ESI-MS	Ion-pair chromatography/electrospray-mass spectrometry
LAS	Linear alkylbenzene sulfonates
LC_{50}	Concentration of a chemical required to kill 50 % of the members of a tested population after a specified test duration
LC/ESI-MS^n	Liquid chromatography-electrospray ionization/multi-stage mass spectrometry
LC/MS	Liquid chromatography-mass spectrometry
LC/MS-MS	Liquid chromatography tandem mass spectrometry
LC/UV	Liquid chromatography with UV detection
MCPE	Chlorophenylchloroethanol
PAHs	Polycyclic aromatic hydrocarbons
PCBs	Polychlorinated biphenyls
PCDDs	Polychlorinated dibenzo-*p*-dioxins
PCDFs	Polychlorinated dibenzofurans
PeBDE	Pentabromodiphenyl ether

POCIS	Polar organic chemical integrative sampler
SPE	Solid phase extraction
SPME	Solid phase microextraction
TLC	Thin layer chromatography
UNEP	United Nations Environment Programme

2.1 Introduction

Chemical contamination was defined as a problem with global relevance, one of nine "planetary boundaries", which should not be transgressed in order to avoid unacceptable global changes (Rockström et al. 2009). In a report on worldwide toxic contamination problems and their implications for human health, mining and industrial activities accounted for nearly all of the documented issues (Blacksmith Institute and Green Cross Switzerland 2011). Overall, industrialization and urbanization in the industrial nations and increasingly also in emerging and developing nations have led to an intensive and still increasing use of water resources (Fig. 2.1). The involved chemical contamination led to a deterioration of rivers, lakes and

Fig. 2.1 Industrial complex in South India, where a tire and rubber production unit, a paper mill, tanneries and chemical production sites discharge the produced wastewaters into public waterways (Photo: Tim Jennerjahn)

coastal waters in many areas. In the industrial nations, these problems were partially solved by source control and by the implementation of technical systems such as sewage treatment. However, there still remain major challenges. The foresight report of the UNEP lists degradation of inland waters in developing nations due to the discharge of untreated wastewaters as one of the 21 most important environmental problems of the twenty-first century (UNEP 2012).

Industries are one of the most important sources for the contamination of aquatic systems. The quantitative composition of the wastewater constituents and their chemical characteristics differ according to the type of the operating production processes. Many organic contaminants are solely discharged by industries which apply certain production processes. Accordingly, high concentrations of such characteristic compounds are only detectable in areas where the related industries are located. Symptomatical are also the high but largely varying concentrations in the wastewaters and the simultaneous occurrence of a large number of chemically heterogenic compounds. A distinct analytical characterization is often hindered due to the large variability of the wastewater quality caused by the discontinuity of many production processes.

The variety of organic compounds which is industrially synthesized is immense. For example, the European Inventory of Existing Commercial Chemical Substances lists about 100,000 synthesized chemicals which were deemed to be on the European Community market in the period 1971–1981. The list is counting for about 99 % of the chemicals' volume on the market. About 30,000–70,000 of the listed chemicals are used on a daily basis (European Inventory of Existing Chemical Substances, Schwarzenbach et al. 2006). The number of byproducts, educts and intermediates though, which might be performed during the synthesis of these regularly used chemicals, is unknown.

The wastewater constituents are discharged into aquatic systems in their dissolved form, associated with colloids or attached to particulate matter. Depending on their physico-chemical properties and the environmental conditions, they undergo different processes after discharge, which include partitioning between phases, transport and/or degradation. A dynamic partitioning is occurring between the following phases: water and atmosphere, water and particulate matter, water and biota as well as particulate matter and biota. Lipophilic organic compounds tend to be bound to particulate matter, so that sediments can be considered as pollution reservoirs (e.g. Baumard et al. 1998). Transport can happen within the dissolved or the particulate phase, sometimes over long distances. This is particularly true for persistent organic pollutants such as DDT, PCDDs/PCDFs and heptachlor.

In the aquatic system, biotic and/or abiotic degradation processes lead to the transformation of organic contaminants to metabolites or to their complete mineralization to H_2O and CO_2. Most organic contaminants – even persistent ones- are degradable under favorable conditions, the latter though within long time periods. The degradation processes depend strongly on the physico-chemical properties of the compounds, e.g. the water solubility. Water soluble organic compounds are often more rapidly degraded than lipophilic compounds (e.g. Berndt 1996).

Dissolved aliphatic alcohols and carboxylic acids were smoothly degraded in an anaerobic milieu, whereas the degradation of the same compound classes with ether bounds or branched hydrocarbon chains took much longer (Kameya et al. 1995). From this it follows that particularly lipophilic organic compounds, possessing functional groups which are not common in nature, rank among the most persistent compounds.

Aquatic biota accumulates particularly such persistent, lipophilic organic contaminants and biomagnification in aquatic food webs has been documented for selected priority pollutants such as HCB, PCBs, PeBDE, dieldrin and mirex (Kelly et al. 2007). The toxicities of several contaminant groups were tested with experimental aquatic species and revealed sublethal to acute toxicity of different parent PAHs, organochlorine pesticides and tributyltin amongst others (e.g. Matthiessen and Law 2002; Thompson et al. 2007). Organic contaminants may have an impact on molecular and cellular processes and also induce changes on the organism, population and community level, which was shown for the antifouling agent tributyltin (Warwick et al. 1990; Fent 2007; Matthiessen and Law 2002). The endocrine disrupting property of this compound has been shown to cause imposex in gastropods (Harding et al. 1999) and to affect not just coastal mollusk populations but also cause reductions in species diversity in coastal invertebrate communities (e.g. Matthiessen and Law 2002). Responses to contaminants are species-specific and help the organisms to cope under various stressors, leading to their acclimation or adaptation to environmental conditions. Sensitive species with a low ability to respond to changes are particularly threatened.

A decline of species holding important ecological functions may negatively affect the whole ecosystem. Many fish and macrobenthic invertebrate species are furthermore exploited as fisheries resources. A reduction of economically important species can therefore cause an impairment of local economies and livelihoods. The contamination of economically important species is moreover posing a threat to the health of humans, as consumers of fish, mussels and crabs.

Accordingly, there have been strong efforts in environmental sciences for a comprehensive characterization of chemical contamination and the related impacts, in order to provide data as a basis for management measures. We present here a comprehensive review of the current state of knowledge about organic contaminants from industrial sources. We present information about (i) the identification of organic contaminants in industrial wastewaters, (ii) how they are traced in aquatic systems, and (iii) their toxicity for aquatic organisms as observed in laboratory experiments and in the field.

Standard search machines were used to find relevant peer reviewed literature from the period 1970–2012. For Table 2.1 also data from two Ph.D. theses and from one report were cited due to the limited available information. Unspecific keywords were used for the literature search such as "industrial wastewaters" and "industrial effluents" in combination with "organic pollutants" and "organic contaminants". From the several 100 matches found with this method, only studies which focus on the identification and chemical/toxicological characterization of low-molecular weight organic compounds in industrial wastewaters and impacted aquatic systems were selected. Studies which focus on treatment techniques, remediation or on bulk parameters to study industrial contamination were neglected. Furthermore, only field studies were considered in which the detected contaminants were unequivocally attributed to industrial sources.

2.2 Chemical Characterization of Industrial Wastewaters

2.2.1 Analytical Approaches for the Identification of Organic Contaminants in Industrial Wastewaters

Improvements in the chemical characterization of industrial effluents are closely linked with the development of appropriate analytical tools. A thorough overview on analytical methods used to characterize complex contaminant mixtures in industrial wastewaters has been given by Castillo and Barceló (1999a). More general reviews not only focusing on industrial but also on municipal effluents have recently been given by Sanchez Rojas and Bosch Ojeda (2005) as well as by Giger (2009). The later compilation described the advances in the detection of polar and amphiphilic contaminants as important constituents in wastewater. A more specific review summarized the application of LC/MS for the analysis of endocrine disrupting agents in wastewater partially considering also industrial effluents (Postigo et al. 2009).

Noteworthy, the enormous contributions of GC/MS and LC/MS techniques with respect to the identification and quantification of organic constituents in different types of industrial wastewater have been summarized in all three reviews. These techniques are responsible for the most important progress in the analytical detection of industrial contaminants, hence nearly all recent studies (reported in this review) are based analytically on well established GC/MS and LC/MS methods.

Recently, only very few studies reported the application of new analytical methods. Some reports represent slight modifications of standard techniques. E.g. Castillo et al. (1999b) reported on the extension of the GC/MS towards higher gas chromatographic temperature (HT-GC/MS) to detect higher boiling substances in industrial effluents. Further studies focused on optimization of sample treatment and extraction procedures. These approaches include sequential solid phase extraction (SSPE) or novel passive sampling devices such as POCIS (Castillo et al. 1999b; Alvarez et al. 2005). Lastly, some publications described the optimization of specific analytical methods for the detection of preselected industrial contaminants. Rodriguez et al. (2004) followed 2-mercaptobenzothiazol in tannery wastewater by liquid chromatography coupled to an amperometric detector. Schriks et al. (2010) quantified glucocorticoids in wastewater using high resolution mass spectrometry.

2.2.2 Organic Contaminants Which Have Been Identified in Different Types of Industrial Wastewaters

Since the 1970s, mainly wastewaters from the textile industry, tanneries, the petrochemical industry, pulp and paper industry, rubber and tire production and from chemical production sites were investigated. The studies were almost exclusively conducted in industrial nations of the northern hemisphere, i.e. Europe, the United States and Japan. Although during the last two decades industrial production sites were more and more shifted to newly industrialized countries, such as China, India and Brazil, studies from these countries are lacking.

In most cases, the effluents were investigated, after treatment and before release to the environment. The sort of applied methods confine the spectrum of identifiable compounds. Using purge techniques or head space SPME for sample preparation (Smith 1990; James and Stack 1997; López-Grimau et al. 2006), only volatile organic compounds were extracted from the sample matrix. A compound separation with liquid chromatography and subsequent application of mass spectrometry (e.g. Castillo and Barceló 2001; Loos et al. 2007; Schriks et al. 2010) allows for the detection of hydrophilic and amphiphilic compounds. Analyses with GC/MS are relevant for the determination of volatile and semivolatile lipophilic organic compounds. Detection of hydrophilic or amphiphilic compounds with GC/MS is possible after derivatization or transformation of the target compounds to more volatile species (e.g. Giger 2009). However, several studies were focused on specific target compounds or compound groups and neglected other substances which were probably also present in the effluents (e.g. Rodriguez et al. 2004; Schriks et al. 2010). Other studies applied a screening approach in order to identify a preferably wide spectrum of organic contaminants (e.g. Ellis et al. 1982; Botalova et al. 2009). Although such approaches allow for the identification of numerous contaminants, there are still many substances which are detectable by GC/MS but remain unidentified because of their unknown mass spectra. The structure elucidation of such compounds is time consuming and includes interpretation of mass spectra and chemical synthesis of reference compounds. Other compounds might have physico-chemical properties which do not allow for the detection with GC/MS, which requires the application of more specific compound extraction, separation and detection methods. The organic compounds identified in different industrial effluents were listed in Table 2.1. It is indicated if only a certain compound group from the respective study was selected. Unspecific organic compounds which might also have a biogenic origin such as *n*-alkanes, fatty acids and fatty acid esters were not included.

In effluents from the ***textile industry***, i.e. dye manufacturing plants and textile mills in the United States, several environmentally relevant compounds like halogenated anilines, benzenes and anthraquinones were found (Games and Hites 1977). Applying headspace techniques and GC/MS, also volatile organic solvents were determined in textile industry effluents (Smith 1990; López-Grimau et al. 2006). Screening analyses of untreated wastewaters revealed the presence of a wide spectrum of contaminants including different alkylated phenols and phthalic acid esters

Table 2.1 Organic compounds identified in industrial wastewaters applying different separation techniques and mass spectrometry

Industry	Identified organic compounds	Region	Wastewater type	Method	Type of listed compounds	Reference
Textile industry	Anisole, aniline, chloroaniline, alkylated anilines, dibromoaniline, tribromoaniline, hydroxyethylbutylbromoaniline, bromonitroaniline, bromodinitroaniline, hydroxyethylbutylaniline, bromodiethylaniline, hydroxyethylcyanoethylaniline, methylmethoxyaniline, nitrochlorobenzene, bromonitrobenzene, methoxyazobenzene, alkylated indolines, trimethylindole, trimethyloxindole, trimethylindolineacetaldehyde, dimethylindolinediacetaldehyde, chlorophenylthiadiazol, aminophenylthiadiazole, dichlorobenzene, hydroxyanthraquinone, dihydroxyanthraquinone, nitrobenzene, aminoanthraquinone, hydroxyaminoanthraquinone, bromoaminohydroxyanthraquinone, caprolactam, dinitrophenol, acetylethylcyanoethylethoxyphenylenediamine	US	Effluents	GC/MS		Games and Hites (1977)
	Perchloroethylene, alkylated cyclohexanes, decahydronaphthalene, xylene, ethylbenzene, trimethylbenzene, methylethylbenzene, trichlorobenzene, chloroisocyanatobenzene, dichloromethane, terpenes, chlorocyclohexanone, dibutylphthalate, chloroform, toluene	US	Effluents	Purge and trap prior to GC/MS		Smith (1990)
	Nitrobenzene sulfonate, methylbenzene sulfonate, chlorobenzene sulfonate, naphthalene sulfonates, alkylated benzene sulfonates	Spain, Portugal	Untreated	IPC-ESI-MS		Alonso et al. (1999)

(continued)

Table 2.1 (continued)

Industry	Identified organic compounds	Region	Wastewater type	Method	Type of listed compounds	Reference
	Ionol, di-*tert*.-butylethylphenol, di-*tert*.-butylmethoxymethylphenol, tetramethylbutylphenols, nonylphenols, methylenebisphenol, di-*tert*.-butylhydroxybenzoic acid, benzoic acid phenyl ester, dimethylphenylacetophenone, hexamine, *tert*.-butylquinone, diethylphtalate, dimethylcarboxymethylexylphtalate, phenol, butylethylhexylphtalate, butyloctylphtalate, di-iso-octylphtalate, benzylbutylphtalate, benzylquinoline, tributylacetylcitrate, diethylaminomethylbenzopyranone, triphenylphosphonic acid	Portugal	Untreated	HT-GC/MS		Castillo and Barceló (2001)
	Polyethylene glycol, carboxylated polyethylene glycols, alkylalcohol polyethoxylate, dibutylphthalate, bis(ethylhexyl)phthalate, LAS, naphthalene sulfonates	Portugal		LC/MS		Castillo and Barceló (2001)
	Xylene, ethylbenzene, toluene	Spain	Effluents	SPME prior to GC/MS		López-Grimau et al. (2006)
	Bisphenol A, nonylphenol, octylphenol, nonylphenol ethoxylates, nonylphenol ethoxycarboxylates, octylphenol ethoxylates, octylphenol ethoxycarboxylates	Belgium, Italy	Effluents	LC/MS-MS		Loos et al. (2007)

Tanneries	Mercaptobenzothiazole, dihydroxypyrimidine, ethylhexylacetate, tris(butoxyethyl)phosphate, trimethylcyclohexanone, trimethylcyclohexenone, trimethylcyclohexanol, benzoic acid, cyclohexanecarboxylic acid, phenylacetic acid, phenylpropionic acid, methoxycinnamic acid, phthalic acid, phenol, cresol, chlorocresol, hexanol, ethylhexanol, hexandiol, glycerol, hydroxyindole, indole acetic acid, hydroxyphenylacetic acid, phenylacetic acid, hydroxyphenylacetic acid, dihydroxyphenylacetic acid, phenylacetic acid methylester, vanilylpropionic acid, dihydroxyphenylpropionic acid, methanoldiethoxylate, butanoldiethoxylate, pentanoldiethoxylate	Germany	Untreated	GC/MS screening	Reemtsma and Jekel (1997)
	Nitrobenzene sulfonate, chlorobenzene sulfonate, naphthalene sulfonates, alkylated benzene sulfonates	Portugal	Untreated	IPC-ESI-MS	Alonso et al. (1999) and Alonso and Barceló (1999)
	Nitrobenzene sulfonate**, chlorobenzene sulfonate*, naphthalene sulfonates	Sweden, Spain	Untreated	IPC-ESI-MS	Alonso and Barceló (1999)
	Polyethylene glycol, alkylalcohol polyethoxylates, nonylphenol ethoxylates, carboxylated polyethylene glycols, bis(ethylhexyl)phthalate, diethylphthalate, nitrophenol, chlorocresol, nitrobenzene sulfonate, chlorobenzene sulfonate, LAS, naphthalene sulfonates, methylthiobenzothiazole, methylsulfinylbenzothiazole, benzothiazolylbenzothiazolethione	Portugal	Untreated	LC/MS	Castillo et al. (1999a)
	LAS, polyethylene glycol, alkylalcohol polyethoxylates, naphthalene sulfonates, naphthalenedisulfonates, chlorobenzene sulfonate	Sweden	Effluents	LC/MS	Castillo et al. (2001)
	Mercaptobenzothiazole	Mexico	Effluents	HPLC-ED	Rodriguez et al. (2004)

(continued)

Table 2.1 (continued)

Industry	Identified organic compounds	Region	Wastewater type	Method	Type of listed compounds	Reference
Petrochemical Industry	Xylene, styrene, methylstyrene, indane, indene, methylindene, alkylated naphthalenes, phenol, cyclooctadiene, isopropylbenzene, ethyltoluene, butoxyethanol, acetophenone, terpineol, acenaphthene, fluorene, diphenylpropanol, cresol, acenaphthylene	US	Petrochemical & petrorefinery effluents	GC/MS		Keith (1974)
	Toluene, C_2-benzene, naphthalene, alkanols, alkanones, phenol, ethylacetate, ethyl-iso-propylether, dioxolane, chlorobenzene, nitrobenzene, dimethyltoluenesulfonamide, azolidinylbutene, pyrrolidinylmethylenebutane, prometone, simazine	US	Refinery effluents	GC/MS screening		Ellis et al. (1982)
	Benzene, toluene, ethylbenzene, xylenes, ethylmethylbenzene, trimethylbenzene, naphthalene	Ireland	Petrochemical effluents	SPME prior to GC/MS		James and Stack (1997)
	Ethylbenzoate, nonylphenol, pentachlorophenol, isothiocyanate-cyclohexane, tetramethylthiourea, bis(ethylhexyl)phthalate, methylpyrrolidinone, di-iso-octylphthalate, dimethylphthalate	Europe	Petrochemical effluents	LC/MS		Castillo et al. (1998)
	Dimethylbenzoic acid, trimethylbenzoic acid, ionol, di-*tert*.-butylphenol, di-*tert*.-butylethylphenol, ethylethoxybenzoate, diethylphthalate, butyloctylpthalate, di-iso-octylpthalate, dimethlyphenylacetophenone, *tert*.-butylhydroxymethylphenylsulfide	Portugal	Petrochemical effluents	HT-GC/MS		Castillo et al. (1999b)
	Chlorinated benzenes, chlorotoluene, chlorinated phenols, trichlorobenzaldehyde	Poland	Coking plant effluents	GC/MS		Czaplicka (2003)

	Indoline, toluamide, alkylated + hydrogenated quinoline derivatives, N-phenylformamide, N-*tert*.-butylbenzamide, tetramethylated piperidinone, triphenylphosphine oxide, alkylindanes, dimethylpyridine, PAHs and alkylated PAHs	Germany	Petrochemical effluents	GC/MS screening	Only source-specific constituents	Botalova et al. (2009) and Botalova and Schwarzbauer (2011)
Pulp and paper industry	Terpenes, methyltrisulfide, hexachloroethane, formylthiophene, acetylthiophene, propionylthiophene, dimethylsulfoxide, ethylcarbamate, nitrotoluene, dimethylsulfone, hydroxybenzaldehyde, hydroxyacetophenone, resin acids, trimethoxyacetophenone, trimethoxybenzaldehyde, methylthiobenzothiazole	US	Effluents	GC/MS screening		Keith (1976)
	Dichlorocatechol, trichlorocatechol, monochlorotrihydroxybenzene, dichlorotrihydroxybenzene, trichlorotrihydroxybenzene, dichlorodihydroxymethoxybenzene	Norway	Chlorination stage effluents	GC/MS		Carlberg et al. (1980)
	Chlorinated phenols, dichloroguaiacol, trichloroguaiacol, tetrachloroguaiacol	Switzerland	Effluents	GC/MS		Leuenberger et al. (1985)
	Trichlorophenol, trichloroguaiacol, dehydroabietic acid, chlorodehydroabietic acid, dichlorodehydroabietic acid	Sweden	Treated and untreated wastewater	GC/MS		Hynning (1996)
	Nonylphenol ethoxycarboxylates	US	Effluents	GC/MS		Field and Reed (1996)
	Nitrilotriacetic acid, EDTA, DTPA	Canada	Effluents	GC/MS		Lee et al. (1996)
	Nitrophenol, chlorinated phenols, catechol	Europe	Effluents	LC/UV		Castillo et al. (1997)
	Bisphenol A, mono-, di-, tri- and tetra-chlorinated bisphenol A	Japan	Effluents	GC/MS		Fukazawa et al. (2001)
	Resin acids, bisphenol A, nonylphenol ethoxycarboxylates, lignin, polychlorinated dibenzothiophene, chlorinated phenols, chloroguaiacols, chlorosyringol, volatile sulphur compounds	Europe	Effluents	Various	Only compounds in effluents	Latorre et al. (2005) and references therein

(continued)

Table 2.1 (continued)

Industry	Identified organic compounds	Region	Wastewater type	Method	Type of listed compounds	Reference
	Ditolylmethane, ditolylethane, di-iso-propylnaphthalene, bis(methylphenoxy)ethane, terphenyl, benzylnaphthyl ether, chloromethylphenoxymethylphenoxyethane, benzylbiphenyl	Japan	Effluents	GC/MS		Terasaki et al. (2008)
	Terpenes, acetylmorpholine, aniline, trimethylpentanedioldi-iso-butyrate, methanol, phenol, alkylated phenols, decalone, acetyloxytrimethylbicycloheptanedione, benzoic acid, abietic acid, dehydroabietic acid, hexahydrodimethyl-iso-propylnaphthalene	Germany	Effluents	GC/MS screening		Botalova and Schwarzbauer (2011)
Rubber and tire production	Benzothiazole, mercaptobenzothiazole, toluene, butoxyethanol, nonylphenol, di-*tert*.- butyl-methylphenol, methylenediphenol, alkylated benzenes, isophorone, methylbiphenyl, cyclohexylamine, diphenylamine, propyldiphenylamine, diethylphthalate, naphthalene, alkylated naphthalenes, phenanthrene, alkylated phenanthrenes, fluoranthene, pyrene, dimethylacridan	US	Effluents	GC/MS		Jungclaus et al. (1976)
	Benzothiazole, methylbenzisothiazole, methylbenzothiazole, methylthiobenzothiazole, benzothiazolone, benzisothiazolol, thiocyanic acid (benzothiazoylthio) methylester, dithiobisbenzothiazol aniline, alkylated anilines, chlorinated anilines, benzoic acid, phenyl(phenylmethyl)thiourea, thiocyanic acid anilinophenylester chlorobenzene, toluene, benzene isothiocyanate, methylamine benzoic acid, (methylthio) methylbenzene, (methylthio)benzoic acid, hydroxybenzoic acid, aminebenzoic acid, cyclohexylamine, benzoic acid hydroxymethylmethylester, diphenyl hydrazinecarboxamide, methanetetraylbis-cyclohexanamine, methylimidazoledione, benzotriazinone	Spain	Partially treated wastewater	GC/MS		Puig et al. (1996)

	Indole, methylindole, methyloxindole, diethylphthalate, dibutylphthalate, dicyclohexylphthalate, methoxyphenol, di-*tert*.-butylphenol, mercaptothiazoline, zinc dimethylthiocarbamate, acetylphenylamine	Thailand	Effluents	GC/MS screening	Frequently occurring compounds	Worawit (2006)
Chemical production sites						
Synthesis of e.g. pharmaceuticals, pesticides, optical brighteners and surfactants	Aminoethyldibenzoazepine, hydroxy-*tert*.-amylphenylbenzotriazole, methylquinoxaline, dimethylquinoxaline, phenylnaphthylamine, *tert*.-butylphenylnaphthylamine, tetramethylbutylphenylnaphthylamine, methylthiobenzothiazole, benzothiadiazole, nitrophenol, *tert*.-amylphenol, di-*tert*.-butylnitrophenol, di-*tert*.-amylcyanophenol, di-*tert*.-amylcyanomethylphenol, propylcarboxymethyldi-*tert*.-butylhydroxyphenylpropionate, trifluoromethylaniline, di-*tert*.-butylbenzoquinonemethideacetic acid, chlorinated diphenylether	US	Effluents	GC/MS, LC/MS		Jungclaus et al. (1978) and Lopez-Avila and Hites (1980)
Synthesis of 2,4,6-Trinitrotoluene	Nitrosomorpholine, nitrobenzonitrile, morpholinoacetonitrile, dinitrobenzene, trinitrobenzene, dimethyldinitrobenzene, methylnitrophenol, dinitromethylphenol, toluene, nitrotoluene, dinitrotoluene, trinitrotoluene, aminonitrotoluene, aminodinitrotoluene, dinitroaniline	US	Effluents	GC/MS		Spanggord et al. (1982)
Synthesized compounds not indicated	Dimethoxymethane, dichloroethane, methylpentanol, styrene, methylbutadiene, propenenitrile, dichloromethane, toluene, methylpentanone, trivinylcyclohexane, divinylbenzene isomers	France	Effluents	SPME prior to GC/MS		Santos et al. (1996)

(continued)

Table 2.1 (continued)

Industry	Identified organic compounds	Region	Wastewater type	Method	Type of listed compounds	Reference
Synthesis of e.g. plastics, lacquers and plant protection agents	LAS, naphthalene sulfonates, hydroxynapthalenedisulfonate, nitrobenzene sulfonate, methylbenzene sulfonate, chlorobenzene sulfonate	Germany	Effluents	LC/MS		Castillo et al. (2001)
Synthesis of industrial intermediates and vitamins	Triphenylphosphinoxide, di-iso-propylidenesorbofuranose, di-iso-propylidenexylohexulofuranosoic acid, bis-ethyl-iso-octanol lactone isomers	Germany	Effluents	GC/MS, LC/ESI-MS^n		Knepper and Karrenbrock (2006)
Synthesis of e.g. plastics, flame-retardants and plant protection agents	Dichlorobenzene, chlorinated anilines, dichloromethylthiobenzene, trifluoromethylacetophenone, tetramethylbutane dinitrile, diphenylpropenenitrile, tetrahydrothiophene dicarboxylic acid, tris(chloropropyl) phosphates, triethylphosphate, dibutylmethylphosphonate	Germany	Effluents	GC/MS screening		Botalova et al. (2011)
Synthesis of various intermediate products	Dimethylpyrazol, tributylamine, tri-iso-octylamines, trioctylamine, tris(ethylhexyl)amine, dibutyldecanamine, trithiolane, methyldiazaadamantanone, tris(chloropropyl) phosphates, trimethylpentanediol-diisobutyrate, hydroxydimethylpropanoic acid hydroxydimethylpropylester	Germany	Effluents	GC/MS screening		Botalova et al. (2011)
Synthesis of e.g. fibers, lacquers and paints	Dimethylpyrazol, dithiolane, dithiane, trithiolane, trithianes, tetrathiane, tetrathiepane, hexathiepane, triacetin	Germany	Effluents	GC/MS screening		Botalova et al. (2011)
Pharmaceutical industry	Cimetidine, bromazepam, diclofenac, decahydroquinoline, metharbital	Germany	Effluents of a pilot treatment plant	FIA-MS-MS, GC/MS, LC/MS		Schröder (1999)

	Hydroxyandrostanone, androstanedione	Germany	Effluents	GC/MS screening	Dsikowitzky (2002)
	Cortisol, cortisone, dexamethasone, prednisolone	Netherlands	Effluents	LC/MS	Schriks et al. (2010)
Food-processing industry					
Pea, fish and porc production	Dichloropropane, toluene, trichloromethane, chlorinated phenols, diethylphthalate, bis(ethylhexyl)phthalate, fluorene, phenanthrene, fluoranthene, pyrene	Denmark	Effluents	Not specified	Maya-Altamira et al. (2008)
Meat production	Dibenzylamine, methylindolinone, tris(chloropropyl) phosphate, dihydrodimethylthiopyranecarboxaldehyde, *tert*.-butylbenzoic acid	Germany	Effluents	GC/MS screening	Botalova and Schwarzbauer (2011)
Cement industry	LAS, naphthalene sulfonates	Germany	Effluents	LC/MS	Castillo et al. (2001)
Automotive industry	Hexamethoxymethylmelamine, butoxyethoxypropanol	Spain	Effluents	GC/MS	Consejo et al. (2005)
Power station	Nitrosomorpholine, benzylidenebenzamine, benzylbenzamide, tris(chloroethyl)phosphate, tris(chloropropyl) phosphate, trimethylphosphate, trimethylpentanedioldi-iso-butyrate, menthol, ethylhexanoic acid ethylhexyl ester, *tert*.-butylbenzoic acid, methyldihydro jasmonate, alkylated naphthalenes	Germany	Effluents	GC/MS screening	Botalova and Schwarzbauer (2011)

(Castillo and Barceló 2001). Phthalic acid esters are used in great quantitities as plasticizers and the annual worldwide production volume was estimated to 5.2 million tons (Mackintosh et al. 2006). In untreated wastewaters, also different naphthalene sulfonates, benzene sulfonates and LAS were detected (Alonso et al. 1999; Castillo and Barceló 2001). Naphthalene sulfonates and anthraquinone sulfonates are employed as starting material for the manufacturing of azo and anthraquinone dyestuffs, and LAS are used as surfactants in detergent formulations (e.g. Fichtner et al. 1995). Whereas LAS are readily biodegradable, naphthalene sulfonates are relatively persistent and were also found in river water samples (Takada et al. 1994; Fichtner et al. 1995; Rieger et al. 2002). Untreated wastewaters and effluents from textile industries also contain compounds which are known as endocrine disrupting chemicals (Castillo and Barceló 2001; Loos et al. 2007). Nonylphenol, nonylphenol ethoxylate, octylphenol, diethylphthalate, bis(ethylhexyl)phthalate and bisphenol A are weakly estrogenic (Petrovic et al. 2004). Azo dyes are the largest group of dyes and account for more than half of the worldwide annual dye production (Stolz 2001). During the degradation of azo dyes, metabolites can be performed, which are themselves problematic for aquatic systems, e.g. 6-Acetylamino-3-amino-naphthalene-2-sulfonic acid, *N*-(bis-hydroxymethyl-phenyl)-acetamide, different aromatic amines and different anilines (Chung and Stevens 1993; Bhaskar et al. 2003; Pinheiro et al. 2004; Bilgi and Demir 2005) (see also Sect. 2.3).

Just like in untreated wastewaters of the textile industry, naphthalene sulfonates, benzene sulfonates and LAS were detectable in effluents and in untreated wastewaters from ***tanneries*** (Table 2.1) (Castillo et al. 1999a, 2001; Alonso et al. 1999; Alonso and Barceló 1999). Untreated wastewaters also contained the endocrine disrupting compounds nonylphenol ethoxylates, diethylphthalate and bis(ethylhexyl)phthalate (Castillo et al. 1999a; Petrovic et al. 2004). The leather industry uses nonylphenol ethoxylates and alkylalcohol polyethoxylates as surfactants to remove skin grease, whereas polyethylene glycol is used as leather lubricant. All these chemicals as well as carboxylated polyethylene glycols, which are very likely a degradation product of polyethylene glycol and alkylalcohol polyethoxylates, were present in the tannery wastewaters (Castillo et al. 1999a, 2001). Benzothiazoles are widely used as fungicides instead of chlorophenols and were also frequently found in tannery wastewaters (Reemtsma and Jekel 1997; Castillo et al. 1999a; Rodriguez et al. 2004). A comprehensive investigation revealed that untreated tannery wastewaters were heavily burdened with organic compounds (DOC 900 mg l^{-1}) (Reemtsma and Jekel 1997). Numerous contaminants from 12 compound classes including cyclohexanes, aromatic carboxylic acids, phenols, indoles and ethoxylates were identified. After anaerobic and biological wastewater treatment, DOC concentration was strongly reduced and the contaminant concentrations of all compound classes decreased. Nevertheless, environmentally relevant compounds were still detectable after the treatment process (Reemtsma and Jekel 1994, 1997). Some of these contaminants were relatively persistent against the treatment process (e.g. benzothiazoles), others were newly performed during treatment like dichlorobenzoic acid and tris (2-butoxyethyl)phosphate (Reemtsma and Jekel 1997).

Several studies focused on the chemical characterization of effluents from the ***petrochemical industry***, i.e. petroleum refining and related industries. Important petroleum products are gaseous fuels, gasoline, solvents, lubricating oil, grease, waxes and asphalt. During the distillation of crude oil in the refineries, fractions with different physico-chemical properties are obtained. Alkylated benzenes, indane and its alkylated derivatives as well as naphthalene were found in several effluent samples (Table 2.1) (Keith 1974; Ellis et al. 1982; James and Stack 1997; Botalova et al. 2009; Botalova and Schwarzbauer 2011). These compounds are e.g. constituents of the gas and naphtha fraction (Speight 2007). Alkylated napthalenes and quinoline derivatives also occurred in effluent samples (Keith 1974; Botalova et al. 2009) and are characteristic constituents of the middle distillate fraction (Speight 2007). Further constituents of crude oil which were also present in effluent samples were indoline, acenaphthene, acenaphthylene and fluorene (Keith 1974; Ellis et al. 1982; Wang et al. 1998; Speight 2007). However, Botalova et al. (2009) reason that indoline and quinoline related compounds may also represent technical byproducts of secondary synthetic processes in petrochemical plants. The presence of xylene, toluene and benzene in effluents was reported in the three earliest studies (Keith 1974; Ellis et al. 1982; James and Stack 1997). These chemicals are primary products of oil refineries and are widely used as educts for a large variety of chemical synthesis, e.g. for the plastic production. By alkylation of toluene with ethylene, followed by hydrogenation, methylstyrene is obtained, which is used for polymerization synthesis (Speight 2007). Styrene and methylstyrene were found in effluent samples, as well as phenol and nitrobenzene (Keith 1974; Ellis et al. 1982). The latter two compounds are e.g. used for the solvent refining of lubricating oils in order to remove undesirable constituents from the charge material (Speight 2007). Phthalic anhydride is a precursor in the synthesis of phthalic acid esters which are used as plasticizers. It is produced in petrochemical plants by oxidation of naphthalene (Speight 2007). The occurrence of different phthalic acid ester derivatives in petrochemical effluents (Castillo et al. 1998, 1999b) could therefore result from the synthesis of starting materials for plastic production in these plants. This is underlined by the fact that ionol and ionol derivatives were also present (Castillo et al. 1999b), which are used as additives in plastic materials and rubber (Yasuhara et al. 1997). Further structurally very diverse compounds were detected in the effluent samples. Summing up, the chemical composition of the petrochemical plant and refinery effluents was found to be very heterogenic. From this it follows that the production processes in the plants and refineries and the wastewater treatment approaches differ strongly which leads to a unique chemical composition of the effluents.

Studies on the chemical characterization and toxicity of wastewaters from the ***pulp and paper industry*** were recently reviewed (Latorre et al. 2005). Contaminants which were identified in pulp and paper production effluents as indicated in this review and information from further studies are presented in Table 2.1. Generally, the effluents contain natural plant constituents such as resin acids (e.g. abietic and dehydroabietic acid), lignin, terpenes, catechol, hydroxybenzaldehyde, and structurally related compounds. The bleaching of chemical pulp with chlorine leads to the presence of the chlorinated derivatives of plant constituents: chlorinated catechols, chlorinated dehydroabietic acid, chlorinated guaiacols and chlorinated syringols. Chlorinated phenols were formerly used as wood preservatives and as a result they have also been encountered in effluent samples (Latorre et al. 2005). Paper mill effluents also contained bisphenol A and nonylphenol ethoxycarboxylates (Field and Reed 1996; Latorre et al. 2005), which were shown to be weakly estrogenic (White et al. 1994; Petrovic et al. 2004). Bisphenol A is a common raw material in the paper production (Fukazawa et al. 2001). The compound is easily chlorinated by the bleaching and disinfection agent sodium hypochlorite, and chlorinated bisphenol A derivatives are performed, which were identified in effluent samples (Fukazawa et al. 2001). During the last decade, the use of chlorine as bleaching agent was reduced and it was replaced by chlorine dioxide, molecular oxygen, peroxide or ozone (e.g. Latorre et al. 2005). The elimination of chlorine as bleaching agent caused the launching of metal binders such as DTPA and EDTA in the paper recycling industry, and their occurrence in the respective effluents. These agents bind metals which would otherwise disturb the bleaching process with hydrogen peroxide (Lee et al. 1996). The replacement of chlorine by other bleaching agents also explains the absence of chlorinated compounds in the two most recent studies from Japan and Germany. The compounds detected in sediment samples from the outfall of the largest paper mill industry area in Japan are used as solvents and color sensitizers in the manufacture of thermal paper (Terasaki et al. 2008). Besides terpenes and resin acids, a photoinitiator (acetyloxytrimethylbicycloheptanedione) and an adsorbent (acetylmorpholine) were found in German effluents from a paper production facility (Botalova and Schwarzbauer 2011). The other effluent constituents were unspecific.

In effluents from ***rubber and tire production*** plants, structurally diverse organic compounds were detected, of which many are known as chemicals frequently used in the rubber industry (Table 2.1). The presence of benzothiazoles and aniline derivatives in effluents was proved in two studies (Jungclaus et al. 1976; Puig et al. 1996) and arises from the use of benzothiazoles and aniline derivatives as catalysts for the vulcanization process (Fishbein 1991). Different phthalic ester derivatives, i.a. diethylphthalate, dibutylphthalate and dicyclohexylphthalate were also detected (Jungclaus et al. 1976; Worawit 2006), being related to the use of phthalic esters as plasticizers in the rubber production (Fishbein 1991). Toluene is used as solvent for rubber processing, di-*tert*.-butylmethylphenol and structurally related compounds are employed as antioxidants. Other important antioxidants are diphenylamines (Fishbein 1991), which can explain the presence of diphenylamine and propyldiphenylamine in the effluent samples from the US. Mineral oils and tar products are widely used in the rubber industry as extenders (Fishbein 1991 and references therein). They contain PAHs, such as naphthalene and phenanthrene and their alkylated derivatives as well as fluoranthene and pyrene, which were all found in the effluents (Jungclaus et al. 1976). Further compounds were detected in the effluents, some of them are structurally related to the compounds described above, but their specific application in the rubber and tire production is unknown.

Due to the wide spectrum of industrial chemicals which are synthesized at ***chemical production sites***, the chemical composition of effluent samples is very heterogenic. For example, different nitrogen containing heterocyclic compounds, sulfur containing compounds and chlorinated compounds were identified in an early study of effluents from a chemical production site in the US (Jungclaus et al. 1978; Lopez-Avila and Hites 1980). During the synthesis of the explosive trinitrotoluene, wastewaters are generated, which are heavily polluted with nitroaromatic compounds, including nitrotoluenes, nitrobenzenes, nitroanilines and nitrobenzonitrile (Spanggord et al. 1982). Effluents from industries which produce modern plastic materials, industrial intermediates, plant production agents and flame retardants contained a completely different spectrum of contaminants (Table 2.1) (Botalova et al. 2011). Many of them are unspecific and also occur in municipal effluents. These are, for example, surfactants such as LAS, flame retardants such as alkyl phosphates and the plasticizer trimethylpentanedioldi-iso-butyrate (e.g. Eganhouse 1986; Dsikowitzky 2002). Other compounds were described to stem exclusively from industrial emission sources, e.g. hexathiepane. Tetramethylbutanedinitrile, several identified tertiary amines and triacetin are potential site-specific markers for the characteristic production processes taking place in three chemical manufacturers, respectively (Botalova et al. 2011). Another study exclusively described byproducts which are performed during specific chemical synthesis. Such compounds are manufactured unintentionally and have no commercial application. Accordingly, little attempts have so far been made to gather information about their properties, their toxicities and their fate in the environment (Knepper and Karrenbrock 2006). Similar to textile industry and tannery wastewaters, naphthalene sulfonates, benzene sulfonates and LAS were detectable in effluents from a manufacturer of plastic materials, lacquers and plant protection agents (Castillo et al. 2001). The analysis of effluents from a chemical manufacturer in France (Santos et al. 1996) revealed the presence of different solvents, which are widely known as water pollutants.

As in the chemical manufacturing, the production processes in the ***pharmaceutical industry*** are very diverse and depending on the manufactured medicines, which is mirrored by complete different chemical compositions of the effluents (Table 2.1). Effluents of a German pilot treatment plant contained the antihistamine cimetidine, the tranquilizer bromazepam, the pain reliever diclofenac and the anticonvulsant metharbital (Schröder 1999). Mainly steroids which are structurally related to androsterone were found in effluents from another German company. They are probably byproducts of the manufacturing of hormone supplements (Dsikowitzky 2002). Further steroid hormones including cortisol, cortisone, dexamethasone and prednisolone were determined in pharmaceutical industry effluents in the Netherlands (Schriks et al. 2010).

We found a rather limited number of studies about effluents from other industrial facilities such as the food-processing industry, cement industry, automotive industry and a power station (Table 2.1). Information about the emissions of the coating and printing industry and of manufacturers of electric devices, toys and computers would be interesting, but are lacking.

Several compound groups appeared in many different industrial effluent types including benzene and naphthalene sulfonates, phthalic acid esters, volatile organic solvents, chlorinated benzenes, phenol and alkylphenols. It illustrates that they are unspecific and implies that their occurrence in the environment does not allow for attributing their presence to certain emission sources. The analytical approaches determine the set of compounds which are detectable in the samples. Compound groups which were addressed frequently were sulfonates and endocrine disrupters. However, screening studies illustrated that emerging environmental contaminants might be generated during wastewater treatment and released to aquatic systems, which are not addressed by monitoring studies. This applies also to byproducts of industrial synthesis. The replacement of environmentally problematic compounds by novel industrial formulation is common and leads to the occurrence of further emerging contaminants in the effluents with unknown impact in the environment.

Overall, the heterogeneity of the chemical composition of effluents from similar industries is striking. The observed differences are related to different industrial production processes, leading to the presence of varying synthesis educts, additives, products and byproducts in the wastewaters. For wastewater treatment, different techniques are applied which exhibit different efficiencies in contaminant removal. Chemical materials are synthesized in batches, leading also to short-term fluctuations in the effluent composition. Many authors addressed these fluctuations and generated composite samples or carried out multiple sampling sessions.

Given the diversity of the production processes and high structural diversity of the effluent constituents, further studies on effluents from chemical production sites, petrochemical and pharmaceutical industry should be carried out. These industries emit a plentitude of structurally diverse organic contaminants. Many of these compounds might be stable in aquatic systems, but there occurrence has not been reported yet and no information is available about their distribution in the environment and their ecotoxicological relevance.

The occurrence of the described compounds in industrial effluents does not indicate their persistence in the environment. The fate of industrial contaminants in aquatic systems is discussed in Sect. 2.3.

2.3 Tracing Industrial Wastewaters in the Environment

In this section, studies on the occurrence and fate of industrial contaminants in aquatic systems are discussed. Only studies are selected, which applied a comprehensive chemical characterization of the samples, and in which the identified contaminants were unequivocally attributed to industrial emissions. Although various studies were published which focus on the occurrence of organic contaminants in industrial wastewaters (Sect. 2.2.2), relatively few studies were found which survey the fate of industrial contaminants in aquatic systems (Table 2.2).

Two early studies were groundbreaking for this topic. Organic contaminants were identified in effluents of a chemical production site (Table 2.1) and subsequently traced in the adjacent river water course and coastal waters (Table 2.2) (Jungclaus et al. 1978 and Lopez-Avila and Hites 1980). The chemical composition

Table 2.2 Organic compounds which were detected in water, sediment or animal samples and could unequivocally be attributed to industrial emissions

Industry	Detected organic compounds	Region	Concentration range	Compartment	Reference
Textile industry	Aminohydroxyphenoxyanthracenedione (CI disperse Red 60), dihydroxybis(methylamino)anthraquinone (CI disperse blue 26), acetamidoacetyloxyethylbromodinitrophenyldiazenylmethoxyanilinoethyl acetate (CI disperse blue 79), bromodinitroaniline	Canada	nq	River water, sediment	Maguire (1992)
	Acetylaminobis(methoxyethyl)aminomethoxyphenylaminobromochlorobenzotriazole, acetylaminocyanoethylethylaminomethoxyphenylaminobromochlorobenzotriazole, acetylaminohydroxyethylaminomethoxyohenylaminobromochlorobenzotriazole	Japan	nq	River water	Nukaya et al. (1997), Oguri et al. (1998), and Shiozawa et al. (2000)
	Aminodichlorobenzothiazole	Korea	nq	River water	Kwon et al. (2003)
	Bromodinitrophenyldiazenyldipropenylaminomethoxyphenylacetamide (CI disperse Blue 373)	Brazil	nq	River water	Umbuzeiro et al. (2005)
Tanneries	Ethylacetate, ethanol, methylpentanone, toluene, methylpropanol, butanol, ethoxyethanol, butoxyethanol, bis(dimethylaminoethyl)ether, hexanediol, ethylhexanol, dimethylbenzylalcohol, butoxyethoxyethanol, hexylglycolmono-iso-butyrate, di-*tert*.-butylmethylphenol, phenol, cresol, di-*tert*.-butyldimethylbicyclohexanone	Japan	9–5,680 μg l^{-1}	River Water	Yasuhara et al. (1981)
Petrochemical industry	Tribromomethane	Greece	98,000 ng/l	Coastal water, water from canals and marshes	Grigoriadou et al. (2008)
	Dimethylpyridine, thioanisole	Germany	<5–130 ng l^{-1}	River water	Botalova and Schwarzbauer 2011
	Alkylated naphthalenes, alkylated phenanthrenes/anthracenes, alkylated dibenzothiophenes, alkylated biphenyls	Indonesia	0.005–20 μg g^{-1} dw	Coastal sediment	Dsikowitzky et al. (2011)

(continued)

Table 2.2 (continued)

Industry	Detected organic compounds	Region	Concentration range	Compartment	Reference
	Alkylated naphthalenes, alkylated phenanthrenes/anthracenes, alkylated dibenzothiophenes, alkylated biphenyls, triphenylphosphine oxide, triphenylphosphine sulfide	Indonesia	nq	Mangrove macrobenthic invertebrates and fish	Dsikowitzky et al. (2011)
Pulp and paper industry	Chlorophenols, chloroguaiacols, chlorocatechols, resin acids	Finland	200–20,400 ng l^{-1}	Lake water	Oikari et al. (1985)
	Chlorophenols, chloroguaiacols, chlorosyringols, chlorocatechols	Sweden	12–3,710 ng l^{-1}	Coastal water	Söderström et al. (1994)
	Chlorophenols, chloroguaiacols, chlorosyringols, chlorocatechols	Sweden	0.027–48.7 μg g^{-1}	Marine fish bile	Söderström et al. (1994)
	Chlorophenols, chloroguaiacols, chlorocatechols	Finland	Not indicated	Coastal sediment	Palm and Lammi (1995)
	Chlorophenols, chloroguaiacols, chlorovanillins, chlorocatechols, chlorinated dehydroabietic acids	New Zealand	<0.01–5.2 μg g^{-1} dw	Lake sediment	Tavendale et al. (1995)
	Nonylphenol ethoxycarboxylates	US	1,700–11,800 ng l^{-1}	River water	Field and Reed (1996)
	Chlorophenols, chloroguaiacols, chlorovanillin, chlorocatechols	New Zealand	<0.01–0.0247 μg g^{-1} dw	River sediment	Judd et al. (1996)
	Chlorophenols, chloroguaiacols, chlorovanillin, chlorocatechols	Finland	2.1 μg ml^{-1}	Fish bile	Leppänen et al. (1998)
	Resin Acids, sitosterol, wood sterols, chlorophenols	Finland	<0.1–3,290 μg g^{-1} dw	Coastal sediment	Meriläinen and Oikari (2008)
	Resin Acids, sitosterol, chlorophenols	Finland	<0.05–7,800 μg g^{-1} dw	Macrobenthic invertebrate and insect tissues	Meriläinen and Oikari (2008)

	Dimethyldiphenylmethane, dimethylphenylethane, di-iso-propyl-naphthalene, bis(methylphenoxy)ethane, terphenyl, benzyl-naphthylether, diphenoxybenzene, benzylbiphenyl, dichloromethylphenoxyethane, dibenzyloxybenzene	Japan	130–5,290 ng l^{-1}	Coastal water	Terasaki et al. (2012)
	Dimethyldiphenylmethane, dimethylphenylethane, di-iso-propyl-naphthalene, bis(methylphenoxy)ethane, terphenyl, benzyl-naphthylether, chloromethylphenoxymethylphenoxyethane	Japan	1.6–190 μg g^{-1} dw	Coastal sediment	Terasaki et al. (2012)
	Dimethyldiphenylmethane, dimethylphenylethane, di-iso-propyl-naphthalene, bis(methylphenoxy)ethane, terphenyl, benzyl-naphthylether, diphenoxybenzene, chloromethylphenoxymethylphenoxyethane, benzylbiphenyl, dichloromethylphenoxyethane, dibenzyloxybenzene	Japan	<0.01–0.0872 μg g^{-1} ww	Marine fish tissue	Terasaki et al. (2012)
Rubber and tire production	Benzothiazole, methylthiobenzothiazole, methylbenz-iso-thiazole, methylbenzothiazole, cyclohexylamine, toluene, aniline	Spain	nq	Irrigation canal water	Puig et al. (1996)
Chemical production sites					
Synthesis of e.g. pharmaceuticals, pesticides, optical brighteners and surfactants	Hydroxy-*tert*.-amylphenylbenzotriazole, hydroxy-*tert*.-butylmethylphenylchlorobenzotriazole, chloroethylamino-iso-propylaminotriazine, chlorobis(ethylamino)triazine, phenyl-naphthylamine, tetramethylbutylphenylnaphthylamine, toluidinesulfonylphenol, diphenyloctylphosphate, *tert*.-amylphenol, methoxydi-*tert*.-butylphenol, di-*tert*.-butylcyanophenol, di-*tert*.-amylcyanophenol, di-*tert*.-amylcyanomethylphenol, chlorophenylisocyanate, chlorotrifluoromethylphenylisocyanate, chlorophenyl-iso-propylcarbamate, chlorotrifluoromethylphenyl-iso-propylcarbamate	US	0.1–1,600 μg g^{-1}	River and estuarine sediments	Jungclaus et al. (1978); Lopez-Avila and Hites (1980)

(continued)

Table 2.2 (continued)

Industry	Detected organic compounds	Region	Concentration range	Compartment	Reference
Synthesis of e.g. pharmaceuticals, pesticides, optical brighteners and surfactants	Aminoethyldibenzoazepine, methylquinoxaline, dimethylquinoxaline, phenylnaphthylamine, *tert.*-butylphenylnaphthylamine, tetramethylbutylphenylnaphthylamine, methylthiobenzothiazole, benzothiadiazole, *tert.*-amylphenol, di-*tert.*-butylnitrophenol, di-*tert.*-amylcyanophenol, di-*tert.*-amylcyanomethylphenol, di-*tert.*-butylbenzoquinonemethideacetic acid, propylcarboxymethyldi-*tert.*-butylhydroxyphenylpropionate	US	100–60,000 ng l^{-1}	River and coastal water	Jungclaus et al. (1978); Lopez-Avila and Hites (1980)
Synthesis of circuit boards	Tetrabromobisphenol A, dimethylated derivative of tetrabromobisphenol A	Sweden	0.27–1.5 μg g^{-1} dw	River sediment	Sellström and Jansson (1995)
Synthesis of chlorinated organic solvents	Chloroform, carbon tetrachloride, trichloroethylene, tetrachloroethylene, pentachlorobenzene, hexachlorobenzene, octachlorostyrene	Spain	11–2,400 ng l^{-1}	River water	Amaral et al. (1996)
Coating industry	Hexamethoxymethylmelamine, pentamethoxymethylmelamine	Netherlands	nq	Surface water	Bobeldijk et al. (2002)
Cloralkali electrolysis	Penta-, hexa- and heptachlorinated naphthalenes	Germany	<0.01–0.55 μg g^{-1} dw	River sediment	Brack et al. (2003)
Chlorinated industrial and agricultural chemicals	Aldrin, biphenyl, chlordane, DDT and DDT-metabolites, endosulfan, fluorine, heptachlor, HCHs, PCBs, PAHs	Azerbaijan	0.002–6.6 μg g^{-1}	Coastal wetland sediment	Swartz et al. (2003)
Chlorinated industrial and agricultural chemicals	Aldrin, dieldrin, endrin, heptachlor, chlordanes, nonachlor, HCHs, tertrachlorobenzene, hexachlorobenzene, pentachloroanisole, pentachlorobenzene, endosulfan, mirex, chlorpyrifos, DDT and DDT-metabolites, PCBs, PAHs	Azerbaijan	0.31–1,717 μg g^{-1} dw	Freshwater turtle tissue	Swartz et al. (2003)
Synthesis of organochlorine compounds	Hexachlorobutadiene	Germany	10–80 ng l^{-1}	River water	Dsikowitzky et al. (2004)
Synthesis of organochlorine compounds	Bis(chloropropyl)ethers, hexachlorobutadiene, octachlorostyrene	Germany	0.0005–0.058 μg g^{-1} dw	River sediment	Kronimus et al. (2004)

Synthesis of industrial intermediates and vitamins	Di-iso-propylidenesorbofuranose, bis-ethyl-iso-octanol lactone isomers, di-iso-propylidenexylohexulofuranosoic acid, triphenylphosphinoxide,	Germany	110–4,000 ng l^{-1}	River water	Knepper and Karrenbrock (2006); Knepper et al. (2000)
Synthesis of e.g. plastics, flame-retardants and plant protection agents	Dichloroaniline, triethylphosphate, tributylphosphate, trimethylpentanedioldiisobutyrate	Germany	<5–160 ng l^{-1}	River water	Botalova and Schwarzbauer (2011)
Synthesis of industrial chemicals, e.g. polymers and solvents	Chloromethyldioxolane, triethylphosphate, trimethylbenzoic acid, *tert.*-butylbenzoic acid	Germany	<5–530 ng l^{-1}	River water	Botalova and Schwarzbauer (2011)
Synthesis of organochlorine pesticides	DDT, DDD, DDE, DDMU, DDA, DBP, DDCN, DDABE, DDAMA, CPE, MCPE, DCPE, HCHs, tetrachlorocyclohexene, pentachlorocyclohexenes, heptachlorocyclohexene, chlorinated benzenes, chlorinated benzoic acids, chlorinated phenols	Germany	<100–75,000,000 ng l^{-1}	Groundwater	Frische et al. (2011)
Pharmaceutical industry	Chlorthalidone, torsemide, zolpidem, azithromycin, warfarin, desmethylazithromycin, erythromycin, dehydrated erythromycin, terbinafine	Croatia	0.8–20.1 μg g^{-1} dw	River sediment	Terzic and Ahel (2011)
Food- processing industry					
Meat production	Dibenzylamine, methylindolinone	Germany	6–18 ng l^{-1}	River water	Botalova and Schwarzbauer (2011)

nq: not quantified dw: dry weight ww: wet weight

The contaminant concentrations in animal tissue and sediments were listed as dry weight or wet weight values, if this information was indicated in the references

of the river water reflected the effluent discharge and the aqueous concentrations of the various compounds followed the rule of simple dilution. The compounds with the highest octanol-water coefficients were found to be strongly associated with particulate matter and were detectable in the sediments, also at the greatest distance from the plant, some of them in high concentrations. Namely benzotriazole, chlorobenzotriazole, diphenylether, trichlorodiphenylether, dibenzoazepine, triazine, phenylbutazone and phenylnaphthylamine occurred throughout the whole environmental system or were present at relatively high concentrations. Some of the compounds evaporated or degraded; for example, the degradation process involved the oxidation of dissolved phenols to quinones. It was concluded that the long-term exposure to this wide variety of contaminants contributed to the lack of biota in the receiving river water course.

Most of the later studies observed solely the fate of industrial contaminants in aquatic systems without the simultaneous characterization of contaminants in the effluents. Many of them were carried out in industrial nations of the northern hemisphere.

Aminodichlorobenzothiazole was identified for the first time in river water near a complex of the ***textile industry*** in Korea (Table 2.2) (Kwon et al. 2003). It is an educt for the synthesis of magenta dyes (EP Patent 0,468,380). Although the Ames test revealed no mutagenic activity of the compound, its presence in public waterways might need further attention due to the unknown effects on the health of humans and aquatic animals. Different dye stuffs and degradation products were also detected in other river systems (Maguire 1992; Nukaya et al. 1997; Oguri et al. 1998; Shiozawa et al. 2000; Umbuzeiro et al. 2005). Another study on riverine contamination by a leather ***tannery*** demonstrated the presence of relatively unspecific non-halogenated organic solvents in water samples (Yasuhara et al. 1981).

High concentrations of tribromomethane, dimethylpyridine and thioanisole were detected in water samples from the vicinity of ***petrochemical industries*** (Grigoriadou et al. 2008; Botalova and Schwarzbauer 2011). However, the origin of tribromomethane and thioanisole could not directly be linked to production processes in the petrochemical plants (Grigoriadou et al. 2008; Botalova and Schwarzbauer 2011). Different alkylated PAHs were detected in water, sediments, mangrove macrobenthos and fish from a tropical lagoon which receives wastewaters from a large oil refinery (Dsikowitzky et al. 2011). Alkylated PAHs are more abundant in crude oil and more persistent in the environment than parent PAHs (Barron and Holder 2003). This suggests their use for the tracing of petrogenic contamination rather than considering only a limited number of parent PAHs. Furthermore, triphenylphosphineoxide, a byproduct of the industrial production of olefins, was detected in many of the animal tissue samples from the tropical lagoon (Dsikowitzky et al. 2011). Its presence in petrochemical effluents was demonstrated (Botalova et al. 2009), but it also occurs in municipal sewage (Rodil et al. 2005).

The bleaching of chemical pulp with chlorine, which is a process frequently applied by the ***pulp and paper industry***, results in the performance of chlorinated derivatives of plant constituents such a chlorophenols, chloroguaiacols, chlorovanillins, chlorosyringols and chlorocatechols (Sect. 2.2.2). Their presence in aquatic systems was demonstrated in a number of studies, and they were detected in water, sediment and animal tissue (Table 2.2). A completely different compound spectrum was found in the vicinity of thermal paper manufacturing plants due to the different production processes. The chemical composition of the water, sediment and animal tissue from the plant vicinity reflected adequately the composition of the wastewaters (Tables 2.1 and 2.2) (Terasaki et al. 2012). Discharges of a ***rubber and tire production facility*** resulted in the presence of different derivatives of benzothiazole in a water irrigation channel (Tables 2.1 and 2.2) (Puig et al. 1996). The compounds remained intact after the purification process in the plant and were traced in the public waterways over more than 100 km distance from the source, indicating their stability.

Tetrabromobisphenol A, a brominated flame retardant, and its dimethylated derivative were found in high concentrations in sediment samples downstream a ***chemical production site*** in Sweden (Sellström and Jansson 1995). It was demonstrated that this particular production site emitted the tetrabromobisphenol A by analyzing circuit boards which are produced by the company and by analyzing sludge from a sewage treatment plant receiving leach water from a landfill with wastes from this plant. The circuit boards and the sludge samples contained high levels of the target compound. In sewage sludge samples from a treatment plant to which no known users of tetrabromobisphenol A were connected, and in sediment samples from upstream the plant, significantly lower levels of the compound were found.

Hexamethoxymethylmelamine was found in surface waters of the Netherlands (Bobeldijk et al. 2002). This compound is used as cross-linking agent in organic coatings and plastics. In Europe, industrial applications include coatings for cans, coils and automobiles (Weiss 1997). HMMM is also used for the formation of melamin nanoparticles and for the production of castings (e.g. Weber et al. 2009). The presence of this compound in surface waters is therefore likely related to its application in specific industrial production processes.

The presence of different chlorinated organic contaminants in Spanish river water reflected the impact of a chlorinated organic solvent factory in the vicinity of the sampling sites (Table 2.2). However, the tap water of the nearby village was not affected and the concentrations of chlorinated compounds were below safety thresholds for drinking water (Amaral et al. 1996). The city of Sumgayit in Azerbaijan was a center for the production of chlorinated industrial and agricultural chemicals of the former Soviet Union. In sediments from wetlands in the vicinity of the industrial complex, high concentrations of toxic organochlorine pesticides were found. In comparison to turtles from reference sites, significantly enhanced levels of all emitted pesticides were measured in two turtle species (*Mauremys caspica*, *Emys orbicularis*) from the wetlands (Swartz et al. 2003). Bitterfeld was an important industrial production complex for chlorinated organic compounds in the former German Democratic Republic. The industrial discharges were released to a brook which flows into the Mulde River, a tributary of the Elbe River. To date chlorinated naphthalenes have been reported in high concentrations in the brook sediments (Table 2.2) (Brack et al. 2003), as well as in Elbe River and Mulde River sediments

(Schwarzbauer, unpublished results). Organochlorine pesticides and other chlorinated organic compounds were also detected in groundwater samples from the vicinity of another production unit of the German Democratic Republic which was closed at the end of the 1980s (Frische et al. 2011). The presence of specific chlorinated contaminants downstream a modern chemical production site in Germany was attributed to the technical synthesis of low volatile chlorinated organic compounds in the plant (Dsikowitzky et al. 2004; Kronimus et al. 2004).

In another study, several types of organic contaminants were identified in industrial wastewaters of the chemical industry which were byproducts of specific production processes (Table 2.1). All of them were also traced in the Rhine River, Germany (Knepper and Karrenbrock 2006; Knepper et al. 2000) (Table 2.2). In river water samples taken downstream at different distances from several industrial point sources in Germany, a set of organic contaminants was identified, which were attributed to the specific discharges. Chloromethyldioxolane only occurred in samples taken downstream a chemical production site and was suggested as site-specific molecular marker (Botalova and Schwarzbauer 2011).

The presence of several pharmaceutical drugs in river sediments was unequivocally attributed to discharges of the ***pharmaceutical industry***. Less known pharmaceutical intermediates and/or transformation products were identified, which had not been reported from freshwater sediments as yet (Terzic and Ahel 2011).

Only in comprehensive studies including industrial wastewaters *and* field samples, the compound spectra found in the wastewater samples could be retrieved in samples from the adjacent aquatic system. In all other cases, except the pulp and paper industry, the compound spectra found in field samples from industrialized areas (Table 2.2) was totally different from those in studies on effluents of the respective industries (Table 2.1). This illustrates again the complexity of the chemical composition of industrial wastewaters, even from the same industry types, and the lack of data about the related environmental impacts (see also Sect. 2.2.2). Screening analyses of the individual contamination profile of industrialized areas are therefore urgently required including the examination of wastewater samples *and* field samples.

Many well-known pollutants such as PCBs, chlorinated benzenes and flame retardants may either origin from industrial or from municipal sources (e.g. Blanchard et al. 2004; Law et al. 2006; Marti et al. 2011). Although their presence in industrialized areas suggests their industrial origin, they exhibit a lot of different applications and therefore cannot unequivocally be attributed to industrial emissions. Studies with a high spatial resolution covering upstream and downstream sampling at potential emission sites are necessary to demonstrate industrial inputs of such unspecific contaminants. It is therefore beneficial to identify industrial organic markers which allow the tracing of specific industrial emissions in the environment. The marker concept has been developed by geochemists to understand the sources, transport and fate of organic compounds. An ideal marker should be source specific and exhibit a conservative behavior. The source specificity refers to the link between a molecular marker and its source. The conservative behavior implies that the marker is stable over time scales relevant to the processes under study (Eganhouse

1997; Takada and Eganhouse 1998). Potential industrial organic markers should be characteristic for wastewaters from different industrial branches and for the specific industrial production processes which are applied.

These markers can then be used to verify the input of specific industrial wastewaters to aquatic systems and to investigate the spatial distribution of the emission. This information is vital to disentangle different emission sources and to further investigate their potential impacts in the environment. The generated knowledge can provide a scientific basis for the development of management measures for a reduction of the contamination. In Table 2.3, potential industrial organic markers for different emission sources are presented as synthesis of Sects. 2.2.2 and 2.3.

The listed compounds are source specific, meaning that their occurrence is related to specific industrial production processes. Due to their structural characteristics it can be excluded that they also occur in nature. Furthermore, they were detectable in field samples indicating their environmental stability. Therefore, they fulfill the criteria for marker compounds. The chemical structures of these potential industrial markers are presented in Fig. 2.2. Some of the compounds which were detected in the environment and were unequivocally attributed to a specific industrial source are known to occur also in municipal sewage such as triphenylphosphineoxide. These compounds were excluded. Future studies should verify the usefulness of the suggested industrial markers and identify more markers for specific industrial production processes.

2.4 Identification of Toxic Organic Contaminants in Industrial Wastewaters

Toxicity tests with microorganisms and algae were successfully conducted to survey the toxic effects of industrial wastewaters before and after wastewater treatment (e.g. Chen et al. 1999; Kungolos 2005a, b). The approach allows to assess the reduction of the toxicity which is achieved by wastewater treatment and to estimate the impact of the treated wastewater on aquatic organisms. The toxicity of numerous single organic contaminants which are commonly found in industrial wastewaters was tested under laboratory conditions for different experimental species (e.g. Nestmann et al. 1980; Drzyzga et al. 1995; Ali and Sreekrishnan 2001).

The mentioned studies give evidence about the toxicity of specific industrial organic contaminants and different industrial wastewaters and are useful to estimate potential impacts in the environment. However, the causative constituents which are responsible for the observed toxicity of whole wastewater samples are not identified. To close this gap of knowledge, mainly two approaches are applied which combine chemical analysis and toxicity evaluation: (1) chemical characterization of wastewaters and toxicity evaluation, and (2) toxicity-based fractionation of the wastewater samples and thereafter identification of the compounds which caused the observed toxic effects. Both approaches are discussed in the following sections.

Table 2.3 Potential marker compounds which could be useful to trace industrial emissions in aquatic systems

Industry	Organic marker	Compartment	Production process
Textile industry	Aminodichlorobenzothiazole	Water	Educt in the synthesis of magenta dyes
	Azo dye stuffs such as CI Disperse Blue 373 and their metabolites	Water	Coloring of textiles
Petrochemical industry	Dimethylpyridine	Water	Refining of crude oil (compound is a constituent of tar)
	Alkylated dibenzothiophenes	Water, sediment, biota	Refining of crude oil (compounds are characteristic constituents of crude oil and therefore characteristic for petrogenic contamination)
Pulp and paper industry	Chloroguaiacols, chlorovanillins, chlorocatechols, chlorinated dehydroabietic acids, chlorosyringols	Water, sediment, biota	Bleaching of chemical pulp with chlorine
	Chloromethylphenoxymethylphenoxyethane, Bis(methylphenoxy)ethane, benzylnaphthyl ether, benzylbiphenyl	Water, sediment and/or biota	Application of sensitizers and sensitizer related compounds for the preparation of thermal paper
Rubber and tire production	Presence of numerous derivatives of benzothiazole	Water	Vulcanization
Chemical production sites	Bis-ethyl-iso-octanol lactone isomers	Water	Butanal synthesis
	Di-iso-propylidenesorbofuranose, di-iso-propylidenexylohexulofuranosoic acid	Water	Byproduct of the vitamin C synthesis
	Hexamethoxymethylmelamine	Water	Coating of metals
	Bis(chloropropyl)ethers, hexachlorobutadiene, octachlorostyrene	Sediment	Synthesis of chlorinated industrial chemicals
	Chloromethyldioxolane, hexachlorobutadiene, octachlorostyrene	Water	Synthesis of chlorinated industrial chemicals

These compounds do not occur in municipal sewage, their origin can be attributed to specific industrial production processes and they were detectable in industrial areas. They can be used to verify the input of specific industrial wastewaters to aquatic systems and to investigate the spatial distribution of the emission

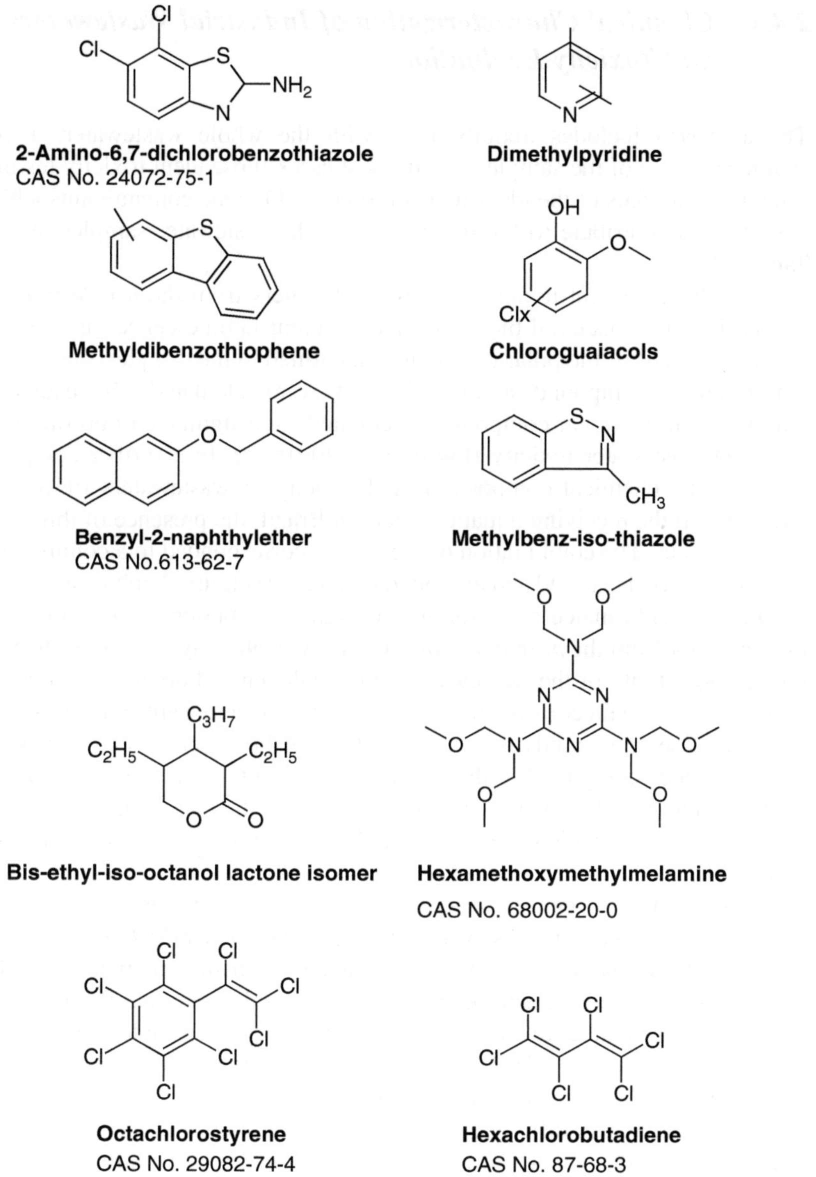

Fig. 2.2 Chemical structures of potential industrial markers; These compounds do not occur in municipal sewage, their origin can be attributed to specific industrial production processes and they were detectable in industrial areas. They can be used to verify the input of specific industrial wastewaters to aquatic systems and to investigate the spatial distribution of the emission

2.4.1 Chemical Characterization of Industrial Wastewaters and Toxicity Evaluation

The approach includes toxicity tests with the whole wastewater, a chemical characterization of the samples and in most cases subsequent toxicity testing using reference materials of the identified compounds. Organic contaminants which were identified to contribute to the toxicity of whole wastewater samples are listed in Table 2.4.

The whole effluent toxicity of raw wastewaters from the ***textile industry*** was assessed using a bacterial biosensor and different luminescence inhibition assays. The EC_{50} values of the polar compounds identified in the samples were determined with reference compounds and toxicity units were calculated. The results revealed that the identified polar compounds accounted for a significant portion of the total observed wastewater toxicity (Farré et al. 2001a, b). In a further comprehensive study on the chemical composition and toxicity of wastewaters from the textile industry and the receiving aquatic system in Brazil, the presence of three prevalent dyes came out. The combination of these dyes corresponded to a commercial product, which was also tested and exhibited mutagenic activity (Umbuzeiro et al. 2005).

Different chlorinated and brominated organic compounds were found in spent bleach liquor from the ***pulp and paper industry*** in Norway. The Ames test revealed mutagenic activity of the wastewater. Three halogenated organic compounds were found to be the main constituents of the wastewaters and were synthesized to investigate if these compounds were responsible for the observed effect. However, all three reference compounds only showed a weak mutagenic activity (Bjørseth et al. 1979; Carlberg et al. 1980). The authors noted that the compounds or the combination of compounds which were responsible for the mutagenic effect could not been identified.

Phenol, dihydroxybenzene and hydroquinone were present in the effluents of a treatment plant receiving wastewaters from a ***chemical production site*** in Italy. The toxicity of the effluent constituents was tested with a battery of experimental species and purchased reference compounds. The test results and data from the literature were used to calculate toxicity units. Effluent toxicity was under- or overestimated by calculating the sum of the toxicity units depending on which toxicity data and test organisms were used (Guerra 2001).

Toxicity tests with different industrial wastewaters from Sweden, Germany and Spain revealed the genotoxic and cytotoxic properties of untreated wastewaters from tanneries and from a chemical production site. The cytotoxic effects correlated with the presence of LAS in the samples (Castillo et al. 2001). The effluents of different industries in South China were tested for their acute toxicity with a battery of test species including green algae (*Pseudokirchneriella subcapitata*), *daphnia*, genetically modified strains of *Escherichia coli* and duckweed (*Lemna minor*) (Fang et al. 2012). In all effluents endocrine disrupting chemicals, PCBs, PAHs and PCDDs/PCDFs were detectable. Due to the complex chemical composition of the effluents, the authors remark that the observed toxicity of the effluents was caused

Table 2.4 Toxic organic contaminants identified in industrial wastewaters by chemical characterization and toxicity evaluation

Industry	Identified toxicant	Wastewater type	Approach	Experimental species	Biological effect	Reference
Textile industry	Alkylalcohol polyethoxylate, naphthalene sulfonates	Untreated wastewater	Toxicity tests with wastewaters, chemical characterization & toxicity tests with single reference compounds	*Escherichia coli*	Reduction of metabolic activity	Farré et al. (2001a)
	Alkylalcohol polyethoxylates, chlorophenols, phenol, nonylphenol polyethoxylate, polyethylene glycol	Untreated wastewater	Toxicity tests with wastewaters, chemical characterization & toxicity tests with single reference compounds	*Vibrio fischeri*	Luminescence inhibition	Farré et al. (2001b)
	Bromodinitrophenyldiazenyldipropenylaminomethoxyphenylacetamide (CI disperse Blue 373), dichloronitrophenylazoethylanilino] propionitrile (CI Disperse Orange 37), bromodinitro-phenyldiazenyldiethylaminopheny-lacetamide (CI Disperse Violet 93)	Effluent	Toxicity tests with effluents, chemical chracterization & toxicity tests with the industrial product	*Salmonella* strains	Mutagenicity	Umbuzeiro et al. (2005)
	Octylphenol, nonylphenol, bisphenol A, estrone, estradiol, triclosan, PCBs, PAHs, PCDD/PCDF	Effluent	Toxicity tests with effluents & examination of wastewater constituents	Battery of different test species	Growth inhibition, mortality	Fang et al. (2012)
Tanneries	LAS, polyethylene glycol, alkylalcohol polyethoxylates, nonylphenol polyethoxylate, benzene sulfonates, naphthalene sulfonates	Untreated wastewater	Toxicity tests with wastewaters & examination of wastewater constituents	*Vibrio fischeri or Photobacterium phosphoreum*, *Salmonella typhimurium*	Luminescence inhibition, mutagenicity	Castillo et al. (2001)

(continued)

Table 2.4 (continued)

Industry	Identified toxicant	Wastewater type	Approach	Experimental species	Biological effect	Reference
Pulp and paper industry	Trichlorotrihydroxybenzenes, bromomethylpropanylbenzene, dichloromethylpropanylbenzene	Spent bleach liquor	Toxicity tests with wastewaters, chemical characterization & toxicity tests with synthesized reference compounds	*Salmonella typhimurium*	Mutagenicity	Bjørseth et al. (1979) and Carlberg et al. (1980)
	Octylphenol, nonylphenol, bisphenol A, estrone, estradiol, triclosan, PCBs, PAHs, PCDD/PCDF	Effluent	Toxicity tests with effluents & examination of wastewater constituents	Battery of different test species	Growth inhibition, mortality	Fang et al. (2012)
Chemical production sites						
Synthesized compounds not indicated	Linear alkylbenzene sulfonates, benzene sulfonates, naphthalene sulfonates	Effluents	Toxicity tests with wastewaters & examination of wastewater constituents	*Vibrio fischeri or Photobacterium phosphoreum*, *Salmonella typhimurium*	Luminescence inhibition, mutagenicity	Castillo et al. (2001)
Synthesis of rubbers, fertilizers and industrial intermediates	Phenol, dihydroxybenzene, hydroquinone	Effluents	Toxicity tests with effluents, chemical characterization & toxicity tests with reference compounds	*Daphnia magna*, *Artemia salina*, *Brachionus plicatilis*	Mobility inhibition, mortality	Guerra (2001)
Coating industry	Octylphenol, nonylphenol, bisphenol A, estrone, estradiol, triclosan, PCBs, PAHs, PCDD/PCDF	Effluent	Toxicity tests with effluents & examination of wastewater constituents	Battery of different test species	Growth inhibition, mortality	Fang et al. (2012)

by a mixture of various substances which makes the identification of the causative constituents very difficult. Moreover, the mixture effects can be synergistic, additive or antagonistic. No further toxicity tests with reference compounds were conducted.

A review of approaches involving chemical analysis and toxicity testing of wastewaters and an overview of toxicity data for phenols, polyethoxylate surfactants, LAS, naphthalene and benzene sulphonates and other organic contaminants was presented by Farré and Barceló (2003). The study of the toxic response of these commonly in industrial wastewaters occurring compounds can help to identify their contribution to the total toxicity of wastewater samples. A broad-range correlation between presence of priority pollutants and whole effluent toxicity across a range of industry types was done by Sarakinos et al. (2000). There were effluent types for which there was one order of magnitude variation in inferred and measured toxicity. Overall, chemical-based assessments tended to overestimate toxicity of effluents containing high metal concentrations and to underestimate the toxicity of pulp mill effluents. Main and interaction effects of contaminants commonly found in textile dyes, pulp and paper mill effluents and oil refinery effluents were assessed by Parvez et al. (2008). Four-component mixtures using reference compounds were prepared and the toxicity of the different combinations were determined using a luminescence inhibition assay with *Vibrio fischeri*. In all tested mixtures, the magnitude of main effects was more significant than the interaction effects. It was furthermore highlighted that the behavior of an individual component changed in presence of other components. Specific physico-chemical properties, i.e. partition coefficient, molecular size and polarity of the compounds partly explained the behavior of the compounds in a mixture.

Although it was demonstrated that toxic effects correlated with the presence of specific contaminants in the investigated wastewater samples, the total toxicity of the samples could not be attributed to the identified compounds. This is due to the complex chemical composition of the samples, which impede a complete chemical characterization of the organic constituents. The presence of compounds which have not been considered as target analytes such as unidentified organic contaminants and heavy metals might also have contributed to the observed toxic activities. Furthermore, also other physico-chemical factors such as salinity, temperature, pH, nutrient concentrations and hardness might influence total wastewater toxicity.

2.4.2 *Effects-Directed Analysis*

This approach includes the extraction of wastewater samples and fractionation applying chromatographic methods such as HPLC and TLC. Subsequently, the obtained fractions are characterized according to their toxicity. The constituents of the toxic fractions are then analyzed by common analytical methods such as GC/MS, LC/MS and LC/MS-MS. As the last step, the toxicity of the identified compounds has to be confirmed (e.g. Schuetzle and Lewtas 1986; Brack 2003; Hewitt and Marvin 2005; Brack et al. 2008).

In the 1980s, the Environmental Protection Agency of the US developed a method for the extraction, fractionation and toxicity testing of unpolar organic contaminants in wastewaters (Burkhard and Ankley 1989; Burkhard et al. 1991). The sample extraction was executed with reversed phase SPE. The extracted compounds were eluted using mixtures of water and methanol. The toxicity of the obtained fractions was tested using *daphnia*. Only the toxic fractions were further separated with HPLC and analyzed with GC/MS. The crucial advantage of this newly developed method was the application of a reversed extraction phase. From this phase, the extracted compounds were eluted using relatively low toxic water–methanol solvent mixtures. The tolerance of most species towards methanol allows a testing of the fractions with up to one percent per volume methanol. Using this approach, effluent samples from the pharmaceutical industry, a textile factory and the pulp and paper industry were investigated and tested with *Vibrio fischeri* (Svenson et al. 1996). Further developments of effects-directed analysis allowed to augment the determinable compound spectrum to more polar and hydrophilic wastewater constituents (Fiehn and Jekel 1996; Fiehn et al. 1997; Reemtsma et al. 1999a, b; Castillo and Barceló 1999b; Reemtsma 2001). The exclusive usage of one test species probably leads to the systematic neglect of toxicants which do not have an effect on the specific cell metabolism of this test organism (Brack 2003). It is therefore a meaningful further methodological development to use a set of test organisms such as daphnia, green algae *and* luminous bacteria (Brack et al. 1999; Brack 2003). Additional to the conventional biotests also the genotoxic, hormonal and bioaccumulative potential of wastewaters should be evaluated (Pessala et al. 2004). The analytical methods applied for the effects-directed analysis of wastewater and field samples were reviewed in detail by Hewitt and Marvin (2005). Toxic organic compounds which were identified in industrial wastewaters applying effects-directed analysis are summarized in Table 2.5.

The acute toxicity of a ***textile industry*** effluent for *Vibrio fischeri* was found to be caused by two unsaturated fatty acids and two tridecanols, which accounted for 84 % of the toxicity of the effluent (Svenson et al. 1996). In the most toxic fractions of untreated textile wastewaters from Spain, a set of organic contaminants was identified including phthalic acid esters and nonylphenol isomers (Castillo and Barceló 2001). Different benzothiazole derivatives, cresol, dichlorobenzoic acid and methylbenzoic acid were identified as toxicants in ***tanneries*** effluents (Reemtsma et al. 1999a, b). In the toxic fractions of tannery wastewater, a variety of anionic and nonionic organic compounds were found. The confirmation step included toxicity tests with artificial water samples which contained the major identified contaminants in similar concentration levels (Castillo and Barceló 1999b).

A number of studies on the identification of toxicants in effluents from the ***pulp and paper industry*** have been conducted in Canada. Moreover, studies of the effects of pulp and paper mill effluents on aquatic organisms were reviewed in detail by

Table 2.5 Toxic organic contaminants identified in industrial wastewaters applying effects-directed analysis

Industry	Identified toxicant	Wastewater type	Experimental species	Biological effect	Reference
Textile industry	Linoleic acid, oleic acid, tridecanol, tridecanol isomer	Effluent	*Vibrio fischeri*	Luminescence inhibition	Svenson et al. (1996)
	Bis(ethylhexyl)phthalate, dibutylphthalate, nonylphenol isomers, nonylphenol ethoxylates, alcohol polyethoxylates	Untreated wastewater	*Daphnia magna*	Immobilisation	Castillo and Barceló (2001)
Tanneries	Thiocyanomethylthiobenzothiazole, mercapto-benzothiazole, Methylthiobenzothiazole	Tanyard wastewater	*Vibrio fischeri*	Luminescence inhibition	Reemtsma et al. (1999a, b)
	Cresol, dichlorobenzoic acid, methylbenzoic acid, methylthiobutyric acid	Beamhouse wastewater	*Vibrio fischeri*	Luminescence inhibition	Reemtsma et al. (1999a, b)
	Polyethylene glycol, nonylphenol polyethoxylate, alkylalcohol polyethoxyl-ates, chlorocresol, pentachlorophenol, diethylphthalate, bis(ethylhexyl)phthalate, LAS, benzene sulfonates, naphthalene sulfonates	Untreated wastewater	*Vibrio fischeri*	Luminescence inhibition	Castillo and Barceló (1999b)
Pulp and paper industry	Trichlorocatechol, tetrachlorocatechol, dichlorohydroxyquinone	Effluent	*Salmonella*	Mortality	McKague (1981)
	Trichlorohydroxyfuranone	Effluent	*Salmonella typhimurium*	Mutagenicity	Holmborn et al. (1984)
	Octadecenoic acid amide, linoleic acid, linoleic acid isomer	Effluent	*Vibrio fischeri*	Luminescence inhibition	Svenson et al. (1996)
	Juvabione, dehydrojuvabione, manool	Effluent	*Oncorhynchus mykiss*	Hepatic mixed-function oxygenase activity	Martel et al. (1997)
	Chlorinated pterostilbene	Effluent	*Oncorhynchus mykiss*	Hepatic mixed-function oxygenase activity	Burnison et al. (1999)
	Methyl dehydroabietate, ethyl dehydroabietate, ethyl abietate, isopimaric acid, dehydroabi-etic acid, abietic acid	Effluent	*Saccharomyces cerevisiae*	Antiestrogenic activity	Terasaki et al. (2009)

(continued)

Table 2.5 (continued)

Industry	Identified toxicant	Wastewater type	Experimental species	Biological effect	Reference
Chemical production sites					
Synthesized compounds not indicated	Hydroxymethylthiobenzothiazole	Effluents	*Ceriodaphnia dubia*, *Pimephales promelas*	Mortality (96h-LC_{50})	Jop et al. (1991)
Synthesized compounds not indicated	Chlorobenzene, nitrobenzene, naphthalene, chloronitrobenzene, naphthol, aniline	Effluent	*Daphnia magna*	Mortality (24h- and 48h-LC_{50})	Yang et al. (1999)
Synthesized compounds not indicated	Benzopyrone, phenol	Effluent	*Daphnia magna*	Mortality (24h-LC_{50})	Jin et al. (1999)
Pharmaceutical industry	Phenylacetic acid methylester, substituted thiazol	Effluent	*Vibrio fischeri*	Luminescence inhibition	Svenson et al. (1996)

Hewitt et al. (2008) and Hewitt (2011). Chlorinated catechols were regularly detected in effluents from pulp mills and were found in water, sediment and fish bile (Tables 2.1 and 2.2). They were also identified as toxicants in chlorination-stage effluents from bleached kraft pulp mills (Table 2.5) (Mckague 1981). Another early study revealed that the mutagenicity of pulp mill effluents for *Salmonella* strains was mainly attributed to trichlorohydroxyfuranone (Holmborn et al. 1984). This compound is known to be performed as byproduct during the chlorination of humic acid (Meier et al. 1986). Effluents from the forest industry contained two unsaturated fatty acids and an unsaturated fatty acid amide as dominant toxicants, and almost the total toxicity of the sample (97 %) was accounted for by these three compounds with a contribution of the interaction of the three (Svenson et al. 1996). Effluents from a thermomechanical pulp mill induced hepatic mixed-function oxygenase activity in the rainbow trout. Organic compounds which naturally occur in the balsam fix *Abies balsamea* were identified as the constituents of the toxic fraction (Martel et al. 1997). Applying an effects-directed analytical approach, also chlorinated pterostilbene was identified as toxicant in pulp mill effluents. The presence of this compound corresponded with the induction of mixed-function oxygenase activity in the rainbow trout. Pterostilbene belongs to the pinosylvin family, a group of naturally occurring substances that are often present in coniferous trees (Burnison et al. 1999). Its chlorinated derivative is performed during the bleaching of the chemical pulp with chlorine. It is interesting that a biogenic compound, modified during bleaching, caused the observed biological response. Resin acids are the main constituents of natural tree resins. Despite their biogenic origin, it was demonstrated that several resin acids which occurred in paper mill effluents caused antiastrogenic effects (Terasaki et al. 2009).

Approximately 70 % of the toxicity of effluents from a ***chemical production site*** in the US was attributed to un-ionized ammonia. However, residual toxicity was assessed applying different fractionation and identification steps and revealed the presence of hydroxymethylthiobenzothiazole in the effluents (Jop et al. 1991). Several organic compounds were identified as the main toxicants in effluents from a chemical production site in China (Yang et al. 1999). A toxicity-based fractionation was also conducted with effluents from another chemical production site in China. A confirmation procedure using standard materials revealed that benzopyrone and phenol accounted for 44.6 % and 32.9 % of the whole effluent toxicity, respectively. It was observed that the two toxicants showed synergistic effects (Jin et al. 1999).

Phenylacetic acid methylester, an educt for the synthesis of pharmaceutical drugs and a substituted thiazol, which was a component of a pharmaceutical drug, were identified as main toxicants in the effluents of the ***pharmaceutical industry*** (Svenson et al. 1996).

The identification of wastewater constituents which are causing recorded toxic effects requires a lot of analytical expertise and has proven to be a "demanding challenge" (Hewitt and Marvin 2005). Although a number of good results were obtained during the early studies in the 1980s and 1990s, only few data were generated during the last decade.

2.5 Chemical Evaluation of Industrial Contamination Linked to Surveys of Environmental Impacts

We present here studies which attempt to correlate a detected industrial contamination in the environment with observed impacts on aquatic organisms. Studies which demonstrated that toxic effects in the field were directly attributed to the occurrence of specific industrial contaminants are discussed in Sect. 2.5.1. Surveys which revealed that the presence of industrial organic contaminants in field samples correlated with changes on the community level of aquatic organisms are discussed in Sect. 2.5.2.

2.5.1 Industrial Contamination and Toxicity Evaluation in the Field

Several studies demonstrated the toxic activity of field samples and related the observed effects to the presence of industrial organic contaminants. In the Nishitakase River, Japan, three mutagens were identified in water samples using *Salmonella typhimurium*. The compounds were identified as phenylbenzotriazole derivatives (Sect. 2.3, Table 2.2). The river is affected by discharges of the ***textile industry*** and because of the structural similarity it was postulated that the compounds are performed during industrial dying or during sewage treatment as metabolites of azo dyes (Nukaya et al. 1997; Oguri et al. 1998; Shiozawa et al. 2000). The mutagenic activity of water samples from the Cristais River, Brazil, for *Salmonella typhimurium* was also attributed to the impact of the textile industry (Umbuzeiro et al. 2004). The combination of three azo dyes, which was also present in the effluents and which corresponded to a commercial product, was identified (Sects. 2.3 and 2.4.1). The dyes were found to contribute to the mutagenic activity of the river water samples, and indirectly also an impact on the quality of the related drinking water was demonstrated. Coastal sediments from the vicinity of plants of the ***petrochemical industry*** in Kuwait were contaminated by heavy metals and PAHs. The fresh wet sediments were acutely toxic for *Vibrio fischeri*. The LC_{50} was negatively correlated with the PAH and heavy metal concentrations suggesting causality. However, no individual contaminant could be singled out as a main toxicant (Beg et al. 2001). Neoplasms and related disorders were found in fish from coastal waters close to a paper mill. Analysis of the sediments revealed the presence of 28 organic contaminants including resin acids, chlorinated phenols, chlorinated guaiacols and PAHs. Among the identified compounds, trichlorophenol, tetrachloroguaiacol, dehydroabietic acid, pyrene and fluoranthene revealed mutagenic activity in tests using *Bacillus subtilis* and *Salmonella typhimurium*. In liver samples of the spotted sea trout *Nibea mitsukurii* from the same area, the mutagen epoxystearic acid was identified (Kinae et al. 1981a, b). The Fenholloway River, US, receives effluents from a paper mill and contains populations of masculinized female eastern mosquitofish *Gambusia holbrooki*. Water samples collected from the river and a control tributary were analysed. Fenholloway River water contained androstenedione. In the river sediments, androstenedione and progesterone were found. The authors concluded that pulp-derived phytosteroids are converted by microbes into progesterone and subsequently into androstenedione and other bioactive steroids. The presence of these compounds in the river might be responsible for the occurrence of the masculinized phenotype of female eastern mosquitofish (Jenkins et al. 2001, 2003). Sediments from a brook which received wastewaters from the chemical production site Bitterfeld were investigated. Bitterfeld was an important industrial production complex for chlorinated organic compounds in the former German Democratic Republic (Sect. 2.3). Organic toxicants were identified applying effects-directed analysis. Methyl parathion, prometryn, phenylnaphthalene amine, PAHs and tributyltin were confirmed as major toxicants using *Vibrio fischeri*, *Daphnia magna* and *Scenedesmus vacuolatus* (Brack et al. 1999).

Another type of studies applied caging experiments in areas affected by industrial discharges to assess whether the contamination had an effect on the exposed experimental species. The procedure included the identification of the causative toxicants. The toxicity of a leather ***tannery*** effluent to a transplanted population of blue mussels *Mytilus edulis* in an Irish estuary was assessed. Additionally, the mussels were exposed to single constituents of the effluents under laboratory conditions. After 1 year, the fitness of the transplanted mussels was at most sites comparable to the control group except those closest to the tannery outfall. The laboratory exposures revealed that the fungicide thiocyanomethylthiobenzothiazole enduced lipid peroxidation in the digestive gland and amoebocyte proliferation in the gills (Walsh and O'Halloran 1997).

Rainbow trouts (*Salmo gairdneri*) were exposed for 48–144 h to effluents from a ***paper mill*** and were examined for the presence of dehydroabietic acid in the whole fish. Dehydroabtiec acid concentrations in the exposed fish were 2–10 times higher than in the control group. The authors supposed that the accumulation of dehydroabietic acid could result in sublethal toxic effects (Fox et al. 1977). However, further toxicity tests to confirm this assumption were not carried out. In the blood plasma of caged rainbow trouts in Lake Saimaa, Finland, chlorinated phenols, chlorinated guaiacols, trichlorosyringol, tetrachlorocatechol and free resin acids were detected. In the control group, none of these compounds were detectable. The lake received bleached pulp mill effluents and the chlorinated contaminants were also present in the lake water (Table 2.2). During the 10-day caging period, increased haemoglobin and decreased plasma protein concentrations as well as a partial inhibition of liver UDP-glucuronosyltransferase was observed, even in a distance of up to 11 km from the source (Oikari et al. 1985). In whitefish *Coregonus lavaretus* exposed for 1 month in the same lake, chlorinated contaminant concentrations in bile and gut lipids correlated with distance from the source. The whitefish responded to the effluent exposure in a dose dependent way seen as a gradient change of the liver enzyme activity (Soimasuo et al. 1995). White sucker *Catostomus commersoni* was caged for 3 days in bleached pulp mill effluent and in a reference stream. The exposed group exhibited a 15–90 fold elevated mixed function oxygenase activity.

For the identification of the causative constituents, liver samples were extracted, fractionated and the fractions were tested for mixed function oxygenase activity in rat cells. Although PCDDs, PCDFs and chlorinated diphenylethers were identified in the toxic fractions, the respective concentrations were too low to be solely responsible for the whole observed effect (Parrott et al. 2000).

2.5.2 Industrial Contamination and Field Surveys of Ecological Changes

Studies on the relationship between chemical contamination and biological diversity in river systems were reviewed by Ricciardi et al. (2009). The authors noted that only a few studies couple chemical and biological analysis to evaluate water quality and the related effects on fluvial diversity. A response of freshwater invertebrates to chemical contamination was demonstrated (e.g. Cao et al. 1997; Day et al. 2006). A loss of sensitive species was observed at the most polluted sites along a pollution gradient in the River Trent, UK. Tolerant species were abundant at intermediately polluted sites and absent at clean sites (Cao et al. 1997). Interestingly, total individuals, total species and species richness of macroinvertebrates was highest near the outfall a sewage treatment plant in Louisiana, US, and declined further away (Day et al. 2006). The faunistic composition in coastal waters receiving sewage was also assessed (e.g. Inglis and Kross 2000; Wake 2005; Simboura et al. 1995; Bigot et al. 2006; Saunders et al. 2007; Calabretta and Oviatt 2008). Univariate and multivariate analysis revealed that the coastal area receiving sewage from Athens, Greece, could be divided into three zones which reflected the degree of pollution affecting the communities (Simboura et al. 1995). At stations closest to an urban center at the US coast, communities with opportunistic taxa were found which persisted in a state of low faunal diversity (Calabretta and Oviatt 2008). However, watershed analysis in these studies did not include a comprehensive characterization of the organic contamination but referred to heavy metal concentrations or standard parameters such as dissolved oxygen content, particulate organic carbon and nutrient concentrations.

In order to link contaminant exposure to community changes in fluviatile systems, a biotic indicator of species at risk was developed and applied in different field studies in Europe and Australia (Schäfer et al. 2007, 2011; Liess et al. 2008). However, multivariate analysis of the contamination of floodplain sediments and community composition in a lake revealed that PAH concentrations did not correlate with changes of the benthic community (De Lange et al. 2009).

The occurrence of tributyltin in coastal waters was found to correlate with changes of the faunal composition of macrobenthos and meiobenthos (Warwick et al. 1990; Matthiessen and Law 2002), whereas oxygen depletion rather than chemical contamination was the dominant anthropogenic influence in an estuary receiving wastewaters (Saiz-Salinas 1997). In the San Francisco Estuary, US, many anthropogenic contaminants occured at levels above known toxicity thresholds. A long term monitoring of associated effects on the estuary's plankton, fish, birds and mammals revealed that no single contaminant was consistently related to effects among the biota considered. This was explained by the wide range of sensitivity to contaminants among the investigated organisms (Thompson et al. 2007). The Todos os Santos Bay in Brazil is a heavily industrialized area and urban center receiving discharges from an oil refinery. Macrobenthic communities in the bay are subject to chronic oil pollution which was documented by a chemical characterization of the sediments. The contamination correlated with a reduction in the number of species and diversity of benthic invertebrates (Venturini et al. 2008). Following the closure of major industries discharging into the Severn Estuary, UK, the concentrations of heavy metals, PCBs and PAHs were lower as compared to 25 years before. The long term monitoring indicated a recovery of faunal diversity in recent years implying that chemical contamination had been a forcing feature for the biota (Langston et al. 2010).

All mentioned studies compared faunal composition in contaminated areas to that in less impacted areas. In most cases, the contamination stemmed from mixed municipal/industrial sources and priority pollutants were considered to characterize the contamination. Ecological surveys of sites contaminated by industrial point sources combined with a comprehensive chemical characterization applying screening methods and toxicity evaluation are missing. Such an approach would be promising to understand the potential impacts of industrial contamination on aquatic diversity.

2.6 Conclusion

Major advancements in the identification of organic contaminants from industrial wastewaters include the development of new analytical techniques during the past decades which allow for the identification of previously unknown, emerging contaminants. Applying LC/MS techniques, polar wastewater constituents were identified which contribute to the whole effluent toxicity but are not detectable applying classical GC/MS methods. The successful application of GC/MS screening techniques led to the identification of structurally diverse organic contaminant groups which are educts, intermediates or byproducts of chemical syntheses conducted in different industrial branches. It enabled also the identification of contaminants which are performed during wastewater treatment and of contaminant metabolites, which might also pose a threat to the environment.

Studies tracing industrial contaminants in aquatic systems and examining the intrinsic underlying processes determining the fate of these compounds were applied. A simplified scheme is presented in Fig. 2.3. A marker concept has been developed by geochemists to understand the sources, transport and fate of organic compounds and was also applied to identify industrial markers.

The advancement of the analytical techniques and the identification of emerging contaminants rendered possible to evaluate the toxicity of these compounds and the potential impacts in the environment applying different toxicological approaches. The combination of chemical and toxicological methods allowed the identification

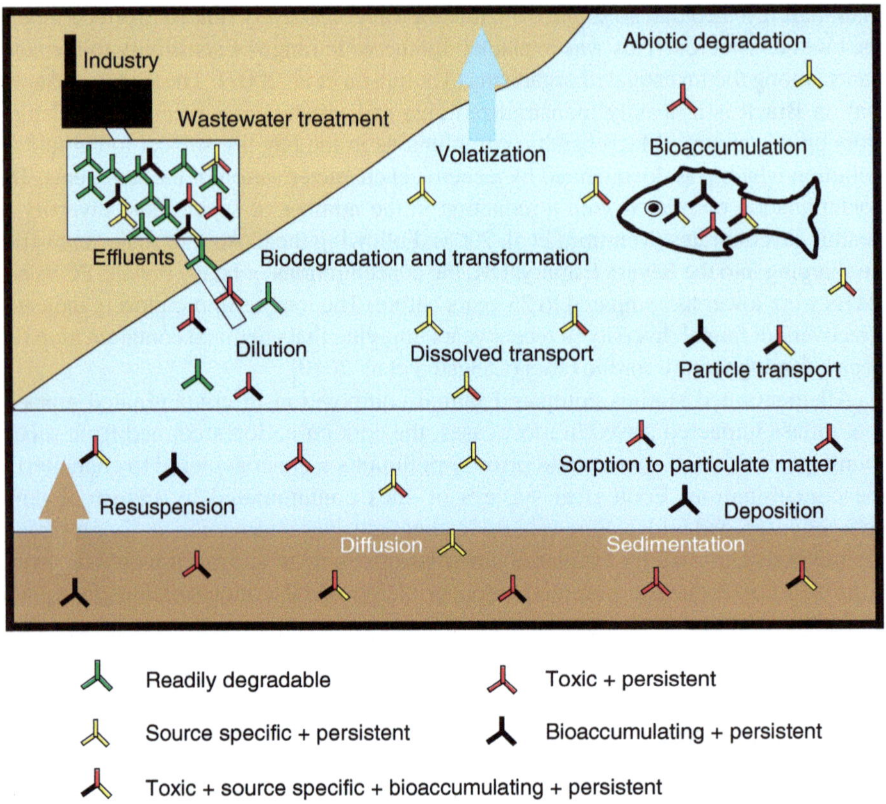

Fig. 2.3 Simplified scheme of the input and fate of industrial contaminants in river systems. The scheme shows the major processes controlling the environmental fate of contaminants which exhibit different physico-chemical properties

of organic constituents which contribute to the toxicity of industrial wastewaters. Toxic activities of different contaminant combinations were determined in order to assess main and interaction effects of contaminants commonly present in industrial wastewaters. The combination of chemical and biological investigations in the field is promising in order to assess the impact of chemical contamination on aquatic diversity.

Nevertheless, this review also shows that there are major gaps in our knowledge on the chemical composition of industrial effluents and their impact in the environment.

Numerous structurally diverse organic contaminants were identified in effluents from chemical production sites, the petrochemical and the pharmaceutical industry. Many of these compounds might be stable in aquatic systems, but as yet only limited attempts were made to trace these compounds in the environment. Only a few studies investigated effluents from other industrial facilities such as the food-processing industry, cement industry, automotive industry and power stations. Information about the emissions of the coating and printing industry and of manufacturers of electric devices, toys and computers would be interesting, but are lacking.

The compound spectra found in field samples from industrialized areas was totally different from those in studies on effluents of the respective industries. This illustrates the complexity of the chemical composition of industrial wastewaters, even from the same industry types. It also shows the difficulty to generate comprehensive information on the chemical composition of industrial wastewaters. Only a limited number of studies focus beyond the identification of organic contaminants also on their fate in the hydrosphere.

Although organic constituents which contribute to the toxicity of industrial wastewaters were identified, the total toxicity of the samples could not be attributed completely to the identified compounds. Unidentified compounds, mixture effects and further physico-chemical factors impede the complete elucidation of the causes for the observed total toxicity. The toxicity of many organic compounds which were detected in industrial wastewaters and which are known to also occur in the environment is as yet unknown and they are not addressed by monitoring studies. Studies which investigate the impact of chemical contamination on aquatic diversity consider only priority pollutants such as PAHs, PCBs and organochlorine pesticides.

It has to be noted that the replacement of environmentally problematic compounds by novel industrial formulations is common but may lead to the presence of further, as yet unidentified contaminant groups in the effluents and in the hydrosphere with unknown impact in the environment. The number of registered substances is rapidly increasing, so does the number of potentially detectable organic contaminants in the environment. An example of an unidentified compound which has, according to the composition of the mass spectrum, at least three chlorine atoms is shown in Fig. 2.4. The compound was detected in effluent samples from a chemical production site.

Because industrial production sites are more and more shifted to emerging nations, such as China, India and Brazil, the use and contamination of water resources in these countries is increasing, but as yet only a few studies about these issues are available.

Research strategies which are in our opinion promising to address these gaps include:

- A comprehensive and industry branch specific chemical characterization of wastewaters, especially from chemical production sites, the petrochemical industry, the pharmaceutical industry and from so far neglected industry branches
- Examination of wastewater samples *and* field samples to assess the individual contamination profiles of industrialized areas
- An extended application of the organic marker concept in order to identify further industrial organic markers which are useful to verify the input of specific industrial wastewaters to aquatic systems, to investigate the spatial distribution of the emission and to disentangle different emission sources
- More information on the chemicals which are used for the different industrial production processes which is valuable for the identification of marker compounds

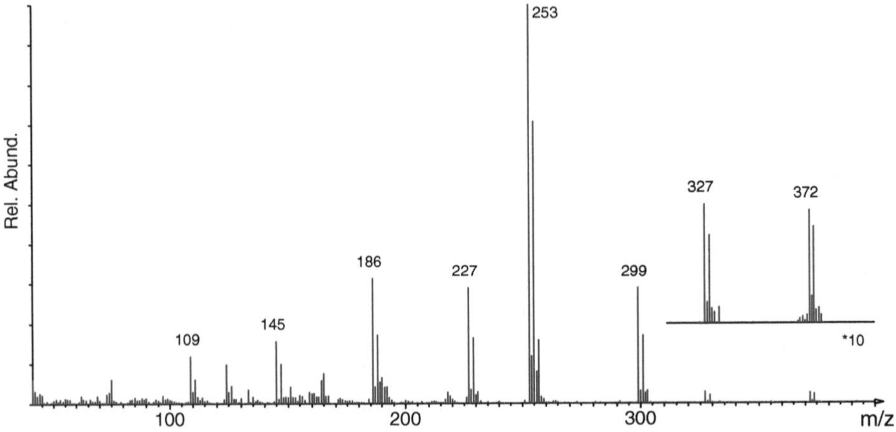

Fig. 2.4 Example of an unidentified compound which has, according to the composition of the mass spectrum, at least three chlorine atoms; The compound was detected in effluent samples from a chemical production site

- Further studies which combine chemical and toxicological methods to identify the constituents which are responsible for the toxicity of wastewaters
- Field studies combining chemical and ecological methods plus toxicity tests to assess the impact of industrial effluents on aquatic diversity
- More studies on industrial emissions and related impacts in emerging and developing nations in order to consider the geographical shift of industrial production

Chemical contamination was defined as a "planetary boundary" due to its global, ubiquitous impact on the physiological development of humans and other organisms with ultimate impacts on ecosystem structure/functions and because it acts as a slow variable affecting other planetary boundaries. Chemical contamination is one of the planetary boundaries, for which no boundary level could be determined. Setting of such a boundary would require knowledge of the critical impacts on organisms of exposure to 10,000 of chemicals and the related threshold concentrations, which is a gigantic task (Rockström et al. 2009). Industries emit a plentitude of structurally very heterogenic organic contaminants. Basic information on the chemical characteristics of these emissions is to some extent lacking but is necessary for the evaluation of associated risks and for the definition of thresholds.

References

Ali M, Sreekrishnan TR (2001) Aquatic toxicity from puklp and paper mill effluents: a review. Adv Environ Res 5:175–196

Alonso MC, Barceló D (1999) Tracing polar benzene- and naphthalene sulfonates in untreated industrial effluents and water treatment works by ion-pair chromatography-fluorescence and electrospray-mass spectrometry. Anal Chim Acta 400:211–231

Alonso MC, Castillo M, Barceló D (1999) Solid-phase extraction procedure of polar benzene- and naphthalene sulfonates in industrial effluents followed by unequivocal determination with ion-pair chromatography/electrospray-mass spectrometry. Anal Chem 71:2586–2593

Alvarez DA, Stackelberg PE, Petty JD, Huckins JN, Furlong ET, Zaugg SD, Meyer MT (2005) Comparison of a novel passive sampler to standard water-column sampling for organic contaminants associated with wastewater effluents entering a New Jersey stream. Chemosphere 61:610–622

Amaral OC, Otero R, Grimalt JO, Albaiges J (1996) Volatile and semi-volatile organochlorine compounds in tap and riverine waters in the area of influence of a chlorinated organic solvent factory. Water Res 30:1876–1884

Barron MG, Holder E (2003) Are exposure and ecological risks of PAHs underestimated at petroleum contaminated sites? Hum Ecol Risk Assess Int J 9:1533–1545

Baumard P, Budzinski H, Michon Q, Garrigues P, Burgeot T, Bellocq J (1998) Origin and bioavailability of PAHs in the Mediterranean Sea from mussel and sediment records. Estuar Coast Shelf Sci 47:77–90

Beg MU, Al-Muzaini S, Saeed T, Jacob PG, Beg KR, Al-Bahloul M, Al-Matrouk K, Al-Obaid T, Kurian A (2001) Chemical contamination and toxicity of sediment from a coastal area receiving industrial effluents in Kuwait. Arch Environ Contam Toxicol 41:289–297

Berndt J (1996) Umweltbiochemie. Gustav Fischer Verlag, Stuttgart, p 278. doi:10.1002/ange.19971090338

Bhaskar M, Gnanamani A, Ganeshjeevan RJ, Chandrasekar R, Sadulla S, Radhakrishnan G (2003) Analyses of carcinogenic aromatic amines released from harmful azo colorants by *streptomyces* SP SS07. J Chromatogr A 1018:117–123

Bigot L, Conand C, Amouroux JM, Frouin P, Bruggemann H, Grémare A (2006) Effects of industrial outfalls on tropical macrobenthic sediment communities in Reunion Island (Southwest Indian Ocean). Mar Pollut Bull 52:865–880

Bilgi S, Demir C (2005) Identification of photooxidation degradation products of CI reactive orange 16 dye by gas chromatography–mass spectrometry. Dyes Pigments 66:69–76

Bjørseth A, Carlberg GE, Møller M (1979) Determination of halogenated organic compounds and mutagenicity testing of spent bleach liquors. Sci Total Environ 11:197–211

Blacksmith Institute and Green Cross Switzerland (2011) The world's worst toxic pollution problems: report 2011. http://www.worstpolluted.org/2011-report.html

Blanchard M, Teil MJ, Ollivon D, Legenti L, Chevreuil M (2004) Polycyclic aromatic hydrocarbons and polychlorobiphenyls in wastewaters and sewage sludges from the Paris area (France). Environ Res 95:184–197

Bobeldijk I, Stoks PGM, Vissers JPC, Emke E, van Leerdam JA, Muilwijk B, Berbee R, Noij THM (2002) Surface and wastewater quality monitoring: combination of liquid chromatography with (geno)toxicity detection, diode array detection and tandem mass spectrometry for identification of pollutants. J Chromatogr A 970:167–181

Botalova O, Schwarzbauer J (2011) Geochemical characterization of organic pollutants in effluents discharged from various industrial sources to riverine systems. Water Air Soil Pollut 221:77–98

Botalova O, Schwarzbauer J, Frauenrath T, Dsikowitzky L (2009) Identification and chemical characterization of specific organic constituents of petrochemical effluents. Water Res 43:3797–3812

Botalova O, Schwarzbauer J, al Sandouk N (2011) Identification and chemical characterization of specific organic indicators in the effluents from chemical production sites. Water Res 45:3653–3664

Brack W (2003) Effect-directed analysis: a promising tool for the identification of organic toxicants in complex mixtures? Anal Bioanal Chem 377:397–407

Brack W, Altenburger R, Ensenbach U, Möder M, Segner H, Schüürmann G (1999) Bioassay-directed identification of organic toxicants in river sediment in the industrial region of Bitterfeld (Germany) – a contribution to hazard assessment. Arch Environ Contam Toxicol 37:164–174

Brack W, Kind T, Schrader S, Möder M, Schüürmann G (2003) Polychlorinated naphthalenes in sediments from the industrial region of Bitterfeld. Environ Pollut 121:81–85

Brack W, Schmitt-Jansen M, Machala M, Brix R, Barceló D, Schymanski E, Streck G, Schulze T (2008) How to confirm identified toxicants in effect-directed analysis. Anal Bioanal Chem 390:1959–1973
Burkhard LP, Ankley GT (1989) Identifying toxicants: NETAC's toxicity-based approach. Environ Sci Technol 23:1438–1443
Burkhard LP, Durhan EJ, Lukasewycz MT (1991) Identification of nonpolar toxicants in effluents using toxicity-based fractionation with gas chromatography/mass spectrometry. Anal Chem 63:277–283
Burnison BK, Comba ME, Carey JH, Parrott J, Sherry JP (1999) Isolation and tentative identification of compound in bleached-kraft mill effluent capable of causing mixed-function oxygenase induction in fish. Environ Toxicol Chem 18:2882–2887
Calabretta CJ, Oviatt CA (2008) The response of benthic macrofauna to anthropogenic stress in Narragansett Bay, Rhode Island: a review of human stressors and assessment of community conditions. Mar Pollut Bull 56:1680–1695
Cao Y, Bark AW, Williams WP (1997) Analysing benthic macroinvertebrate community changes along a pollution gradient: a framework for the development of biotic indices. Water Res 31:884–892
Carlberg GE, Gjøs N, Møller M, Gustavsen KO, Tveten G, Renberg L (1980) Chemical characterization and mutagenicity testing of chlorinated tryhydroxybenzenes identified in spent bleach liquors from a sulphite plant. Sci Total Environ 15:3–15
Castillo M, Barceló D (1999a) Determination of organic pollutants in industrial wastewater. In: Barceló D (ed) Sample handling and trace analysis of pollutants: techniques, applications and quality assurance. Elsevier Science, Amsterdam. ISBN 13: 978-0-444-82831-6
Castillo M, Barceló D (1999b) Identification of polar toxicants in industrial wastewaters using toxicity-based fractionation with liquid chromatography/mass spectrometry. Anal Chem 71:3769–3776
Castillo M, Barceló D (2001) Characterisation of organic pollutants in textile wastewaters and landfill leachate by using toxicity-based fractionation methods followed by liquid and gas chromatography coupled to mass spectrometric detection. Anal Chim Acta 426:253–264
Castillo M, Puig D, Barceló D (1997) Determination of priority phenolic compounds in water and industrial effluents by polymeric liquid–solid extraction cartridges using automated sample preparation with extraction columns and liquid chromatography – use of liquid – solid extraction cartridges for stabilization of phenols. J Chromatogr A 778:301–311
Castillo M, Oubiña A, Barceló D (1998) Evaluation of ELISA kits followed by liquid chromatography-atmospheric pressure chemical ionization-mass spectrometry for the determination of organic pollutants in industrial effluents. Environ Sci Technol 32:2180–2184
Castillo M, Alonso MC, Riu J, Barceló D (1999a) Identification of polar, ionic, and highly water soluble organic pollutants in untreated industrial wastewaters. Environ Sci Technol 33:1300–1306
Castillo M, Barceló D, Pereira AS, Aquino Neto FR (1999b) Characterization of organic pollutants in industrial effluents by high-temperature gas chromatography–mass spectrometry. TrAC Trends Anal Chem 18:26–36
Castillo M, Alonso MC, Riu J, Reinke M, Klöter G, Dizer H, Fischer B, Hansen PD, Barceló D (2001) Identification of cytotoxic compounds in European wastewaters during a field experiment. Anal Chim Acta 426:265–277
Chen C-Y, Chen J-N, Chen S-D (1999) Toxicity assessment of industrial wastewater by microbial testing method. Water Sci Technol 39:139–143
Chung K-T, Stevens B Jr (1993) Degradation of azo dyes by environmental microorganisms and helminthes. Environ Toxicol Chem 12:2121–2132
Consejo C, Ormad MP, Sarasa J, Ovelleiro JL (2005) Treatment of wastewater coming from painting processes: application of conventional and advanced oxidation technologies. Ozone Sci Eng 27:279–286
Czaplicka M (2003) Qualitative and quantitative determination of halogenated derivatives in wastewatere from coking plant. J Sep Sci 26:1067–1071

Day JW Jr, Westphal A, Pratt R, Hyfield E, Rybczyk J, Kemp GP, Day JN, Marx B (2006) Effects of long-term municipal effluent discharge on the nutrient dynamics, productivity, and benthic community structure of a tidal freshwater forest wetland in Louisiana. Ecol Eng 27:242–257

de Aragão Umbuzeiro G, Roubicek DA, Rech CM, Sato MIZ, Claxton LD (2004) Investigating the sources of the mutagenic activity found in a river using the Salmonella assay and different water extraction procedures. Chemosphere 54:1589–1597

De Lange HJ, Peeters ETHM, Harmsen J, Maas H, De Jonge J (2009) Seasonal variation of total and biochemically available concentrations of PAHs in a floodplain lake sediment has no effect on the benthic invertebrate community. Chemosphere 75:319–326

Drzyzga O, Jannsen S, Blotevogel KH (1995) Toxicity of diphenylamine and some of its nitrated and aminated derivatives to the luminescent bacterium *Vibrio fischeri*. Ecotoxicol Environ Saf 31:149–152

Dsikowitzky L (2002) Umweltgeochemische Charakterisierung der niedermolekularen organischen Fracht des Flusssystems Lippe. PhD thesis, RWTH Aachen University, 131 pp

Dsikowitzky L, Schwarzbauer J, Littke R (2004) The anthropogenic contribution to the organic load of the Lippe river (Germany), part II: quantification of specific organic contaminants. Chemosphere 57:1289–1300

Dsikowitzky L, Nordhaus I, Jennerjahn T, Khrycheva P, Sivatharshan Y, Yuwono E, Schwarzbauer J (2011) Anthropogenic organic contaminants in water, sediments and benthic organisms of the mangrove-fringed Segara Anakan Lagoon, Java, Indonesia. Mar Pollut Bull 62:851–862

Eganhouse RP (1986) Long-chain alkylbenzenes: their analytical chemistry, environmental occurrence and fate. Int J Environ Anal Chem 26:241–263

Eganhouse RP (1997) Molecular markers and environmental organic geochemistry: an overview. In: Eganhouse RP (ed) Molecular markers in environmental chemistry. American Chemical Society, Washington, pp 1–21

Ellis DD, Jone CM, Larson RA, Schaeffer D (1982) Organic constituents of mutagenic secondary effluents from wastewater treatment plants. Arch Environ Contam Toxicol 11:373–382

Fang YX, Ying GG, Zhang LJ, Zhao JL, Su HC, Yang B, Liu S (2012) Use of TIE techniques to characterize industrial effluents in the Pearl River Delta region. Ecotoxicol Environ Saf 76:143–152

Farré M, Barceló D (2003) Toxicity testing of wastewater and sewage sludge by biosensors, bioassays and chemical analysis. TrAC Trends Anal Chem 22:299–310

Farré M, Pasini O, Alonso MC, Castillo M, Barceló D (2001a) Toxicity assessment of organic pollution in wastewaters using a bacterial biosensor. Anal Chim Acta 426:155–165

Farré M, García MJ, Tirapu L, Ginebreda A, Barceló D (2001b) Wastewater toxicity screening of non-ionic surfactants by Toxalert® and Microtox® bioluminescence inhibition assays. Anal Chim Acta 427:181–189

Fent K (2007) Ökotoxikologie. Georg Thieme Verlag, Stuttgart/New York, p 338. ISBN 978-3-13-109993-8

Fichtner S, Lange FT, Schmidt W, Brauch H-J (1995) Determination of aromatic sulfonates in the river Elbe by on-line ion-pair extraction and ion-pair chromatography. Fresenius J Anal Chem 353:57–63

Fiehn O, Jekel M (1996) Comparison of sorbents using semipolar to highly hydrophilic compounds for a sequential solid-phase extraction procedure of industrial wastewaters. Anal Chem 68:3083–3089

Fiehn O, Vigelahn L, Kalnowski G, Reemtsma T, Jekel M (1997) Toxicity-directed fractionation of tannery wastewater using solid-phase extraction and lumenescence inhibition microtiter plates. Acta Hydrochim Hydrobiol 25:11–16

Field JA, Reed RL (1996) Nonylphenol polyethoxy carboxylate metabolites of nonionic surfactants in U.S. paper mill effluents, municipal sewage treatment plant effluents, and river waters. Environ Sci Technol 30:3544–3550

Fishbein L (1991) Chemicals used in the rubber industry. Sci Total Environ 101:33–43

Fox ME, Whittle DM, Kaiser KLE (1977) Dehydroabietic acid accumulation by rainbow trout (*Salmo gairdneri*) exposed to kraft mill effluent. J Gt Lake Res 3:155–158

Frische K, Schwarzbauer J, Ricking M (2011) Structural diversity of organochlorine compounds in groundwater affected by an industrial point source. Chemosphere 81:500–508

Fukazawa H, Hoshino K, Shiozawa T, Matsushita H, Terao Y (2001) Identification and quantification of chlorinated bisphenol a in wastewater from wastepaper recycling plants. Chemosphere 44:973–979

Games LM, Hites RA (1977) Composition, treatment efficiency, and environmental significance of dye manufacturing plant effluents. Anal Chem 49:1433–1440

Giger W (2009) Hydrophilic and amphiphilic water pollutants: using advanced analytical methods for classic and emerging contaminants. Anal Bioanal Chem 393:37–44

Grigoriadou A, Schwarzbauer J, Georgakopoulos A (2008) Molecular indicators for pollution source identification in marine and terrestrial water of the industrial area of Kavala city, North Greece. Environ Pollut 151:231–242

Guerra R (2001) Ecotoxicological and chemical evaluation of phenolic compounds in industrial effluents. Chemosphere 44:1737–1747

Harding MJC, Davies IM, Bailey SK, Rodger GK (1999) Survey of imposex in dogwhelks (*Nucella lapillus*) from North Sea coasts. Appl Oranomet Chem 13:521–538

Hewitt LM (2011) Effects-directed studies of pulp and paper mill effluents. In: Brack W (ed) Effect-directed analysis of complex environmental contamination, vol 15, Hdb Env chem., pp 267–284. doi:10.1007/978-3-642-18384-3_11

Hewitt LM, Marvin CH (2005) Analytical methods in environmental effects-directed investigations of effluents. Mutat Res 589:208–232

Hewitt LM, Kovacs TG, Dubé MG, MacLatchy DL, Martel PH, McMaster ME, Paice MG, Parrott JL, van den Heuvel MR, van der Kraak GJ (2008) Altered reproduction in fish exposed to pulp and paper mill effluents: roles of individual compounds and mill operating conditions. Environ Toxicol Chem 27:682–697

Holmborn B, Voss RH, Mortimer RD, Wog A (1984) Fractionation, isolation, and characterization of Ames mutagenic compounds in kraft chlorination effluents. Environ Sci Technol 18:333–337

Hynning P-A (1996) Separation, identification and quantification of compounds of industrial effluents with bioconcentration potential. Water Res 30:1103–1108

Inglis GJ, Kross JE (2000) Evidence for systemic changes in the benthic fauna of tropical estuaries as a result of urbanization. Mar Pollut Bull 41:367–376

James KJ, Stack MA (1997) Rapid determination of volatile organic compounds in environmentally hazardous wastewaters using solid phase microextraction. Fresenius J Anal Chem 358:833–837

Jenkins RL, Angus RA, McNatt H, Howell WM, Kemppainen JA, Kirk M, Wilson EM (2001) Identification of androstenedione in a river containing paper mill effluent. Environ Toxicol Chem 20:1325–1331

Jenkins RL, Wilson EM, Angus RA, Howell WM, Kirk M (2003) Androstenedione and progesterone in the sediment of a river receiving paper mill effluent. Toxicol Sci 73:53–59

Jin H, Yang X, Yin D, Yu H (1999) A case study on identifying the toxicant in effluent discharged from a chemical plant. Mar Pollut Bull 39:122–125

Jop KM, Kendall TZ, Askew AM, Foster RB (1991) Use of fractionation procedures and extensive chemical-analysis for toxicity identification of a chemical-plant effluent. Environ Toxicol Chem 10:981–990

Judd MC, Stuthridge TR, McFarlane PN, Anderson SM, Bergman I (1996) Bleached kraft pulp mill sourced organic chemicals in sediments from a New Zealand river: Part II Tarawera river. Chemosphere 33:2209–2220

Jungclaus GA, Games LM, Hltes RA (1976) Identification of trace organic compounds in tire manufacturing plant wastewaters. Anal Chem 48:1894–1896

Jungclaus GA, Lopez-Avila V, Hites RA (1978) Organic compounds in an industrial wastewater: a case study of their environmental impact. Environ Sci Technol 12:88–96

Kameya T, Murayama T, Urano K, Kitano M (1995) Biodegradation ranks of priority organic compounds under anaerobic conditions. Sci Total Environ 170:43–51

Keith LH (1974) Chemical characterization of industrial wastewaters by gas chromatography–mass spectrometry. Sci Total Environ 3:87–102

Keith LH (1976) Identification of organic compounds in unbleached treated kraft paper mill wastewaters. Environ Sci Technol 10:555–564

Kelly BC, Ikonomou MG, Blair JD, Morin AE, Gobas FAPC (2007) Food web-specific biomagnification of persistent organic pollutants. Science 317:236–239

Kinae N, Hashizume T, Makita T, Tomita I, Kimura I, Kanamori H (1981a) Studies on the toxicity of pulp and paper mill effluents-I. Mutagenicity of the sediment samples derived from kraft paper mills. Water Res 15:17–24

Kinae N, Hashizume T, Makita T, Tomita I, Kimura I, Kanamori H (1981b) Studies on the toxicity of pulp and paper mill effluents-II. Mutagenicity of the extracts of the liver from spotted sea trout (*Nibea mitsukurii*). Water Res 15:25–30

Knepper TP, Karrenbrock F (2006) Pollutants as byproducts and degradation products of chemical synthesis. In: The Encyclopedia of environmental analysis Rhine, Hdb Env Chem vol 5, Part L., pp 235–254. doi:10.1007/698_5_033

Knepper TP, Muller J, Wulff T, Maes A (2000) Unknown bisethylisooctanollactone isomers in industrial waste water: isolation, identification and occurrence in surface water. J Chromatogr A 889:245–252

Kronimus A, Schwarzbauer J, Dsikowitzky L, Heim S, Littke R (2004) Anthropogenic organic contaminants in sediments of the Lippe River, Germany. Water Res 38:3473–3484

Kungolos A (2005a) Application of microbiotests and activated sludge respirometry for the evaluation of industrial wastewater toxicity. Bull Environ Contam Toxicol 74:801–808

Kungolos A (2005b) Evaluation of toxic properties of industrial wastewater using on-line respirometry. J Environ Sci Health Toxic Hazard Subst Environ Eng 40:869–880

Kwon J-H, Kwon J-W, Kim K, Kim Y-H (2003) Identification of a water pollutant 2-amino-6, 7-dichlorobenzothiazole, at a river near a textile industrial complex in Korea. Bull Environ Contam Toxicol 70:54–62

Langston WJ, Pope ND, Jonas PJC, Nikitic C, Field MDR, Dowell B, Shillabeer N, Swarbrick RH, Brown AR (2010) Contaminants in fine sediments and their consequences for biota of the Severn Estuary. Mar Pollut Bull 61:68–82

Latorre A, Rigol A, Lacorte S, Barceló D (2005) Organic compounds in paper mill wastewaters. In: The handbook of environmental chemistry vol 5, Part O. Springer Berlin Heidelberg, pp 25–51. doi:10.1007/b98606

Law RJ, Allchin CR, de Boer J, Covaci A, Herzke D, Lepom P, Morris S, Tronczynski J, de Wit CA (2006) Levels and trends of brominated flame retardants in the European environment. Chemosphere 64:187–208

Lee H-B, Peart TE, Kaiser KLE (1996) Determination of nitrilotriacetic, ethylenediaminetetraacetic and diethylenetriaminepentaacetic acids in sewage treatment plant and paper mill effluents. J Chromatogr A 738:91–99

Leppänen H, Marttinen S, Oikari A (1998) The use of fish bile metabolite analyses as exposure biomarkers to pulp and paper mill effluents. Chemosphere 36:2621–2634

Leuenberger C, Giger W, Coney R, Graydon JW, Molnar-Kubica E (1985) Persistent chemicals in pulp mill effluents. Water Res 19:885–894

Liess M, Schäfer RB, Schriever CA (2008) The footprint of pesticide stress in communities-species traits reveal community effects of toxicants. Sci Total Environ 406:484–490

Loos R, Hanke G, Umlauf G, Eisenreich SJ (2007) LC–MS–MS analysis and occurrence of octyl-and nonylphenol, their ethoxylates and their carboxylates in Belgian and Italian textile industry, waste water treatment plant effluents and surface waters. Chemosphere 66:690–699

Lopez-Avila V, Hites RA (1980) Organic compounds in an industrial wastewater their transport into sediments. Environ Sci Technol 14:1382–1390

López-Grimau V, Guadayol JM, Griera JA, Gutiérrez MC (2006) Determination of non halogenated solvents in industrial wastewater using solid phase microextraction (SPME) and GC-MS. Lat Am Appl Res 36:49–55

Mackintosh CE, Maldonado JA, Ikonomou MG, Gobas FAPC (2006) Sorption of phthalate esters and PCBs in a marine ecosystem. Environ Sci Technol 40:3481–3488

Maguire RJ (1992) Occurrence and persistence of dyes in a Canadian river. Water Sci Technol 25:264–270

Martel PH, Kovacs TG, O'Connor BI, Voss RH (1997) Source and identity of compounds in a thermomechanical pulp mill effluent inducing hepatic mixed-function oxygenase activity in fish. Environ Toxicol Chem 16:2375–2383

Marti N, Aguado D, Segovia-Martinez L, Bouzas A, Seco A (2011) Occurrence of priority pollutants in WWTP effluents and Mediterranean coastal waters of Spain. Mar Pollut Bull 62:615–625

Matthiessen P, Law RJ (2002) Contaminants and their effects on estuarine and coastal organisms in the United Kingdom in the late twentieth century. Environ Pollut 120:739–757

Maya-Altamira L, Eriksson E, Baun A (2008) Source analysis and hazard screening of xenobiotic organic compounds in wastewater from food-processing industries. Water Air Soil Pollut 8:505–517

McKague AB (1981) Some toxic constituents of chlorination-stage effluents from bleached kraft pulp mills. Can J Fish Aquat Sci 38:739–743

Meier JR, Ringhand HP, Coleman WE, Schenck KM, Munch JW, Streicher RP, Kaylor WH, Kopfler FC (1986) Mutagenic by-products from chlorination of humic acid. Environ Health Perspect 69:101–107

Meriläinen P, Oikari A (2008) Uptake of organic xenobiotics by benthic invertebrates from sediment contaminated by the pulp and paper industry. Water Res 42:1715–1725

Nestmann ER, Lee EGH, Matula TI, Douglas GR, Mueller JC (1980) Mutagenicity of constituents identified in pulp and paper mill effluents using the salmonella/mammalian-microsome assay. Mutat Res 79:203–212

Nukaya H, Yamashita J, Tsuji K, Terao Y, Ohe T, Sawanishi H, Katsuhara T, Kiyokawa K, Tezuka M, Oguri A, Sugimura T, Wakabayashi K (1997) Isolation and chemical-structural determination of a novel aromatic amine mutagen in water from the Nishitakase River in Kyoto. Chem Res Toxicol 10:1061–1066

Oguri A, Shiozawa T, Terao Y, Nukaya H, Yamashita J, Ohe T, Sawanishi H, Katsuhara T, Sugimura T, Wakabayashi K (1998) Identification of a 2-phenylbenzotriazole (PBTA)-type mutagen, PBTA-2, in water from the Nishitakase River in Kyoto. Chem Res Toxicol 11:1195–1200

Oikari A, Holmbom B, Ånäs E, Miilunpalo M, Kruzynski G, Castrén M (1985) Ecotoxicological aspects of pulp and paper mill effluents discharged to an inland water system: distribution in water, and toxicant residues and physiological effects in caged fish (*Salmo gairdneri*). Aquat Toxicol 6:219–239

Palm H, Lammi R (1995) Fate of pulp mill organochlorines in the Gulf of Bothnia sediments. Environ Sci Technol 29:1722–1727

Parrott JL, van den Heuwel MR, Hewitt LM, Baker MA, Munkittrick KR (2000) Isolation of MFO inducers from tissues of white suckers caged in bleached kraft mill effluent. Chemosphere 41:1083–1089

Parvez S, Venkatamaran C, Mukherji S (2008) Toxicity assessment of organic contaminants: evaluation of mixture effects in model industrial mixtures using 2nd full factorial design. Chemosphere 73:1049–1055. The handbook of environmental chemistry, Volume 5J. Springer, Berlin/Heidelberg. doi:10.1007/698_2009_18

Pessala P, Schultz E, Nakari T, Joutti A, Herve S (2004) Evaluation of wastewater effluents by small-scale biotests and a fractionation procedure. Ecotoxicol Environ Saf 59:263–272

Petrovic M, Eljarrat E, Lopez de Alda MJ, Barceló D (2004) Endocrine disrupting compounds and other emerging contaminants in the environment: a survey of new monitoring strategies and occurrence data. Anal Bioanal Chem 378:549–562

Pinheiro HM, Touraud E, Thomas O (2004) Aromatic amines from azo dye reduction: status review with emphasis on direct UV spectrophotometric detection in textile industry wastewaters. Dyes Pigments 61:121–139

Postigo C, Kuster M, Villagrasa M, Rodriguez-Mozaz S, Brix R, la Farré M, Lopez de Alda M, Barceló D (2009) Liquid chromatography – mass spectrometry methods for analysis of endocrine-disrupting chemicals in wastewaters. HdB env Chem 5J, pp 227–272. The handbook of environmental chemistry, Volume 5J. Springer, Berlin/Heidelberg. doi:10.1007/698_2009_18

Puig A, Ormad P, Roche P, Sarasa J, Gimeno E, Ovelleiro JL (1996) Wastewater from the manufacture of rubber vulcanization accelerators: characterization, downstream monitoring and chemical treatment. J Chromatogr A 733:511–522
Reemtsma T (2001) Prospects of toxicity-directed wastewater analysis. Anal Chim Acta 426:279–287
Reemtsma T, Jekel M (1994) Improving GC-MS screening analysis by functional-group-related silica gel fractionation of industrial wastewater extracts. J High Resolut Chromtogr 17:589–592
Reemtsma T, Jekel M (1997) Dissolved organics in tannery wastewaters and their alteration by a combined anaerobic and aerobic treatment. Water Res 31:1035–1046
Reemtsma T, Fiehn O, Jekel M (1999a) A modified method for the analysis of organics in industrial wastewater as directed by their toxicity to *Vibrio fischeri*. Fresenius J Anal Chem 363:771–776
Reemtsma T, Putschew A, Jekel M (1999b) Industrial wastewater analysis: a toxicity-directed approach. Waste Manag 19:181–188
Ricciardi F, Bonnineau C, Faggiano L, Geiszinger A, Guasch H, Lopez-Doval J, Muñoz I, Proia L, Ricart M, Romaní A, Sabater S (2009) Is chemical contamination linked to the diversity of biological communities in rivers? Trends Anal Chem 28:592–602
Rieger P-G, Meier H-M, Gerle M, Vogt U, Groth T, Knackmuss H-J (2002) Xenobiotics in the environment: present and future strategies to obviate the strategies to obviate the problem of biological persistence. J Biotechnol 94:101–123
Rockström J, Steffen W, Noone K, Persson Å, Chapin FS, Lambin E, Lenton TM, Scheffer M, Folke C et al (2009) Planetary boundaries: exploring the sage operating space for humanity. Ecol Soc 14:32–63
Rodil R, Quintana JB, Reemtsma T (2005) Liquid chromatography-tandem mass spectrometry determination of nonionic organophosphorus flame retardants and plasticizers in wastewater samples. Anal Chem 77:3083–3089
Rodriguez DM, Wrobel K, Jimenez MGG (2004) Determination of 2-mercaptobenzothiazole (MBT) in tannery wastewater by high performance liquid chromatography with amperometric detection. Bull Environ Contam Toxicol 73:818–824
Saiz-Salinas JI (1997) Evaluation of adverse biological effects induced by pollution in the Bilbao Estuary (Spain). Environ Pollut 96:351–359
Sanchez Rojas F, Bosch Ojeda C (2005) Effluent analysis in analytical chemistry: an overview. Anal Bioanal Chem 382:978–991
Santos FJ, Galceran MT, Fraisse D (1996) Application of solid-phase microextraction to the analysis of volatile organic compounds in water. J Chromatogr A 742:181–189
Sarakinos HC, Bermingham N, White PA, Rasmussen JB (2000) Correspondence between whole effluent toxicity and the presence of priority substances in complex industrial effluents. Environ Toxicol Chem 19:63–71
Saunders JE, Zahed KMA, Paterson DM (2007) The impact of organic pollution on the macrobenthic fauna of Dubai Creek (UAE). Mar Pollut Bull 54:1715–1723
Schäfer RB, Caquet T, Siimes K, Mueller R, Lagadic L, Liess M (2007) Effects of pesticides on community structure and ecosystem functions in agricultural streams of three biogeographical regions in Europe. Sci Total Environ 382:272–285
Schäfer RB, Kefford BJ, Metzeling L, Liess M, Burgert S, Marchant R, Pettigrove V, Goonan P, Nugegoda D (2011) A trait database of stream invertebrates for the ecological risk assessment of single and combined effects of salinity and pesticides in South-East Australia. Sci Total Environ 409:2055–2063
Schriks M, Van Leerdam JA, Van der Linden SC, Van der Burg B, Van Wezel A, De Voogt P (2010) High-resolution mass spectrometric identification and quantification of glucocorticoid compounds in various wastewaters in the Netherlands. Environ Sci Technol 44:4766–4774
Schröder HF (1999) Substance-specific detection and pursuit of non-eliminable compounds during biological treatment of waste water from the pharmaceutical industry. Waste Manag 19:111–123

Schuetzle D, Lewtas J (1986) Bioassay-directed chemical analysis in environmental research. Anal Chem 58:1060A–1075A

Schwarzenbach RP, Escher BI, Fenner K, Hofstetter TB, Johnson CA, von Gunten U, Wehrli B (2006) The challenge of micropollutants in aquatic systems. Science 313:1072–1077

Sellström U, Jansson B (1995) Analysis of tetrabromobisphenol a in a product and environmental samples. Chemosphere 31:3085–3092

Shiozawa T, Tada A, Nukaya H, Watanabe T, Takahashi Y, Asanoma M, Ohe T, Sawanishi H, Katsuhara T, Sugimura T, Wakabayashi K, Terao Y (2000) Isolation and identification of a new 2-phenylbenzotriazole-type mutagen (PBTA-3) in the Nikko River in Aichi, Japan. Chem Res Toxicol 13:535–540

Simboura N, Zenetos A, Panayotidis P, Makra A (1995) Changes in benthic community structure along an environmental pollution gradient. Mar Pollut Bull 30:470–474

Smith B (1990) Identification and reduction of toxic pollutants in textile mill effluents. Office of Waste Reduction, North Carolina State University, N.C. Department of Environment, Health, and Natural Resources, Raleigh, p 108

Söderström M, Wachtmeister CA, Förlin L (1994) Analysis of chlorophenolics from bleach kraft mill effluents (BKME) in bile of perch (*Perca fluviatlis*) from the Baltic Sea and development of an analytical procedure also measuring chlorocatechols. Chemosphere 28:1701–1719

Soimasuo R, Jokinen I, Kukkonen J, Petänen T, Ristola T, Oikari A (1995) Biomarker responses along a pollution gradient: effects of pulp and paper mill effluents on caged whitefish. Aquat Toxicol 31:329–345

Spanggord RJ, Gibson BW, Keck RG, Thomas DW (1982) Effluent analysis of wastewater generated in the manufacture of 2,4,6-trinitrotoluene 1 characterization study. Environ Sci Technol 16:229–232

Speight JG (2007) The chemistry and technology of petroleum, 3rd edn. CRC Press, Boca Raton, p 945. ISBN 0-8493-9067-2

Stolz A (2001) Basic and applied aspects in the microbial degradation of azo dyes. Appl Microbiol Biotechnol 56:69–80

Svenson A, Norin H, Hynning P-A (1996) Toxicity-directed fractionation of effluents using the bioluminescence of *Vibrio fischeri* and gas chromatography/mass spectroscopy. Identification of toxic components. Environ Toxicol Water Qual Int J 11:277–284

Swartz CD, Donnelly KC, Islamzadeh A, Rowe GT, Rogers WJ, Palatnikov GM, Mekhtiev AA, Kasimov R, McDonald TJ, Wicklife JK, Presley BJ, Bickham JW (2003) Chemical contaminants and their effects in fish and wildlife from the industrial zone of Sumgayit, Republic of Azerbaidjan. Ecotoxicology 12:509–521

Takada H, Eganhouse RP (1998) Molecular markers of anthropogenic waste. In: Meyers RA (ed) Encyclopedia of environmental analyses and remediation. Wiley, New York, pp 2883–2940

Takada H, Mutoh K, Tomita N, Miyadzu T, Ogura N (1994) Rapid removal of linear alkylbenzene sulfonates (LAS) by attached biofilm in an urban shallow stream. Water Res 28:1953–1960

Tavendale MH, Wilkins AL, Langdon AG, Mackie KL, Stuthridge TR, McFarlane PN (1995) Analytical methodology for the determination of freely available bleached kraft mill effluent-derived organic constituents in recipient sediments. Environ Sci Technol 29:1407–1414

Terasaki M, Fukazawa H, Tani Y, Makino M (2008) Organic pollutants in paper-recycling process water discharge areas: first detection and emission in aquatic environment. Environ Pollut 151:53–59

Terasaki M, Shiraishi F, Fukazawa H, Makino M (2009) Development and validation of chemical and biological analyses to determine the antiestrogenic potency of resin acids in paper mill effluents. Environ Sci Technol 43:9300–9305

Terasaki M, Jozuka K, Makino M (2012) Identification and accumulation of aromatic sensitizers in fish from paper recycling in Japan. Environ Toxicol Chem 31(1202):1208

Terzic S, Ahel M (2011) Nontarget analysis of polar contaminants in freshwater sediments influenced by pharmaceutical industry using ultra-high-pressure liquid chromatography-quadrupole time-of-flight mass spectrometry. Environ Pollut 159:557–566

Thompson B, Adelsbach T, Brown C, Hunt J, Kuwabara J, Neale J, Ohlendorf H, Schwarzbach S, Spies R, Taberski K (2007) Biological effects of anthropogenic contaminants in the San Francisco Estuary. Environ Res 105:156–174

Umbuzeiro GA, Freeman HS, Warren SH, Oliveira DP, Terao Y, Watanabe T, Claxton LD (2005) The contribution of azo dyes to the mutagenic activity of the Cristais River. Chemosphere 60:55–64

UNEP (2012) In: Alcamo J, Leonard SA (eds) 21 Issues for the 21st century: result of the UNEP foresight process on emerging environmental issues. United Nations Environment Programme (UNEP), Nairobi, p 56

Venturini N, Muniz P, Bícego MC, Martins CC, Tommasi LR (2008) Petroleum contamination impact on macrobenthic communities under the influence of an oil refinery: integrating chemical and biological multivariate data. Estuar Coast Shelf Sci 78:457–467

Wake H (2005) Oil refineries: a review of their ecological impacts on the aquatic environment. Estuar Coast Shelf Sci 62:131–140

Walsh AR, O'Halloran J (1997) The toxicity of leather tannery effluent to a population of the blue mussel *Mytilus edulis* (L.). Ecotoxicology 6:137–152

Wang Z, Fingas M, Blenkinsopp S, Sergy G, Landriault M, Sigouin L, Foght J, Semple K, Westlake DWS (1998) Comparison of oil composition changes due to biodegradation and physical weathering in different oils. J Chromatogr A 809:89–107

Warwick RM, Platt HM, Clarke KR, Agard J, Gobin J (1990) Analysis of macrobenthic and meiobenthic community structure in relation to pollution and disturbance in Hamilton Harbor, Bermuda. J Exp Mar Biol Ecol 138:119–142

Weber B, Bremser W, Hiltrop K (2009) Creating new materials with melamine resins. Prog Org Coat 64:150–155

Weiss KD (1997) Paint and coatings: a mature industry in transition. Prog Polym Sci 22:203–245

White R, Jobling S, Hoare SA, Sumpter JP, Parker MG (1994) Environmentally persistent alkylphenolic compounds are estrogenic. Endocrinology 135:175–182

Worawit W (2006) Characterization and fenton treatment of hazardous organic substances in para rubber industrial wastewater in Southern Thailand. Ph.D. thesis, Chulalongkorn University, doi:123456789/8327

Yang L, Yu H, Yin D, Jin H (1999) Application of the simplified toxicity identification evaluation procedures to a chemical works effluent. Chemosphere 38:3571–3577

Yasuhara A, Shiraishi H, Tsuji M, Okuno T (1981) Analysis of organic substances in highly polluted river water by mass spectrometry. Environ Sci Technol 15:570–573

Yasuhara A, Shiraishi H, Nishikawa M, Yamamoto T, Uehiro T, Nakasugi O, Okumura T, Kenmotsu K, Fukui H, Nagase M, Ono Y, Kawagoshi Y, Baba K, Noma Y (1997) Determination of organic components in leachates from hazardous waste disposal sites in Japan by gas chromatography–mass spectrometry. J Chromatogr A 774:321–332

Chapter 3
Fly Ash Pollutants, Treatment and Recycling

Alessandra Gianoncelli, Annalisa Zacco, Rudolf P.W.J. Struis, Laura Borgese, Laura E. Depero, and Elza Bontempi

Contents

A. Gianoncelli • A. Zacco • L. Borgese • L.E. Depero • E. Bontempi (✉)
INSTM and Chemistry for Technologies Laboratory, Brescia University,
Via Branze 38 25023 Brescia, Italy
e-mail: elza.bontempi@ing.unibs.it

Rudolf P.W.J. Struis
Laboratory for Bioenergy and Catalysis (LBK), Paul Scherrer Institute, General Energy Research (ENE), CH 5232 Villigen, Switzerland

École Polytechnique Fédérale de Lausanne (EPFL-ENAC-IIE), 1015 Lausanne, Switzerland

E. Lichtfouse et al. (eds.), *Pollutant Diseases, Remediation and Recycling*,
Environmental Chemistry for a Sustainable World 4, DOI 10.1007/978-3-319-02387-8_3,

Abstract This chapter reviews fly ash typology, composition, treatment, deposition, recycling, functional re-use, and metals and organic pollutants abatement. Fly ash is a by-product of power and incineration plants operated either on coal and biomass, or municipal solid waste. The growing of environmental awareness and increasing energy and material demand will foster recycling. Recycling will help to reuse valuable materials which would otherwise be wasted, and reduce energy consumption and greenhouse gas emissions from extraction and processing. Fly ash is world's fifth largest material resource because of the large amount of ash produced in the world. Fly ash can be classified into several categories: coal fly ash obtained from power plant burning coal; flue gas desulphurisation fly ash, that is the byproduct generated by the air pollution control equipment in coal-fired power plants to prevent (reduce) the release of SO_2; biomass fly ash produced in the thermal conversion of biomass; and municipal solid waste incineration (MSWI) fly ash, that is the finest residue obtained from the scrubber system in a municipal solid waste incineration plant.

Fly ash often contains pollutants such as heavy metals and organic compounds. The composition of fly ash is very variable, depending on their origins, then also the pollutants can be very different. For example, MSWI fly ash are the most problematic ash in terms of contaminant content. We review existing techniques for fly ash inertization, separately considering heavy metals entrapment or organic abatement. We show that fly ash is a valuable resource with potential use in several applications like agriculture, synthesis of zeolite and geopolymer, adsorbent and building materials. Finally all advantages in fly ash recovery and re-use are discussed. It is shown that fly ash recycling will reduce landfilling disposal, raw materials employ, greenhouse gas emission and water consumption.

Keywords Fly ash • Heavy metals • Organic pollutants • Recycling • FGD • Coal • MSWI • RHA

List of Abbreviations

AES Acid Extraction Sulphide stabilisation process
APC Air Pollution-Control
CDD ChloroDibenzo-p-Dioxin
CDF ChloroDibenzo-p-Furan

COSMOS	Colloidal Silica Medium to Obtain Safe inert
DC	Direct Current
DDT	DichloroDiphenylTrichloroethane
EDTA	EthyleneDiamineTetraAcetate
FA	Fly Ash
FGD	Flue Gas Desulphurisation
HpCDD	HeptaChloroDibenzo-p-Dioxin
HpCDF	HeptaChloroDibenzo-p-Furan
HRGC/HRMS	High Resolution Gas Chromatography/High Resolution Mass Spectrometry
HxCDD	HexaChloroDibenzo-p-Dioxin
HxCDF	HexaChloroDibenzo-p-Furan
LCA	Life Cycle Assessments
LOI	Loss of Ignition
MSWI	Municipal Solid Waste Incineration
OCDD	OctaChloroDibenzo-p-Dioxin
OCDF	OctaChloroDibenzo-p-Furan
PAH	Polycyclic Aromatic Hydrocarbon
PCB	Polychlorinated Bifenyl
PCDD	PolyChloroDibenzo-p-Dioxin
PCDF	PolyChloroDibenzo-p-Furan
PeCDD	PentaChloroDibenzo-p-Dioxin
PeCDF	PentaChloroDibenzo-p-Furan
RHA	Rice Husk Ash
S/S	Solidification/Stabilisation
TCDD	TetraChloroDibenzo-p-Dioxin
TCDF	TetraChloroDibenzo-p-Furan
TEQ	Toxic Equivalent
VOC	Volatile Organic Compound
WHO-TEQ	World Health Organisation-Toxic Equivalent

3.1 Introduction

This review addresses fly ash (FA) typology, composition, treatment, deposition, recycling, functional re-use, and metals and organic pollutants abatement of FAs resulting from power and incineration plants operated either on coal and biomass, or, municipal solid waste. The emphasis of the review lies with outlining major insights, developments, and trends in FA treatment during the past four decades (>1976) and the survey cites about 500 publications. From the three sources (coal, biomass, MSW), the use of coal predominates. Recent estimates put global coal FA production in the region of 750 million tons/year (Blissett and Rowson 2012), which means a raise by about 50 % compared to the last decade. This is attributable to the increasing energy demand from economically uprising countries, such as India, China, and Brazil. This trend may be enforced in the future by countries seeking interim solutions to nuclear power free-energy production. But, so far, European countries put many efforts in generating heat, energy, and power from renewable biomass-based sources in both, small- and large-scale, grate furnaces (domestic boilers, stoves, district heating plants) and by diversification of its energy supply from biomass (Rajamma et al. 2009; Yin et al. 2008). With time, growing environmental awareness and increasing energy and material demand turned the initial goals for FA treatment (MSW volume reduction, inhibiting metal leaching, inertization, stabilization, landfill deposition, or, deriving secondary raw ores and additives for building materials) towards also exploiting the chemical or physical potentials of FA constituents, thus, rendering them second-life functionality. This had been made possible by profound knowledge on FA composition, morphology, and changes between different sources. The treatment, reuse, or, final destiny of FA depends highly, but not only, on its mineral composition. They are outlined below briefly, because they set the framework for developments and trends in FA treatment.

Coal FA is one of the most complex anthropogenic materials that can be characterized, because not only it can comprise up to about 300 individual minerals (Vassilev and Vassileva 2005), but also the amorphous aluminosilicate nature of coal FA makes its structure difficult to characterize. Other determining factors are coal source, method of combustion power plants (Scheetz and Earle 1998; Helmuth 1987), and, typology of emission control device, storage, and, handling (Jala and Goyal 2006).

Flue gas desulphurisation (FGD) FA is the byproduct generated by the air pollution control equipment in coal-fired power plants to prevent (reduce) the release of SO_2 into the air, as it is an atmospheric pollutant and a precursor to acid rain. The term flue gas desulphurisation generally refers to wet scrubbers that remove sulphur dioxide emissions from large electric utility boilers (mainly coal combustion).

Biomass FA produces problems in the thermal conversion process itself (for example, slagging and fouling phenomena) and, depending on quantity and quality, can also pose environmental management problems. The composition of ash produced in a biomass power plant depends strongly on the biomass source (Mukherjee et al. 2008; Burke 2007). Moreover, some biosorbents are selective for certain types of heavy metals, whereas others are not (Garg et al. 2007; Tarley and Arruda 2004). This is also apparent with element compositions of FA from different biomass (Thy et al. 2006; Rajamma et al. 2009), showing that they vary widely with biomass type.

MSWI FA is a collective name for FA recovered from the gas stream and the APC residue from the scrubber system in a municipal solid waste incineration plant (De Boom and Degrez 2012). Depending upon waste type, combustion unit, melting conditions, and air-pollution-control device, the heterogeneous composition of MSWI FA can vary substantially (Kirby and Rimstidt 1993; Song et al. 2004). Usually, they comprise highly volatile elements, such as Cl and S, among larger amounts of other elements like Si, Al, K, Ca, plus traces of potentially toxic heavy metals (Pb, Zn, etc.), incompletely combusted carbon residues, and possibly also hazardous persistent organic pollutants (POP), like PAH, PCB, chlorinated dioxins, etc. Potentially leachable heavy metals concentrate more in glassy than in crystalline phases (Zevenbergen and Comans 1994) due to MSWI-specific conditions promoting volatilization, melting, condensation, and, precipitation processes (cf. Eighmy et al. 1995). The formation of (more) stable secondary minerals may occur naturally (by aging) or brought about purposely by several reactions (oxidation, carbonation, neutralization of pH, dissolution, precipitation, sorption), whereas degradation of POPs by aging either is unlikely or involves impractically long deposit times, therefore, demanding special treatment.

This review addresses FA topologies, FA treatments, organic pollutants abatement, and FA recycling in detail one after the other in Sects. 3.2, 3.3, 3.4 and 3.5 and it ends with conclusions regarding FA treatments and temporal trends.

3.2 Fly Ash Typologies

3.2.1 Coal Fly Ash

Coal FA is a predominantly inorganic residue obtained from the flue gases of furnaces at pulverised coal power plants. When coal is burnt, the minerals entrained in the coal are thermally transformed into chemical species that are reactive or could be chemically activated, for example, by the addition of calcium hydroxide (Vassilev and Vassileva 1996). Most estimates in the current literature put global coal FA production in the region of 500 million tons/year (Ahmaruzzaman 2010). Figure 3.1 shows the estimated coal FA amount production in 2005, in some countries.

These estimates are based on data at least 10 years old. In the last years, coal consumption has increased up to 50 %, mainly due to the economic growth of emerging countries, like China and India. Considering this, a more reliable estimate would mean that 750 million tonnes of coal FA is generated on a global basis each year (Blissett and Rowson 2012; Izquierdo and Querol 2012). Coal FA is generally produced at 1,200–1,700 °C from the various inorganic and organic constituents of the parent coal. The formation mechanism of FA particles is schematically represented in Fig. 3.2. The first step of the combustion makes the coal mineral matter conversion in char. The char obtained from coal burns out at much higher temperatures. Always coal includes also extraneous minerals, which gradually decompose and are released within the char as its fragments. After the minerals convert to gases and eventually condense to form solid ash particles. Homogeneous condensation results in ash particles with dimensions ranging between 0.01 and 0.2 μm. Fragmentation of included minerals produces particles with dimensions ranging between 0.2 and 10 μm. Residual mineral matter undergoes a series of complex

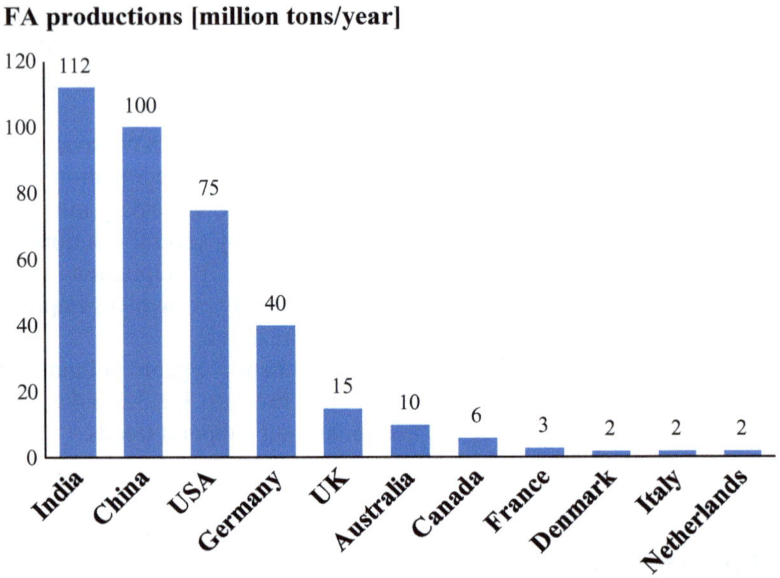

Fig. 3.1 Fly ash production (million tons/year) in different countries (Pandey and Singh 2010) in 2005

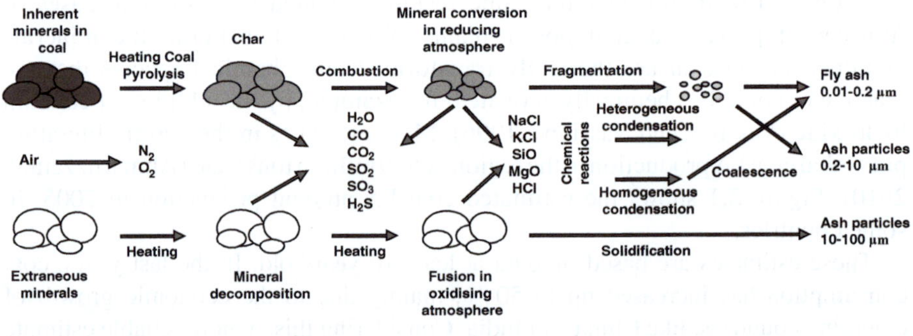

Fig. 3.2 Simplified mechanism of coal fly ash formation from pulverised fuel combustion (Tomeczek and Palugniok 2002)

transformations to form predominantly spherical particles with a size range between 10 and 100 μm (Sarkar et al. 2005).

The characteristics of produced FA differ depending on the coal source, the method of combustion of power plants (Scheetz and Earle 1998; Helmuth 1987), and typologies of emission control devices, storage and handling (Jala and Goyal 2006). Therefore, FA shows a wide variation in their physicochemical and mineralogical properties. For example, the pH of FA can vary from 4.5 to 12.0 depending largely on the sulphur content of the parent coal. About 300 individual minerals

Table 3.1 Bulk chemical composition of coal FA by region (Blissett and Rowson 2012). Amount of carbon was measured by a Loss of Ignition (LOI) test (Kruger 1998)

	Range (mass%)				
Component	Europe	US	China	India	Australia
SiO_2	28.5–59.7	37.8–58.5	35.6–57.2	50.2–59.7	48.8–66.0
Al_2O_3	12.5–35.6	19.1–28.6	18.8–55.0	14.0–32.4	17.0–27.8
Fe_2O_3	2.6–21.2	6.8–25.5	2.3–19.3	2.7–14.4	1.1–13.9
CaO	0.5–28.9	1.4–22.4	1.1–7.0	0.6–2.6	2.9–5.3
MgO	0.6–3.8	0.7–4.8	0.7–4.8	0.1–2.1	0.3–2.0
Na_2O	0.1–1.9	0.3–1.8	0.6–1.3	0.5–1.2	0.2–1.3
K_2O	0.4–4	0.9–2.6	0.8–0.9	0.8–4.7	1.1–2.9
P_2O_5	0.1–1.7	0.1–0.3	1.1–1.5	0.1–0.6	0.2–3.9
TiO_2	0.5–2.6	1.1–1.6	0.2–0.7	1.0–2.7	1.3–3.7
MnO	0.03–0.2	nd	nd	0.5–1.4	nd
SO_3	0.1–12.7	0.1–2.1	1.0–2.9	nd	0.1–0.6
LOI	0.8–32.8	0.2–11.0	nd	0.5–5.0	nd

have been identified in different coal FAs (Vassilev and Vassileva 2005). The main detected crystalline phases are mullite, quartz, magnetite, haematite and anhydrite (Helmuth 1987; Wang 2008). Glass is also found. Quartz and mullite are the major crystalline constituents of low-calcium ash, whereas high-calcium FA consists of quartz, tricalcium aluminate, calcium silicate, and tetracalcium aluminosilicate (Ahmaruzzaman 2010). The less predominant minerals include calcite, pyrite and hematite. Because of this structural variety, coal FA is one of the most complex anthropogenic materials that can be characterised.

Table 3.1 reports bulk chemical composition of coal FA depending of their geographical region. In general it can be seen that coal FA has a bulk chemical composition containing a variety of metal oxides in the order $SiO_2 > Al_2O_3 > Fe_2O_3 > CaO > K_2O > MgO$, Na_2O, TiO_2. However, it is evident that there are significant differences in composition, not only among regions, but also inside the same region. In addition, a large amount of amorphous content was also found (Blissett and Rowson 2012). The amorphous aluminosilicate nature of FA, that makes its structure difficult to characterize, makes coal FA also very versatile, since the glassy component of FA reacts first before the crystalline phases. This is very important to decide their final use. Coal FA mineralogy also depends on the coal source. Coal can be generally divided in four typologies, which are anthracite, bituminous, sub-bituminous, and lignite. Therefore, final coal FA composition is also determined by the type of coal burned to produce it. Generally it is shown that coal FA from sub-bituminous and lignite coals is characterised by higher CaO and MgO and lower Al_2O_3 with respect to the higher grade fuels such as bituminous and anthracite coals. On the contrary, bituminous coal FA contains less than 10 % CaO (and usually does not contain any crystalline compounds of Calcium) in total and often mainly consist of aluminosilicate glass. Coal FA that contains more than 15 % CaO is composed of calcium aluminosilicate glass in addition to

Table 3.2 Range of chemical compounds of fly ash produced from different coal types

Component (wt%)	Bituminous	Sub–bituminous	Lignite	Antracite
SiO_2	20–60	40–60	15–45	28–57
Al_2O_3	5–35	20–30	10–25	18–36
Fe_2O_3	10–40	4–10	4–15	3–16
CaO	1–12	5–30	15–40	1–27
MgO	0–5	1–6	3–10	1–4
SO_3	0–4	0–2	0–10	0–9
Na_2O	0–4	0–2	0–6	0–1
K_2O	0–3	0–4	0–4	0–4
LOI	0–15	0–3	0–5	1–8

Ahmaruzzaman (2010) and Moreno et al. (2005)

Table 3.3 Classification systems of the US standards bodies for fly ash use in concrete (ASTM C618)

Class ASTM C618	$SiO_2+Al_2O_3+Fe_2O_3$ (%)	SO_3 (%)	Moisture (%)	LOI (%)
C	>50	<5	<3	<6
F	>70	<5	<3	<12

substantial proportions of crystalline Calcium compounds including tricalcium aluminate, tetracalcium aluminosilicate, calcium silicate and calcium oxide (Manz 1999).

Table 3.2 shows the normal range of chemical constituents of bituminous coal FA with those of lignite coal FA and sub-bituminous and antracitre coal FA. Lignite and sub-bituminous coal FA have a higher calcium oxide content and lower loss of ignition (LOI, that is an indication of the amount of carbon in FA) than FA from bituminous coals. Lignite coal FA may have a higher concentration of sulphite compounds than bituminous coal FA.

Based on its mineralogy, American Society for Testing and Materials (ASTM) grouped coal FA into two classes, designated C and F, respectively (Blissett and Rowson 2012).

Table 3.3 shows that in the ASTM classification Class F coal FA has a combined SiO_2, Al_2O_3, and Fe_2O_3 content of greater than 70 % compared to greater than 50 % for Class C coal FA. Another difference between Class F and Class C is that the amount of alkalis (combined sodium and potassium), and sulphites, are generally higher in Class C FA than in the Class F FA. On the basis of this classification, class C coal FA is derived from lignite and sub-bituminous coals and class F coal FA is derived from bituminous and anthracite coals.

Coal FA classifications, widely reported in literature, are fundamental to relate its mineralogy to its applications. The aim is to use a system designed to distinguish coal FA types that will be suitable to choose appropriate FA for use in building applications (Kamseu et al. 2011). Indeed, class F ash is regarded as a pozzolanic material. A pozzolan is a siliceous, or a siliceous and aluminous material that has no

Table 3.4 Approach for classification of coal FA based on chemical composition

Class	$SiO_2+Al_2O_3+K_2O+TiO_2+P_2O_5$ (%)	$CaO+MgO+SO_3+Na_2O+MnO$ (%)	Fe_2O_3 (%)
Sialic	>77	<11.5	<11.5
Calsialic	<89	>11.5	<11.5
Ferrisialic	<89	<11.5	>11.5
Ferricalcsialic	<77	>11.5	>11.5

Vassilev and Vassileva (2007)

intrinsic cementitious property. Moreover, when it is in a very finely divided form, it can chemically react with calcium hydroxide at ordinary temperatures and in presence of moisture, and form compounds exhibiting cementitious properties. The nature of lignite and sub-bituminous FA, that contain high calcium oxide quantity, can induce the formation of cementitious materials also in the absence of $Ca(OH)_2$ (Manz 1999). However, this classification is prescriptive, rather than based on performance, so it can be controversial because, in practice, many class C FAs can meet the performance requirements of class F ashes (Manz 1999). Vassilev and Vassileva (2007) proposed a new classification system based on an analysis of 41 European coal FAs. It groups the main bulk oxides together to create the four tiers classification system presented in Table 3.4.

The new proposed approach is based on a classification system where a further distinction is made between the phase and mineralogical composition of the coal FA (Vassilev and Vassileva 2007).

The chemical composition of coal FA has also been extensively investigated. The major matrix elements are Si, Al, and Fe together with significant percentages of Ca, K, Na and Ti (Matti et al. 1990). FA contains essential macro-nutrients like P, K, Ca, Mg and S and micro-nutrients including Fe, Mn, Zn, Cu, Co, B and Mo. The type of coal used for combustion affects the sulphur content of FA. Coal FAs contain many elements at concentrations greater than 50 mg/kg. Some of these may produce environmental concern (Izquierdo and Querol 2012). For example, coal FA is often rich in trace elements like Mercury, Cobalt and Chromium (Izquierdo and Querol 2012), that are concentrated in smaller FA particles (Adriano et al. 1980). Al in FA is mostly bound in insoluble aluminosilicate structures, and this considerably limits its biological toxicity. The ratio between Ca and S is strictly related to the pH of the water-ash system and so plays the dominant role in the leachability of most elements contained in FA (Izquierdo and Querol 2012). In pH range from 7 to 10, Be, Cd, Co, Cu, Fe, Mg, Mn, Ni, Pb, Si, Sn, Th, Tl, U, and Zn have the minimum solubility; by contrast, oxyanionic-forming species, i.e. As, B, Cr, Mo, Sb, Se, V, and W, display their maximum leachability in the same pH range (Izquierdo and Querol 2012).

Table 3.5 shows the concentrations of trace elements in 23 European coal FAs. These values are in agreement with the elemental composition found for coal FA from other areas (Jankowski et al. 2006). It is noticeable that elements, such as As, Cr, Pb, and Se, may be present in coal FA. Moreover, coal FA generally

Table 3.5 Trace element content in 23 European fly ashes from coal

	Trace element composition (ppm)		
Element	25th Perc.	Median	75th Perc.
As	40	55	97
B	135	259	323
Ba	639	1,302	1,999
Be	6	8	12
Cd	1	2	2
Co	30	35	48
Cr	137	148	172
Cu	73	86	118
Ge	3	7	15
Hg	0.2	0.2	0.3
Li	150	185	252
Mo	7	11	13
Ni	87	96	144
Pb	59	80	109
Rb	50	108	147
Sb	4	4	8
Se	6	7	13
Sn	7	8	10
Sr	384	757	1,647
Th	25	30	37
U	9	12	18
V	202	228	278
Zn	123	154	175

Moreno et al. (2005)

shows a relatively low content of environmentally available toxic metals, mainly in the leachable fraction. The leachability of toxic elements is closely related to the phase with which they are associated and to leaching conditions such as pH (Jankowski et al. 2006). It is very recently reported (Świetlik et al. 2012) that high combustion temperature leads to a reduction in the concentration of Cd and Pb in FA, because the most volatile metals are concentrated in the finest particles that are less efficiently collected by the electrostatic precipitator system. Consequently, the level of toxic metals leached with water is generally low. Additionally, in the presence of Ca, the leachability of As and V is attenuated due to the occurrence of low solubility phases (Izquierdo and Querol 2012). From the morphological point of view, coal FA consists of fine, glass-like particles, with dimensions that fall in the range from 0.01 to 100 μm. The morphology of coal FA particles is controlled primarily by the combustion temperature and subsequent cooling rate. The particles are predominantly spherical in shape. Scanning electron microscopy analysis has revealed that coal FA samples consist of solid spheres, hollow spheres (cenospheres), and irregular unburned carbon. Mineral aggregates containing corundum, quartz, and magnetite particles have also been observed (Benezet et al. 2008; Kutchko and Kim 2006) (Fig. 3.2). Cenospheres are believed to be formed by the expansion of CO_2 and H_2O gasses, evolved from minerals within the coal being burnt (Fisher et al. 1976). The predominant forces are, however, the pressure and surface tension on the melts as well as gravity (Fisher et al. 1976). The predominant spherical microscopic structure of fine FA is a consequence of equilibrium of forces on the molten inorganic particle as it is forced up the furnace or smoke stack against gravity. The molten inorganic particles cool down rapidly, maintaining their equilibrium shape. Generally, the percentage of cenospheres increases with the ash content in coal, and decrease with Fe_2O_3 concentration (Ghosal and Self 1995). Because cenospheres are hollow, they have a low bulk density. Therefore, FA has low bulk density (normally less than 1 g/cm^3), high surface area and light texture (Asokan et al. 2005).

3.2.2 Flue Gas Desulphurisation (FGD) Fly Ash

The coal burning can convert sulphur-bearing impurities to gaseous SO_2, an atmospheric pollutant and a precursor to acid rain. In the past two decades, restrictions on the emission of sulphur dioxide from coal-fired power plants have been increasingly stringent. In response to these requirements, power plants have had to remove increasing amounts of SO_2 from the flue gases before releasing them to the atmosphere. This problem has spurred the development of various types of scrubbing processes to convert SO_2 from flue gases into solid products for disposal or beneficial reuse. Flue gas desulphurisation (FGD) FA is the byproduct generated by the air pollution control equipment in coal-fired power plants, to reduce sulphur amount in air. The term flue gas desulphurisation generally refers to wet scrubbers that remove sulphur dioxide emissions from large electric utility boilers (mainly coal combustion). These systems are capable of removing approximately 70–90 % of the SO_2 in the flue gas, depending on the operating conditions of the system.

Since the end of the 1970s, more than 100 million tons of FGD FA has been produced in hard coal and lignite power plants in Europe (Leiva et al. 2010). Many types of FGD systems have been developed. These can be classified into two major groups, wet and dry processes, which are subdivided in several types depending on the specific chemical reactions taking place and the employed flow conditions (Leiva et al. 2010). Among them, the wet limestone FGD process is the most widely employed because of its high desulphurisation performance, reliability and low utility consumption (Kikkawa et al. 2002). In wet FGD scrubbing systems, the scrubbing liquid contains an alkali reagent to enhance the absorption of SO_2 and eventually other acid gases. More than a dozen different reagents can be used, with lime and limestone being the most diffused.

Sodium-based solutions (sometimes referred to as clear solutions) provide better SO_2 solubility and less scaling problems than lime or limestone. However, sodium reagents are much more expensive. Generally, SO_2 is removed from the flue gas by absorption into limestone slurry, which in turn is extracted from the system as gypsum slurry and is dewatered. FDG FA chemical composition is mainly gypsum, so the finally produce FA are called FGD FA and/or FGD gypsum. The other impurities include about 10 % of extra water, organic carbon, and limestone (Miao et al. 2012). FGD FA volumes and quality were essentially influenced by power station parameters, such as mode of operation, type of fuel/sulphur content, technology of the desulphurisation unit and absorbents type. Concerning coal FA, libraries and web sites can be found (see for example the American Coal Ash Association) with much additional information. In contrast, much less data are available on the properties of FGD FA. Kost et al. (2005) published a study reporting chemical physical properties of FGD samples collected from 13 different locations. The composition of all samples was dominated by Ca, S, Al, Fe, and Si and strong preferential partitioning into the acid insoluble residue (i.e., coal ash residue) was observed for Al, Ba, Be, Cr, Fe, Li, K, Pb, Si, and V. Deionised water leachates and diluted acetic acid leachates had mean pH values of >11.2 and high mean concentrations of S (primarily as SO_4^{2-}) and Ca. Concentrations of Ag, As, Ba, Cd, Cr, Hg, Pb, and Se, were below drinking water standards in leachates. FGD FA is mostly uniform in particle size. Specific surface area was related to particle size and it varied between 1.3 m^2/g with bed ash and 9.5 m^2/g with spray dryer material. Many of the chemical and physical properties of these FGD FAs were associated with the quality of the coal rather than the combustion and SO_2 scrubbing used processes.

3.2.3 Municipal Solid Waste Incineration (MSWI) Fly Ash

Municipal solid waste incineration (MSWI) is a highly efficient technique for waste management, because it reduces significantly the waste volume. The main economic advantage of this system is that the energy produced when burning waste is recovered. However, MSWI technology causes environmental problems, mainly related to the disposal of FA produced during combustion. There are different types of ash residues that result from cleaning emissions from MSW incinerators and it is important to differentiate between FA and air pollution-control (APC) residues. FA is the finer fraction of the particulate (<200 μm) which is entrained in the flue gas and subsequently trapped in an electrostatic precipitator or fabric filter and later removed before any further treatment of the gaseous effluents. FA can be recovered separately from the gas stream, or it can be incorporated into the APC residues. APC residues are produced from a scrubber systems that involve the injection of an alkaline material (for example lime) to remove acid gases, particulates and flue gas condensation/reaction products. FAs and APC residues are often considered together, as a unique output from the MSWI plant. This residue is often called MSWI FA (De Boom and Degrez 2012). Depending upon types of solid waste, combustion unit, melting conditions and air pollution-control device, MSWI FA can have a heterogeneous composition (Kirby and Rimstidt 1993; Song et al. 2004). The mineralogy of MSWI FA is an assemblage of crystalline and amorphous phases (amorphous can be higher than 50 %) (Bontempi et al. 2010b). MSWI FA is preferentially composed of sulphate and aluminosilicate glass as well as silicate minerals (Eighmy et al. 1995; Forestier and Libourel 1998; Fermo et al. 1999, 2000). MSWI FA is also rich in soluble salts (as for example halite and sylvite). The toxic elements in MSWI FA are more concentrated in the glassy phases rather than in the crystalline phases (Zevenbergen and Comans 1994). MSWI FA has a higher fraction of water-soluble compounds than bottom ash. This factor is considered an important aspect for treatment, disposal and possible utilisation of MSWI residue (Hjelmar 1996). In fact, 15–25 % of the total mass of FA and 30–40 % of the total mass of the dry/semidry acid gas scrubbing APC residues containing salts are readily dissolved in water.

Ranges of chemical elements that can be found in MSWI FA are reported in Table 3.6. MSWI FA comprises highly volatile elements, such as Cl and S, in addition to some constituents like Si, Al, K and Ca. The high calcium content is due to excess lime used in the scrubbing process, while the high levels of chloride originate from the significant volumes of polyvinyl chloride present in MSW. As a result of volatilisation and subsequent condensation of particles as well as concentration phenomena occurring during combustion, MSWI FA contain high concentrations of heavy metals (i.e., As, Cd, Hg, Pb and Zn) (Forestier and Libourel 1998; Sabbas et al. 2003) and can contain organic micro-pollutants. The division of metals into solid and gas phases (Fermo et al. 1999) may be due to the effect of incineration temperature and chemical environment in the combustion chamber. In the combustion zone, metals or metal species behave differently related to their thermal behaviours. Consequently, the metal entrapped in a glassy phase or adsorbed into FA particles depends upon a wide range of processes such as volatilisation, melting, subsequent condensation and precipitation (Eighmy et al. 1995).

A recent paper showed extensive characterisation of MSWI FA collected in a German incinerator. (Bayuseno and Schmahl 2011). As an example, the composition data of this ash is given in Table 3.7. As shown in Table 3.7, the composition of MSWI FA change with aging. Several studies (Speiser et al. 2000; Sabbas et al. 2003) demonstrated that a number of weathering reactions can promote mineralogical changes of MSWI FA. For example, because of natural carbonation, and several other factors such as pH, redox potential, temperature and humidity conditions, some mineralogical change can occur over time. The mineralogical changes in FA by aging (both hydrated and naturally aged fly ash) may include reactions as oxidation, carbonation, neutralisation of pH, dissolution, precipitation and sorption – towards formation of stable secondary minerals, such as ettringite, hydrocalumite, and a C–S–H phase. In particular, the natural carbonation produces a change in the mineralogical of FA and formation of new phases that eventually

Table 3.6 Ranges of chemical elements in fly ash from municipal waste incineration

Element	Fly ash (wt. %)
Si	9.5–19
Al	4.9–7.8
Fe	1.8–3.5
Ca	7.4–13
Mg	1.1–1.9
K	2.3–4.7
Na	2.2–5.7
Ti	0.75–1.2
Si	0.11–0.32
Cl	4.5–10.1
P	0.48–0.96
Mn	0.08–0.17
Element	**Fly ash (ppm)**
Ag	31–95
As	49–320
Ba	920–1,800
Be	nd
Cd	250–450
Co	29–69
Cr	140–530
Cu	860–400
Hg	0.8–7
Mo	15–49
Ni	92–240
Pb	7,400–19,000
Sc	6.1–33
Sn	1,400–1,900
Sr	<80–250
V	32–150
W	nd
Zn	19,000–41,000

Hjelmar (1996)

impacts on its leaching behaviour (Sabbas et al. 2003). Then, such structural changes may also be beneficial, leading to significant reductions in trace elements for leaching (Meima and Comans 1999).

It is possible to schematize the distribution of heavy metals in FA as follows (Bayuseno 2006): the most lithophilic (e.g., Cr, Mn, Na, Ti and V) and siderophilic elements (e.g., Co, Mo, Ni, Fe and Sn) are more concentrated in the larger particles (bottom ash). Some elements (W, Mo, Zr, Ti, Co, Al, Si, Ba, Sr), which are hardly vaporised during incineration, would preferentially melt and thereby tend to be

Table 3.7 Mineralogy of MSWI FA as collected and after natural aging of 1 year

Mineral species	Fresh fly ash (wt%)	Aged fly ash (wt. %)
Amorphous	44.2(6)	45.3(6)
Alunite	1.8(1)	0.9(1)
Anhydrite	8.5(1)	5.6(1)
Anorthite-albite (Plagiocase)	1.4(1)	1.6(2)
Apatite	2.3(2)	1.3(1)
Bassanite	–	1.3(1)
$Ca_2S(Ca_2SiO_4)$	2.2(2)	1.4(1)
$Ca_3S(Ca_3SiO_5)$	2.6(2)	2.1(2)
$Ca_3A(Ca_3Al_2O_6)$	0.9(1)	1.3(1)
Calcite	1.5(1)	1.4(1)
Caracolite	0.2(0)	0.2(0)
Cristobalite	0.4(0)	–
Digenite	0.2(0)	0.2(0)
Diopside	2.3(2)	1.5(2)
Dolomite	1.1(1)	0.6(1)
Enstatite	2.3(1)	2.5(2)
Garnet	0.4(1)	0.6(1)
Halite	4.6(1)	5.7(0)
Hematite	0.5(1)	0.6(0)
Hercynite	0.6(1)	1.4(1)
Hydrocalumite	–	0.6(1)
Iron	0.4(0)	–
Kalsilite	0.4(1)	0.4(1)
Lime	0.3(0)	0.3(0)
Magnetite	0.5(1)	0.9(1)
Massicotite	0.2(0)	–
Melilite	3.3(1)	4.4(2)
Minium	1.0(0)	0.2(0)
Muscovite/ilite	–	0.6(1)
Nepheline	1.4(1)	0.8(1)
Portlandite	0.2(0)	–
Potassium tetrachlorozincate	9.3(2)	–
Quartz	2.1(0)	2.9(1)
Rutile	0.6(1)	0.6(1)
Sanidine feldspar	1.3(0)	2.4(1)
Sodalite	–	0.3(1)
Sylvite	–	0.6(0)
Syngenite	–	9.6(2)
Tobermorite 14-Å	–	0.3(1)
Ulvoespinel	0.7(0)	0.8(0)
Wurtzite	0.5(0)	0.3(0)

Bayuseno and Schmahl (2011)

Values in parentheses indicate the least-squares estimated standard deviation referring to the least significant digit to its left, a zero indicates an estimated standard deviation <0.05 %

tightly fixed in phases inherited from MSWI FA (Hjelmar 1996). On the contrary, the chalcophilic elements (e.g., As, Cd, Cu, Pb and Zn) are readily vaporised at an intermediate temperature and are the major constituents of FA (Forestier and Libourel 1998; Belevi and Moench 2000). Hg, that is extremely volatile, may be emitted almost totally in the vapour phase and therefore may be not found in the MSWI FA. Finally, the atmophilic elements (C, S, Cl, and F) generally remain in the vapour or enter in the minute particles of condensed-phases (e.g., Cu, Fe, and Sn) after electro filtration. MSWI FA may also contain persistent organic pollutants such as polychlorinated dibenzo-dioxins (PCDD) and polychlorinated dibenzo-furans (PCDF) (Rani et al. 2008).

3.2.4 Fly Ash from Biomass Matter Burning

Combustion of biomass is a common technique to generate heat and power in both small- and large-scale grate furnaces (domestic boilers, stoves, district heating plants). In recent years, pressures on global environment and sustainability have led to an increasing interest on renewable energy sources, and diversification of Europe's energy supply. Among the available resources, the biomass plays an important role, since it is considered a renewable and CO_2 neutral energy resource. Nowadays, biomass is ranked as fourth after oil, gas and coal, providing approximately 14 % of the world's energy needs (Shen et al. 2009). The Seventh Framework Programme of the European Commission addresses research priorities on biomass. The objective is to develop and implement technologies for electricity, heating and cooling from biomass (Rajamma et al. 2009; Yin et al. 2008). Moreover, one of the problems associated to biomass combustion is related with the ash production, in the thermal conversion process itself (for example, slagging and fouling phenomena), and also its environmental management.

The quantity and quality of ashes produced in a biomass power plant are strongly influenced by the characteristics of the biomass source (Mukherjee et al. 2008; Burke 2007). For example, combustion of wood generates low amounts of ashes to be managed; on the contrary herbaceous biomass, agriculture wastes and rice husk have higher ash content. Agricultural residues are usually mainly composed of cellulose, hemicelluloses, and lignin. Other components are extractives, lipids, proteins, simple sugars, starches, water, hydrocarbons, ash, and many more compounds that contain a variety of functional groups present in the binding process (Sun 2010). The functional groups present in biomass molecules include acetamido, carbonyl, phenolic, amido, amino, sulfhydryl carboxyl, ester, and hydroxyl groups (Beveridge and Murray 1980; Gupta and Ali 2000). These groups have the affinity for metal complexation. Some biosorbents are selective for certain types of metals depending on their chemical composition, whereas other are no specific and can bind a wide range of heavy metals (Garg et al. 2007; Tarley and Arruda 2004). Biomass ashes differ from coal FA, in particular in what concerns its chemical composition and mineralogy and exhibit more variation in both composition and amount of inorganic material than is typical of coal. The characteristics of ashes from biomass combustion are influenced by biomass source (for example, herbaceous material, wood or rice husk), and combustion technology (for example, fixed bed or fluidised bed) (Obernberger et al. 1997). The ashes from biomass burning are quite different also depending on the type of soil and harvesting (ASTM 2005). Furthermore, even for the same type of biomass, the properties of its FA depend also on some growth and production factors including weather, season, storage and geographic origins (Bridgeman et al. 2007; Wiselogel et al. 1996). In addition, the temperature in the furnace and its hydrodynamics influence the degree of slagging and fouling in the furnace and boiler, because of the relatively low melting temperature characteristic of biomass ashes and consequently the composition of formed final ash (Werther et al. 2000).

Variations in the inorganic fraction of fuels are directly reflected in ash compositions. The different kinds of installations result in multiple kind of ash flow, each with differences in morphology, composition and leaching behaviour. There are several paper in literature that reports chemical composition of biomass FA (Thy et al. 2006; Rajamma et al. 2009), showing how these compositions are widely variable.

Generally, the main inorganic elements in biomass ash are Ca, K, Na, Si and P. Some of these act as important nutrients for the biomass (Masia et al. 2007). Moreover, some biomass fuels have high silicon content (e.g. rice husk) while some have high alkali metal content (wood). Whereas the elemental composition of the ash is determined by the inorganic constituents in the parent biomass, the phase composition and the crystallinity is strictly dependent on the used combustion technique. Typically, FA from neat biomass combustion has more alkali (Na and K) and less alumina (Al_2O_3) than coal FA (Thy et al. 2006; Llorente and Garcia 2006). The solid byproducts that can be obtained from the thermochemical processing of biomass (e.g., gasification and pyrolysis) are generally a mixture of char and ash. Whereas the ash typically recovered from biomass combustion is mostly inorganic, that from biomass gasification can have unburned carbon contents ranging from 10 to 60 % (Leiva et al. 2007). Despite that, generally, biomass ash does not contain toxic metals like coal ash; sometimes the concentration of toxic heavy metals (for example, Zn, Cd, Pb, Hg) may be problematic in the FA (Demirbas 2005; Tulonen et al. 2012). In addition to its high carbon content, this ash can also contain organic contaminants, like PCDD/F and polyaromatic hydrocarbons (PAHs). This makes the management of gasification FA a challenging task, especially for contaminated fuels (Gómez-Barea et al. 2009). Because ash with a silica content of >50 wt% can be used as a substitute source for silica in the manufacturing process of a silicate glass (Park et al. 2009), or as constituent in the cement production (Girón et al. 2013) the silica content is the most evaluated component of FA from biomass.

Table 3.8 Silica content of some biomass fly ash

	Temperature (°C)			
Biomass	400	600	800	1,000
Rice husk	85	85.9	90.4	92.1
Sugarcane leaves	68	69.7	65.6	67.9
Straw	69.9	71.6	81.2	85
Eucalyptus wood	2.1	1.5	1.7	1.5
Bagasse	47.3	49.7	53.1	63.7
Palm petiole	65.7	67.7	60.3	65.2
Palm bunches	46.8	47.7	44.4	48
Coconut shell	12.1	9.3	12.9	25.4
Corn Stubble	34.6	34.1	33.6	39.8
Cassava	1.4	0.9	5.3	9.5
Rubber tree wood	10.9	8	7.7	5.4

Ruangtaweep et al. (2011)

Table 3.8 reports as a reference, silica content of some biomass FAs (Ruangtaweep et al. 2011) at different sintering temperatures. It has been found that, rice husk ash, sugarcane leaves ash, straw ash, and palm petiole ash at all sintering temperatures, and bagasse ash at 800 and 1,000 °C, have silica content of >50 wt%.

Rice husk ash, in particular, is the main studied ash from biomass. Rice is one of the major crops grown throughout the world. Covering 1 % of the earth's surface, rice is now a way of life, being grown on every continent and is deeply embedded in cultures, rituals and myths.

World rice production is approximately 645 million tons. Asian farmers produce rice about 90 % of total production of 100,000 tons or more, with two countries, China and India, growing more than half of the total crop (Ruangtaweep et al. 2011). For every 4 tons of rice, 1 ton of rice husk is produced. Rice husk is generally employed as biomass for energy recovery. The energy generating from rice husk has received great support (Wang et al. 1999), particularly in those countries that are primarily dependent on imported oil as their energy needs. In heat generation, rice husk has a calorific value of 12.1–15.2 MJ/kg (Beagle 1981). Rice husk has a chemical composition varying from sample to sample, depending upon the varieties of paddy sown, geographical conditions, soil chemistry, proportion of irrigated area, fertilizer used, climatic variation, timeliness of crop production operations and agronomic practices in the paddy growth process (Chaudhary and Jollands 2004; DSIRI 1992; Rao 1980; Patel et al. 1987). Rice husk major constituents comprising of cellulose, hemicellulose, lignin, hydrated silica (Chandrasekhar et al. 2005), which the exterior composed a large amount of silica (Mahin 1990; Ismail and Waliuddin 1996). After combustion, 15–30 % of this results in rice hush ash (RHA) (Mehdinia et al. 2011), that is collected from the particulate collection equipment attached upstream to the stacks of rice husk-fired boilers (Naiya et al. 2009). It contains 80–95 % of silica and minor amounts of

Table 3.9 Chemical analysis of rice husk ash

Constituent	Amount (wt %)
Silica (SiO_2)	80
Alumina (Al_2O_3)	3.93
Sulfur trioxide (SO_3)	0.78
Iron oxide (Fe_2O_3)	0.41
Calcium oxide (CaO)	3.84
Magnesium oxide (MgO)	0.25
Sodium oxide (Na_2O)	0.67
Potassium oxide (K_2O)	1.45
Loss on ignition (LOI) at 850 °C	8.56

Ismail and Waliuddin (1996)

metallic elements (Chandrasekhar et al. 2005). Chemical composition and crystallinity of RHA vary considerably according to the burning technique, depending on the temperature regime and gasification structures. For example, during the combustion at 550–800 °C, amorphous silica tends to be formed; on the contrary, at greater temperatures, crystalline silica is generated, both having different properties and specification for the possible final use. The chemical composition of rice husk ash is given in Table 3.9.

3.3 Fly Ash Treatments

Air pollution is a major problem for incineration. In modern incinerators, an advanced pollution control system is designed to minimize the pollution and ensure compliance with environmental standards. A dry/wet scrubber is used to spray fine atomised slurry or lime powder into the hot exhaust gas in order to neutralize the acidic gases such as sulphur oxides and hydrogen chloride. An activated carbon column or injected activated carbon spray is used to absorb the heavy metals and organic pollutants, such as polychlorinated biphenyl's and volatile organic compound, in the exhaust gas. A selective non-catalytic reduction system is used to remove nitrogen oxides by adding urea or ammonia for reaction to nitrogen, carbon dioxide and water. The bag filter system acts to filter and remove the fine particulates and dust particles (Lam et al. 2010). A selective catalytic reduction (SCR) is also used to convert nitrogen oxides into diatomic nitrogen (N2) and water (H2O) with through a catalyst. In particular this techniques utilized a gaseous reductant, for example anhydrous ammonia, aqueous ammonia or urea, that is added to a stream of flue or exhaust gas and is absorbed onto a catalyst. This technique reduce nitrogen oxides by 70–95 % and so, comparing with selective non-catalytic system, this is more efficient (Finocchio et al. 2006).

Concerning the presence of contaminants, MSWI FAs are the most problematic ashes. The composition of municipal solid waste varies over time and from country

Table 3.10 Quantification of elements found in the fly ash mixture

Element	Concentration (mg/l)
S	62
Cl	309
K	94
Ca	1,846
Ti	21
Cr	1
Mn	2
Fe	69
Cu	9
Zn	47
Br	7
Sr	5
Pb	12

Bontempi et al. (2010b)

to country, due to the differences in lifestyle and waste recycling processes of a country; the ash content will vary too. Generally, the chemical and physical characterisation of ash will depend on the compositions of the raw MSW, the operational conditions, the type of incinerator and air pollution control system design. Chemical composition of MSWI FA is quite variable because it depends on the incinerated waste composition, which varies by season. The chemical composition of ashes shows that the major common oxides present in ash are SiO_2, Al_2O_3, CaO, Fe_2O_3, Na_2O, and K_2O. CaO is the most abundant compound that exists in MSWI FA (Alba et al. 1997). The most abundant elements are Si, Al, Fe, Ca, Mg, K, Na and Cl. With regard heavy metals, Cd, Cr, Cu, Hg, Ni, Pb, and Zn are the most abundant, with Zn and Pb generally found in the largest amounts (see Table 3.10) (Bontempi et al. 2010b). These metals may cause leaching problems and are harmful to the environment without proper treatment. Generally, the heavy metals content in FA is high due to the vaporisation of metals during combustion and the process of metals adsorption on the surface of FA particles. MSWI FA contains high chloride content. This may be due to the lime scrubber in the air pollution control system, which removes acidic gases such as HCl, thus resulting in a high amount of chloride content remaining in FA after the air pollution control system.

Another problem of FA is the content of organic pollutant like dioxin. The international emission standard limit for dioxin concentration in flue gas is 0.1 ng I-TEQ/m^3. In order to minimize the flue gas dioxin concentration, dioxin is removed from vapor phase to solid phase by lime scrubbing and activated carbon injection to adsorb dioxin (Hinton and Lane 1991; Lee et al. 2008). There are a number of ways of minimising dioxin/furan formation during MSW incineration which can significantly reduce the dioxin/furan. A two-stage approach system was designed to get complete combustion and minimize formation. In the combustion system, the combustion temperature should be above 1,000 °C, the combustion residence time should be greater than 1 s and the combustion chamber turbulence should be

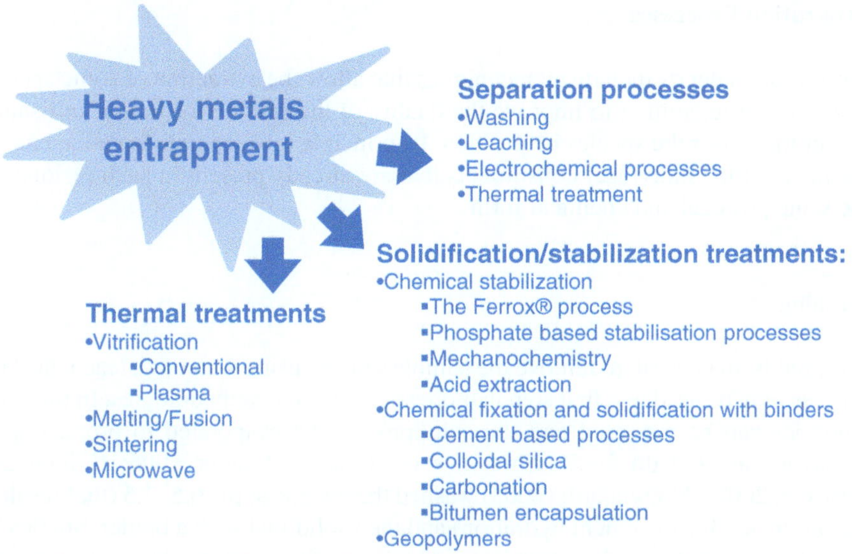

Fig. 3.3 Classification of several techniques for heavy metals entrapment

represented by a Reynolds number greater than 50,000, with good MSW feed preparation and controlled feed rate. In the end-of-pipe treatment systems, very rapid gas cooling from 400 to 250 °C should be achieved. Semi-dry lime scrubbing and bag filtration coupled with activated carbon injection adsorption as end-of-pipe treatments can all play a role in prevention or minimisation of dioxins in the final flue gas emission to the atmosphere (McKay 2002). Due to the volatilisation and subsequent condensation as well as concentration phenomena acting during combustion, FA and air pollution control (APC) residues bear high concentrations of heavy metals, salts as well as organic micro-pollutants. There are two possible ways of handling FA residues: landfilling after adequate treatment or recycling as a secondary material after inertisation. There are two classes of proposed technologies: one about heavy metals entrapment and one for the organics pollutants abatement.

3.3.1 Heavy Metal Entrapment

In order to utilize the waste and reduce the environmental impact, the treatments for FA can be grouped into three classes (IAWG 1997): (1) separation processes; (2) solidification/stabilisation (S/S); (3) thermal methods. In practice, it may be useful to set up the treatment with separation techniques (generally washing or leaching with fluid solutions that are more aggressive than water), followed by thermal treatments or stabilisation/solidification techniques. In Fig. 3.3 it is reported the classification of these treatments.

Separation Processes

Separation methods include all techniques that allow the extraction of some species from the waste, aiming to improve the quality of the residue for further utilisation and/or to recover the species in question. Its principle is to separate from the bulk of the material the individual constituents that are already present in such residues in the same physical and chemical form.

Washing

The goal is, in general, to remove the soluble salts by using water as a leachant. This process may be used as a first step in more sophisticated methodologies. In this way, chlorides can be extracted and the solutions can be evaporated to produce salts (Francois and Criado 2007; Chimenos et al. 2005; Mulder 1996; Nzihou and Sharrock 2002). Mangialardi (2003) washed the residue at pH 6.5–7.5 (that resulted less unfavourable to cement hydration) and then solidified with a binder. Studies by the author have shown that washing methods, involving water alone, are not very appropriate since large quantities of heavy metals (e.g. Pb and Zn) are also released simultaneously with the soluble salts (Quina et al. 2008). Rani et al. (2008) explained several washing treatments. In particular for incinerator FA, they showed that it possible washing FA with only water or with water and acid. In the first case, only 1 % of Pb and other heavy metals were removed. Instead, in the second case, a significant amount of chlorine and other heavy metals were removed to such a point that the acid washed ask was utilised as a secondary raw materials in road construction. Some chemical additives, such as soluble phosphates, may be used to overcome this problem. After washing to remove soluble salts, Francois and Criado (2007) utilised phosphate to trap heavy metals in stable crystalline phases as apatite (calcium phosphate).

Leaching

Leaching can be defined as the dissolution of a soluble constituent from a solid phase into a solvent and usually the leaching is the first important operation for the whole treatment process. Leaching arises because of the chemical reactions taking place at the scale of the individual waste particles as well as of the contaminant transport processes via the fluid moving through the solid particles. Aspiring at the extraction of heavy metals for further recovery, solutions for leaching other than water may be used. Pb and Zn are the most important elements in this process, and the target is to obtain a solution in which concentrations are high enough to facilitate further separation and recovery.

Two possibilities can be implemented for improving leaching efficiency, when the released quantities are too low: adding chemical additives or restart with the leaching process (Laethem et al. 1994)

In the literature, many studies use chemical agents to develop heavy metal leaching. Sabbas et al. (2003) showed the influence of pH on leaching. In fact, pH is strongly related to the nature of the particular contaminant under concern as well as the mineral phase(s) in which this is bound. Three main typical leaching behaviours for solubility-controlled leaching have been identified: (1) Cation-forming species and non-amphoteric metal ions (e.g. Cd) that displays fairly constant high values at $pH<4$, and decreases strongly up to $pH=8–9$, remaining approximately constant or slightly increasing for higher pH values. (2) Amphoteric metals (including Al, Pb, Zn) increase solubility under both strongly acidic and strongly alkaline conditions. And, (3), Oxyanion-forming elements (e.g. As, Cr, Mo, V, B, Sb); usually their solubility decreases in alkaline ranges ($pH>10$) (Sabbas et al. 2003). van der Bruggen et al. (1998) and Katsuura et al. (1996) operated by acid extraction processes; van der Bruggen et al. (1998) utilised acid solution with ethylenediaminetetraacetate (EDTA) to increase the solubility of Pb and Katsuura et al. (1996) described that it is possible to recover solid salts containing close to 99 % of NaCl. Ecke et al. (2000) explain Japan practice in which nine acid extraction units for treating the MSWI residues are used. Nagib and Inoue (2000) carried out a series of leaching experiments using three kind of acid: sulphuric, hydrochloric acid and acetic acid. The advantage of employing sulphuric acid was to dissolve Zn and other metals from the FA leaving Pb, which is insoluble in sulphuric acid solution. The advantage of hydrochloric acid and acetic acid is the ability to dissolve all heavy metals generating leach solutions that contain both Pb and Zn, which can be separated by means of solvent extraction. Fedje et al. (2010), due to the alkalinity of the ash, investigated some alternative leaching media such as EDTA, ammonium nitrate, ammonium chloride and a number of organic acids and to compare them with the usual mineral acids and water. The use of mineral acids and ethylene-diaminetetraacetate (EDTA) mobilised many elements, especially Cu, Zn and Pb, whereas the organic acids generally were not very efficient as leaching agents for metals. Leaching using ammonium nitrate was especially effective for the release of Cu. Hong et al. (2000), utilised chelating agents to recover Cr, Cu, Pb and Zn.

It is clear that the efficiency of the extraction depends greatly upon the heavy metal in question, the extracting agent, the pH and the liquid/solid ratio used. Huang et al. (2011) utilised lixiviant extraction method at room temperature that seems more attractive from the point of view of recovery and recycling of metals. Many lixiviants can be chosen for the extraction of heavy metals from FA, such as inorganic mineral acids like sulphuric acid, hydrochloric acid, and nitric acid, organic acids like citric acid, oxalic acid, acetic acid, tartaric acid, or, chelating reagents like nitrilotriacetic acid (NTA), ethylenediaminetetraacetate (EDTA), and diethylenetriaminepentaacetate (DTPA), and, in some cases, alkaline solutions like ammonium and sodium hydroxides.

The leaching of heavy metals can also be improved through bioleaching, by using bacteria leaching processes. Krebs et al. (2001) through bioleaching by *Thiobacilli thiooxidans* made an extraction of higher than 80 % for Cd, Cu and Zn but there was a low efficiency for Pb; while Bosshard et al. (1996), through bioleaching by *Aspergillus Niger* explained as the environmental quality of the residues, that contain heavy metals, can be improved with respect to re-use.

Due to the mobility of the water-leachable compounds, a successful removal of the metals is expected applying supercritical fluid extraction with CO_2 in which traces of water and extractants are dissolved. The use of supercritical CO_2 is particularly attractive, because it is an environmentally benign solvent with a high diffusivity and a low viscosity allowing easy penetration into the smallest pores of particles such as FA. Supercritical fluid extraction does not require an expensive solvent removal step. The extractant can be rinsed out with CO_2, while releasing the pressure after supercritical fluid extraction results in the entire evaporation of the residual CO_2. Furthermore, CO_2 can be recycled without problems (Kersch et al. 2004).

Nam and Namkoong (2012) explained the use of electron beam irradiation on FA that significantly increased the leaching potential of heavy metals from FA. When the intensity of electron beam irradiation increased gradually up to 210 kGy, concentration of Pb and Zn in the leachate increased linearly as absorbed dose increased, while that of Cu went through no significant change. Notable change in metal form in FA was observed when FA was irradiated and it was related to leaching potential of the metals. Change in Pb form was much greater than in Cu form.

Electrochemical Processes

Electrochemical processes are methodologies that involve reduction/oxidation reactions at the interface of electrodes and electrolyte, induced by a potential difference in the electrodes. During the electrochemical process, hydrogen is generated at the cathode, metals are deposited on its surface (due to the reduction reaction) and oxygen is released at the anode (through the oxidation reaction). The main advantage of these methods is that they do not require the addition of chemicals, but removing metals when the concentration is low is a major problem (Yang and Tsai 1998). The variables that significantly affect this process are current density, temperature, mixing conditions, distance between the electrodes and the pH of the solution. Pedersen (2002) showed a kind of electrochemical process: the electrodialytic remediation. The electrodialytic process is a remediation technique used for the removal of heavy metals and chlorides from FAs. It has been introduced as a possible new extraction method. In electrodialytic remediation, the FA is suspended in an aqueous phase and placed between two electrodes (Fig. 3.4).

When a low voltage current is applied, the ions in the ash slurry (in compartment III) begin to migrate in the electric field either toward the positive electrode (anode) or toward the negative electrode (cathode), depending on their speciation. As a result, the ions are transported out of the ash slurry and into the electrolyte solutions in compartment II or IV, from where they can be separated by conventional methods (e.g. electroplating, ion exchange, or precipitation). Pedersen (2002) and Pedersen et al. (2005) demonstrated that the electrodialytic remediation method enhanced the extractability of heavy metals in MSWI fly ashes compared to more traditional extraction methods. Above all when combining the electrodialytic treatment with the addition of an appropriate assisting agent to the ash slurry before and during

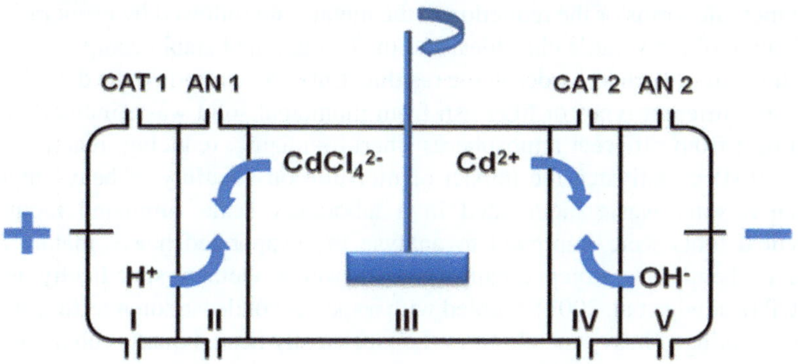

Fig. 3.4 The principle of electrodialytic remediation of fly ash (Pedersen 2002)

remediation, it was possible to improve the desorption of one or more heavy metals specifically, thereby improving the extraction efficiency considerably for targeting metals compared to basis experiments using only water or acid as assisting agents. Also Lima et al. (2012) utilised electrodialytic process: they washed the ash before the electrodialytic remediation period started. Then the suspension was decanted to grossly separate the solid from the liquid. The remaining suspension was filled up to L/S of 4 using distilled water. They concluded that the best compromise for removal of Cd, Pb, Zn, Cu, and Cr was obtained with a solution of 0.25 M ammonium citrate and 1.25 % of NH_3 as assisting agent (Lima et al. 2012; Pedersen 2002; Pedersen et al. 2005).

Thermal Treatment

This method may be used for removing heavy metals through evaporation processes (using temperatures lower than melting point), while at the same time enabling the recycling of the metals (decontamination). In general, the heavy metals of interest are Zn, Pb, Cu, and Cd. In the literature, some studies have given up promising results regarding the separation of heavy metals by thermal treatment: Auer et al. (1999) utilised HCl gas to recycle heavy metals by fractionated condensation. It is a promising method, because heavy metals in presence of HCl (gas) are volatile. Jakob et al. (1995, 1996) analysed evaporation of heavy metals through heat treatment; they showed that with temperatures between 1,000 and 1,100 °C, the evaporation efficiencies for Pb, Cd, and Cu are 98 %, and higher, and that the evaporated amounts as well as the rates of evaporation were strongly influenced by chlorides. Stucki and Jakob (1997) showed that performing thermal treatment under chloride containing and reducing conditions, more than 99 % of the Cd, Cu, Pb, and Zn present in FA predominantly evaporated as chlorides and they can be collected separately as heavy metal condensates. They also asserted that the principal thermochemical reactions that control the evaporation of heavy metals were the evaporation of the metallic forms or the reduction at the metal state followed by evaporation, the evaporation of the volatile chlorides, and the formation of stable compounds within the solid matrix, such as oxides in the residue. Lutz (2002) studied the detoxification of several different types of filter ash from municipal solid waste incinerators and the impact from different atmospheres (inert, oxidizing, reducing, reactive). Zhao et al. (2004) investigated the impact of moisture on volatility of heavy metals in municipal solid waste incinerated in a laboratory scale simulated incinerator. Analytical tools were improved to analyse the evaporated heavy metals during thermo(−chemico) treatment quantitatively as well as element-specifically by using TG-ICP (Ludwig et al. 2007) coupled with a special condensation interface to transfer the volatilised heavy metals from the electrically heated quartz tubular often to the TG-ICP (Ludwig et al. 2000a, 2001). Synchrotron-based X-ray absorption spectroscopy allowed gaining insights in the fate of Zn (Struis et al. 2004) and Pb (Struis et al. 2009) by studying MSWI FA at room temperature, as a function of treatment temperature and atmosphere (inert gas or air). With time, the questions regarding heavy metal partitioning expanded also to other wastes, such as electronic scrap (Scharnhorst et al. 2007). Very recent studies concern the evaluation of health risks related to waste incineration of polymer nanocomposites (Roes et al. (2012) and its possible persistence in a MSWI plant (Walser et al. 2012).

Solidification/Stabilisation Treatments

Solidification/stabilisation (S/S) treatments are among the most prevalent processes used for waste incineration residues. The principal function of solidification/stabilisation is to produce a material with physical (specific surface area, porosity, etc.), mechanical (durability, mechanical strength, etc.) and chemical properties capable to immobilised hazardous components originally present in waste. The solidification/stabilisation process utilizes chemically reactive formulations that, together with the water and other components in sludges and other aqueous hazardous wastes, form stable solids. The material used for solidification/stabilisation (S/S) not only solidifies the hazardous waste by chemical means but also insolubilises, immobilises, encapsulates, destroys, sorbs, or otherwise interacts with selected waste components. The result of these interactions is solids that are non-hazardous or less hazardous than the original waste. The solidification process reduces the mobility of the pollutants in the treated material through encapsulation, as an effect of the reduced surface area and low permeability. On the other hand, the major goal of stabilisation is to convert the pollutants into less soluble or less toxic forms, with or without solidification. Chemical stabilisation and then solidification of the waste is the best approach to solidification/stabilisation (S/S) technology. The solidification/stabilisation processes (S/S) use additives or binders, such as cements or pozzolanic materials, in order to immobilize hazardous components initially present in waste. It consists of mixing the waste with water and binders and sometimes with chemical additives such as sodium silicate, soluble phosphates and carbonates. Even if the S/S treatments are the most used methods, they are no completely efficiency for the treatment of soluble salts; in fact because of FA residues contain high levels of salts, the leaching of these salts from the S/S matrix over a long time has a poor performance. The consequent increase in metals released is due to the loss of physical properties and durability of the treated product. However, the S/S treatments are used in order to observe landfill criteria (Sabbas et al. 2003; Quina et al. 2008). Although the stabilisation technology for MSWI FA has advanced, certain problems with conventional stabilisation technologies cannot be ignored, such as long-term stability and the cost of stabilisation chemicals. In addition, carbonation is an important topic in recent research on the long-term leaching of heavy metals.

Chemical Stabilisation

Chemical stabilisation processes have been proposed, which basically involve chemical precipitation of heavy metal-incorporating insoluble compounds and/or heavy metal substitution/adsorption into various mineral species. The principal forms of chemical agents used include sulphides (IAWG 1997; Katsuura et al. 1996), soluble phosphates (Derie 1996; Eighmy et al. 1997; Hjelmar et al. 1999; Nzihou and Sharrock 2002), ferrous iron sulphate (Lundtorp et al. 1999) and carbonates (Hjelmar et al. 1999). As a result either of hydration, carbonation, or oxidation/reduction, these changes may lead to considerable reductions in trace elements (including heavy metals as Cd, Cu, Pb, Zn and Mo) and leaching. These can give go up to pH decrease, contaminant sorption processes and formation of more stable mineral species (Sabbas et al. 2003).

The Ferrox® Process

The Ferrox®-process was developed in Denmark as an on-site treatment for FA residues prior to landfill disposal (Lundtorp et al. 2002; Jensen et al. 2002). The process consisted in mixing of three mayor units that were ferrous sulphate solution, recycled water and FA residues; suddenly the oxidation of ferrous iron took place followed by pH adjustment with either additional $FeSO_4$ or H_2SO_4 and finally, in a plate-and-frame filter press, the treated residue was dewatered. Aeration oxidised the precipitated ferrohydroxide to ferrihydrite and the suspension was then separated into treated residue and wastewater. This process was reported to leach out large amounts of salts (Cl, Na, K and Ca) but low concentrations of trace metals (Pb and Cd) from FA residues (Lundtorp et al. 2002). Hu (2005) studied the effectiveness of mixing $FeSO_4$ and $Fe_2(SO_4)_3$ solutions with FA residues in order to stabilize heavy metals and they concluded that the solution reacted with the FA to form $Ca_4Fe_9O_{17}$, which coated the surface of ash particles, and for that reason the heavy metal releasing was decreased. Other authors (Christensen et al. 2001) have used iron sulphate to stabilize FA residues. The wastewater generated contained low levels of heavy metals, but high concentrations of salts. The process reduced the leaching of Pb by more than two orders of magnitude at fixed pH. Leaching of Cd, Zn and

Cu were also significantly reduced. However, the effect on elements that forms oxyanions (for example Cr) was marginal and there was no reduction in leaching of Hg. Unfortunately the stability of iron oxides and hydroxides can change with time; so some authors (Sorensen et al. 2000), proposed an additional heat treatment at 600–900 °C after the Ferrox® process. This thermal treatment of Ferrox stabilised FA residues produced a more stable product with respect to leaching of heavy metals. In fact Ferrihydrite was formed initially but converted upon thermal treatment into more stable and crystalline compounds (maghemite and hematite). In general since most of the salts are removed by this process, the treated residues are expected to be less affected by physical disintegration than the cement solidified product.

Phosphate Based Stabilisation Processes

Stabilisation by the addition of phosphate (particularly phosphoric acid) is one of the most promising and intensively investigated techniques. The phosphate phases likely to precipitate are extremely numerous, particularly calcium phosphates and heavy metal phosphates such as Copper, Lead and Zinc phosphate. Since FA is rich in Calcium, that is generally present in the form of calcium carbonates, most of the leachable metals are caught during the precipitation of calcium phosphates including traces of metals. Above all apatite is one of the most thermodynamically stable phosphate species. Piantone et al. (2003) proposed an approach for monitoring the phosphation of FA produced by Municipal Solid Waste (MSW) incineration. The phosphate process includes the Neutrec® process developed by Solvay S.A. (scrubbing of flue gases by contact with sodium bicarbonate), and it has the objective of minimising the disposal of FA residues in hazardous landfills, obtaining a no-hazardous residue.

This treatment consisted in four steps: (1) filtration of the flue gases producing mixtures of soluble and insoluble mineral particles; (2) ash washing at basic pH to remove salts; (3) phosphation of the solid residues to trap metals, and, (4), drying and calcination of the solid phosphate residues to destroy organic compounds (Fig. 3.5). Piantone et al. (2003) washed the FA with a solution with a pH close to 10.5 (by acidification with hydrochloric acid) to permit the extraction of soluble salts with a low dissolution of the metals, particularly Zinc and Lead, and an increase hydrated lime ($Ca(OH)_2$), as excess lime would tend to increase phosphoric acid consumption. To make phosphation, a solution phosphoric acid (H_3PO_4 10 %) was added to FA. The pH ranges from 6.5 to 8. They obtained a precipitation of calcium phosphate that trapped Lead and Zinc by destruction of just formed calcite. After drying, they used calcination process between 600 and 900 °C for 1 h that is important to destroy the dioxins (PCDD/F) always associated with this kind of ash. They obtained fine sand that was creamy white to grey in colour. They extracted soluble salts (NaCl, Na_2SO_4, and Na_2CO_3) after adsorption of the organic compounds by activated charcoal and then extracted metals by ion exchange resins in the Solvay process that are then recycled (Piantone et al. 2003).

Rani et al. (2008) shows the VKI® process. This process has been developed by a consortium of Danish companies and it is explained by Hjelmar et al. (Hjelmar and Birch 1997; Hjelmar et al. 2001; Hjelmar and Hansen 2006).

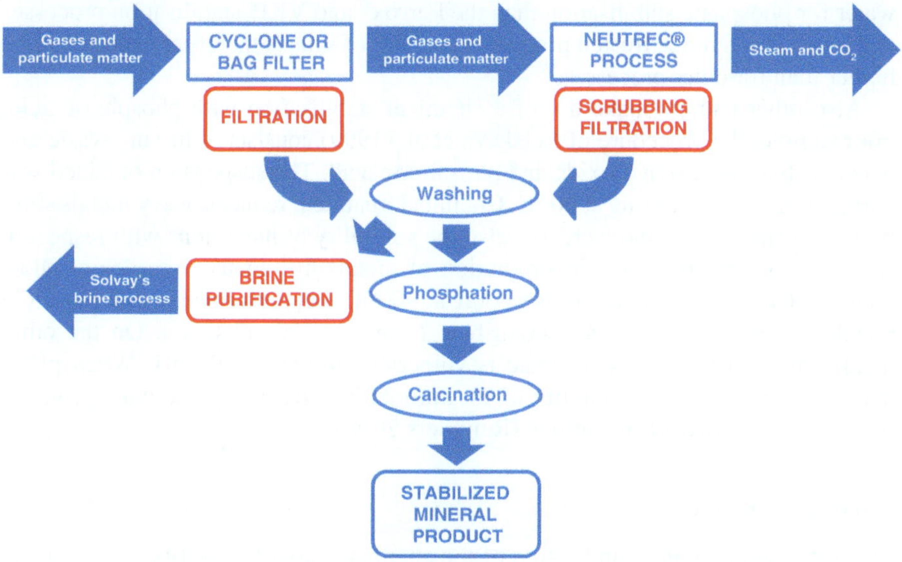

Fig. 3.5 Fly ash stabilisation process (Piantone et al. 2003)

It is now a one-stage treatment process that improves the leaching characteristics of FA residues. Such process involves the removal of soluble salts by aqueous extraction, chemical stabilisation of the washed product with carbon dioxide (CO_2) or phosphoric acid (H_3PO_4), washing the product and final dewatering. pH buffering capacities of both carbon dioxide and phosphoric acid are very important; in fact this mechanism reduces the pH of the residues without totally removing the alkalinity which protects against future acidification. The wastewater from the process must be treated to reduce the content of trace elements and heavy metals before to discharge to a treatment plant. For this reason it could be treated to comply with the EU leaching criteria for acceptance at hazardous waste landfill, and in most cases also with the criteria for landfilling of wastes at a site for non-hazardous waste which accepts stable, non-reactive hazardous (Rani et al. 2008). In many aspects, the VKI® process is like to the Ferrox® stabilisation process with the exception of the use of chemical agents that in the VKI® process are CO_2 and/or H_3PO_4. The VKI® stabilised products are expected to demonstrate low leaching because heavy metals as carbonates and phosphates have low solubility.

Eighmy et al. (1997) made a stabilisation of FA by using 1.2 mol of phosphoric acid/Kg of residue with water at an L/S ratio of 0.4 (WES-PHix® process). They found that in case of FA from an incinerator of municipal waste, treatment with phosphate reduced the fraction available for leaching of divalent heavy metals, by 38 % for Cd, 58 % for Cu, 99 % for Pb and 28 % for Zn. Stabilisation occurs due to precipitation of metal phosphates rather than by adsorption of metals onto phosphates. The products obtained were crystalline and amorphous calcium phosphates, tertiary metal phosphates and apatite family minerals. This process consumes less water for phosphate stabilisation than the Ferrox® and VKI® stabilisation processes and retains salts in the treated product. Leaching of salts is therefore expected to be higher than for other processes.

Also other authors make use of a chemical stabilisation with phosphoric acid. For example, the procedure of Iretskaya et al. (1999) consists of mixing waste and water with a calcium hydroxide and phosphoric acid. The suspension obtained was stirred, filtering and drying at 60 °C. Chemical treatment reduces heavy metals solubility. Thermal treatment decreases also the solubility of metal ions with respect to chelating agents due to the incorporation of metals in a matrix of calcium phosphate. After this, a calcinations procedure at 900 °C was performed in order to avoid metal loss by evaporation and contribute to apatite crystallisation. On the other hand, immobilisation by phosphate is vulnerable to changes in pH. When pH is decreased from 7 to 4, solubility of Pb immobilised by phosphate may go up by roughly three orders of magnitude (Reijnders 2005).

Mechanochemistry

Mechanochemistry is a subdivision of the chemistry used to describe the chemical and physicochemical transformation of substances caused by the mechanical energy during aggregation. Mechanochemical effects have been utilised in many fields, such as metallurgy, crystal engineering, materials engineering, coal industry, building industry, agriculture, pharmacy, and waste treatment, and they are environmentally friendly. It is a high potential technology for heavy metals stabilisation in FA.

In recent years, Nomura et al. (2010) have developed a mechanochemical process for inhibition of heavy metal elution from FA and used the residues as cement materials. For the mechanochemical treatment procedure, they used a planetary ball mill with a couple of pots composed of stainless steel (45 cm^3). Seven steel balls were located in each of the pots; the pots were positioned on a rotating disk so that the pots and the disk rotated in opposite directions. In another work, Nomura et al. (2006) treated FA mechanochemically and they showed the elution property of lead (Pb) and alkali from a cemented solid made of FA treated by mechanochemical process with blast furnace (BF) slag and an additive. The process of mechanochemical treatment and cementation completely inhibited Pb elution from FA. A possible inhibition mechanism might be that Pb compounds were changed into different forms and stabilised during the treatment. The pH value in elution from mechanochemical cemented solid was suppressed to 8–9. These results suggest that FA treated with BF slag mechanochemically can be use as cement material.

Lately, Nomura et al. (2008) used lead oxide (PbO), a typical Pb compound in FA, as a model substance to explore the lead immobilisation in mechanochemical FA recycling. In their testing, they added PbO and calcium oxide (CaO) mixture to FA. The planetary ball mill was rotated at 700 rpm for no less than 2 hours to ensure dioxin degradation, with a 15 min cooling interval every 15 min. After mechanochemical treatment, the residue was removed from the pots for Pb leaching tests and also used as a constituent in cement. They showed that mechanochemical treatment of FA inhibited Pb leaching by 93 %, and additional inhibition (99.9 %)

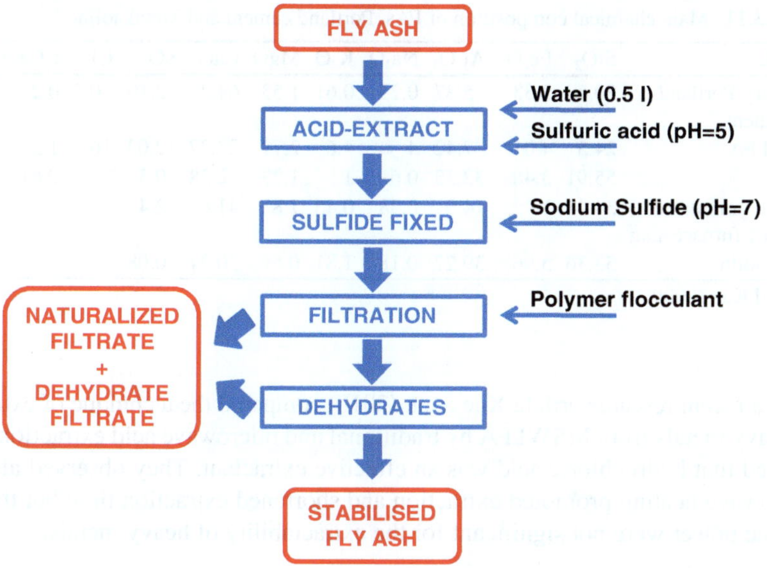

Fig. 3.6 Flow diagram of the acid extraction sulphide stabilisation (AES) process (Katsuura et al. 1996)

was observed on the production of cement from the treated FA. The decrease of pH, observed by the production of cement, is one of the reasons for the further inhibition of Pb leaching (Nomura et al. 2008). Mechanochemical process necessitates no heating and produces insignificant gas emission. Therefore, a large plant with sophisticated facilities is not needed, and simple closed-system designs are more likely to be accepted by local residents. The immobilisation of heavy metals in FA by mechanochemical treatment may consequently be a practical pretreatment technique for the not dangerous recycling of the FA as a component in cement (Guo et al. 2010).

Acid Extraction

Acid extraction sulphide stabilisation (AES) has been used for treating incinerator residues. In the AES process, incinerator FA is acid-extracted and stabilised by NaHS to transform residual heavy metals into insoluble heavy metal sulphides. The process flow scheme is shown in Fig. 3.6. Katsuura et al. (1996) explained that the acidic agent extracted soluble heavy metals and added sodium sulphide stabilised the remaining heavy metals. Concentrations of Pb leaching (0.2 mg/l) from the AES treated electrostatic precipitator FA were below the required values and the wastewater was processed by evaporation/crystallisation. The FA was discharged as stabilised cake for final landfill disposal and the solid salt residue was recycled for soda production. Condensed water was recycled and used for cooling; the plant did not discharge wastewater.

Table 3.11 Main chemical composition of FAs, Portland cement and Metakaoline

Sample	SiO_2	Fe_2O_3	Al_2O_3	Na_2O	K_2O	MgO	CaO	SO_3	Cl^-	f-CaO	Loss
Ordinary Portland cement	20.6	3.33	5.37	0.23	0.61	1.53	64.2	2.19	0.5	0.2	1.85
MSWI FA	24.5	4.01	7.42	4	4.6	2.72	23.37	12.03	10	1.5	22.04
FA	55.91	3.48	32.33	0.6	1	1.23	2.28	0.37	–	0.61	0.73
Ground granulated blast-furnace slag	34.4	1	14.9	0.38	0.45	7.8	41.6	2.4	–	–	−1.77
Metakaolin	53.36	3.96	39.27	0.18	1.81	0.59	0.24	0.08	–	–	1.1

Shi and Kan (2009)

In a recent research article Xue et al. (2010) compared the technique of extraction of heavy metals from MSWI FA by traditional and microwave acid extraction. They showed that hydrochloric acid was an effective extractant. They observed also that microwave heating promoted extraction and shortened extraction time but the time and the power were not significant for the extractability of heavy metals.

Chemical Fixation and Solidification with Binders

Treatments with hydraulic or chemical binders generally yield good leaching properties at relatively low costs (Sabbas et al. 2003). The aim is to incorporate the waste in the binder system and produce a monolithic solid with structural integrity and long-term stability. Solidification inhibits leaching by reducing waste/leachant contact and by forming a stable pH environment in which many heavy metals remain insoluble.

Cement Based Processes

S/S system of cement-based materials is the most frequently applied approach to minimize the environmental impact of hazardous wastes, compared with other technologies. It provides also the possibility of reusing the final materials. This technology involves mixing FA with cement to form solidified cement immobilisation body of a durable, monolithic material after hydration, which will not easily leach hazardous components under the disposal conditions, thereby reducing the dangerous leaching composition of the waste by a large margin. American State Bureau of Environmental Protection refers to cement immobilisation as the best technology to dispose the poisonous and harmful wastes (Landreth 1986). Shi and Kan (2009) showed that MSWI FA was mainly composed of silica, calcium, alumina oxide, and iron oxide, which were quite similar to those of complementary cementitious materials commonly used, such as granulated blast-furnace slag and coal FA (see Table 3.11).

This suggested that MSWI FA could have pozzolanic or hydraulic behaviour and its addition to the cement could have a beneficial role in the development of the hydrated cement paste. Thus, MSWI FA does not only imply pollution, but also allows resource reusing. They evaluated the flexural and compressive strength of the mortar specimens and they concluded that they decreased gradually with increasing of MSWI FA content (Shi and Kan 2009).

Cement is the most frequently used and widely studied binder. Solidification with cement is widely understood and simple, easily available, and results in a stable product (Roy et al. 1991). Kawakami et al. (1996) made a comparison of treatments with commercially stabilising agents and cement. They tested a solidifying powder SFPA (Sumitomo fine powder agent) that was a mixture of inorganic agents and they concluded that by using Sumitomo fine powder agent, the heavy metals can be immobilised and no toxic gases were formed during the chemical treatment. While Kamon et al. (2000) utilised cement, $Al_2(SO_4)_3$, $Ca(OH)_2$ and coal FA, and they concluded that the stabilised residue could be used for embankments to minimising the geo-environmental impact. Youcai et al. (2002) made stabilisation with sodium sulphide, thiourea, sodium hydroxide, and ethylenediaminetetraacetic acid and they compared the solidification effects of these chemicals with those of cement and asphalt. They concluded that chemical stabilisation processes, in particular by using sodium sulphide as the chemical agent, were strongly recommended, having advantages in terms of the low volume expansion and environmental safety of the stabilised products and cost balances. Geysen et al. (2004) showed a comparison of immobilisation of FA residues with different cements and with different silica containing materials: pyrogenic silica-aerosil, microsilica, waterglass, aluminosilicate, bottom ash and glass. They observed that the Pb leachability was reduced 3–50 times on the type and amount of cement used and on the curing time and that the physical immobilisation of Pb is better in case of the cement-treatment than with micro-silica.

Colloidal Silica

Bontempi et al. (2010a) have recently proposed a new method for MSWI FA inertisation mainly based on the use of colloidal silica as metal stabilizer. FA from MSWI, Flue Gas Desulphurisation (FGD), and coal, were used in the process together with aqueous colloidal dispersions of SiO_2 particles with high specific surface area. The experiments showed that the quantity of metals decreased in the leachate of the inertised MSWI FA. In particular, the Zn contribution decreased with respect to the leachate of raw MSWI FA and the Pb peak disappeared completely (Fig. 3.7). The TXRF data given in Table 3.12 also show that hardly no metals leached from FGD FA, whereas only some V, Cr, As, and Se were released from coal FA after treatment. So, it appeared that the procedure is effective in the inertisation of FA not only for Pb, but also for V, As, and Se.

Considering the legislation of Italy, cf. DM 03/08/2005 (DECRETO 3 agosto 2005 "Definizione dei criteri di ammissibilità dei rifiuti in discarica"), the treated FAs are suitably inertised for landfill deposition. From Table 3.12, it is also seen that there is a high content of chlorine and bromine that can be recovered by washing prior to obtaining the final material with the lowest quantity of salt, which Bontempi

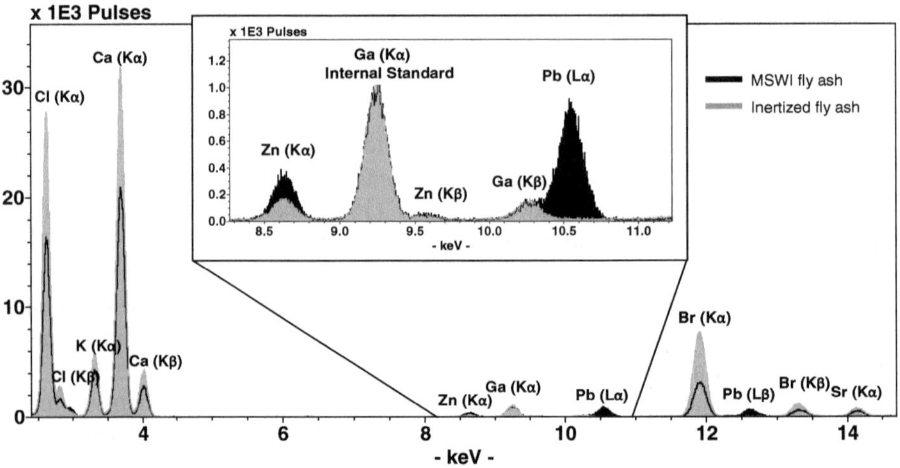

Fig. 3.7 TXRF spectra collected with the leaching solution of MSWI fly ash (*black*) and inertised material (*gray*) (Bontempi et al. 2010a)

Table 3.12 Results of the TXRF quantitative analysis performed with the leaching solution of MSWI FA, FGD FA, coal FA, and the corresponding inertised mixture before washing

Element	MSWI fly ash mg/l	FGD fly ash mg/l	Coal fly ash mg/l	Inertized mixture mg/l
P	n.d.	n.d.	0.3±0.1	n.d.
S	n.d.	1.1±0.9	67±6	98±9
Cl	2,477±234	980±93	n.d.	3,522±288
K	433±46	6.6±0.8	6.6±0.6	479±41
Ca	1,850±171	794±97	173±14	2,318±177
V	n.d.	n.d.	0.29±0.02	n.d.
Cr	n.d.	n.d.	0.031±0.004	n.d.
Fe	0.18±0.08	n.d.	0.03±0.01	1.1±0.4
Zn	3.7±0.3	0.03±0.01	0.013±0.002	1.6±0.1
As	n.d.	n.d.	0.048±0.004	n.d.
Se	n.d.	n.d.	0.96±0.07	n.d.
Br	33±2.8	4.4±1	0.012±0.001	71±6
Rb	n.d.	n.d.	0.012±0.002	n.d.
Sr	n.d.	1.8±0.1	2.4±0.19	n.d.
Ba	n.d.	n.d.	0.19±0.02	n.d.
W	n.d.	n.d.	0.053±0.005	n.d.
Pb	13±1	n.d.	n.d.	n.d.

Bontempi et al. (2010a)
n.d. not detected

et al. (2010a) designated as COSMOS (Colloidal Silica Medium to Obtain Safe inert). Comparing XRD results of MSWI FA with the corresponding COSMOS material (Fig. 3.8) showed that FA inhibits a significant amount of portlandite that is not found with the COSMOS sample. On the contrary, COSMOS showed the

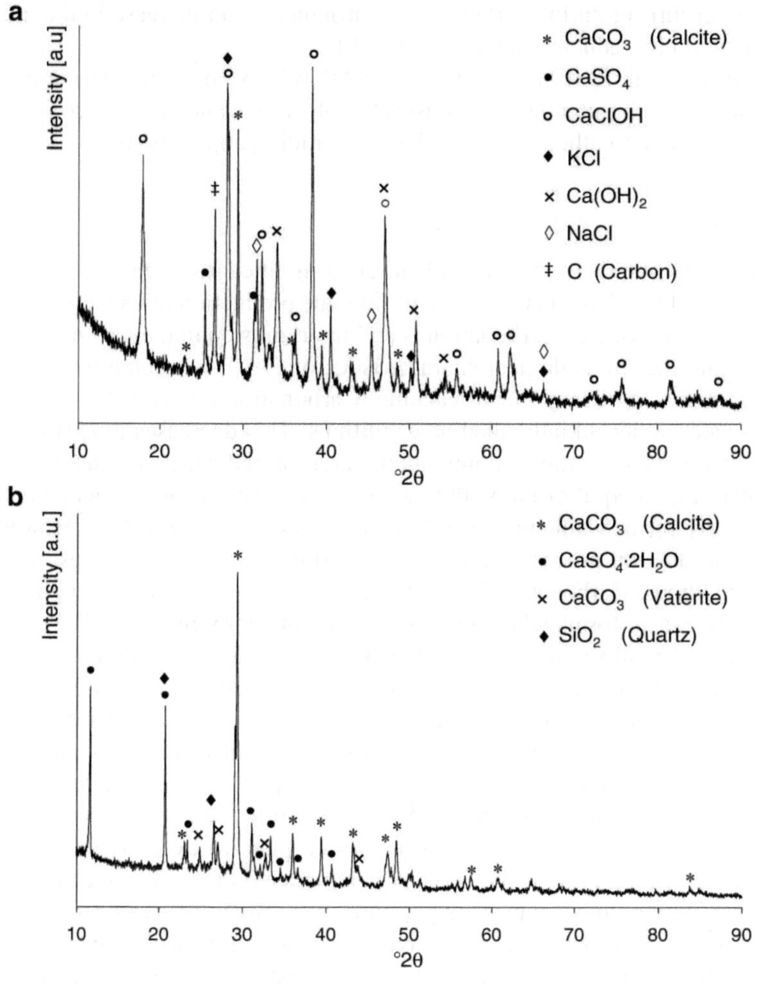

Fig. 3.8 XRD pattern collected of MSWI fly ash (**a**) and the final COSMOS material (**b**) (Bontempi et al. 2010a)

presence of a small number of crystalline phases, with ~25 % calcium carbonates as the main phase, together with calcium sulphate, and silicon oxide, alongside ~65 % of amorphous material, suggesting that only a small amount of colloidal silica crystallised. This may be because the inertisation process was performed at room temperature. Because calcite was the predominant newly formed phase after the FA inertisation process with colloidal silica, the XRD results with COSMOS may be seen to point at a carbonation reaction similar to that of portlandite with CO_2, viz., $Ca(OH)_2 + CO_2 \rightarrow CaCO_3 + H_2O$. Similarly, $CaClOH + CO_2 \rightarrow CaCl_2 + H_2O$, thus both calcium forms could disappear after carbonation, while the $CaCO_3$ peak increases. Notwithstanding these, it is also easy to suppose that the chemical reactions occurring during carbonation contributed to an increase in the efficiency of inertisation by reducing the metals mobility.

Bontempi et al. (2010a) concluded that MSWI FA could be considered a valuable resource with potential for industrial applications and could become an interesting alternative to other inert materials for building applications.

Carbonation

The emissions of CO_2 from MSWI plants are a significant source of greenhouse gas. In order to reduce CO_2 emissions, controlling the production of CO_2 inside incineration plants is important. Carbonation is an important step in the weathering process by changing the mineralogical characteristics of FA. It has some positive effects, such as decreasing heavy metal leaching. Carbonation involves the dissolution of CO_2 in water under initially alkaline conditions. This decreases the pH by precipitating calcite, which is the predominant mineral formed during FA maturation, until the material is in equilibrium with CO_2. Moreover, heavy metals can be entrapped in the newly formed minerals. The CO_2 source is atmospheric CO_2. However, the natural carbonation process can be accelerated using different sources of CO_2 and various methods. Carbonation is capable of inducing setting and strength development in the waste forms where hydration is significantly retarded. The carbonated solidification/stabilisation (S/S) product develops higher strength in comparison to non-carbonated products. Waste containing heavy metals has increased susceptibility to carbonation. Incorporating waste species into a carbonate phase by a mechanism involving solid solution reactions with calcite also improves metal fixation (Malviya and Chaudhary 2006). Li et al. (2007) proposed an accelerated carbonation treatment for the treatment of MSWI FA. They used two reactors: one was a stainless steel chamber incorporating a cooling plate to examine the influence of temperature. Water, used as the coolant, counteracted the heat of reaction and controlled the temperature. Changes in pressure and temperature were monitored using digital gauges. The second carbonation reactor was a closed chamber containing 100 % CO_2 at a relative humidity (RH) of 75 % operated at ambient laboratory temperature. It was used to ascertain the optimum water/solid (w/s) ratio for ash carbonation. Li et al. (2007) explained the mechanism of carbonation by the following equations:

$$H_2O + CO_2 \rightarrow H_2CO_3$$
$$H_2CO_3 \rightarrow H^+ + HCO_3^- \rightarrow 2H^+ + CO_3^{2-}$$
$$Ca^{2+} + CO_3^{2-} \rightarrow CaCO_3$$

They affirmed that water is necessary to promote the reaction of CO_2 but that too much water can limit the reaction due to the blockage of pores in the solid. Moreover, whereas the original pH of the ash is around 12–12.5, it is 7–10 after carbonation. Li et al. (2007) reported differences between the diffractograms of original FA from the MSWI plant and the carbonated samples (Table 3.13). Untreated ash contains

Table 3.13 Main mineral phases in ash untreated and carbonated

Mineral phases	Formula	Untreated ash	Carbonated ash
Lime	CaO	×	
Portlandite	$Ca(OH)_2$	×	
Calcium chloride hydroxide	$Ca(OH)Cl$	×	
Quartz	SiO_2	×	×
Anhydrite	$CaSO_4$	×	×
Halite	$NaCl$	×	×
Sylvite	KCl	×	
Friedels salt	$Ca_2Al(OH)_6Cl\ (H_2O)_2$	×	×
Nordstrandite	$Al(OH)_3$		×
Calcite/vaterite	$CaCO_3/Al_2O_3 \cdot 3CaO \cdot 2SiO_2$	×	×
Gehlenite	$Ca_2Al_2SiO_7Ca_2Al$ $(Al \cdot 5B \cdot 5Si \cdot 5Cr \cdot O_7)$		×

Li et al. (2007)

lime, portlandite, and calcium chloride hydroxide (Ca(OH)Cl) due to lime addition during gas treatment in the MSWI plant. They showed that in carbonated samples the calcite peak intensity increased, while those of lime/portlandite/Ca(OH)Cl decreased and finally disappeared. They also identified many Si–Al–Ca salts in the carbonated ashes, including gehlenite, braunite, and larnite.

Although Li et al. (2007) measured a decrease of ion levels with leachates; the decreased mobility of chlorides was still higher than the acceptance value. Ecke (2003) showed that carbonation of FAs from municipal waste incinerators has been found to reduce leachability of Pb and Zn by two orders of magnitude, but to increase leaching of Cd with a factor 10, and that leaching of metals from cement, such as Cr, Ni and Sb, may well increase because of carbonation in the long run.

Bitumen Encapsulation

Bitumen is a residue from petroleum or coal-tar refining and it consists of a mixture of high molecular weight hydrocarbons. A waste encapsulation process using bitumen was developed in France in the 1960s for radioactive waste disposal. It has a number of benefits as an encapsulation binder, such as chemical inertness, impermeability to water, resistance to breaking, and relative simplicity of processing. Based on the asphalt manufacturing process (Rani et al. 2008), a stabilisation technique for the treatment of FA residues was developed by Shell Bitumen in the early 1990s. FA residues were stirred under addition of about 50 wt% hot bitumen for several hours to guarantee absolute covering. The properties of bitumen, including durability, resistance to micro-organisms and many chemical substances, make it a stable and effective binder for encapsulating hazardous waste. The resultant material either was put into steel beats or transferred directly into moulds at a storage site for waste. Hard grade bitumen was employed to guarantee that the stabilised waste preserved structural integrity for a long time (Rani et al. 2008).

Geopolymer

Geopolymerisation technology has been proposed to stabilize and solidify FA. Unlike a glass, geopolymers are obtained by the reaction between a solid alumino-silicate, where Ca is absent or contained in low amount, and a highly concentrated alkaline solution, that is the chemical activator, as for example aqueous alkali (Na, K or Li) hydroxide or silicate solution. Geopolymers are formed at room or low (40–90 °C) temperatures (Duxson et al. 2005). In particular, the reaction is a geo-synthetic reaction whose mechanism can be characterised by the following reaction steps (Davidovits 1994):

$$\left(Si_2O_{5,}Al_2O_2\right)_n + nH_2O \xrightarrow{NaOH,KOH} n\left(OH\right)_3 - Si - O - Al^- - \left(OH\right)_3$$

$$n\left(OH\right)_3 - Si - O - Al^- - \left(OH\right)_3 \xrightarrow{NaOH,KOH} \{Na,K\} - \left(SiO - O - Al^-O\right)_n - 3nH_2O$$

The technology has received increasing attention, because of its low cost, flexibility, low processing temperature, and its capability to immobilize and stabilize hazardous wastes. Geopolymers used in solidification/stabilisation (S/S) systems to immobilize toxic metals have a function similarly to cement binders in terms of encapsulation. The approach is the same as for S/S of hazardous wastes by pozzolan-based binders, which also do not remove heavy metals from the polluted waste, but bind them physically and chemically in the solid matrix in order to reduce their mobility. Depending on the source material and the processing conditions, geopolymers in construction can manifest improved chemical and physical properties, such as acid attack resistance, fire resistance, structural integrity, low permeability, high compressive strength and durability, low shrinkage, fast or slow setting, and low thermal conductivity. The mechanism of heavy metals immobilisation is believed to be either physical, or, chemical, such that metals are either fixed in the geopolymer network, or, possibly bound into the structure for charge balancing roles or remain physically trapped by the surrounding network.

Lancellotti et al. (2010) introduced 20 % of incinerator FAs, deriving from electrofilter and fabric filter into a geopolymeric matrix. They focused on the evaluation of the possibility of inertizing MSWI FA in a metakaolin geopolymer matrix. They observed a chemical efficiency in terms of heavy metals immobilisation; in fact, the release of heavy metals from geopolymeric materials, after 1 or 7 days, was lower than the values of the as-received FA (see Table 3.14). They evaluated also the stabilisation of the new geopolymeric matrix and they observed that there were no traces of degradation or fragmentation, even not under a certain pressure applied on the samples.

They concluded that geopolymers, based on metakaolin and sodium hydroxide/silicate compositions, stabilize the FA against leaching, showing release of heavy metals which allows the disposal in a landfill for not dangerous wastes.

Galiano et al. (2011) studied the use of geopolymeric agents in the S/S of hazardous metals with MSWI FA. They demonstrated the stabilisation of metals, such as Pb, Cd, Cr, Zn, and Ba, using coal FA-based geopolymerisation technology.

Table 3.14 Heavy metals release values for the geopolymer samples after 1 and 7 days and the as-received fly ashes after 1 day, according to EN 12457 (mg/l), compared to the limit values set by Italian legislation (DM 03/08/2005)

Sample	Cr	Cd	Ni	Cu	Pb
Limit value	1 ppm	0.02 ppm	1 ppm	5 ppm	1 ppm
GPEF20(1)	0.02	0	0	0.04	0
GPEF20(7)	0.09	0	0	0.07	0.03
GPFF20(1)	0.53	0	0	0.03	0.1
GPFF20(7)	0.64	0	0	0.03	0.1
GP(1)	0	0	0	0.03	0.01
GP(7)	0.03	0	0	0.03	0.02
EF (1)	2.41	3.55	0.06	0.01	0.12
FF (1)	2.31	3.22	0	2.63	3.94

Lancellotti et al. (2010)

GP geopolymer, *EF* fly ash by electrofilter, *FF* Fly ash by fabric filter, *GPEF20(1)* leachate of geopolymer with 20 % of fly ash by electro filter after 1 day, *GPEF20(7)* leachate of geopolymer with 20 % of fly ash by electrofilter after 7 days, *GPFF20(1)* leachate of geopolymer with 20 % of fly ash by fabric filter after 1 day, *GPFF20(7)* leachate of geopolymer with 20 % of fly ash by fabric filter after 7 days

They compared the geopolymer S/S solids with similar solids obtained after the stabilisation of the MSWI waste with classical hydraulic binders. They observed that the compressive strength developed by the S/S solids increased with time with all treated samples and that Zn, Sb, and Sn were excellently immobilised in the S/S solids, while Mo, V, and Cr (which are preferentially present as oxyanions) showed the worst results with most of the mixtures. With regard to the speciation of metals, they concluded that Zn, Pb, and Cd exhibited typical hydroxide-amphoteric behaviour and that the leachability of Cd, Cr, Pb, and Zn was generally lower with the geopolymer S/S solids at neutral and acidic pHs. Moreover, some authors (Zheng et al. 2011) made also a water-wash pretreatment in order to promote the early strength of geopolymer and this process resulted in a higher final strength compared to the equivalent without water-wash.

Thermal Treatments

The thermal treatment of FA is used extensively in some countries to obtain reduced leaching from the residues and reduced volume, as well as, to obtain a treated material that is suitable for reuse. Thermal treatment techniques include vitrification, melting, sintering, and microwave employment. Regardless of the process, in most cases, thermal treatment results in a more homogeneous, denser product with improved leaching properties. Vitrification also adds the benefits of physical encapsulation of contaminants in the glass matrix. A major drawback to these methods, however, is that they require substantial amounts of energy and especially vitrifying and melting result in the mobilisation of volatile elements, such as Hg, Pb, Zn, during the thermal treatment process (Sabbas et al. 2003).

Vitrification

Vitrification is a process whereby residues are mixed with glass precursor materials and then combined at high temperatures into a single-phase amorphous, glassy product. Typical vitrification temperatures are at 1,000–1,500 °C and the material is cooled to form a single solid phase (amorphous and homogeneous). The waste material is melted with additives (glass precursors) in order to fix the contaminants in the final matrix (alumina-silicates). The retention mechanisms are chemical bonding of inorganic species in the residues with glass-forming materials, such as silica, and encapsulation of residue constituents by a layer of glassy material (Sabbas et al. 2003). This technique has the advantage of stabilising hazardous waste, reducing the amount of waste to be landfilled and producing potentially re-usable materials. After vitrification, there is an obvious change in microstructure and mineralogy, an excellent resistance against leaching of heavy metal ions, and a significant decomposition rate (99.95 %) of PCDD/Fs in Toxic EQuivalent (TEQ) in the produced slag. In fact, thermal methods are very efficient at destroying dioxins, furans, and other toxic organic compounds. Vitrification to produce glass or glass–ceramics varies from using conventional electric furnaces to plasma arc technology.

Conventional

Vitrification of FA requires appropriate SiO_2 content to achieve a suitable glass composition as demonstrated by many authors (Bernardo et al. 2010; Ferreira et al. 2003; Izumikawa 1996; Park and Heo 2002; Sabbas et al. 2003). Research has demonstrated that vitrification of FA immobilizes hazardous metals, destroys organic pollutants, and results in an 80–90 % reduction in waste volume. It also showed that it reduced leaching of Cr and Pb, but leaching of Zn and Cu increased. In particular, Ito (1996) made the vitrification of FA by swirling-flow furnace and concluded that the treatment was effective on reducing the volume, detoxification, and stabilisation of FA from fabric filter. Izumikawa (1996) made a metal recovery from FA generated from vitrification processes (at 1,350 °C). The process consisted of pre-treatment, leaching, sulphurisation, and waste water treatment. They concluded that almost 100 % of toxic heavy metals in vitrification could be recovered by chemical treatment and a detoxified FA was obtained. Frugier et al. (2002) studied the influence of composition variations of FA on the degree and kinetics of the vitrification and crystallisation in the vitrified material. They concluded that the composition of FA was very important on the degree and kinetics of the process of vitrification. The vitrification decreased when the alkali content increased and the silicon content decreased.

Plasma

Plasma arc treatment is a high-energy technology able to treat a range of scheduled wastes. In plasma arc treatment, a thermal plasma field is created by directing an electric current through a low pressure gas stream. A generic plasma process flow

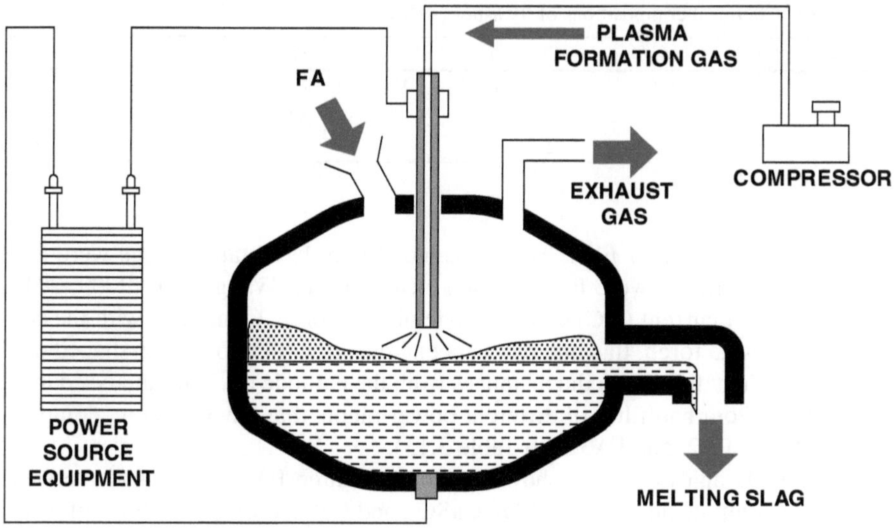

Fig. 3.9 Process flow diagram for the plasma vitrification of fly ash (Sakai and Hiraoka 2000)

diagram for treating hazardous wastes is shown in Fig. 3.9. The thermal plasma system consists of six parts including a steel reactor, a plasma torch, power supply system, working gas unit, cooling water unit, and exhaust gas cleaning unit. The plasma torch installed inside the reactor is the core of this system. The plasma system consists of a direct current (DC) hollow graphite cathode installed through the top of a furnace and supported by a vertical manipulator column. An inert gas, in particular argon, is introduced down the centre of the cathode to generate a stable plasma arc that is transferred to the furnace melt. The anode consists of conductive elements built into the furnace hearth. The process is able to manage varying particle size and chemical composition ashes. The furnace operates under controlled reducing conditions and runs approximately at temperatures between 1,500 and 1,600 °C; therefore even residues containing heavy metals can be melted uniformly with a short time. Water cooling elements are used to improve the lifetime of the furnace refractory coatings. The plasma power is modulated to maintain the melt temperature. The ash is rapidly melted and molten slag phase continuously overflows the converter where it is either water granulated or cast. The melted slag is removed continuously via the outlet port. The atmosphere in this furnace is oxidative. Exhaust gases exiting the unit are treated in a thermal oxidizer to fully oxidize residual combustible gas species. The gas is cooled and particulates removed in a bag filter prior to venting the cleaned gas to atmosphere. The particle free off-gas enters a wet scrubber dosed with a caustic solution to remove acid gases and is subsequently vented through the stack (Sakai and Hiraoka 2000; Rani et al. 2008).

Sakai and Hiraoka (2000) demonstrated the recovery of Pb and Zn during the vitrification of FA. Also Mugica et al. (1995) analysed the vitrification process with plasma and they explained that this technology was adequate for FA from

Table 3.15 Chemical compositions of fly ashes (wt%)

Sample	Fe_2O_3	MgO	CaO	TiO_2	Al_2O_3	SiO_2	K_2O	Na_2O	Cl
FA1	5.9	1.4	10.3	0.9	17.8	48.9	1.8	1.9	2.4
FA2	1.5	2.4	28.4	1.6	8.6	32.8	4.2	3.9	9.1
FA3	4.9	3.3	21.2	0.8	7.4	34.8	4.9	4.1	13.6

Wang et al. (2010)

electrostatic precipitator filter. They concluded that the obtained slags were inert and the heavy metals were fixed in the silicate matrix. Wang et al. (2009, 2010) utilised a direct current (DC) double anode plasma torch. Compared with the conventional plasma torch, this one has a special design with two nozzle-shaped copper anodes set at different axial distances from the cathode tip. They analysed three kinds of FA from bag filter of different power plants that have different composition (see FA1, FA2, and FA3 with Table 3.15).

From XRD analyses, they showed that these three FA samples had the same crystalline components, such as SiO_2, $CaSO_4$, and $CaCO_3$. Concentrations of heavy metals in FA were high and included Cd, Cr, Cu, Ni, Pb, and Zn. The concentrations and residual fractions of heavy metals in slag were significantly enriched. The residual fractions of Cr and Ni were approximately 84–96 % and 90–94 %, respectively. It showed that Cr and Ni were stable and remained almost completely in the slag. However, the slag contained extremely fewer Cd and Pb due to visibly observable volatilisation during the thermal plasma treatment process. They observed that the residual fractions of heavy metals in slag presumably decreased in the following order: Cr > Ni > Cu > Zn > Pb > Cd. Overall, the residual fraction of each heavy metal in slag from sample FA1 (which was rich in chloride ions) was higher than those of sample FA2 and FA3, illustrating that chloride in FA could accelerate the volatilisation of heavy metals with low boiling points through vitrification (Wang et al. 2010).

Melting (or Fusion)

Melting is similar to vitrifying, but this process does not include the addition of glass materials. The glass formed may not be homogeneous and results in a multiple-phased product. Often, several molten metal phases are produced. It is possible to separate specific metal phases from the melted product and recycle these metals perhaps after refinement. Temperatures are similar to those used in vitrifying (Sabbas et al. 2003).

The melting processes are divided into two groups:

A: Fuel-burning melting system
B: Electric melting system

The group A has been subdivided in five subgroups:

- surface melting furnaces;
- swirling-flow melting furnaces;

- coke-bed melting furnaces;
- rotary kiln melting furnaces;
- internal melting furnaces.

While group B has been subdivided in four subgroups:

- electric-arc melting furnaces;
- electric resistance melting furnaces;
- plasma melting furnaces;
- induction melting furnace (high-frequency, low-frequency).

It is not possible to say what the best technique is. Moreover, it is fundamental to choose the most appropriate process for the particular condition and situation. For example, in the case of a big incinerator with a power generator, the electric-arc melting furnaces (which is used and explained before in vitrification technique) is generally the best technique, because they can use the recovery electric power. However, in the case of a smaller incinerator without power generator, one of the fuel-burning melting systems is preferred (Sakai and Hiraoka 2000). The fuel-burning melting systems utilize heavy oil, kerosene, and gas as a fuel. The structure of the furnace consists of an inner body, burner, and surface for melting. The surface melting furnace works in such a way that continuously supplied incinerator residues melt at its surface by the heat of the fuel burning and leave the furnace via the outlet port. Therefore, the melted slag touches the furnace body directly and the incinerator deposits operate as an insulator to protect the furnace body. This kind of furnace has a quite large exhaust gas amount and it is more appropriate for a small capacity range of wastes.

Sakai and Hiraoka (2000) utilised melting technology for FA. They observed that inorganic elements were reorganised after melting treatment according to their boiling temperature. Particularly, Si, Al, Ca, Fe, Mg, Mn, P, Cr, and Sn have high boiling points and they were converted into slag, while elements such as Cd and Pb with a low boiling point, were found in exhaust gas. Jung et al. (2005) studied the metal behaviour in MSWI FA after melting treatments and they concluded that the performance depended on the volatility of the metals. In particular, the behaviour of Zn was also influenced by the oxidising condition in the furnace and the presence of chlorine in the mixture could have a significant effect on Cu, but not on Pb. Nowak et al. (2012) tested different amounts and types of chlorides (NaCl, $MgCl_2$, $CaCl_2$) on heavy metals removal from MSWI FA by thermal treatments. They performed heavy metals removal by mixing the ash with chlorides and treating the mixture thermally at $T = 1{,}000 \pm 100$ °C. They showed that NaCl chlorinated heavy metals directly, but also that it vaporised without reacting in significant amounts. On the other hand, $CaCl_2$ and $MgCl_2$ first formed HCl and/or Cl_2 by reacting with H_2O and/or O_2 (indirect chlorination). $CaCl_2$ and $MgCl_2$ were generally more effective for removal of heavy metals, such as Cd, Cr, Cu, Pb, and Zn, than NaCl. With respect to resource conservation, Nowak et al. (2012) concluded that $MgCl_2$ seemed to be the best choice due to high heavy metal removals (especially Zn) and low remaining chloride, which is also necessary for further use of the treated ash.

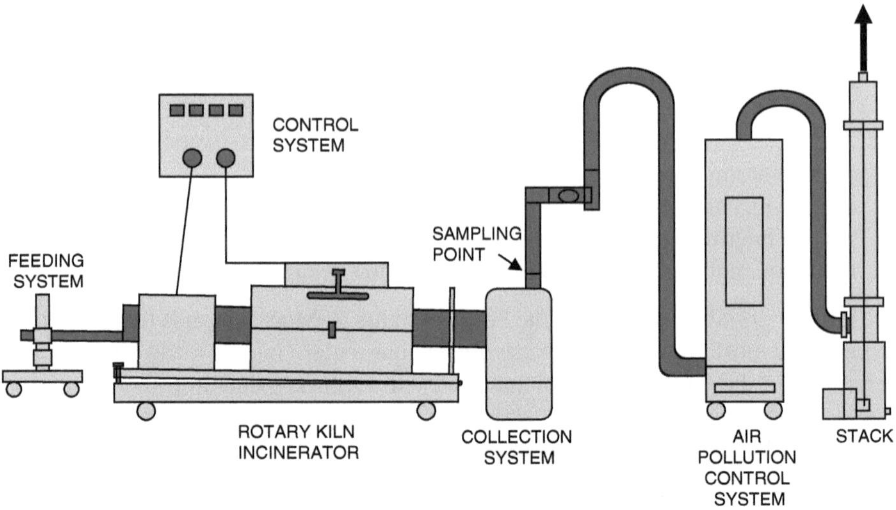

Fig. 3.10 The rotary kiln incinerator (Wey et al. 2006)

Sintering

Sintering is an alternative method to recycling FA and it is a viable method to produce glasses and glass-ceramics from incinerator residues. Sintering involves heating the residues to a point where particles bonding take place and chemical phases in the residues reorganize. This leads to a denser product with less porosity and a higher strength. Usually the temperature involved in this process is in the range of 900–1,000 °C. When MSW is incinerated, some level of sintering will typically occur in the incineration furnace. This is particularly the case when a rotary kiln is used as component of the incineration process (Sabbas et al. 2003). In general, sintered incinerator FA produces ceramic materials with low metals leachability.

Some authors utilised water-washing before sintering as a pre-treatment to lower the sintering temperature. In particular, Mangialardi (2001) utilised water to wash the FA and concluded that water-washing greatly improved the chemical and mechanical characteristics of sintering products. Wey et al. (2006) investigated the sintering behaviour of FA with a rotary kiln and looked for a solution to reduce the concentrations of heavy metals to acceptable levels by considering also the partitions of the heavy metals in the process. The rotary kiln incinerator (Fig. 3.10) consisted of a continuous feeder system, a main sintering chamber, and a collection system of sintered ash. The temperature in the chamber was measured by a thermocouple and an air pollution control device cleaned the exhaust gas before coming out into the atmosphere. The temperature and rotary speed of the chamber was controlled by a control system. They also utilised water-washing treatment for decreasing the sintering temperature and time. They explained that water-washing removed the chlorides of the FA to improve the quality of sintering products between

700 and 900 °C and they concluded that concentration of Pb decreased in sintering products only under 900 °C.

Karamanov et al. (2003a, b) analysed the sintering behaviour of a glass obtained from MSWI FA. They obtained glass-ceramics containing gehlenite by melting a mixture of 70 % of MSWI FA and 30 % of waste from feldspar production at 1,350 °C. They quenched the melt in water to form a glass frit and sintering between 950 and 1,050 °C. They showed that crystallisation was influenced by the heating rate. At high rates, surface crystallisation was the predominant mechanism, while, at low heating rates, bulk crystallisation occurred. Heating at low rates formed glass–ceramics containing open and closed porosity. Nonporous sintered samples were obtained using isothermal treatment at 1,120 °C. They concluded that the properties of the glass–ceramics formed were similar to those of marble and granite.

Also Wang et al. (1998) studied the sintering process at different conditions with respect to compact pressure, sintering temperature, and time. They analysed the strengths of the materials produced during sintering process. They showed that the strengths increased as the temperature arose at 1,120 and 1,140 °C, but decreased with an increasing ignition loss of FA. Moreover, they analysed heavy metals and concluded that these metals were fixed in the sintered materials and rendered low metals leachability with the ceramic-like solid obtained by the sintering process.

Microwave

Microwave energy is a non-ionising electromagnetic radiation with frequencies in the range of 300 MHz to 300 GHz. The use of microwave heating in the treatment of FA is relatively new even if this technique was developed over 50 years ago. Compared with other conventional thermal treatment technologies, the microwave technique offers the advantage of selective, uniform, and rapid heating, because of its characteristics of polar oscillation and effect of dielectric losses. The application of microwave radiation is generally used in many fields: for example in materials joining (Siores and Rego 1995), sintering and fusion of ceramics (Agrawal 1998), treatment of hazardous waste (Sedhom et al. 1992; Jones et al. 2002), digestion procedures for environmental materials (Lamble and Hill 1998), and treatment of minerals (Kingman and Rowson 1998; Appleton et al. 2005).

Rawlings et al. (2006) and Fukui et al. (2006) immobilised coal FA into a glass-ceramic material by microwave processing in a much shorter time and they compared the physical and mechanical properties to those obtained via conventional heating.

In a second work, Fukui et al. (2007) compared the hydrothermal treatment with microwave heating and conventional heating of coal FA with an aqueous solution of NaOH and a mixture of rice husk ash as a silica source at 100 °C. Figure 3.11 shows their experimental scheme, which comprised two steps: In the first step, they dissolved FA into aqueous NaOH solution at T = 100 °C with an electric heater for 1 hour and then separated the insoluble FA from the solution by filtration. In the

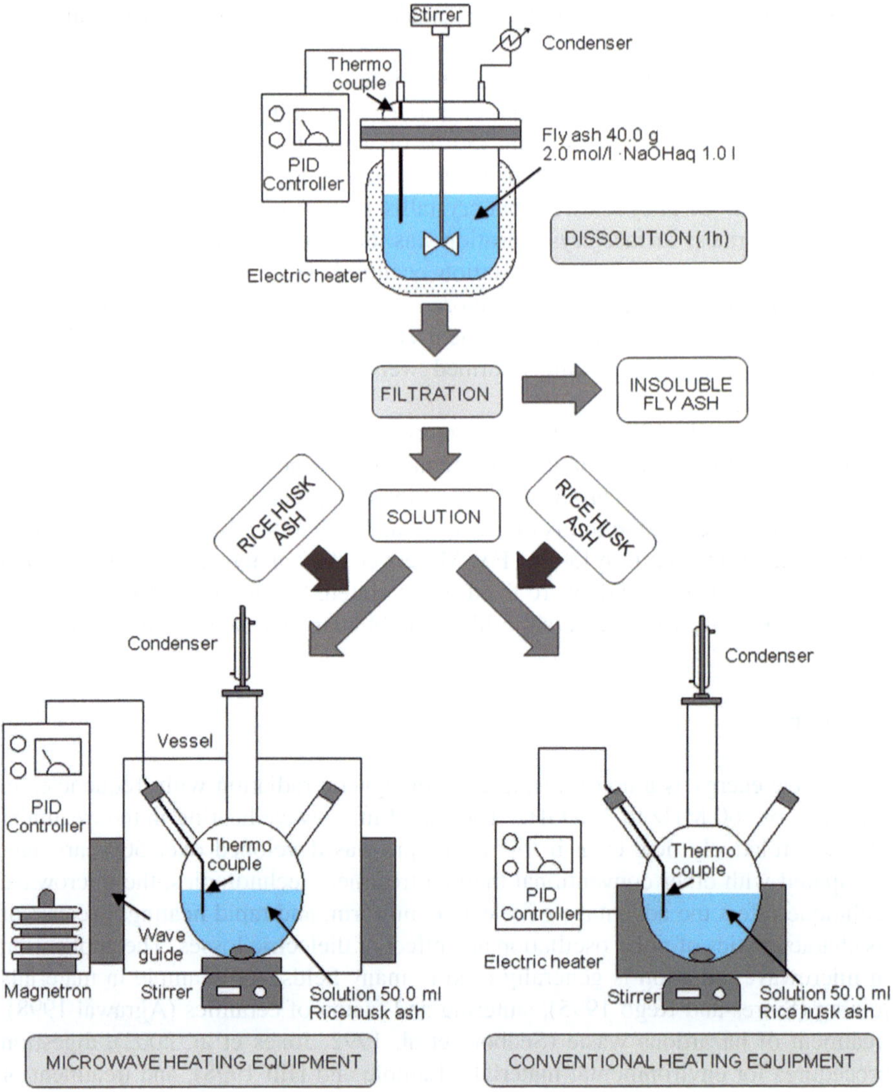

Fig. 3.11 Schematic diagram of the experimental scheme (Fukui et al. 2007)

second step, they added rice husk ash as a silica source to the solution and heated up the slurry by microwaves and by a conventional electric heater for purpose of comparison (Fig. 3.11).

They explained that the microwave heating shorted the reaction time respect to conventional electric heater and that the rice husk ash (RHA) was added in order to change the crystalline phase of the obtained zeolite. They also analysed the method to synthesize phillipsite, which is a mineral of the zeolite group and showed that the crystalline phase of the zeolite produced from FA by hydrothermal treatment depends on the heating method, but microwave heating requires more rice husk ash than conventional heating in order to synthesize phillipsite selectively. They found that the particle size of phillipsite obtained by microwave was smaller than that obtained by conventional process.

Chou et al. (2009) studied the sintering of MSWI FA by microwave heating. They investigated the effects of process time, aging time, and material and salt contents. They observed that MSWI FA could be sintered under microwave processing to form a glass–ceramic material. This process was able to immobilize heavy metals. In particular, Zn, Cr, Pb, Ca, and Cu with FA were mainly in oxide phases, but, after thermal treatment, the leaching concentration of Pb was higher than allowed by legislation. They also observed that the sintering efficiency increased as the microwave time was longer. As soluble salts decreased the efficiency of sintering, they washed them out before microwave processing. They also made aging tests during 0–30 days and concluded that aging did not increase the leaching of heavy metals, but increased the weights of the materials obtained by sintering of MSWI FA by 40–60 % by absorbing water from air.

3.3.2 Organic Pollutants Abatement

FA is often considered as a hazardous waste, because sometimes it has high percentage of heavy metals and organics. Besides a high content of inorganic compounds, incineration residues also contain abundant carbon compounds resulting from incomplete combustion, unburned organic matter, and carbon compounds formed during the incineration process. Chlorinated dioxins and benzofurans and PAH are theoretically degradable, but their full degradation in soils may take very long time. Ashes often contain organic compounds, in particular, persistent hazardous organic compounds, also called persistent organic pollutants (POP), such as polycyclic aromatic hydrocarbons (PAH), polychlorinated biphenyl's (PCB), methyl sulphates, and chlorinated dioxins and benzofurans. Dioxins are a class of structurally and chemically related polyhalogenated aromatic hydrocarbons that mainly includes polychlorinated dibenzo-p-dioxins (PCDDs or simply "dioxins"), poly-chlorinated dibenzofurans (PCDFs or simply "furans"), and polychlorinated 'dioxin-like' biphenyls (PCBs). They constitute a group of persistent environmental chemicals and usually occur as a mixture of congeners. Dioxins had come to public attention in the year 1982 when an explosion at ICMESA factory in Seveso, Italy, deposited these chemicals over a wide area. USEPA (1994a, b) reported that only 7 of the 75 possible PCDD congeners and 10 of the 135 possible PCDF congeners have chlorine substitution in the 2,3,7,8 positions and they render dioxin-like toxicity. Similar toxicity applies with only 12 of 209 possible PCB congeners having four or more chlorine atoms. Figure 3.12 shows the basic structure of PCDDs, PCDFs, and PCBs, respectively.

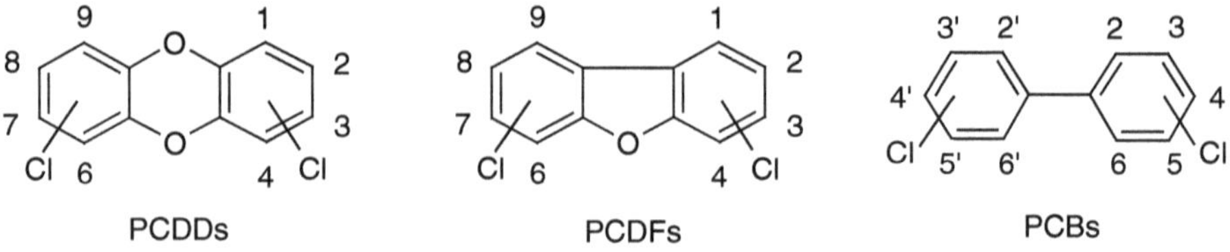

Fig. 3.12 Chemical structure of PCDD/Fs and PCBs with the numbering convention at the positions on benzene rings where chlorine or other halogen atoms can be substituted

PCDDs and PCDFs are synthesised in MSWI after combustion by reactions between the resulting FA and flue gas at low temperature (350–250 °C). FA produced in well-controlled combustion of coal has been found to contain quite low levels of chlorinated dioxins and benzofurans, together with low amounts of PAH, while wood combustion may produce ashes with considerable amounts of PAH, chlorinated dioxins and benzofurans (Reijnders 2005). The two most important mechanisms proposed in the literature thus either concern the formation of PCDDs and PCDFs from precursors, such as chlorinated phenols or benzenes, or, de novo-synthesis reactions between more or less complex, non-chlorinated, compounds (and FA in general) and a chlorine source, such as Cl_2, HCl, or, inorganic chlorides. In both cases, FA can act as a catalyst and it was demonstrated that the presence of oxygen is essential (Lasagni et al. 2009).

Several different kinds of organics pollutant abatements have been studied over the past three decades. Although thermal treatment has produced the mayor results, the manufacturing and maintenance of the melting facilities are expensive, the exhaust gases must be treated, and high-temperature treatment carries a risk of accidents. Therefore, other approaches for degrading organo-chlorinated compounds had been developed as well. Krishnamurthy and Brown (1980) showed e.g. that the dehalogenation methods mostly involve the use of low-valent metal, such as alkali metal in alcohol, or, Mg and Zn in acidic or basic solution. It is an effective alternative procedure for decomposing chlorinated organic wastes under relatively mild conditions without the formation of the toxic by-products and it is a high potential technology for degrading PCDDs and PCDFs. Figure 3.13 shows some of the other techniques developed so far.

Thermal Treatment

There exist thermal applications for decomposing dioxins and organics in general. In the literature, several studies focused on the thermal treatment of dioxins. For example, Vogg and Stieglitz (1986) decomposed 95 % of dioxins by thermal treatment at 600 °C for 2 h, but they also observed the formation of new dioxins at a lower temperature. Thermal technology can also be used for FA treatment to decompose dioxins under suitable conditions to render recyclable, fused-slag byproducts.

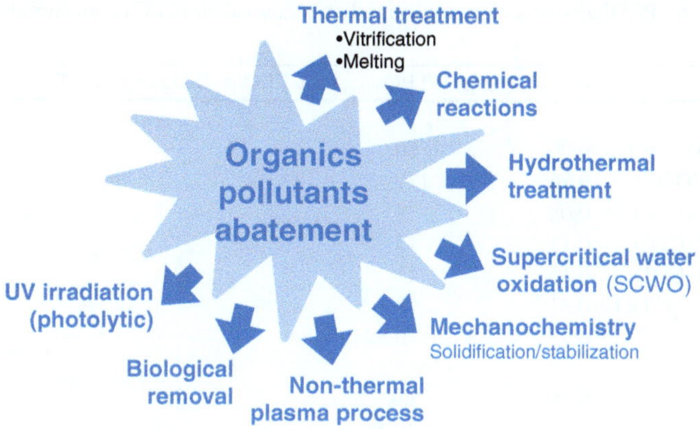

Fig. 3.13 Classification of several techniques for organic pollutants abatement

Vitrification

Thermal vitrification technology is a more satisfactory technique to treat organic pollutant contents in FA than other thermal treatment. Until now, various types of melting furnaces have been developed and used to vitrify incineration residues into non-toxic glassy slag, for example, electric ovens, coke-bed melting furnaces, rotary kiln with electric heaters, sintering in burning furnaces, plasma melting furnaces, surface melting furnaces, and swirling-flow melting furnaces (Wang et al. 2009; Kulkarni et al. 2008).

Wang et al. (2009) utilised in particular a thermal plasma reactor and they showed that after vitrification the decomposition was 99.3 % in PCDD/Fs and 99.95 % in Toxic Equivalent (TEQ) (see Table 3.16). They explained that unlike FA, the thermal glassy slag obtained was amorphous, compact, and homogeneous.

Kuo et al. (2003) found for PAH in ash a reduction of over 99.9 % during vitrification. However, volatilisation was also significant, requiring provisions for reducing PAH emissions into ambient air. Ito (1996) found that over 98 % of chlorinated dioxins and benzofurans decomposed in a vitrification process operating at 1,300–1,400 °C. Significant decomposition of hazardous organics, such as PAH, chlorinated dioxins and benzofurans, is not to be expected for immobilisation processes operating at much lower temperatures and also neither by reaction with cement (temperature up to 100 °C) nor by 'natural weathering' (ambient temperature).

Melting

Non-toxic materials can be obtained through melting processes, the melted slag can be used as a resource, and the PCDDs/PCDFs in FA are decomposed at the typical melting operating temperatures.

Table 3.16 The PCDD/Fs concentrations and Toxic Equivalent (TEQ) concentrations in the fly ash and slag

	Fly ash (ng/g)	$TEQ_{fly\ ash}$ pg I-TEQ/g	Slag (ng/g)	TEQ_{slag} pg I-TEQ/g
PCDDs				
2,3,7,8-TCDD	0.06	60	–	–
1,2,3,7,8-PeCDD	0.26	130	–	–
1,2,3,4,7,8-HxCDD	0.08	8	–	–
1,2,3,6,7,8-HxCDD	0.15	15	–	–
1,2,3,7,8,9-HxCDD	0.11	11	–	–
1,2,3,4,6,7,8-HpCDD	0.81	8.1	–	–
OCDD	0.32	0.32	–	–
PCDFs				
2,3,7,8-TCDF	0.69	69	–	–
1,2,3,7,8-PeCDF	0.53	26.5	–	–
2,3,4,7,8-PeCDF	0.83	415	–	–
1,2,3,4,7,8-HxCDF	0.47	47	–	–
1,2,3,6,7,8-HxCDF	0.49	49	–	–
1,2,3,7,8,9-HxCDF	0.12	12	–	–
2,3,4,6,7,8-HxCDF	0.47	47	–	–
1,2,3,4,6,7,8-HpCDF	1.08	10.8	0.05	0.5
1,2,3,4,7,8,9-HpCDF	0.12	1.2	–	–
OCDF	0.34	0.34	–	–
Total PCDDs	1.79	232.4	–	–
Total PCDFs	5.14	677.8	0.05	0.5
Total PCDD/Fs	6.93	910.2	0.05	0.5
PCDD/PCDFs	0.35	0.34		

Wang et al. (2009)

Sakai and Hiraoka (2000) utilised a surface melting furnace for thermal treatment of MSWI residue containing organics pollutant at the high temperature of approximately 1,400 °C. The operation of this kind of furnace was explained under Thermal treatments (Sect. 3.3.1.3). They found that from 320 ng/g (=3.7 ng TEQ/g) dioxins in the MSWI FA studied by them, only 0.012 ng/g (0.00049 ng TEQ/g) remained in the slag after thermal treatment. They concluded that the decomposition rate of 99.93 % for PCDDs/PCDFs (99.94 % in TEQ) and the slag produced by the melting process was environmentally compatible.

Nishida et al. (2001) developed a technology to crystallize melted slag (which had low strength) into higher strength stones to produce good quality stones. Other effective applications for the produced stones could be using them as aggregates in pavement blocks or asphalt. The technology was also capable to destroy 99 % of dioxins in ash.

Non-thermal Plasma Process

Non-thermal plasma is a very economical process, because it is performed at room temperatures, involves low maintenance, does not require auxiliary fuel, and eliminates disposal problems or sensitivity to poisoning by sulphur or halogen containing compounds.

Table 3.17 Detailed content of PCDDs and PCDFs, before and after non-thermal plasma treatment

		Dioxin content (ng/g)		
		Before	After discharge	
Type of dioxin			Ash	Air
PCDDs	T_4CDD_S	35	10	–
	P_5CDD_S	71	29	–
	H_6CDD_S	97	44	–
	H_7CDD_S	82	45	–
	O_8CDD	78	47	–
	Total	363	175	–
PCDFs	T_4CDF_S	52	18	–
	P_5CDF_S	66	32	–
	H_6CDF_S	89	47	–
	H_7CDF_S	61	34	–
	O_8CDF	28	17	–
	Total	296	148	–
PCDDs + PCDFs		659	323	–

Zhou et al. (2003)

Zhou et al. (2003) demonstrated the utility of non-thermal plasma technology on the destruction of dioxins from MSWI FA by inducing a surface chemical transformation of the dioxin-containing FA by pulse discharging using a different polarity of pulse voltage. They showed that the concentration of dioxin congeners influenced their destruction; in particular, higher concentrations showed higher destruction efficiency, and, vice versa (see Table 3.17).

They concluded that this technology was able to destroy dioxins and, in particular, that of TCDD. The destruction efficiency with the PCDD-type dioxins decreased in the order: TCDDs > PeCDDs > HxCDDs > HpCDDs > OCDD. With furan congeners, the order was: TCDFs > PeCDFs > HxCDFs > HpCDFs > OCDF.

UV Irradiation (Photolytic)

Another promising method for degradation of dioxins is photocatalytic reactions by semiconductors films, such as TiO_2, ZnO, CdS, and Fe_2O_3, under UV or solar light to generate conduction band electrons and valence band holes (e^- and h^+) that are able to initiate oxidoreduction-chemical reactions on semiconductors. This reaction takes place at room temperature and ambient pressure with low energy photons. TiO_2 has been used predominantly as semiconductor photocatalyst.

Kim et al. (2006) showed that valence band holes were powerful oxidants that initiate the degradation reactions of a wide variety of organic compounds. They reported that a complete degradation of 2-chlorordibenzo-p-dioxin and 2,7-dichlorodibenzo-p-dioxin was observed after 2 and 90 h, respectively, in UV illuminated aqueous suspension with no significant intermediates detection. With

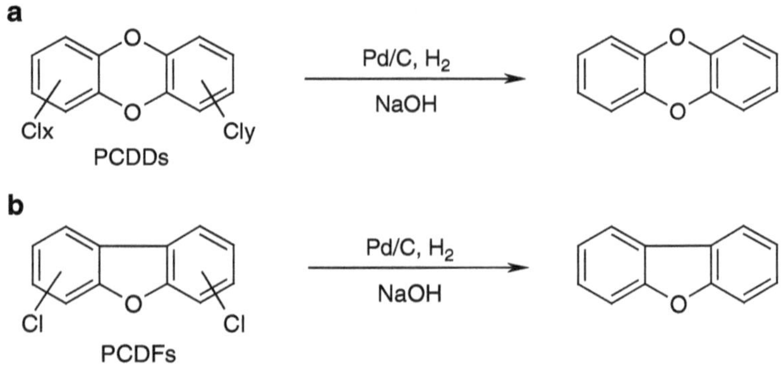

Fig. 3.14 The equation of the catalytic dechlorination of PCDDs (**a**) and PCDFs (**b**) (Zhijun et al. 2007)

photocatalytic degradation of highly chlorinated dioxin compounds, Choi et al. (2000) found that degradation rates of dioxins decreased with the number of chlorine and increased with the intensity of light and the TiO_2 coating weight. Additionally, Everaert and Baeyens (2004) found that oxides of Pt and Au supported on silica-boria-alumina were effective for the destruction of dioxins at 200 °C. Andersson et al. (1998) explained that to avoid blockage of the catalyst with coarse FA particles and ammonium sulphate, the catalyst for the destruction of dioxins functions better after preliminary cleaning. The advantage of applying selective catalytic reduction (SCR) over the other method is the elimination of complicated disposal problems with the residual matter, whereas the catalyst lacked the capacity of removing contaminants as wide as with activated coke.

Chemical Reaction

A chemical reaction involves use of a reagent for the decomposition of polychlorinated aromatic compounds. Catalytic dechlorination with noble-metal and transition-metal catalysts is an especially simple and efficient method. Molecular hydrogen is often used as a hydrogen source in the catalytic dechlorination. In recent years, catalytic dechlorination methods have been developed including transfer hydrogenation by means of hydrogen donors, such as metal hydrides formic acid and its salts, and alcohols that need water or methanol as solvents.

Hagenmaier et al. (1991) demonstrated that, under oxygen deficiency, the destruction of PCDD, PCDF, and other chlorinated aromatics was a general property with FA from waste incineration processes. The authors proved that dechlorination/hydrogenation of PCDD/PCDF and other chlorinated aromatic compounds with FA were catalysed under certain conditions.

Zhijun et al. (2007) studied an efficient method for the detoxification of dioxins and furan using 5 % of palladium on carbon (Pd/C) as reaction promoting catalyst and which can be regenerated and recycled (Fig. 3.14). They showed that the

Table 3.18 Conversions of 2,3,7,8-chlorinated PCDD/Fs with Pd/C-catalysed dechlorination

Congeners	Initial concentration (pg/ml)	Final concentration (pg/ml)	Conversion (%)
2.3.7.8-TCDD	4,029	11	99.73
1.2.3.7.8-PeCDD	9,756	19	99.81
1.2.3.4.7.8-HxCDD	11.359	13	99.89
1.2.3.6.7.8-HxCDD	23.08	18	99.92
1.2.3.6.7.8-HxCDD	16.501	20	99.88
1.2.3.4.6.7.8-HpCDD	193.708	73	99.96
OCDD	521.638	88	99.98
Total CDDs	780.071	242	99.97
I-TEQ	16.46	27	99.84
2.3.7.8-TCDF	10.449	n.d.	100
1.2.3.7.8-PeCDF	23.525	n.d.	100
2.3.4.7.8-PeCDF	19.322	10	99.95
1.2.3.4.7.8-HxCDF	31.259	9	99.97
1.2.3.6.7.8-HxCDF	36.84	6	99.98
1.2.3.7.8.9-HxCDF	2,914	n.d.	100
2.3.4.7.8.9-HxCDF	30.914	11	99.96
1.2.3.4.6.7.8-HpCDF	112.908	68	99.94
1.2.3.4.7.8.9-HpCDF	14.3	n.d.	100
OCDF	34.638	n.d.	100
Total CDFs	317.069	104	99.97
I-TEQ	23.382	8	99.97
Total I-TEQ	39.841	35	99.91

Zhijun et al. (2007)
Concentrations: determined by isotopic dilution. High resolution gas chromatographs/High resolution mass spectrometer; n.d. means not detected

detoxification reaction of PCDDs and PCDFs using commercial Pd/C catalyst had high activity in water/isopropanol solution at 40 °C under normal pressure. They included water for economical and ecological reasons, because they found no environmental pollutants after treatment.

They observed that the dechlorination efficiencies were higher for octa- and hepta-chlorinated congeners than for tetra- and penta-chlorinated ones and that it worked well in water, rendering over 99 % efficiencies with almost all of the PCDD/Fs congeners studied (Table 3.18) within short reaction times. In particular, the concentration of 2,3,7,8-TCDD, which is one of the most persistent congeners in dioxins, decreased from 4,029 to 11 pg/ml, thus, with a 99.73 % efficiency. They also found that the catalytic degradation of PCDFs was easier than that of PCDDs.

Mitoma et al. (2004) studied detoxification of polychlorinated aromatic compounds by means of metallic calcium in ethanol. They found that metallic calcium could be kept stable under atmospheric conditions for a long time compared with metallic sodium, because the surface was coated with $CaCO_3$ formed from contact with air. Moreover, ethanol, which was one of the safe solvents for humans, acted not only as a solvent, but also as an accelerator due to its ability to remove the carbonated coating. They showed that PCDDs, PCDFs, and PCBs were reduced in 98–100 % conversions by treatment in ethanol at room temperature and that the TEQ for the total residues of isomers was reduced from 22,000 to 210 pg-TEQ at room temperature. Regarding costs and safety of the reagents, they concluded that this decomposition method was one of the most environmental and economical promising detoxification methods for dioxins.

Hydrothermal Treatment

A hydrothermal treatment is a physicochemical process based on the relation between temperature, relative humidity, and time. FAs were put into water (or other solutions) and subjected to hydrothermal treatment at high pressure and temperature. Hydrothermal condition is relatively mild compared with high temperature refractory conditions and it is more energy saving, because the steam produced in the incineration heat recovery boiler can be used as heating source. In the literature, there are some works about the elimination of the PCDDs and PCDFs under hydrothermal condition.

Ma and Brown (1997) found that NaOH containing methanol was an effective solution for dioxins decomposition. They showed that treating FAs containing 1,100 ng/g total dioxins by hydrothermal treatment using this solution at 300 °C for 20 min the level of dioxins decreased to 0.45 ng/g. They concluded that the process was superior to purely thermal treatment at the same temperature and the regenerated FAs could be used in the cement industries.

Also Yamaguchi et al. (1996) showed that PCDDs/PCDFs decompose completely when the hydrothermal reactions took place when using a 1 N NaOH solution containing 10 vol. % methanol as solvent at $T=300$ °C for 20 min and they observed that the treated FA toxicity of PCDDs/PCDFs decreased to 0.03 ng-TEQ/g. Although in both works a good decomposition of PCDDs/PCDFs was obtained, the reagent methanol that they used was poisonous and the alkaline environment would result in caustic brittleness of the reaction wall which works under high pressure.

In order to enhance decomposing PCDDs/PCDFs during hydrothermal process, Xie et al. (2010) studied carbohydrazide as a strong reductant at temperatures of 245 and at 260 °C and achieved by total concentration decomposition efficiencies with PCDDs and PCDFs higher than 80 % and 90 %, respectively.

In a recent work, Hu et al. (2012) studied raw FA containing a total PCDDs/PCDFs amount of 11,463.3 ng/kg (TEQ=628.8 ng/kg). To improve the decomposition of PCDDs/PCDFs during hydrothermal treatment, they used a mixture of ferric and ferrous sulphates with the ratio Fe(III)/Fe(II)=2 as reactant. They used an autoclave at two different temperatures and corresponding pressures ($T=245$ °C/$P=3.46$ MPa; $T=290$ °C/7.44 Mpa). After processing, the resulting cake was dried at $T=80$ °C for 2–3 days. Figure 3.15 shows the scheme of the process.

They found that without the ferric/ferrous sulphate-mixture, the decomposition efficiency and the destruction efficiency of the PCDDs/PCDFs with MSWI FA was

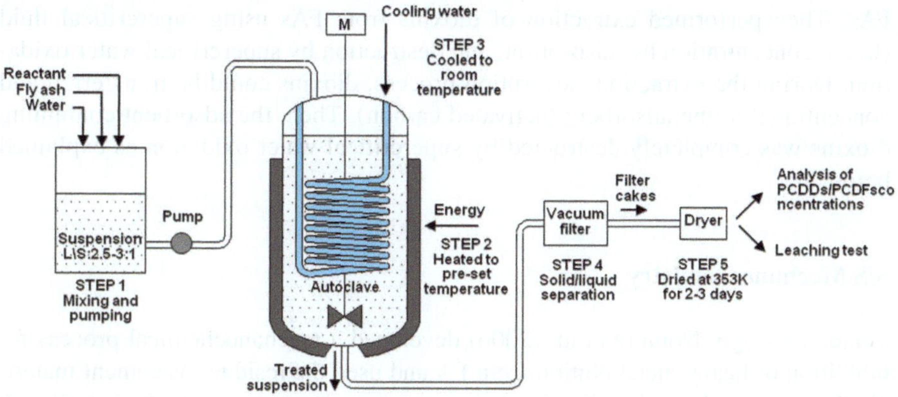

Fig. 3.15 Process scheme of hydrothermal process (Hu et al. 2012)

80.03 % and 46.17 %, respectively. The decomposition efficiency was defined as the ratio of total PCDDs/PCDFs concentration reduction by hydrothermal processing to total original PCDDs/PCDFs concentrations in experimental fly ashes, while the destruction efficiency was defined as the ratio of the reduced I-TEQ value to original I-TEQ value. With the Fe(II)/Fe(III) sulphate mixture, the decomposition efficiency of PCDDs/PCDFs raised a little to a range of 81.88–92.30 %, whereas the destruction efficiency of PCDDs/PCDFs increased dramatically to a range of 80.82–89.59 %. It thus appeared that the beneficial effect from the added Fe compounds on the decomposition of PCDDs/PCDFs in MSWI FAs was limited, but also that it could enhance the destruction efficiency greatly. The authors also reported that the decomposition and destruction efficiencies of dioxin were higher at T=290 °C than those at T=245 °C and they attributed it to the change in the water properties with temperature and pressure under hydrothermal conditions. They concluded that the temperature was likely the most significant parameter for destructing PCDDs and PCDFs and that a higher reaction temperature is to be preferred when possible.

Supercritical Water Oxidation (SCWO)

Supercritical water oxidation is another way for an effective dioxin remediation. It consists in a waste treatment process using supercritical water, which exists as a phase above the critical temperature (647.3 K) and critical pressure (22.12 MPa).

Sako et al. (1997) utilised the process for the decomposition of dioxins in MSWI FAs with air, pure oxygen gas and hydrogen peroxide as oxidizer. They performed a reaction under conditions of temperature (673 K), pressure (30 MPa), and time 30 min. They found that the decomposition yield of dioxins is 99.7 % with the use of supercritical water and hydrogen peroxide. Recently, the same group (Sako et al. 2004) studied a hybrid process for the destruction of dioxins in

FAs. They performed extraction of dioxins from FAs using supercritical fluid (CO_2), concentration by adsorption, and destruction by supercritical water oxidation. During the extraction–adsorption process, dioxins could be transferred and concentrated to the adsorbent (activated carbon). Then, the adsorbent containing dioxins was completely destructed by supercritical water oxidation as explained before.

S/S Mechanochemistry

Some years ago, Nomura et al. (2006) developed a mechanochemical process for inhibition of heavy metal elution from FA and used the residues as cement materials. For the mechanochemical treatment process, they used a planetary ball mill with a pair of stainless steel pots (45 cm^3). Seven steel balls were placed in each of the pots. The pots were placed on a rotating disk in such a way that the pots and the disk rotated in opposite directions. In particular, they mechanochemically treated FA with blast furnace slag to use FA as a cement resource. They reported the degradation behaviour of dioxins in FA caused by mechanochemical treatment. The concentration of dioxins in FA decreased from 9 ng-TEQ/g to 30 pg-TEQ/g after mechanochemical treatment for 4 h, and no dioxins was detected after 8 h.

Yan et al. (2007) also investigated the potential of mechanochemical treatment to degrade PCDD/Fs contained in FA via grounding with calcium oxide (CaO) under atmospheric pressure. The results indicated that PCDD/Fs included in real FA could be degraded by mechanochemical treatment. The degradation efficiency of PCDD/Fs increased with increasing ratio of CaO and that it might increase with increasing rotational speed. Although the destruction of chlorinated compounds like DDT and PCDD/Fs by milling with calcium oxide was successful, the authors deemed it uneconomical on an industrial scale due to the long milling time required. However, using a hydride donor compound ($NaBH_4$, $LiAlH_4$) or metal Na/Mg and a hydrogen donor species involved much shorter milling times (ranging from minutes to a few hours) with contaminated materials as well as with highly concentrated or pure chlorinated compounds and their mixtures at room temperature. Therefore, the authors considered the latter approach fast and economic.

Biological Removal

Bioremediation is an alternative waste management method that overcomes some of the shortcomings of existing methods. It is economical and has minimal impact on the environment. And, it is one of the most important areas in environmental microbiology and biotechnology. Several microbial species were found to catabolise dibenzo-p-dioxin (PCDD) and dibenzofuran (PCDF), and their congeners, and also allows removing dioxin from FA.

Nam et al. (2005) utilised a particular microorganism (*Sphingomonas wittichii RW1*), because they found it to be potential dioxin degrader. They analysed MSWI FA in which the total PCDD concentration was 28,852 pg/g-ash (WHO-TEQ=877.4 pg/g-ash).

Table 3.19 Removal of PCDD/Fs from incinerator fly ash by microbial agent

Homolog	Congener	WHO-TEQ conc. (ng/kg-ash) Control (Autoclaved control.)	Sample	Decrease (%)
TCDD	2,378	70.4	31.4	55.4
PeCDD	12,378	96.3	40.9	57.53
HxCDD	123,478	12.2	5.7	52.79
HxCDD	123,678	104.4	45.9	56.03
HxCDD	123,789	69.6	31.1	55.32
HpCDD	1,234,678	83.6	33.8	59.57
OCDD	12,346,789	10.8	3.6	66.67
TCDF	2,378	18.5	8.7	52.97
PeCDF	12,378	9.4	4	57.45
PeCDF	23,478	353.6	66.7	81.14
HxCDF	123,478	38	12.9	66.05
HxCDF	123,678	39.2	13.6	65.31
HxCDF	123,789	73.8	26.2	64.5
HxCDF	234,678	19	6.8	64.21
HpCDF	1,234,678	15.1	5.2	65.56
HpCDF	1,234,789	2.3	0.9	60.87
OCDF	12,346,789	1.4	0.4	71.43
SUM		1,018.3	338.5	66.76

Nam et al. (2008)

They observed that the removal efficiency with PCDDs congeners increased with the degree of chlorination from 48.4 % (with TCDD) to 86.4 % (with OCDD). For example, they reduced PCDDs from 28.85 to 7.08 ng/g-ash (i.e. with 75.5 % removal efficiency). They concluded that *Sphingomonas wittichii RW1* could decompose PCDD, PCDF, and their metabolites, in MSWI FA and they might serve as a potential candidate for the industrial elimination of PCDDs/PCDFs in FA. In a more recent work, Nam et al. (2008) treated MSWI FA with a biocatalyst for 21 days using a solid-state fermentation system. They obtained a reduction of 69 % of total toxic PCDD/PCDF present in MSWI FA. In particular, 55.4 % of the most toxic 2,3,7,8-TCDD and 57.5 % of the second most toxic 1,2,3,7,8-PeCDD were removed by the microbial biocatalyst (see Table 3.19).

The results shown in Table 3.19 were achieved using a newly manufactured microbial agent consisting of five fungi and four bacteria. They authors concluded that this microbial biocatalyst could be a potential candidate for the development of a biological treatment technology to eliminate PCDD/Fs in FA.

3.4 Fly Ash Recycling

FA is considered contaminating mainly due to its enrichment in potentially toxic trace elements that condensed from the flue gas. Therefore, most of the FA produced was disposed off as landfill in earlier times, but this practice has become increasingly more under examination with time for environmental concerns, because of

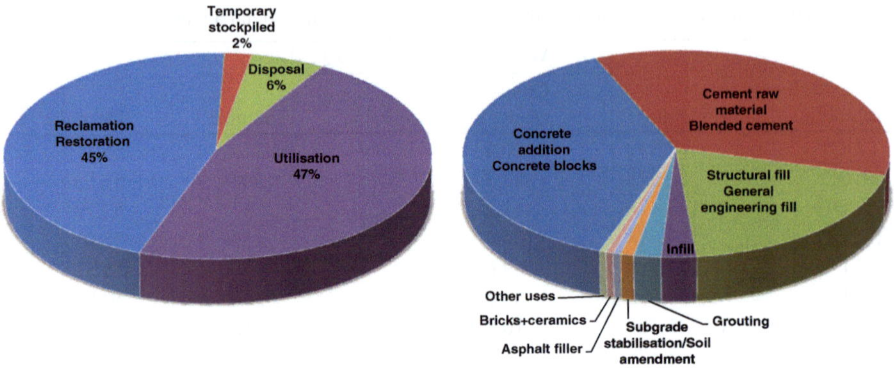

Fig. 3.16 Main fields of use in Europe (ECOBA 2008)

environmental costs that are incurred so that a company's activities do not damage the environment or that any such damage is put right. Figure 3.16 shows the main fields of FA employ in the EU nowadays (ECOBA 2008).

Generally, the most re-used FA is coal FA. Recent estimates put global coal FA production in the region of 750 million tonnes per year (Blissett and Rowson 2012). Current coal FA utilisation amounts to about 39 % in the USA and 47 % in Europe (ECOBA 2008; ACAA 2010). The global average is estimated to be close to 25 % (Wang 2008). On the other hand, FA from Municipal Solid Waste Incineration (MSWI FA) arising from flue gas treatment are partially recovered in some cases only (i.e. some industrial processes are used to recover certain fractions of residues), but, in general, the residues are still sent for disposal, often by landfill and commonly following specific treatments. In some cases, MSWI FA residues are placed in salt mines where they are used as backfill. Its possible employ is limited to some applications, as for example as geopolymer (Galiano et al. 2011) or inert materials (Zacco et al. 2012), after the stabilisation of leachable metals. However, certain problems with stabilisation technologies cannot be ignored, such as long-term stability and the cost of stabilisation chemicals.

Disposal of FA may soon be too costly due to stricter legislative requirements, less land being destined for disposal, and ecological concerns, probably, even forbidden. An economically and environmentally viable solution to this problem should include an increased utilisation of FA and the search for new products rather than land disposal. Other incentives are e.g. financial returns from the sale of the by-product, or, at least, an offset of the processing and disposal costs, and the replacement of some scarce or expensive natural resources by the by-products. Considerable research had been conducted worldwide on the re-use of FA materials and making them more affordable. Utilisation of FA can be in the form of an alternative to another industrial resource, process, or, application. Figure 3.17 presents possible uses of FA based on its properties (Wang 2006). These processes and applications include addition of FA to cement and other concrete products or its addition to construction materials as a light-weight aggregate, roadway and pavement

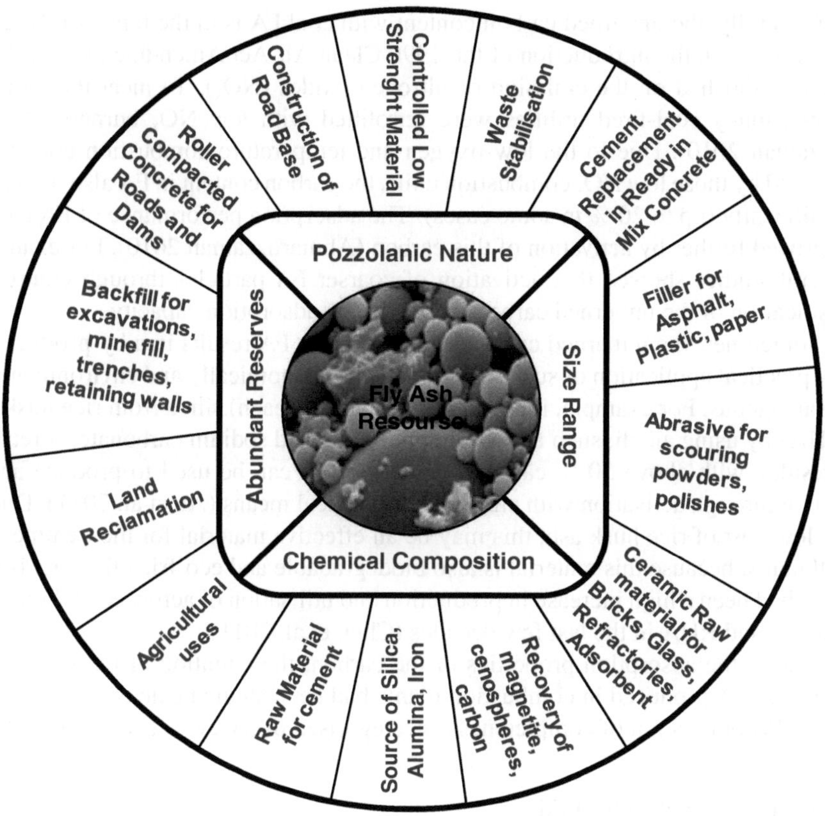

Fig. 3.17 Possible uses of fly ash based on its properties (Wang 2006)

utilisation, employ as infiltration barrier and underground void filling, adsorber in soil and water treatment for environmental improvement, and so on.

The following six Sects. (3.4.1, 3.4.2, 3.4.3, 3.4.4, 3.4.5 and 3.4.6) describe alternative uses of FA and associated research at full length.

3.4.1 Adsorbent

Thanks to its good adsorption capacities, activated carbon produced by carbonising organic materials is today's most widely used adsorbent (Gaikwad 2004). However, the high cost of the activation process limits its employ. On the other hand, the adsorbent properties of FA (and in particular those from coal and rice husk) are attributed to their unburned carbon content and specific surface area. This carbon can be a precursor for activated carbons because it has gone through devolatilisation during combustion in the power station furnace (Maroto-Valer et al. 1999).

Generally, the unburned carbon content with coal FA is in the range of 2–12 %. However, with the introduction of the 1990 Clean Air Act Amendments, caps have been established on the emission of nitrogen oxides (NO_x). To meet the requirements, many coal-fired utilities were retrofitted with low NO_x burners (Ahmaruzzaman 2010). Due to the low oxygen and temperature combustion conditions required by those low NO_x combustion units, the carbon content in FA also increased significantly (up to 20 % in some cases). The adsorption performance of FA can be improved further by activation of this carbon (Ahmaruzzaman 2010). For example, several studies showed that activation of coarser FA particles through controlled gasification of the unburned carbon increased the adsorption capacity.

Sometimes, the unburned carbon separated from FA results in a by-product and any practical application of such material results economically and environmentally advantageous. For example, it is possible to extract (leach) silica from rice husk ash (RHA) by using alkali, such as sodium hydroxide and sodium carbonate. It renders a residue with above 50 % carbon content, which can be used to produce active carbon through activation with chemical or physical means (An et al. 2011). Due to the low cost of rice husk ash, this may be an effective material for the treatment of pollutants, because this material is also biodegradable and eco-friendly. Fact is that there had been a high increase in production and utilisation of activated carbon from coal FA and RHA in the last few decades (Chen et al. 2011).

Due to the adsorption properties of the carbon, the retention of hazardous elements by FA produced in combustion plants had been studied extensively in recent years. Without claim of completeness, a categorised overview is given below.

Nitrogen and Sulphur Oxides

The adsorption of NO_x using activated chars recovered from FA was reported (Rubel et al. 2005). Carbon-rich fractions from a gasifier adsorbed one-third of the NO_x compared with a commercial carbon. Activated carbon from unburned carbon in coal FA had also been used for removal of NO (Rubio et al. 2007). FA treated with calcium hydroxide had been tested as a reactive adsorbent for SO_2 removal (Al-Shawabkeh et al. 1995). It was reported that $Ca(OH)_2$-FA mixtures were a low-cost SO_2 control option (Davini 1995, 1996). Davini (2002) also tested a process using activated carbon derived from FA plus $Ca(OH)_2$ for SO_2 and NO_x adsorption from industrial flue gas and this mixture exhibited characteristics similar to that of non-FA-derived activated carbons used with flue gases normally. It was also recommended that the mineral matter must be removed efficiently from unburned carbon of FA before activation; in this manner an efficient activated carbon for NO_x adsorption can be obtained (Rubio et al. 2007).

Mercury

Due to its high removal efficiency, activated carbon can also be used to remove mercury from flue gas and which is otherwise emitted into the atmosphere as well.

However, due to its high cost, large-scale applications in utility boilers had not been developed (EPA 1997). It has also been observed that some coal FA can capture mercury. Although the role of inorganic components of FA in this capture remains unclear, several studies showed that unburned carbon from FA shows certain capabilities for adsorbing elemental mercury (López-Antón et al. 2007; Suárez-Ruiz et al. 2007; Suárez-Ruiz and Parra 2007). Researchers at the Pennsylvania State University have proposed a method to separate unburned coal from FA (Skillings Mining Review 1999). To understand the role of different types of unburned carbons with FA in mercury capture, some studies associated types of particles and textural properties of FA with the amount of captured Hg (Pavlish et al. 2003; Hower et al. 2000; Suárez-Ruiz et al. 2007; Suárez-Ruiz and Parra 2007). It was demonstrated that mercury retention can be dependent on the mode of occurrence of this element in the gas phase (metallic Hg is retained in FA in a greater proportion than $HgCl_2$). Several studies made to understand the entrapment mechanism as a function of the physical and chemical interactions between the carbon surface and mercury showed that the carbon surface contains some adsorption centers, called primary sites. When a molecule of the adsorbate adsorbs on a primary site, the adsorbed molecule can act as a secondary center for the adsorption of more molecules (Dubini 1965).

Organic Gas

In addition to the adsorption of NO_x, SO_x, and Hg, in flue gas, FA had also been shown to be effective in adsorption of organic gas. Peloso et al. (1983) studied the adsorption of toluene vapours. In accord with other researchers (Rovatti et al. 1988), coal FA obtained after thermal activation showed satisfactory adsorption performance for toluene vapours. Good results were also reported for aromatic hydrocarbon and m-xylene (Rotenberg et al. 1991).

PAH

Not only heavy metals, but also a complex mixture of organic compounds, originated during combustion process, was associated with FA particles from all origins, but mainly with MSWI FA. The nature of these compounds has only partly been elucidated, but it is evident that FA can contain persistent hazardous organic compounds, also called persistent organic pollutants, such as polycyclic aromatic hydrocarbons (PAH), polychlorinated biphenyl's (PCB), methyl sulphates, and chlorinated dioxins and benzofurans (Fiedler 2003; Johansson and Van Bavel 2003a, b; Kim et al. 2003; Mininni et al. 2004; Schmid et al. 2000; Tame et al. 2003; Van Gerven et al. 2004; Wheatley and Sadhra 2004). The organic compounds identified in FA extracts include polar mutagenic organic chemicals (Yoshino and Urano 1998). This can create a problem of secondary environmental pollution.

Heavy Metals

The adsorption of heavy metals by FA was found to depend on several factors, such as the initial concentration of heavy metals and contact time. In particular, the initial concentration of heavy metal has a great effect on the adsorption capacity of the FA. The adsorption capacity of FA was also found to depend on the surface activities, such as available specific surface area for solute surface interaction, equilibrium between competitive adsorption from all the cations, ionic size, and stability of bonds between heavy metals and FA. The last important factor governing the adsorption of heavy metal by FA is pH, because, with increasing pH in a certain pH range, most metal adsorption increases up to a certain pH value, before decreasing with further increase in pH (Ahmaruzzaman 2010). Among the investigated metals, Pb, As, Cu, Cd, Ni, Cr, and Hg are the most often studied. Coal FA was found to be good adsorbent for removal Pb, Zn (Prabhu et al. 1981), and Cd (Gashi et al. 1988), from effluents in the battery and fertilizer industries. Removal efficiencies were greater than 70 %. FA was also effective for the removal of arsenic. Coal FA was also tested for removal of As (V) from water (Diamadopoulos et al. 1993). Coal FA was employed for the removal of Cu (Panday et al. 1985; Gashi et al. 1988) and Hg (Kapoor and Viraraghavan 1992) from aqueous solution. The lime content in the FA resulted to be a significant factor influencing the adsorption of Cr(VI) and Cd(II) and FA was found to have a higher adsorption capacity for Cd(II) as compared to Cr(VI) (Grover and Narayanaswamy 1982; Dasmahapatra et al. 1996). Coal FA was also shaped into pellets and used for the removal of Cu and Cd ions (Papandreou et al. 2007). To improve removal efficiencies and adsorption capacities, chemical modifications of FA can be also done (Zhu et al. 2012). For example, FA impregnated with Cl, Fe, and Cu salts facilitate a more porous structure, which is much more efficient in the metal adsorption. Finally, Apak et al. (1996) reported that coal FA is a good adsorbent for both radionuclides of ^{137}Cs and ^{90}Sr. Table 3.20 resumes the metals adsorption capacity using coal FA. The utilisation of RHA was studied for the adsorption of Pb(II) and Hg(II) (Feng et al. 2004). Selective adsorption of various metal ions (Na, K, Mg, Ca, Cu, Cd, Mn, Hg, Cr, Pb, and Fe) by FA was also reported (Papachristou et al. 1985). Heavy metal-specific adsorption capacities using fly ash can be found with Table 3.20.

Water Purification

Several studies focused on the use of FA for water depuration. The mainly employed fly ashes for this application are coal FA and RHA. Concerning water pollution, heavy metals are most serious pollutants and the contamination of water by toxic heavy metals through the discharge of industrial wastewater is a worldwide environmental problem. Heavy metal toxicity provides severe health concerns: it can reduce mental and central nervous functions, lower energy levels, damage blood composition, lungs, kidneys, liver, and other vital organs. Table 3.21 reports permissible limits and health effects of various toxic heavy metals.

Table 3.20 Metal type-specific adsorption capacity using coal fly ash summarised by Ahmaruzzaman (2010)

Metals	Adsorbent	Adsorption capacity [mg/g]	References
Zn^{2+}	Coal fly ash	0.07–13.3	Banerjee et al. (2003), Bayat (2002a, c)
	Fly ash	0.27–11.11	Weng and Huang (1994), Querol et al. (2002)
	Rice husk ash	5.88–14.30	Bhattacharya et al. (2006), Srivastava et al. (2008)
	Bagasse fly ash	2.34–13.21	Srivastava et al. (2006), Gupta and Ali (2000), Gupta and Sharma (2003)
Cd^{2+}	Coal fly ash	18.98	Papandreou et al. (2007)
	Fly ash	0.01–207.3	Apak et al (1998), Ayala et al. (1998), Bayat (2002a), Srivastava et al. (2006), Weng and Huang (1994), Apak et al (1998), Panday et al. (1985)
	Rice husk ash	3.04	Srivastava et al. (2008)
	Bagasse fly ash	1.24–6.19	Srivastava et al. (2006), Gupta and Sharma (2003)
Pb^{2+}	Fly ash	18.0–483.4	Apak et al (1998); Ricou et al. (1999);Wang et al. (2008)
	Treated rice husk ash	12.61	Feng et al. (2004)
	Bagasse fly ash	285–566	Gupta et al. (1998a)
Cu^{2+}	Coal fly ash	20.92	Papandreou et al. (2007)
	Fly ash	0.76–207.3	Panday et al. (1985), Ayala et al. (1998), Apak et al. (1998), Lin and Chang (2001), Rao et al. (2003), Ricou et al. (1999), Wang et al. (2008)
	Bagasse fly ash	2.26–2.36	Gupta and Ali (2000)
Ni^{2+}	Fly ash	0.03–14.0	Banerjee et al. (2003), Bayat (2002a), Ricou et al. (1999), Rao et al. (2002)
	Bagasse fly ash	1.12–6.48	Gupta et al. (2003), Srivastava et al. (2006)
Cr^{3+}	Fly ash	52.6–106.4	Kelleher et al. (2002)
Cr^{6+}	Fly ash	0.31–23.86	Panday et al. (1984), Bhattacharya et al. (2008), Banerjee et al. (2004), Srivastava et al. (2006)
	Rice husk ash	25.64	Bhattacharya et al. (2008)
	Bagasse fly ash	4.25–4.35	Gupta et al. (1999)
Hg^{2+}	Fly ash	0.63–11.0	Sen and De (1987), Banerjee et al. (2004), Kapoor and Viraraghavan (1992)
	Treated rice husk ash	6.72	Feng et al. (2004)
As^{5+}	Fly ash	7.7–27.8	Diamadopoulos et al. (1993)

Heavy metal and metalloid removal from aqueous solutions is commonly performed by several processes, such as ion exchange, reverse osmosis, adsorption, chemical precipitation, solvent extraction, etc. Among these processes, the adsorption process should be a simple and effective technique for heavy metals removal from wastewater, also, because it usually produces few volumes of sludge. FA has potential application in wastewater treatment mainly, because of its chemical composition, as it contains alumina, silica, ferric oxide, calcium oxide, magnesium oxide, and carbon. Also, its physical properties, such as porosity, particle size

Table 3.21 Permissible limits and health effects of various toxic heavy metals

Metals	Permissible limits for industrial effluent discharge – Inland surface water (mg/L)	Permissible limits for potable water (mg/L)		Health hazards
	WHO	WHO	EU Standard	
Nickel	–	0.02	0.02	Causes chronic bronchitis, reduced lung function, cancer of lungs
Zinc	5.0–15.0	3	–	Causes short-term illness called "metal fume fever" and restlessness
Copper	0.05–1.5	2	2	Long term exposure causes stomachache, irritation of nose, mouth, eyes, headache
Cadmium	0.1	0.003	0.005	Carcinogenic, cause lung fibrosis, dyspnoea
Lead	0.1	0.01	0.01	Suspected carcinogen, anemia, muscle and joint pains, kidney problem and high blood pressure
Total chromium	–	0.05	0.05	Suspected human carcinogen, producing lung tumors
Arsenic	–	0.01	0.01	Carcinogenic, producing liver tumors, and gastrointestinal effects
Mercury	–	0.001	0.001	Excess dose may cause headache, abdominal pain, and diarrhea, paralysis, and gum inflammation, loosening of teeth, loss of appetite, etc.
Iron	0.1–1.0	0.2	0.2	Excess amounts cause rapid pulse rates, congestion of blood vessels, hypertension
Manganese	0.05–0.5	0.5	0.05	Excess amounts toxic, and causes growth retardation, fever, sexual impotence, muscles fatigue, eye blindness
Vanadium	–	1.4	–	Very toxic, and may cause paralysis

Ahmaruzzaman (2011)

distribution, and surface area, are in accord with this employ. Finally, in order to maximize metal adsorption by hydrous oxides, it is generally necessary to adjust the pH of wastewater using lime and sodium hydroxide. And, here, the alkaline nature of FA could also be a good neutralising agent (Bayat 2002a, b).

Apart from heavy metals in wastewater, other inorganic compounds can principally also be dangerous for human health, such as those containing phosphor, fluoride, boron, and nitrogen. Another incentive for recovering nitrogen and phosphor is their re-use as fertilizer. Serious wastewater polluting contaminants also concern organic compounds, such as phenols and dyes.

Phosphor

Phosphorous presence on surface and groundwater is a consequence of concentrated agricultural activities (such as soil fertilisation, feed lots, diaries, and pig and poultry farms) and is the origin of water quality problems in rivers and lakes. Since aluminum, iron, and calcium can adsorb or precipitate phosphates, FA emerges as a potential candidate to treat phosphate-laden effluents thanks to its composition that is rich in silica and aluminum, iron, and calcium oxides. The addition of coal FA to wastewater with phosphate was found to produce insoluble or low solubility salt, as for example calcium phosphate (Cheung and Venkitachalam 2000).

Fluoride

Phosphate fertilizer can also induce high fluoride concentration in ground water resource. Because of an excessive amounts of fluoride may cause adverse health effects to humans and animals, great attention was made by using coal FA for the removal of fluoride from waters (Chaturvedi et al. 1990). The good condition for favourable FA application was found at low concentration, high temperature, and acidic pH.

Boron

On the contrary, only few studies were reported on boron adsorption using coal FA (Hollis et al. 1988).

Nitrogen

The removal of nitrogen compounds from wastewater is of increasing importance (Joshi and Lothia 1997; Mukherjee et al. 2008). Nitrogen contamination is due to intensive use of nitrogen fertilizers, and uncontrolled on-land discharges of raw and treated wastewater. Because of the presence of numerous impurities in the wastewater, many kinds of absorption materials resulted not suitable to remove nitrate. Chemical reduction involve production of ammonia (Querol et al. 2002) and physical adsorption is not good for removing excess nitrate, due partly to their high cost. The majority of biological denitrification treatments rely on bacteria, which require organic carbon substrate. The carbon added source is generally methanol, ethanol, or, acetic acid (Bhattacharjee and Kandpal 2002; Thy et al. 2006). The disadvantages of this source are the need of a close and costly process control, the risk of overdosing and deep knowledge about the operation of the biological system. Moreover, traditional liquid carbon sources are expensive and potentially hazardous. In a recent study, rice husk was chosen as a carbon source and the only physical support for microorganisms (Shao et al. 2009). It was shown that it is safer to use rice husk as substrate in wastewater denitrification when compared with traditional liquid carbon sources and a rapid increase of nitrate removal efficiency was found.

Phenols

Khanna and Malhotra (1997) proved that coal FA has a good potential also for phenolic compounds adsorption. Phenol removal was found strongly dependent on temperature and pH value of treatment solution (Alemany et al. 1996). Experiments were made also on wastewater (Singh et al. 1994; Singh and Nayak 2004). The potential of rice husk ash (RHA) for phenol adsorption from aqueous solution was investigated (Mahvi et al. 2004) also by using lagooning mixture of FA and wastewater (Lorenz 1954). Rice husk ash resulted very effective than rice husk for phenol removal. Rice husk ash obtained from a rice mill in Kenya was used for removal of some phenolic compounds in water (Damaris et al. 2002).

Dyes

The potential of coal FA for removal of dyes from wastewater was investigated by Ramakrishna and Viraraghavan (Viraraghavan and Ramakrishna 1999; Ramakrishna and Viraraghavan 1997; Mohan et al. 2002). The dye removal was proven inversely proportional to the size of FA particle: better adsorption was observed at lower particle sizes, because of increased external surface area available for adsorption. Brown coal FA was also examined as low-cost adsorbents for removal of synthetic dyes from waters (Janos et al. 2003). FA can be regenerated after the adsorption, using suitable reagents. For example, Batabyal et al. (1995) regenerated the used, saturated, coal FA with 2 % aqueous H_2O_2 solution. Coal FA has been also employed for its potential for use as a photocatalyst for the removal of dyes from water under visible light (Bayat 2002c). The rice husk ash was shown to be a potential cost effective adsorbent for the removal of methylene blue dye from aqueous systems (Chandrasekhar and Pramada 2006).

Table 3.22 presents a summary of adsorption capacity of various organic compounds on FA.

Table 3.22 Adsorption capacity of various organic compounds by fly ash

Organic compounds	Adsorbent	Capacity (mg/g)	References
Phenol	FA	67	Akgerman and Zardkoohi (1996)
	Sugar FA	0.47–0.66	Gupta et al. (1998b)
	Wood FA	5.4	Daifullah and Gad (1998)
Ortho-chloro phenol	Coal-FA	0.8–1.0	Kao et al. (2000)
	FA	98.7	Aksu and Yener (2001)
2,4-Dichloro phenol	FA	22	Kao et al. (2000)
	Coal FA	1.5–1.7	Daifullah and Gad (1998)
3-Chloro phenol	FA	20	Akgerman and Zardkoohi (1996)
Para-chloro phenol	FA	118.6	Aksu and Yener (2001)
2-Nitro phenol	Wood FA	143.8	Singh and Nayak (2004)
	FA	5.80–6.44	Gupta et al. (1998b)
3-Nitro phenol	FA	6.52–8.06	Daifullah and Gad (1998)
4-Nitro phenol	Sugar FA	0.76–1.15	Singh and Nayak (2004)
Wood	FA	134.9	Gupta et al. (1998b)
	FA	7.80–9.68	Daifullah and Gad (1998)
Para-nitro phenol Bagasse	FA	8.3	Gupta et al. (1998b)
Cresol Coal	FA	85.4–96.4	Dutta et al. (2003)
m-Cresol Wood	FA	34.5	Hollis et al. (1988)
p-Cresol Wood	FA	52.5	Hollis et al. (1988)
2,4 Dimethyl phenol	FA	1.39	Batabyal et al. (1995)
DDD Sugar	FA	$(7.5–7.7) \times 10^{-3}$	Gupta and Ali (2001)
DDE Sugar	FA	$(6.5–6.7) \times 10^{-3}$	Gupta and Ali (2001)
Lindane Bagasse	FA	$(2.4–2.5) \times 10^{-3}$	Gupta et al. (2002)
Malathion Bagasse	FA	$(2.0–2.1) \times 10^{-3}$	Gupta et al. (2002)
Carbofuran	FA	1.54–1.65	Kumari and Saxena (1988)
TCB	FA	0.35	Nollet et al. (2003)
HeCB	FA	0.15	Nollet et al. (2003)

Closing Remarks

Despite the great potential in the absorption applications of FA, the effectiveness of adsorption process depends on the conditions and variables used for the adsorption process, such as temperature, pH, ionic strength, existence of competing organic or inorganic compounds in solution, initial adsorbent concentration, contact time, etc. In addition, some problems can arise because of the competitive adsorption of organic compounds on FA: there are few data for the adsorption of organic compounds in presence of metals and other contaminants. Finally, there are a lack of data concerning the reproducibility of the adsorption properties and the equilibrium data. Other problems limit the actual FA employ as adsorber. Because FA is a waste material, its utilisation involves the potential leaching of some elements into water: the surface layer of FA particles (probably microns in thickness) contains a significant amount of leachable material, which is deposited during cooling after combustion.

3.4.2 *Agriculture*

Generally, FA contains potentially toxic elements, such as Cd, Pb, Ni, As, Cu, Zn, Cr, Se, etc. (Rautaray et al. 2003; Lee et al. 2006; Tiwari et al. 2008; Adriano et al. 2002), which may contaminate soil (Pandey and Singh 2010). Nearly 5–30 % of toxic elements present in FA, especially, Cd, Cu, and Pb, is leachable (Natusch and Wallace 1974). Other elements are toxic to plants or animals if they are present in high concentrations: for example, B causes growth depressions (Adriano et al. 2002) and Mo toxicity due to FA employ in soil had also been reported (Tolle et al. 1983). The metals may also percolate and contaminate ground water or they may contaminate the nearby water body. However, leachability of heavy metals from coal FA is relatively low and leaching extent depends on the conditions of the system. Trace metal concentration in the leachate depends on FA weight/solution, concentration of elements, pH, temperature, pressure, and time (Pandey and Singh 2010). Besides potentially toxic heavy metals, FA can also contain radioactive elements like ^{238}U, ^{232}Th, ^{40}K, ^{226}Ra, ^{210}Pb, ^{228}Ra, ^{222}Ru, ^{220}Ru (Sharma 1989; Mandal and Sengupta 2003; Papastefanou 2008), and other radionuclides, such as uranium (U) and thorium (Th) series (Tadmore 1986). However, most FAs are not significantly enriched in radioactive elements compared to common soils or rocks (Zielinski and Finkelman 1997). It was also found that the ground water quality due to disposal of FA remained unaffected with respect to radionuclide contamination (Cothern and Smith 1987; Zielinski and Finkelman 1997). A complex mixture of toxic organic compounds can also be associated with FA particles. Mainly depending on thermal treatment, the organic compounds identified with FA extracts can include mutagens and carcinogens as well.

Many applications for the use of mainly coal FA in agriculture and soil remediation had been proposed and studied in the past decades. A broad selection is presented below and it is categorised by keywords for ease of presentation.

Metal Immobilisation and Soil Remediation

FA is an alkaline waste material with high adsorption capacity and high surface area available for element adsorption, rendering many cationic metals less mobile (Ciccu et al. 2003). For this reason, increased attention was paid to the use of coal FA for metal-contaminated soil stabilisation, aiming to minimize the mobility of toxic heavy metals. When coal FA is added to the soil, it reduces metals mobility and availability and it helps to restore soil properties. The mechanism of immobilisation involves the phenomenon of adsorption, complexation, and (co)precipitation (Pandey and Singh 2010). Chemical immobilisation by means of soil amendments has been recognised as a valuable innovative technique for a wide range of contaminated sites (Vangronsveld and Cunningham 1998). Chemical immobilisation not only inhibits the transport of contaminants, such as heavy metals into deeper soil layers and into groundwater, but also promotes the revegetation (Vangronsveld et al. 1996). In a study conducted on soils contaminated with heavy metals in an Italian mine site, the mixing with coal FA showed decreased levels of heavy metals content in percolating water (Ciccu et al. 2003). To solve problems related to metal solubility and acid mine drainage, Iyer and Scott (2001) and Xenidis et al. (2002) suggested the use of coal FA. It was employed for land stabilisation in mining areas (Jarvis and Brooks 1996), in pyritic mine tailing (Sonderegger and Donovan 1984), and to reduce Cu and Pb mobility and bioavailability in soil (Kumpiene et al. 2007). According to the literature, FA used for metal immobilisation represents an effective, eco-friendly management option for reclamation of metal contaminated soil. Its employ had been recognised as a valuable emerging technique for a wide range of contaminated sites. It is a fast-working alternative compared with phytoextraction, which takes several years to remove metals from contaminated soils due to its low bioavailability (Denuex-mustin et al. 2003). Similarly, inducing metal bioavailability by the use of synthetic chelates, such as EDTA, are also not considered anymore, because of their very low biodegradability compared with microbial chelates (White 2001). However, the surface charge of some phases (for example iron oxides) present in FA is pH dependent, therefore, the metal adsorption capacities can change with changing pH (Cornell and Schwertmann 2003). Synthesised zeolites from coal FA or zeolites mixed with FA had also been used in the immobilisation of heavy metals in polluted soils (Belviso et al. 2010).

Impact on Soil Properties, Plant Nutrients, Plant Growth

There are many studies dedicated on the impact of coal FA on soil properties (cf. Tejasvi and Kumar 2012). Indeed, FA not only has good adsorption properties, but it is also rich in many macro and micro plant nutrients, showing interesting properties for agricultural purposes. Due to the large amounts of FA materials produced, they can be considered a potential resource not only for heavy metal adsorption, but also for improving problematic soil systems. Some review papers discuss this application (cf. Pandey and Singh 2010). The use of coal combustion residue in agriculture was initially due to the presence of essential nutrients (except organic C and N), which can promote plant growth and alleviated the soils nutrient deficiency (Mittra et al. 2005). FA was employed as a soil amendment in agriculture to buffer the soil pH (Phung et al. 1978), improving soil texture (Chang et al. 1977), raising the nutrient status of the soil (Rautaray et al. 2003), etc. It was found that FA amendment enhanced the plant growth when using 10–25 % FA (Dwivedi et al. 2007), but it rendered adverse (toxic) impacts with concentrations above 50 %, because of the reduction in growth parameters, such as plant height, root biomass, number of tillers, grain, and straw weight. For example, Srivastava et al. (1995) reported that 10 % utilisation of FA in soil resulted in better growth, dry matter production, and increased photosynthetic pigments, in Lactuca sativa. FA was also found to improve the flowering and fruiting of tomato plant (Khan and Khan 1996). Campbell et al. (1983) found that FA addition at the rate of 10 % increased the water holding capacity for fine and coarse sands. On the other hand, there are several factors that can restrict the ash disposal in soils, such as the content of potentially toxic elements, high salinity, and reduced solubility of some nutrients due to the high pH with some FAs (Page et al. 1979). In addition, FAs are poor in both phosphorous (because excessive Fe and Al convert soluble P to insoluble P compounds and these are not readily available to plants; Adriano et al. 1980) and nitrogen (N_2 is absent, because it is oxidised into gaseous constituents during the combustion).

Value of Preliminary Predictive Tests

FA being a combustion residue shows a wide variation in physico-chemical and mineralogical characteristics depending on the conditions of combustion, nature of combustion source, type of emission control devices, storage, and handling methods (Jala and Goyal 2006). In some cases, laboratory leaching tests appeared to be a poor predictor of what happened in reality. This possibility was exemplified by the unexpectedly very high leaching of Zn, W, Be, and Cd, from a coal ash dump in Poland (Twardowska and Szczepanska 2002). Indeed, external factors may be poorly reflected in leaching tests in the form as they are commonly used. Wallander et al. (1997) found that heavy metals immobilised as apatites (characterised by a calcium phosphate lattice) may be released due to the activity of fungi. The presence of humic acids is important for leaching of persistent organic pollutants (Comans et al. 2003; Kim et al. 2002; Osako et al. 2002; Sakai et al. 2000; Van der Sloot et al. 2001). However, humic acids tend to be absent in leaching tests. In view of environmental impact, this may be a significant problem when the leachability is much higher than predicted by leaching tests (Reijnders 2005). As a consequence, the commercialisation of coal FA as a fertilizer in agricultural sector for crop production is generally uncommon. To promote the use of this potential resource in large volumes, it is very important to evaluate the impact of FA on different soil characteristics, such as soil fertility, soil health, soil microbes, soil bio-chemical activity, and soil nitrogen cycling. For effective FA incorporation in soil, a good understanding of the possible changes in soil biota, especially by earthworms, to amendment of agricultural soil with FA is also necessary.

Impact on Soil Fertility

FA can improve the physical properties of soil and its nutrient status (Rautaray et al. 2003). Recently, pot culture experiments were conducted to verify the effect of coal FA for soil amendment to improve the fertility of soil leading to higher productivity of the test crop. Forty percent FA was found most suitable for growth and yield of test crop (Tejasvi and Kumar 2012). Moreover, the effect of FA addition to soil fertility largely depends upon the properties of original FA and soil. The majority of crops need optimum pH values of between 6.5 and 7.0, because in this range the availability of most nutrients to plants is maximised. FAs are generally alkaline. The initial increase in soil pH after FA amendment is explained by the rapid release of Ca, Na, Al, and OH^- ions (Wong and Wong 1990). Lime in FA reacts with acidic components in soil and releases nutrients in the form and amount beneficial to crop plants. Indeed, FA has been used for correction of sulphur and boron deficiency in acid soils (Chang et al. 1977). FA applied on acidic strip mine spoils at different places increased the yield of many crops, thanks to increased availability of Ca^{2+}, Mg^{2+}, and preventing toxic effects of Al^{3+} and Mn^{2+} and other metallic ions by neutralising the soil acidity (Fail and Wochok 1977). On the contrary, fertility strongly decreases at very low pH levels as dissolution and bioavailability of some metals that are toxic to plants (as for example Mn) increases. The activity of certain metals may increase with an increase in pH. For example, Al is the most abundant metal in FA. Aluminium is relatively insoluble as $Al(OH)_3$ at neutral pH, but it exists predominantly as highly soluble and toxic aluminate anions above a soil pH of 8.0.

Impact on Soil Biota

There are only few works about the effects of coal FA amendment on soil biological properties. Numerous short-term laboratory incubation studies showed that the addition of FA to soils can inhibit microbial respiration, numbers, size, enzyme activity, and soil nitrogen cycling processes, such as nitrification and nitrogen mineralisation (Arthur et al. 1984; Cerevelli et al. 1986; Wong and Wong 1986; Pitchtel 1990; Pitchtel and Hayes 1990; Garau et al. 1991). Survival rate and function of earthworms are recognised as valuable indices of soil health and fertility in agriculture. Structural features of burrows are known to significantly influence hydrology, gas diffusion, and nutrient distribution (Bouma 1991; Bouché and Al-Addan 1997; Bastardie et al. 2002). However, there are only few studies regarding the effects of FA amended soil on the earthworm (Nahmani et al. 2005; Muir et al. 2007). Only very recently, Markad et al. (2012) evaluated toxicological effects of FA on some earthworms. The risk assessment and management of heavy metals suggested that leachable toxicant from FA can induce harmful effects, such as oxidative stress, destabilisation of lysosomal membranes, and DNA damage.

Impact of FA Amended with Organic Wastes

Other studies suggest that coal FA can find better application in agriculture or improvement of degraded soils when combined with organic amendments, such as cow manure, farmyard manure, sewage sludge, press mud, paper factory sludge, crop residues, or, organic compost (Kumpiene et al. 2007; Sajwan et al. 2003; Shen et al. 2008). Certain inhibitory effects to soil microbes by toxic components of FA may also be attenuated by the application of organic materials. For example, the toxicity of boron is the major limiting factor in agricultural use of FA, because it induces inhibition of microbial respiration; the boron release can be prevented by the co-application of a readily oxidisable organic substrate (Page et al. 1979). A considerable change in the physicochemical properties of soil, rising of pH, and increased rice crop yield was obtained by mixed application of FA and paper factory sludge and farmyard manure in different combinations (Mittra at al. 2005). The toxicity of heavy metals from FA, such as Pb, Ni, Cr, Zn, Cu, and Cd, were reduced with the increase of organic matter content in the FA dump by building complexes with the heavy metals (Juwarkar and Jambhulkar 2008). Other beneficial effects for combining FA with organic matter on soil are e.g. enhanced biological activity (Kumpiene et al. 2007), reduced heavy-metal availability, killing pathogens in the sludge (Wong 1995); reduced leaching of major nutrients (Sajwan et al. 2003); improved soils through higher nutrient concentrations, better texture, lower bulk density, higher porosity and mass moisture content, higher content of fine-grained minerals (Shen et al. 2008), and being beneficial for vegetation (Rautaray et al. 2003).

Use of FA as Container Substrate

Another possible employ of coal FA in agriculture is to use it as container substrates for ornamental plant production (Chen and Li 2006; Jayasinghe et al. 2009, 2010). Any material formulated to provide support, water retention, aeration, and/or nutrient retention for plant production is known as container substrate. Numbers of studies have demonstrated the feasibility and advantages of utilising synthetically produced aggregates by combining different wastes together with FA (Jayasinghe and Tokashiki 2006; Jayasinghe et al. 2005, 2007, 2008, 2009). Since ornamental plants are grown for aesthetic purposes only, element accumulation in plant of toxic elements is generally not of much concern.

Use of FA as Insecticide

Several studies have revealed that coal FA can be used as insecticide in agricultural areas against a range of lepidopterous and coleopterous pests infesting rice, vegetables, greens, and certain other field crops (Narayanasamy and Gnanakumar 1989; Basu et al. 2009; Sankari and Narayanasamy 2007).

The application of FA dusts strongly increased the plant resistance mechanism and enhanced the fruit yield (Eswaran and Manivannan 2007). FA could effectively control various pests infesting several vegetables both under laboratory and field conditions.

Final Remarks

In conclusion, the potential of coal FA as a resource material in agriculture is now well established: it was shown that FA can be directly used as a soil conditioner in different degraded soils, mainly for the purpose of forestry and floriculture. It can be used in combination with chemical fertilizer to obtain additional benefits in terms of improvement in soil physical characteristics, increased yields, etc. However, the recommendation for a large FA application to agricultural soils cannot be made, unless extensive trials are done to establish a proper combination of FA with each type of soil to found its quality and safety. In the near future, much more attention must be paid to some fundamental aspects related to FA incorporation to soil, like long-term studies of soil quality, soil fertility, soil health, and continuous monitoring of the properties of soil and FA to verify the long-term impact of FA employ. For example, although coal FA is used directly as a soil amelioration agent, there are issues regarding potentially mobile toxic elements connected with the soil that is being used to grow crops. In this case, Pandey and Singh (2010) suggest pre-treatment steps.

3.4.3 Building Materials

Cement is the most cost and energy intensive component of concrete: apart from water, humans do not use any other material in a greater quantity than concrete. In 2000, global cement production was about 1,537 million tonnes and the associated CO_2 emissions were estimated to be 1579 Mt (van Oss and Padovani 2003). Portland cement is the most common type of cement in general use. The Portland cement industry generates an estimated 7 % of global anthropogenic CO_2 emissions (Mehta 2004). Thus, even small reductions in the embodied CO_2 emissions of concrete can make significant global impacts. Thanks to the presence of cementitious compounds like calcium and reactive glass, coal FA is quite suitable in Portland cement products and its employ as partial replacement of cement had been studied extensively in the last decade, because of economical and environmental reasons.

There are essentially three applications for coal FA in cement: (1) Replacement of cement in Portland cement concrete, (2) Use as pozzolanic material in the production of pozzolanic cements, and, (3) Use as retardant ingredient with cement and replacement of gypsum ($CaSO_4 \cdot 2H_2O$).

Use of FA as Replacement of Cement in Portland Cement Concrete

The utilisation of FA is mainly due to its beneficial effects, such as lower water demand for similar workability, reduced bleeding, and lower evolution of heat. In addition, FA is employed because of its economical advantage as pozzolana for partial replacement of cement. FA employ reduces the cost of construction and increase its durability. Indeed, FA concrete provides much strong and stable protective cover to the steel against natural weathering action. FA can also improve the concrete workability without compromising strength; for this reason it was employed in fiber reinforced concrete that is characterised by low workability. Addition of FA as an admixture increased early-age compressive strength and long-term corrosion resistance characteristics of concrete (Maslehuddin 1989). The employ of FA in concrete produces less permeability, because the spherical FA particles lead to improved packing, i.e., they facilitate a more dense paste, and further pozzolanic reaction. FA reaction with the soluble calcium hydroxide in cement produces the more stable (less-soluble) cementitious compound of calcium silicate hydrate, hereby also increasing the resistance to corrosion and ingress of corrosive liquids. In addition, the reaction products also tend to fill capillary voids in the concrete mixture, thereby, reducing permeability of the concrete.

Chemical activation of FA reactivity was demonstrated to be a method to increase the use of FA in concrete. To accelerate pozzolanic reactions, chemical activators, such as Na_2SO_4 or $CaCl_2$, were employed. They changed pozzolanic reaction mechanisms between FA and lime (Shi and Qian 2003), resulting in decreased setting time, accelerated strength development, and increased strength of materials, especially with a high percentage of FA.

High-volume FA concrete has emerged as a construction material in its own right. This concrete can contain more than 50 % of FA by mass of total cementitious materials. FA has been used extensively in large volume placement and mass concrete applications (Reiner and Rens 2006), because the lower heat of hydration produced compared with straight Portland cement concrete allows controlling volume expansion and reducing cracking at early ages.

Replacing cement with FA by less than 50 % rendered a 5–7 % reduction in the water requirement for the designated slump and the rate and volume of bleeding water was either higher, or, about the same, compared with the control mixture (Ravina and Mehta 1986). Concrete containing high volume of FA exhibited excellent mechanical properties (Volz 2012), very low permeability to chloride ions, and good durability with regard to repeated freesing and thawing, (Giaccio and Malhotra 1988; Langley et al. 1989; Malhotra 1990); it also showed no adverse expansion when reactive aggregates were incorporated into concrete (Malhotra 1990). On the contrary, FA is generally under-utilised (McCarthy and Dhir 1999), particularly, in relation to its application as a binder.

Use of FA as Pozzolanic Material in the Production of Pozzolanic Cements

The pozzolanic reaction, which is normally a slow process, can be accelerated by using coal FA with reduced particle size (Ahmaruzzaman 2010). Typical FA has a mean particle diameter ranging from 20 to 30 μm, but ultra-fine FA with a mean particle diameter of 1–5 μm can also be produced. The finer FA can react more completely than the coarser particles and renders durability and strength benefits at a much earlier age (e.g. less than 90 days compared with more than 1 year with typical FA), even, with a smaller dosage of ultra-fine FA. Concrete durability properties, such as resistance to alkali sulphate reaction, sulphate attack, and corrosion, are also enhanced by ultra-fine FA. Generally, ultra-fine FA is used as a replacement in a ratio of 5–15 % of the cement weight. Moreover, there is a considerable potential for increasing the utilisation of FA in cementitious materials, either, by simple adjustment of the concrete constituent materials, or, using a more refined concrete-mix-proportioning techniques (McCarthy and Dhir 1999).

Use of FA as Retardant Ingredient with Cement and Replacement of Gypsum

Coal FA can also be combined with binder material(s) other than normal Portland cement. These routes are of importance for achieving more sustainable, high-performance, concrete. It has been shown that the use of FA at levels of about 30 % of the binder allows obtaining similar or enhanced performance compared with that of Portland cement concrete of equivalent 28 day strength. Higher levels of FA are also used, but only in special situations where properties other than strength, as for example control of heat, had been important (Ahmaruzzaman 2010). FA use at higher levels can have a detrimental effect on early-concrete strength, because the use of high levels of this relatively slower reacting material reduces the mechanical properties of Portland cement. A possible way of overcoming this potential restriction is to combine FA with cement exhibiting high early-strength characteristics, or, with cement that is reacting faster. High early strength can be achieved by grounding the cement to a greater fineness, or, the use of chemically composed cement (Ahmaruzzaman 2010). The advantages of combining FA with these materials are several. Indeed, it is possible to increase the use of FA contents in concrete for mainstream structural applications. In addition, the dilution of cement with slow reacting FA will reduce the rate of setting, promote the development of a better hydrate structure, boost the early strength of the FA-enriched concrete, and potentially allows using larger amounts of material.

To reduce CO_2 emissions and improve cement quality, reactive amorphous silica is often used as additive in modern cement production to reduce clinker consumption. Successful examples include silica fume produced from silicone industry. Reactive silica can also be produced from combustion of organic residues. Because of the organic nature, reactive silica so produced is often called bio-silica. Plant by-products as for example rice husk can produce bio-silica. When it is burned and then blended with Portland cement, the final product is called biocement. The use of biocement has showed environmental, economic, and technical benefits (Hosseini et al. 2011). Under controlled incineration, rice husk ash (RHA) contains amorphous silica. Literature reports that up to 30 % (by weight) of RHA could be blended with Portland cement without adversely affecting the strength and permeability properties of the concrete (Ganesan et al. 2008). It was reported that reduced porosity in RHA concrete also increased resistance to chloride ion penetration, which is an advantage in prevention of corrosion of reinforcement in marine environment (Zhang and Mohan 1996). For self-consolidating, high-performance concrete, the hardening properties were improved with higher RHA content due to the microfilling and pozzolanic effects of rice husk ash (Safiuddin 2008).

Biocement with interesting properties can also be made with ash obtained after biomass burning. The demand of renewable energy sources has increased the use of biomass in the production of heat and electricity. This means a higher FA quantity from biomass. FAs from biomass have a predominantly inorganic fraction and a minor organic fraction (i.e. unburned carbon). The composition depends on the different parameters and conditions of the combustion process, such as the type of biomass, the load, and the combustion and operating conditions in which the process develops (Girón et al. 2013). A variety of plant residues contain sufficient Si to generate amorphous bio-silica for biocement production.

Table 3.23 shows the chemical composition of ash produced from organic residues. Values for Portland cement are included for reference (Hosseini et al. 2011).

With respect to the use of organic residue ashes, the following comments can be made:

- Biocement made with vetiver grass ash considerably improved mortar resistance against acid attack (Nimityongskul et al. 2003).
- The corn cob ash blended cement reduced water absorption in concrete and increase its resistance to chemical attack (Adesanya and Raheem 2010).
- Tangchirapat et al. (2007) showed that ground oil palm ash is an excellent pozzolanic material to be used as a Portland cement replacement.
- Singh et al. (2007) showed that a biocement with 20 % of bamboo leaf ash, that contains amorphous bio-silica compressive strength comparable to Portland cement after 28 days of curing.
- Sewage sludge has appreciable quantities of silicon that makes it suitable for biocement production. Sewage sludge can be used in biocement production via two different processes: (1), by blending its incinerated ash with Portland cement or, (2), by co-combustion of sewage sludge with limestone before its addition to Portland cement (Hosseini et al. 2011). However, the chemical properties of sewage sludge ash are expected to be highly variable, depending on what materials enter the waste stream and the wastewater treatment process.
- Studies on the use of sawdust ash for biocement production have shown that it is cost effective and offers a large potential market for this ash (Sumaila and Job 1999; Udoeyo and Dashibil 2002). Sawdust ash increases the consistency of the cement paste; however, the workability and strength of this biocement declines when a high sawdust ash content is selected.
- The potential of paper mill sludge ash as 'clinker' in cement manufacture was also proven (García et al. 2008). There is little information regarding wheat straw ash, but an increase in sodium sulphate resistance of the concrete was observed with an increasing percentage of ash addition (Binici et al. 2008). Moreover, despite promising results, only few works have been devoted to biocement production and additional experiments are necessary to uncover whether there is a problem in using these ashes. Combustion of other alternative fuels must also consider the effect on the FA quality.

The disposal of metal-laden FA, like MSWI FA, is a problem. Generally, this FA is disposed off as landfill. FA containing heavy metals can be incorporated into the cementitious mixture to reduce the heavy metals leachability from solidified products and without the risks of unacceptable delay of cement setting. The feasibility of a washing pretreatment of MSWI FA with water as a means of maximising the incorporation of FA in cementitious matrices (ash content up to 75–90 % by weight of total solid) had been proven (Mangialardi et al. 1999), without the risk of heavy

Table 3.23 Chemical composition of ash produced from organic residues following combustion at the specified temperature and duration

Organic residue	SiO_2 %	Al_2O_3 %	Fe_2O_3 %	CaO%	MgO%	Na_2O%	K_2O%	Loss on ignition [%]	Combustion Temperature [°C]	Combustion time [h]	References
Sawdust	67.20	4.09	2.26	9.98	5.80	0.08	0.11	4.67	NDa	ND	Elinwa and Mahmood (2002)
Paper mill sludge	25.70	18.86	0.87	43.51	5.15	1.56	1.31	ND	ND	ND	Mozaffari et al. (2006)
Rice husk	93.20	0.40	0.10	1.10	0.10	0.10	1.30	3.70	650	ND	Chindaprasirt et al. (2008)
Vetiver grass	57.48	3.73	1.71	5.45	1.24	0.12	15.49	11.76	900	9	Nimityongskul et al. (2003)
Corn cob	66.38	7.48	4.44	11.57	2.06	0.41	4.92	1.30–1.47	650	8	Adesanya and Raheem (2009)
Sugar cane straw	59.06	4.75	3.18	19.59	2.25	0.73	4.75	2.05	800	1	Martirena Hernándezet al. (1998)
Oil palm shell	63.60	1.60	1.40	7.60	3.90	0.10	6.90	9.60	ND	ND	Chindaprasirt et al. (2008)
Wheat straw	54.24	4.55	1.05	12.54	2.39	ND	ND	7.22	670	5	Biricik et al. (2000)
Bamboo leaf	75.90	4.13	1.22	7.47	1.85	0.21	5.62	ND	600	2	Dwivedi et al. (2006)
Sewage sludge	50.60	12.80	7.21	1.93	1.48	0.32	1.70	ND	700	3	Pan et al. (2003)
Portland cement	22.00	6.00	3.00	65.00	1.00	0.20	0.80	1.29	1,400	ND	Nimityongskul et al. (2003)

Hosseini et al. (2011)

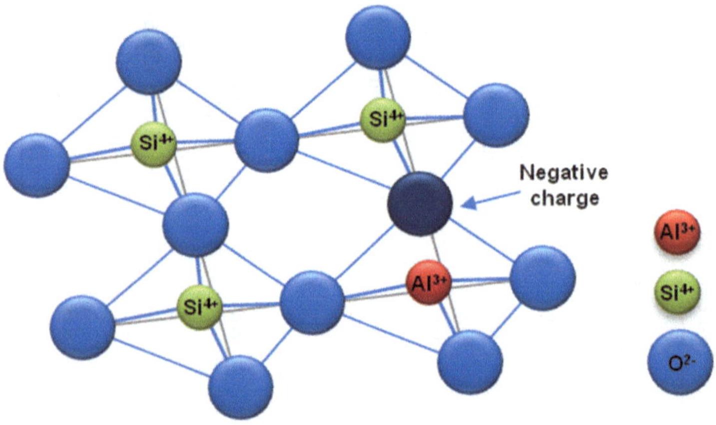

Fig. 3.18 Structure of zeolite framework of tetrahedral $[SiO_4]^{4-}$ with a Si/Al substitution ($[AlO_4]^{5-}$) yielding a negative charge, and consequently a cation exchange capacity (Querol et al. 2002)

metals leaching from solidified products. Sintering procedure was also applied to MSWI FA with the aim of manufacturing sintered products for reuse as concrete aggregates (Mangialardi 2001). The addition of up to 50 % treated FA will not affect the strength and hardness, and the leaching property is acceptable for the use in road construction. However, the long-term durability has not yet been determined (Aubert et al. 2004, 2006).

It was shown that FA can be used as adsorbent with the aim of wastewater treatment (see Sect. 3.4.1). After effluent treatment, FA can be disposed off ultimately in cementitious materials. For example, coal FA, which is generally used as immobilisation agents for heavy metal in aqueous solution, was employed to entrap Pb(II), Cd(II), and Cu(II), and then solidified by cement-based technology to hard concrete blocks and they should not pose any risk to the environment (Kilinçkale et al. 1997). The setting and hardening properties of the obtained mortars and the flexural and mechanical strengths of the solidified samples were optimised with respect to the dosage of natural and metal-loaded solid wastes. Weng and Huang (1994) showed that for an extended curing time, it is possible to prepare mortars with metal-laden FA that have strength comparable to cement only. However, a final recommendation on the use of this mortar for construction purposes necessitates the evaluation of its properties on a long-term basis regarding strength, durability, sulphate resistance, wetting/drying and freesing/thawing cycles, and volume change of the metal-laden FA cement.

3.4.4 *Synthesis of Zeolite*

Zeolites are crystalline aluminium–silicates. Their structure is a network of $[SiO_4]^{4-}$ and $[AlO_4]^{5-}$ tetrahedral linked to each other at the corners by sharing their oxygens (Fig. 3.18) (Ahmaruzzaman 2010)

The tetrahedral connection makes a three-dimensional network with lots of voids and open spaces, producing many useful properties of zeolites, such as adsorption of molecules in the huge internal channels. The characteristic that identify zeolites from silica minerals is due to the substitution of aluminum into the crystalline framework: the replacement of $[SiO_4]^{4-}$ tetrahedra by the $[AlO_4]^{5-}$ tetrahedra causes excess negative charges and cations are needed to neutralize them. Compensation of the negative charge is done by associated cations such as H^+, Na^+, K^+, Ca^{2+}, NH^{4+}. As a consequence of their structural properties, zeolites have a wide range of industrial applications (Breck 1984) mainly based on ion exchange, and gas and water adsorptions.

Zeolites exist as natural minerals, due to alkaline activation of glassy volcanic rocks, or they can be synthesised from a wide variety of high-Si and Al starting materials. The similarity of the FA composition to some volcanic material was the reason for the synthesis of zeolites from this waste product. The conversion of FA in zeolites is one of the most promising approaches to improve its utilisation. Zeolite can be obtained by hydrothermal treatment of FA (Querol et al. 1995). In particular, due to the presence of high content of reactive phases, such as aluminosilicate glass and its high specific surface, FA is a suitable starting material for zeolite synthesis. In addition, FA has a high Si/Al ratio, which allows the synthesis of low-Si zeolites with a high ion exchange capacity, a high selectivity for polar molecules, and a large pore volume (Höller and Wirsching 1985). All developed methodologies for zeolites synthesis from coal FA (Querol et al. 2002) are based on the dissolution of Al–Si-bearing FA phase with alkaline solutions (mainly NaOH and KOH solutions) and subsequent precipitation of zeolitic material. Coal, lignite, and rice husk, ashes were used as raw materials for zeolite synthesis (Chareonpanich et al. 2004). Zamaow and Schultze (1993) showed that the obtained materials have important potential industrial applications, mainly as adsorbents and as replacement for phosphate in detergents. Zeolites synthesised from FA were employed for phosphate and ammonia immobilisation from wastewater (Takami et al. 2000; Murayama et al. 2003). Therefore, simultaneous removal of ammonium and phosphate from aqueous solution by these materials was also shown to be possible (Takami et al. 2000; Murayama et al. 2003). Surface modified FA zeolites have been developed for monitoring of VOCs in indoor air and remove organic compounds (Rayalu et al. 2006). In addition to above usages, zeolite was proven to be a suitable matrices for slow release of the biofertiliser. Surface modified zeolites have been used to adsorb biofertiliser and other negatively charged microbes (Bowman 2003).

Table 3.24 shows the zeolite that can be synthesised from coal FA. The potential industrial application of these zeolitic materials is varied.

Zeolite synthesised from FA was employed also in the removal of heavy metals from wastewater (Wu et al. 2012). Somerset et al. (2008) showed that it is effective in reducing the Pb(II) and Hg(II) concentrations by 95 % and 30 %, respectively. The removal of arsenic, from drinking water was also reported (Goswami and Das 2000). FA was converted to zeolites by also for the removal of metal ions, radiocesium (Mimura et al. 2001), Lead and Cadmium (Shawabkeh et al. 2004). Surface modified zeolite from FA was employed for the adsorption of phenol and

Table 3.24 Zeolites and other neomorphic phases synthesised from fly ash and Joint Committee of Powder Diffraction Standard (JCPDS) codes for XRD identification

Zeolitic product	Chemical formula	JCPDS
(Na)P1 zeolite	$Na_6Al_6Si_{10}O_{32} \cdot 12H_2O$	39-0219
(Na)P zeolite	$Na_{3.6}Al_{3.6}Si_{12.4}O_{32} \cdot 12H_2O$	44-0052
Phillipsite	$K_2Al_2Si_3O_{10} \cdot H_2O$	30-0902
K-chabazite	$K_2Al_2SiO_6 \cdot H_2O$	12-0194
Zeolite F linde	$KAlSiO_{4.1} \cdot 5\ H_2O$	25-0619
Herschelite	$Na_2Al_2Si_{1.68}O_{7.44} \cdot 1.8H_2O$	31-1271
Faujasite	$Na_2Al_2Si_{3.3}O_{8.8} \cdot 6.7H_2O$	12-0228
Zeolite A	$NaAlSi_{1.1}O_{4.2} \cdot 2.25HO_2$	43-0142
Zeolite X	$NaAlSi_{1.23}O_{4.46} \cdot 3.07H_2O$	39-0218
Zeolite Y	$NaAlSi_{2.43}O_{6.86} \cdot 4.46H_2O$	38-0239
Perlialit	$K_9NaCaAl_{12}Si_{24}O_{72} \cdot 15H_2O$	38-0395
Analcime	$NaAlSi_2O_6 \cdot H_2O$	19-1180
Hydroxy-sodalite	$Na_{1.08}Al_2Si_{1.68}O_{7.44} \cdot 1.8H_2O$	31-1271
Hydroxy-cancrinite	$Na_{14}Al_{12}Si_{13}O_{51} \cdot 6H_2O$	28-1036
Kalsilite	$KalSiO_4$	33-0988
Tobermorite	$Ca_5(OH)_2Si_6O_{16} \cdot 4H_2O$	19-1364

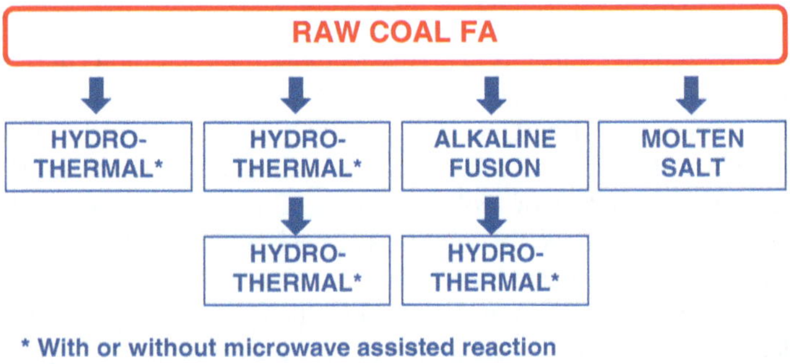

Fig. 3.19 Summary of approaches for the synthesis of zeolites from coal fly ash (Blissett and Rowson 2012)

o-chlorophenol (Kamble et al. 2008) and thiophene and benzothiophene (Ngamcharussrivichai et al. 2008). Although MSWI FA has a low content of SiO_2 and Al_2O_3 (15–30 %), different types of zeolites had been synthesised from MSWI FA by fusion or hydrothermal process (Yang and Yang 1998; Miyake et al. 2002). However, it was found that the residual liquid after the synthesis contains high concentrations of heavy metals, such as Pb and Zn, thus, necessitating further treatment before discharge (Yang and Yang 1998). The main limitation for zeolites synthesis from FA is that the reaction rate requires relatively high temperatures (125–200 °C) to dissolve Si and Al from the coal FA particles. Figure 3.19 summarises the different approaches that had been studied for the synthesis of zeolites from coal FA (Blissett and Rowson 2012). Moreover, because of the variability of FA composition, it is very difficult to suppose the employ of FA in high-value applications, but it can be used for some of the low value applications that utilize the good ion exchange capacity (Ahmaruzzaman 2010).

3.4.5 Geopolymers

A geopolymeric reaction is a geosynthetic reaction of aluminosilicate minerals in the presence of an alkali solution at low temperatures and first developed in the 1970s (Davidovits 1994). The reactions are already reported in Sect. 3.3.1.2. The resulting matrix poses good physical, chemical, and mechanical properties: these include low density, micro and nano-porosity, low shrinkage, high mechanical strength, good thermal stability, durability, surface hardness, fire, and chemical resistance, such as compressive strength (van Jaarsveld and van Deventer 1999). The major focus of these materials is that they can replace the use of Portland cement as a binder in concrete applications.

Several studies were made to use FA as source for geopolymer synthesis regarding the dissolution reaction of FA, rate and thermodynamic properties of the reaction, and the mechanism of the hardening process involved in geopolymerisation (Mustafa Al Bakri et al. 2011). For example, it was shown that geopolymers made from FA can harden rapidly at room temperature and that the mortar develops compressive strengths in the range of 80–120 MPa (Kumar et al. 2007a, b; Blissett and Rowson 2012). Moreover, geopolymer was considered a valid technology for MSWI FA inertisation, because of its strong fixing capacity for heavy metals (Jin et al. 2011; Lancellotti et al. 2010). Heavy metal immobilisation occurring in geopolymerisation process was explained by the following mechanisms: (1) metal ions are taken into the geopolymer network, (2) metal ions are bound into the structure for charge balancing roles, and, (3), the heavy metals-containing precipitate is physically encapsulated (Zheng et al. 2011). One disadvantage of the geopolymerisation is that the inorganic salts, which are common with MSWI FA, affects the process negatively and may have to be eliminated first by water-washing pretreatment of the MSWI FA. The use of geopolymer binder was proposed for making pervious concrete (also known as porous concrete, or, water-permeable concrete). It is a special concrete with high void content to allow air or water to move through the concrete. This concrete can be used in the construction of park areas, areas with light traffic, pedestrian walkways, tennis courts, greenhouse, and other civil engineering and architectural works. The voids of pervious concrete make it also suitable for use in other applications, such as thermally insulating and acoustic absorption. Pervious concrete was also successfully prepared from alkali-activated lignite with high-calcium FA binder, rendering a material with density 30 % lower than that of the conventional concrete (Tho-in et al. 2012). Moreover, several factors mainly related to the chemistry of the raw materials can affect the performance of the final geopolymer products, such as compressive strength, setting time, strength development, composition of the gel phase, and the microstructure of the alkali-activated material. For example, Rickard et al. (2010) showed that high iron contents appear to have a negative effect on the high temperature performance of geopolymers. Bakharev (2006) showed that crystallinity of FA is a key factor for final product performance. And, when different coal FAs are used as source material for geopolymers production, a FA with high crystalline fraction renders a material with a much lower early-mechanical strength than with a predominantly amorphous FA.

Coal FA characteristics are highly variable between power stations and strongly dependent on the coal source and burning conditions. Therefore, various FA characteristics must be considered (clarified) before as geopolymers (Rickard et al. 2012). The most important specifics are:

- particle size [as it influences the reactivity] (Chen-Tan et al. 2009),
- glass content [as it sets the amount of reactive material] (Williams and van Riessen 2010),
- Si:Al ratio in the glass [as it controls mechanical strength and thermal resistance] (Rickard et al. 2011),
- iron content [as it is fundamental in high temperature applications] (Rickard et al. 2010),
- Calcium content [as it affects setting time] (Dombrowski et al. 2007).

As with concrete, FA geopolymer can have the problems of low workability and rapid setting time owing to the high viscosity of sodium silicate solution. To improve its workability, additional water can be incorporated into the mixture, resulting in a geopolymer with a lower compressive strength. On the contrary, for some applications, a short setting time of geopolymer can be necessary and in those cases accelerators must be used. The widely used accelerator is calcium chloride, but its use controversial, because it can induce corrosion in concrete reinforcement. Rattanasak et al. (2011) showed that also calcium sulphate can be used for reactions acceleration. Instead, Boonserm et al. (2012) used 5 % FGD FA as it is rich in calcium sulphate and they attributed the strength improvement of the geopolymer mortar to the sulphate ions by increasing the dissolution of Al ions, thus, enhancing the geopolymerisation process and strength of the geopolymer. Although large-scale application of geopolymers is very complex, mainly due to the variability of the employed FA for synthesis, high performance geopolymer concretes are starting to become commercialised (van Deventer et al. 2012).

3.4.6 Other Applications

Irrespective of origin, FAs have a good potential for use in ceramic products and their employ in this field had been reported extensively: there are many studies (Barbieri and Lancellotti 2004; Yilmaz 2012; Andreola et al. 2008) about glass–ceramics obtained from FA. The results indicated that FA, when mixed with traditional materials, had the necessary requirements to be used as a raw material for production of ceramic tiles. Ceramic processing route has several advantages for FA re-use: (1) ceramics can be processed at reasonably low temperatures, (2), a large number of ceramics product is tolerant to variations in FA composition, (3), ceramics exhibit reasonable chemical durability, are radiation resistant, and can endure changes occurring at different environments, such as humidity, temperature, sun, salt, and wind. In addition, (4), ceramic heat treatment process is a valid technology for the inertisation of FA.

Tiles

For tiles production, addition of 20 wt% MSWI FA had been shown to produce high compressive strength and low water absorption after sintering at T = 960 °C (Haiying et al. 2007). The amount of leachable toxic heavy metals reduced to one-tenth compared with the FA itself. Acceptable durability was also found possible using 50 wt% MSWI ash. (Also see roofing tiles under "Bricks".)

Glass Materials

As FA is generally composed of oxides, such as SiO_2, Al_2O_3, CaO, and Fe_2O_3, it had also received attention as a low cost material for the manufacture of glass–ceramics and glass materials (Chu et al. 2013). FA with a silica content of >50 wt% can be used as a substitute for silica in the manufacturing process of silicate glass, because the glass product can immobilize and stabilize the heavy metals in the matrix (Park et al. 2009). Several studies also reported on the recycling of MSWI FA as glass-based materials (Boccaccini et al. 1997; Romero et al. 1999). Physical and mechanical properties of these materials were found very promising compared with literature values for glasses and ceramics derived from other waste and non waste sources (Erol et al. 2007).

Bricks

Although FA is commonly used in cements, it has rarely been applied for bricks. Recently, Cicek and Tanriverdi (2007) showed that FA might improve the compressive strength of bricks and make them more resistant to frost (Lingling et al. 2005). It was found that coal FA bricks are as durable as clay bricks and, in certain aggressive environments, perform better than clay bricks. In the brick production, FA can be used in the range of 40–70 wt%. In the field of cultural heritage, there is almost no research about FA addition in the manufacture of replacement bricks for use in restoration work. A mandatory criterion for bricks employ in this field is that the bricks must have the same physical and mechanical parameters as the original bricks used in the building being restored and that is not easy when employing FA. In addition to brick products, Shulman (2007) included FA in the manufacture of lightweight roofing products, such as rigid roofing tiles. These roofing tiles have the advantage to be lighter than clay products and providing a good fire resistance, making them an excellent replacement for cedar shake roofing in high fire danger areas.

Other Products

The manufacture of sintered lightweight aggregates from FA is a good way to utilize a large quantity of FA. FA had been employed for roadway construction, embankment soil stabilisation, subgrade-based course material, aggregate filler, bituminous pavement additive, and mineral filler for bituminous concrete (Kim et al. 2012). Use of coal FA as aggregates in road works was estimated to reduce construction costs by about 10–20 % (Ahmaruzzaman 2010). Very recently, waste paper sludge FA, which contains more than 20 % lime (CaO), was employed as clay soil stabilizer. It was shown that addition of 10 % of this FA increased the value of compressive strength from 0 to 28 days of curing periods compared with control (unstabilised) soil (Khalid et al. 2012). Regardless of composition, coal FA is a raw material suitable for use as lightweight aggregate in the production of concrete with good compressive strength (Sarabèr et al. 2012).

Filling Material

Mine back filling was shown to be an attractive option for FA recycling for the plants located near the coal mine. Good potential for this application was demonstrated for those areas where sand is scarce. FAs of all origin were employed also for mine void filling. During 1999–2000, India used about 60,000 tonnes of ash for backfilling underground mines (Mathur 2000).

Catalyst Constituent

The application of coal FA as heterogeneous catalyst material had received a great deal of attention (Wang 2008). Heterogeneous catalysis is appealing, because it is often easier to recover the catalyst after reaction completion than with homogeneous catalysis. With heterogeneous catalysis, catalytic materials (usually a metal) can be supported on other materials. The activity of the supported catalyst depends on both the active component and its interaction with the support matter. The most used catalyst support material covers a range of metal oxides, such as Al_2O_3, SiO_2, TiO_2, and MgO (Wang and Lu 2007). Thanks to its composition, coal FA can also be employed as support. For example, Wang and Lu (2007) used FA as support material for nickel in the application of CO_2 methane reforming and found them stable and capable of high conversion with activities close to those achieved by the well-known systems. However, coal FA had also been employed as the catalytically active component. Recently, the synthesis of a highly active, nano-crystalline, thermally stabilised, solid acid catalyst had been realised using a sol–gel technique for loading sulphated zirconia onto coal FA. The catalyst demonstrated high conversions of benzene (87 %) and toluene (93 %) (Khatri et al. 2010). FA had also been used as the active component in the catalytic oxidation of volatile organic compounds (Born et al. 1993). In addition, acid activation of coal FA was performed to produce a nano-crystalline, solid acid catalyst for the esterification of salicylic acid with acetic anhydride and methanol to produce acetylsalicylic acid (aspirin) and methyl salicylate (oil of wintergreen) in a single step without the need for solvents (Khatri and Rani 2008).

Mesoporous Silica

Coal FA and RHA had been investigated for their potential as silica source for the manufacture of mesoporous silica. Porous solids are materials that can be used in a variety of separation and catalytic processes. Due to high cost and toxicity of conventional reagents, there is a great demand for new materials for the manufacture of mesoporous materials (Kumar et al. 2001). The methods of synthesis are very similar to those employed for zeolite manufacture. The main difference is that, during or prior to crystallisation, structure templates are added. Moreover, there are a few studies on the potential applications for such materials. In addition, nearly pure amorphous silica can be obtained from RHA (Kalapathy et al. 2000).

Polymer Composite Material

Different studies had been conducted to find suitable applications for FA with various polymers (Yang et al. 2006; Chaowasakoo and Sombatsompop 2007) for various reasons: plastics production requires a remarkable consumption of oil-based resources, which are notoriously non-renewable. For polymers, many filler materials had been developed in the past. The main fillers for polymers are carbon black, calcium carbonate, talc, kaolin, mica, and wollastonite, but the use of highly dispersible silica could improve the mechanical performance of the composite further. Rigid polyurethane foams are used in several industrial areas, such as thermal insulation and automotive sectors, due to their interesting properties, such as low thermal conductivity, low density, high abrasion resistance, and good shock absorption. However, these materials are very combustible; their flames spread fast, and they exhibit high heat release rates. Rubbers have immense potential to englobe FA materials compared with other rubber cheapeners. Clearly, developing a new class of environment-friendly structural materials by including an inert produced from waste also enhances the sustainability of the final composite.

FA-filled polymer composites are becoming attractive, because of their wide applications and low cost. In particular, Sreekanth et al. (2011) showed that coal FA increases mechanical and thermal performance of the final composite. For example, Bosio et al. (2012) showed that COSMOS material, which is inertised MSWI FA formed by silica, calcium carbonate, and sulphate, has an interesting composition for use as polymer filler. The inert inserted into polymer matrices integrated very well and it was found to disperse homogeneously with nylon. Deepthi et al. (2010) used FA as filler in high-density polyethylene to develop lightweight, reinforced composites. Wood FA was used as reinforcing material for polymers. It gives rise to vastly improved mechanical properties over that of the base polymer (Ayswarya et al. 2012). Oil FA (produced from oil fuel power plants) showed interesting properties to be used as a polymer-filler composite (Khan et al. 2011). FA can be added directly to polyurethane foams to increase its fire resistance. Usta (2012) showed that FA significantly enhanced fire resistance and thermal stability of polyurethane foams due to the inert nature of FA with fire. A similar effect was found with polycarbonate (Soyama et al. 2007).

Rubber Enforcement

Rubbers have immense potential to englobe FA materials compared with other rubber cheapeners. FA behaves uniquely as it does not show any property deterioration, even, at high loading where conventional cheapeners fail. It thus provides a novel route to its reutilisation in significant quantities. RHA was employed as a filler in rubbers (Arayapranee et al. 2005), especially in epoxidised natural rubber, to ameliorate the tensile strength, tear strength, hardiness, and cure characteristics, of the rubber. Generally, RHA is not as good as fumed silica and carbon black in improving some rubber properties, but it can replace them in some occasions (Zeng et al. 2012).

3.4.7 Closing Remarks

It is evident that FA has a great potential in the environmental applications. FA is an interesting alternative to replace activated carbon, building materials, zeolites and so on. However, despite the large number of published laboratory data, FA has not been applied at an industrial scale in Europe and mainly for two reasons. Firstly, because the variability in chemistry, mineralogy, and morphology, of different FA hinders its use directly. Secondly, because FA may create a problem of secondary environmental pollution in that the surface layer of FA particles may contain significant amounts of readily leachable materials. Europe is currently facing a social and economical crisis, which is partly also due to the cost of raw materials that is growing continuously due to their scarcity in the EU. On the other hand, environmental consideration and public concern request the society to move toward sustainable development and cleaner technologies. In order to maximize productivity and to minimize waste, it is important to recognize when secondary products can be recovered from a processing step. The potentialities to use FA as raw material must be given intensive attention abroad the nations. Resource preservation must be focused with priority on the recovery of input resources, also offering new economical opportunities.

3.5 Environmental Advantages in Fly Ash Reuse

The current use of new resources and the disposal of waste materials must be reconsidered in view of a sustainable development, because the manufacturing industries require enormous amounts of different natural resources, like raw materials (ores), energy, water, etc. In particular, the high amount of waste generated is still far from being used in its totality, neither as product nor as by-product, making technological alternatives needed to reduce its possible impact on the environment. Despite sometimes posing environmental pollution, FA can be considered an important raw material for various applications. The utilisation of FA in building, removal of toxic

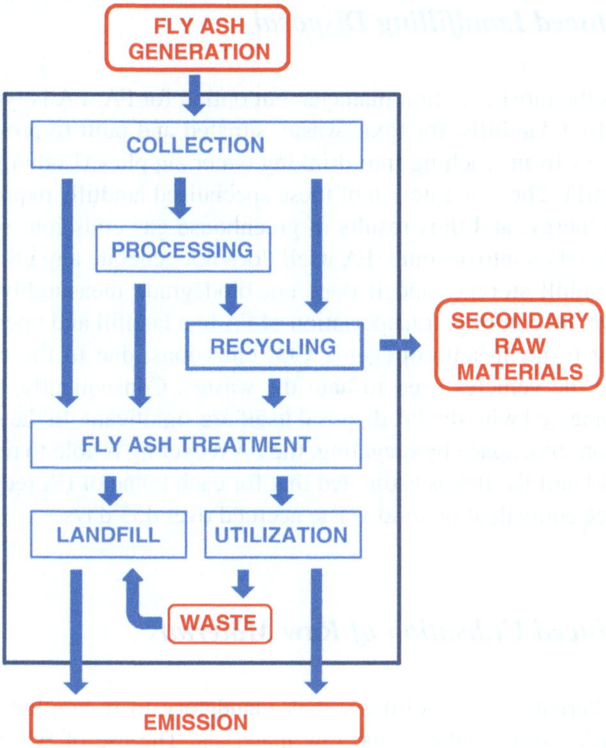

Fig. 3.20 Environmentally and economically sustainable fly ash management (Sabbas et al. 2003)

materials like organics or heavy metals, dyes, zeolite synthesis, and so on, can help a great deal in preserving natural resources and reduction of environmental pollution.

As depicted in Fig. 3.20, an environmentally and economically sustainable FA management includes all the reported processes, starting from FA generation to landfilling (Sabbas et al. 2003; Ludwig et al. 2003), i.e.:

- FA generation: all combustion process generates FA
- FA collection, including source separation into different material streams
- Processing, either to prepare FA for reuse or to modify FA characteristics toward final land disposal
- Recycling: production of secondary materials from FA
- FA treatment, including several technologies (diffusely addressed in this review)
- FA utilisation, covering all the utilisation options of FA after processing
- Landfilling

The recycling and reuse of FA have a number of environmental benefits, including reduced landfilling disposal, reduced utilisation of virgin resources, and reduced amounts of greenhouse gases. A few of the environmental benefits of FA employ are described below.

3.5.1 Reduced Landfilling Disposal

Landfilling is the most common management option for FA. FA is typically placed in specialised FA landfills, for toxic waste, situated and built to prevent trace elements in the FA from leaching into drinking water supplies (Ludwig et al. 2000b; Baur et al. 2001). The construction of these specialised landfills requires additional material and energy, and thus results in greenhouse gas emission, which actually also should be taken into account. FA itself does not generate any utility emissions, because of landfill storage, and it does not biodegrade measurably in anaerobic conditions. On the contrary, transportation of FA to a landfill and operation of landfill equipment result in anthropogenic CO_2 emissions, due to the combustion of fossil fuels in the vehicles used to haul the wastes. Consequently, the landfilling emissions connected with the FA disposal itself are significant. In the same way that households conserve space by recycling, the FA recycling is able to reduce the need for additional landfills. It was estimated that for each tonne of FA recycled, it saves a landfill space equivalent of solid waste accrued over 455 days.

3.5.2 Reduced Utilisation of Raw Materials

To give an alternative to landfilling, it is mandatory to re-employ the otherwise deposited wastes as secondary-hand raw materials. The use of FA in construction results in significant savings of the natural resources, due to savings in cost for aggregates. FA ash has chemical and physical properties that make if suitable for use in building applications. It is mainly due to its composition that includes silica, alumina, iron, and other oxides. These characteristics allow FA to replace Portland cement in concrete products, or, to be used in the production of bricks. Often, FA needs no processing and it can be delivered from the power plants directly to the ready-mix producer. This means that the natural materials – like sand and fine aggregates – can be saved for other uses. This extends the service life of current virgin sources and delays the need to find new sources. This may be very important for Europe, where more than 20 % of the resources used are imported (Bruckner et al. 2012). Green building emphasizes the use of "recycled content" as a part of the components needed with many different structures. The green building initiative encourages using FA in concrete and other products that contain recycled materials.

3.5.3 Reduction in Greenhouse Gases

Agricultural lime contributes a prime role in the global fluxes of the greenhouse gases, such as carbon dioxide, nitrous oxide, and methane. Its contribution through emission of CO_2 to the atmosphere was estimated to be 9 Tg (1 Tg = 10^{12} g = 106 metric tonnes) of CO_2 from an approximate 20 Tg of applied agricultural lime in

2001 (West and McBride 2005). Many researchers discussed opportunities for reducing potential global warming by altering the agronomic practices (Robertson et al. 2000). One recommendation was to the utilisation of FA instead of agricultural lime for minimising global warming (Bernoux et al. 2003; West and McBride 2005). An experimental study revealed that 1 tonne of FA could sequester up to 26 kg of CO_2, i.e., 38.18 tonnes of FA per tonne of CO_2 sequestered (Montes-Hernandez et al. 2009). Therefore, the use of FA instead of lime as soil ameliorant can reduce the net CO_2 emission and thereby lessen global warming.

In the building materials sector, the use of FA to replace cement can decrease cement in concrete mixture and results in decreasing CO_2 from the cement production. During the production of Portland cement, energy is consumed and carbon dioxide is being released. Decomposition of limestone into equimolar quantities of CaO and CO_2 is an essential process in the production of cement clinker. Through stoichiometry, it can be shown that for every tonne of CaO produced, 0.44 tonnes of carbon dioxide are generated. This CO_2 is thought to be a major contributor to the greenhouse effect and the global warming of the planet (Ferreira et al. 2003; Tietenberg 2003). In addition, the decomposition of limestone is an energy-consuming (endothermic) process and the high energy consumption of the kiln adds an emission load depending on the nature of the fuel and its location. However, using FA as a substitute for virgin materials reduces the emissions of greenhouse gases. By using recycled FA in cement production, carbon dioxide emissions are decreased, because FA doesn't need to be "preheated," since it was already heated during the combustion process. According to one estimate, use of 1 tonne of FA in concrete will avoid 2 tonnes of CO_2 emitted from cement production and reduces green-house effect and global warming (Krishnamoorthy 2000; Naik and Tyson 2000). It was calculated that by using a mix of 25 % coal FA, a reduction in overall concrete greenhouse gas emissions of 13–15 % is possible with typical concrete mixes (Flower and Sanjayan 2007). In addition, MSWI FA can be made to react with atmospheric CO_2 to form materials with improved properties. Accelerated carbonation imparts chemical and mineralogical changes to the FA, which reduce their environmental impact through cementation by means of carbonate precipitation (Jiang et al. 2009). Thus, accelerated carbonation can be used not only for FA treatment, but also for the capture of CO_2.

3.5.4 Reduction in Water Consumption

Using FA as partial replacement for cement reduces the amount of water required in the concrete mixing process. It was calculated (Mehta 2001) that the concrete industry uses one trillion liters of water every year in mixing alone. It was also estimated that concrete made with coal ash requires at least 10 % less water to produce a long lasting product. Moreover, the mixing water amount may be cut in half by using better aggregate grading practices and by expanding the use of mineral admixtures, including FA (Mehta 2001).

3.5.5 Other Cost Saving Aspects

Coal and FGD FAs and RHA are generally available practically for free at power plant stations and hence only costs related with its transportation have to be considered when using FA as fill material in construction works. For example, use of coal FA in road works results in reduction of construction cost by about 10–20 % (ACAA 2011).

Use of coal FA as by-product aggregate in the manufacture of lightweight construction products presents its main advantage in economic saving when saving weight. This is especially noticeable when products, such as bricks, made from FA, because FA-based bricks weigh, on average, one-third less than conventional clay-fired bricks, thus, also lower shipping costs and raising profit margins.

For FA bricks manufacturing units, the main raw material can be FA that is generated by thermal power stations in large quantities as for example MSWI FA (Zhang and Qi 2012). Then, another economical advantage in the use of FA is the abundance of low-cost FA available to make the bricks, yielding an excellent product. Only transportation charges for FA to the manufacturing unit and cost of the raw material must be considered.

3.5.6 Closing Remarks

FA can be considered a valuable resource. Possible applications need to be studied to facilitate its employ to all areas of economic interest. To increase the acceptance and use of FA in all available applications, more analysis of long-term economic and environmental impacts, possibly through employing life cycle assessments (LCAs), is needed. Indeed, there is a scarcity in the knowledge on the environmental impact of FA as a raw material (Blissett and Rowson 2012).

3.6 Conclusion

During past decades, several treatments have been developed to deal with FA-specific problems. With time, the conventional palette of FA treatments (separation, solidification, stabilization, immobilization, heating) have been widened enormously by employing electrochemistry, mechanochemistry, chemical reaction, chemical fixation, thermal plasma, microwave energy, up to, organic pollutants abatement using e.g. non-thermal plasma, UV, hydrothermal processing, or, supercritical water oxidation. Nowadays, research also focuses on FA destinies other than landfill deposition, or, use as compensatory building material, or, upgrading towards secondary raw ores, by exploiting the chemical or physical potential of certain FA constituents (cf. aluminosilicate glass is a suitable material for zeolite synthesis, carbon in coal or rice husk absorbs well). Also worth pursuing is the recovery of agriculture-relevant nutrients (N, P, S) from biomass- or coal-based FAs. The sophisticated reuse of FA promises more environmental advantages and furthers a sustainable utilization of raw material sources for future societies.

Acknowledgement The authors from the Brescia University (Italy) acknowledge LIFE+ financial instrument of the European Community (LIFE+ 2008 project ENV/IT/434 and LIFE+ 2011 project ENV/IT/256) and RPWJ Struis thanks Prof. Chr. Ludwig (EPFL, Switzerland) for valuable discussions.

References

ACAA (2010) American Coal Ash Association: Coal Combustion Product (CCP) Production & Use ACAA, Aurora CO

ACAA (2011) American Coal Ash Association: ash at work. Applications, science and sustainability of coal ash. Lesser Pubblications, New York

Adesanya DA, Raheem AA (2009) Development of corn cob ash blended cement. Constr Build Mater 23:347–352

Adesanya DA, Raheem AA (2010) A study of the permeability and acid attack of corn cob ash blended cements. Constr Build Mater 24:403–409

Adriano DC, Page AL, Elseewi AA, Chang AC, Straugham I (1980) Utilization and disposal of fly-ash and coal residues in terrestrial ecosystem; a review. J Environ Qual 9:333–344

Adriano DC, Weber J, Bolan NS, Paramasivan S, Koo Bon-Jun, Sajwan KS (2002) Effects of high rates of coal fly ash on soil, turfgrass, and groundwater quality. Water Air Soil Pollut 139:365–385

Agrawal DK (1998) Microwave processing of ceramics. Curr Opin Solid State Mater Sci 3:480–485

Ahmaruzzaman M (2010) A review on the utilization of fly ash. Prog Energy Combust Sci 36:327–363

Ahmaruzzaman M (2011) Industrial wastes as low-cost potential adsorbents for the treatment of wastewater laden with heavy metals. Adv Colloid Interface Sci 166:36–59

Akgerman A, Zardkoohi M (1996) Adsorption of phenolic compounds on fly ash. J Chem Eng Data 41:185–187

Aksu Z, Yener J (2001) A comparative adsorption/biosorption study of mono-chlorinated phenols onto various sorbents. Waste Manag 21:695–702

Alba N, Gasso S, Lacorte T, Baldasano JM (1997) Characterization of municipal solid waste incineration residues from facilities with different air pollution control systems. J Air Waste Manage Assoc 47:1170–1179

Alemany LJ, Jimenez MC, Larrubia MA, Delgado F, Blasco JM (1996) Removal of phenol from aqueous solution by adsorption on to coal fly ash. Adsorpt Sci Technol 13:527–536

Al-Shawabkeh A, Matsuda H, Hasatani M (1995) Comparative reactivity of treated FBC- and PCC-fly ash for SO2 removal. Can J Chem Eng 73:678–685

An DM, Guo YP, Zou B, Zhu YC, Wang ZC (2011) A study on the consecutive preparation of silica powders and active carbon from rice husk ash. Biomass Bioenerg 35:1227–1234

Andersson P, Rappe C, Maaskant O, Unsworth JF, Marklund S (1998) Low temperature catalytic destruction of PCDD/F in flue gas from waste incineration. Organohalog Compd 36:109–112

Andreola F, Barbieri L, Hreglich S, Lancellotti I, Morselli L, Passarini F, Vassura I (2008) Reuse of incinerator bottom and fly ashes to obtain glassy materials. J Hazard Mater 153:1270–1274

Apak R, Atun G, Guclu K, Tutem E (1996) Sorptive removal of cesium-137 and strontium-90 from water by unconventional sorbents. II. Usage of coal fly ash. J Nucl Sci Technol 33:396–402

Apak R, Tutem E, Hugul M, Hizal J (1998) Heavy metal cation retention by unconventional sorbents (red muds and fly ashes). Water Res 32:430–440

Appleton TJ, Gloder RI, Kingman SW, Lowndes IS, Read AG (2005) Microwave technology for energy-efficient processing of waste. Appl Energy 81:85–113
Arayapranee W, Naranong N, Rempel GL (2005) Application of rice husk ash as fillers in the natural rubber industry. J Appl Polym Sci 98:34–41
Arthur MA, Zwick TC, Tolle DA, Voris PV (1984) Effects of fly ash on microbial CO2 evolution from an agricultural soil. Water Air Soil Pollut 22:209–216
Asokan P, Saxena M, Asolekar SR (2005) Coal combustion residues environmental implications and recycling potentials. Resour Conserv Recycl 43:239–262
ASTM (2005) Standard specification for coal fly ash and raw or calcined natural pozzolan for use in concrete (C618-05). In: Annual book of ASTM standards, concrete and aggregates, vol 04.02. American Society for Testing Materials, Philadelphia
Aubert JE, Husson B, Vaquier A (2004) Use of municipal solid waste incineration fly ash in concrete. Cem Concr Res 34:957–963
Aubert JE, Husson B, Sarramone N (2006) Utilization of municipal solid waste incineration (MSWI) fly ash in blended cement. Part 1: Processing and characterization of MSWI fly ash. J Hazard Mater 136:624–631
Auer PO, Eichler B, Ludwig C, Stucki S, Wochele J (1999) Recycling heavy metals by the method of fractionated condensation. In: 4th world congress – recovery, recycling, re-integration, vol IV, Geneve, pp 328–333
Ayala J, Blanco F, Garcia P, Rodriguez P, Sancho J (1998) Austrian fly ash as a heavy metals removal material. Fuel 77:1147–1154
Ayswarya EP, Abraham BT, Thachil ET (2012) HDPE-ash nanocomposites. J Appl Polym Sci 124:1659–1667
Bakharev T (2006) Thermal behaviour of geopolymers prepared using class F fly ash and elevated temperature curing. Cem Concr Res 36:1134–1147
Banerjee SS, Jayaram RV, Joshi MV (2003) Removal of nickel(II) and zinc(II) from wastewater using fly ash and impregnated fly ash. Sep Sci Technol 38:1015–1032
Banerjee SS, Joshi MV, Jayaram RV (2004) Removal of Cr(VI) and Hg(II) from aqueous solutions using fly ash and impregnated fly ash. Sep Sci Technol 39:1611–1629
Barbieri L, Lancellotti I (2004) Incinerator waste as secondary raw material: examples of applications in glasses, glass-ceramics and ceramics. Geol Soc Spec Pub 236:423–433
Bastardie F, Cannavacciuolo M, Capowiez Y, de Dreuzy J-R, Bellido A, Cluzeau D (2002) A new simulation for modeling the topology of earthworm burrow systems and their effects on macropore flow in experimental soils. Biol Fertil Soils 36:161–169
Basu M, Pande M, Bhadoria PBS, Mahapatra SC (2009) Potential fly-ash utilization in agriculture: a global review. Progr Nat Sci 19:1173–1186
Batabyal D, Sahu A, Chaudhuri SK (1995) Kinetics and mechanism of removal of 2,4- dimethyl phenol from aqueous solutions with coal fly ash. Sep Technol 5:179–186
Baur I, Ludwig C, Johnson CA (2001) The leaching behavior of cement stabilized air pollution control residues: a comparison of field and laboratory investigations. Environ Sci Technol 35:2817–2822
Bayat B (2002a) Comparative study of adsorption properties of Turkish fly ashes. I. The case of nickel(II), copper(II) and zinc(II). J Hazard Mater 3897:1–23
Bayat B (2002b) Comparative study of adsorption properties of Turkish fly ashes. II. The case of chromium(VI) and cadmium(II). J Hazard Mater 3898:1–16
Bayat B (2002c) Combined removal of zinc (II) and cadmium (II) from aqueous solutions by adsorption onto high-calcium Turkish fly ash. Water Air Soil Pollut 136:69–92
Bayuseno AP (2006) Mineral phases in raw and processed municipal waste incineration residues-towards a chemical stabilisation and fixation of heavy metals. PhD thesis, Bochum, Germany
Bayuseno AP, Schmahl WW (2011) Characterization of MSWI fly ash through mineralogy and water extraction original research article resources. Conserv Recycl 55:524–534
Beagle EC (1981) Rice husk conversion to energy. FAO, Rome
Belevi H, Moench H (2000) Factors determining the element behavior in municipal solid waste incinerator. 1. Field studies. Environ Sci Technol 34:2501–2506

Belviso C, Cavalcante F, Ragone P, Fiore S (2010) Immobilization of Ni by synthesising zeolite at low temperatures in a polluted soil. Chemosphere 78:1172–1176
Benezet JC, Adamiec P, Benhassaine A (2008) Relation between silico-aluminous fly ash and its coal of origin. Particuology 6:85–92
Bernardo E, Bonomo E, Dattoli A (2010) Optimisation of sintered glass–ceramics from an industrial waste glass. Ceram Int 36:1675–1680
Bernoux M, Volkoff B, Carvalho MS (2003) CO2 emissions from liming of agricultural soils in Brazil. Global Biogeochem Cycles 17:18–21
Beveridge TJ, Murray RGE (1980) Sites of metal deposition in the cell wall of Bacillus subtilis. J Biotechnol 141:876–887
Bhattacharjee U, Kandpal TC (2002) Potential of fly utilization in India. Energy 27:151–166
Bhattacharya AK, Mandal SN, Das SK (2006) Adsorption of Zn(II) from aqueous solution by using different adsorbents. Chem Eng J 123:43–51
Bhattacharya AK, Naiya TK, Mandal SN, Das SK (2008) Adsorption, kinetics and equilibrium studies on removal of Cr(VI) from aqueous solutions using different low-cost adsorbents. Chem Eng J 137:529–541
Binici H, Yucegok F, Aksogan O, Kaplan H (2008) Effect of corncob, wheat straw, and plane leaf ashes as mineral admixtures on concrete durability. J Mater Civ Eng 20:478–484
Biricik H, Akoz F, Turker F, Berktay I (2000) Resistance to magnesium sulfate and sodium sulfate attack of mortars containing wheat straw ash. Cem Concr Res 30:1189–1197
Blissett RS, Rowson NA (2012) A review of the multi-component utilisation of coal fly ash. Fuel 97:1–23
Boccaccini AR, Bucker M, Bossert J, Marszalek K (1997) Glass matrix composites from coal flyash and waste glass. Waste Manage 17:39–45
Bontempi E, Zacco A, Borgese L, Gianoncelli A, Ardesi R, Depero LE (2010a) A new method for municipal solid waste incinerator (MSWI) fly ash inertization, based on colloidal silica. J Environ Monit 12:2093–2099
Bontempi E, Zacco A, Borgese L, Gianoncelli A, Ardesi R, Depero LE (2010b) A new powder filler, obtained by applying a new technology for fly ash inertisation procedure. Adv Sci Technol 62:27–33
Boonserm K, Sata V, Pimraksa K, Chindaprasirt P (2012) Improved geopolymerization of bottom ash by incorporating fly ash and using waste gypsum as additive. Cem Concr Compos 34:819–824
Born JGP, Mulder P, Louw R (1993) Fly ash mediated reactions of phenol and monochlorophenols: oxychlorination, deep oxidation, and condensation. Environ Sci Technol 27:1849–1863
Bosio A, Ardesi R, Zacco A, Brisotto M, Gianoncelli A, Gelfi M, Depero LE, Bontempi E (2012) Reuse of COSMOS filler, obtained by Municipal Solid Waste Incineration fly ash inertization, in plastic materials. In: Proceedings of third international conference on "industrial and hazardous waste management", Chania, Crete, Greece, 12–14 Sept 2012
Bosshard PP, Bachofen R, Brandl H (1996) Metal leaching of fly ash from municipal waste incineration by Aspergillus. Niger Environ Sci Technol 30:3066–3070
Bouché MB, Al-Addan F (1997) Earthworms, water infiltration and soil stability: some new assessments. Soil Biol Biochem 29:441–452
Bouma J (1991) Influence of soil macroporosity on environmental quality. Adv Agron 46:1–37
Bowman RS (2003) Applications of surfactant-modified zeolites to environmental remediation. Microporous Mesopor Mater 61:43–56
Breck DW (1984) Ion exchange reactions in zeolites. In: Zeolite molecular sieves, structure, chemistry, and use, vol 7. Robert E. Krieger Publishing, Malabar, p 245213
Bridgeman TG, Darvell LI, Jones JM, Williams PT, Fahmi R, Bridgewater AV, Barraclough T, Shield I, Yates N, Thain SC, Donnison IS (2007) Influence of particle size on the analytical and chemical properties of two energy crops. Fuel 86:60–72
Bruckner M, Giljum S, Lutz C, Wiebe KS (2012) Materials embodied in international trade – global material extraction and consumption between 1995 and 2005. Global Environ Change 22:568–576

Burke M (2007) CCP experts gather in India. In: Ash at work, vol 2. American Coal Ash Association, Aurora, pp 17–19
Campbell DJ, Fox WE, Aitken RL, Bell LC (1983) Physical characteristics of sands amended with fly ash. Aust J Soil Res 21:147–154
Cerevelli S, Petruzzelli G, Perna A, Menicagli R (1986) Soil nitrogen and fly ash utilization: a laboratory investigation. Agrochemica 30:27–33
Chandrasekhar S, Pramada PN (2006) Rice husk ash as an adsorbent for methylene blue-effect of ashing temperature. Adsorption 12:27–43
Chandrasekhar S, Pramada PN, Praveen L (2005) Effect of organic acid treatment on the properties of rice husk silica. J Mat Sci 40:6535–6544
Chang AC, Lund LJ, Page AL, Warneke JE (1977) Physical properties of fly ash amended soils. J Environ Qual 6:267–270
Chaowasakoo T, Sombatsompop N (2007) Mechanical and morphological properties of fly ash/epoxy composites using conventional thermal and microwave curing methods. Compos Sci Technol 67:2282–2291
Chareonpanich M, Namto T, Kongkachuichay P, Limtrakul J (2004) Synthesis of ZSM- 5 zeolite from lignite fly ash and rice husk ash. Fuel Process Technol 85:1623–1634
Chaturvedi AK, Yadava KP, Pathak KICK, Singh VN (1990) Defluoridation of water by adsorption of fly ash. Water Air Soil Pollut 49:51–61
Chaudhary DS, Jollands MS (2004) Characterization of rice hull ash. J Appl Polym Sci 93:1–8
Chen J, Li Y (2006) Coal fly ash as an amendment to container substrate for Spathiphyllum production. Bioresour Technol 97:1920–1926
Chen Y, Zhu Y, Wang Z, Li Y, Wang L, Ding L, Gao X, Ma Y, Guo Y (2011) Application studies of activated carbon derived from rice husks produced by chemical-thermal process – a review. Adv Colloid Interface Sci 163:39–52
Chen-Tan NW, van Riessen A, Chi VLY, Southam DC (2009) Determining the reactivity of a fly Ash for production of geopolymer. J Am Ceram Soc 92:881–887
Cheung KICK, Venkitachalam TH (2000) Improving phosphate removal of sand infiltration system using alkaline fly ash. Chemosphere 41:243–249
Chimenos J, Fernández A, Cervantes A, Miralles L, Fernández M, Espiell F (2005) Optimizing the APC residue washing process to minimize the release of chloride and heavy metals. Waste Manage 25:686–693
Chindaprasirt P, Rukzon S, Sirivivatnanon V (2008) Resistance to chloride penetration of blended Portland cement mortar containing palm oil fuel ash, rice husk ash and fly ash. Constr Build Mater 22:932–938
Choi W, Hong SJ, Chang YS, Cho Y (2000) Photocatalytic degradation of polychlorinated dibenzo-p-dioxins on TiO2 film under UV or solar light irradiation. Environ Sci Technol 34:4810–4815
Chou SY, Lo SL, Hsieh CH, Chen CL (2009) Sintering of MSWI fly ash by microwave energy. J Hazard Mater 163:357–362
Christensen TH, Lundtrop K, Jensen L, Sorensen A, Mogensen EPB (2001) Stabilization of waste incinerator APC-residues with FeSO4. In: Proceedings of eighth international landfill symposium, Sardinia, Cagliari, Italy
Chu TC, Wang KS, Lin KL, Chien CC, Chen JH (2013) Synthesis of waste-derived glass-ceramics from MSWI fly ash and EAF dust: kinetics of nucleation and crystallization. Environ Prog Sustain Energy 32(3):480–488. doi:10.1002/ep.11647
Ciccu R, Ghiani M, Serci A, Fadda S, Peretti R, Zucca A (2003) Heavy metal immobilization in the mining-contaminated soils using various industrial wastes. Miner Eng 16:187–192
Cicek T, Tanriverdi M (2007) Lime based steam autoclaved fly ash bricks. Constr Build Mater 21:1295–1300
Comans R, Zuiver E, Geelhoed P, Hoede D (2003) Characterisation of leaching properties of C-fix products and components. ECN, Petten
Cornell RM, Schwertmann U (2003) The iron oxides: structure, properties, reactions, occurrences and uses, 2nd edn. Wiley-VCH, Weinheim

Cothern CR, Smith JE (1987) Environmental radon. Plenum Press, New York, p 363
Daifullah AEH, Gad H (1998) Sorption of semi-volatile organic compounds by bottom and fly ashes using HPLC. Adsorpt Sci Technol 16:273–283
Damaris NM, Paul MS, Rachel MN, Geoffrey NK (2002) Adsorption and detection of some phenolic compounds by rice husk ash of Kenyan origin. J Environ Monit 4:978–984
Dasmahapatra GP, Pal TK, Bhadra AK, Bhattacharya B (1996) Studies on separation characteristics of hexavalent chromium from aqueous solution by fly ash. Sep Sci Technol 31:2001–2009
Davidovits J (1994) Geopolymers: inorganic polymeric new materials. J Mater Educ 16:91–139
Davini P (1995) Investigation of flue gas desulphurization by fly ash and calcium hydroxide mixtures. Resour Conserv Recycl 15:193–201
Davini P (1996) Investigation of the SO2 adsorption properties of Ca(OH)2-fly ash systems. Fuel 75:713–716
Davini P (2002) Flue gas treatment by activated carbon obtained from oil-fired fly ash. Carbon 40:1973–1979
De Boom A, Degrez M (2012) Belgian MSWI fly ashes and APC residues: a characterisation study. Waste Manage 32:1163–1170
Deepthi MV, Sharma M, Sailaja RRN, Anantha P, Sampathkumaran P, Seetharamu S (2010) Mechanical and thermal characteristics of high density polyethylene-fly ash cenospheres composites. Mater Des 31:2051–2060
Demirbas A (2005) Potential applications of renewable energy sources, biomass combustion problems in boiler power systems and combustion related environmental issues. Prog Energy Combust Sci 31:171–192
Denuex-mustin S, Roussel-Debet S, Mustin C, Henner P, Munier-Lamy C, Colle C, Berthein J, Garnier-Laplace J, Leyval C, Mobilite et al (2003) Transfer des elements en traces: in.uence des microorganisms du so. Tec et Doc, Paris p 282
Derie R (1996) A new way to stabilise fly ash from municipal incinerators. Waste Manage 16:711–716
Diamadopoulos E, Loannidis S, Sakellaropoulos GP (1993) As(V) removal from aqueous solutions by fly ash. Water Res 27:1773–1777
Dombrowski K, Buchwald A, Weil M (2007) The influence of calcium content on the structure and thermal performance of fly ash based geopolymers. J Mater Sci 42:3033–3043
DSIRI (1992) Department of Scientific and Industrial Research India. Technology evaluation in rice milling industry, New Delhi, India
Dubini MM (1965) J Phys Chem 39:697–704
Dutta B, Basu JK, Das Gupta S (2003) Removal of cresol from aqueous solution using fly ash as adsorbent: experiments and modeling. Sep Sci Technol 38:1345–1360
Duxson P, Provis JL, Lukey GC, Mallicoat SW, Kriven WM, van Deventer JSJ (2005) Understanding the relationship between geopolymer composition, microstructure and mechanical properties. Colloid Surf A 269:47–58
Dwivedi VN, Singh NP, Das SS, Singh NB (2006) A new pozzolanic material for cement industry: bamboo leaf ash. Int J Phys Sci 1:106–111
Dwivedi S, Tripathi RD, Srivastava S, Mishra S, Shukla MK, Tiwari KK, Singh R, Rai UN (2007) Growth performance and biochemical responses of three rice (Oryza sativa L.) cultivars grown in fly-ash amendment soil. Chemosphere 67:140–151
Ecke H (2003) Sequestration of metals in carbonated municipal solid waste incineration (MSWI) fly ash. Waste Manage 23:631–640
Ecke H, Sakanakura H, Matsuto T, Tanaka N, Lagerkvist A (2000) State-of-the-art treatment processes for municipal solid waste incineration residues in Japan. Waste Manag Res 18:41–51
ECOBA (2008) European Coal Combustion Products Association: production and utilisation of CCPs in 2008 in Europe (EU 15)
Eighmy TT, Eusden JD, Krzanowski JE, Dominggo DS, Stampfli D (1995) Comprehensive approach toward understanding element speciation and leaching behaviour in municipal solid waste incineration electrostatic precipitator ash. Environ Sci Technol 29:629–646

Eighmy TT, Crannell BS, Butler LC, Cartledge FK, Emery EF, Oblas D, Krzanowski JE, Eusden JD, Shaw EL, Francis CA (1997) Heavy metal stabilisation in municipal solid waste combustion dry scrubber residue using soluble phosphate. Environ Sci Technol 37:3330–3338

Elinwa AU, Mahmood YA (2002) Ash from timber waste as cement replacement material. Cem Concr Compos 24:219–222

EPA (1997) Mercury study report to Congress EPA-452/R-97-10

Erol M, Küçükbayrak S, Ersoy-Meriçboyu A (2007) Characterization of coal fly ash for possible utilization in glass production. Fuel 86:706–714

Eswaran A, Manivannan K (2007) Effect of foliar application of lignite fly ash on the management of papaya leaf curl disease. Acta Hort (ISHS) 740:271–275

Everaert K, Baeyens J (2004) Catalytic combustion of volatile organic compounds. J Hazard Mater 109:113–139

Fail JL Jr, Wochok ZS (1977) Soyabean growth on fly ash amended strip mine spoils. Plant Soil 48:473–484

Fedje KK, Ekberg C, Skarnemark G, Steenari BM (2010) Removal of hazardous metals from MSW fly ash – an evaluation of ash leaching methods. J Hazard Mater 173:310–317

Feng Q, Lin Q, Gong F, Sugita S, Shoya M (2004) Adsorption of lead and mercury by rice husk ash. J Colloid Interface Sci 278:1–8

Fermo P, Cariati F, Pozzi A, Demartin F, Tettamanti M, Collina E, Lasagni M, Pieta D, Puglisi O, Russo U (1999) The analytical characterization of municipal solid waste incinerator fly ash: methods and preliminary results. Fresenius J Anal Chem 365:666–673

Fermo P, Cariati F, Pozzi A, Tettamanti M, Collina E, Pieta D (2000) Analytical characterization of municipal solid waste incinerator fly ash: Part II. Fresenius J Anal Chem 366:267–272

Ferreira C, Ribeiro A, Ottosen L (2003) Possible applications for municipal solid waste fly ash. J Hazard Mater B 96:201–216

Fiedler H (2003) Persistent organic pollutants. Springer, Berlin

Finocchio E, Busca G, Notaro M (2006) A review of catalytic processes for the destruction of PCDD and PCDF from waste gases. Appl Catal B Environ 62:12–20

Fisher GL, Chang DPY, Brummer M (1976) Fly-Ash collected from electrostatic precipitators-microcrystalline structure and mystery of spheres. Science 192:553–555

Flower D, Sanjayan J (2007) Green house gas emissions due to concrete manufacture. Int J Life Cycle Assess 12:282–288

Forestier LL, Libourel G (1998) Characterization of flue gas residues from municipal solid waste combustors. Environ Sci Technol 32:2250–2256

Francois D, Criado C (2007) Monitoring of leachate at a test road using treated fly ash from municipal solid waste incinerator. J Hazard Mater B139:543–549

Frugier P, Godon N, Vernaz E, Larché F (2002) Influence of composition variations on the initial alteration rate of vitrified domestic waste incineration fly-ash. Waste Manage 22:137–142

Fukui K, Arai K, Fukui K, Yoshida H (2006) Phillipsite synthesis from fly ash prepared by hydrothermal treatment with microwave heating. Adv Powder Technol 17:369–382

Fukui K, Fukui K, Yamamoto T, Yoshida H (2007) Effects of microwave irradiation on the crystalline phase of zeolite synthesized from fly ash by hydrothermal treatment. Adv Powder Technol 18:381–393

Gaikwad RW (2004) Removal of Cd (II) from acqueous solution by activated charcoal derived from coconut shell. Electron J Environ Agric Food Chem 3:702–709

Galiano LY, Fernández Pereira C, Vale J (2011) Stabilization/solidification of a municipal solid waste incineration residue using fly ash-based geopolymers. J Hazard Mater 185:373–381

Ganesan K, Rajagopal K, Thangavel K (2008) Rice husk ash blended cement: assessment of optimal level of replacement for strength and permeability properties of concrete. Constr Build Mater 22:1675–1683

Garau MA, Dalmau JL, Felipo MT (1991) Nitrogen mineralization in soil amended with sewage sludge and fly ash. Biol Fertil Soils 12:199–201

García R, Vigil de la Villa R, Vegas I, Frías M, Sánchez de Rojas MI (2008) The pozzolanic properties of paper sludge waste. Constr Build Mater 22:1484–1490

Garg UK, Kaur MP, Garg VK, Sud D (2007) Removal of hexavalent Cr from aqueous solutions by agricultural waste biomass. J Hazard Mater 140:60–68
Gashi ST, Daci NM, Ahmeti XM, Selimi TJ, Hoxha EM (1988) Removal of heavy metals from industrial wastewaters, chemistry for protection of the environment. Elsevier, Amsterdam
Geysen D, Vandecasteele C, Jaspers M, Wauters G (2004) Comparison of immobilisation of air pollution control residues with cement and with silica. J Hazard Mater 107:131–143
Ghosal S, Self SA (1995) Particle size-density relation and cenosphere content of coal fly-ash. Fuel 74:522–529
Giaccio GM, Malhotra VM (1988) Concrete incorporating high volumes of ASTM Class F fly ash. Cem Concr Aggreg 10:88–95
Girón RP, Ruiz B, Fuente E, Gil RR, Suárez-Ruiz I (2013) Properties of fly ash from forest biomass combustion. Fuel 114(3):71–77
Gómez-Barea A, Vilches LF, Leiva C, Campoy M, Fernández-Pereira C (2009) Plant optimisation and ash recycling in fluidised bed waste gasification. Chem Eng J 146:227–236
Goswami D, Das AK (2000) Removal of arsenic from drinking water using modified fly-ash bed. Int J Water 1:61–70
Grover M, Narayanaswamy MS (1982) Removal of hexavalent chromium by adsorption on flay ash. J Inst Eng (India), Part EN: Environ Eng Div 1:36–39
Guo X, Xiang D, Duan G, Mou P (2010) A review of mechanochemistry applications in waste management. Waste Manage 30:4–10
Gupta VK, Ali I (2000) Utilisation of bagasse fly ash (a sugar industry waste) for the removal of copper and zinc from wastewater. Sep Purif Technol 18:131–140
Gupta VK, Ali I (2001) Removal of DDD and DDE from wastewater using bagasse fly ash, a sugar industry waste. Water Res 35:33–40
Gupta VK, Sharma S (2003) Removal of zinc from aqueous solutions using bagasse fly ash – a low cost adsorbent. Ind Eng Chem Res 42:6619–6624
Gupta VK, Mohan D, Sharma S (1998a) Removal of lead from wastewater using bagasse fly ash – a sugar industry waste material. Sep Sci Technol 33:1331–1333
Gupta VK, Sharma S, Yadav IS, Mohan D (1998b) Utilization of bagasse fly ash generated in the sugar industry for the removal and recovery of phenol and p-nitrophenol from wastewater. J Chem Technol Biotechnol 71:180–186
Gupta VK, Mohan D, Sharma S, Park KT (1999) Removal of chromium (VI) from electroplating industry wastewater using bagasse fly ash – a sugar industry waste material. Environmentalist 19:129–136
Gupta VK, Jain CK, Ali I, Chandra S, Agarwal S (2002) Removal of lindane and malathion from wastewater using bagasse fly ash – a sugar industry waste. Water Res 36:2483–2490
Gupta VK, Ali I, Jain CK, Sharma M, Saini VK (2003) Removal of cadmium and nickel from wastewater using bagasse fly ash – a sugar industry waste. Water Res 37:4038–4044
Hagenmaier H, Horch K, Fahlenkamp H, Schetter G (1991) Destruction of PCDD and PCDF in refuse incineration plants by primary and secondary measures. Chemosphere 23:1429–1437
Haiying Z, Youcai Z, Jingyu Q (2007) Study on use of MSWI fly ash in ceramic tile. J Hazard Mater 141:106–114
Helmuth R (1987) Fly ash in cement and concrete. Portland Cement Association, Skokie, pp 36–61
Hinton WS, Lane AM (1991) Characteristics of municipal solid waste incinerator fly ash promoting the formation of polychlorinated dioxins. Chemosphere 22:473–483
Hjelmar O (1996) Disposal strategies for municipal solid waste incineration residues. J Hazard Mater 47:345–368
Hjelmar O, Birch H (1997) Treatment of air pollution control residues from MSW incinerators prior to landfilling. In: Proceedings of Sardinia 97, sixth international landfill symposium, S. Margharita di Pula, Cagliari, Italy, pp 535–544
Hjelmar O, Hansen JB (2006) A one-stage treatment process for improving the leaching characteristics of APC residues from MSW incinerators. In: Proceedings of WASCON 2006, sixth international conference on the environmental and technical implications of construction with

alternative materials – science and engineering of recycling for environmental protection. ISCOWA, Belgrade, pp 97–105. ISBN 86–9088 15–0–6
Hjelmar O, Birch H, Hansen JB (1999) Development of a process for treatment of APC residues from MSW incinerators prior to landfilling. In: Proceedings of Sardinia 99, seventh international landfill symposium I, pp 543–548
Hjelmar O, Birch H, Hansen JB (2001) The VKI process for treatment of APC residues from MSW incinerators: process development and optimisation in pilot scale. In: Proceedings of the second ISWA seminar on waste-to-energy – state of the art and latest news, Malmö, Sweden, 25–26 October
Höller H, Wirsching U (1985) Zeolites formation from fly ash. Foftschritte der Mineralogist 63:21–43
Hollis JF, Keren R, Gal M (1988) Boron release and sorption by fly ash as affected by pH and particle size. J Environ Qual 17:181–184
Hong KJ, Tokunaga S, Kajiuchi T (2000) Extraction of heavy metals from MSW incinerator fly ashes by chelating agents. J Hazard Mater 75:57–73
Hosseini MM, Shao Y, Whalen JK (2011) Biocement production from silicon-rich plant residues: perspectives and future potential in Canada. Biosyst Eng 110:351–362
Hower JC, Maroto-Valer MM, Taulbee DN, Sakulpitakphon T (2000) Mercury capture by distinct fly ash carbon forms. Energy Fuel 14:224–226
Hu SH (2005) Stabilization of heavy metals in municipal solid waste incineration ash using mixed ferrous/ferric sulfate solution. J Hazard Mater 123:158–164
Hu Y, Zhang P, Chen D, Zhou B, Li J, Li XW (2012) Hydrothermal treatment of municipal solid waste incineration fly ash for dioxin decomposition. J Hazard Mater 207–208:79–85
Huang K, Inoue K, Harada H, Kawakita H, Ohto K (2011) Leaching behavior of heavy metals with hydrochloric acid from fly ash generated in municipal waste incineration plants. Trans Nonferrous Met Soc China 21:1422–1427
IAWG (International Ash Working Group), Chandler AJ, Eighmy TT, Hartlén O, Kosson D, Sawell SE, van der Sloot H, Vehlow J (1997) Municipal solid waste incinerator residues, vol 67, Studies in environmental science. Elsevier, Amsterdam
Iretskaya S, Nzihou A, Zahraoui C, Sharrock P (1999) Metal leaching from MSW fly ash before and after chemical and thermal treatments. Environ Prog 18:144–148
Ismail MS, Waliuddin AM (1996) Effect of rice husk ash on high strength. Concr Const Build Mater 10:521–526
Ito T (1996) Vitrification of fly ash by swirling-flow furnace. Waste Manage 16:453–460
Iyer RS, Scott JA (2001) Power station fly ash e a review of value-added utilization outside of the construction industry. Resour Conserv Recycl 31:217–228
Izquierdo M, Querol X (2012) Leaching behaviour of elements from coal combustion fly ash: an overview. Int J Coal Geol 94:54–66
Izumikawa C (1996) Metal recovery from fly ash generated from vitrification process for MSW ash. Waste Manage 16:501–507
Jakob A, Stucki S, Kuhn P (1995) Evaporation of heavy metals during the heat treatment of municipal solid waste incinerator fly ash. Environ Sci Technol 29:2429–2436
Jakob A, Stucki S, Struis RW (1996) Complete heavy metal removal from fly ash by heat treatment – influence of chlorides an evaporation rates. Environ Sci Technol 30:3275–3283
Jala S, Goyal D (2006) Fly ash as a soil ameliorant for improving crop production: a review. Bioresour Technol 97:1136–1147
Jankowski J, Ward CR, French D, Groves S (2006) Mobility of trace elements from selected Australian fly ashes and its potential impact on aquatic ecosystems. Fuel 85:243–256, Special issue: the 21st annual International Pittsburgh Coal Conference
Janos P, Buchtova H, Ryznarova M (2003) Sorption of dyes from aqueous solutions onto fly ash. Water Res 37:4938–4944
Jarvis ST, Brooks TG (1996) The use of PFA: cement pastes in the stabilization of abundant mine workings. Waste Manage 16:135–144
Jayasinghe GY, Tokashiki Y (2006) Utilization of coal fly ash for the production of artificial aggregates as a crop growth medium with acidic "Kunigami Mahji" soils in Okinawa, Japan. J Solid Waste Manag Technol 32:1–9

Jayasinghe GY, Tokashiki Y, Kitou M (2005) Development of synthetic lightweight soil aggregates utilizing coal fly ash and mine clay as waste materials. Tropic Agric Res Ext 8:1–12

Jayasinghe GY, Tokashiki Y, Kitou M (2007) Assessment and utilization of synthetic soil aggregates developed by using coal fly ash and paper waste. In: Conference proceedings of Agronomy Society of America (ASA), Crop Science Society of America (CSSA), and Soil Science Society of America (SSSA), 4–8 November, New Orleans, LA, USA. A Century of Integrating Crops, Soils and Environment, Madison

Jayasinghe GY, Tokashiki Y, Kitou M (2008) Characteristics of synthetic soil aggregates produced by mixing acidic "Kunigami Mahji" soil with coal fly ash and their utilization as a medium for crop growth. Soil Sci Plant Nutr 54:264–276

Jayasinghe GY, Tokashiki Y, Kitou M, Kinjo K (2009) Coal fly ash based synthetic aggregates as potential alternative container substrates for ornamentals. J Plant Nutr Soil Sci 172:720–728

Jayasinghe GY, Tokashiki Y, Kitou M (2010) Use of synthetic soil aggregates as a containerized growth medium component to substitute peat in the ornamental plant production. Arch Agron Soil Sci 56(2):183–199

Jensen DL, Lundtrop K, Christensen TH (2002) Treatment of waste incineration air-pollution control residues with FeSO4: laboratory investigation of design parameters. Waste Manage Res 20:80–89

Jin M, Huang C, Chen L, Sun X, Wang L (2011) Immobilization of MSWI fly ash with geopolymers. Adv Mater Res 150–151:1564–1569

Jiang J-g, Du X-j, Chen M-z, Zhang C (2009) Continuous CO(2) capture and MSWI fly ash stabilization, utilizing novel dynamic equipment. Environ Pollut 157:2933–2938

Johansson I, Van Bavel B (2003a) Levels and patterns of polycyclic aromatic hydrocarbons in incineration ashes. Sci Total Environ 311:221–231

Johansson I, Van Bavel B (2003b) Polycyclic aromatic hydrocarbons in weathered bottom ash from incineration of municipal solid waste. Chemosphere 53:123–128

Jones DA, Lelyveld TP, Mavrofidis SD, Kingman SW, Miles NJ (2002) Microwave heating applications in environmental engineering – a review. Resour Conserv Recycl 34:75–90

Joshi RC, Lothia RP (1997) Fly ash in concrete: production, properties and uses, vol 2, Advances in concrete technology. Gordon and Breach Science Publishers, New York

Jung CH, Matsuto T, Tanaka N (2005) Behavior of metals in ash melting and gasification-melting of municipal solid waste (MSW). Waste Manage 25:301–310

Juwarkar AA, Jambhulkar HP (2008) Restoration of fly ash dumps through biological interventions. Environ Monit Assess 139:355–365

Kalapathy U, Proctor A, Shultz J (2000) Simple method for production of pure silica from rice hull ash. Bioresour Technol 73:257–262

Kamble SP, Mangrulkar PA, Bansiwal AK, Rayalu SS (2008) Adsorption of phenol and o-chlorophenol on surface altered fly ash based molecular sieves. Chem Eng J 138:73–83

Kamon M, Katsumi T, Sano Y (2000) MSW fly ash stabilized with coal ash for geotechnical application. J Hazard Mater 76:265–283

Kamseu E, Leonelli C, Obonyo E (2011) Int Ceram Rev 60:221–225

Kao PC, Tzeng JH, Huang TL (2000) Removal of chlorophenols from aqueous solution by fly ash. J Hazard Mater 76:237–249

Kapoor A, Viraraghavan T (1992) Adsorption of mercury from wastewater by fly ash. Adsorpt Sci Technol 9:130–147

Karamanov A, Pelino M, Hreglich A (2003a) Sintered glass–ceramics from municipal solid waste-incinerator fly ashes. Part I: The influence of the heating rate on the sinter-crystallisation. J Eur Ceram Soc 23:827–832

Karamanov A, Pelino M, Salvo M, Metekovits I (2003b) Sintered glass–ceramics from incinerator fly ashes. Part II. The influence of the particle size and heat-treatment on the properties. J Eur Ceram Soc 23:1609–1615

Katsuura H, Inoue T, Hiraoka M, Sakai S (1996) Full-scale plant study on fly ash treatment by the acid extraction process. Waste Manage 16:491–499

Kawakami I, Esaki M, Tetsuyama I, Sumitomo M (1996) Immobilization of fly ash from msw incinerators and ash-melting furnaces. Waste Manage 16:483–489

Kelleher BP, O'Callaghan MN, Leahy MJ, O'Dwyer TF, Leahy JJ (2002) The use of fly ash from the combustion of poultry litter for the adsorption of chromium(III) from aqueous solution. J Chem Technol Biotechnol 77:1212–1218

Kersch C, Ortiz SP, Woerlee GF, Witkamp GJ (2004) Leachability of metals from fly ash: leaching tests before and after extraction with supercritical CO2 and extractants. Hydrometallurgy 72:119–127

Khalid N, Mukri M, Kamarudin F, Arshad MF (2012) Clay soil stabilized using Waste Paper Sludge Ash (WPSA) mixtures. Electron J Geotech Eng 17:1215–1225

Khan MR, Khan MW (1996) The effect of fly-ash on plant growth and yield of tomato. Environ Pollut 92:105–111

Khan MJ, Al-Juhani AA, Ul-Hamid A, Shawabkeh R, Hussein IA (2011) J Appl Polym Sci 122:2486–2496

Khanna SK, Malhotra P (1997) Kinetics and mechanism of phenol adsorption on fly ash. Ind J Environ Health 19:224–237

Khatri C, Rani A (2008) Synthesis of a nano-crystalline solid acid catalyst from fly ash and its catalytic performance. Fuel 87:2886–2892

Khatri C, Mishra MK, Rani A (2010) Synthesis and characterization of fly ash supported sulfated zirconia catalyst for benzylation reactions. Fuel Process Technol 91:1288–1295

Kikkawa H, Nakamoto T, Morishita M, Yamada K (2002) New wet FGD process using granular limestone. Ind Eng Chem Res 41:3028–3036

Kilinçkale F, Aythan S, Apak R (1997) Solidification/stabilization of heavy metal- loaded red muds and fly ashes. J Chem Technol Biotechnol 69:240–246

Kim Y, Lee D, Osako M (2002) Effect of dissolved humic matters of the leachability of PCDD/F from fly ash. Chemosphere 47:599–605

Kim E, Oh J, Chang Y (2003) Effects of forest fire on the level and distribution of PCDD/Fs and PAH in soil. Sci Total Environ 311:177–189

Kim SH, Kwak SY, Suzuki T (2006) Photocatalytic degradation of flexible PVC/TiO2 nanohybrid as an eco-friendly alternative to the current waste landfill and dioxin-emitting incineration of post-use PVC. Polymer 47:3005–3016

Kim Y-R, Pinto I, Park S-W (2012) Experimental evaluation of anti-stripping additives in bituminous mixtures through multiple scale laboratory test results. Constr Build Mater 29:386–393

Kingman FSW, Rowson NA (1998) Microwave treatment of minerals – a review. Miner Eng 11:1081–1087

Kirby CS, Rimstidt JD (1993) Mineralogy and surface properties of municipal solid waste ash. Environ Sci Technol 27:652–660

Kost DA, Bigham JM, Stehouwer RC, Beeghly JH, Fowler R, Traina SJ, Wolfe WE, Dick WA (2005) Chemical and physical properties of dry flue gas desulfurization products. J Environ Qual 34:676–686

Krebs W, Bachofen R, Brandl H (2001) Growth stimulation of sulfur oxidizing bacteria for optimization of metal leaching efficiency of fly ash from municipal solid waste incineration. Hydrometallurgy 59:283–290

Krishnamoorthy R (2000) Ash utilisation in India prospect and problems. Barrier and utilisation option for large volume application of fly ash in India. In: Hajela V (ed) Proceedings of the Workshop on USAID/India Greenhouse Gas Pollution Prevention Project. pp 63–67

Krishnamurthy S, Brown HC (1980) Selective reductions. 27. Reaction of alkyl halides with representative complex metal hydrides and metal hydrides. Comparison of various hydride reducing agents. J Org Chem 45:849–856

Kruger JE (1998) An overview. In: Kruger JE (ed) Guides on the use of South African fly ash as a cement extender guide. The South African Coal Fly Ash Association (SACAA), Cresta, p 2

Kulkarni PS, Crespo JC, Afonso CAM (2008) Dioxins sources and current remediation technologies – a review. Environ Int 34:139–153

Kumar P, Mal N, Oumi Y, Yamana K, Sano T (2001) Mesoporous materials prepared using coal fly ash as the silicon and aluminium source. J Mater Chem 11:3285–3290

Kumar R, Kumar S, Mehrotra S (2007a) Towards sustainable solutions for fly ash through mechanical activation, resources. Conserv Recycl 52:157–179

Kumar S, Kumar R, Alex TC, Bandopadhyay A, Mehrotra SP (2007b) Influence of reactivity of fly ash on geopolymerisation. Adv Appl Ceram 106:120–127

Kumari K, Saxena SK (1988) Adsorption thermodynamics of carbofuran of fly ash. Colloid Surf 33:55–61

Kumpiene J, Lagerkvist A, Maurice C (2007) Stabilization of Pb and Cu contaminated soil using coal fly ash and peat. Environ Pollut 145:365–373

Kuo Y, Lin T, Tsai P, Lee W, Lin H (2003) Fate of polycyclic aromatic hydrocarbons during vitrification of incinerator ash in a coke bed furnace. Chemosphere 51:313–319

Kutchko BG, Kim AG (2006) Fly ash characterization by SEM-EDS. Fuel 85:2537–2544

Laethem B, van Herck P, Geuzens P, Vandecasteele C (1994) Integrated treatment of MSW-residues: treatment of fly ash in view of metal recovery. WASCON, pp 525–538

Lam CHK, Ip AWM, Barford JP, McKay G (2010) Use of incineration MSW ash: a review. Sustainability 2:1943–1968

Lamble KJ, Hill SJ (1998) Critical review – microwave digestion procedures for environmental materials. Analyst 123:103R–133R

Lancellotti I, Kamseu E, Michelazzi M, Barbieri L, Corradi A, Leonelli C (2010) Chemical stability of geopolymers containing municipal solid waste incinerator fly ash. Waste Manag 30:673–679

Landreth RE (1986) Guide to the disposal of chemically stabilized and solidified waste. EPA SW-72 U.S. Environmental Protection Agency, Cincinnati

Langley WS, Carette GG, Malhotra VM (1989) Structural concrete incorporating high volumes of ASTM Class F fly ash. ACI Mater J 86:507–514

Lasagni M, Collina E, Grandesso E, Piccinelli E, Pitea D (2009) Kinetics of carbon degradation and PCDD/PCDF formation on MSWI fly ash. Chemosphere 74:377–383

Lee H, Ha HS, Lee CS, Lee YB, Kim PJ (2006) Fly ash effect on improving soil properties and rice productivity in Korean paddy soil. Bioresour Technol 97:1490–1497

Lee VKC, Cheung WH, McKay G (2008) PCDD/PCDF reduction by the co-combustion process. Chemosphere 70:682–688

Leiva C, Gomez-Barea A, Vilches LF, Ollero P, Vale J, Fernandez-Pereira C (2007) Use of biomass gasification fly ash in lightweight plasterboard. Energy Fuel 21:361–367

Leiva C, Arenas CG, Vilches LF, Vale J, Gimenez A, Ballesteros JC, Fernández-Pereira C (2010) Use of FGD gypsum in fire resistant panels. Waste Manage 30:1123–1129

Li X, Fernández Bertos M, Hills CD, Carey PJ, Simon S (2007) Accelerated carbonation of municipal solid waste incineration fly ashes. Waste Manage 27:1200–1206

Lima AT, Ottosen LM, Ribeiro AB (2012) Assessing fly ash treatment: remediation and stabilization of heavy metals. J Environ Manage 95:110–115

Lin CJ, Chang JE (2001) Effect of fly ash characteristics on the removal of Cu(II) from aqueous solution. Chemosphere 44:1185–1192

Lingling X, Wei G, Tao W, Nanru Y (2005) Study on fired bricks with replacing clay by fly ash in high volume ratio. Constr Build Mater 19:243–247

Llorente MJF, Garcia JEC (2006) Concentration of elements in woody and herbaceous biomass as a function of the dry ashing temperature. Fuel 85:1273–1279

López-Antón MA, Díaz-Somoano M, Martínez-Tarazona MR (2007) Mercury retention by fly ashes from coal combustion: influence of the unburned coal content. Ind Eng Chem Res 46:927–931

Lorenz K (1954) Secondary treatment of power-plant phenol waste with fly ash and cinders. Gesundh Ing 75:189

Ludwig C, Schuler AJ, Wochele J, Stucki S (2000a) Source: measuring heavy metals by quantitative thermal vaporization. Water Sci Technol 42:209–216

Ludwig C, Johnson CA, Kappeli M, Ulrich A, Riediker S (2000b) Hydrological and geochemical factors controlling the leaching of cemented MSWI air pollution control residues: a lysimeter field study. J Contam Hydrol 42:253–272

Ludwig C, Lutz H, Wochele J, Stucki S (2001) Studying the evaporation behavior of heavy metals by thermo-desorption spectrometry. Fresenius J Anal Chem 371:1057–1062
Ludwig C, Hellweg S, Stucki S (2003) Municipal solid waste management. Springer, Berlin/Heidelberg/New York. ISBN 3-540-44100-X
Ludwig C, Wochele J, Jorimann U (2007) Measuring evaporation rates of metal compounds from solid samples. Anal Chem 79:2992–2996
Lundtorp K, Jensen DL, Sørensen MA, Mogensen EPB, Christensen TH (1999) Stabilization of APC-Residues with FeSO4. In: Proceedings of Sardinia 99, seventh international landfill symposium I, pp 549–556
Lundtorp K, Jensen DL, Sorensen MA, Christensen TH, Mogensen EPB (2002) Treatment of waste incinerator air-pollution-control residues with FeSO4: concept and product characterisation. Waste Manage Res 20:69–79
Lutz H (2002) Detoxification of filter ashes from municipal solid waste incinerators, ETH Zürich, PhD thesis no 14653, Switzerland
Ma WP, Brown PW (1997) Hydrothermal reactions of fly ash with Ca(OH)2 and CaSO4.2H2O. Cem Concr Res 27:1237–1248
Mahin DB (1990) Energy from rice residue, biomass energy and technology report. Winrock International Institute for Agricultural Development, Arlington
Mahvi AH, Maleki A, Eslami A (2004) Potential of rice husk and rice husk ash for phenol removal in aqueous systems. Am J Appl Sci 1:321–326
Malhotra VM (1990) Durability of concrete incorporating high-volume of low- calcium (ASTM class F) fly ash. Cem Concr Compos 12:271–277
Malviya R, Chaudhary R (2006) Factors affecting hazardous waste solidification/stabilization: a review. J Hazard Mater B 137:267–276
Mandal A, Sengupta D (2003) Radioelemental study of Kolaghat, thermal power plant, West Bengal, India: possible environmental hazards. Environ Geol 44:180–186
Mangialardi T (2001) Sintering of MSW fly ash for reuse as a concrete aggregate. J Hazard Mater 87:225–239
Mangialardi T (2003) Disposal of MSWI fly ash through a combined washing-immobilisation process. J Hazard Mater 98:225–240
Mangialardi T, Paolini AE, Polettini A, Sirini P (1999) Optimization of the solidification/stabilization process of MSW fly ash in cementitious matrices. J Hazard Mater 70:53–70
Manz OE (1999) Coal fly ash: a retrospective and future look. Fuel 78:133–136
Markad VL, Kodam KM, Ghole VS (2012) Effect of fly ash on biochemical responses and DNA damage in earthworm, dichogaster curgensis. J Hazard Mater 215–216:191–198
Maroto-Valer MM, Taulbee DN, Schobert HH, Hower JC, Andersen JM (1999) Use of unburned carbon in fly ash as precursor for the development of activated carbons. In: International ash utilization symposium, paper #19
Martirena Hernández JF, Middendorf B, Gehrke M, Budelmann H (1998) Use of wastes of the sugar industry as pozzolana in lime-pozzolana binders: study of the reaction. Cem Concr Res 28:1525–1536
Masia AAT, Buhre BJP, Gupta RP, Wall TF (2007) Use of TMA to predict deposition behaviour of biomass fuels. Fuel 86:2446–2456
Maslehuddin M (1989) Effect of sand replacement on the early-age strength gain and long-term corrosion-resisting characteristics of fly ash concrete. ACI Mater J 86:58–62
Mathur AK (2000) Ash utilisation in NTPC. In: Proceedings of the workshop on fly ash utilisation: issues and strategies, pp 41–45
Matti SS, Mukhopadhyay TM, Gupta SK, Banerjee SK (1990) Evaluation of fly ash as a useful material in agriculture. J Ind Soc Soil Sci 38:342–344
McCarthy MJ, Dhir RK (1999) Towards maximising the use of fly ash as a binder. Fuel 78:121–132
McKay G (2002) Dioxin characterization, formation and minimization during municipal solid waste (MSW) incineration: review. Chem Eng J 86:343–368
Mehdinia SM, Latif PA, Abdullah AM, Taghipour H (2011) Synthesize and characterization of rice husk silica to remove the hydrogen sulfide through the physical filtration system. Asian J Sci Res 4:246–254

Mehta PK (2001) Reducing the environmental impact of concrete. Concrete International, October 2001, pp 61–66
Mehta PK (2004) High-volume fly ash concrete for sustainable development. In: Wank K (ed) International workshop on sustainable development and concrete technology. Iowa State University Publication, Beijing
Meima JA, Comans RNJ (1999) The leaching of trace elements from municipal solid waste incinerator bottom ash at different stages of weathering. Appl Geochem 14:159–171
Miao Z, Yang H, Wu Y, Zhang H, Zhang X (2012) Experimental studies on decomposing properties of desulfurization gypsum in a thermogravimetric analyzer and multiatmosphere fluidized beds. Ind Eng Chem Res 51:5419–5423
Mimura H, Yokota K, Akiba K, Onodera Y (2001) Alkali hydrothermal synthesis of zeolites from coal fly ash and their uptake properties of cesium ion. J Nucl Sci Technol 38:766–772
Mininni G, Sbrilli A, Guerriero E, Rotatori M (2004) Dioxins and furans formation in pilot incineration tests of sewage sludge spiked with organic chlorine. Chemosphere 54:1337–1350
Mitoma Y, Uda T, Egashira N, Simion C, Tashiro H, Tashiro M, Fan X (2004) Approach to highly efficient dechlorination of PCDDs, PCDFs, and coplanar PCBs using metallic calcium in ethanol under atmospheric pressure at room temperature. Environ Sci Technol 38:1216–1220
Mittra BN, Karmakar S, Swain DK, Ghosh BC (2005) Fly ash a potential source of soil amendment and a component of integrated plant nutrient supply system. Fuel 84:1447–1451
Miyake M, Tamura C, Matsuda M (2002) Resource recovery of waste incineration fly ash: synthesis of zeolites A and P. J Am Ceram Soc 85:1873–1875
Mohan D, Singh KP, Singh G, Kumar K (2002) Removal of dyes from wastewater using fly ash, a low-cost adsorbent. Ind Eng Chem Res 41:3688–3695
Moreno N, Querol X, Andrés JM, Stanton K, Towler M, Nugteren H, Janssen-Jurkovicová M, Jones R (2005) Physico-chemical characteristics of European pulverized coal combustion fly ashes. Fuel 84:1351–1363
Mozaffari E, O'Farrell M, Kinuthia JM, Wild S (2006) Improving strength development of wastepaper sludge ash by wet-milling. Cem Concr Compos 28:144–152
Mugica J, Aguirre P, Fresnillo P (1995) Vitrification de cenizas volantes con plasma. Ing Quim Abril 137–142
Muir MA, Yunusa IAM, Burchett MD, Lawrie R, Chan KY, Manoharan V (2007) Short-term responses of contrasting species of earthworms in an agricultural soil amended with coal flyash. Soil Biol Biochem 39:987–992
Mukherjee AB, Zevenhoven R, Bhattacharya P, Sajwan KS, Kikuchi R (2008) Mercury flow via coal and coal utilization by-products: a global perspective. Resour Conserv Recycl 52:571–591
Mulder E (1996) Pre-treatment of MSWI fly ash for useful application. Waste Manag 16:181–184
Murayama N, Yoshida S, Takami Y, Yamamoto H, Shibata J (2003) Simultaneous removal of NH4 and PO43- in aqueous solution and its mechanism by using zeolite synthesized from coal fly ash. Sep Sci Technol 38:113–130
Mustafa Al Bakri AM, Kamarudin H, Bnhussain M, Khairul Nizar I, Rafiza AR, Izzat AM (2011) Chemical reactions in the geopolymerisation process using fly ash-based geopolymer: a review. Aust J Basic Appl Sci 5:1199–1203
Nagib S, Inoue K (2000) Recovery of lead and zinc from fly ash generated from municipal incineration plants by means of acid and/or alkaline leaching. Hydrometallurgy 56:269–292
Nahmani J, Capowiez Y, Lavelle P (2005) Effects of metal pollution on soil macroinvertebrate burrow systems. Biol Fertil Soils 42:31–39
Naik TR, Tyson SS (2000) Environmental benefits from the use of coal combustion products (CCP). In: Verma CVJ, Rao SV, Kumar V, Krishnamoorthy R (eds) Proceedings of the second international conference on fly ash disposal, utilisation, pp 40–43
Naiya TK, Bhattacharya AK, Mandal S, Das SK (2009) The sorption of Lead(II) ions on rice husk ash. J Hazard Mater 163:1254–1264
Nam S, Namkoong W (2012) Irradiation effect on leaching behavior and form of heavy metals in fly ash of municipal solid waste incinerator. J Hazard Mater 199–200:440–447

Nam IH, Hong HB, Kim YM, Kim BH, Murugesan K, Chang YS (2005) Biological removal of polychlorinated dibenzo-p-dioxins from incinerator fly ash by Sphingomonas wittichii RW1. Water Res 39:4651–4660
Nam IH, Kim YM, Murugesan K, Jeon JR, Chang YY, Chang YS (2008) Bioremediation of PCDD/Fs-contaminated municipal solid waste incinerator fly ash by a potent microbial biocatalyst. J Hazard Mater 157:114–121
Narayanasamy P, Gnanakumar D (1989) Lignite fly-ash: a nonpolluting substance for tackling pest problems. In: Devaraj KV (ed) Progress in pollution research. University of Agricultural Sciences, Bangalore, pp 201–206
Natusch DFS, Wallace JR (1974) Urban aerosol toxicity: the influence of particle size. Science 186:695–699
Ngamcharussrivichai C, Chatratananon C, Nuntang S, Prasassarakich P (2008) Adsorptive removal of thiophene and benzothiophene over zeolites from Mae Moh coal fly ash. Fuel 87: 2347–2351
Nimityongskul P, Panichnava S, Hengsadeekul T (2003) Use of vetiver grass ash as cement replacement materials. ICV- Held in Guangzhou, China, pp 6–9
Nishida K, Nagayoshi Y, Ota H, Nagasawa H (2001) Melting and stone production using MSW incinerated ash. Waste Manag 21:443–449
Nollet H, Roels M, Lutgen P, Van Der Meeren P, Verstraete W (2003) Removal of PCBs from wastewater using fly ash. Chemosphere 53:655–665
Nomura Y, Fujiwara K, Takada M, Nakai S, Hosomi M (2006) Detoxification of fly ash by mechanochemical treatment with blast furnace slag and the usability of the residues as cement materials. J Jpn Soc Waste Manag Expert 17:355–360
Nomura Y, Fujiwara K, Takada M (2008) Lead immobilization in mechanochemical fly ash recycling. J Mater Cycle Waste Manag 10:14–18
Nomura Y, Fujiwara K, Terada A, Nakai S, Hosomi M (2010) Prevention of lead leaching from fly ashes by mechanochemical treatment. Waste Manag 30:1290–1295
Nowak B, Frías Rocha S, Aschenbrenner P, Rechberger H, Winter F (2012) Heavy metal removal from MSW fly ash by means of chlorination and thermal treatment: influence of the chloride type. Chem Eng J 179:178–185
Nzihou A, Sharrock P (2002) Calcium phosphate stabilization of fly ash with chloride extraction. Waste Manag 22:235–239
Obernberger I, Biedermann F, Widmann W, Riedl R (1997) Concentrations of inorganic elements in biomass fuels and recovery in the different ash fractions. Biomass Bioenerg 12:211–224
Osako M, Kim Y, Lee D (2002) A pilot and field investigation on mobility of PCDDs/ PCDFs in landfill site with municipal solid waste incineration residue. Chemosphere 48:849–856
Page AL, Elseewi AA, Straugham I (1979) Physical and chemical properties of fly- ash from coal fired plants with reference to environmental impacts. Residue Rev 71:83–120
Pan SC, Tseng DH, Lee CC, Lee C (2003) Influence of the fineness of sewage sludge ash on the mortar properties. Cem Concr Res 33:1749–1754
Panday KK, Prasad G, Singh VN (1984) Removal of Cr(VI) from aqueous solutions by adsorption on fly ash-wollastonite. J Chem Technol Biotechnol 34A:367–374
Panday KK, Prasad G, Singh VN (1985) Copper (II) removal from aqueous solutions by fly ash. Water Res 19:869–873
Pandey VC, Singh N (2010) Impact of fly ash incorporation in soil systems. Agric Ecosyst Environ 136:16–27
Papachristou E, Vasilikiotis G, Alexiades C (1985) Selective adsorption of heavy metal cations by using fly ash. In: Proceedings of the first international symposium on environmental technology for developing countries, Istanbul, Turkey
Papandreou A, Stournaras CJ, Panias D (2007) Copper and cadmium adsorption on pellets made from fired coal fly ash. J Hazard Mater 148:538–547
Papastefanou C (2008) Radioactivity of coals and fly ashes. J Radioanal Nucl Chem 275:29–35
Park YJ, Heo J (2002) Vitrification of fly ash from municipal solid waste incinerator. J Hazard Mater 91:83–93

Park JS, Taniguchi S, Park YJ (2009) Alkali borosilicate glass by fly ash from a coal-fired power plant. Chemosphere 74:320–324
Patel M, Karera A, Prasanna P (1987) Effect of thermal and chemical treatments on carbon and silica contents in rice husk. J Mater Sci 22:2457–2464
Pavlish JH, Sondreal EA, Mann MD, Olson ES, Galbreath KICK, Laudal DL (2003) Status review of mercury control options for coal-fired power plants. Fuel Process Technol 82:89–165
Pedersen AJ (2002) Evaluation of assisting agents for electrodialytic removal of Cd, Pb, Zn, Cu and Cr from MSWI fly ash. J Hazard Mater B 95:185–198
Pedersen AJ, Ottosen LM, Villumsen A (2005) Electrodialytic removal of heavy metals from municipal solid waste incineration fly ash using ammonium citrate as assisting agent. J Hazard Mater 122:103–109
Peloso A, Rovatti M, Ferraiolo G (1983) Fly ash as adsorbent material for toluene vapours. Resour Conserv Recycl 10:211–220
Phung HT, Lund LJ, Page AL (1978) In: Adriano DC, Brisbin IL (eds) Potential use of fly-ash as a liming material in environmental chemistry and cycling processes. CONF-760429. U.S. Department of Energy, p 504
Piantone P, Bodenan F, Derie R, de Pelsenaire G (2003) Monitoring the stabilisation of municipal solid waste incineration fly ash by phosphation. Waste Manag 23:225–243
Pitchtel JR, Hayes JM (1990) Influence of fly ash on soil microbial activity and populations. J Environ Qual 19:593–597
Pitchtel JR (1990) Microbial respiration in fly-ash sewage sludge-amended soils. Environ Pollut 63:225–237
Prabhu PVSS, Narayanaswamy MS, Narasa Raju TSS (1981) Adsorption of zinc from aqueous solutions by fly ash. IAWPC Tech Annu 8:46–52
Querol X, Alastuey A, Jose LFT, Angel LS (1995) Synthesis of zeolites by alkaline activation of ferro-aluminous fly ash. Fuel 74:1226–1231
Querol X, Moreno N, Umana JC, Alastuey A, Hernandez E, Lopez-Soler A, Plana F (2002) Synthesis of zeolites from coal fly ash: an overview. Int J Coal Geol 50:413–423
Quina MJ, Bordado JC, Quinta-Ferreira RM (2008) Treatment and use of air pollution control residues from MSW incineration: an overview. Waste Manag 28:2097–2121
Rajamma R, Ball RJ, Tarelho LAC, Allen GC, Labrincha JA, Ferreira VM (2009) Characterisation and use of biomass fly ash in cement-based materials. J Hazard Mater 172:1049–1060
Ramakrishna KR, Viraraghavan T (1997) Dye removal using low cost adsorbents. Water Sci Technol 36:189–196
Rani DA, Boccaccini AR, Deegan D, Cheeseman CR (2008) Air pollution control residues from waste incineration: current UK situation and assessment of alternative technologies. Waste Manag 28:2279–2292
Rao VMHG (1980) Utilization of rice husk- a preliminary analysis. J Sci Ind Res 39:495–515
Rao M, Parwate AV, Bhole AG (2002) Removal of Cr6þ and Ni2þ from aqueous solution using bagasse and fly ash. Waste Manag 22:821–830
Rao M, Parwate AV, Bhole AG, Kadu PA (2003) Performance of low-cost adsorbents for the removal of copper and lead. J Water Suppl: Res Technol-AQUA 52:49–58
Rattanasak U, Pankhet K, Chindaprasirt P (2011) Effect of chemical admixtures on properties of high-calcium fly ash geopolymer. Int J Miner Metall Material 18:364–369
Rautaray SK, Ghosh BC, Mittra BN (2003) Effect of fly ash, organic wastes and chemical fertilizers on yield, nutrient uptake, heavy metal content and residual fertility in a rice–mustard cropping sequence under acid lateritic soils. Bioresour Technol 90:275–283
Ravina D, Mehta PK (1986) Properties of fresh concrete containing large amounts of fly ash. Cem Concr Res 16:227–238
Rawlings RD, Wu JP, Boccaccini AR (2006) Glass-ceramics: their production from wastes – a review. J Mater Sci 41:733–761
Rayalu SS, Bansiwal AK, Meshram SU, Labhsetwar N, Devotta S (2006) Fly ash based zeolite analogues: versatile materials for energy and environment conservation. Catal Surv Asia 10:74–88

Reijnders L (2005) Disposal, uses and treatments of combustion ashes: a review. Resour Conserv Recycl 43:313–336

Reiner M, Rens K (2006) High-volume fly ash concrete: analysis and application. Pract Period Struct Des Constr 11:58–64

Rickard WDA, van Riessen A, Walls P (2010) Thermal character of geopolymers synthesized from Class F fly ash containing high concentrations of iron and alpha-quartz. Int J Appl Ceram Technol 7:81–88

Rickard WDA, Williams RP, Temuujin J, van Riessen A (2011) Assessing the suitability of three Australian fly ashes as an aluminosilicate source for geopolymers in high temperature applications. Mater Sci Eng A 528:3390–3397

Rickard WDA, Temuujin J, van Riessen A (2012) Thermal analysis of geopolymer pastes synthesised from five fly ashes of variable composition. J Non-Cryst Solid 385:1830–1839

Ricou P, Lecuyer I, Le Cloirec P (1999) Removal of Cu2+, Zn2+ and Pb2+ by adsorption onto fly ash and fly ash/lime mixing. Water Sci Technol 39:239–247

Robertson GP, Paul EA, Harwood RR (2000) Greenhouse gases in intensive agriculture: contributions of individual gases to the radiative forcing of the atmosphere. Science 289:1922–1925

Roes L, Patel MK, Worrell E, Ludwig C (2012) Preliminary evaluation of risks related to waste incineration of polymer nanocomposites. Sci Total Environ 417:76–86

Romero M, Rawlings RD, Rincon JM (1999) Development of a new glass-ceramic by means of controlled vitrification and crystallisation of inorganic wastes from urban incineration. J Eur Ceram Soc 19:2049–2058

Rotenberg SJ, Mettzler G, Poliner J, Bechtold WE, Eidson AF, Newton GJ (1991) Adsorption kinetics of vapor-phase m-xylene on coal fly ash. Environ Sci Technol 25:930–935

Rovatti M, Peloso A, Ferraiolo G (1988) Susceptibility to regeneration of fly ash as an adsorbent material. Resour Conserv Recycl 1:137–143

Roy A, Eaton HC, Cartledge FK, Tittlebaum ME (1991) Solidification/stabilization of heavy metal sludge by a Portland cement/fly ash binding mixture hazard. Waste Hazard Mater 833–839

Ruangtaweep Y, Kaewkhao J, Kedkaew C, Limsuwan P (2011) Investigation of biomass fly ash in Thailand for recycle to glass production. Procedia Eng 8:58–61

Rubel A, Andrews R, Gonzalez R, Groppo J, Robl T (2005) Adsorption of Hg and NOx on coal by-products. Fuel 84:911–916

Rubio B, Izquierdo MT, Mayoral MC, Bona MT, Andres JM (2007) Unburnt carbon from coal fly ashes as a precursor of activated carbon for nitric oxide removal. J Hazard Mater 143:561–566

Sabbas T, Polettini A, Pomi R, Astrup T, Hjelmar O, Mostbauer P, Cappai G, Magel G, Salhofer S, Speiser C, Heuss-Assbicher S, Klein R, Lehner P (2003) Management of municipal solid waste incinerator residues. Waste Manag 23:61–88

Safiuddin M (2008) Development of self-consolidating high performance concrete incorporating rice husk ash, a Thesis presented to the University of Waterloo in fulfillment of the thesis requirement for the degree of doctor of philosophy in Civil Engineering, Waterloo, ON, Canada

Sajwan KS, Paramasivam S, Alva AK, Adriano DC, Hooda PS (2003) Assessing the feasibility of land application of fly ash, sewage sludge and their mixtures. Adv Environ Res 8:77–91

Sakai S, Hiraoka M (2000) Municipal solid waste incinerator residue recycling by thermal processes. Waste Manag 20:249–258

Sakai S, Urano S, Takatsuki H (2000) Leaching behavior of PCBs and PCDDs/DFs from some waste materials. Waste Manag 20:241–247

Sako T, Sugeta T, Otake K, Sato M, Tsugumi M, Hiaki T, Hongo M (1997) Decomposition of dioxins in fly ash with supercritical water oxidation. J Chem Eng Jpn 30:744–747

Sako T, Kawasaki S, Noguchi H, Kimura T, Sato H (2004) Destruction of dioxins and PCBs in solid wastes by supercritical fluid treatment. Organohalog Compd 66:1187–1193

Sankari SA, Narayanasamy P (2007) Bio-efficacy of fly-ash based herbal pesticides against pests of rice and vegetables. Curr Sci 92:811–816

Sarabèr A, Overhof R, Green T, Pels J (2012) Artificial lightweight aggregates as utilization for future ashes – a case study. Waste Manag 32:144–152

Sarkar A, Rano R, Mishra K, Sinha I (2005) Particle size distribution profile of some Indian fly ash – a comparative study to assess their possible uses. Fuel Process Technol 86:1221–1238
Scharnhorst W, Ludwig C, Wochele J, Jolliet O (2007) Heavy metal partitioning from electronic scrap during thermal end-of-life treatment. Sci Total Environ 373:576–584
Scheetz BE, Earle R (1998) Utilization of fly ash. Curr Opin Solid State Mater Sci 3:510–520
Schmid J, Elser A, Ströbel R, Crowe M (2000) Dangerous substances in waste. European Environment Agency, Copenhagen
Sedhom E, Dauerman L, Ibrahim N, Windgasse G (1992) Microwave treatment of hazardous wastes: "fixation" of chromium in soil. J Microw Power Electromagn Energy 27:81–86
Sen AK, De AK (1987) Adsorption of mercury (II) by coal fly ash. Water Res 21:885–888
Shao L, Xu ZX, Jin W, Yin HL (2009) Rice husk as carbon source and biofilm carrier for water denitrification. Pol J Environ Stud 18:693–699
Sharma S (1989) Fly ash dynamics in soil water systems. Crit Rev Environ Control 19:251–275
Shawabkeh R, Al-Harahsheh A, Hami M, Khlaifat A (2004) Conversion of oil shale ash into zeolite for cadmium and Lead removal from wastewater. Fuel 83:981–985
Shen JF, Zhou XW, Sun DS, Fang JG, Liu ZJ, Li Z (2008) Soil improvement with coal ash and sewage sludge: a field experiment. Environ Geol 53:1777–1785
Shen DK, Gu S, Luo KH, Bridgewater AV, Fang MX (2009) Kinetic study on thermal decomposition of woods in oxidative environment. Fuel 88:1024–1030
Shi HS, Kan LL (2009) Characteristics of municipal solid wastes incineration (MSWI) fly ash–cement matrices and effect of mineral admixtures on composite system. Constr Build Mater 23:2160–2166
Shi C, Qian J (2003) Increasing coal fly ash use in cement and concrete through chemical activation of reactivity of fly ash. Energy Source 25:617–628
Shulman DM (2007) An aggregate replacement. A power plant byproduct may be ideal for concrete production. Concr Producer 25:41–42
Singh BK, Nayak PS (2004) Sorption equilibrium studies of toxic nitro-substituted phenols on fly ash. Adsorpt Sci Technol 22:295–309
Singh BK, Misran NM, Rawatn NS (1994) Sorption characteristics of phenols on fly ash and impregnated fly ash. Ind J Environ Health 36:1–7
Singh NB, Das SS, Singh NP, Dwivedi VN (2007) Hydration of bamboo leaf ash blended Portland cement. Indian J Eng Mater Sci 14:69–76
Siores E, Rego DD (1995) Microwave applications in materials joining. J Mater Process Technol 48:619–625
Skillings Mining Review (1999) Penn State researchers study use of coal by-products. Skill Min Rev 88:9
Somerset V, Petrik L, Iwuoha E (2008) Alkaline hydrothermal conversion of fly ash precipitates into zeolites 3: the removal of mercury and Lead ions from wastewater. J Environ Manage 87:125–131
Sonderegger JL, Donovan JJ (1984) Laboratory simulation of fly ash as an amendment to pyrite-rich tailings. Ground Water Monit Rev 4:75–80
Song GJ, Kim KH, Seo YC, Kim SC (2004) Characteristics of ashes from different locations at the MSW incinerator equipped with various air pollution control devices. Waste Manag 24:99–106
Sorensen MA, Bender-Koch C, Stackpoole MM, Bordia RK, Benjamin MM, Christensen TH (2000) Effects of thermal treatment on mineralogy and heavy metal behaviour in iron oxide stabilised APC residues. Environ Sci Technol 34:4620–4627
Soyama M, Inoue K, Iji M (2007) Flame retardancy of polycarbonate enhanced by adding fly ash. Polym Adv Technol 18:386–391
Speiser C, Baumann T, Niessner R (2000) Morphological and chemical characterisation of calcium-hydrate phases formed in alteration processes of deposited municipal solid waste incinerator bottom ash. Environ Sci Technol 34:5030–5037
Sreekanth MS, Joseph S, Mhaske ST, Mahanwar PA, Bambole VA (2011) Effects of mica and fly ash concentration on the properties of polyester thermoplastic elastomer composites. J Thermoplast Compos Mater 24:317–331

Srivastava K, Farooqui A, Kulshreshtha K, Ahmed KJ (1995) Effect of fly ash amended soil on growth of Lactuca sativa L. J Environ Biol 16:93–96

Srivastava VC, Mall ID, Mishra IM (2006) Modelling individual and competitive adsorption of cadmium(II) and zinc(II) metal ions from aqueous solution onto bagasse fly ash. Sep Sci Technol 41:2685

Srivastava VC, Mall ID, Mishra IM (2008) Removal of cadmium(II) and zinc(II) metal ions from binary aqueous solution by rice husk ash. Colloid Surf A 312:172–184

Struis RPWJ, Ludwig C, Lutz H (2004) Speciation of zinc in municipal solid waste incineration fly ash after heat treatment: an X-ray absorption spectroscopy study. Environ Sci Technol 38:3760–3767

Struis RPWJ, Nachtegaal M, Mattenberger H, Ludwig C (2009) The fate of lead in MSWI-fly ash during heat treatment: an X-ray absorption spectroscopy study. Adv Eng Mater 11:507–512

Stucki S, Jakob A (1997) Thermal treatment of incinerator fly ash: factors influencing the evaporation of ZnCl2. Waste Manag 17:231–236

Suárez-Ruiz I, Parra JB (2007) Relationship between the textural properties, fly ash carbons and Hg capture in fly ashes derived from the combustion of anthracitic pulverized feed blends. Energy Fuel 21:1915–1923

Suárez-Ruiz I, Hower JC, Thomas GA (2007) Hg and Se capture and fly ash carbons from combustion of complex pulverized feed blends mainly of anthracitic coal rank in Spanish power plants. Energy Fuel 21:59–70

Sumaila SA, Job OF (1999) Properties of SDA-OPC concrete: a preliminary assessment. J Environ Sci 3:155–159

Sun RC (2010) Cereal straw as a resource for sustainable biomaterials and biofuels. Elsevier, Amsterdam. ISBN 978-0-444-53234-3

Świetlik R, Trojanowska M, Jóźwiak Małgorzata A (2012) Evaluation of the distribution of heavy metals and their chemical forms in ESP-fractions of fly ash. Fuel Process Technol 95:109–118

Tadmore J (1986) Radioactivity from coal-fired power plants: a review. J Environ Radioact 4:77–244

Takami Y, Murayama N, Ogawa K, Yamamoto H, Shibata JJ (2000) Water purification property of zeolite synthesized from coal fly ash. Min Mater Process Inst Jpn 116:987–991

Tame NW, Dlugogorski BZ, Kennedy EM (2003) Increased PCDD/F formation in the bottom ash from fires of CCA-treated wood. Chemosphere 50:1261–1263

Tangchirapat W, Saeting T, Jaturapitakkul C, Kiattikomol K, Siripanichgorn A (2007) Use of waste ash from palm oil industry in concrete. Waste Manag 27:81–88

Tarley CRT, Arruda MAZ (2004) Biosorption of heavy metals using rice milling byproducts, characterization and application for removal of metals from aqueous effluents. Chemosphere 54:987–995

Tejasvi A, Kumar S (2012) Impact of fly ash on soil properties. Natl Acad Sci Lett 35:13–16

Tho-in T, Sata V, Chindaprasirt P, Jaturapitakkul C (2012) Pervious high-calcium fly ash geopolymer concrete. Constr Build Mater 30:366–371

Thy P, Jenkins BM, Grundvig S, Shiraki R, Lesher CE (2006) High temperature elemental losses and mineralogical changes in common biomass ashes. Fuel 85:783–795

Tietenberg T (2003) Environmental and natural resource economics, 6th edn. Addison Wesley, Boston, pp 404–412

Tiwari S, Kumari B, Singh SN (2008) Evaluation of metal mobility/immobility in fly ash induced by bacterial strains isolated from the rhizospheric zone of Typha latifolia growing on fly ash dumps. Bioresour Technol 99:1305–1310

Tolle DA, Arthur MF, Van Voris P (1983) Microcosm/field comparison of trace element uptake in crops grown in fly ash amended soil. Sci Total Environ 31:243–263

Tomeczek J, Palugniok H (2002) Kinetics of mineral matter transformation during coal combustion. Fuel 81:1251–1258

Tulonen T, Arvola L, Strömmer R (2012) Cadmium release from afforested peatlands and accumulation in an aquatic ecosystem after experimental wood ash treatment. Eur J Forest Res 131(5):1529–1536

Twardowska I, Szczepanska J (2002) Solid waste: terminological and long term environmental risk assessment problems exemplified in a power plant fly ash study. Sci Total Environ 285:29–51
Udoeyo FF, Dashibil PU (2002) Sawdust ash as concrete material. J Mater Civ Eng 14:173–177
USEPA (1994a) Health assessment document for 2, 3, 7, 8-tertachlorodibenzo-p-dioxin (TCDD) and related compounds. EPA/600/Bp-92/001c estimating exposure to dioxin-like compounds, epa/600/6-88/005cb. Office of Research and Development, Washington, DC
USEPA (1994b) Combustion emission technical resource document (CETRED) report no EPA 530-R-94-014. Office of Solid Waste Emergency Response, Washington, DC
Usta N (2012) Investigation of fire behavior of rigid polyurethane foams containing fly ash and intumescent flame retardant by using a cone calorimeter. J Appl Polym Sci 124:3372–3382
van der Bruggen B, Vogels G, Van Herck P, Vandecasteele C (1998) Simulation of acid washing of municipal solid waste incineration fly ashes in order to remove heavy metals. J Hazard Mater 57:127–144
Van der Sloot HA, Kosson DS, Hjelmar O (2001) Characteristics, treatment and utilization of residues from municipal waste incineration. Waste Manag 21:753–765
van Deventer JS, Provis JL, Duxson P (2012) Technical and commercial progress in the adoption of geopolymer cement. Miner Eng 29:89–104
Van Gerven T, Geysen D, Vandercasteele C (2004) Estimation of the contribution of a municipal waste incinerator to the overall emission and human intake of PCBs in Wilrijk, Flanders. Chemosphere 54:1303–1308
van Jaarsveld JGS, van Deventer JSJ (1999) Effect of the alkali metal activator on the properties of fly ash-based geopolymers. Ind Eng Chem Res 38:3929–3941
van Oss HG, Padovani AC (2003) Cement manufacture and the environment Part II: environmental challenges and opportunities. J Ind Ecol 7:93–126
Vangronsveld J, Cunningham SD (1998) Metal-contaminated soils: in situ inactivation and phytorestoration. Springer/R.G. Landes Company, Berlin/Heidelberg/Georgetown
Vangronsveld J, Colpaert JV, van Tichelen KK (1996) Reclamation of a bare industrial area contaminated by nonferrous metals: physicochemical and biological evaluation of the durability of soil treatment and revegetation. Environ Pollut 94:131–140
Vassilev SV, Vassileva CG (1996) Mineralogy of combustion wastes from coal-fired power stations. Fuel Process Technol 47:261–280
Vassilev SV, Vassileva CG (2005) Methods for characterization of composition of fly ashes from coal-fired power stations: a critical overview. Energy Fuel 19:1084–1098
Vassilev SV, Vassileva CG (2007) A new approach for the classification of coal fly ashes based on their origin, composition, properties, and behaviour. Fuel 86:1490–1512
Viraraghavan T, Ramakrishna KR (1999) Fly ash for colour removal from synthetic dye solutions. Water Qual Res J Can 34:505–517
Vogg H, Stieglitz L (1986) Thermal-behavior of PCDD/PCDF in fly-ash from municipal incinerators. Chemosphere 15:1373–1378
Volz JS (2012) High-volume fly ash concrete for sustainable construction. Adv Mater Res 512–515:2976–2981
Wallander H, Wickman T, Jacks G (1997) Apatite as a P source in mycorrhizal and non-mycorrhizal Pinus sylvestris. Plant Soil 196:123–131
Walser T, Limbach KL, Brogioli R, Erismann E, Flamigni L, Hattendorf B, Juchli M, Krumeich F, Ludwig C, Prikopsky K, Rossier M, Saner D, Stark WJ, Sigg A, Hellweg S, Günther D (2012) Persistence of engineered nanoparticles in a municipal solid-waste incineration plant. Nat Nanotechnol 7:520–524. doi:10.1038/NNANO.2012.64
Wang S (2008) Application of solid ash based catalysts in heterogeneous catalysis. Environ Sci Technol 42:7055–7063
Wang S, Lu GQ (2007) Effect of chemical treatment on Ni/fly-ash catalysts in methane reforming with carbon dioxide. In: Fbio Bellot Noronha MS, Eduardo Falabella S-A (eds) Studies in surface science and catalysis, vol 167. Elsevier, New York, pp 275–280
Wang KS, Chiang KY, Perng JK, Sun CJ (1998) The characteristics study on sintering of municipal solid waste incinerator ashes. J Hazard Mater 59:201–219

Wang XK, Sun CQ, Cai HW (1999) Origin of the Chinese cultivated rice (Oryza sativa L.). Chin Sci Bull 44:295–304
Wang S, Terdkiatburana T, Tade MO (2008) Single and co-adsorption of heavy metals and humic acid on fly ash. Sep Purif Technol 58:353–358
Wang Q, Yan J, Tu X, Chi Y, Li X, Lu S, Cen K (2009) Thermal treatment of municipal solid waste incinerator fly ash using DC double arc argon plasma. Fuel 88:955–958
Wang Q, Yan JH, Chi Y, Li XD, Lu SY (2010) Application of thermal plasma to vitrify fly ash from municipal solid waste incinerators. Chemosphere 78:626–630
Weng CH, Huang CP (1994) Treatment of metal industrial wastewater by fly ash and cement fixation. J Environ Eng 120:1470–1487
Werther J, Saenger M, Hartge EU, Ogada T, Siagi Z (2000) Combustion of agricultural residues. Prog Energy Combust Sci 26:1–27
West TO, McBride AC (2005) The contribution of agricultural lime to carbon dioxide emissions in the United States: dissolution, transport, and net emissions. Agric Ecosyst Environ 108:145–154
Wey MY, Liu KY, Tsai TH, Chou JT (2006) Thermal treatment of the fly ash from municipal solid waste incinerator with rotary kiln. J Hazard Mater B137:981–989
Wheatley AD, Sadhra S (2004) Polycyclic aromatic hydrocarbons in solid residues from waste incineration. Chemosphere 55:743–749
White PJ (2001) Phytoremediation assisted by microorganism. Trends Plant Sci 6:502
Williams RP, van Riessen A (2010) Determination of the reactive component of fly ashes for geopolymer production using XRF and XRD. Fuel 89:3683–3692
Wiselogel AE, Agblevor FA, Johnson DK, Deutch S, Fennell JA, Sanderson MA (1996) Compositional changes during storage of large round switch grass bales. Bioresour Technol 56:103–109
Wong JWC (1995) The production of artificial soil mix from coal fly ash and sewage sludge. Environ Technol 16:741–751
Wong MH, Wong JWC (1986) Effects of fly ash on soil microbial activity. Environ Pollut Ser A 40:127–144
Wong JWC, Wong MH (1990) Effects of fly-ash on yields and elemental composition of two vegetables, Brassica parachinensis and B. chinensis. Agric Ecosyst Environ 30:251–264
Wu C, Tang Y, Tang L (2012) Removal of heavy metal from wastewater using zeolite from fly ash. Adv Mater Res 518–523:2736–2739
Xenidis A, Mylona E, Paspaliaris I (2002) Potential use of lignite fly ash for the control of acid generation from sulphidic wastes. Waste Manag 22:631–641
Xie JL, Hu YY, Chen DZ, Zhou B (2010) Hydrothermal treatment of MSWI fly ash for simultaneous dioxins decomposition and heavy metal stabilization. Front Environ Sci Eng China 4:108–116
Xue J, Wang W, Wang Q, Liu S, Yang J, Wui T (2010) Removal of heavy metals from municipal solid waste incineration (MSWI) fly ash by traditional and microwave acid extraction. J Chem Technol Biotechnol 85:1268–1277
Yamaguchi H, Shibuya E, Kanamaru Y, Uyama K, Nishioka M, Yamasaki M (1996) Hydrothermal decomposition of PCDDs/PCDFs in MSWI fly ash. Chemosphere 32:203–208
Yan JH, Peng Z, Lu SY, Li XD, Ni MJ (2007) Degradation of PCDD/Fs by mechanochemical treatment of fly ash from medical waste incineration. J Hazard Mater 147:652–657
Yang G, Tsai C (1998) A study on heavy metal extractability and subsequent recovery by electrolysis for a municipal incinerator fly ash. J Hazard Mater 58:103–120
Yang GCC, Yang TY (1998) Synthesis of zeolites from municipal incinerator fly ash. J Hazard Mater 62:75–89
Yang Y-F, Gai G-S, Cai Z-F, Chen Q-R (2006) Surface modification of purified fly ash and application in polymer. J Hazard Mater 133:276–282
Yilmaz G (2012) Structural characterization of glass-ceramics made from fly ash containing SiO 2-Al 2O 3-Fe 2O 3-CaO and analysis by FT-IR-XRD-SEM methods. J Mol Struct 1019:37–42

Yin C, Rosendahl LA, Kaer SK (2008) Grate-firing of biomass fort heat and power production. Progr Energy Combust 34:725–754

Yoshino H, Urano K (1998) Mutagenic activities of exhaust gas from sludge incineration plants. Sci Total Environ 215:41–49

Youcai Z, Lijie S, Guojian L (2002) Chemical stabilization of MSW incinerator fly ashes. J Hazard Mater 95:47–63

Zacco A, Gianoncelli A, Ardesi R, Sacrato S, Guerini L, Bontempi E, Tomasoni G, Alberti M, Depero LE (2012) Use of colloidal silica to obtain a new inert from municipal solid waste incinerator (MSWI) fly ash: first results about reuse. Clean Technol Environ Policy 14:291–297

Zamaow MJ, Schultze LE (1993) In: Zeolite-93, Idaho pp 2200–2211

Zeng Z, Liu H, Yu H, Peng Z (2012) Structure and thermal properties of RHA/NR composite. Adv Mater Res 482–484:1275–1280

Zevenbergen C, Comans RNJ (1994) Geochemical factors controlling the mabilization of major elements during weathering of MSWIash. In: Goumans JJM, van der Sloot HA, Aalbers TG (eds) Environmental aspects of construction with waste materials, vol 60, Studies in environmental sciences. Elsevier, Amsterdam, pp 179–194

Zhang MH, Mohan V (1996) High-performance concrete incorporating rice husk ash as a supplementary cementing material. ACI Mater J 93:629–636

Zhang H, Qi J (2012) Reuse of air pollution control ash. Environ Eng Sci 29:902–906

Zhao YC, Stucki S, Ludwig C, Wochele J (2004) Impact of moisture on volatility of heavy metals in municipal solid waste incinerated in a laboratory scale simulated incinerator. Waste Manag 24:581–587

Zheng L, Wang C, Wang W, Shi Y, Gao X (2011) Immobilization of MSWI fly ash through geopolymerization: effects of water-wash. Waste Manag 31:311–317

Zhijun Y, Chuanhai X, Qing Z, Jiping C, Xinmiao L (2007) Catalytic detoxification of polychlorinated dibenzo-p-dioxins and polychlorinated dibenzofurans in fly ash. Waste Manag 27:588–592

Zhou YX, Yan P, Cheng ZX, Nifuku M, Liang XD, Guan ZC (2003) Application of non-thermal plasmas on toxic removal of dioxin-contained fly ash. Powder Technol 135–136:345–353

Zhu T, Kuang J, Xu W, Ye M, Guo Y, Liu W (2012) Study on mercury adsorption performance of modified fly ash. Adv Mater Res 343–344:246–249

Zielinski RA, Finkelman RB (1997) Radioactive elements in coal and fly ash: abundance, forms and environmental significance. US Geological Survey Fact Sheet FS-163-97

Chapter 4
Organotin Compounds from Snails to Humans

Ana C.A. Sousa, M. Ramiro Pastorinho, Shin Takahashi, and Shinsuke Tanabe

Contents

A.C.A. Sousa (✉)
CICS-UBI – Health Sciences Research Centre, University of Beira Interior,
Av. Infante D. Henrique, 6200-506 Covilhã, Portugal

Department of Biology & CESAM, University of Aveiro,
Campus Universitário Santiago, 3810-193 Aveiro, Portugal

Center for Marine Environmental Studies (CMES), Ehime University,
2-5 Bunkyo-cho, Matsuyama 790-8577, Japan
e-mail: anasousa@ua.pt; anasousa@fcsaude.ubi.pt

E. Lichtfouse et al. (eds.), *Pollutant Diseases, Remediation and Recycling*,
Environmental Chemistry for a Sustainable World 4, DOI 10.1007/978-3-319-02387-8_4,

Abstract Human activities are producing and transforming more than 140,000 chemical compounds according to the European Union Regulation REACH (registration, evaluation, authorization and restriction of chemicals). Humans and wildlife can be exposed to chemicals at all stages of the chemical life cycle: raw material extraction, manufacturing and processing, use, recycling or disposal. Some chemicals can mimic or antagonize hormone action, affecting the normal functioning of organisms. These chemicals, defined as endocrine disruptors, include among many other organotin compounds.

Organotins are the most widely used organometallic compounds for agricultural, industrial and biomedical applications with an annual production of about 60,000 tons. The most notorious organotin is tributyltin, a potent biocide used in antifouling paints. Tributyltin has adverse and even bizarre effects, such as imposex, on aquatic organisms. Tributyltin, derivatives and to a lesser extent other classes of organotins are found in all compartments of the aquatic environment due to the intensive use of organotins for the protection of ships hulls and other immersed structures. The lowest concentrations are found in the water column, whereas the highest are detected in sediments and biota, particularly in top predators. Humans are highly exposed because they consume top predators and other fishes and shellfish. Tributyltin became thus a well-known example of how chemicals can have a global detrimental effect on the environment, even to the eyes of public opinion.

Organotins include also tributyltin derivatives (dibutyltin, monobutyltin), phenyltins (triphenyltin, diphenyltin and monophenyltin), octyltins (trioctyltin, dioctyltin and monooctyltin) and methyltins: trimethyltin, dimethyltin and monomethyltin. They are employed as biocides, polyvinyl chloride (PVC) stabilizers and industrial catalysts for the manufacture of silicone and polyurethane foams. Such diverse industrial applications imply their presence on our everyday lives in clothes, pillows, toys, wallpaper, plastic food containers, pipes including drinking water mains and household piping, medical devices and breast implants. Hence, humans are exposed to organotins not solely through ingestion of contaminated food but also by inhalation and ingestion of dust and through direct contact with treated products. As a consequence, organotins have already been detected in human samples and toxicity data reveals that they are endocrine disruptors, imunotoxicants and carcinogens. Moreover, a link to obesity was recently established.

We review the levels, fate and effects of organotin compounds towards wildlife and humans, starting with a description of organotin applications, with particular incidence in antifouling paints. The need to develop antifouling strategies is presented and the history on the use of tributyltin as the active biocide described. The global contamination of the marine environment and the deleterious effects of tributyltin towards non-target organisms are addressed. Particular attention is drawn to the imposex phenomenon, with an historical perspective and a brief explanation on the induction

M.R. Pastorinho
Department of Biology & CESAM, University of Aveiro,
Campus Universitário Santiago, 3810-193 Aveiro, Portugal

S. Takahashi • S. Tanabe
Center for Marine Environmental Studies (CMES), Ehime University,
2-5 Bunkyo-Cho, Matsuyama 790-8577, Japan

mechanisms proposed. A case study of imposex in a sentinel species – *Nassarius reticulatus* – is also presented. The restrictions on tributyltin use in antifouling paints are also described alongside with the new regulations for other organotins used in consumer products. The sources and pathways of organotins in the environment are briefly addressed and an overview of marine environmental levels is provided alongside with references to regional and global monitoring surveys. At last, studies in human exposure to organotins are presented, a thorough review on human tissues burdens is performed and future research is proposed.

Keywords AFS Convention • Antifouling paints • Bioaccumulation • Body burden • Environmental levels • Exposure pathways • Gastropods • Human exposure • Imposex • Invasive/Noninvasive matrices • Mussels • Organotin compounds • Tributyltin (TBT)

List of Abbreviations

9CRA	9-*cis*-retinoic acid
AF	Antifouling
AFS Convention	International Convention on the Control of Harmful Antifouling Systems on Ships
AOS	Average Oviduct Stage
ATSDR	Agency for Toxic Substances and Disease Registry
BW	Body Weight
BTs	Butyltins
DBT	Dibutyltin
DDT	Dichlorodiphenyltrichloroethane
DL	Detection Limit
DMT	Dimethyltin
DOT	Dioctyltin
DPT	Diphenyltin
dw	dry weight
EC	European Commission
EFSA	European Food Safety Authority
FPLI	Female Penis Length Index
IMO	International Maritime Organization
LOQ	Limit of Quantification
MBT	Monobutyltin
MeTs	Methyltins
MMT	Monomethyltin
MPT	Monophenyltin
MOT	Monooctyltin
OcTs	Octyltins
OSPAR	Convention for the Protection of the Marine Environment of the North-East Atlantic

OTs	Organotin Compounds
PPAR	Peroxisome Proliferator-Activated Receptor
PTs	Phenyltins
PVC	Polyvinyl Chloride
REACH	Registration, Evaluation, Authorization and Restriction of Chemicals
RPLI	Relative Penis Length Index
RPSI	Relative Penis Size Index
RXR	Retinoic X Receptor
T	Testosterone
TBT	Tributyltin
TBT-SPC	Tributyltin Self-Polishing Copolymer paint
TDI	Tolerable Daily Intake
TeBT	Tetrabutyltin
TMT	Trimethyltin
TPT	Triphenyltin
TOT	Trioctyltin
VDSI	*vas deferens* sequence index
WHO	World Health Organization
ww	wet weight

4.1 Introduction

Human activities are responsible for the production and transformation of thousands of chemical compounds. These chemicals, 143,835 according to the pre-register at the European Union Regulation on Chemicals, REACH (Registration, Evaluation, Authorization and Restriction of Chemicals), are fundamental in modern life, totally or partially constituting the materials, preparations and products inherent to our daily life. With the global output (produced and shipped) valued at $5 trillion as of 2011, they have a relevant role in the world economy (American Chemistry Council 2012). Release to the environment due to deficient management is widespread and the potential for accumulation by wildlife and humans is very high.

Some of those chemicals have the ability to mimic or antagonize the action of hormones, thus affecting the normal functioning of organisms leading to deleterious effects (Diamanti-Kandarakis et al. 2009). These chemicals, defined as endocrine disrupting chemicals became ubiquitous in the environment and include, among many others, organotin compounds.

Organotin compounds are used in marine activities and agriculture, as pesticides and fungicides; and in industry, for the manufacture of polyvinyl chloride (PVC) and silicone amongst other products where they act as catalysts (Sect. 4.2). Despite numerous applications in consumer products and household items (e.g. textiles, footwear, medical implants, childcare articles, pillows, etc.), their notoriety is due to tributyltin, a potent biocide used until 2008 in antifouling paints formulations, which caused well known negative effects on aquatic organisms, as for example imposex in gastropods (Sects. 4.3 and 4.4). This phenomenon is considered the best example of endocrine disruption in wildlife and has received tremendous attention from the scientific community over the past several years. The extensive contamination of the marine environment by tributyltin and its metabolites, alongside with the deleterious effects towards non target organisms lead to the introduction of restrictions on their use that culminated with a global ban in 2008 (Sect. 4.5). Despite their ban, organotins are still of concern in the marine environment mainly due to the sequestration by sediments that occurred during the pre-ban period, which leads them to act nowadays as a secondary source.

Due to its widespread use, organotins can be found almost everywhere, from soils and sediments to water, biota and the atmosphere (Sect. 4.6). In general terms, the lowest concentrations are detected in the water column, whereas the highest are detected in sediments and biota, particularly in top predators. Since organisms provide a better integration of contaminant exposure over long periods of time, several monitoring programs were developed using different taxonomic groups in order to assess contamination. Several of those monitoring surveys reveal that fish and shellfish frequently consumed by humans have high levels of butyltin compounds. Such evidences demonstrated that tributyltin used in antifouling paints (and its metabolites) are reaching humans trough the consumption of contaminated seafood.

The hugely diverse industrial application grants organotins a ubiquitous presence in every aspect of human life through clothes, pillows, toys, wallpaper, plastic food containers, pipes (including drinking water mains and household piping), medical devices and breast implants (Sect. 4.7). This scenario creates frequent exposure opportunities by dermal contact with treated products or through ingestion of contaminated dust, which are cumulative with the ingestion of contaminated food (particularly seafood). Organotins have already been detected in human samples and toxicity data reveals that they are endocrine disruptors, imunotoxicants, and carcinogens and they were recently linked to obesity.

The review here presented was based in a digest made from peer reviewed literature available through specialized search engines, e.g. SCOPUS. Simultaneously, an attempt was made to obtain and integrate as much grey literature as possible, as the authors recognize its importance in the particular case of organotins and its environmental, societal and economic impacts. Given the intrinsic nature of this literature, no claim of comprehensiveness is made by the authors.

4.2 Organotin Compounds: Chemical Profile, Production and Applications

4.2.1 Chemical Structure

Organotin compounds (OTs) are organometals characterized by a tin (Sn) atom covalently bound to one or more organic substituents (e.g. methyl, ethyl, butyl, propyl, phenyl, octyl) (Hoch 2001; WHO 1990). Chemically these compounds are represented by the general formula RSnX, in which R is an organic alkyl or aryl group and X is an inorganic or organic ligand as, for instance, chloride, fluoride, oxide, hydroxide, carboxylate or thiolate (Hoch 2001; Sekizawa et al. 2003; WHO 1990). According to the number of organic groups, organotins can be classified in four distinct classes: monoorganotins ($RSnX_3$), diorganotins (R_2SnX_2), triorganotins (R_3SnX), and tetraorganotins (R_4Sn) (Table 4.1).

The chemical and physical properties of organotin compounds vary significantly, depending mainly upon the number and nature of the R groups, but also upon the type of ligand (X). Organotins solubility in water, for instance, tends to decrease with both the increase in the number and length of the organic substituents; however the nature of the ligand can also play an important role (Hoch 2001; WHO 1990). The toxicity of organotins also varies greatly, being strongly influenced by the number and nature of the organic groups. In general, inorganic tin is non-toxic whereas tri-substituted compounds have maximum toxicological activity (Hoch 2001; Sekizawa et al. 2003).

4.2.2 Production and Industrial Applications

> Organotins belong to the most important anthropogenic organometallic compounds.
>
> (Fent 2006)

With an estimated annual production of about 60,000 tons, organotins belong to the most widely used organometallic compounds family (Mala 2008). Excluding methyltins that can be produced naturally by bacteria (Guard et al. 1981; Hallas et al. 1982), all organic forms of tin are of anthropogenic origin.

The first organic tin compounds were synthesized in the mid nineteenth century firstly by Löwig, in 1852, closely followed by Frakland, in 1853 (de Carvalho Oliveira and Santelli 2010; Omae 2003). However, their industrial production started only a century later, during the 1940s, when the plastic industry, especially the one involved in the production of polyvinyl chloride (PVC), began to expand. Organotin compounds are used as additives in PVC manufacturing process to prevent its discoloration and embrittlement under the influence of light and heat (Hoch 2001). Besides PVC stabilization, organotins are also used as catalysts in the production of polyurethane foams and silicones.

The finding of the biocidal properties of tri-substituted organotins in the late 1950s broadened their applications to become the active ingredient in antifouling paint (AF) formulations (Sekizawa et al. 2003; WHO 1990) which led to its ubiquitous presence in the marine environment.

Less significant (in terms of production volume) applications comprise: agricultural pesticides and fungicides; biocides in construction materials and household items; surface disinfectants (including those used in hospitals and veterinary practice); laundry sanitizers; rodent-repellents; scintillation detectors for γ- and X-rays; ballistic

Table 4.1 Examples of selected organotin compounds with their respective formula, molecular weight and chemical structure. Adapted from PubChem (www.pubchem.ncbi.nlm.nih.gov)

Class	Compound	Abbreviation	Formula	Molec. weight	Chemical structure
Butyltins (BTs)	Monobutyltin	MBT	C_4H_9Sn	175.82426	
	Dibutyltin	DBT	$C_8H_{18}Sn$	232.93852	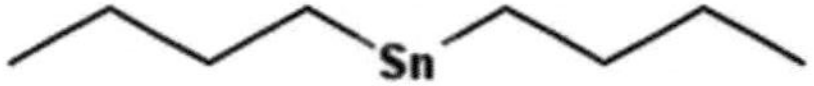
	Tributyltin	TBT	$C_{12}H_{27}Sn$	290.05278	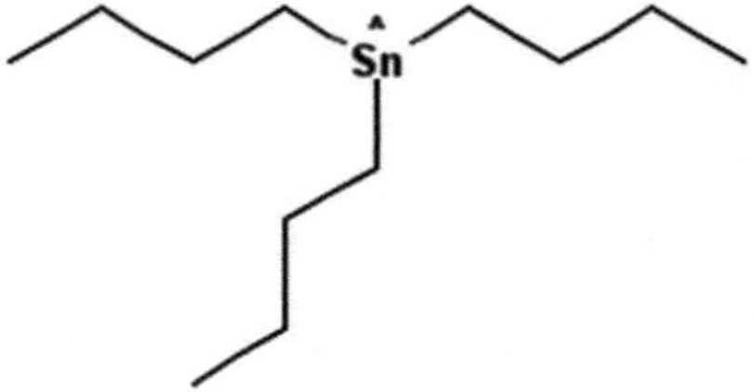
Phenyltins (PTs)	Monophenyltin	MPT	C_6H_7Sn	97.82978	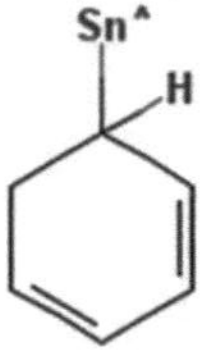
	Diphenyltin	DPT	$C_{12}H_{10}Sn$	272.9178	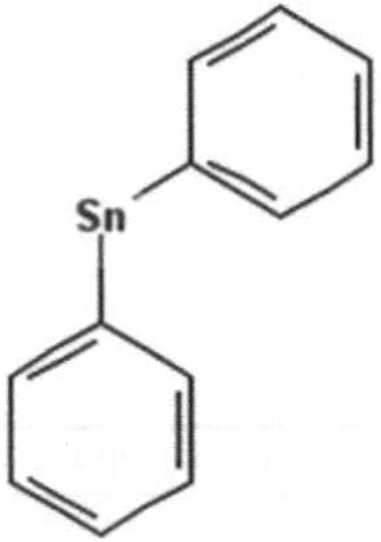

(continued)

Table 4.1 (continued)

Class	Compound	Abbreviation	Formula	Molec. weight	Chemical structure
	Triphenyltin	TPT	$C_{18}H_{15}Sn$	350.0217	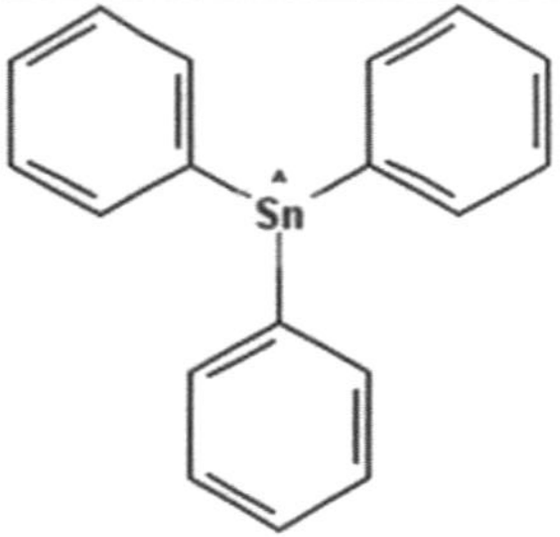
Octyltins (OcTs)	Monooctyltin	MOT	$C_8H_{20}Sn$	234.9544	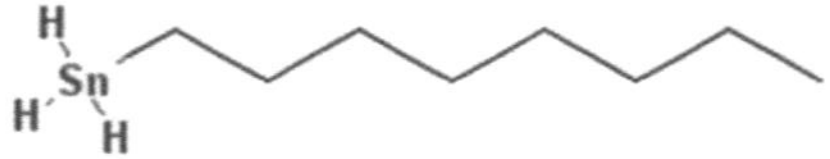
	Dioctyltin	DOT	$C_{16}H_{34}Sn$	345.15116	
	Trioctyltin	TOT	$C_{24}H_{51}Sn$	458.37174	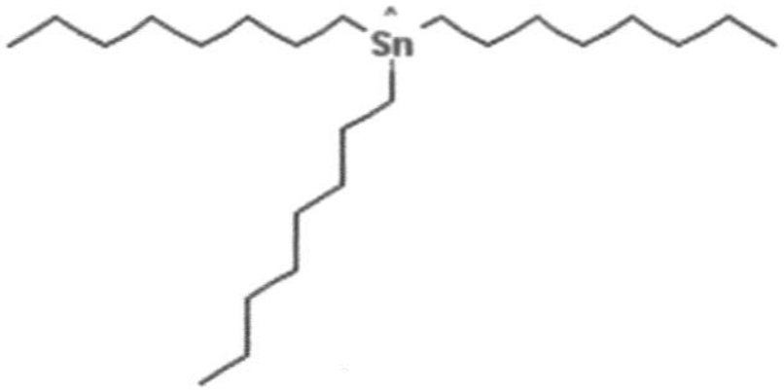

additives for solid-rocket engine fuels; ionophores in liquid membrane ion-selective electrodes and pharmaceuticals (e.g. anti-inflammatory and cancer treatments drugs) (Allsopp et al. 2001; Mala 2008; Valla and Bakola-Christianopoulou 2007; WHO 1990) (Table 4.2).

Despite the incredible amount of applications (a reflex of the high number of organotin compounds available – more than 800, according to Hoch (2001)), the majority of organotins production goes to the PVC (polyvinyl chloride) industry (mainly dibutyltins, dioctyltins and mono substituted organotins), and to the manufacture of polyurethane foams, and silicones (Fig. 4.1). Less than 20 % of the organotins annual production is used as biocides and pesticides (EVISA 2009).

4.3 Antifouling Paints: The Organotin Family Flag Ship

4.3.1 The Need to Develop Antifouling Strategies

Biofouling can be described as the settlement and growth of marine organisms on immersed structures. It begins with a complex community of bacteria, diatoms, protozoa and algal spores; and evolves to an even more complex community that includes algae, hydrozoans, crustaceans, tubeworms and mollusks (Omae 2003). The development of such community on ships poses serious problems to the shipping industry (Figs. 4.2 and 4.3).

According to the International Maritime Organization, six months is the necessary time for an unprotected ship hull at sea to gather 150 kg m^2 of fouling organisms. Considering for example a supertanker with 40,000 m^2, the amount of biofouling organisms may reach 6,000 tones. The consequent increase in the hull roughness and weight leads to a decrease in the ship performance due to enhanced drag. In order to maintain the same range of average speed an increase of 40–50 % in the fuel consumption is expected (IMO 1999). Alongside an increase in fuel consumption, costs associated with cleaning, repainting and dry docking services have also to be considered (Hoch 2001). Furthermore, biofouling also facilitates the transport of non-indigenous species (Dafforn et al. 2011). The use of antifouling technologies becomes, thus, unavoidable.

4.3.2 The Use of Tributyltin as Biocide in Antifouling Paints

Diverse antifouling strategies were developed throughout naval history. However, it was only after the 1950s when the outstanding biocidal properties of trialkyltins were discovered that the antifouling industry gained a new balance, reaching today the overall annual value of €3 billion (Dafforn et al. 2011).

The discovery of the biocidal activity of triorganotins was attained by a team of Dutch scientists led by van der Kerk (see Omae 2003 and references therein).

Table 4.2 Summary of organotin compounds applications

Organotin classes	General formula	Applications	Selected references
Monosubstituted	$RSnX_3$	Stabilizers in PVC films Glass treatment (Limited applications when compared to other OTs)	WHO (1990) and ATSDR (2005)
Disubstituted	R_2SnX_2	Stabilizers in plastics industry (particularly PVC) Catalysts in the production of polyurethane foams and in room-temperature vulcanization of silicones Glass treatment processes as precursors for SnO_2 film Dewormers in poultry farming Water-proofing agents for cellulosic materials (e.g. cotton textiles, paper and wood) Flame retardants for wool fabrics Binder in water-based varnishes	WHO (1990), Hoch (2001), de Carvalho Oliveira and Santelli (2010) and Antizar-Ladislao (2008)
Trisubstituted	R_3SnX	Biocides in antifouling paint formulations Fungicides, insecticides, miticides, and antifeedants in agrochemical industry Pesticides for ornamental plants Miticides in citrus fruits Acaricides in vineyards Insecticide and fungicide in wood preservation Biocide in construction materials Disinfectants and biocides for cooling systems in power stations, pulp and paper mills, textile mills, breweries, tanneries Insecticides and antifeedants in textiles Biocides in allergic pillows Biocides in insoles for shoes Biocides in cycling shorts padding Biocides in sprays for athlete's foot treatment	WHO (1990), Hoch (2001) and RPA (2005)
Tetrasubstituted	R_4Sn	Intermediates in the preparation of other organotin compounds Oil stabilizers	WHO (1990) and de Carvalho Oliveira and Santelli (2010)

*OT*s organotin compounds, *PVC* polyvinyl chloride, *SnO_2* tin dioxide

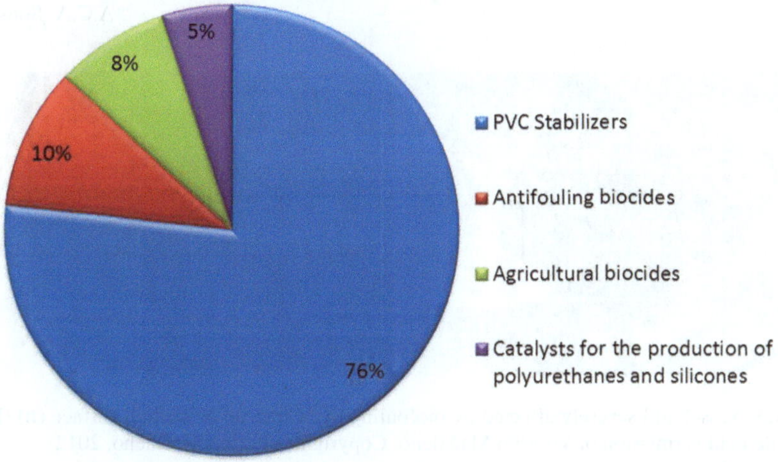

Fig. 4.1 Major industrial applications of organotins as relative percentage of total production. *PVC* polyvinyl chloride (Data cited from EVISA 2009)

Fig. 4.2 Biofouling. When submerged into water, a community of algae and invertebrates develops in the ship's hull (**a**). Such community enhances corrosion in ships' essential structures (**b**) and increases roughness which in turn leads to an increase in hydrodynamic drag and fuel consumption. In order to overcome such problem the naval industry developed antifouling systems that prevent the attachment and growth of organisms in the submerged surfaces. The more effective the system, the larger periods a ship can spend at sea without need for dry-docking (**c**, **d**)

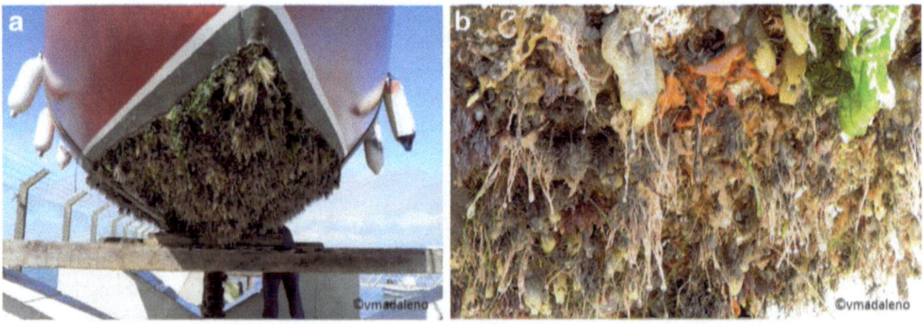

Fig. 4.3 Vessel hull severely affected by biofouling (**a**); Close up of the hull surface (**b**) (Printed with the kind permission of Vitorino Madaleno. Copyright: Vitorino Madaleno, 2012)

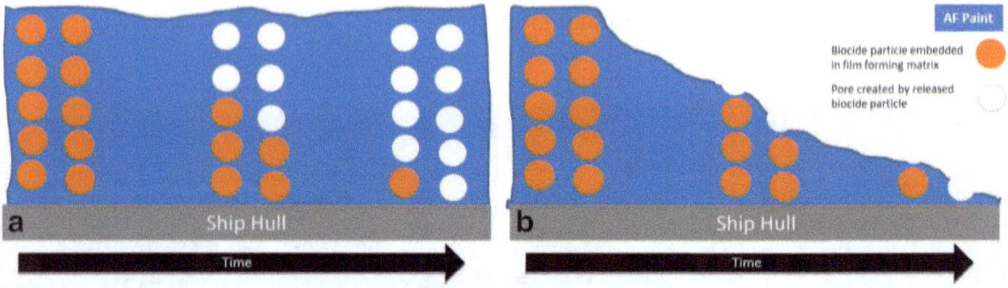

Fig. 4.4 Simplified schematic representation of biocide release in traditional antifouling paints. (**a**) Insoluble matrix; (**b**) Soluble matrix (Adapted from Chambers et al. 2006). *AF paint*: Antifouling paint

Following this discovery, tributyltin was tested as molluscicide in African countries, in order to prevent the spread of schistosomiasis, a disease caused by a trematod parasite that has several freshwater snails as intermediate hosts (WHO 1990). After this first success, the antifouling properties of tributyltin were recognized and this compound was registered as wood preservative and afterwards it was also used as biocide in antifouling paints (WHO 1990).

The first tributyltin based antifouling paints (Fig. 4.4) had the biocide dispersed in a resinous matrix, causing the release to be highly uneven, with very fast leaching in the first months becoming much slower afterwards, until the complete removal of the biocide from the matrix 18–24 months after application (IMO 1999). The late 1960s saw the development of a new technology: instead of being dispersed in the paint matrix, tributyltin was chemically bonded to the polymer base (Fig. 4.5). With these self-polishing paints the leaching rate was controlled and constant throughout the life of the paint, which could reach up to 60 months (IMO 1999).

As a consequence of these enhanced properties, by the 1970s most seagoing vessels used tributyltin based antifouling paints (IMO 1999). At that time, tributyltin

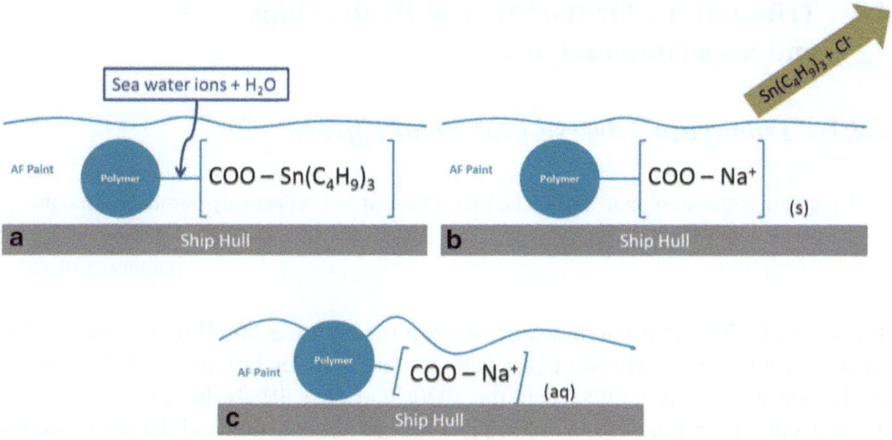

Fig. 4.5 Schematic representation of a tributyltin self-polishing copolymer paint (TBT-SPC) (**a**) The ester linkage bonding tributyltin onto the polymer backbone suffers "hydrolysis" by sea water ions and water molecules; (**b**) as a consequence tributyltin moieties are released; (**c**) the loss of tributyltin moieties causes fundamental changes in the copolymer, leading to its erosion by the moving sea water (the self-polishing effect). *AF paint* Antifouling paint, *aq* aqueous, *s* solid (Adapted from Yebra et al. 2004)

was considered as the ideal antifouling agent, firstly due to their effectiveness and long duration, which prevented costly dry dock operations; secondly because it was thought that tributyltin was easily degraded by UV light and microorganisms in the aquatic environment, making them more environmental friendly than the antifouling compounds used up until then: arsenic and dichlorodiphenyltrichloroethane (DDT) (IMO 1999; Omae 2003). In fact, most of the experimental data available at that time supported the belief that "organotin compounds were ideal compounds because they change to nontoxic inorganic tin compounds after use", as a scientist of the time would explain to his co-workers (Matsuda cited from Omae 2006).

However, this belief was rapidly and dramatically denied by hard facts. Several reports demonstrated that tributyltin was persistent in the marine environment being responsible for several deleterious effects on non-target organisms (see Sect. 4.4). Due to its ubiquitous application for decades, tributyltin was released into the aquatic environment in an unprecedented scale for a man-made chemical. The magnitude of this phenomenon can be appreciated if one considers the data reviewed by Hoch (2001): a commercial ship during a 3-day stay in a harbor can release as much as 600 ng L^{-1} (in terms of Sn). Once in the water column, and due to its strong affinity to suspended particles, tributyltin will accumulate in sediments that act primarily as sinks and reservoirs, and secondarily as sources of the compound (Hoch and Schwesig 2004).

4.4 Tributyltin – Or the Story of Boats, Ships and Superimposed Sex

4.4.1 Tributyltin Adverse Biological Effects

> Tributyltin is the most toxic xenobiotic ever produced and deliberately introduced into the marine environment
>
> (Goldberg 1986)

The harmful effects of tributyltin were first recorded in Arcachon Bay, an enclosed bay located in the French Atlantic coast with abundant oyster farming facilities. Oyster production was severely affected in the 1970s and early 1980s due to severe growth problems that were followed by a complete lack of reproduction and the emergence of shell calcification anomalies (see review by Alzieu 2000). Such problems were associated with the existence of numerous marinas on the bay. The high numbers of vessels docked there were releasing high amounts of tributyltin into the aquatic environment.

In the beginning of the 1980s, tributyltin was linked to a sexual disorder recorded in some gastropod species in the United States and England (Blaber 1970; Bryan et al. 1987; Smith 1971, 1981a, b). Females of *Nassarius obsoletus* (Say), in the USA; and *Nucella lapillus* (L.), in England; were affected by imposex – a phenomenon characterized by the superimposition of male characters (penis and *vas deferens*) into functionally normal prosobranch females (Smith 1971). In advanced stages, the development of the *vas deferens* can block the vulva leading to female sterilization. Such was the case of *N. lapillus* females that experienced dramatic population declines that ultimately led to its extinction at heavily tributyltin polluted locations (Bryan et al. 1986). Thus, imposex is regarded as an example of endocrine disruption whose effects are notorious at the individual and also at the community levels. It has been reported to occur in more than 200 gastropod species worldwide (Shi et al. 2005) and it has been used as a fairly specific biomarker of tributyltin pollution. A more detailed description of imposex and its relevancy as a tributyltin pollution monitor is provided in the next section.

Despite the particular sensitivity patented by molluscs (with the most notorious examples mentioned above) tributyltin is extremely toxic towards a wider range of organisms, from bacteria to mammals, including humans (Table 4.3). Publications regarding the toxicity of tributyltin towards different taxa vary widely in its focus and extension. As an example of the discrepancy in the number of published works between the different taxa, a survey in Scopus database in October 2012, retrieves 28 results when searching for "tributyltin" and "phytoplankton" in "Title, abstract and keywords" field, whereas for "tributyltin" and "fish" it retrieves 430 results. Moreover, for certain taxa most of the publications available are not directly related to toxic effects. As an example, for bacteria, most of the current research focus on tributyltin resistance and the possibility of using these organisms in bioremediation strategies (see Cruz (2012) from a complete review on the subject). The objective is

Table 4.3 Overview of tributyltin effects, by taxonomic group

Group	Effects	Selected references
Bacteria	Toxic (particularly to Gram positive) Inhibits growth and metabolism Affects respiration Decreases productivity Inhibits solute transport Inhibits macromolecules biosynthesis and transhydrogenase	WHO (1990), Wuertz et al. (1991), Gadd (2000), Mendo et al. (2003), Martins et al. (2005) and Cruz et al. (2012)
Phytoplankton	Reduces marine and freshwater microalgae growth at low concentrations Reduces respiration and photosynthetic activity Alters photosynthetic pigment content Induces drastic changes in biochemical composition Reduces primary productivity Induces changes in the community structure	WHO (1990), Beaumont and Newman (1986), Bryan and Gibbs (1991), Fargašová (1996, 1997), Fargašová and Kizlink (1996), Nudelman et al. (1998), Petersen and Gustavson (1998), White et al. (1999), Gadd (2000) and Sidharthan et al. (2002)
Plants	Impairs the development of motile spores of some macroalgae Reduces growth in several marine angiosperms Decreases growth and transpiration rates Reduces photosynthetic activity Bioaccumulates and induces stress in plants used for human consumption	WHO (1990), Trapp et al. (2004), Jensen et al. (2004), Caratozzolo et al. (2007), Lespes et al. (2009) and Carvalho et al. (2010)
Crustaceans	Reduces reproductive performance Reduces neonate survival Inhibits larvae developmental ratio Decreases juvenile growth rates Induces changes in the community structure	WHO (1990), Kusk and Petersen (1997), Waldock et al. (1999), Dahllof et al. (2001), Takeuchi et al. (2001) and Aono and Takeuchi 2008
Molluscs	Shell thickening Larval mortality Inhibits egg development Reduces survival ship of hatchlings Decreases growth Alters sex ratio Induces heat-shock proteins Induces DNA damage Imposex[a] Sterility	Langston (1990), WHO (1990), Bryan and Gibbs (1991), Leung et al. (2007), Horiguchi (2009b), Sousa et al. (2005a, 2009b), Barroso et al. (2002), Hagger et al. (2006) and Park et al. 2012

(continued)

Table 4.3 (continued)

Group	Effects	Selected references
Fish	Inhibits growth Masculinization Induces sperm abnormalities Reduces fecundity Inhibits ovarian development Embryo abnormalities Larval malformations Liver vacuolation Inhibits Cytochrome P450 Hematopoietic tissue hyperplasia Neurotoxic (modulation of the glutamate signalling pathway) Induces DNA damage Induces severe damage in the thyroid gland Lipotoxic	WHO (1990), Fent (1996a), McAllister and Kime (2003), Shimasaki et al. (2003), Zheng et al. (2005), Zhang et al. (2007, 2009, 2008, 2011, 2013) and Zuo et al. (2009, 2012)
Mammals	Reproductive anomalies, including reduced spermatogenesis and embryo malformations Induces adipose tissue differentiation and obesity Neurobehavioral alterations Immunological disorders Cardiovascular toxicity Suppression of osteoclastogenesis	WHO (1990), Grün and Blumberg (2006), Ohtaki et al. (2007), Yonezawa et al. (2007), Antizar-Ladislao (2008), Chen et al. (2008) and Mala (2008)

[a]For a complete description on imposex see Sect. 4.4.2

being pursued with vascular plants, particularly in saltmarshes (for further information see Carvalho 2010).

4.4.2 Imposex as a Biomarker of Tributyltin Pollution

> Imposex is the best documented example of endocrine disruption in wildlife.
>
> (Matthiessen and Gibbs 1998)

In 1970, Blaber recorded for the first time the appearance of a penis in females of the gastropod *Nucella lapillus* collected in Plymouth Sound, United Kingdom. The severity of this condition was higher inside harbors and lower farther away, but the causative agent was unidentified. One year later, Smith (1971) also noticed the appearance of a penis and *vas deferens* in *Nassarius obsoletus* females in the United States. He coined the phenomenon as imposex but at that time the cause was still unknown. In order to unveil what was causing this bizarre phenomenon, Smith conducted an extensive field survey in which over 3,000 *N. obsoletus* specimens were examined (Smith 1981a). He studied nine pairs of marinas and control sites and concluded that the imposex levels were significantly higher inside marinas (Smith 1981a). Such results suggested that this phenomenon was caused by a chemical compound present in marinas. To test this hypothesis, Smith performed a series of experiments in which he transferred snails between a yacht basin and a distant 'clean' locality and vice-versa (Smith 1981b). He also exposed in the laboratory, snails not affected by imposex to seawater containing organotin based antifouling paints (Smith 1981b). The results clearly demonstrated that snails exposed both *in situ* and in the laboratory developed imposex and that imposex was suppressed (but not lost) when affected females were transferred to clean seawater (Smith 1981b). With Smith's work, the causal relationship between imposex and tributyltin was finally established (Smith 1981a, b). Later, this relationship was confirmed by Bryan and co-workers through a series of laboratory and field experiments using *N. lapillus* (Bryan et al. 1987). The same team proposed the use of imposex in *N. lapillus* to monitor tributyltin pollution (Gibbs et al. 1987). At that time, several indices were proposed to evaluate the degree of imposex within a population: percentage of affected females (%I); relative penis size index (RPSI) defined as the ratio between female penis size (FPS) and male penis size (MPS): RPSI=(FPS*100/MPS) and the *vas deferens* sequence index (VDSI). For *N. lapillus*, Gibbs and co-workers (1987) proposed a VDSI scheme with 6 stages with stage 0 corresponding to unaffected females and stage 6 corresponding to sterile ones.

Besides these two species (*N. obsoletus* and *N. lapillus*), imposex is now known to affect more than 200 gastropod species (Shi et al. 2005). It is currently a worldwide phenomenon, being reported in Europe; North and South America; Asia; Africa; Australia and even in such remote locations as the Arctic and Antarctic (see e.g. Gibson and Wilson 2003; Horiguchi 2012; Negri and Marshall 2009; OSPAR 2010; Shi et al. 2005; Strand et al. 2006; Tallmon 2012), where it is used as a biomarker of tributyltin pollution. The robustness of imposex as a specific biomarker led the OSPAR Commission to recommend the use of imposex to monitor tributyltin pollution levels across Europe (OSPAR 2003).

Taking a Closer Look into Penis Bearing Females – *N. Reticulatus* Case Study

The marine netted whelk *Nassarius reticulatus* is a ubiquitous nassarid gastropod in European coastal areas from the Canaries and Azores to Norway and throughout the Mediterranean and Black Seas (Fretter and Graham 1994). It has been successfully used in several tributyltin biomonitoring programs along European coastline (Table 4.4). This species is commonly found at sheltered and shallow (including intertidal) sites, being particularly abundant in estuaries and inside harbors (Barroso et al. 2002). It is also widespread in offshore waters (Barroso et al. 2011; Rato et al. 2006, 2008).

Table 4.4 List of studies reporting the occurrence of imposex in *N. reticulatus* throughout Europe

Country	Study area	References
England	Coastal areas in south-west England	Bryan et al. (1993)
Germany	Coastal areas	Martina et al. (2007) and references therein
France	Coastal areas in Brittany	Stroben et al. (1992a, b), Oehlmann et al. (1993), Huet et al. (1995) and Wirzinger et al. (2007)
Spain	Coastal areas North Atlantic: Galician and Basque coast	Barreiro et al. (2001, 2004); Ruiz et al. (2005), Quintela et al. (2006), Couceiro et al. (2009) and Rodríguez et al. (2009, 2010)
Portugal	Coastal areas along the continental Portuguese coast	Gibbs et al. (1997); Barroso et al. (2000, 2001, 2002, 2005); Pessoa et al. (2001); Sousa et al. (2005b, 2007b, 2009b), Rato et al. (2009) and Laranjeiro et al. (2010)
Portugal	Offshore areas, Portugal	Santos et al. (2004), Rato et al. (2006, 2008) and Barroso et al. (2011)

The use of *N. reticulatus* to monitor tributyltin pollution was proposed by Stroben and co-workers in the early 1990s (Stroben et al. 1992b). This species was latter recommended by the OSPAR Commission to monitor the effects of tributyltin in coastal areas where *N. lapillus* is absent (OSPAR 2003). The imposex development in *N. reticulatus* occurs at tributyltin seawater concentrations of 1 ng Sn L^{-1} being thus less sensitive than *N. lapillus* for which imposex development starts at seawater levels as low as 0.4 ng Sn L^{-1} (Bryan et al. 1993; Stroben et al. 1992a).

The following indices are used to evaluate imposex in *N. reticulatus*:

(%I) Percentage of affected females, i.e., number of affected females*100/total females. This index is useful at low polluted locations; at highly polluted sites its usefulness is dubious since usually all females are affected by imposex and no distinction can be attained.

FPL Female Penis Length (mm). Important parameter in low, medium and highly polluted locations as it enables the distinction between sites with similar pollution levels.

RPLI Relative Penis Length Index; RPLI = (FPL*100)/MPL.[1] As for FPL this parameter is very useful in the entire pollution spectrum.

AOS Average Oviduct Stage. This index proposed by Barreiro et al. (2001) assesses the degree of oviduct masculinization. In normal females the oviduct is a straight line (AOS = 0) whereas in highly affected specimens it gets convoluted resembling the male *vesicula seminalis* (AOS = 2). Useful at moderate/highly polluted locations since at low polluted sites AOS value is usually zero.

[1] MPL: male penis length (mm).

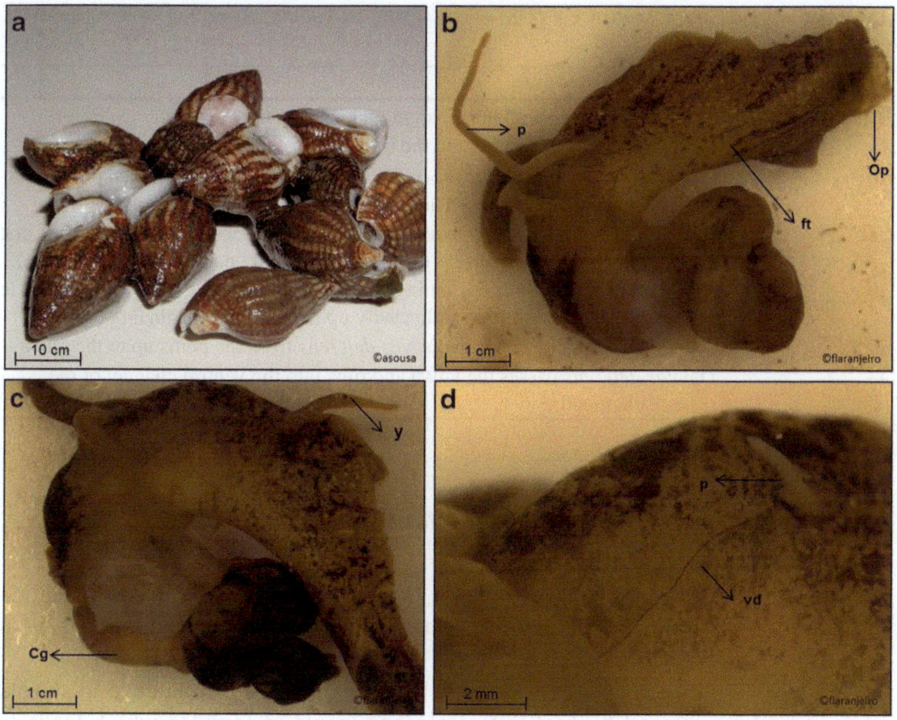

Fig. 4.6 *Nassarius reticulatus*. (**a**) Adult specimens collected at Ria de Aveiro, Portugal and (**b**–**d**) Shell stripped individuals: (**b**) Male specimen; (**c**) Imposex affected female; (**d**) Close up of the female penis area. *Op* operculum, *p* penis, *ft* foot, *y* eye, *Cg* capsule gland, *vd*: vas deferens (Printed with the kind permission of Filipe Laranjeiro. Copyright: Filipe Laranjeiro, 2008)

VDSI *Vas deferens* sequence index. Classifies the degree of females' masculinization and is considered as the most valid index to assess tributyltin pollution as, in extreme cases, it can express the reproductive capacity of the population (Stroben et al. 1992b) (Fig. 4.6).

Table 4.5 summarizes the four different VDS stages proposed by Stroben et al. (1992b) for *N. reticulatus*. Stage 4+ was considered as the end of imposex development in the nassarid in the initial scheme proposed by Stroben et al. (1992b), however some authors proposed the computation of stage 4+ to 4.5 (Barreiro et al. 2001) or to 5 (Barroso et al. 2002) in order to better discriminate between highly polluted locations. The majority of recent reports use the latter (Barroso et al. 2005; Laranjeiro et al. 2010; Rato et al. 2008, 2009; Rodríguez et al. 2009; Ruiz et al. 2005; Sousa et al. 2005b, 2007a, 2009b).

Table 4.5 *Vas deferens* sequence (VDS) stages in *N. reticulatus* according to Stroben et al. (1992b)

VDS stage	Description
0	Unaffected female with no apparent male characters.
1a	Tiny penis without duct behind the right ocular tentacle.
1b	No penis, but a short, distal *vas deferens* tract behind the right ocular tentacle.
2	Penis with a closing or closed duct behind the right ocular tentacle.
3a	Penis with duct continuing in an incomplete distal tract of the *vas deferens* that is growing out successively towards the vaginal opening.
3b	Penis lacking; *vas deferens* running continuously from the right ocular tentacle over the bottom of the mantle cavity up to the vulva (vaginal opening).
4	Penis with duct and a continuous *vas deferens* from the penis up to the vulva.
4+	The *vas deferens* passes the vulva and runs into the ventral channel of the capsule gland.

4.4.3 Mechanisms Underlying Imposex Induction

Several hypotheses have been postulated in order to explain imposex induction mechanism and three possible pathways have been identified: (A) steroid; (B) neuroendocrine, and (C) retinoic.

The steroid pathway postulates that tributyltin causes an imbalance in steroid hormones characterized by higher testosterone levels in females which leads to the imposex development; this disruption in steroid homeostasis can be promoted either through interferences with the steroid biosynthesis (see Bettin et al. 1996; Spooner et al. 1991) or with steroid excretion (see Gooding et al. 2003; Ronis and Mason 1996).

The neuroendocrine pathway suggests that the homeostasis of certain neuroendocrine factors can be disrupted by tributyltin, leading to the formation of accessory sex organs in females (Féral and Le Gall 1983; Oberdörster and McClellan-Green 2000, 2002).

The retinoic pathway proposes that tributyltin and triphenyltin mimic the endogenous ligand of retinoid X receptor (RXR) and thus activate the signaling cascades which are retinoic acid dependent (Castro et al. 2007; Coelho et al. 2012; Horiguchi et al. 2007a, b, 2010; Lima et al. 2011; Nishikawa et al. 2004; Sousa et al. 2010).

So far, both neuroendocrine and steroidal theories could not provide unequivocal experimental results as they were unable to promote the increase in female penis length to the same extent as tributyltin whilst in the RXR theory a single injection of 9-*cis*-retinoic acid (9CRA), the natural ligand for human RXRs, induced the development of imposex (percentage of affected females) as well as the substantial growth in female penises to the same extent as tributyltin or triphenyltin, when administrated at similar concentrations (Horiguchi 2006).

A complete and more detailed description of imposex induction mechanisms can be found in the following reviews: Castro et al. (2007), Horiguchi (2009a), (2012), Nishikawa (2006).

Recently, Pascoal and co-workers (2013), suggested the involvement of a novel pathway. Those authors studied the gene expression profile of *N. lapillus* when exposed to tributyltin, and their results support the involvement of the three previously suggested imposex induction mechanisms (steroid, neuroendocrine and retinoid). Furthermore, they reported the involvement of putative peroxisome proliferator-receptor (PPAR) pathway. In a series of laboratory experiments, they demonstrated that, when injected with rosiglitazone, a well-known PPARγ ligand, *Nucella lapillus* females' developed a penis over a two month period. In face of such results, the authors proposed an alternative mechanism in which tributyltin acts in the PPARγ signaling pathway, probably via the RXR/PPAR heterodimer that in turn triggers imposex development. The activation of this heterodimer could be *via* RXR or PPAR, which explains the imposex induction with 9CRA and also with rosiglitazone. This was the first time that the PPAR signaling cascade was implicated in endocrine disruption in invertebrates. Considering that, in vertebrates the PPAR signaling pathway is implicated in endocrine disruption (e.g. Grün and Blumberg 2006), Pascoal et al. (2013) pioneer study highlighted once again the existence of a common signaling pathway between invertebrates and vertebrates species (Castro et al. 2005; Santos et al. 2012).

Despite the recent advances in the field (e.g. Coelho et al. 2012; le Maire et al. 2009; Lima et al. 2011; Pascoal et al. 2013; Sternberg et al. 2010), many questions remain unsolved, particularly the ones concerning the interplay between all pathways. It remains therefore a challenge for the future the identification of all the intervenients in this bizarre and complex phenomenon, which arguably became the best biomarker ever used to monitor a specific type of pollution.

4.5 Regulating Organotin Compounds

4.5.1 Antifouling Paints International Regulations

The first regulations were adopted by those countries that experienced negative outcomes due to tributyltin pollution. Therefore, the first restrictive initiative was adopted by France in 1982, motivated by the collapse of the oyster farming industry due to tributyltin pollution in Arcachon Bay, with the prohibition to use tributyltin on small boats (<25 m). Four years later, the United Kingdom banned the sale of tributyltin based antifouling paints for small boats and aquaculture structures due to the severe ecological impact of this pollutant on mollusk populations (see review by Alzieu 1998). Similar restrictions were adopted by other countries, namely USA and New Zealand in 1988; Australia, Norway and Canada in 1989 and Japan in 1990.

The European Union introduced in 1989 the Directive 89/677/EEC banning the use of tributyltin and triphenyltin on small boats (<25 m). This directive was afterwards implemented by each member state. In Portugal, for example, the Directive was transposed to the internal law in 1993. However, the first initiative was adopted 1 year before, when tributyltin and triphenyltin based antifouling paints were banned from the Navy fleet.

Following the introduction of the tributyltin partial ban, some recovery of severely affected gastropod and oyster populations, as well as other marine communities was visible (Alzieu 1998; Blanck and Dahl 1998; Evans et al. 1995, 1996; Harding et al. 1997; Miller et al. 1999; Minchin et al. 1995; Rees et al. 2001) however this recovery was not generalized. Several reports described that imposex and tributyltin levels around ports were not decreasing and that those locations were hotspots of tributyltin pollution (Bailey and Davies 1989; Davies and Bailey 1991; Gibson and Wilson 2003; Minchin et al. 1995, 1996; Morgan et al. 1998; Smith 1996). In Portugal, for example, an increase on *Nucella lapillus* imposex levels was reported between 1995 and 2000 (Barroso and Moreira 2002; Santos et al. 2002), indicating that the partial ban was ineffective in Portuguese coastal waters.

Simultaneously, other studies demonstrated that tributyltin pollution was not restricted to harbor areas, and populations from the open ocean as well as the ones from offshore areas were also affected (Barreiro et al. 2001; Mensink et al. 1996; Michel et al. 2001; Ten Hallers-Tjabbes et al. 1994). Tributyltin was even detected in remote and supposedly undisturbed ecosystems such as Antarctica (Negri et al. 2004) and the Great Barrier Reef in Australia (Haynes et al. 2002; Haynes and Loong 2002; Marshall et al. 2002).

Given all the scientific evidences that a partial ban was not effective on reducing tributyltin environmental levels and recognizing the need to protect the marine environment, the International Maritime organization (IMO) adopted, in 2001, a total ban on the use of tributyltin based antifouling paints through the 'International Convention on the Control of Harmful Antifouling Systems on Ships' (AFS Convention) from 2003 onwards. This convention, however, could only enter into force one year after 25 states, representing 25 % of the world's merchant shipping tonnage, ratified it.

Due to the slow ratification process and to promote the implementation of the AFS Convention in the member states, the European Union moved to the adoption of Directive 2002/62/EC and Regulation 782/2003 and the for the total interdiction of organotins antifouling paints application on European Union ships after 1st of July 2003 and the prohibition of organotins antifouling paints on ships' hulls from 1st of January 2008. From this date onwards the prohibition was extended to all ships entering European Union ports.

In September 2007, Panama became the 25th state ratifying the Convention and therefore the necessary conditions to the implementation of the Convention were achieved (Gipperth 2009). One year after, on the 1st September 2008, the AFS convention entered into force and from that date onwards organotins were banned from antifouling paints formulations (Fig. 4.7). However, the production and application of organotins based antifouling paints is still possible in some developing countries which are not members of the International Maritime Organization (Yi et al. 2012).

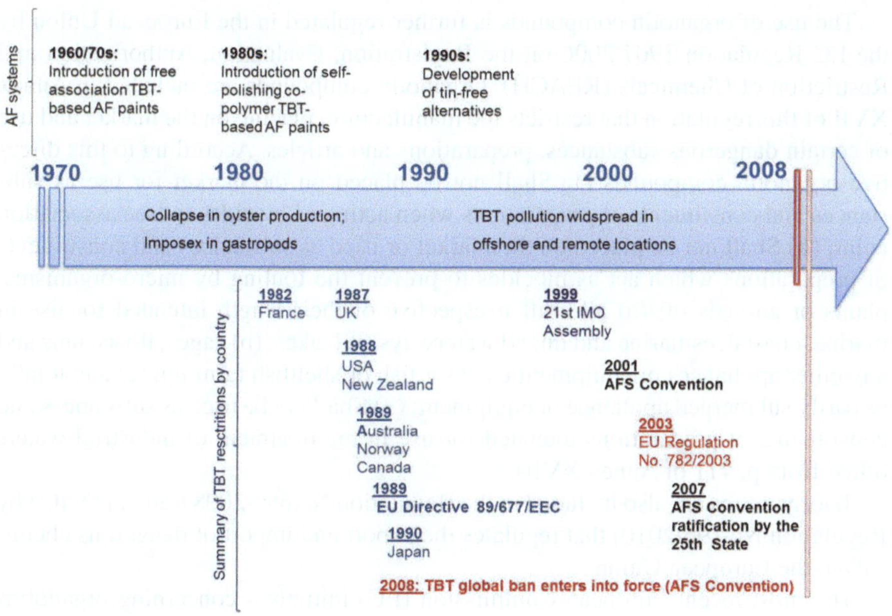

Fig. 4.7 Tributyltin-based antifouling paints timeline, with the indication of major developments in the antifouling (*AF*) industry; the negative environmental impacts and the summary of international regulations adopted by country. For detailed descriptions on worldwide regulations see reviews by Champ (2000) and Sonak et al. (2009) (Adapted from IMO 2002). AF paints Antifouling paints, AFS Convention: International Convention on the Control of Harmful Antifouling Systems on Ships; IMO International Maritime Organization, TBT Tributyltin

4.5.2 European Legislation for Organotin Compounds

As previously mentioned, in 1989, the European Union adopted the Directive 89/677/EEC that banned the use of tributyltin and triphenyltin as biocides in antifouling paints formulations applied to small boats (<25 m). The use of tri-substituted organotins was further restricted in 1998, under the Directive 98/8/EC of the European Parliament and of the Council of 16 February 1998, concerning the placing of biocidal products on the market (cited from Decision 2009/425/EC). Organotin compounds are also included as priority hazardous substances in the European Water Framework Directive (Directive 2000/60/EC). The use of triphenyltin as pesticide in agriculture was, in 2002, banned in Europe through Decision 2002/478/EC and Decision 2002/479/EC.

Since 2003 the use of organotin compounds in antifouling paints was completely banned in the European Union through the Directive 2002/62/EC and Regulation 782/20036 (see previous section for more details).

The use of organotin compounds is further regulated in the European Union by the EC Regulation 1907/2006 on the Registration, Evaluation, Authorization and Restriction of Chemicals (REACH). Organotin compounds are included in Annex XVII of this regulation that restricts the manufacture, placing on the market and use of certain dangerous substances, preparations and articles. According to this directive organotin compounds (1) Shall not be placed on the market for use as substances and constituents of preparations when acting as biocides in free association paint; (2) Shall not be placed on the market or used as substances and constituents of preparations which act as biocides to prevent the fouling by micro-organisms, plants or animals of: (a) all craft irrespective of their length intended for use in marine, coastal, estuarine and inland waterways and lakes; (b) cages, floats, nets and any other appliances or equipment used for fish or shellfish farming; (c) any totally or partly submerged appliance or equipment; (3) Shall not be used as substances and constituents of preparations intended for use in the treatment of industrial waters (cited from p. 411 of Annex XVII).

Triorganotins are also included in the Regulation No 689/2008 (later amended by Regulation No 196/2010) that regulates the export and import of dangerous chemicals in the European Union.

The most recent European Commission (EC) initiatives concerning organotins deals with the restriction of these substances in consumer products. In May 2009, the EC adopted Decision 2009/425/EC. In 2010, this Decision was incorporated into Annex XVII of REACH Regulation (EC Regulation No 1907/2006) *via* Regulation No 276/2010 (Table 4.6).

4.6 The Widespread Occurrence of Organotin Compounds

4.6.1 Sources and Pathways of Organotins in the Environment

The extensive usage of organotin compounds over the last decades is responsible for an extraordinary input of these contaminants into the environment. Regardless of the innumerous applications of organotins, it was the use of tributyltin, and to a lesser extent triphenyltin, as biocides in antifouling paints the main responsible for the introduction of these substances in the aquatic environment. Tributyltin (or triphenyltin) leaches directly from the antifouling paints into the water column where it is degraded into less toxic compounds by light or microorganisms through a series of progressive removal or organic groups according to the following scheme (de Carvalho Oliveira and Santelli 2010; Omae 2003):

$$R_3Sn^+ \rightarrow R_2SnX^{2+} \rightarrow RSnX^{3+} \rightarrow Sn(IV)$$

This stepwise dealkyllation or dearylation ultimately forms inorganic tin, a non-toxic element. Several factors can influence the degradation rate of organic tin

Table 4.6 Summary of organotin restrictions in consumer products according to Decision 2009/425/EC and Regulation No 276/2010. The restrictions are for mixtures, articles or part thereof when the concentration is greater than the equivalent of 0.1 % by weight of tin

Compound	Products	Effective from
Tri-substituted OTs, including TBT and TPT	All articles or part of articles	1st July 2010
Dibutyltin (DBT) compounds	Mixtures	1st January 2012
	Articles or part of articles (except food contact materials)[a]	
	One-component and two-component room temperature vulcanisation sealants (RTV-1 and RTV-2 sealants) and adhesives	1st January 2015
	Paints and coatings containing DBT compounds as catalysts when applied on articles	
	Soft polyvinyl chloride (PVC) profiles whether by themselves or coextruded with hard PVC	
	Fabrics coated with PVC containing DBT compounds as stabilizers when intended for outdoor applications	
	Outdoor rainwater pipes, gutters and fittings, as well as covering material for roofing and facades	
Dioctyltin (DOT) compounds	Textile articles intended to come into contact with the skin	1st January 2012
	Gloves	
	Footwear or part of footwear intended to come into contact with the skin	
	Wall and floor coverings	
	Childcare articles	
	Female hygiene products	
	Nappies	
	Two-component room temperature vulcanization moulding kits (RTV-2 moulding kits)	

OTs organotins, *PVC* polyvinyl chloride, *TBT* Tributyltin, *TPT* Triphenyltin
[a]Food containing materials are regulated by Regulation No 1935/2004

compounds into inorganic tin, namely: temperature, light intensity, pH, salinity, and the nature and density of microbial communities (de Carvalho Oliveira and Santelli 2010; Hoch 2001). This last factor is probably the most important one as abiotic transformation is of minor importance when compared to biological mediated degradation (Fent 2006).

However, and despite the fact that tributyltin can be degraded into less toxic forms (dibutyltin, monobutyltin and inorganic tin) in the water column under aerobic conditions, its low aqueous solubility and strong affinity to particulate matter and suspended particles, favors its deposition in sediments. Once in sediments, and under anaerobic conditions, the degradation of tributyltin is very slow with reported half-lives of several decades (see Fent 2006 and references therein). For those

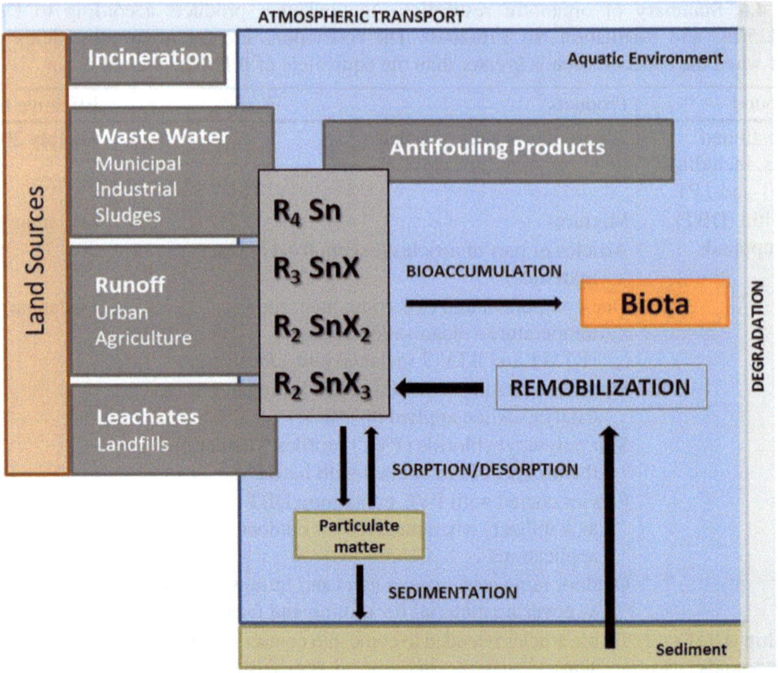

Fig. 4.8 Schematic representation of the main sources and pathways of organotin compounds in the environment. The use of tributyltin as biocide in antifouling paints is still considered the main source of organotins into the aquatic environment; tributyltin can be directly released from the vessel's surface, or it can be introduced into the aquatic environment through dry docking operations or even during the manufacturing processes via industrial effluents. Other important routes include the runoff from agriculture fields; the discharge of contaminated effluents, the deposition of sewage sludge and the leachates from landfills (for references see text)

reasons, sediments are considered tributyltin ultimate sink (Hoch 2001; Omae 2006). However, tributyltin might be bioavailable again as a consequence of its re-suspension, diffusion into the water column and/or decomposition (Fig. 4.8) (Burton et al. 2005; Díez et al. 2002; Hoch 2001).

As previously mentioned, organotins have several applications besides its usage as biocides in antifouling paints and therefore they are released into the environment through other routes. Those routes include, for example, the runoff from agricultural fields where organotins (mainly triphenyltin), are used as pesticides (Fent 1996a; Yi et al. 2012). They also include the leaching from consumer products that enters the environment through the discharge of effluents and sewage sludge from wastewater treatment plants (Fent 1996a, 2006). The disposal of products that contain organotins in landfills can also lead to the contamination of soils and ground waters, although at a small scale (Fent 1996a, b). Recently, Krupp et al. (2010) demonstrated that organotins can be released from landfill sites and that in the atmosphere they are stable enough to travel over long distances.

4.6.2 Organotins Levels in the Environment

Due to its widespread usage, organotins can be found almost everywhere, from soils and sediments to water, biota and the atmosphere. The distribution of organotins in the environment has been extensively addressed over the last decades and several comprehensive reviews on the subject have been published (Table 4.7), therefore only a general overview on the topic is provided here.

Table 4.7 Summary of recent reviews addressing environmental levels of organotin compounds

Reference	Geographical location	Compounds	Compartments: Water	Compartments: Biota	Compartments: Sediments
ATSDR (2005)	Several regions around the world	MBT DBT TBT			Lake sediments River sediments Marine/coastal sediments
Fent (2006)	Several regions around the world	MBT DBT TBT ∑BTs	Freshwater	Mussels Other bivalves Gastropods Amphipods	Marine/coastal sediments
		MPT DPT TPT ∑PTs	Seawater	Fish Marine mammals	Sediment cores
Omae (2006)	USA Canada Germany	MeTs	Rain water Freshwater Estuarine water Seawater	Algae Seaweed Invertebrates Fish	Marine/coastal sediments
Antizar-Ladislao (2008)	Several regions around the world	MBT DBT TBT	Seawater	Mussels Other bivalves Echinoderms Fishes	Marine/coastal sediments
Harino et al. (2008)	Japan	MBT DBT TBT ∑BTs MPT DPT TPT ∑PTs	Seawater	Plankton Mussels Fishes	Marine/coastal sediments Sediment cores
Cao et al. (2009)	China	MBT DBT TBT ∑BTs	Seawater Freshwater	Gastropods	Lake sediments River sediments Marine/coastal sediments

(continued)

Table 4.7 (continued)

Reference	Geographical location	Compounds	Compartments Water	Biota	Sediments
Kannan and Tanabe (2009)	Polar regions	MBT		Mussels	
	Coastal areas (mainly Asia-Pacific)	DBT TBT ∑BTs		Squids	
	Inshore and offshore areas			Coastal and pelagic fishes	
	Global marine environment			Coastal and oceanic birds	
				Marine mammals	
Yi et al. (2012)	Several regions around the world	MPT DPT TPT ∑PTs	Seawater	Plankton Mussels	Coastal sediments
				Other bivalves	
				Gastropods	
				Echinoderms	
				Arthropods	
				Crustaceans	
				Fishes	
				Birds	
				Mammals	

MBT Monobutyltin, *DBT* Dibutyltin, *TBT* Tributyltin, *TeBT* Tetrabutyltin, ∑*BTs* Sum of butyltins (∑BTs=MBT+DBT+TBT), ∑*PTs*, Sum of phenyltins (∑PTs=MPT+DPT+TPT), *MPT* Monophenyltin, *DPT* Diphenyltin, *TPT* Triphenyltin, *MeTs* Methyltins, *Sn*: Inorganic tin

In general terms, the lowest concentrations are detected in the water column whereas the highest ones are detected in sediments and biota (Fig. 4.9), particularly in top predators (Fent 2006; Kannan and Tanabe 2009). From a geographical perspective, levels tend to be higher close to point sources and therefore coastal areas where main ports, dockyards and marinas are located exhibit the highest levels (see e.g. Fent 2006). Even after the introduction of the tributyltin global ban, levels inside harbors remained extremely high, probably due to the paint removal operations in dry docks and the remobilization from sediments, that are now acting as secondary emission sources (Kotrikla 2009; Ruiz et al. 2008; Sousa et al. 2009b).

Although the highest levels of organotins have been reported for cosmopolitan coastal areas, reports on the occurrence of organotins in remote locations such as the Arctic and Antarctic are also available (e.g. Negri and Marshall 2009; Strand and Asmund 2003). Organotins, particularly butyltins were detected even in deep sea organisms and sediments at depths as high as 4 km (see review by Harino et al. (2009b) and Fig. 4.10).

While relatively low when compared to the organotins concentrations found in coastal sediments, the levels detected in deep sea areas are high enough to pose risk for

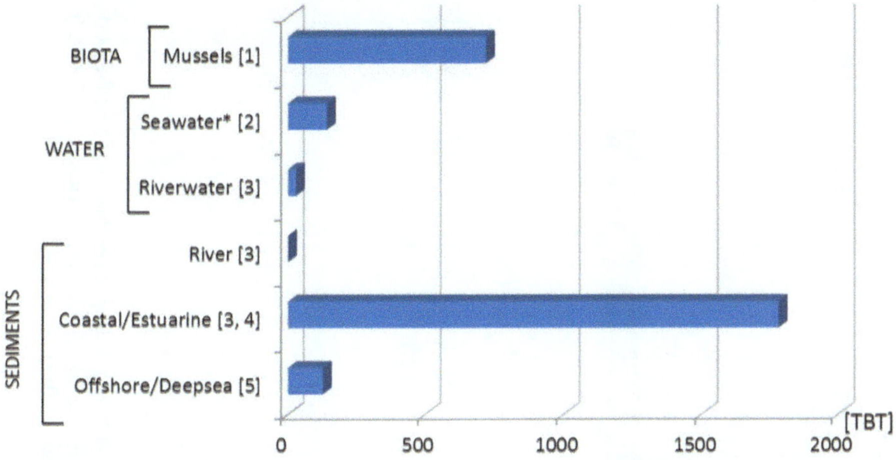

Fig. 4.9 Tributyltin (*TBT*) concentrations in sediments (ng Sn g^{-1} dw), water (ng Sn L^{-1}) and biota (ng Sn g^{-1} dw). Since tributyltin levels exhibits huge variations amongst different geographical locations, only reports concerning the levels in the same country (Portugal) are depicted. Cited from: [1] Sousa et al. (2009a); [2] Galante-Oliveira et al. (2011); [3] Díez et al. (2005); [4] Sousa et al. (2007b); [5] Sousa et al. (2012). Seawater (*) concentrations represents hexane extractable tin

those ecosystems (see Antizar-Ladislao 2008; Garg et al. 2011; Sousa et al. 2012 and references therein). In Nazaré Canyon (one of the largest submarine canyons of the world), for example, organotins levels are generally lower than the ones reported for other offshore and deep sea areas (Fig. 4.10). Even so, tributyltin concentrations at some sampling sites are higher than the Environmental Assessment Criteria (EAC) set by the OSPAR Commission for tributyltin in sediments (0.004 ng Sn g^{-1} dw) (OSPAR 2008), implying that at those locations negative impacts on ecosystems are likely to occur.

4.6.3 Regional and Global Biomonitoring Surveys

Since organisms provide a better indication of contamination than water samples, by integrating contaminant exposure over long periods of time (Fent 2006), several monitoring programs were developed using organisms from different taxonomic groups in order to assess contamination in the world's oceans.

Those programs report in great measure the use of invertebrates, such as crustaceans as in the "Caprella watch program" (Takeuchi et al. 2001, 2004). Yet, mussels' unique characteristics (e.g. sessile lifestyle, wide geographical distribution, tolerance to a considerable range of salinity, resistance to stress and high accumulation of a wide range of contaminants) together with the easy sampling, make this bivalve a favorite bioindicator for the marine environment (Goldberg 1975; Tanabe and Subramanian 2006). The "Mussel watch program" was first implemented in the United States (Goldberg 1975) and was subsequently used in several geographical

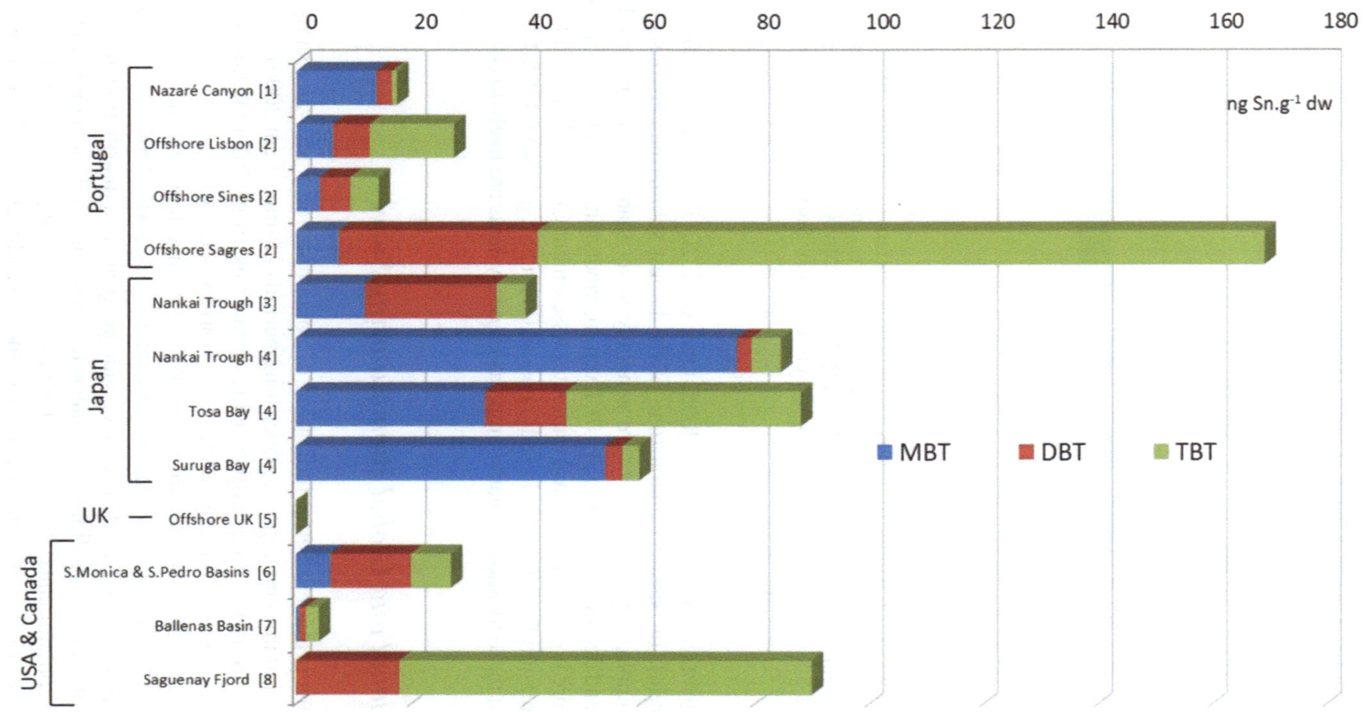

Fig. 4.10 Monobutyltin (*MBT*), Dibutyltin (*DBT*) and Tributyltin (*TBT*) concentrations (ng Sn g^{-1} dw) in offshore and deep sea sediments around the world. Data cited from: [1] Sousa et al. (2012); (2) Goméz-Ariza et al. (2006); [3] Harino et al. (2005); [4] Harino et al. (2009c); [5] Thomas et al. (2000); [6] Venkatesan et al. (1998); [7] Stewart and Thompson (1994); [8] Viglino et al. (2004) (Adapted from Sousa et al. 2012)

locations including Asia and Europe. In the Asia-Pacific region, for example, a mussel watch program was undertaken in the late 1990s by Tanabe and co-workers (Tanabe 2000). In their work, butyltin levels in mussels collected from coastal areas of Cambodia, China (Hong Kong and southern China), Malaysia, India, Indonesia, the Philippines, and Vietnam were quantified and the results obtained demonstrated that butyltin contamination is widespread in those areas (Sudaryanto et al. 2000, 2002). Several other surveys were undertaken in the Asia-Pacific coastal areas (e.g. Harino et al. 2009a; Hong et al. 2002; Hwang et al. 1999; Kan-Atireklap et al. 1998; Prudente et al. 1999; Sudaryanto et al. 2004, 2005) and results suggest that the concentrations of butyltins in mussels are significantly correlated with the per capita gross national product (GNP) of the respective countries (Kannan and Tanabe 2009).

In Europe, no extensive project like the one undertaken by Tanabe and co-workers exists however several regional surveys were undertaken by different researchers. Figure 4.11 illustrates the wide range of butyltin concentrations detected in soft tissues of the Mediterranean mussel, *Mytillus galloprovincialis*, across several European Countries. Furthermore, the OSPAR and the Helsinki Commissions in their regular monitoring programs assesses the levels of several pollutants, including organotins, in mussel's soft tissues (e.g. HELCON 2010; OSPAR 2010).

Concentrations of organotins in mussels' tissues are dependent on several factors such as water temperature, salinity, oxygen content, biological activity, and distance from point sources (Díez et al. 2005; Sousa et al. 2009a). For those reasons mussels are particularly suitable in regional monitoring programs, particularly the ones that integrate temporal pollution trends. The efficacy of the tributyltin ban, for example, has been monitored in several countries using mussels as bioindicators (e.g. Choi et al. 2010; Harino et al. 1999; Nemanič et al. 2009; Zanon et al. 2008). Generally, pre-tributyltin ban levels are higher than the ones after ban implementation (see Sousa et al. 2009a and references therein).

Regardless of their appropriateness for regional monitoring surveys, mussels are not the ideal organisms to understand global contamination of the marine environment. Being sessile, they solely reflect local contamination levels. Global contamination is generally evaluated using long lived top predators species, as for instance, skipjack tuna or marine mammals (Kannan and Tanabe 2009; Tanabe 1999; Ueno et al. 2004). Several reports described the use of squids, fishes, birds and marine mammals to monitor global organotins pollution patterns (e.g. Guruge et al. 1996; Harino et al. 2007; Kannan et al. 1998; Yamada et al. 1997). Recently, all the data from those reports was compiled and reviewed by Kannan and Tanabe (2009). For this reason no further discussion on this subject will be held here. Readers are encouraged to follow the thorough digest therein.

4.7 On the Way to Humans

Humans are exposed to organotin compounds in multiple ways. Figure 4.12 depicts the most common exposure pathways. Generally, ingestion of contaminated food is considered to be the main source (Guérin et al. 2007). Among food items, shellfish

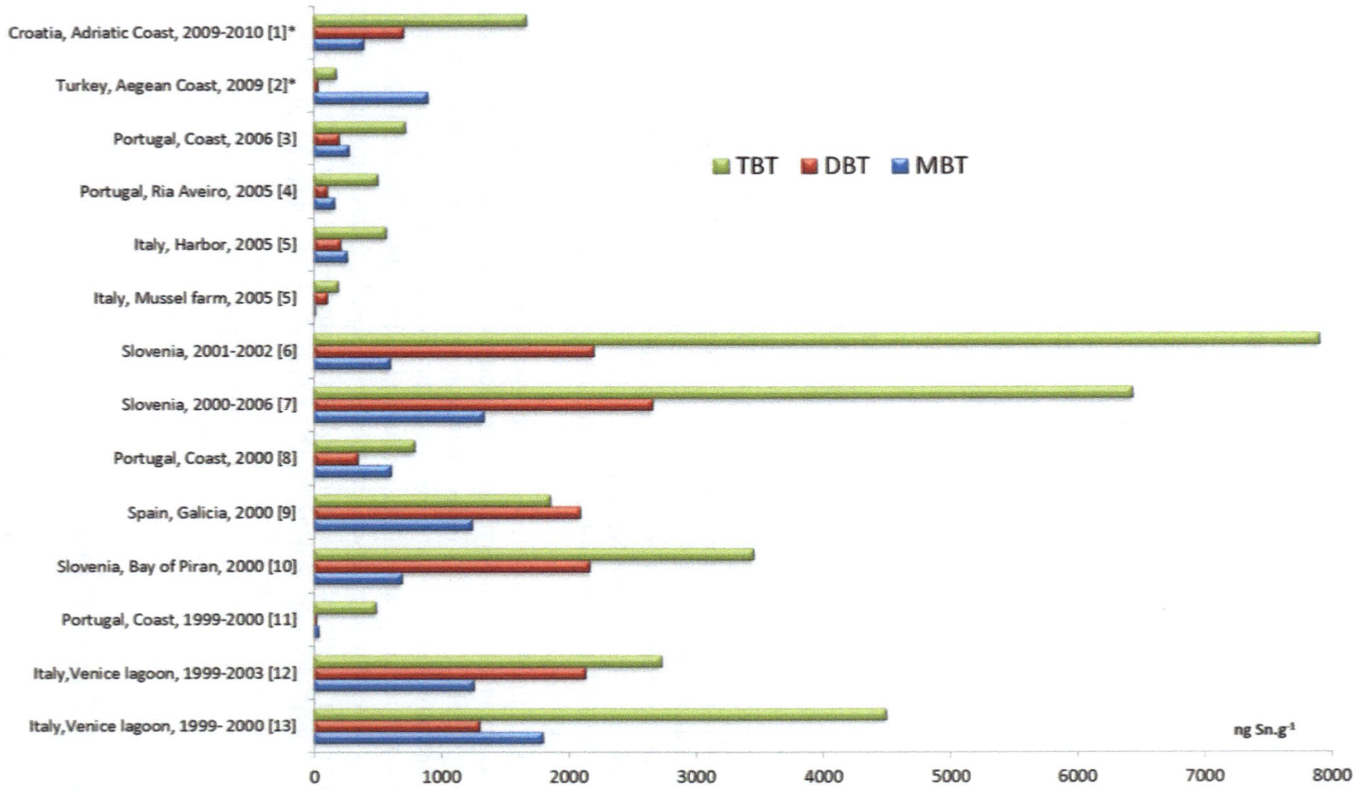

Fig. 4.11 Maximum concentrations of Tributyltin (*TBT*), Dibutyltin (*DBT*) and Monobutyltin (*MBT*) in *Mytilus galloprovincialis* tissues collected around Europe between 1999 and 2010. All concentrations are reported on a dry weight basis, except for (*) that are expressed on a wet weight basis. Data cited from: [1] Furdek et al. (2012); [2] Kucuksezgin et al. (2011); [3] Sousa et al. (2009a); [4] Sousa et al. (2007b); [5] Magi et al. (2008); [6] Ščančar et al. (2007); [7] Nemanič et al. (2009); [8] Barroso et al. (2004); [9] Ruiz et al. (2005); [10] Nemanič et al. (2002); [11] Díez et al. (2005); [12] Zanon et al. (2008); [13] Bortoli et al. (2003)

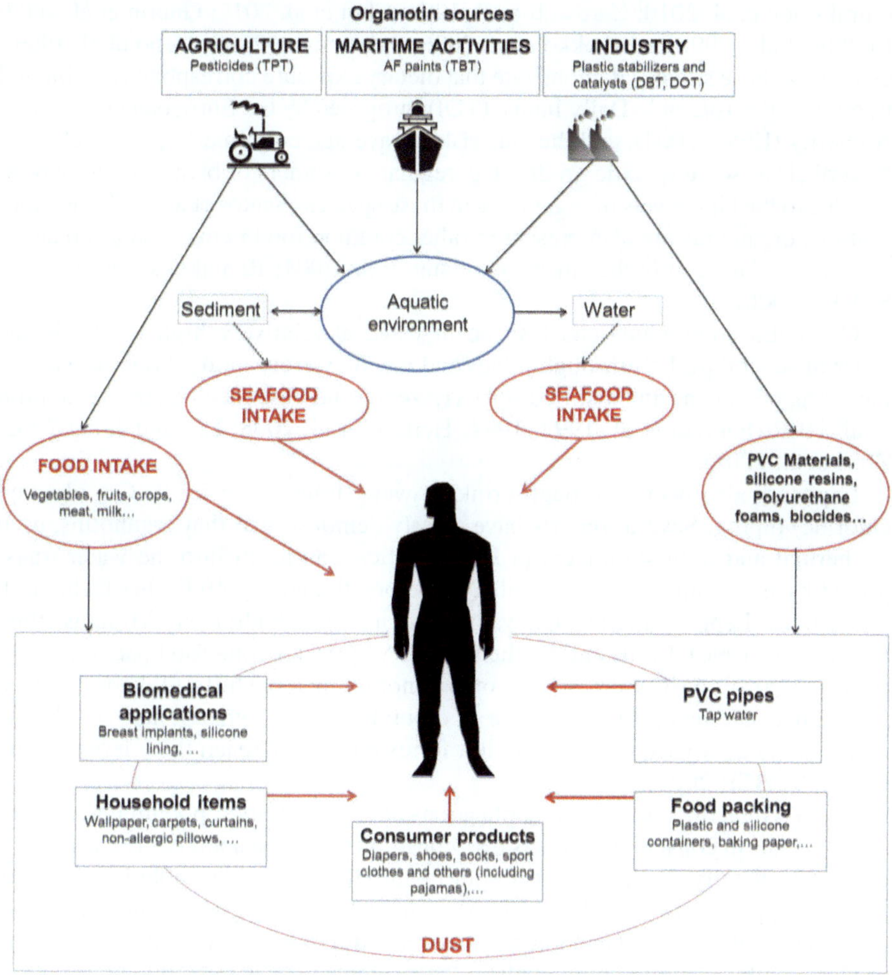

Fig. 4.12 Schematic representation of possible sources and exposure pathways of organotin compounds towards humans. The main exposure pathway is considered to be the ingestion of contaminated seafood. Other dietary sources include vegetables, crops, fruits and alcoholic beverages. PVC (polyvinyl chloride) piping is responsible for the intake of organotins *via* drinking water. The generalized presence of organotins in many consumer products also constitutes an important source through decay to the indoor environment and consequently to humans (via dust ingestion, for instance). Biomedical applications such as breast implants may also contribute to the total organotins burden. See text for details. *TBT* Tributyltin, *TPT*, Triphenyltin, *DBT* Dibutyltin, *DOT* Dioctyltin

and fish products tend to present the highest levels, due to the global contamination of the marine environment by tributyltin. Several studies described the levels of organotins in seafood and addressed the risk of its consumption towards humans

(Airaksinen et al. 2010; Cardwell et al. 1999; Choi et al. 2012; Guérin et al. 2007; Keithly et al. 1999; Rantakokko et al. 2006; Santos et al. 2009; Ueno et al. 1999). Generally, those surveys demonstrate that dietary exposure corresponds to a limited fraction of the Tolerable Daily Intake (TDI)[2] proposed by the European Food Safety Authority (EFSA 2004), and thus the risk for average consumers is relatively low. Nevertheless, some specific groups, e.g. regular consumers of bivalves, might be at risk due to the high levels of organotins in those species (Santos et al. 2009). Besides seafood, organotins are also present in other common food items such as potatoes, vegetables, fruits, milk and meat (Qunfang et al. 2004; Rantakokko et al. 2006; SCHER 2006).

Organotin compounds can also be ingested at relatively high levels (in the order of several μg L^{-1}) through juices and a wide variety of alcoholic beverages, including beer, wine and distilled spirits (Azenha and Vasconcelos 2002; Campillo et al. 2012; Forsyth et al. 1992, 1994; Heroult et al. 2008; Liu and Jiang 2002; Zhao et al. 2008).

Exposure also occurs through drinking water transported by PVC (polyvinyl chloride) piping. Several reports have clearly demonstrated that organotins, used as thermal and light stabilizers in PVC plastics, can leach into the water transported by these pipes (Adams et al. 2011; Forsyth and Jay 1997; Fristachi et al. 2009; Jones-Lepp et al. 2001; Quevauviller et al. 1991; Sadiki and Williams 1999; Sadiki et al. 1996). Furthermore, the use of PVC and silicone food packing materials may also constitute a source of organotins in foodstuffs (Takahashi et al. 1999). It is estimated that exposure to dibutyltin from domestic use of silicone molds commonly used to bake cookies, for example, can reach the tolerable daily intake (SCHER 2006).

Besides diet, the innumerous applications of PVCs, plastics, polyurethane foams and silicones in consumer products also constitute potential sources. The exposures can occur directly through contact with the product, as for example in clothes or medical applications (SCHER 2006) or indirectly through contaminated dust (Fromme et al. 2005; Kannan et al. 2010; Santillo et al. 2001, 2003). The list of products with organotin compounds in their composition is tremendous, including such diverse items as childrens' toys, clothes, sanitary items (e.g. diapers and sanitary pads), wallpaper, carpets, and pillows (SCHER 2006). Upon degradation, the majority of these articles will contribute to the total load of organotins in the indoor environment. Levels of organotin compounds, including octyltins, were recently analyzed in house dust samples revealing that this is an important exposure pathway that cannot be neglected when addressing human exposure issues (Kannan et al. 2010 and references therein).

[2]The Scientific Panel on Contaminants in the Food Chain (EFSA 2004) established a group Tolerable Daily Intake (TDI) for tributyltin, dibutyltin, triphenyltin and dioctyltin of 0.25 μg kg^{-1} body weight.

4.7.1 *What Goes Around, Comes Around: Unraveling Organotins in Humans*

The occurrence of organotin compounds in humans was first registered by Kannan and Falandysz (1997) during a monitoring survey conducted in the Polish Coast of the Baltic Sea. Those authors analyzed butyltin levels in sediments and in liver samples from fish, birds, marine mammals and humans (collected by autopsy from nine male corpses) (Table 4.8). The results (2.4–11.0 ng $\sum$BTs g^{-1} wet weight (ww)) demonstrated, for the first time, that humans were, in fact, exposed to butyltins. Motivated by these findings, this team decided to further investigate human exposure to organotins in a series of studies where blood analysis had a central role (Kannan et al. 1999; Whalen et al. 1999). Levels of monobutyltin, dibutyltin and tributyltin were quantified in whole blood samples collected from 32 adults in Central Michigan, USA (Kannan et al. 1999). Monobutyltin was detected in 53 % of the samples, with the highest value of 27 ng mL^{-1}; dibutyltin was detected in 81 % of the samples with the highest value of 16 ng mL^{-1}, while the highest tributyltin concentration registered was of 85 ng mL^{-1}, with the compound being detected in 70 % of the samples. Kannan's team also detected butyltin levels in blood samples from six healthy volunteers at unexpectedly high concentrations (64–155 ng $\sum$BTs mL^{-1}) (Whalen et al. 1999). Such findings pointed towards frequent human exposure to butyltins. Two possible exposure scenarios were then suggested: (i) direct exposure through the ingestion of contaminated food (mainly seafood); and (ii) indirect exposure from household products containing butyltins (Kannan et al. 1999, 2000).

In order to unravel the potential exposure pathways, Takahashi et al. (1999) quantified the levels of monobutyltin, dibutyltin and tributyltin in several plastic products alongside with liver samples from terrestrial mammals and humans. Results demonstrated that butyltins were present at extremely high concentrations in several household items and that concentrations found in human livers (n=4) were higher than those previously reported by Kannan and Falandysz (1997) in the Baltic study (59–96 ng $\sum$BTs g^{-1} ww *versus* 2.4–11.0 ng $\sum$BTs g^{-1} ww, respectively). The higher levels found in Japanese individuals might be a consequence of the higher fish consumption by the Japanese population, however any conclusions are limited by the reduced number of samples analyzed.

The levels of organotins in human liver were later investigated by Nielsen and Strand (2002) by quantification of monobutyltin, dibutyltin, tributyltin and also triphenyltin in 18 Danish men that "died unexpectedly". Tributyltin and triphenyltin were below detection limit (0.3 ng g^{-1} ww and 3 ng g^{-1} ww, respectively) in all samples, whereas monobutyltin and dibutyltin levels varied between 0.3 and 4.7 ng g^{-1} ww and 0.8 and 28.3 ng g^{-1} ww, respectively. A significant association between age and dibutyltin/monobutyltin ratios was observed and the authors suggested that it could be a consequence either of a more recent exposure in younger men or a lower capacity to debutylate dibutyltin (Nielsen and Strand 2002).

The occurrence of dibutyltin, the main tributyltin metabolite, in the human liver (Nielsen and Strand 2002; Takahashi et al. 1999) indicates that organotins are

Table 4.8 Summary of organotin levels detected in human biological tissues and fluids

Study	Target population	Date	Type of sample	No	Target OTs	Results	Units
Kannan and Falandysz (1997)	Male adults Gdansk, Poland	1994	Liver	9	∑BTs	2.4–11.0	ng g^{-1} ww
Takahashi et al. (1999)	Adults, Ehime, Japan	1997–1998	Liver	4	MBT	14–22	ng g^{-1} ww
					DBT	45–78	
					TBT	<2.0	
					∑BTs	59–96	
Nielsen and Strand (2002)	Male adults, Denmark	1999–2000	Liver	18	MBT	0.3–4.7	ng g^{-1} ww
					DBT	0.8–28.3	
					TBT	<0.3	
					∑BTs	1.1–33.0	
					TPT	<3	
Kannan et al. (1999)	Adults Michigan, USA	1998	Whole blood	32	MBT	<7–27	ng mL^{-1}
					DBT	<2.5–16	
					TBT	<1–85	
					∑BTs	<DL–101	
Whalen et al. (1999)	Healthy adults, USA	na	Whole blood	6	MBT	17–31	ng mL^{-1}
					DBT	12–94	
					TBT	4.9–31	
					∑BTs	64–155	
Rüdel and Steinhanses (2001) in Rudel (2003)	Students, Germany	na	Whole blood	30	MBT	<0.4–1.6	ng mL^{-1}
					DBT	<0.3–1.4	
					TBT	<0.3	
					MPT	<1	
					DPT	<0.3	
					TPT	<0.4	

Lo et al. (2003)	Healthy adults, Germany	na	Serum	8	MBT	<0.02	ng mL^{-1}
					DBT	<0.02	
					TBT	<0.02–0.05	
					TeBT	<0.02	
					TPT	0.18–0.67	
					MOT	<0.02	
					DOT	<0.02	
Peters (2004)	Indiscriminate individuals, The Netherlands	na	Whole blood	91	MBT	≤0.1	ng g^{-1}
					DBT	<0.1	
					TBT	≤0.1	
					MPT	0.1–0.5	
					DPT	0.2–2.4	
					TPT	<0.4	
					MOT	<0.2	
					DOT	<0.2	
Rantakokko et al. (2008)	Fishermen and their families, Finland	2004–2005	Whole blood	300	MBT	<0.20	ng mL^{-1}
					DBT	<0.18–0.38	
					TBT	<0.32	
					MPT	<0.35	
					DPT	<0.03	
					TPT	<0.04–0.56	
					DOT	<0.72	
Nagase et al. (1995)	na	na	Hair	na	TBT	≤15	ng g^{-1} (as chlorides)
					TPT	<10	
Uhl (1986) in Appel (2004)	One volunteer that ingested TBT	na	Urine	1	MBT[a]	na	na
					DBT[a]		
					TBT		
Gui-bin et al. (2000a)	Intoxicated individuals, Jiangxi, China	1999	Urine	2	DMT	20.4–79.7	ng mL^{-1}
					TMT	42.0–83.3	

(continued)

Table 4.8 (continued)

Study	Target population	Date	Type of sample	No	Target OTs	Results	Units
Gui-bin et al. (2000a)	Intoxicated individuals, Jiangxi, China	1999	Blood	1	DMT	nd	ng g^{-1}
					TMT	70.3	
Gui-bin et al. (2000b)	One intoxication victim, Jiangxi, China	1999	Heart	1	DMT	0.100	μg Sn g^{-1}
			Liver			1.93	
			Stomach			0.104	
			Kidney			1.05	
Gui-bin et al. (2000b)	One intoxication victim, Jiangxi, China	1999	Heart	1	TMT	1.48	μg Sn g^{-1}
			Liver			1.42	
			Stomach		TMT	0.304	
			Kidney			0.47	
Yoo et al. (2007)	One individual exposed to DMT at the workplace	na	Urine	1	MMT	nd	μg g^{-1} creatinine adjusted
					DMT	up to 774	
					TMT	up to 946	
Yoo et al. (2007)	One individual exposed to DMT at the workplace	na	Blood	1	DMT	5.3–47.6	μg L^{-1}
					TMT	10.9–266.9	
Zachariadis and Rosenberg (2009a)	Male individuals	na	Urine	4	MBT	<13–34	ng Sn L^{-1}
					DBT	<10	
					TBT	<12–49	
					TeBT	<15	
					MPT	<17	
					DPT	<50	

Cui et al. (2011)	Occupationally exposed workers	na	Urine	13	MMT	<8.1–261.9	ng Sn L^{-1}
					DMT	26.0–6,353	
					TMT	<5.6–1,277	
					∑MeTs	26.0–7,892	
Cui et al. (2011)	General population	na	Urine	2	MMT	<8.1	ng Sn L^{-1}
					DMT	<2.5–154.7	
					TMT	<5.6	
					∑MeTs	<DL–7,892	
Mino et al. (2008)	Breastfeeding women, Japan	2002	Breast milk	67	MBT	<2.5	ng mL^{-1} (as chlorides)
					DBT	2.5–9.5	
					TBT	<1.3	
					TPT	<1.3	

OTs organotins, *MBT* Monobutyltin, *DBT* Dibutyltin, *TBT* Tributyltin, *TeBT* Tetrabutyltin, ∑*BTs:* sum of butyltins (MBT+DBT+TBT); *MPT* Monophenyltin, *DPT* Diphenyltin, *TPT* Triphenyltin, *MOT* Monooctyltin, *DOT* Dioctyltin, *MMT* Monomethyltin, *DMT* Dimethyltin, *TMT* Trimethyltin, ∑*MeTs* sum of methyltins (MMT+DMT+TMT), *<DL* below detection limit, *ww* wet weight, *na* information not available

[a]Referred by Appel (2004) as TBT metabolites

bioavailable after exposure (Appel 2004). Furthermore, the fact that dibutyltin is the main butyltin species detected, suggests that the tributyltin hepatic debutylation is a fast process in humans (Nielsen and Strand 2002; Takahashi et al. 1999).

Studies on organotins' published after 2000 disclose relatively low levels of butyltins (Lo et al. 2003; Peters 2004; Rantakokko et al. 2008; Rudel 2003) when compared to those reported in previously cited publications (Kannan et al. 1999; Whalen et al. 1999). A survey conducted with 30 students from two German cities, for example, revealed that monobutyltin and dibutyltin were quantifiable only in 3.3 and 17 % of the samples, respectively; whereas the remaining compounds were always below detection limit (see Table 4.8 for details) (Rüdel and Steinhanses (2001) in Rudel 2003).

In another study, Lo et al. (2003) quantified the levels of organotins in human plasma from eight healthy adults and demonstrated that monobutyltin, dibutyltin, tetrabutyltin, monooctyltin and dioctyltin were always below detection (Detection Limit=0.02 ng mL^{-1}). However, tributyltin was detected in four samples (with values up to 0.5 ng mL^{-1}) and triphenyltin in all, revealing for the first time triphenyltin accumulation in humans. Low concentrations of organotins were also detected by Peters (2004) when analyzing the blood of 91 Dutch individuals. In fact, dibutyltin, monophenyltin, diphenyltin and triphenyltin, were always below detection limit whereas octyltins compounds – monooctyltin and dioctyltin – were detected in 13 and 14 % of the samples analyzed, respectively. Monobutyltin was quantified only in 3 % of the samples.

The occurrence of organotin compounds in human blood was further investigated in a Finnish population of fishermen and their families (Rantakokko et al. 2008). Monobutyltin, dibutyltin, tributyltin, monophenyltin, diphenyltin, triphenyltin and dioctyltin were quantified in 300 whole blood samples collected between 2004 and 2005, in order to understand the health benefits and risks of high fish consumption. The results disclosed relatively low levels for most of the organotins analyzed (Table 4.8). In fact, monobutyltin, tributyltin, monophenyltin, diphenyltin and dioctyltin were always below the limit of quantification (LOQ). Dibutyltin was detected in two of the samples, whereas triphenyltin was detected in 37. Of these samples, triphenyltin median concentrations were only 1.5*LOQ (LOQ=0.04 ng mL^{-1}). For these reasons, the authors couldn't establish associations between triphenyltin levels and possible health outcomes. However, by dividing triphenyltin data in two different categories (<LOQ and >LOQ), a statistically significant association between age, fish consumption and triphenyltin levels in whole blood could be established. The results confirmed that fish consumption was associated with triphenyltin levels in whole human blood.

Overall, the levels of organotins detected in blood samples are relatively low; particularly in surveys conducted after the implementation of restrictions on the use of tributyltin based antifouling paints. Furthermore, in recent surveys tributyltin and its metabolites are less important than triphenyltin (Lo et al. 2003; Rantakokko et al. 2008) or octyltin compounds (Peters 2004). Such evidences suggest that organotins sources have changed over the last years. Most probably, in the late 1990s organotins were incorporated through food ingestion, particularly fish and shellfish highly contaminated with butyltins due to tributyltin widespread usage in antifouling paints. Recently, tributyltin levels in seafood have decreased and therefore the exposure to organotins from other sources became more important. Such sources include, for example, the ingestion of food contaminated with triphenyltin (used as agriculture pesticide) and widespread exposure to certain groups of organotins, particularly octyltins, derived from the leaching of many consumer products (Kannan et al. 2010).

The global ban on tributyltin represented a tremendous change in the marine ecosystems. That change inevitably reflected in the human exposure pathways. Since the amount of studies focused on the evaluation of body burdens of each compound is, as we demonstrated, so scarce, it becomes impossible to establish the real dimensions of such a paradigmatic event in the human population.

4.7.2 In Search for the "Ideal" Matrix to Estimate Organotin Body Burdens

Organotins exhibit both lipophilic and ionic properties; therefore they can accumulate in lipids and they can also bind to proteins such as glutathione and α-keratins (Appel 2004; Kannan and Falandysz 1997). Due to this dual behavior, organotins detection is expected in such distinct matrices such as liver, kidney, blood, hair and nails, urine, or breast milk.

The selection of the matrices to be analyzed is dependent on the target chemical itself, the amount necessary and the analytical technique to be used, particularly its detection limit (Esteban and Castaño 2009). It is also largely conditioned by practical and ethical reasons. The use of invasive matrices such as internal organs or blood, for example, requires highly trained personnel for drawing blood or perform a biopsy (NRC 2006). On the other hand the use of noninvasive matrices, such as urine tends to increase participation in a study, because people are more likely to provide urine or hair samples than undergo blood draw (NRC 2006). Besides, the use of noninvasive matrices usually allows the collection of a larger sample volume, many times critical for the quantification (due to detection limit issues) of the target chemical.

Invasive Matrices

Liver and kidneys are rich in glutathione (Kannan and Falandysz 1997). Considering that butyltins tend to bind to this tripeptide, these two organs are good candidates to assess organotins burdens in humans. In fact, liver and kidney proved to be good matrices both in wildlife and human studies (Sect. 4.7.1). However, due to the above noted reasons kidney and liver samples are difficult to obtain and this is the likely reason why most of the available literature focus on other organs/tissues.

Hair and Nails

The use of hair and nails exhibits a series of advantages such as low costs and easiness of collection, transport and storage (Esteban and Castaño 2009). Furthermore, hair samples may provide information on temporal exposure patterns by segmental analysis (Esteban and Castaño 2009). The main disadvantage on the use of hair, and to a lesser extent nails, is related with the difficulty to differentiate internal and external exposures. These difficulties are further exacerbated by the use of perms, hair dyes and nail polish (ATSDR 2001; Schramm 2008). The use of hair and nails has already been validated to biomonitor organometals like methylmercury (ATSDR 2001; Esteban and Castaño 2009).

α-keratins are major structural components of hair, nails and feathers. It is, thus, expectable that organotins might be deposited in those matrices. Previous reports described that in birds, for example, butyltins levels tend to be higher in feathers than in the rest of the analyzed tissues (Guruge et al. 1996, 1997; Senthilkumar et al. 1998, 1999). In fact, feathers are considered to be an important excretion route for butyltins in birds and they have been proposed as good bioindicators to monitor butyltins in avian species (Guruge et al. 1997). In marine mammals, levels of butyltins in hair are also very high when compared with other tissues; and therefore the use of hair as an indicator of butyltins contamination for these animals has been proposed (Kim et al. 1996; Takahashi et al. 2000). In humans, hair has been routinely used to monitor a wide variety of chemicals, such as mercury (ATSDR 2001; Schramm 2008). Similarly to what happens with methylmercury, that has a high affinity for reduced sulfhydryl groups, organotins accumulation occurs through their binding to the –SH and =NH groups of selected proteins (Kannan and Falandysz 1997). Despite such lines of evidence, no hard data corroborating the usefulness of organotins levels analysis in human hair exists. To our best knowledge, no surveys were conducted using human hair and only one study, dealing solely with the analytical intricacies of organotins extraction and quantification, revealed the levels of tributyltin and triphenyltin in human hair (Nagase et al. 1995). However information concerning the analyzed samples' characteristics, e.g. target population or number of samples, were not disclosed by the authors (Nagase et al. 1995) (see Table 4.8 for details). Without further studies on human hair, its suitability as a non-invasive matrix for organotins monitoring remains merely speculative.

Similarly, nails are also potential bioindicators of organotins exposure, as they are rich in α-keratins. Yet, to our best knowledge organotin levels in human nails was never examined.

Urine

Urine is one of the most commonly used matrix for human biomonitoring, particularly for water soluble chemicals such as metals (Esteban and Castaño 2009). Urine samples are easy to collect; furthermore large sample volumes can be easily obtained (Dirtu et al. 2012; Esteban and Castaño 2009).

One of the first reports of organotins in human urine arose from a controversial experiment conducted in the 1980s: one volunteer ingested a mixture of 3 mL of cherry brandy and 7 mL of ethanol, in which tributyltin oxide had been dissolved. The results revealed that only 5 % of the dose was found in the urine collected afterwards (mainly as dibutyltin) and that the tributyltin metabolites decreased rapidly on the first days after exposure (Uhl 1986 cited from Appel 2004).

Recently, Zachariadis and co-workers developed and validated speciation methods to quantify organotins in urine (Zachariadis and Rosenberg 2009a, b, 2012). Initially, the extraction method was validated and checked through the quantification of a suite of organotins in four samples collected from male donors (Zachariadis and Rosenberg 2009a). Levels of dibutyltin, tetrabutyltin, monophenyltin and diphenyltin were always below detection limit. Monobutyltin and tributyltin were detected in 25 and 50 % of the analyzed samples, respectively whereas triphenyltin couldn't be quantified with the proposed technique (Zachariadis and Rosenberg 2009a). In their subsequent work, new methods were described and validated in order to quantify additional organic tin compounds, including less water soluble ones, such as tributyltin and triphenyltin (Zachariadis and Rosenberg 2009b; Zachariadis and Rosenberg 2012). Despite such analytical advances, no comprehensive survey on the levels of butyltins and phenyltins in human urine is available.

Methylated organotins are more water soluble and have lower boiling points than other organotins due to their smaller organic functional groups being thus excreted through urine (Cui et al. 2011). Consequently, urine samples constitute an ideal matrix to characterize methyltin exposure in humans. Recently, Cui et al. (2011) investigated methyltins levels, including monomethyltin, dimethyltin and trimethyltin in urine samples from the employees of an organotin production plant. Dimethyltin was the major compound detected. Furthermore, relatively high levels of trimethyltin were detected in 92 % of the workers (Cui et al. 2011). According to those authors, and considering that dimethyltins were the gross majority of the organotins produced on the plant, the fact that trimethyltin was detected may imply that methylation occurs in the organism, and therefore more research on the levels and associated risks with methyltins should be performed. To the best of our knowledge, this is the only epidemiological study that addressed the levels of methyltins in human urine. The few remaining studies refer to acute poisoning events (Gui-bin et al. 2000a, b; Suzuki et al. 2008; Yoo et al. 2007).

Breast Milk

Breast milk is a commonly used matrix to biomonitor lipophilic chemicals due to its high fat content (Esteban and Castaño 2009; Tanabe and Kunisue 2007). Lipophilic chemicals such as Persistent Organic Pollutants (POPs) are stored and equilibrated in the body in fat rich tissues and can pass into the breast milk to be excreted (Esteban and Castaño 2009). Therefore, breast milk may contain chemicals from exposure events that occurred earlier in a woman's life, thus reflecting cumulative exposures (NRC 2006). Breast milk can simultaneously be used to estimate the infant's intake of selected chemicals (NRC 2006).

To the best of our knowledge, only one study addressed the presence of organotins in human breast milk samples. Mino et al. (2008) quantified the levels of monobutyltin, dibutyltin, tributyltin and triphenyltin in breast milk from 62 Japanese lactating women living in Okayama and Kagawa prefectures. The results revealed that only dibutyltin had detectable levels, and solely in a small percentage (16 %) of the surveyed women (see Table 4.8 for details). Such results prompted the researchers to conclude that human breast milk was largely uncontaminated by organotins (Mino et al. 2008). Nevertheless, the fact that milk organotin levels are quite low doesn't imply that the same is true for other matrices.

Some Remarks Regarding Ideal Matrices

The limited information on total human body burdens and the distribution of the different organotin classes in the different compartments renders the interpretation of levels in a single matrix incomplete. The use of internal organs such as liver and kidney in conjunction with blood in representative target groups would constitute the ideal framework to cover these knowledge lacune. However, the already pointed ethical, technical and social constraints turn this approach into a difficult task.

In realistic terms, an integrated study in which organotins levels are simultaneously quantified in several matrices including urine, blood, breast milk and hair from the same individual is thus necessary in order to better understand the distribution of organotins in the human body, but especially to form a correct perception on which is the best noninvasive matrix to use in future monitoring surveys. Without such type of information the estimation of organotins burdens in humans will keep on being an elusive goal.

4.8 Conclusion

Organotins are the most widely used organometallic compounds for agricultural, industrial and biomedical applications with an annual production of about 60,000 tons. Despite an overwhelming presence in society and the environment, and of some well-studied cases, (e.g. tributyltin and its derivatives in the marine environment), to this day there is a huge gap of knowledge concerning the importance of each exposure pathway, the internal distribution of each class inside organisms and joint effects of organotins, being this particularly true for humans. The use of internal organs in conjunction with blood in representative target groups would constitute the ideal framework to tackle this issue. However, ethical, technical and social constraints render this approach unfeasible, with special incidence in the human case. We propose integrated studies where organotins levels are quantified simultaneously in several matrices, e.g. urine, blood, and hair to understand tissue distribution and to select the most suitable noninvasive matrix to use with respect to the target population, organotin compound of interest and selected analytical technique. Only this approach will allow accurate estimates of total organotin burdens.

Acknowledgments This work was financially supported by Portuguese Foundation for Science and Technology FCT trough program COMPETE and the projects PEst-C/SAU/UI0709/2011 and PEst-C/MAR/LA0017/2011. Ana C. A. Sousa and M. Ramiro Pastorinho acknowledge FCT for the postdoctoral grants SFRH/BPD/65884/2009 and SFRH/BPD/26689/2006. The authors also thank Filipe Laranjeiro and Vitor Madaleno for the kind permission to use their pictures.

References

Adams WA, Xu Y, Little J, Fristachi AF, Rice G, Impellitteri CA (2011) Predicting the migration rate of dialkyl organotins from PVC pipe into water. Environ Sci Technol 45:6902–6907. http://dx.doi.org/10.1021/es201552x

Airaksinen R, Rantakokko P, Turunen AW, Vartiainen T, Vuorinen PJ, Lappalainen A, Vihervuori A, Mannio J, Hallikainen A (2010) Organotin intake through fish consumption in Finland. Environ Res 110:544–547. http://dx.doi.org/10.1016/j.envres.2010.06.004

Allsopp M, Santillo D, Johnston P (2001) Hazardous chemicals in carpets. Greenpeace Research Laboratories Technical Note 01/2001. University of Exeter. http://www.greenpeace.to/publications/carpet.pdf

Alzieu C (1998) Tributyltin: case study of a chronic contaminant in the coastal environment. Ocean Coast Manag 40:23–26. http://dx.doi.org/10.1016/S0964-5691(98)00036-2

Alzieu C (2000) Environmental impact of TBT: the French experience. Sci Total Environ 258:99–102. http://dx.doi.org/10.1016/S0048-9697(00)00510-6

American Chemistry Council (ACC) (2012) Global business of chemistry: global chemical shipments by country/region. http://www.americanchemistry.com/Jobs/EconomicStatistics/Industry-Profile/Global-Business-of-Chemistry

Antizar-Ladislao B (2008) Environmental levels, toxicity and human exposure to tributyltin (TBT)-contaminated marine environment. A review. Environ Int 34:292–308. http://dx.doi.org/10.1016/j.envint.2007.09.005

Aono A, Takeuchi I (2008) Effects of tributyltin at concentrations below ambient levels in seawater on *Caprella danilevskii* (Crustacea: Amphipoda: Caprellidae). Mar Pollut Bull 57:515–523. http://dx.doi.org/10.1016/j.marpolbul.2008.02.031

Appel KE (2004) Organotin compounds: toxicokinetic aspects. Drug Metab Rev 36:763–786. http://dx.doi.org/10.1081/DMR-200033490.

ATSDR (2001) Summary Report, Hair Analysis Panel discussion: exploring the state of the science. The Agency for Toxic Substances and Disease Registry, Division of Health Assessment and Consultation and Division of Health Education and Promotion, Atlanta, p 199. http://www.atsdr.cdc.gov/HAC/hair_analysis/index.html

ATSDR (2005) Toxicological profile for tin and tin compounds. U.S. Department of Health and Human Services: Public Health Service Agency for Toxic Substances and Disease Registry, Atlanta. http://www.atsdr.cdc.gov/toxprofiles/tp55.pdf

Azenha M, Vasconcelos MT (2002) Butyltin compounds in portuguese wines. J Agric Food Chem. 50:2713–2716. http://dx.doi.org/abs/10.1021/jf0115544

Bailey SK, Davies IM (1989) The effects of tributyltin on dogwhelks (*Nucella lapillus*) from Scottish coastal waters. J Mar Biol Assoc UK 69:335–354. http://dx.doi.org/10.1017/S0025315400029453

Barreiro R, González R, Quintela M, Ruiz JM (2001) Imposex, organotin bioaccumulation and sterility of female *Nassarius reticulatus* in polluted areas of NW Spain. Mar Ecol Prog Ser 218:203–212. http://dx.doi.org/10.3354/meps218203

Barreiro R, Quintela M, Ruiz JM (2004) TBT e imposex en Galicia: los efectos de un disruptor endocrino en poblaciones de gasterópodos marinos. Ecosistemas 13:13–29. http://www.revistaecosistemas.net/pdfs/45.pdf

Barroso CM (2001) The use of imposex for monitoring TBT pollution in the Portuguese Coast, with emphasis on the gastropod *Nassarius (Hinia) reticulatus* (L.). PhD, University of Aveiro, Aveiro, pp 1–215

Barroso CM, Moreira MH (2002) Spatial and temporal changes of TBT pollution along the Portuguese coast: inefficacy of the EEC directive 89/677. Mar Poll Bull 44:480–486. http://dx.doi.org/10.1016/S0025-326X(01)00260-0

Barroso CM, Moreira MH, Gibbs PE (2000) Comparison of imposex and intersex development in four prosobranch species for TBT monitoring of a southern European estuarine system (Ria de Aveiro, NW Portugal). Mar Ecol Prog Ser 201:221–232. http://dx.doi.org/10.3354/meps201221

Barroso CM, Moreira MH, Bebianno MJ (2002) Imposex, female sterility and organotin contamination of the prosobranch *Nassarius reticulatus* from the Portuguese coast. Mar Ecol Prog Ser 230:127–135. http://dx.doi.org/10.3354/meps230127

Barroso CM, Mendo S, Moreira MH (2004) Organotin contamination in the mussel *Mytilus galloprovincialis* from Portuguese coastal waters. Mar Pollut Bull 48:1149–1153. http://dx.doi.org/10.1016/j.marpolbul.2004.03.006

Barroso CM, Reis-Henriques MA, Ferreira M, Gibbs PE, Moreira MH (2005) Organotin contamination, imposex and androgen/oestrogen ratios in natural populations of *Nassarius reticulatus* along a ship density gradient. Appl Organomet Chem 19:1141–1148. http://dx.doi.org/10.1002/aoc.955

Barroso CM, Rato M, Verissimo A, Sousa A, Santos JA, Coelho S, Gaspar MB, Maia F, Galante-Oliveira S (2011) Combined use of *Nassarius reticulatus* imposex and statolith age determination for tracking temporal evolution of TBT pollution in the NW Portuguese continental shelf. J Environ Monit 13:3018–3025. http://dx.doi.org/10.1039/C1EM10508F

Beaumont AR, Newman PB (1986) Low levels of tributyl tin reduce growth of marine micro-algae. Mar Pollut Bull 17:457–461. http://dx.doi.org/10.1016/0025-326X(86)90835-0

Bettin C, Oehlmann J, Stroben E (1996) TBT-induced imposex in marine neogastropods is mediated by an increasing androgen level. Helgol Meeresunters 50:299–317. http://dx.doi.org/10.1007/BF02367105

Blaber SJM (1970) The occurrence of a penis-like outgrowth behind the right tentacle in spent females of *Nucella lapillus* (L.). Proc Malacol Soc Lond 39:231–233. http://mollus.oxfordjournals.org/content/39/2-3/231.extract

Blanck H, Dahl B (1998) Recovery of marine periphyton communities around a Swedish marina after the ban of TBT use in antifouling paints. Mar Pollut Bull 36:437–443. http://dx.doi.org/10.1016/S0025-326X(97)00209-9

Bortoli A, Troncon A, Dariol S, Pellizzato F, Pavoni B (2003) Butyltins and phenyltins in biota and sediments from the Lagoon of Venice. Oceanologia 45:7–23. http://www.iopan.gda.pl/oceanologia/451borto.pdf

Bryan GW, Gibbs PE (1991) Impact of low concentrations of tributyltin (TBT) on marine organisms: a review. In: Newman MC, McIntosh AW (eds) Metal ecotoxicology: concepts and applications. Lewis Publishers, Ann Arbor/Boca Raton/Boston, pp 323–361

Bryan GW, Gibbs PE, Hummerstone LG, Burt GR (1986) The decline of the gastropod *Nucella lapillus* around the south-west England: evidence for the effect of tributyltin from antifouling paints. J Mar Biol Assoc UK 66:611–640. http://dx.doi.org/10.1017/S0025315400042247

Bryan GW, Gibbs PE, Burt GR, Hummerstone LG (1987) The effects of tributyltin (TBT) accumu lation on adult dog-whelks, *Nucella lapillus*: long-term field and laboratory experiments. J Mar Biol Assoc UK 67:525–544. http://dx.doi.org/10.1017/S0025315400027272

Bryan GW, Burt GR, Gibbs PE, Pascoe PL (1993) *Nassarius reticulatus* (Nassariidae: Gastropoda) as an indicator of tributyltin pollution before and after TBT restrictions. J Mar Biol Assoc UK 73:913–929. http://dx.doi.org/10.1017/S0025315400034809

Burton ED, Phillips IR, Hawker DW (2005) In-situ partitioning of butyltin compounds in estuarine sediments. Chemosphere 59:585–592. http://dx.doi.org/10.1016/j.chemosphere.2004.10.067

Campillo N, Viñas P, Peñalver R, Cacho JI, Hernández-Córdoba M (2012) Solid-phase microextraction followed by gas chromatography for the speciation of organotin compounds in honey and wine samples: a comparison of atomic emission and mass spectrometry detectors. J Food Compos Anal 25:66–73. http://dx.doi.org/10.1016/j.jfca.2011.08.001

Cao D, Jiang G, Zhou Q, Yang R (2009) Organotin pollution in China: an overview of the current state and potential health risk. J Environ Manag. http://dx.doi.org/10.1016/j.jenvman.2008.06.007

Caratozzolo R, Bellini E, Melati MR, Pellerito C, Fiore T, Agati PD, Scopelliti M, Pellerito L (2007) Interference of tributyltin(IV)chloride on the vascular plant cells. Appl Organomet Chem 21:66–72. http://dx.doi.org/10.1002/aoc.1175

Cardwell RD, Keithly JC, Simmonds J (1999) Tributyltin in U.S. market-bought seafood and assessment of human health risks. Hum Ecol Risk Assess 5:317–335. http://dx.doi.org/10.1080/10807039991289464

Carvalho P (2010) Halophytes in technologies for TBT remediation in sediments. LAP Lambert Academic Publishing, Colne. ISBN 978-3-8433-5708-1

Carvalho P, Basto M, Silva M, Machado A, Bordalo A, Vasconcelos M (2010) Ability of salt marsh plants for TBT remediation in sediments. Environ Sci Pollut Res 17:1279–1286. http://dx.doi.org/10.1007/s11356-010-0307-1

Castro LF, Santos M, Reis-Henriques M (2005) The genomic environment around the Aromatase gene: evolutionary insights. BMC Evol Biol 5:43. http://www.biomedcentral.com/1471-2148/5/43

Castro LFC, Lima D, Machado A, Melo C, Hiromori Y, Nishikawa J, Nakanishi T, Reis-Henriques MA, Santos MM (2007) Imposex induction is mediated through the Retinoid X Receptor signalling pathway in the neogastropod *Nucella lapillus*. Aquat Toxicol 85:57–66. http://dx.doi.org/10.1016/j.aquatox.2007.07.016

Chambers LD, Stokes KR, Walsh FC, Wood RJK (2006) Modern approaches to marine antifouling coatings. Surf Coat Technol 201:3642–3652. http://dx.doi.org/10.1016/j.surfcoat.2006.08.129

Champ MA (2000) A review of organotin regulatory strategies, pending actions, related costs and benefits. Sci Total Environ 258:21–71. http://dx.doi.org/10.1016/S0048-9697(00)00506-4

Chen Y, Zuo Z, Chen S, Yan F, Chen Y, Yang Z, Wang C (2008) Reduction of spermatogenesis in mice after tributyltin administration. Toxicology 251:21–27. http://dx.doi.org/10.1016/j.tox.2008.06.015

Choi M, Moon H-B, Yu J, Eom J-Y, Choi H-G (2010) Temporal trend of butyltins in seawater, sediments, and mussels from Busan Harbor of Korea between 2002 and 2007: tracking the effectiveness of tributylin regulation. Arch Environ Contam Toxicol 58:394–402. http://dx.doi.org/10.1007/s00244-009-9428-2

Choi M, Moon H-B, Choi H-G (2012) Intake and potential health risk of butyltin compounds from seafood consumption in Korea. Arch Environ Contam Toxicol 62:333–340. http://dx.doi.org/10.1007/s00244-011-9688-5

Coelho I, Lima D, André A, Melo C, Ruivo R, Reis-Henriques MA, Santos MM, Castro LFC (2012) Molecular characterization of Adh3 from the mollusc *Nucella lapillus*: tissue gene expression after tributyltin and retinol exposure. J Mollusc Stud 78:343–348. http://dx.doi.org/10.1093/mollus/eys018

Couceiro L, Díaz J, Albaina N, Barreiro R, Irabien JA, Ruiz JM (2009) Imposex and gender-independent butyltin accumulation in the gastropod *Nassarius reticulatus* from the Cantabrian coast (N Atlantic Spain). Chemosphere 76:424–427. http://dx.doi.org/10.1016/j.chemosphere.2009.03.035

Cruz ASH (2012) Tributyltin (TBT) resistance in *Aeromonas molluscorum* Av27. PhD thesis, University of Aveiro, p 225

Cruz A, Oliveira V, Baptista I, Almeida A, Cunha Â, Suzuki S, Mendo S (2012) Effect of tributyltin (TBT) in the metabolic activity of TBT-resistant and sensitive estuarine bacteria. Environ Toxicol 27:11–17. http://dx.doi.org/10.1002/tox.20605

Cui Z, Zhang K, Zhou Q, Liu J, Jiang G (2011) Determination of methyltin compounds in urine of occupationally exposed and general population by in situ ethylation and headspace SPME coupled with GC-FPD. Talanta 85:1028–1033. http://dx.doi.org/10.1016/j.talanta.2011.05.018

Dafforn KA, Lewis JA, Johnston EL (2011) Antifouling strategies: history and regulation, ecological impacts and mitigation. Mar Pollut Bull 62:453–465. http://dx.doi.org/10.1016/j.marpolbul.2011.01.012

Dahllof I, Agrenius S, Blanck H, Hall P, Magnusson K, Molander S (2001) The effect of TBT on the structure of a marine sediment community – a boxcosm study. Mar Pollut Bull 42:689–695. http://dx.doi.org/10.1016/S0025-326X(00)00219-8

Davies IM, Bailey SK (1991) The impact of tributyltin from large vessels on the dogwhelk *Nucella lapillus* (L.) populations around Scottish oil ports. Mar Environ Res 32:202–211. http://dx.doi.org/10.1016/0141-1136(91)90042-7

de Carvalho Oliveira R, Santelli RE (2010) Occurrence and chemical speciation analysis of organotin compounds in the environment: a review. Talanta 82:9–24. http://dx.doi.org/10.1016/j.talanta.2010.04.046

Diamanti-Kandarakis E, Bourguignon JP, Giudice LC, Hauser R, Prins GS, Soto AM, Zoeller RT, Gore AC (2009) Endocrine-disrupting chemicals: an Endocrine Society Scientific statement. Endocr Rev 30:293–342. http://dx.doi.org/10.1210/er.2009-0002

Díez S, Ábalos M, Bayona JM (2002) Organotin contamination in sediments from the Western Mediterranean enclosures following 10 years of TBT regulation. Water Res 36:905–918. http://dx.doi.org/10.1016/S0043-1354(01)00305-0

Díez S, Lacorte S, Viana P, Barceló D, Bayona JM (2005) Survey of organotin compounds in rivers and coastal environments in Portugal 1999–2000. Environ Pollut 136:525–536. http://dx.doi.org/10.1016/j.envpol.2004.12.011

Dirtu A, Van den Eede N, Malarvannan G, Ionas A, Covaci A (2012) Analytical methods for selected emerging contaminants in human matrices – a review. Anal Bioanal Chem 404:2555–2581. http://dx.doi.org/10.1007/s00216-012-6053-0

EFSA (2004) Opinion of the scientific panel on contaminants in the food chain on a request from the commission to assess the health risks to consumers associated with exposure to organotins in foodstuffs. EFSA J 102: 1–119. http://www.efsa.eu.int

Esteban M, Castaño A (2009) Non-invasive matrices in human biomonitoring: a review. Environ Int 35:438–449. http://dx.doi.org/http://dx.doi.org/10.1016/j.envint.2008.09.003

Evans SM, Leksono T, McKinnel PD (1995) Tributyltin pollution: a diminishing problem following legislation limiting the use of TBT-based antifouling paints. Mar Pollut Bull 30:14–21. http://dx.doi.org/10.1016/0025-326X(94)00181-8

Evans SM, Evans PM, Leksono T (1996) Widespread recovery of dogwhelks, *Nucella lapillus* (L.), from tributyltin contamination in the North Sea and Clyde Sea. Mar Pollut Bull 32:263–269. http://dx.doi.org/10.1016/0025-326X(95)00127-9

EVISA (2009) European Virtual Institute for Speciation Analysis – EU bans certain organotin compounds. http://www.speciation.net/News/EU-bans-certain-organotin-compounds-;~/2009/06/05/4298.html

Fargašová A (1996) Inhibitive effect of organotin compounds on the chlorophyll content of the green freshwater alga *Scenedesmus quadricauda*. Bull Environ Contam Toxicol 57:99–106. http://dx.doi.org/10.1007/s001289900161

Fargašová A (1997) The effects of organotin compounds on growth, respiration rate, and chlorophylla content of *Scenedesmus quadricauda*. Ecotoxicol Environ Saf 37:193–198. http://dx.doi.org/10.1006/eesa.1997.1538

Fargašová A, Kizlink J (1996) Effect of organotin compounds on the growth of the freshwater alga *Scenedesmus quadricauda*. Ecotoxicol Environ Saf 34:156–159. http://dx.doi.org/10.1006/eesa.1996.0057

Fent K (1996a) Ecotoxicology of organotin compounds. Crit Rev Toxicol 26:1–117. http://dx.doi.org/10.3109/10408449609089891

Fent K (1996b) Organotin compounds in municipal wastewater and sewage sludge: contamination, fate in treatment process and ecotoxicological consequences. Sci Total Environ 185:151–159. http://dx.doi.org/10.1016/0048-9697(95)05048-5

Fent K (2006) Worldwide occurrence of organotins from antifouling paints and effects in the aquatic environment. In: Konstantinou IK (ed) Antifouling paint biocides. Springer, Berlin/Heidelberg, pp 71–100. http://dx.doi.org/10.1007/698_5_050
Féral C, Le Gall S (1983) The influence of a pollutant factor (tributyltin) on the neuroendocrine mechanism responsible for the occurrence of a penis in the females of *Ocenebra erinacea*. In: Lever J, Boer HH (eds) Molluscan neuroendocrinology. North Holland, Amsterdam, pp 173–175
Forsyth DS, Jay B (1997) Organotin leachates in drinking water from chlorinated poly(vinyl chloride) (CPVC) pipe. Appl Organomet Chem 11:551–558. http://dx.doi.org/10.1002/(SICI)1099-0739(199707)11:7<551::AID-AOC606>3.0.CO;2–0
Forsyth DS, Weber D, Cleroux C (1992) Determination of butyltin, cyclohexyltin and phenyltin compounds in beers and wines. Food Addit Contam 9:161–169. http://dx.doi.org/10.1080/02652039209374058
Forsyth DS, Sun WF, Dalglish K (1994) Survey of organotin compounds in blended wines. Food Addit Contam 11:343–350. http://dx.doi.org/10.1080/02652039409374233
Fretter V, Graham A (1994) British prosobranch molluscs, their functional anatomy and ecology. The Ray Society, Dorset Press, Dorchester
Fristachi A, Xu Y, Rice G, Impellitteri CA, Carlson-Lynch H, Little JC (2009) Using probabilistic modeling to evaluate human exposure to organotin in drinking water transported by polyvinyl chloride pipe. Risk Anal 29:1615–1628. http://dx.doi.org/10.1111/j.1539-6924.2009.01307.x
Fromme H, Mattulat A, Lahrz T, Rüden H (2005) Occurrence of organotin compounds in house dust in Berlin (Germany). Chemosphere 58:1377–1383. http://dx.doi.org/10.1016/j.chemosphere.2004.09.092
Furdek M, Vahčič M, Ščančar J, Milačič R, Kniewald G, Mikac N (2012) Organotin compounds in seawater and *Mytilus galloprovincialis* mussels along the Croatian Adriatic Coast. Mar Pollut Bull 64:189–199. http://dx.doi.org/10.1016/j.marpolbul.2011.12.009
Gadd GM (2000) Microbial interactions with tributyltin compounds: detoxification, accumulation, and environmental fate. Sci Total Environ 258:119–127. http://dx.doi.org/10.1016/S0048-9697(00)00512-X
Galante-Oliveira S, Oliveira I, Ferreira N, Santos JA, Pacheco M, Barroso C (2011) *Nucella lapillus* L. imposex levels after legislation prohibiting TBT antifoulants: temporal trends from 2003 to 2008 along the Portuguese coast. J Environ Monit 13:304–312. http://dx.doi.org/10.1039/C0EM00140F
Garg A, Meena RM, Jadhav S, Bhosle NB (2011) Distribution of butyltins in the waters and sediments along the coast of India. Mar Pollut Bull 62:423–431. http://dx.doi.org/10.1016/j.marpolbul.2010.12.003
Gibbs PE, Bryan GW, Pascoe PL, Burt GR (1987) The use of the dog-whelk, *Nucella lapillus*, as an indicator of tributyltin (TBT) contamination. J Mar Biol Assoc UK 67:507–523. http://dx.doi.org/10.1017/S0025315400027260
Gibbs PE, Bebianno MJ, Coelho MR (1997) Evidence of the differential sensitivity of neogastropods to tributyltin (TBT) pollution, with notes on a species (*Columbella rustica*) lacking the imposex response. Environ Technol 18:1219–1224. http://dx.doi.org/10.1080/09593331808616643
Gibson CP, Wilson SP (2003) Imposex still evident in eastern Australia 10 years after tributyltin restrictions. Mar Environ Res 55:101–112. http://dx.doi.org/10.1007/s00128-011-0482-x
Gipperth L (2009) The legal design of the international and European Union ban on tributyltin antifouling paint: direct and indirect effects. J Environ Manag 90:S86–S95. http://dx.doi.org/10.1016/j.jenvman.2008.08.013
Goldberg ED (1975) The mussel watch – a first step in global marine monitoring. Mar Pollut Bull 6:111–114. http://dx.doi.org/10.1016/0025-326X(75)90271-4
Goldberg ED (1986) TBT: an environmental dilemma. Environment 28:17–44. http://dx.doi.org/10.1080/00139157.1986.9928814
Goméz-Ariza JL, Santos MM, Morales E, Giráldez I, Sánchez-Rodas D, Vieira N, Kemp JF, Boon JP, Ten Hallers-Tjabbes CC (2006) Organotin contamination in the Atlantic Ocean off the Iberian Peninsula in relation to shipping. Chemosphere 64:1100–1108. http://dx.doi.org/10.1016/j.chemosphere.2005.11.068

Gooding MP, Wilson VS, Folmar LC, LeBlanc GA (2003) The biocide tributyltin reduces the accumulation of testosterone as fatty acid esters in the mud snail (*Ilyanassa obsoleta*). Environ Health Perspect 111:426–430. http://dx.doi.org/10.1289/ehp.5779

Grün F, Blumberg B (2006) Environmental obesogens: organotins and endocrine disruption via nuclear receptor signaling. Endocrinology 147:s50–s55. http://dx.doi.org/10.1210/en.2005-1129

Guard HE, Cobet AB, Coleman Iii WM (1981) Methylation of trimethyltin compounds by estuarine sediments. Science 213:770–771. http://dx.doi.org/10.1126/science.213.4509.770

Guérin T, Sirot V, Volatier JL, Leblanc JC (2007) Organotin levels in seafood and its implications for health risk in high-seafood consumers. Sci Total Environ 338:66–77. http://dx.doi.org/10.1016/j.scitotenv.2007.08.027

Gui-bin J, Qun-fang Z, Bin H (2000a) Tin compounds and major trace metal elements in organotin-poisoned patient's urine and blood measured by gas chromatography-flame photometric detector and inductively coupled plasma-mass spectrometry. Bull Environ Contam Toxicol 65:277–284. http://dx.doi.org/10.1007/s001280000125

Gui-bin J, Qun-fang Z, Bin H (2000b) Speciation of organotin compounds, total tin, and major trace metal elements in poisoned human organs by gas chromatography-flame photometric detector and inductively coupled plasma-mass spectrometry. Environ Sci Technol 34:2697–2702. http://dx.doi.org/10.1021/es0008822

Guruge KS, Tanabe S, Iwata H, Taksukawa R, Yamagishi S (1996) Distribution, biomagnification, and elimination of butyltin compound residues in common cormorants (*Phalacrocorax carbo*) from Lake Biwa, Japan. Arch Environ Contam Toxicol 31:210–217. http://dx.doi.org/10.1007/BF00212368

Guruge KS, Tanabe S, Fukuda M, Yamagishi S, Tatsukawa R (1997) Comparative tissue distribution of butyltin compounds in common cormorants (*Phalacrocorax carbo*) from Japan. Toxicol Environ Chem 58:197–208. http://dx.doi.org/10.1080/02772249709358409

Hagger JA, Depledge MH, Oehlmann J, Jobling S, Galloway TS (2006) Is there a causal association between genotoxicity and the imposex effect? Environ Health Perspect 114:20–26. http://dx.doi.org/10.1289/ehp.8048

Hallas LE, Means JC, Cooney JJ (1982) Methylation of tin by estuarine microorganisms. Science 215:1505–1507. http://dx.doi.org/10.1126/science.215.4539.1505

Harding MJC, Rodger GK, Davies IM, Moore JJ (1997) Partial recovery of the dogwhelk (*Nucella lapillus*) in Sullom Voe, Shetland from tributyltin contamination. Mar Environ Res 44:285–304. http://dx.doi.org/10.1016/S0141-1136(97)00008-1

Harino H, Fukushima M, Kawai S (1999) Temporal trends of organotin compounds in the aquatic environment of the Port of Osaka, Japan. Environ Pollut 105:1–7. http://dx.doi.org/10.1016/S0269-7491(98)00217-6

Harino H, Iwasaki N, Arai T, Ohji M, Miyazaki N (2005) Accumulation of organotin compounds in the deep-sea environment of Nankai Trough, Japan. Arch Environ Contam Toxicol 49:497–503. http://dx.doi.org/10.1007/s00244-005-7014-9

Harino H, Ohji M, Wattayakorn G, Adulyanukosol K, Arai T, Miyazaki N (2007) Accumulation of organotin compounds in tissues and organs of stranded whales along the coasts of Thailand. Arch Environ Contam Toxicol 53:119–125. http://dx.doi.org/10.1007/s00244-006-0128-x

Harino H, Eguchi S, Yamamoto Y, Kurokawa Y, Kawai S, Arai T, Ohji M, Yamato S, Kobayashi N, Miyazaki N (2008) Distribution of organotin compounds in representative coastal areas from Japan: a review. Coast Mar Sci 32:88–95. http://repository.dl.itc.u-tokyo.ac.jp/dspace/bitstream/2261/20993/1/CMS320113.pdf

Harino H, Arai T, Ohji M, Ismail A, Miyazaki N (2009a) Contamination profiles of antifouling biocides in selected coastal regions of Malaysia. Arch Environ Contam Toxicol 56:468–478. http://dx.doi.org/10.1007/s00244-008-9252-0

Harino H, Arai T, Ohji M, Miyazaki N (2009b) Organotin contamination in deep sea environments. In: Arai T, Harino H, Ohji M, Langston W (eds) Ecotoxicology of antifouling biocides. Springer Japan, Tokyo, pp 95–108. http://dx.doi.org/10.1007/978-4-431-85709-9_6

Harino H, Iwasaki N, Arai T, Ohji M, Miyazaki N (2009c) Occurrence of antifouling biocides and fluorinated alkyl compounds in sediment core from deep sea: Suruga Bay, Tosa Bay, and Nankai Tough, Japan. Arch Environ Contam Toxicol 57:661–669. http://dx.doi.org/10.1007/s00244-009-9374-z

Haynes D, Loong D (2002) Antifoulant (butyltin and copper) concentrations in sediments from the Great Barrier Reef World Heritage Area, Australia. Environ Pollut 120:391–396. http://dx.doi.org/10.1016/S0269-7491(02)00113-6

Haynes D, Christie C, Marshall P, Dobbs K (2002) Antifoulant concentrations at the site of the Bunga Teratai Satu grounding, Great Barrier Reef, Australia. Mar Pollut Bull 44:968–972. http://dx.doi.org/10.1016/S0025-326X(02)00114-5

HELCON (2010) Hazardus substances in the Baltic Sea: an integrated thematic assessment of hazardous substances in the Baltic Sea Baltic Sea. Baltic Sea Environ Proc 120B:119. http://www.helcom.fi/stc/files/Publications/Proceedings/bsep120B.pdf

Heroult J, Bueno M, Potin-Gautier M, Lespes G (2008) Organotin speciation in French brandies and wines by solid-phase microextraction and gas chromatography – pulsed flame photometric detection. J Chromatogr A 1180:122–130. http://dx.doi.org/10.1016/j.chroma.2007.11.084

Hoch M (2001) Organotin compounds in the environment – an overview. Appl Geochem 16:719–743. http://dx.doi.org/10.1016/S0883-2927(00)00067-6

Hoch M, Schwesig D (2004) Parameters controlling the partitioning of tributyltin (TBT) in aquatic systems. Appl Geochem 19:323–334. http://dx.doi.org/10.1016/s0883-2927(03)00131-8

Hong HK, Takahashi S, Min BY, Tanabe S (2002) Butyltin residues in blue mussels (*Mytilus edulis*) and arkshells (*Scapharca broughtonii*) collected from Korean coastal waters. Environ Pollut 117:475–486. http://dx.doi.org/10.1016/S0269-7491(01)00185-3

Horiguchi T (2006) Masculinization of female gastropod mollusks induced by organotin compounds, focusing on mechanism of actions of tributyltin and triphenyltin for development of imposex. Environ Sci 13:77–87. http://www.ncbi.nlm.nih.gov/pubmed/16788559

Horiguchi T (2009a) Mechanism of imposex induced by organotins in gastropods. In: Takaomi A, Harino H, Ohji M, Langston WJ (eds) Ecotoxicology of antifouling biocides. Springer Japan, Tokyo, pp 111–123. http://dx.doi.org/10.1007/978-4-431-85709-9

Horiguchi T (2009b) The endocrine-disrupting effect of organotin compounds for aquatic organisms. In Takaomi A, Harino H, Ohji M, Langston WJ (eds) Ecotoxicology of antifouling biocides. Springer Japan, Tokyo, pp 125–146. http://dx.doi.org/10.1007/978-4-431-85709-9_8

Horiguchi T (2012) Ecotoxicological impacts of organotins: an overview. In Pagliarani A, Trombetti F, Ventrella V (eds) Biochemical and biological effects of organotins. Bentham Science, Sharjah, pp 3–24. http://dx.doi.org/0.2174/97816080526531120101 0vii

Horiguchi T, Nishikawa T, Ohta Y, Shiraishi H, Morita M (2007a) Retinoid X receptor gene expression and protein content in tissues of the rock shell *Thais clavigera*. Aquat Toxicol 84:379–388. http://dx.doi.org/10.1016/j.aquatox.2007.06.019

Horiguchi T, Ohta Y, Nishikawa T, Shiraishi F, Shiraishi H, Morita M (2007b) Exposure to 9-cis retinoic acid induces penis and vas deferens development in the female rock shell, *Thais clavigera*. Cell Biol Toxicol 24:553–562. http://dx.doi.org/10.1007/s10565-007-9051-9

Horiguchi T, Urushitani H, Ohta Y, Iguchi T, Shiraishi H (2010) Establishment of a polyclonal antibody against the retinoid X receptor of the rock shell Thais clavigera and its application to rock shell tissues for imposex research. Ecotoxicology 19:571–576. http://dx.doi.org/10.1007/s10646-009-0447-6

Huet M, Fioroni P, Oehlmann J, Stroben E (1995) Comparison of imposex response in three prosobranch species. Hydrobiologia 309:29–35. http://dx.doi.org/10.1007/BF00014469

Hwang HM, Oh JR, Kahng SH, Lee KW (1999) Tributyltin compounds in mussels, oysters and sediments of Chinhae Bay, Korea. Mar Environ Res 47:61–70. http://dx.doi.org/10.1016/S0141-1136(98)00023-3

IMO (1999) Anti-fouling systems moving towards the non-toxic solution. International Maritime Organization, London. http://www.imo.org/includes/blastDataOnly.asp/data_id%3D7986/FOULING2003.pdf

IMO (2002) Focus on IMO: antifouling systems. International Maritime Organization, London, p 31. http://www.imo.org/blast/blastDataHelper.asp?data_id=7986&filename=FOULING2003.pdf

Jensen HF, Holmer M, Dahllöf I (2004) Effects of tributyltin (TBT) on the seagrass *Ruppia maritima*. Mar Pollut Bull 49:564–573. http://dx.doi.org/10.1016/j.marpolbul.2004.03.010

Jones-Lepp TL, Varner KE, Hilton BA (2001) Speciation and detection of organotins from PVC pipe by micro-liquid chromatographyelectrospray-ion trap mass spectrometry. Appl Organomet Chem 15:933–938. http://dx.doi.org/10.1002/aoc.246

Kan-Atireklap S, Yen NTH, Tanabe S, Subramanian A (1998) Butyltin compounds and organochlorine residues in green mussel (*Perna viridis* L.) from India. Toxicol Environ Chem 67:409–424. http://dx.doi.org/10.1080/02772249809358631

Kannan K, Falandysz J (1997) Butyltin residues in sediment, fish, fish-eating birds, harbour porpoise and human tissues from the Polish coast of the Baltic Sea. Mar Pollut Bull 34:203–207. http://dx.doi.org/10.1016/S0025-326X(96)00146-4

Kannan K, Tanabe S (2009) Global contamination by organotin compounds. In Arai T, Harino H, Ohji M, Langston WJ (eds) Ecotoxicology of antifouling biocides. Springer, Tokyo, pp 39–60. http://dx.doi.org/10.1007/978-4-431-85709-9_3

Kannan K, Guruge KS, Thomas NJ, Tanabe S, Giesy JP (1998) Butyltin residues in southern sea otters *Enhydra lutris nereis* found dead along California coastal waters. Environ Sci Technol 32:1169–1175. http://dx.doi.org/10.1021/es970914u

Kannan K, Senthilkumar K, Giesy JP (1999) Occurrence of butyltin compounds in human blood. Environ Sci Technol 33:1776–1779. http://dx.doi.org/abs/10.1021/es990011w

Kannan K, Senthilkumar K, Giesy JP (2000) Response to comment on "Occurrence of butyltin compounds in human blood". Environ Sci Technol 34:1879–1880. http://dx.doi.org/10.1021/es002003h

Kannan K, Takahashi S, Fujiwara N, Mizukawa H, Tanabe S (2010) Organotin compounds, including butyltins and octyltins, in house dust from Albany, New York, USA. Arch Environ Contam Toxicol 58:901–907. http://dx.doi.org/10.1007/s00244-010-9513-6

Keithly JC, Cardwell RD, Henderson DG (1999) Tributyltin in seafood from Asia, Australia, Europe, and North America: assessment of human health risks. Hum Ecol Risk Assess 5:337–354. http://dx.doi.org/10.1080/10807039991289473.

Kim GB, Lee JS, Tanabe S, Iwata H, Tatsukawa R, Shimazaki K (1996) Specific accumulation and distribution of butyltin compounds in various organs and tissues of the Steller sea lion (Eumetopias jubatus): comparison with organochlorine accumulation pattern. Mar Pollut Bull 32:558–563. http://dx.doi.org/10.1016/0025-326X(96)84576-0

Kotrikla A (2009) Environmental management aspects for TBT antifouling wastes from the shipyards. J Environ Manag 90:S77–S85. http://dx.doi.org/10.1016/j.jenvman.2008.07.017

Krupp EM, Merle JK, Haas K, Foote G, Maubec N, Feldmann J (2010) Volatilization of organotin species from municipal waste deposits: novel species identification and modeling of atmospheric stability. Environ Sci Technol 45:943–950. http://dx.doi.org/10.1021/es102512u

Kucuksezgin F, Aydin-Onen S, Gonul LT, Pazi I, Kocak F (2011) Assessment of organotin (butyltin species) contamination in marine biota from the Eastern Aegean Sea, Turkey. Mar Pollut Bull 62:1984–1988. http://dx.doi.org/10.1016/j.marpolbul.2011.06.020

Kusk KO, Petersen S (1997) Acute and chronic toxicity of tributyltin and linear alkilbenzene sulfonate to the marine copepod *Acartia tonsa*. Environ Toxicol Chem 16:1629–1633. http://dx.doi.org/10.1002/etc.5620160810

Langston WJ (1990) Toxic effects of metals and the incidence of metal pollution in marine ecosystems. In: Furness RW, Rainbow PS (eds) Heavy metals in the marine environment. CRC Press, Boca Raton, pp 101–122

Laranjeiro F, Sousa A, Takahashi S, Tanabe S, Barroso CM (2010) Combination of field monitoring and laboratory bioassays for the assessment of TBT pollution in Ria de Aveiro. In: Hamamura N, Suzuki S, Mendo S, Barroso CM, Iwata H, Tanabe S (eds) Interdisciplinary studies on environmental chemistry – biological responses to contaminants: from molecular to community level. TERRAPUB, Tokyo, Japan, pp 175–188. http://www.terrapub.co.jp/onlineproceedings/ec/03/pdf/BR_03175.pdf

le Maire A, Grimaldi M, Roecklin D, Dagnino S, Vivat-Hannah V, Balaguer P, Bourguet W (2009) Activation of RXR-PPAR heterodimers by organotin environmental endocrine disruptors. EMBO Rep 10:367–373. http://dx.doi.org/10.1038/embor.2009.8

Lespes G, Marcic C, Heroult J, Le Hecho I, Denaix L (2009) Tributyltin and triphenyltin uptake by lettuce. J Environ Manag 90:S60–S68. http://dx.doi.org/10.1016/j.jenvman.2008.07.019

Leung KMY, Grist EPM, Morley NJ, Morritt D, Crane M (2007) Chronic toxicity of tributyltin to development and reproduction of the European freshwater snail *Lymnaea stagnalis* (L.). Chemosphere 66:1358–1366. http://dx.doi.org/10.1016/j.chemosphere.2006.06.051

Lima D, Reis-Henriques MA, Silva R, Santos AI, Filipe C. Castro L, Santos MM (2011) Tributyltin-induced imposex in marine gastropods involves tissue-specific modulation of the retinoid X receptor. Aquat Toxicol 101:221–227. http://dx.doi.org/10.1016/j.aquatox.2010.09.022

Liu JY, Jiang GB (2002) Survey on the presence of butyltin compounds in Chinese alcoholic beverages, determined by using headspace solid-phase microextraction coupled with gas chromatography – flame photometric detection. J Agric Food Chem 50:6683–6687. http://dx.doi.org/10.1021/jf025712i

Lo S, Alléra A, Albers P, Heimbrecht J, Jantzen E, Klingmüller D, Steckelbroeck S (2003) Dithioerythritol (DTE) prevents inhibitory effects of triphenyltin (TPT) on the key enzymes of the human sex steroid hormone metabolism. J Steroid Biochem Mol Biol 84:569–576. http://dx.doi.org/10.1016/S0960-0760(03)00074-8

Magi E, Liscio C, Pistarino E, Santamaria B, Di Carro M, Tiso M, Scaloni A, Renzone G, Cosulich M (2008) Interdisciplinary study for the evaluation of biochemical alterations on mussel *Mytilus galloprovincialis* exposed to a tributyltin-polluted area. Anal Bioanal Chem 391:671–678. http://dx.doi.org/10.1007/s00216-008-2055-3

Mala N (2008) Toxicity and the cardiovascular activity of organotin compounds: a review. Appl Organomet Chem 22:598–612. http://dx.doi.org/10.1002/aoc.1436

Marshall P, Christie C, Dobbs K, Green A, Haynes D, Brodie J, Michalek-Wagner K, Smith A, Storrie J, Turak E (2002) Grounded ship leaves TBT-based antifoulant on the Great Barrier Reef: an overview of the environmental response. Spill Sci Technol Bull 7:215–221. http://dx.doi.org/10.1016/S1353-2561(02)00040-3

Martina D, Claudia S, Jean B, Cornelius B, Ulrike S-O, Jörg O (2007) Prosobranch snails as test organisms for the assessment of endocrine active chemicals – an overview and a guideline proposal for a reproduction test with the freshwater mudsnail *Potamopyrgus antipodarum*. Ecotoxicology 16:169–182. http://dx.doi.org/10.1007/s10646-006-0106-0

Martins JD, Jurado AS, Moreno AJM, Madeira VMC (2005) Comparative study of tributyltin toxicity on two bacteria of the genus Bacillus. Toxicol In Vitro 19:943–949. http://dx.doi.org/10.1016/j.tiv.2005.06.019

Matthiessen P, Gibbs PE (1998) Critical appraisal of the evidence for tributyltin-mediated endocrine disruption in mollusks. Environ Toxicol Chem 17:37–43. http://dx.doi.org/10.1897/1551-5028(1998)017<0037:CAOTEF>2.3.CO;2

McAllister BG, Kime DE (2003) Early life exposure to environmental levels of the aromatase inhibitor tributyltin causes masculinisation and irreversible sperm damage in Zebrafish (*Danio rerio*). Aquat Toxicol 63:309–316. http://dx.doi.org/10.1016/S0166-445X(03)00154-1

Mendo S, Nogueira PR, Ferreira SCN, Silva RG (2003) Tributyltin and triphenyltin toxicity on benthic estuarine bacteria. Fresenius Environ Bull 12:1361–1367

Mensink BP, Ten Hallers-Tjables CC, Kralt J, Freriks IL, Boon J (1996) Assessment of imposex in the commun whelk, *Buccinum undatum* (L.) from the Eastern Scheldt, the Netherlands. Mar Environ Res 41:315–325. http://dx.doi.org/10.1016/0141-1136(95)00022-4

Michel P, Averty B, Andral B, Chiffoleau JF, Galgani F (2001) TBT along the coast of Corsica (Western Meditterranean): a persistent problem. Mar Pollut Bull 42:1128–1132. http://dx.doi.org/10.1016/S0025-326X(01)00101-1

Miller KL, Fernandes TF, Read PA (1999) The recovery of populations of dogwhelks suffering from imposex in the Firth of Forth 1987–1997/98. Environ Pollut 106:183–192. http://dx.doi.org/10.1016/S0269-7491(99)00076-7

Minchin D, Oehlmann J, Duggan CB, Stroben E, Keatinge M (1995) Marine TBT antifouling contamination in Ireland, following legislation in 1987. Mar Pollut Bull 30:633–639. http://dx.doi.org/10.1016/0025-326X(95)00016-G

Minchin D, Stroben E, Oehlmann J, Bauer B, Duggan CB, Keatinge M (1996) Biological indicators used to map organotin contamination in Cork Harbour, Ireland. Mar Pollut Bull 32:188–195. http://dx.doi.org/10.1016/0025-326X(95)00120-C

Mino Y, Amano F, Yoshioka T, Konishi Y (2008) Determination of organotins in human breast milk by gas chromatography with flame photometric detection. J Health Sci 54:224–228. http://dx.doi.org/10.1248/jhs.54.224

Morgan E, Murphy J, Lyons R (1998) Imposex in *Nucella lapillus* from TBT contamination in South and South-West Wales: a continuing problem around ports. Mar Pollut Bull 36:840–843. http://dx.doi.org/10.1016/S0025-326X(98)00087-3

Nagase M, Kondo H, Hasebe K (1995) Determination of tributyltin and triphenyltin compounds in hair and fish using a hydrolysis technique and gas chromatography with flame photometric detection. Analyst 120:1923–1926. http://dx.doi.org/10.1039/AN9952001923

Negri A, Marshall P (2009) TBT contamination of remote marine environments: Ship groundings and ice-breakers as sources of organotins in the Great Barrier Reef and Antarctica. J Environ Manag 90:S31-S40. http://dx.doi.org/10.1016/j.jenvman.2008.06.009

Negri AP, Hales LT, Battershill C, Wolff C, Webster NS (2004) TBT contamination identified in Antarctic marine sediments. Mar Pollut Bull 48:1142–1144. http://dx.doi.org/10.1016/j.marpolbul.2004.03.004

Nemanič TM, Leskovšek H, Horvat M, Vrišer B, Bolje A (2002) Organotin compounds in the marine environment of the Bay of Piran, Northern Adriatic Sea. J Environ Monit 4:426–430. http://dx.doi.org/10.1039/b111582k

Nemanič MT, Milačič R, Ščančar J (2009) A survey of organotin compounds in the Northern Adriatic Sea. Water Air Soil Pollut 196:211–224. http://dx.doi.org/10.1007/s11270-008-9770-4

Nielsen JB, Strand J (2002) Butyltin compounds in human liver. Environ Res 88:129–133. http://dx.doi.org/10.1006/enrs.2001.4321

Nishikawa J (2006) Imposex in marine gastropods may be caused by binding of organotins to retinoid X receptor. Mar Biol 149:117–124. http://dx.doi.org/10.1007/s00227-005-0210-3

Nishikawa J, Mamiya S, Kanayama T, Nishikawa T, Shiraishi F, Horiguchi T (2004) Involvement of the retinoid X receptor in the development of imposex caused by organotins in gastropods. Environ Sci Technol 38:6271–6276. http://dx.doi.org/abs/10.1021/es049593u

NRC (2006) Human biomonitoring for environmental chemicals. The National Academies Press, Washington, DC. ISBN:9780309102728. http://www.nap.edu/openbook.php?record_id=11700

Nudelman MA, Carro C, Nudelman NS (1998) Effects of tin(IV) chloride and of organotin compounds on aquatic micro-organisms. Appl Organomet Chem 12:67–75. http://dx.doi.org/10.1002/(SICI)1099-0739(199801)12:1<67::AID-AOC673>3.0.CO;2-#

Oberdörster E, McClellan-Green P (2000) The neuropeptide APGWamide induces imposex in the mud snail, *Ilyanassa obsoleta*. Peptides 21:1323–1330. http://dx.doi.org/10.1016/S0196-9781(00)00274-6

Oberdörster E, McClellan-Green P (2002) Mechanisms of imposex induction in the mud snail, *Ilyanassa obsoleta*: TBT as a neurotoxin and aromatase inhibitor. Mar Environ Res 54:715–718. http://dx.doi.org/10.1016/S0141-1136(02)00118-6

Oehlmann J, Stroben E, Fioroni P (1993) Fréquence et degré d'expression du pseudohermaphrodisme chez quelques prosobranches sténoglosses des côtes françaises (surtout de la baie de Morlaix et de la Manche) 2. Situation jusqu'au printemps de 1992. Cah Biol Mar 34:343–362

Ohtaki K, Aihara M, Takahashi H, Fujita H, Takahashi K, Funabashi T, Hirasawa T, Ikezawa Z (2007) Effects of tributyltin on the emotional behavior of C57BL/6 mice and the development of atopic dermatitis-like lesions in DS-Nh mice. J Dermatol Sci 47:209–216. http://dx.doi.org/10.1016/j.jdermsci.2007.05.001

Omae I (2003) Organotin antifouling paints and their alternatives. Appl Organomet Chem 17:81–105. http://dx.doi.org/10.1002/aoc.396

Omae I (2006) Chemistry and fate of organotin antifouling biocides in the environment. In: Hutzinger O (ed) Water pollution. Handbook of environmental chemistry, vol 5. Springer, Berlin/Heidelberg, pp 17–50. http://dx.doi.org/10.1007/698-5-048

OSPAR (2003) Joint assessment and monitoring program – guidelines for contaminant-specific biological effects monitoring: Technical annex 3 (TBT-specific biological effects monitoring). Ospar Commission, London. http://www.ospar.org/content/content.asp?menu=00900301400135_000000_000000

OSPAR (2008) 2007/2008 CEMP assessment – trends and concentrations of selected hazardous substances in the marine environment. Ospar Commission, London. http://www.ospar.org/documents/dbase/publications/p00378_2007-2008_CEMP_assessment.pdf

OSPAR (2010) Quality status report 2010. OSPAR Commission, London, p 177. http://qsr2010.ospar.org/en/media/chapter_pdf/QSR_complete_EN.pdf

Park K, Kim R, Park JJ, Shin HC, Lee JS, Cho HS, Lee YG, Kim J, Kwak IS (2012) Ecotoxicological evaluation of tributyltin toxicity to the equilateral venus clam, *Gomphina veneriformis* (Bivalvia: Veneridae). Fish Shellfish Immunol 32:426–433. http://dx.doi.org/10.1016/j.fsi.2011.11.031

Pascoal S, Carvalho G, Vasieva O, Hughes R, Cossins A, Fang Y, Ashelford K, Olohan L, Barroso C, Mendo S, Creer S (2013) Transcriptomics and in vivo tests reveal novel mechanisms underlying endocrine disruption in an ecological sentinel, *Nucella lapillus*. Mol Ecol 22:1589–1608. http://dx.doi.org/10.1111/mec.12137

Pessoa MF, Fernando A, Oliveira JS (2001) Use of imposex (Pseudohermaphroditism) as indicator of the occurrence of organotin compounds in Portuguese coastal waters – Sado and Mira estuaries. Environ Toxicol 16:234–241. http://dx.doi.org/10.1002/tox.1029

Peters RJB (2004) Man-made chemicals in human blood. TNO-report R 2004/493. Netherlands Organisation for Applied Scientific Research, Apeldoorn, p 48

Petersen S, Gustavson K (1998) Toxic effects of tri-butyl-tin (TBT) on autotrophic pico-, nano-, and microplankton assessed by a size fractionated pollution-induced community tolerance (SF-PICT) concept. Aquat Toxicol 40:253–264. http://dx.doi.org/10.1016/S0166-445X(97)00049-0

Prudente M, Ichihashi H, Kan-atireklap S, Watanabe I, Tanabe S (1999) Butyltins, organochlorines and metal levels in green mussel, *Perna viridis* L. from the coastal waters of the Philippines. Fish Sci 65:441–447

Quevauviller P, Donard OFX, Bruchet A (1991) Leaching of organotin compounds from poly(vinyl chloride) (PVC) material. Appl Organomet Chem 5:125–129. http://dx.doi.org/10.1002/aoc.590050210

Quintela M, Couceiro L, Ruiz JM, Barreiro R (2006) Discovery of imposex in the gastropod *Cyclope neritea* now invading Galicia (north-west Spain). J Mar Biol Assoc UK 86:1171–1173. http://dx.doi.org/10.1017/S0025315406014159

Qunfang Z, Guibin J, Zhongyang L, Lina L, Chungang Y, Yongning W (2004) Survey of butyltin compounds in 12 types of foods collected in China. Food Addit Contam 21:1162–1167. http://dx.doi.org/10.1080/02652030400019364

Rantakokko P, Kuningas T, Saastamoinen K, Vartiainen T (2006) Dietary intake of organotin compounds in Finland: a market-basket study. Food Addit Contam 23:749–756. http://dx.doi.org/10.1080/02652030600779908

Rantakokko P, Turunen A, Verkasalo PK, Kiviranta H, Männistö S, Vartiainen T (2008) Blood levels of organotin compounds and their relation to fish consumption in Finland. Sci Total Environ 399:90–95. http://dx.doi.org/10.1016/j.scitotenv.2008.03.017

Rato M, Sousa A, Quintã R, Langston W, Barroso C (2006) Assessment of inshore/offshore tributyltin pollution gradients in the Northwest Portugal continental shelf using *Nassarius reticulatus* as a bioindicator. Environ Toxicol Chem 25:3213–3220. http://dx.doi.org/10.1897/06-167R.1

Rato M, Gaspar MB, Takahashi S, Yano S, Tanabe S, Barroso C (2008) Inshore/offshore gradients of imposex and organotin contamination in *Nassarius reticulatus* (L.) along the Portuguese coast. Mar Pollut Bull 56:1323–1331. http://dx.doi.org/10.1016/j.marpolbul.2008.04.004

Rato M, Ferreira N, Santos J, Barroso CM (2009) Temporal evolution of imposex in *Nassarius reticulatus* (L.) along the Portuguese coast: the efficacy of EC regulation 782/2003. J Environ Monit 11:100–107. http://dx.doi.org/10.1039/b810188d

Rees CM, Brady BA, Fabris GJ (2001) Incidence of imposex in *Thais orbita* from Port Phillip Bay (Victoria, Australia), following 10 years of regulation on use of TBT. Mar Pollut Bull 42:873–878. http://dx.doi.org/10.1016/S0025-326X(01)00041-8

Rodríguez JG, Tueros I, Borja Á, Franco J, Ignacio García Alonso J, Garmendia JM, Muxika I, Sariego C, Valencia V (2009) Butyltin compounds, sterility and imposex assessment in *Nassarius reticulatus* (Linnaeus, 1758), prior to the 2008 European ban on TBT antifouling paints, within Basque ports and along coastal areas. Cont Shelf Res 29:1165–1173. http://dx.doi.org/10.1016/j.csr.2009.01.005

Rodríguez JG, Rouget P, Franco J, Garmendia JM, Muxika I, Valencia V, Borja Á (2010) Evaluation of the use of transplanted *Nassarius reticulatus* (Linnaeus, 1758), in monitoring TBT pollution, within the European Water Framework Directive. Ecol Indic 10:891–895. http://dx.doi.org/10.1016/j.ecolind.2009.12.005

Ronis MJJ, Mason AZ (1996) The metabolism of testosterone by the periwinkle (*Littorina littorea*) in vitro and in vivo: Effects of tributyltin. Mar Environ Res 42:161–166. http://dx.doi.org/10.1016/0141-1136(95)00069-0

RPA (2005) Risk assessment studies on targeted consumer applications of certain organotin compounds. Prepared for the European Commission by Risk & Policy Analysts Limited (RPA). Dated Sept 2005. http://ec.europa.eu/enterprise/sectors/chemicals/files/studies/organotins_3rd_report_16_sept_2005_en.pdf

Rudel H (2003) Case study: bioavailability of tin and tin compounds. Ecotoxicol Environ Saf 56:180–189. http://dx.doi.org/10.1016/S0147-6513(03)00061-7

Ruiz JM, Barreiro R, González JJ (2005) Biomonitoring organotin pollution with gastropods and mussels. Mar Ecol Prog Ser 287:169–176. http://dx.doi.org/10.3354/meps287169

Ruiz JM, Barreiro R, Couceiro L, Quintela M (2008) Decreased TBT pollution and changing bioaccumulation pattern in gastropods imply butyltin desorption from sediments. Chemosphere 73:1253–1257. http://dx.doi.org/10.1016/j.chemosphere.2008.07.029

Sadiki A-D, Williams DT (1999) A study on organotin levels in Canadian drinking water distributed through PVC pipes. Chemosphere 38:1541–1548. http://dx.doi.org/10.1016/S0045-6535(98)00374-9

Sadiki A-I, Williams DT, Carrier R, Thomas B (1996) Pilot study on the contamination of drinking water by organotin compounds from PVC materials. Chemosphere 32:2389–2398. http://dx.doi.org/10.1016/0045-6535(96)00134-8

Santillo D, Johnston P, Brigden K (2001) The presence of brominated flame retardants and organotin compounds in dusts collected from Parliament buildings from eight countries. Greenpeace Research Laboratories Technical Note 03/2001. Greenpeace Research Laboratories, Department of Biological Sciences, University of Exeter, Exeter. http://archive.greenpeace.org/toxics/reports/eudust.pdf

Santillo D, Labunska I, Davidson H, Johnston P, Strutt M, Knowles O (2003) Consuming chemicals: hazardous chemicals in house dust as an indicator of chemical exposure in the home: Part I – UK. Greenpeace Research Laboratories Technical Note 01/2003, Greenpeace Research Laboratories, Exeter, p 74. http://www.greenpeace.to/greenpeace/?p=900

Santos MM, Ten Hallers-Tjables CC, Santos AM, Vieira N (2002) Imposex in *Nucella lapillus*, a bioindicator for TBT contamination: re-survey along the Portuguese coast to monitor the effectiveness of EU regulation. J Sea Res 48:217–223. http://dx.doi.org/10.1016/S1385-1101(02)00166-1

Santos MM, Vieira N, Reis-Henriques MA, Santos AM, Gómez-Ariza JL, Giráldez I, Ten Hallers-Tjabbes CC (2004) Imposex and butyltin contamination off the Oporto Coast (NW Portugal): a possible effect of the discharge of dredged material. Environ Int 30:793–798. http://dx.doi.org/10.1016/j.envint.2004.01.005

Santos MM, Enes P, Reis-Henriques MA, Kuballa J, Castro LFC, Vieira MN (2009) Organotin levels in seafood from Portuguese markets and the risk for consumers. Chemosphere 75:661–666. http://dx.doi.org/10.1016/j.chemosphere.2008.12.066

Santos MM, Reis-Henriques MA, Castro LFC (2012) Lipid homeostasis perturbation by organotins: effects on vertebrates and invertebrates. In Pagliarani A, Trombetti F, Ventrella V (eds)

Biochemical and biological effects of organotins. Bentham Science, Sharjah, pp 3–24. http://dx.doi.org/0.2174/978160805265311201010vii.
Ščančar J, Zuliani T, Turk T, Milačič R (2007) Organotin compounds and selected metals in the marine environment of Northern Adriatic Sea. Environ Monit Assess 127:271–282. http://dx.doi.org/10.1007/s10661-006-9278-6
SCHER (2006) Scientific Committee on Health and Environmental Risks (SCHER). Revised assessment of the risks to health and the environment associated with the use of the four organotin compounds TBT, DBT, DOT and TPT. Health and Consumer Protection Directorate. Opinion adopted by the SCHER during the 14th plenary of 30 Nov 2006. European Commission, EC. http://ec.europa.eu/health/ph_risk/committees/04_scher/docs/scher_o_047.pdf
Schramm KW (2008) Hair-biomonitoring of organic pollutants. Chemosphere 72:1103–1111. http://dx.doi.org/10.1016/j.chemosphere.2008.04.017
Sekizawa J, Sutur G, Birnbaum L (2003) Integrated human and ecological risk assessment: a case study of tributyltin and triphenyltin compounds. Hum Ecol Risk Assess 9:325–342. http://dx.doi.org/10.1080/718990536
Senthilkumar K, Kannan K, Tanabe S, Prudente M (1998) Butyltin compounds in resident and migrant birds collected from Philippines. Fresenius Environ Bull 7:561–571
Senthilkumar K, Tanabe S, Kannan K, Subramanian A (1999) Butyltin residues in migratory and resident birds collected from South India. Toxicol Environ Chem 68:91–104. http://dx.doi.org/10.1080/02772249909358648
Shi HH, Huang CJ, Zhu SX, Yu XJ, Xie WY (2005) Generalized system of imposex and reproductive failure in female gastropods of coastal waters of mainland China. Mar Ecol Prog Ser 304:179–189. http://dx.doi.org/10.3354/meps304179
Shimasaki Y, Kitano T, Oshima Y, Inoue S, Imada N, Honjo T (2003) Tributyltin causes masculinization in fish. Environ Toxicol Chem 22:141–144. http://dx.doi.org/10.1897%2F1551-5028%282003%29022%3C0141%3ATCMIF%3E2.0.CO%3B2
Sidharthan M, Young KS, Woul LH, Soon PK, Shin HW (2002) TBT toxicity on the marine microalga *Nannochloropsis oculata*. Mar Pollut Bull 45:177–180. http://dx.doi.org/10.1016/S0025-326X(01)00283-1
Smith BS (1971) Sexuality in the American mud snail, *Nassarius obsoletus* Say. Proc Malacol Soc Lond 39:377–378. http://mollus.oxfordjournals.org/content/39/5/377.extract#
Smith BS (1981a) Reproductive anomalies in stenoglossan snails related to pollution from marinas. J Appl Toxicol 1:15–21. http://dx.doi.org/10.1002/jat.2550010105
Smith BS (1981b) Male characteristics on female mud snails caused by antifouling bottom paints. J Appl Toxicol 1:22–25. http://dx.doi.org/10.1002/jat.2550010106
Smith PJ (1996) Selective decline in imposex levels in the dogwhelk *Lepsiella scobina* following a ban on the use of TBT antifoulants in New Zealand. Mar Pollut Bull 32:362–365. http://dx.doi.org/10.1016/0025-326X(96)84830-2
Sonak S, Pangam P, Giriyan A, Hawaldar K (2009) Implications of the ban on organotins for protection of global coastal and marine ecology. J Environ Manag 90:S96–S108. http://dx.doi.org/10.1016/j.jenvman.2008.08.017
Sousa A, Génio L, Mendo S, Barroso CM (2005a) Comparison of the acute toxicity of tributyltin and copper to veliger larvae of *Nassarius reticulatus* (L.). Appl Organomet Chem 19:324–328. http://dx.doi.org/10.1002/aoc.886
Sousa A, Mendo S, Barroso CM (2005b) Imposex and organotin contamination in *Nassarius reticulatus* (L.) along the Portuguese coast. Appl Organomet Chem 19:315–323. http://dx.doi.org/10.1002/aoc.856
Sousa A, Laranjeiro F, Takahashi S, Tanabe S, Barroso CM (2007a) Usefulness of *Nassarius reticulatus* imposex levels to monitor decreasing levels in TBT pollution: field and laboratory studies. Organohalogen Compd 69:1075–1078
Sousa A, Matsudaira C, Takahashi S, Tanabe S, Barroso C (2007b) Integrative assessment of organotin contamination in a southern European estuarine system (Ria de Aveiro, NW Portugal): tracking temporal trends in order to evaluate the effectiveness of the EU ban. Mar Pollut Bull 54:1645–1653. http://dx.doi.org/10.1016/j.marpolbul.2007.07.005

Sousa A, Ikemoto T, Takahashi S, Barroso CM, Tanabe S (2009a) Distribution of synthetic organotins and total tin levels in *Mytilus galloprovincialis* along the Portuguese coast. Mar Pollut Bull 58:1130–1136. http://dx.doi.org/10.1016/j.marpolbul.2009.04.007

Sousa A, Laranjeiro F, Takahashi S, Tanabe S, Barroso CM (2009b) Imposex and organotin prevalence in a European post-legislative scenario: temporal trends from 2003 to 2008. Chemosphere 77:566–573. http://dx.doi.org/10.1016/j.chemosphere.2009.06.049

Sousa A, Barroso CM, Tanabe S, Horiguchi T (2010) Involvement of retinoid X receptor in imposex development in *Nucella lapillus* and *Nassarius reticulatus* – preliminary results. In: Hamamura N, Suzuki S, Mendo S, Barroso CM, Iwata H, Tanabe S (eds) Interdisciplinary studies on environmental chemistry – biological responses to contaminants: from molecular to community level, vol 3. TERRAPUB, Tokyo, Japan, pp 189–196. http://www.terrapub.co.jp/onlineproceedings/ec/03/pdf/BR_03189.pdf.

Sousa ACA, Oliveira IB, Laranjeiro F, Takahashi S, Tanabe S, Cunha MR, Barroso CM (2012) Organotin levels in Nazaré canyon (west Iberian Margin, NE Atlantic) and adjacent coastal area. Mar Pollut Bull 64:422–426. http://dx.doi.org/10.1016/j.marpolbul.2011.11.013

Spooner N, Gibbs PE, Bryan GW, Goad LJ (1991) The effect of tributyltin upon steroid titres in the female dogwhelk, *Nucella lapillus*, and the development of imposex. Mar Environ Res 32:37–49. http://dx.doi.org/10.1016/0141-1136(91)90032-4

Sternberg R, Gooding M, Hotchkiss A, LeBlanc G (2010) Environmental-endocrine control of reproductive maturation in gastropods: implications for the mechanism of tributyltin-induced imposex in prosobranchs. Ecotoxicology 19:4–23. http://dx.doi.org/10.1007/s10646-009-0397-z

Stewart C, Thompson JAJ (1994) Extensive butyltin contamination in southwestern coastal British Columbia, Canada. Mar Pollut Bull 28:601–606. http://dx.doi.org/10.1016/0025-326X(94)90361-1

Strand J, Asmund G (2003) Tributyltin accumulation and effects in marine molluscs from West Greenland. Environ Pollut 123:31–37. http://dx.doi.org/10.1016/S0269-7491(02)00361-5

Strand J, Glahder CM, Asmund G (2006) Imposex occurrence in marine whelks at a military facility in the high Arctic. Environ Pollut 142:98–102. http://dx.doi.org/10.1016/j.envpol.2005.09.014

Stroben E, Oehlmann J, Fioroni P (1992a) *Hinia reticulata* and *Nucella lapillus*. Comparison of two gastropod tributyltin bioindicators. Mar Biol 114:289–296. http://dx.doi.org/10.1007/BF00349532

Stroben E, Oehlmann J, Fioroni P (1992b) The morphological expression of imposex in *Hinia reticulata* (Gastropoda: Buccinidae): a potential indicator of tributyltin pollution. Mar Biol 113:625–636. http://dx.doi.org/10.1007/BF00349706

Sudaryanto A, Takahashi S, Tanabe S, Muchtar M, H R (2000) Occurrence of butyltin compounds in mussels from Indonesian coastal waters and some Asian countries. Water Sci Technol 42:71–79. http://www.iwaponline.com/wst/04207/wst042070071.htm

Sudaryanto A, Takahashi S, Monirith I, Ismail A, Muchtar M, Zheng J, Richardson BJ, Subramanian A, Prudente M, Hue ND, Tanabe S (2002) Asia-Pacific mussel watch: monitoring of butyltin contamination in coastal waters of Asian developing countries. Environ Toxicol Chem 21:2119–2130. http://dx.doi.org/10.1002/etc.5620211015

Sudaryanto A, Takahashi S, Iwata H, Tanabe S, Ismail A (2004) Contamination of butyltin compounds in Malaysian marine environments. Environ Pollut 130:347–358. http://dx.doi.org/10.1016/j.envpol.2004.01.002

Sudaryanto A, Takahashi S, Iwata H, Tanabe S, Muchtar M, Razak H (2005) Organotin residues and the role of anthropogenic tin sources in the coastal marine environment of Indonesia. Mar Pollut Bull 50:226–235. http://dx.doi.org/10.1016/j.marpolbul.2004.11.041

Suzuki Y, Endo Y, Ogawa M, Kim Y, Onda N, Yamanaka K (2008) Development of an analytical method to confirm toxic trimethylated tin in human urine. J Chromatogr B 868:116–119. http://dx.doi.org/10.1016/j.jchromb.2008.04.032

Takahashi S, Mukai H, Tanabe S, Sakayama K, Miyazaki T, Masuno H (1999) Butyltin residues in livers of humans and wild terrestrial mammals and in plastic products. Environ Pollut 106:213–218. http://dx.doi.org/10.1016/S0269-7491(99)00068-8

Takahashi S, Le LTH, Saeki H, Nakatani N, Tanabe S, Miyazaki N, Fujise Y (2000) Accumulation of butyltin compounds and total tin in marine mammals. Water Sci Technol 42:97–108

Takeuchi I, Takahashi S, Tanabe S, Miyazaki N (2001) Caprella watch: a new approach for monitoring butyltin residues in the ocean. Mar Environ Res 52:91–113. http://dx.doi.org/10.1016/S0141-1136(00)00265-8

Takeuchi I, Takahashi S, Tanabe S, Miyazaki N (2004) Butyltin concentrations along the Japanese coast from 1997 to 1999 monitored by Caprella spp. (Crustacea: Amphipoda). Mar Environ Res. http://dx.doi.org/10.1016/j.marenvres.2003.11.005

Tallmon D (2012) Tributyltin contamination and imposex in Alaska harbors. Bull Environ Contam Toxicol 88:245–249. http://dx.doi.org/10.1007/s00128-011-0482-x

Tanabe S (1999) Butyltin contamination in marine mammals – a review. Mar Pollut Bull 39:62–72. http://dx.doi.org/10.1016/S0025-326X(99)00064-8

Tanabe S (2000) Asia-Pacific mussel watch progress report. Mar Pollut Bull 40:651–651. http://dx.doi.org/10.1016/S0025-326X(00)00019-9

Tanabe S, Kunisue T (2007) Persistent organic pollutants in human breast milk from Asian countries. Environ Pollut 146:400–413. http://dx.doi.org/10.1016/j.envpol.2006.07.003

Tanabe S, Subramanian A (2006) Bioindicators in developing countries – monitoring persistent organic pollutants. Kyoto University Press/Trans Pacific Press, Kyoto/Melbourne. ISBN:1-920901-11-6, 978-920901-11-0.

Ten Hallers-Tjabbes CC, Kemp JF, Boon JP (1994) Imposex in whelks (*Buccinum undatum*) from the open North Sea: relation to shipping traffic intensities. Mar Pollut Bull 28:311–313. http://dx.doi.org/10.1016/0025-326X(94)90156-2

Thomas KV, Blake SJ, Waldock MJ (2000) Antifouling paint booster biocide contamination in UK marine sediment. Mar Pollut Bull 40:739–745. http://dx.doi.org/10.1016/S0025-326X(00)00010-2

Trapp S, Ciucani G, Sismilich M (2004) Toxicity of tributyltin to willow trees. Environ Sci Pollut Res 11:327–330. http://dx.doi.org/10.1065/espr2004.05.202

Ueno S, Susa N, Furukawa Y, Komatsu Y, Koyama S, Suzuki T (1999) Butyltin and phenyltin compounds in some marine fishery products on the Japanese market. Arch Environ Health 54:20–25. http://dx.doi.org/10.1080/00039899909602232

Ueno D, Inoue S, Takahashi S, Ikeda K, Tanaka H, Subramanian AN, Fillmann G, Lam PKS, Zheng J, Muchtar M, Prudente M, Chung K, Tanabe S (2004) Global pollution monitoring of butyltin compounds using skipjack tuna as a bioindicator. Environ Pollut 127:1–12. http://dx.doi.org/10.1016/S0269-7491(03)00261-6

Valla V, Bakola-Christianopoulou M (2007) Chemical aspects of organotin derivatives of beta-diketones, quinonoids, steroids and some currently used drugs: a review of the literature with emphasis on the medicinal potential of organotins. Synth React Inorg Met-Org Nano-Met Chem 37:507–525. http://dx.doi.org/10.1080/15533170701558087

Venkatesan MI, Chalaux N, Bayona JM, Zeng E (1998) Butyltins in sediments from Santa Monica and San Pedro basins, California. Environ Pollut 99:263–269. http://dx.doi.org/10.1016/S0269-7491(97)00136-X

Viglino L, Pelletier É, St.-Louis R (2004) Highly persistent butyltins in northern marine sediments: a long-term threat for the Saguenay Fjord (Canada). Environ Toxicol Chem 23:2673–2681. http://dx.doi.org/10.1897/03-674

Waldock R, Rees HL, Mattiessen P, Pendle MA (1999) Surveys of the benthic infauna of the Crouch Estuary (UK) in relation to TBT contamination. J Mar Biol Assoc UK 79:225–232. http://dx.doi.org/10.1017/S0025315497000258

Whalen MM, Loganathan BG, Kannan K (1999) Immunotoxicity of environmentally relevant concentrations of butyltins on human natural killer cells in vitro. Environ Res 81:108–116. http://dx.doi.org/10.1006/enrs.1999.3968

White JS, Tobin JM, Cooney JJ (1999) Organotin compounds and their interactions with microorganisms. Can J Microbiol 45:541–554. http://dx.doi.org/abs/10.1139/w99-048#.ULK3i4d3aFA

WHO (1990) Environmental health criteria for tributyltin compounds. Environmental Health Criteria. World Health Organization, Geneva. http://www.inchem.org/documents/ehc/ehc/ehc116.htm

Wirzinger G, Vogt C, Bachmann J, Hasenbank M, Liers C, Stark C, Ziebart S, Oehlmann J (2007) Imposex of the netted whelk *Nassarius reticulatus* (Prosobranchia) in Brittany along a transect from a point source. Cah Biol Mar 48:85–94

Wuertz S, Miller CE, Pfister RM, Cooney JJ (1991) Tributyltin-resistant bacteria from estuarine and freshwater sediments. Appl Environ Microbiol 57:2783–2789. http://aem.asm.org/content/57/10/2783.full.pdf+html

Yamada H, Takayanagi K, Tateishi M, Tagata H, Ikeda K (1997) Organotin compounds and polychlorinated biphenyls of livers in squid collected from coastal waters and open oceans. Environ Pollut 96:217–226. http://dx.doi.org/10.1016/S0269-7491(97)00019-5

Yebra DM, Kiil S, Dam-Johansen K (2004) Antifouling technology – past, present and future steps towards efficient and environmentally friendly antifouling coatings. Prog Org Coat 50:75–104. http://dx.doi.org/10.1016/j.porgcoat.2003.06.001

Yi AX, Leung KMY, Lam MHW, Lee J-S, Giesy JP (2012) Review of measured concentrations of triphenyltin compounds in marine ecosystems and meta-analysis of their risks to humans and the environment. Chemosphere 89:1015–1025. http://dx.doi.org/10.1016/j.chemosphere.2012.05.080

Yonezawa T, Hasegawa S-I, Ahn J-Y, Cha B-Y, Teruya T, Hagiwara H, Nagai K, Woo J-T (2007) Tributyltin and triphenyltin inhibit osteoclast differentiation through a retinoic acid receptor-dependent signaling pathway. Biochem Biophys Res Commun 355:10–15. http://dx.doi.org/10.1016/j.bbrc.2006.12.237

Yoo CI, Kim Y, Jeong KS, Sim CS, Choy N, Kim J, Eum JB, Nakajima Y, Endo Y, Kim YJ (2007) A case of acute organotin poisoning. J Occup Health 49:305–310. http://dx.doi.org/10.1539/joh.49.305

Zachariadis GA, Rosenberg E (2009a) Determination of butyl- and phenyltin compounds in human urine by HS-SPME after derivatization with tetraethylborate and subsequent determination by capillary GC with microwave-induced plasma atomic emission and mass spectrometric detection. Talanta 78:570–576. http://dx.doi.org/10.1016/j.talanta.2008.12.007

Zachariadis GA, Rosenberg E (2009b) Speciation of organotin compounds in urine by GC–MIP-AED and GC–MS after ethylation and liquid–liquid extraction. J Chromatogr B 877:1140–1144. http://dx.doi.org/10.1016/j.jchromb.2009.02.065

Zachariadis GA, Rosenberg E (2012) Speciation analysis of triethyl-lead and tributyl-tin compounds in human urine by liquid–liquid extraction and gas chromatography microwave-induced plasma atomic emission detection. J Sep Sci 35:1132–1137. http://dx.doi.org/10.1002/jssc.201101041

Zanon F, Rado N, Centanni E, Zharova N, Pavoni B (2008) Time trend of butyl- and phenyl-tin contamination in organisms of the Lagoon of Venice (1999–2003). Environ Monit Assess 152:35–45. http://dx.doi.org/10.1007/s10661-008-0294-6

Zhang J, Zuo Z, Chen Y, Zhao Y, Hu S, Wang C (2007) Effect of tributyltin on the development of ovary in female cuvier (*Sebastiscus marmoratus*). Aquat Toxicol 83:174–179. http://dx.doi.org/10.1016/j.aquatox.2007.03.018

Zhang J, Zuo Z, Chen R, Chen Y, Wang C (2008) Tributyltin exposure causes brain damage in *Sebastiscus marmoratus*. Chemosphere 73:337–343. http://dx.doi.org/10.1016/j.chemosphere.2008.05.072

Zhang J, Zuo Z, He C, Cai J, Wang Y, Chen Y, Wang C (2009) Effect of tributyltin on testicular development in *Sebastiscus marmoratus* and the mechanism involved. Environ Toxicol Chem 28:1528–1535. http://dx.doi.org/10.1897/08-347.1

Zhang J, Zuo Z, Wang Y, Yu A, Chen Y, Wang C (2011) Tributyltin chloride results in dorsal curvature in embryo development of *Sebastiscus marmoratus* via apoptosis pathway. Chemosphere 82:437–442. http://dx.doi.org/10.1016/j.chemosphere.2010.09.057

Zhang J, Zuo Z, Xiong J, Sun P, Chen Y, Wang C (2013) Tributyltin exposure causes lipotoxicity responses in the ovaries of rockfish, *Sebastiscus marmoratus*. Chemosphere 90:1294–1299. http://dx.doi.org/10.1016/j.chemosphere.2012.10.078

Zhao K, Zhao Y, Ling Y, Chu X (2008) Analysis the pollution level of organotins in aquatic food and port wine. Wei Sheng Yan Jiu 37:327–331. http://www.ncbi.nlm.nih.gov/pubmed/18646535

Zheng R, Wang C, Zhao Y, Zuo Z, Chen Y (2005) Effect of tributyltin, benzo(a)pyrene and their mixture exposure on the sex hormone levels in gonads of cuvier (*Sebastiscus marmoratus*). Environ Toxicol Pharmacol 20:361–367. http://dx.doi.org/10.1016/j.etap.2005.03.005

Zuo Z, Cai J, Wang X, Li B, Wang C, Chen Y (2009) Acute administration of tributyltin and trimethyltin modulate glutamate and N-methyl-d-aspartate receptor signaling pathway in *Sebastiscus marmoratus*. Aquat Toxicol 92:44–49. http://dx.doi.org/10.1016/j.aquatox.2009.01.008

Zuo Z, Wang C, Wu M, Wang Y, Chen Y (2012) Exposure to tributyltin and triphenyltin induces DNA damage and alters nucleotide excision repair gene transcription in *Sebastiscus marmoratus* liver. Aquat Toxicol 122–123:106–112. http://dx.doi.org/10.1016/j.aquatox.2012.05.015

Chapter 5
Surfactants: Chemistry, Toxicity and Remediation

Sharrel Rebello, Aju K. Asok, Sathish Mundayoor, and M.S. Jisha

Contents

Abstract Surfactant toxicity has aroused a worldwide alert leading to various regulations on its usage and disposal. In this context much concern arises regarding the biodegradability and eco-friendliness of surfactants. Various reviews on surfactant and its toxicity are available; however there is a lack of a concise review covering surfactant types, primary and secondary toxicity of surfactants, evaluating the level of surfactant pollution worldwide. This chapter describes the safety concerns of surfactants on the aquatic system, terrestrial ecosystem and particularly on

S. Rebello • A.K. Asok • M.S. Jisha (✉)
School of Biosciences, Mahatma Gandhi University, Kottayam, Kerala 685 560, India
e-mail: jishams@mgu.ac.in

S. Mundayoor
Rajiv Gandhi Centre for Biotechnology, Thiruvananthapuram, Kerala India

E. Lichtfouse et al. (eds.), *Pollutant Diseases, Remediation and Recycling*, 277
Environmental Chemistry for a Sustainable World 4, DOI 10.1007/978-3-319-02387-8_5,

humans. We present remediation methods to solve surfactant contamination, such as ozonation, UV radiation, catalyst coupled autooxidation. Biological degradation of surfactants is highlighted with reference to the most commonly used anionic detergents sodium dodecyl sulfate (SDS) and linear alkyl benzene sulfonate (LAS). Biodegradation pathways and mechanism are discussed. Green surfactants are presented. Finally the relevance and role of biosurfactants as alternatives to synthetic detergents is described.

Keywords Bioremediation • Biosurfactants • Sodium Dodecyl Sulfate • Linear Alkyl Benzene Sulfonate • Pollution • Anionic • Methylene Blue active agents • Synthetic • Toxicity • Sewage treatment • Green surfactants • Alkylsulfatase

5.1 Introduction

"Cleanliness is next to Godliness" the English proverb reveals the relevance of cleanliness for human beings from the time immemorial. Traced back from the ancestral Babylonian ash- oil soap formula to the currently available soaps, cleansers and detergents, surfactants appear in various forms. Of these detergents indeed have become indispensable elements of man's life all along his steps aiming cleanliness and tidiness. Apart from serving as cleansing agents, surfactants find many industrial applications as additives in paints, as textile softeners, as antistatic agents, in metal processing and in oil drilling operations. Some surfactants have antimicrobial properties which provide the basis for their utility as biocides (Ginkel 1989). The significance of surfactants in pharmaceuticals (Sebag and Vanlerberghe 1990; Zhong 1999; Hari et al. 2005; Ghodrat 2006; Florence and Attwood 2011), agrochemicals (Beigel et al. 1998; Chen et al. 2008; Li et al. 2009; Shu-de 2009; Huang et al. 2010), food industry (Kralova and Sjoblom 2009), textile industry (Jing-xin 2004), petroleum industry (Bhardwaj and Hartland 1993; Frazier Jr et al. 1993; Schramm 2000; Torres et al. 2003), soil remediation (Medha and Lee 1996; Volkering et al. 1997; Chu and Kwan 2003; Lopez et al. 2005; Vreysen and Maes 2005; Yuan et al. 2006; Zhang and Lo 2006; Giannis et al. 2007; Khalladi et al. 2009), contaminated aquifer remediation (Fountain et al. 1991, 1996), gas and coal recovery (Chistyakov 2001), microelectronics (Ilardi et al. 1995; Carswell et al. 2003; Schroeder et al. 2007), and many other industries (Karsa 1990), give an idea on the prospects of surfactant utility and disposal. Applications of chemical surfactants and biosurfactants in various industries along with its microbiological and biotechnological implications are critically reviewed (Singh et al. 2007).

Basically, surfactants are any organic substance, intentionally added to achieve cleaning, rinsing and/or fabric softening due to its surface active properties. A single molecule of surfactant contains a strong hydrophobic group linked to a

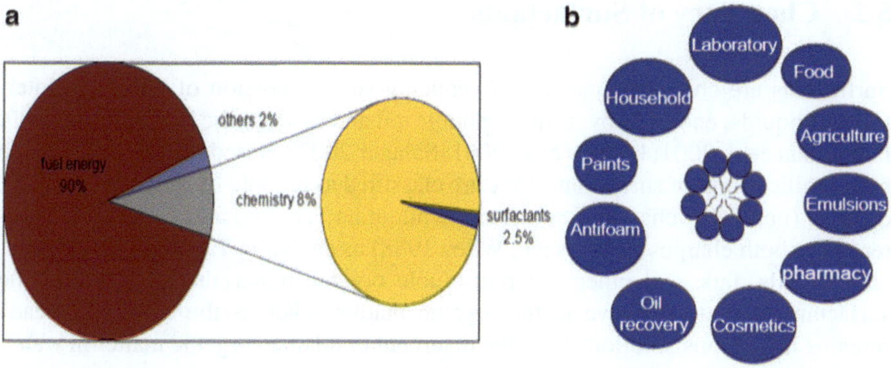

Fig. 5.1 (**a**) Crude oil utilization in the year 2011 for surfactant synthesis, 90 % for fuel energy, 2 % others, 8 % for chemistry (of which only 2.5 % used for surfactant synthesis). (**b**) Multiple roles of surfactants forming micelles in various industries

hydrophilic one. Such molecules tend to congregate at the interfaces between the aqueous medium and the other phases of the system such as air, oily liquids, and particles, thus imparting properties such as foaming, emulsification, and particle suspension (Eaton et al. 1995). Surfactants are especially noted for their wetting qualities and their effectiveness as emulsifiers. More over some surfactants readily adsorb into surfaces which lead to surface modification. These properties account for the exploitation of the surfactants in many product areas as shown in Fig. 5.1.

Industries worldwide discharge a wide range of surfactants, or surface-active agents, to their wastewater treatment facilities. Once used, surfactants enter the water bodies, where they can cause problems if they persist long, leading to the accumulation of potentially toxic or otherwise harmful substances (Deschenes et al. 1996) and cause serious environmental problems (Takada et al. 1992; Abd-Allah 1995). Water pollution caused by synthetic surfactants has been increasing during the past few years due to their extensive use in household, agriculture and other cleaning operations. Surfactants are ubiquitous and in untreated effluents, certain classes of surfactants can be present in sufficient concentrations to constitute toxicity problems to aquatic organisms (Ankley and Burkhard 1992), even between 0.4 and 40 mg/L (Abel 1974).

Extensive research on the surfactant toxicity do exist (Woltering 1984; Lewis 1991; Schweigert et al. 2000; Chaturvedi and Kumar 2010a), however, assessment of contemporary pollution profile of surfactants is relevant. The toxicity and biodegradation of the most commonly used anionic surfactants SDS and LAS is detailed in this review, with special reference to its biodegradability and safe disposal. Information available on the metabolic pathway and molecular mechanism of surfactant degradation, methods and alternatives to combat the problem of surfactant contamination are also discussed. The relevance and utility of biosurfactants as an alternative to current synthetic detergents are also reviewed.

5.2 Chemistry of Surfactants

Surfactants are chemicals capable of reducing surface tension of liquids or interfaces of liquids, endowed by its hydrophobic tail and hydrophilic head (Rosen 1989; Fainerman et al. 2001; Pletnev et al. 2001; Salager 2002). Based on the charge of the hydrophilic group of surfactants, they are classified as anionic (negatively charged), cationic (positively charged), non-ionic (without any charge) and ampholytic/zwitter ionic (both charges) (Myers and Wiley 1988) as shown in Table 5.1. In the presence of oils, fats, and other water insoluble organic materials, the "tail" of the surfactant tends to dissolve in the organic matter, whereas the carboxyl "head" remains in aqueous solution. Thus the detergent emulsifies organic matter in water.

Surfactants do occur as simple monomers, but sometimes exist as more complex polymers (Holland and Rubingh 1992; Lindman and Thalberg 1992). Commonly used detergents appear as a mixture of surfactants, abrasives, enzymes, preservatives,

Table 5.1 Classification of surfactants based on the charge of their hydrophobic and hydrophilic groups, their prospective uses and examples

Category		Used in	Examples
Anionic	Sulfates ($ROSO_3^-M^+$)	Shampoo, cleanser, liquid hand soap, acne treatment, hair color, bleaching, relaxer, exfoliant/scrub, dandruff/scalp treatment, bubble bath, foot treatment, shaving cream, anti-aging	Ammonium lauryl sulfate, Sodium Dodecyl sulfate, sodium lauryl ether sulfate
	Sulfonates ($RSO_3^-M^+$)	Fabric protectors, stain repellents, metal plating and fire-fighting foams	Perfluorooctanesulfonic acid, linear alkylbenzene sulfonates (LAS), perfluorobutanesulfonate
	Phosphates ($ROPO_3^-M^+$)	Detergents	sodium tripolyphosphate
	Carboxylates ($RCOO^-M^+$)	Soaps, fluorocarboxylates (in teflon, wire insulators, fire fighting foam, wax papers, water and oil repellent fabrics)	Sodium stearate, perfluorooctanoate (PFO), perfluorononanoate (PFOA),
Cationic ($R_4N^+X^-$)		Mouthwashes, toothpastes, hair conditioners, lozenges, throat sprays, breath sprays, cosmetics, nasal sprays, corrosion inhibitors, antiseptics	Cetyl trimethyl ammonium bromide (CTAB), benzalkonium chloride, cetylpyridinium chloride
Zwitter ionic or Amphoteric		Increases amplification of DNA in PCR	Betaines $\mathbf{RN^+(CH_3)_2CH_2CH_2SO_3^-}$
		In non-stick cooking spray	Lecithin
		Protein purification	CHAPS
Non ionic		Shampoos, creams, lotions	Cetyl alcohol, stearyl alcohol, alkylpolyglucosides, tween 80, triton X 100

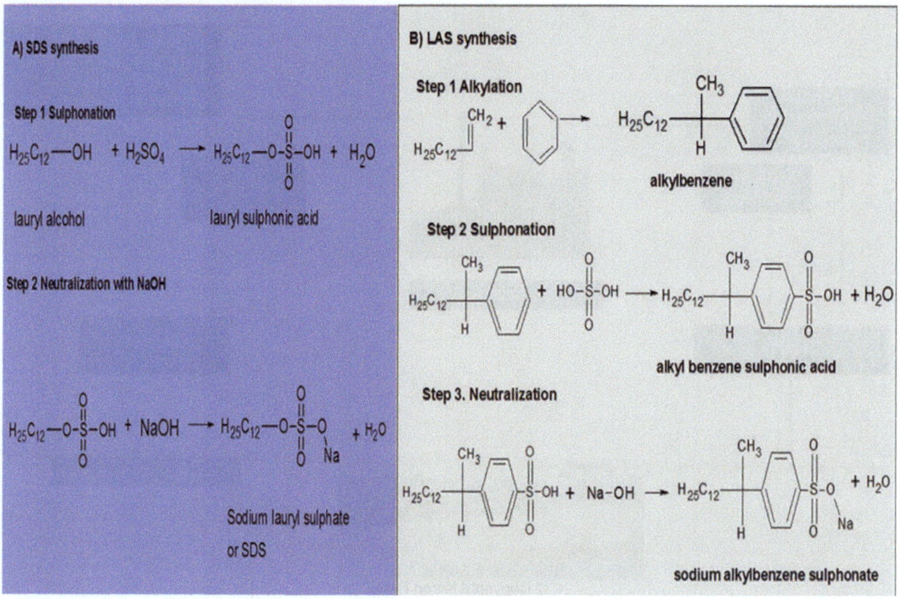

Fig. 5.2 Chemical synthesis of anionic surfactants (**a**) sodium dodecyl sulfate (SDS) synthesis involving sulfonation and subsequent neutralisation (**b**) linear alkyl benzene sulfonate (LAS) synthesis by alkylation, sulfonation and subsequent neutralisation (Constructed based on Behler et al. 2001)

corrosion inhibitors, foaming agents, optical brighteners, fabric softeners, colours, perfumes etc.

Surfactants are industrially synthesised either using non-renewable petroleum based substrates or renewable oleo chemical based substrates followed by various chemical modifications (Behler et al. 2001). The anionic surfactant SDS is synthesised by sulfonation of petrochemical or oleo chemical based lauryl alcohols, whereas, LAS, is exclusively synthesised from petrochemical by-products as depicted in Fig. 5.2. SDS is linear molecule with an alkyl tail of 12 carbon atoms, attached to a sulfate group giving the molecule the amphiphilic properties required of a surfactant. LAS have a hydrophobic alkyl chain and a hydrophilic head with a benzene ring and a sulfonate group. It's not a single compound, but ideally, a mixture of 20 compounds, of closely related homologues and isomers.

5.3 Surfactant Pollution – Worldwide and the Indian Scenario

The detergent consumption in India is less in comparison to other Asian countries. The per capita detergent consumption in India is around 2.7 kg/year, whereas in places like Philippines and Malaysia, it is 3.7 kg and in USA it is around 10 kg. The

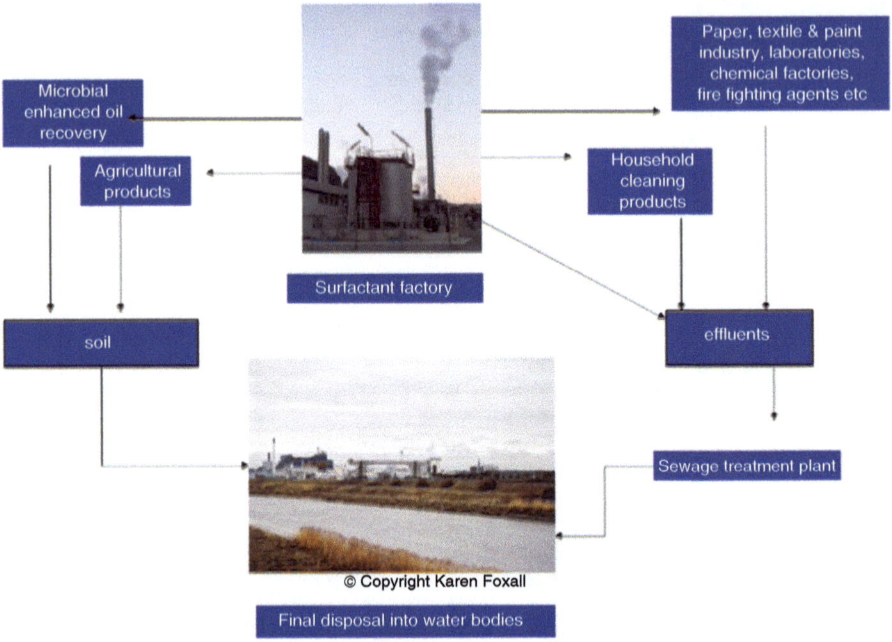

Fig. 5.3 Flow chart depicting course of surfactant pollution

high consumption rates of detergents also develop a high detergent concentration in our water bodies as depicted in Fig. 5.3. According to European Union wastewater quality criteria, reference level of methylene blue active agents is ≤ 0.3 g/m^{-3} and according to the World Health Organization, anionic detergent limit of drinking water is under 0.2 g/m^{-3} (Minareci et al. 2009).

There have been some detailed research papers and review articles on the occurrence of various surfactants and their degradation products in the environment (Cosovic et al. 1985; Holt et al. 1989; Jensen 1999; Matthijs et al. 1999; Carlsen et al. 2002; Gonzalez-Mazo et al. 2002; Sanderson et al. 2006; Gonzalez et al. 2012). According to broadly accepted risk assessment schemes for chemical substances, environmental compatibility requires proof that the use of the chemical will not result in environmental concentration levels higher than the ecotoxicological no effect concentration. The accumulation of surfactants in river sediments (Rico-Rico et al. 2009), marine water and sediments (Petrovic et al. 2002), infiltrated ground water (Field et al. 1992) and sewage effluents with concentrations up to 1,090 μg/L (Holt et al. 1989) also has been observed.

The presence of LAS in sewage works varies depending on the type of sewage treatment and their use in industrial processing in addition to domestic activities. An average LAS concentration of 1–10 mg/L can be found in municipal wastewater treatment dealing with only domestic wastewater (Field et al. 1992) but this range is visibly increased when industrial wastes from washing processes are also treated (Beltran et al. 2000a). LAS was found in treated sludge at high concentrations of up to 30,200 mg/kg dry weight (Berna et al. 1989) and in surface waters at concentrations of up to 416 μg/L (Fox et al. 2000), which is quite above its predicted no effect concentration of 250 μg/L (van de Plassche et al. 1999). Incomplete degradation of LAS surfactants was observed in Brazilian surface waters and their polar metabolites persisted in drinking waters, which questions its portability (Eichhorn et al. 2002). The raw discharge of drainage into the rivers, increase the level of surfactants and other contaminants in these water bodies (Eichhorn et al. 2002; Cirelli and Ojeda 2008), posing a threat to residing macro and micro populations.

The extensive use of phosphate based surfactants in 1980s and subsequent problem of eutrophication in water bodies (Hoffman and Bishop 1994; Stow et al. 2001) was resolved by a ban of phosphate detergents in various states of US, Europe and many western countries giving noteworthy results. However, in various developing countries the usage of phosphate based detergents continue at alarmingly high rates (Khurana 2002), leading to excessive growth of algae. According to Chris Knud Hansen one pound of phosphorus can grow 700 lb of algae (Knud-Hansen 1994). In India, phosphorus is recognised as a pollutant chemical as per norms of the Law of Environment Protection (1989), yet its usage is on the rise (Khurana 2002). Most laundry detergents in India are phosphate-based, but there is no control or regulation for phosphate use in detergents. The Bureau of Indian Standards has laid down the standards for eco-labelling of detergents in India, which could help to develop a phosphate free environment. But such ecomarking still remain ineffective. In such a scenario studies based on detergent toxicity and degradation gains much relevance.

Most of the ponds in Varanasi city have become eutrophic and are highly contaminated especially with high amount of detergents (Chaturvedi and Kumar 2010b). A Study on chemical contamination of water due to household laundry detergents at Ludhiana revealed that with the use of powder detergents, there was a significant increase in the level of pH, TDS, chlorides, sulfate, carbonate and bicarbonate in wash water, whereas very negligible change was found in all the above chemical parameters with the use of liquid detergents (Goel and Kaur 2012). Detergent pollution resulting from textile industry of Tirupur of Tamil Nadu also contributes to environmental pollution (De Neve 2009). The extent of detergent contamination of Sasthamkotta lake of Kerala during a period of 10 years has drastically decreased the portability of the water (Prakasam and Joseph 2000; Krishnakumar 2010). Surfactants were detected in water samples collected from surface and ground (2–62 μg/L), bore wells and open wells (22–427 μg/L) in Tirupati of Andhra Pradesh, and open municipal drainage waste waters (50–720 μg/L) (Kanchi et al. 2012).

The dumping of sewage – domestic and industrial as well as various human activities like bathing and washing of cloths aggravates the surfactant pollution in many artificial lakes like Fateh sagar, Pichhola lake of Udaipur (Rajasthan) (Chaudhury and Meena 2007).

5.4 Safety Concerns on Surfactants

The magnitude of disposal of surfactants and its subsequent toxic effects, quite often, are left unnoticed or remain masked, unless drastic effects are observed. The surfactant pollution of Yarkon river of Israel in 2008 as a result of chemical factory fire, gained so much attention only with the massive death of fish. Surfactants caused pooled in toxicity affecting the ecosystem and its inhabitants, the toxicity increasing down the trophic levels. The extent of surfactant toxicity is concentration dependant and it varies with season (Marcomini et al. 2000), prevalence of anaerobic environment (Mungray and Kumar 2009), predominance of urban and industrial sewage outlets (Lara-Martin et al. 2008), distance of sampling points from port of discharge (Gonzalez-Mazo et al. 2002), occurrence of offshore oil spill or gas exploration, diurnal discharge of sewage (Kantin et al. 1981), salinity (Lewis 1992), hardness of water (Oya et al. 2008) and presence of adsorbents (Casellato and Negrisolo 1989; Marin et al. 1994). The extend and mode of toxicity is also influenced by the chemistry of the pollutant surfactant, its alkyl chain length (Lewis 1991) and ethylene oxide (EO) molar ratio (Hall et al. 1989). Most toxic surfactants have EO ratio ≤ 15, while EO ratio 30–50 is found to be less toxic. A surfactant is considered toxic if the EC_{50}/LC_{50} is below 1 mg/L after 96 h testing on fish and algae and 48 h on Daphnia. Environmentally benign surfactants should preferably be above 10 mg/L (Myers and Wiley 2005; Warson and Finch 1998).

The cycle of surfactant toxicity starts from its very synthesis, disposal and subsequent exposure to the environment. Surfactant synthesis critically affects the environment aggravating the problems related to global warming, climate change, ozone layer depletion and greenhouse gas emission which cannot be totally avoided. Both petrochemical or oleo chemical based surfactant production result in atmospheric emission (NOx, CO_2, SO_2, hydrocarbons), water borne wastes and solid wastes capable of causing eutrophication and acidification of rivers and lakes (Pittinger et al. 1993; Stalmans et al. 1995). Comprehensive evidence has shown that surfactants can injure several vital functions in the body of an organism including inactivation of important enzymes such as esterase and phosphatase (Allen et al. 1965), disruption of normal cell function by alteration of membrane permeability (Braaten et al. 1972), interruption of cellular respiration (Abel and Skidmore 1975) or membrane lysis (Partearroyo et al. 1990).

Different classes of surfactants or mixtures impart toxicity to different extend, with cationic the most toxic, followed by anionic, zwitterionic and non-ionic in decreasing order of toxicity. The impact of anionic surfactants, the predominantly used class, on the environment and its inhabitants are assessed.

5.4.1 Toxicity on Microbial World

The impacts of surfactants on the microbial world vary with each species and extend of pollution. The counts of the micro-floral populations (bacteria, fungi, algae) of the river Nile were inhibited by detergent treatments, while some varieties of algae

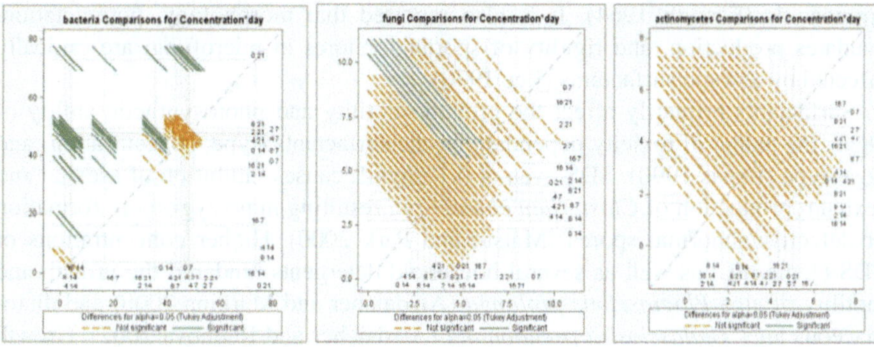

Fig. 5.4 Influence of linear alkyl benzene sulfonate (LAS) on counts of bacteria, fungi and actinomycetes of soil statistically analysed by Tukey's method (Asok and Jisha 2012b)

were promoted (Issa and Ismail 1994). Bacterial degradation of surfactants is well established (which will be discussed in biodegradation of surfactants). However surfactants are also found to have deleterious effects on various other bacteria such as phosphate solubilising *Acinetobacter junii* (Ivankovic et al. 2009), autotrophic ammonia oxidizing *Nitrosomonas* and *Nitrosospira* Strains (Brandt et al. 2001) and bioluminescent *Vibrio fischeri* (Lima et al. 2011). SDS and Triton X 100 (nonionic) affected the growth and nitrogen fixing ability of cyanobacteria *Gloeocapsa* even at concentrations of 50 and 500 ppm respectively (Tozum-Calgan and Atay-Guneyman 1994). Surfactant based membrane lysis, DNA damage and starvation is found as counterparts even in bacteria capable of utilizing SDS as sole carbon source (Klebensberger et al. 2006). Effect of LAS on inherent populations of bacteria, fungi and actinomycetes in soil gave statistically significant results for bacteria and fungi as depicted in Fig. 5.4 (Asok and Jisha 2012b). It was found that LAS apart from its known capacity for destabilizing proteases (Russell and Britton 2002), was also capable of significantly decreasing α-amylase activity of *Bacillus licheniformis*, even at concentrations lower than its CMC. This noteworthy loss in enzymatic activity is most likely due to the electrostatic interactions inherent to the anionic character of LAS (Bravo Rodriquez et al. 2006).

The effect of surfactants on microorganisms can be due to simple reduction of surface tension or by germicidal action (Parr and Norman 1965). Toxic action of surfactants may also be due to reactions at the cell surface like depolarisation of cell membrane, resultant decrease in absorption of essential nutrients and oxygen consumption. Another effect may be delayed release of toxic metabolic products from the cell leading to a build up, both effects ultimately resulting in the death of the organism. Microarray analysis of SDS and LAS treated *Sacharomyces cerevisiae* cells revealed that these detergents caused damage to membranes, alterations in carbon metabolism and induced an oxidative stress response (Sirisattha et al. 2004). Exposure of *Saccharomyces cerevisiae* to critical concentration (0.5 g/L) of SDS caused 10 % increase in the extra cellular GSH concentration (Wei et al. 2003); an enzyme involved in various physiological processes including detoxification. The fungicidal effect of several anionic surfactants could be due to leakage of aminoacids (Forsyth 1964). It is also reported that morphology, pigmentation, exudates production, and rigidity of sporangiophores in microfungi are critically affected by anionic surfactants (Lee 1970).

Surfactants primarily affect the growth, motility and photosynthetic ability of algae, the extent of toxicity dependant on the surfactants (type, concentration) and algal type (Lewis 1990). SDS even at 0.1 mg/mL causes inhibition of asexual and sexual reproduction of *Closterium ehrenbergii* resulting in no zygospore formation or defective/abnormal spores (Matsui and Park 2000). Higher concentrations of SDS (1 mg/mL) as well as several household detergents hindered the growth and motility of alga *Plagioselmis prolonga* (Aizdaicher and Markina 2006) and diatomaceous alga *Thalassiosira pseudonana* (Aizdaicher and Reunova 2002). Growth inhibition and self accumulating effects were observed in duck weeds exposed to SDS (Diriligen and Ince 1995). Long-term (7-days) exposure to the detergent Ariel caused significant toxicity to *Euglena gracilis* affecting its cell density, motility, swimming velocity and chlorophyll a content at a concentrations above 1 mg/L (Azizullah et al. 2012). The photosynthetic ability of algae decreased exponentially with increasing surfactant concentration and decrease in algal biomass (Maksimov and Parshikova 2006), with cationic surfactants causing the most potent inhibition. The resistance to surfactants and unapparent photosynthetic inhibition in some algae like *Chlorella emersonii* was associated to the presence of a trilaminar outer wall composed of nonhydrolyzable macromolecules (Corre et al. 1996), while *Chlorella vulgaris* lacking trilaminar outer wall showed sensitiveness to LAS. A study of the influence of Ca^{2+} to the toxicity of LAS on algae (*Daphnia magna*) showed that the toxicity of LAS increased with alkyl chain length and water hardness (Verge et al. 2001). Comparison of the environmental response of surfactants on photosynthetic response of lake plankton with that of laboratory showed that the laboratory effect concentrations were typically either similar to or lower than those in the field (Lewis and Hamm 1986).

Generally reports on mutagenic or genotoxic activity of anionic surfactants are not available, except for one stating that saponified coconut oil based anionic surfactant (used in cosmetic and hygienic products) was found to have cytotoxic, mutagenic or genotoxic activity on prokaryotes depending on its aggregation state (Petta et al. 2004). However, such genotoxic effect of saponified coconut oil was not found to effect humans as per a recent study (Burnett et al. 2011).

5.4.2 *Toxicity on Soil and Plants*

Water bodies such as lakes and rivers the common source of water for irrigating agricultural purposes, sometimes are also contaminated with surfactants. Irrigation of agricultural land with surfactant polluted water and use of sewage sludge as modes of fertilization, leaves behind surfactant residues in the terrestrial soil and cultivated food products. Even at low concentrations, surfactants seem to alter soil physics, soil chemistry and soil biology significantly, whereby sorption processes

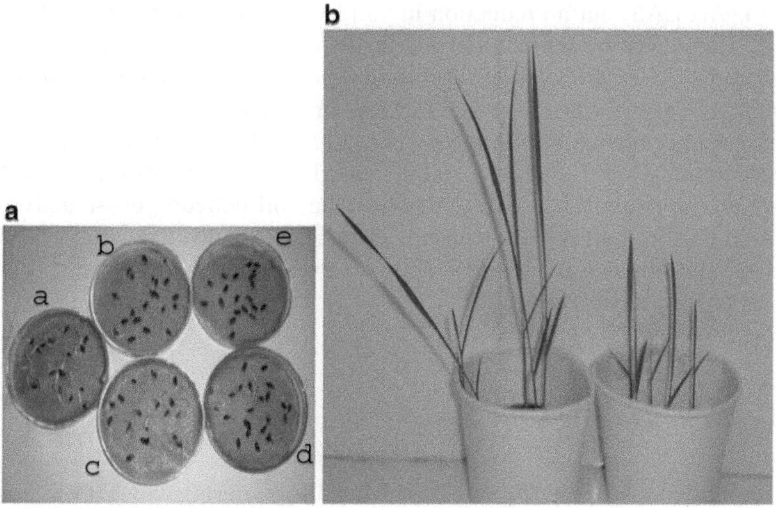

Fig. 5.5 Effect of the surfactant linear alkyl benzene sulfonate (LAS) on germination of rice seedlings (**a**) germination inhibition: *a* (control – 0 mg/L LAS), *b* (10 mg/L), *c* (1 mg/L), *d* (1,000 mg/L), *e* (100 mg/L); (**b**) plant growth of surfactant untreated and treated rice plant respectively (Unpublished data from our laboratory)

play a dominant role (Kuhnt 1993). Surfactants primarily affect the roots of plants suppressing or killing the roots, with comparatively less inhibition in the shoots of wheat seedlings (Rinallo et al. 1988). However, the damage to roots were found to reverse in time in the case of lettuce (Bubenheim et al. 1997), which was due to the microbial degradation of surfactants associated with the roots. Figure 5.5 illustrates the effect of LAS on germination of rice seedlings.

Use of detergent contaminated water for cultivation reduces the photosynthetic rate and chlorophyll content in bean plants (Jovanic et al. 2010) and sunflower leaves (Gadallah 1996). SDS under various concentrations affected the activity of invertase and SOD of soybean leaf in vivo and in vitro, indicating the oxidative stress imparted in plants on surfactant exposure (Xiaoli et al. 2000). The short-term effects of aqueous LAS on soil microbiology were modified by the dosage of LAS with sewage sludge and by a prolonged incubation time, which may allow for microbial recovery (Elsgaard et al. 2001). However, the functional diversity of the aerobic, heterotrophic bacterial community was rather insensitive to LAS (Vinther et al. 2003).

A 14 day series of toxicity tests with LAS on oat (*Avena sativa*), turnip (*Brassica rapa*) and mustard (*Sinapis alba*) in a sandy loam at different concentrations showed that estimated EC_5 value was in the range 50–200 mg kg–1 and EC_{50} value was in the range 200–300 mg kg-1 (Gunther and Pestemer 1992). During the study the lowest EC_5 value was recorded for oat (5 mg kg–1) and highest EC_5 value was for mustard (200 mg kg–1), whereas the EC_{50} value of both oat and mustard was noted as 300 mg kg–1). Considerable short-term acute physiological damage was observed on ryegrass in a field experiment using an application rate of 500 kg/ha LAS, but no reduction in yield was found after harvest (Litz et al. 1987). As reviewed by Jensen the 14th day EC_{50} of LAS on six wild plant species as well as on the cultivated plant *Brassica rapa*- turnip ranged between 90.1 and 204 mg/kg (Jensen 1999). In soils, LAS also impede microbial processes, such as bacterial iron reduction (Kristiansen et al. 2003). The continuous application of the anionic surfactant LAS to the soil increased the acid and alkaline phosphatase activity and arylsulfatase activity, whereas the soil dehydrogenase activity was decreased on continuous LAS exposure (Sanchez-Peinado et al. 2009). The application of high concentrations of LAS also exerted a selective pressure on the heterotrophic bacterial diversity.

5.4.3 Toxicity to Aquatic System

Surfactant toxicity has been reported since late 1960s, during which it was observed that the exposure of fish *Ictalurus natalis* to detergent levels (0.5 ppm) even much lower than its sublethal concentrations caused disruption of chemoreceptors (Bardach et al. 1965), generating various interesting reviews to date (Lewis 1991; Cserhati et al. 2002; Ying 2006; Chaturvedi and Kumar 2010c; Konnecker et al. 2011). Chronic and sublethal toxicities of anionic and nonionic surfactants to aquatic animals occur at concentrations usually greater than 0.1 mg/L (Lewis 1991). Adverse biological effects on aquatic organisms occur especially when surfactants occur at relatively high concentrations (Romanelli et al. 2004). As per a more detailed evaluation of toxic surfactant concentrations, anionic surfactants are harmful (LC_{50} between 10 and 100 mg/L^{-1}), whereas nonionic ones are toxic (LC_{50} between 1 and 10 mg/L^{-1}) or even highly toxic (LC_{50} below 1 mg/L^{-1}) (Liwarska-Bizukojc et al. 2005).

Pollutants are taken up by aquatic organisms via the general body surface, across the gills and through the gut lining following ingestion of food. The surfactant toxicity is primarily a function of the ability of the surfactant to adsorb and penetrate the cell membrane of aquatic organisms (Rosen et al. 2001). The ability of surfactants to bind to various bioactive macromolecules, including starch (Merta and Stenius 1999) and proteins (Nielsen et al. 2000), peptides and DNA (Bhattacharya and Mandal 1997), further augments its toxic potential.

The aftermath of surfactant exposure can be clearly visualized in various organs like gills (Mallatt 1985), liver, kidney (Rosety-Rodríguez et al. 2002), spleen and intestines (Ribelles et al. 1995) of fishes. Exposure to high surfactant concentration results in gill epithelial disruption causing subsequent asphyxiation or osmoregulatory failure (Abel 1976; Mallatt 1985), while exposure to sublethal concentrations of surfactants causes gill epithelial hyperplasia (Mallatt 1985; Susmi et al. 2010) as depicted in Fig. 5.6 and subsequent respiratory stress by increasing the diffusion distance of oxygen. High concentrations of SDS cause death of fishes primarily due to three modes: decrease in surface tension, destruction of tissue and alteration of biomacromolecules (Ribelles et al. 1995).

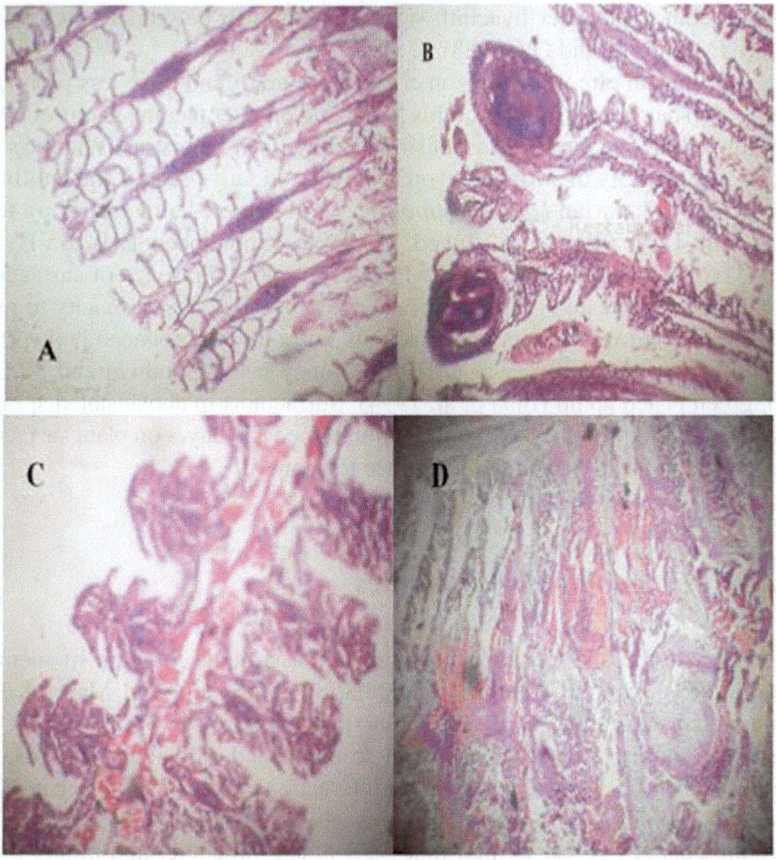

Fig. 5.6 Histopathologic analysis of grass carp gills exposed to sodium dodecyl sulfate (SDS) (**a**) Control 30 days. (**b**) 2 mg/1 SDS 30 days. (**c**) 2.5 mg/1 SDS 30 days. (**d**) 3 mg/1 SDS 30 days (Susmi et al. 2010)

Studies on the effect of SDS on the fish *Cyprinus carpio* revealed that its swimming capacity was reduced 5 times and oxygen consumption was increased 2.8 times in relation to the control at a concentration of 10 ppm. In general, toxic effects of SDS on swimming activity are more pronounced in smaller fish whereas the effects on oxygen consumption is more pronounced in larger ones (Barbieri et al. 1998).

Exposure to SDS and LAS concentrations of 0.3, 0.6, 1.5, 3 and 6 mg/L for 60 min caused a significant inhibitory effect on fertilization success in gilthead *Sparus aurata* L sperm (Rosety et al. 2001) and LAS exposure affected the sperm motility (Rosety et al. 2003). SDS exposure cause acute inflammatory reactions, oxidative stress and damage to mucus layer of fishes, which predispose them to microbial attack (Susmi et al. 2010).

Surfactant exposure on coastal vegetation indirectly result in uptake of marine chloride – sodium and subsequent reduction of water surface tension, primarily caused by erosion of the epicuticular wax of plants (Badot et al. 1995). Inherent salinity tolerance of water hyacinth was also found to be reduced by surfactant exposure (Muramoto and Oki 1988).

The detrimental effects of SDS on crustaceans are found to be less pronounced, yet reports on detrimental effects of this surfactant on filter feeding habits of bivalves (Ostroumov 2003) and mussel suspension feeding (Ostroumov and Widdows 2006) do exist. The toxicity of surfactants on survival, oxidative stress, and cholinesterase activity of planarian *Dugesia japonica* was evaluated and according to estimated 48-h LC_{50} they were ranked as SDS > 4-nonylphenol > LAS > Hyamine 1622 > CTAB > Triton X-100 > PFOS > PFOA (Li 2008). Presence of surfactants in fungicide, insecticide and pesticide sprays could possibly cause toxicity to insects, as per reports on toxic effects of surfactants on bollworm, tobacco budworm and pink bollworm (Wolfenbarger et al. 1967), honey bees (Goodwin and McBrydie 2000), green peach aphid (Imai et al. 1994). Surfactants in above stated sprays aid the penetration of the active compound through the waxy layer on plant surfaces and insect exoskeletons by reducing the surface tension.

5.4.4 *Effect on Higher Vertebrates*

In humans anionic surfactants target mainly the stratum corneum of membrane bilayer of sensitive skin resulting in dermatitis (Marrakchi and Maibach 2006), aphthous ulcers (Chahine et al. 1997). In spite of its use in various fields safety concerns on its usage do exist based on resultant and carcinogenic nitrosamines formed during SDS synthesis or by its interactions with other nitrogen bearing ingredients within a formulation utilizing this ingredient (CIRSL 1983). The American Cancer Society denied that SDS is carcinogenic, and points out that the substance, while undoubtedly a skin irritant, a strong detergent intended to remove oil and soil, but there is no link between use of this product and cancer risk (Doyle 2010). The toxicological study for LAS indicates no severe phenotypic or genotypic toxic effects after repeated exposure in experimental animals (IPCS 1996). Thus, it is likely that the toxicity of these anionic surfactants are relatively low in human and wild animals as the molecular weight increases, probably due to lower adsorption in the intestine. An acute toxic effect by anionic surfactants is therefore not to be likely but a chronic effect, can however, be more possible since a regular dosage of a human is about 5 mg/person from drinking waters, detergents, toothpaste and food (IPCS 1996).

A quantitative structure – activity relationship (QSAR) study revealed that the hydration capacity of n-alkyl sulfates was closely correlated with the irritating potential; the maximum was found at C_{12} analogue (Wilhelm et al. 1993). C_{18} compounds caused cell injury whereas C_{10} and C_{16} compounds caused more severe membrane destruction and protein denaturation (Kotani et al. 1994). SDS causes more severe skin dehydration than dodecyl trimethyl ammonium bromide; complete repair of the irritant reaction was achieved 17 days after surfactant exposure (Wilhelm et al. 1994). The cutaneous toxicity of surfactants in normal human keratinocytes assessed by cytotoxicity, arachidonic acid release and regulation of interleukin-1a mRNA revealed that the effect of SDS was higher than that of the nonionic surfactants Triton-X-100 and Tween 20 (Shivji et al. 1994).

The bio-concentration potential of anionic surfactants was higher than other surfactants (Tolls et al. 1994) that could result in modification or inactivation of proteins and destabilising lipoprotein membranes of various cells. Surfactants induced DNA damage and membrane damage to human lymphocytes was also observed, with cationics causing intense damage as opposed by anionics and non-ionics (Hrabak et al. 1982). A similar singular effect of cationics (but not by anionic and amphoterics), to induce apoptosis in normal and cancer cells of mammals is also reported (Enomoto et al. 2007). Intraperitoneal incorporation of synthetic surfactants SDS, LAS and Emulgen 913 depressed the content of microsomal cytochrome P450, while enhanced the activity of heme oxygenase and metallothionein in liver of rats (Ariyoshi et al. 1990; Ariyoshi et al. 1991). The exposure to clinical concentrations (1 μg/mL) of LAS decreased the viability of rat thymocytes exposed to H_2O_2, an oxidative stress increasing the [Ca^{2+}] levels (Yamaguchi et al. 2006). Anionic surfactants are also found to modify the IgE production in mice (Clausen et al. 2000).

The probable exposure of eyes to surfactants present in shampoos, ocular lens cleaners and other hair cleaning products, instigated studies assessing the toxicity of surfactants to eyes using mouse/rabbit models (Furrer et al. 2000), corneal cell cultures (Vian et al. 1995; Xu et al. 2000; Cooper et al. 2001; Cater and Harbell 2006) and various other animal alternative approaches (Kapoor et al. 2009; McNamee et al. 2009). Several cytotoxicity tests reveal that nonionics have the least toxicity in the order as cationic > anionic > amphoteric > non-ionic (Grant et al. 1992). Some non-ionic surfactants are found to anaesthetise the eye ball and thus combinations of non-ionic surfactants with anionics would make many shampoos gentle to eye (Conry 1980). Interestingly, nonionic surfactants were able to reduce the damaging effect of anionic surfactants; however, the molecular basis of the phenomenon has not been elucidated (Eagle et al. 1992). Loss of eye lens transparency, significant increase in lens wet weight and axial length was seen at 24 h post exposure in cultured bovine lens cells treated with 0.1–0.025 % SDS (Bantseev et al. 2003).

5.4.5 Secondary Toxicity of Surfactants

Generally the presence of surfactants helps in the degradation of polycyclic hydrocarbons, but the degradation of PAH was inhibited by SDS because this surfactant was preferred as a growth substrate (Tiehm 1994). This suggests that the presence of this detergent in the water bodies would indirectly lead to bioaccumulation of other hydrocarbons. Reports are also available that adding SDS and *Pseudomonas aeruginosa* UG2 biosurfactants inhibit polycyclic aromatic hydrocarbon biodegradation in a weathered creosote-contaminated soil (Deschenes et al. 1996).

The presence of surfactants may be important for the fate of pesticides at effluent-irrigated sites because they may increase the apparent solubility of hydrophobic pesticides (Vigon and Rubin 1989). Concentrations of LAS in untreated municipal effluents have been reported to range between 1 and 10 mgL^{-1} (IPCS 1996). The surfactants are able to form micelles by incorporating hydrophobic compounds within it at the individual critical micelle concentration (CMC) of a surfactant (Rosen et al. 2001), thereby increasing the solubility of hydrophobic pesticides (Vigon and Rubin 1989).

Another study revealed that LAS exposure caused more than double cadmium transfer through the gills from Rainbow trout when tested in concentrations expected to be found in a polluted recipient (0.9 μg/L Cd + 0.05 mg/L LAS) (Part et al. 1985). Thus presence of detergents in water bodies could lead to secondary accumulation of other pollutants.

5.5 Tackling Surfactant Pollution

Surfactant toxicity has aroused worldwide attempts to reduce after effects of these silent toxicants. The strict regulations on usage of phosphate free surfactants, remediation of waste water before disposal and promotion of green surfactants are manifestations of these attempts. The usage of liquid detergents than powder forms results in less surfactants toxicity according to a case study at Ludhiana (Goel and Kaur 2012). Interesting suggestions for hands on preparation of laundry and liquid detergents for hard and soft water could also add to our attempts to reduce surfactant toxicity (Khurana 2002). The use of non-toxic-biodegradable natural soaps and soapnuts is yet another promising approach (Ghai 2011).

5.5.1 Remediation Before Disposal

The influence and relevance of surfactants in mans life is too immense, that totally avoiding them from our day to day life seems to be impossible and unpractical. Better management of surfactant use and disposal has become the need of the hour, both at industrial and domestic level. Strict regulations in the effective remediation of surfactants before disposal should be done. This section describes the various methods of surfactant remediation and various steps that could reduce surfactant pollution at domestic level.

Physical and Chemical Methods

Various physical, chemical and biological methods of surfactant detoxification are reported. Physical treatment of surfactants by ozonation and advanced oxidation using various combinations of ozone, hydrogen peroxide, ultraviolet light irradiation, and iron salts were found effective in degrading recalcitrant surfactants, including LAS, alkylphenol ethoxylates and quaternary ammonium surfactants (Ikehata and El-Din 2004). Various other techniques like electrocoagulation (Yuksel et al. 2009), nanofiltration (Korzenowski et al. 2012), sonochemical degradation, foam fractionation and wet air oxidation are also used.

Oxidation Based Methodologies

Detoxification of micro pollutants such as surfactants, pesticides, herbicides and microtoxins from drinking water mainly relies on high oxidizing capacity of ozone (Beltran et al. 2000a). LAS ozonation in surface waters intended for human consumption demonstrated that combinatorial use of O_3 and powdered activated carbon approach is the most efficacious than traditional O_3 or H_2O_2 based oxidation systems, considerably increasing the LAS removal rate and also reducing the concentration of dissolved organic carbon (Rivera-Utrilla et al. 2006). As per a comparative study on various oxidative LAS remediation strategies based on ClO^-, ClO_2, $KMnO_4$, O_3, O_3/H_2O_2, O_3/activated carbon, the latter was the only technique that brought about substantial LAS removal and reduction of dissolved organic carbon (Mendez-Diaz et al. 2009). The chloride based oxidant (ClO^-) of the above study gave no LAS removal but also formed trihalomethanes increasing the toxicity of the water. The oxidative auto-degradation of SDS catalyzed by alumina-supported Co-Zn composite has also been described (Usman et al. 2001). Wet air oxidation of organic pollutants in high temperatures (174–320 °C) and pressures (2.17–20.7 MPa) (Dietrich et al. 1985), using compressed air or pure oxygen is also attempted, but such treatment of LAS resulted in non-biodegradable compounds (Mantzavinos et al. 2001).

Photocatalytic Degradation

The photocatalytic degradation of surfactants in water by solar fenton like oxidation reaction and the influence of various chemical parameters on degradation is critically studied (Bhatkhande et al. 2002; Amat et al. 2004; Venhuis and Mehrvar 2004; Bandala et al. 2008a; Bandala et al. 2008b). Comparison of photo-fenton reaction and cobalt/peroxymonosulfate/ultraviolet (Co/PMS/UV) process showed that the use of solar radiation increased the degradation rate in one magnitude order when compared with dark experiments and further increase of Co/PMS reagent concentration increased reaction rate twice to achieve almost complete surfactant degradation in 5 min (Bandala et al. 2008a). The photo degradation of SDS and LAS in reactors in presence of TiO_2 catalyst and UV light is also well studied (Hidaka 1998). According to this study the surfactant competitively binds to TiO_2 and on exposure to UV light, radical attack on the aromatic ring and alkyl chain brings about degradation.

Foam Fractionation

Foam fractionation is a chemical process in which hydrophobic molecules are preferentially separated from a liquid solution using rising columns of foam. It has been shown to be an effective method of removing anionic or cationic surfactants from effluent streams. Cationic surfactants were easily removed from water by foam fractionation than the anionic surfactants studied (Tharapiwattananon et al. 1996). Foam fractionation process is controlled by many process and material parameters such as: airflow rate, foam column geometry, feed concentration and added salt (Sonc and Grilc 2004; Boonyasuwat et al. 2005). The surfactant concentration in the collapsed foam is 21.5 times the feed concentration.

Sonochemical Degradation

The utility of sonochemical reactors for anionic surfactant degradation from waste waters is well studied the rate of degradation proportionally increasing with sonication time, but decreasing with surfactant increase (Dehghani et al. 2010). The potential of using ultrasonic irradiation for the removal of LAS revealed that it increases with the frequency of radiation, the addition of NaCl or H_2O_2 to this system had adverse effect on LAS conversion, while addition of Fe^{2+} either alone or in conjunction with H_2O_2 (fenton reagent) had a positive effect on degradation of LAS (Manousaki et al. 2004). Usage of 20 kHz ultrasound at 40 °C, pH at 2.5 throughout and addition of extra amounts of zero valent iron and H_2O_2 during the degradation resulted in 93 % reduction of LAS (Naldoni et al. 2011). In another study, the use of low frequency ultrasonic irradiation caused 80 % LAS conversion after 120 min of sonication at 125 W and 30 °C; nonetheless, LAS and its degradation intermediates proved difficult to oxidise as only about 20–25 % of the initial carbon content was transformed to carbon dioxide (Abu-Hassan et al. 2006).

Electrochemical Degradation

Electrocoagulation can be addressed as a method of wastewater treatment when electric current goes through an electrolysis cell supplied with soluble electrodes (Sequeira 1994). This technique is also effective in the treatment of the strongly acidic effluents arising from electrokinetic surfactant-aided soil-flushing of polluted soils using aluminium electrodes (anodes and cathodes) (Lopez-Vizcaino et al. 2012).

Biological Methods

Biotransformation of organic contaminants in the natural environment has been extensively studied to understand microbial ecology, physiology and evolution for their potential in bioremediation (Mishra and Lal 2001; van Hylckama Vlieg and Janssen 2001; Watanabe 2001). Microbial biodegradation provides a safer, environmentally benign and cost effective alternative to physico-chemical methods for surfactant remediation (Oya and Hisano 2010). Biodegradation of surfactants is most often performed by diverse soil or aquatic microorganisms leading to generation of water and carbon dioxide gas (Schleheck et al. 2000). Surfactant degradation is predominantly carried out by various species of *Pseudomonas*, yet many other bacterial species are also reported to participate in surfactant remediation as listed in Table 5.2.

Biodegradation of anionic surfactants was initiated in early 1960s (Payne and Feisal 1963), with the isolation of two unknown bacterial soil isolates capable of degrading short- or long-chained organic acids and alcohols of SDS and three of five phenyl-placement isomers of dodecyl benzene sulfonate. But isomers with phenyl placement at carbon 4 or 5 were toxic and killed the bacteria. The primary SDS splitting enzyme was reported soon (Hsu 1963) and identified as a primary alcohol sulfatase. Anionic surfactant degradation is initiated by alkylsulfatases which convert them to corresponding alcohol by removing the sulfate/sulfonate moiety. Growth of the bacteria on SDS as the sole carbon source induced glyoxylate bypass enzymes, isocitrate lyase and malate synthetase, in addition to alkylsulfatases (Williams and Payne 1964). The degradation studies of SDS, LAS and three commercial laundry detergents in Chesapeake Lake revealed that they were biodegradable only with initial concentrations below 25 ppm (Cook and Goldman 1974). However, continuous research and optimisation of surfactant degradation has resulted in the isolation of *Pseudomonas* capable of removing almost up to 4 % level of SDS in 48 h (Jovcic et al. 2010). Bacteria capable of degrading aromatic sulfonates have been isolated from industrial sewage treatment plants (Zimmermann et al. 1982; Thurnheer et al. 1990) or obtained through continuous adaptation (Kuhm et al. 1991).

Metabolism of sodium dodecyl sulfate (SDS) by the detergent-degrading bacterium *Pseudomonas* C12B using a ^{14}C radiotracer which showed that 70 % of the radiolabel was released as $^{14}CO_2$ at completion, where as the remaining isotope was incorporated cells (Thomas and White 1989). As depicted in Fig. 5.7, SDS was degraded yielding with the sequential production from [^{1-14}C] SDS of 1-dodecanol, dodecanal, and dodecanoic acid. At this point the pathway diverged leading either to formation of $^{14}CO_2$ via beta-oxidation or to elongation to C_{14}, C_{16}, and C_{18} fatty acyl residues with rapid incorporation into lipid fractions such as phospholipids. Biodegradation of LAS begins at the terminus of the alkyl chain with an omega-oxidation and is followed by successive cleavage of C_2 fragments (ß-oxidation) (Huddleston and Allred 1963; Swisher 1963). These intermediates are further biodegraded by oxidative removal of the aromatic ring and cleavage of the sulfonate group (Setzkorn and Huddleston 1965; Swisher 1967). Core research aiming optimized surfactant degradation has been carried out in the case of anionic surfactants SDS and LAS (Abboud et al. 2007; Asok and Jisha 2012a). SDS is found be highly degradable both in aerobic and anaerobic conditions. LAS biodegradation in turn is found to be inhibited by anaerobic conditions (Mungray and Kumar 2009).

Table 5.2 Microorganisms involved in bioremediation of anionic surfactants SDS and LAS

Sl. No.	Organism	Surfactant	References
1.	*Pseudomonas* C12B	SDS, LAS	Payne (1963) and Payne and Feisal (1963)
2.	*Hansenula and Candida*	LAS	Standard and Ahearn (1970)
3.	*Vibrio* sp.	LAS	Bird and Cain (1972)
4.	*Pseudomonas aeruginosa strain W51D*	LAS	Soberon-Chavez et al. (1996)
5.	*Nocardia amarae MB-11*	LAS	Bhatia and Singh (1996)
6.	*Bacillus cereus*	SDS	Singh and Kumar (1998)
7.	*Spongia officinalis*	LAS	Perez et al. (2002)
8.	*Phanerochaete chrysosporium*	LAS	Yadav et al. (2001)
9.	*Parvibaculum lavamentivorans DS-1'and Comamonas Testosteroni* sp. *B-2 and KF-1*	LAS	Schleheck et al. (2004)
10.	*Pseudomonas aeruginosa*	LAS	Lijun et al. (2005)
11.	*Parvibaculum lavamentivorans*^{T}	LAS	Schleheck and Cook (2005)
12.	*Pseudomonas* sp.	LAS	Prats et al. (2006)
13.	*Parvibaculum lavamentivorans strain DS-1*	LAS	Schleheck and Cook (2005) and Schleheck et al. (2007)
14.	*Acinetobacter calcoaceticus, Pantoea agglomerans*	SDS, LAS	Abboud et al. (2007)
15.	*Pseudomonas beteli, Acinetobacter johnsoni*	SDS	Hosseini et al. (2007)
16.	*Klebsiella oxytoca*	SDS	Shukor et al. (2009)
17.	*Stenotrophomonas maltophilia*	LAS	Farzaneh et al. (2010)
18.	*Parvibaculum lavamentivorans DS-1*^{T} *and Comamonas testosteroni* sp. *KF-I*	LAS	Schleheck et al. (2010)
19.	*P. alcaligenes and P. mendocina*	SDS	Chaturvedi and Kumar (2010b)
20.	*Pseudomonas putida*	SDS	Chaturvedi and Kumar (2011b)
21.	*Pseudomonas aeruginosa* S7	SDS	Yeldho et al. (2011)
22.	*P. aeruginosa, P. mendocina, P. stutzeri, P. alcaligenes, P. pseudoalcaligenes, P. putida* and *P. otitidis*	SDS	Chaturvedi and Kumar (2011a)
23.	*P. aeruginosa* MTCC 10311	SDS	Ambily and Jisha (2011) and Ambily and Jisha (2012)
24.	*P. nitroreducens* (MTCC 10463) and *P. aeruginosa* (MTCC 10462)	LAS	Asok and Jisha (2012a)
25.	*Alcaligenes odorans, Citrobacter diversus, Micrococcus luteus* and *P. putida* consortium	LAS	Eniola (2012)

The studies on detergent degradation took a new phase by characterisation of the enzymes involved in degradation. Multiple alkylsulfatases are produced by each surfactant degrading bacteria. For example *Pseudomonas* C12B can produce five such enzymes (Bateman et al. 1986) and *Pseudomonas putida* FLA six (Lillis et al. 1983).

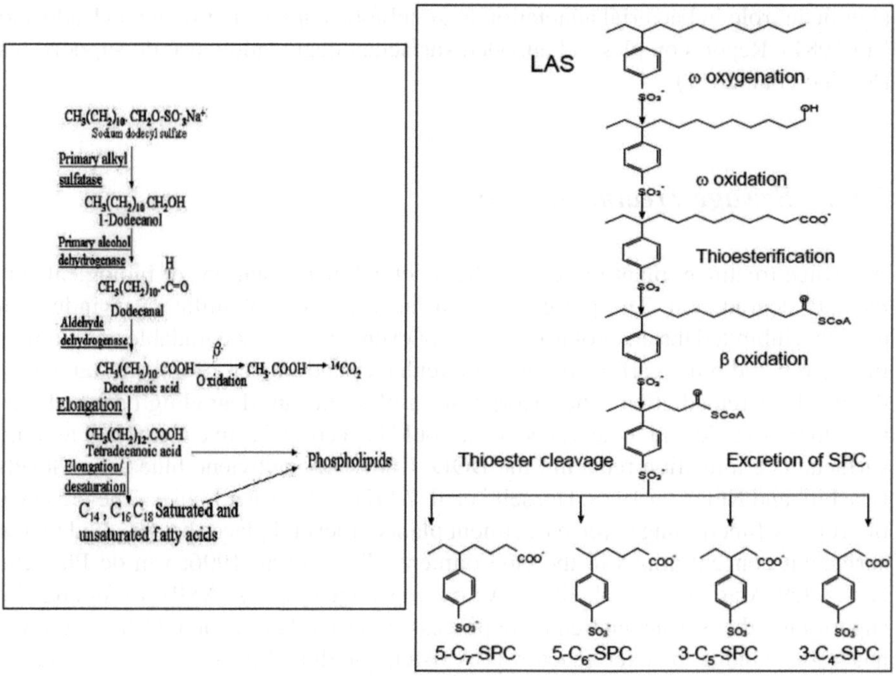

Fig. 5.7 Pathway of sodium dodecyl sulfate (SDS) and linear alkyl benzene sulfonate (LAS) biodegradation as per Thomas and White (1989) and constructed as per Huddleston and Allred (1963), Swisher (1963), and Asok (2011) respectively

Of the five alkyl sulfatases produced by *Pseudomonas* C12B, two (designated P1 and P2) are active towards primary alkyl sulfates, whereas the other three (SI, S2 and S3) act on secondary alkyl sulfates (Cloves et al. 1980). *Pseudomonas* sp. is capable of producing a multiplicity of alkyl sulfatases, in contrast *C. terrigena* produced only two secondary alkylsulfatases irrespective of the culture conditions employed (Fitzgerald 1975). Alkyl sulfatases also have use in white biotechnology as they can be employed for the enantio-convergent transformation of racemic sulfate esters into a single stereo – isomeric secondary alcohol, with a theoretical yield of 100 %. This is a major improvement over traditional kinetic resolution processes, which yield both enantiomers, each at 50 % (Gadler and Faber 2007).

Analysis of alkyl sulfatase in parent and cured strains of *Pseudomonas* confirmed that both enzymes are encoded by the chromosome. The nucleotide sequence of two chromosomally located genes sdsA and sdsB, coding for alkylsulfatase and its transcriptional regulator respectively, were identified to play significant role in SDS degradation (Davison et al. 1992). Evidence to the transcriptional regulation of sdsA gene by sdsB protein was further proved (Jovcic et al. 2010). The ability of *Pseudomonas* C12 B to utilize alkyl benzene sulfonate also appears to be coded by the chromosome (Kostal et al. 1998). A novel alkylsulfatase gene, sdsAP, was cloned from a newly isolated bacterium *Pseudomonas* sp. S9 and expressed in heterologous host of *E. coli* (Long et al. 2011). Plasmid encoded character often plays significant role in bacterial adaptation to xenobiotics in the environment (Kado and Liu 1981). Reports on plasmid encoded surfactant degradation also do support this (Yeldho et al. 2011).

5.5.2 Sewage Treatment Plants

A sewage treatment plant represents the practical manifestations of biological surfactants degradation. The presence of surfactants (alcohol sulfates) in industrial effluents inhibited the anaerobic digestion of even readily biodegradable compounds like starch and other carbohydrates (Feitkenhauer and Meyer 2002), substantiating the need for remediation. The incorporation of surfactant degrading bacterial cultures in household and industrial sewage could be a cost effective method of anionic surfactants elimination reducing the BOD, COD and methylene blue active levels (Hashim and Kulandai 1988; Hosseini et al. 2007) in the water bodies. The presence of properly functioning sewage treatment plants in several places has resulted in low surfactant concentrations in the environment (Trehy et al. 1996; van de Plassche et al. 1999; Marcomini et al. 2000). Activated sludge systems (ASP) are found to be superior to other surfactant treatment processes like oxidation pond (OP) and upflow anaerobic sludge blanket reactor (UASBR), yielding 99 % anionic surfactant removal and detoxifying the surfactants (Mungray and Kumar 2008). As per this comparative study treated effluents from upflow anaerobic sludge blanket reactor and oxidation pond based sewage treatment plants when discharged to aquatic ecosystems are likely to cause substantial risk to aquatic environment due to the presence of anionic surfactants while effluents from activated sludge process are not supposed to pose risk.

The immobilisation of surfactant degrading bacteria in bioreactors gave promising results in various attempts to combat surfactant pollution. Effective use of immobilized cells of *Pseudomonas* C12 B on porous glass beads with either unmodified or silanized surface in a bioreactor showed 85 % SDS removal efficiency (Roig et al. 1998). The removal of SDS from synthetic wastewater in an Intermittent Cycle Extended Aeration System (ICEAS) resulted in more than 98 % removal of SDS (Mortazavi et al. 2008). The use fluidized bed reactors enabled the anaerobic degradation of LAS by microbial consortia (*Bacteroidetes, Firmicutes, Actinobacteria* and *Proteobacteria*) in different support material giving 99 ± 2 % removal rates (de Oliveira et al. 2010).

The combinatorial use of ozonation along with activated sludge system could further reduce the foam ability, residual surfactant concentration and COD in the treated effluents (Beltran et al. 2000b). Acclimation of a mixed culture to ozonated products was beneficial to highly improve biodegradation rates after preozonation. However, photo-degradation along with biodegradation of LAS, didn't promote the biodegradability of the surfactant, whereas presence of some intermediates of photo degradation reduced LAS biodegradability (Mehrvar and Tabrizi 2006).

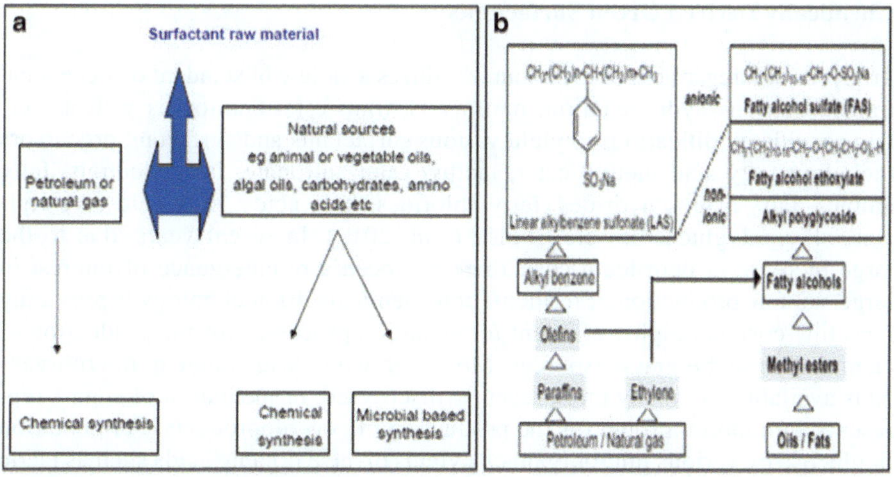

Fig. 5.8 Surfactant synthesis based on their raw material and subsequent modification. (**a**) Strategies possible (**b**) some examples adapted from Hill (2000)

5.5.3 Green Surfactants

The idea of going green has launched the use of renewable materials for surfactant synthesis resulting in so called green surfactants. This new class of biodegradable and biocompatible products is a response to the increasing consumer demand for products that are both "greener", milder and more efficient (Benvegnu and Sassi 2010). The use of renewable resources for surfactant synthesis rather than petrochemicals would reduce the liberated CO_2 levels by 37 % in EU (Patel et al. 1999). A critical review on synthesis of renewable surfactants explains clearly the various strategies involved in green surfactant synthesis (Foley et al. 2012). Green surfactants are defined as biobased amphiphilic molecules obtained from nature or synthesized from renewable raw materials. Various renewable raw materials particularly triglycerides, carbohydrate sources and organic acids (produced by fermentation) serve as starting materials in surfactant synthesis, of which, triglycerides/sterols contribute to the hydrophobic part while sugars/amino acids contribute to the hydrophilic part of green surfactants. Occurrence of multiple renewable raw materials including triglycerides such as palm kernel oil, coconut oil (Hill 2000), cashew nut shell liquid (Tyman 1979), algal oils (Foley et al. 2011), vegetable oil (Guo et al. 2000; Infante et al. 2004), carbohydrates such as sugar beet, molasses, hemicelluloses and inulin (Hill and Rhode 1999) help to generate diverse biobased surfactants. Green surfactants can be synthesized from renewable raw materials either by chemical modification or utilizing the biosynthetic machinery of biotic community (plants, microbes, yeast etc.) yielding biosurfactants as depicted in Fig. 5.8.

Chemically Derived Green Surfactants

Triglycerides, regardless of their source, utilizes a variety of standard oleochemical transformations-hydrogenation, hydrolysis, trans-esterification as well as certain specific modifications to yield various surfactants and surfactant precursors including fatty acid methyl ester, methyl ester sulfonate, fatty alcohols, fatty amines, fatty acid anhydrides, fatty chlorides, fatty acids, fatty acid carboxylates, alkylpolyglucosides etc. (Foley et al. 2012). In recent years, due to the large increase in petroleum cost, there has been a re-emergence of interest in large-volume production of fermentation chemicals. Biotechnology is providing new, low-cost and highly efficient fermentation processes for the production of chemicals from biomass resources. Moreover, with a wide range of microorganisms available and many more recently discovered, fermentation of sugars represents an important route for the production of new bioproducts. Fermentation of glucose by various microorganisms yields diverse organic acids such as lactic acid, citric acid, ascorbic acid, itaconic acid, glutamic acid and lysine derivatives (Corma et al. 2007). Table 5.3 summarizes the various classes of green surfactants available and their uses. Among these alkylpolyglucoside is readily biodegradable and it shows very good skin compatibility in dermatological investigations causing less skin irritability than SDS. On account of the performance and the high quality, regarding the light colour and the good odour, this self-emulsifying compound is particularly appropriate to cosmetic lotions and creams (Weuthen et al. 1995). Alkylpolyglucosides are found to be superior to various carbohydrate based surfactants and are extensively produced on account of its good performance, mildness and completely renewable based nature. Methyl ester sulfonate offers an environmentally friendly and viable alternative to the currently used workhorse surfactant LAS due to its high biodegradability (Ghazali 2002).

As noted in Fig. 5.8 surfactants could be synthesized either from petrochemicals or natural oleochemicals, but the use of green surfactants does not always bring a solution to ecotoxicity as the surfactants chemically remains the same irrespective of the mode of its synthesis. For, example while surfactants like linear alkyl sulfate and secondary alkyl sulfonates are purely petrochemical based, surfactants like alcohol sulfates, alcohol ether sulfates and alkyl ethoxylates are partly fossil fuel based or partly oleo chemical based as depicted in Fig. 5.8b. The European life cycle inventory of selected surfactants (Stalmans et al. 1995), described in detail the resource requirements over the environmental release during petrochemical and oleo chemical based surfactant synthesis. A close analysis of this data revealed that no technical or scientific basis exists to support a general environmental superiority claim, either for an individual surfactant or for various based on sourcing from petrochemical, oleochemical, or agricultural feedstocks and minerals thereby suggesting that environmental acceptability is not associated with raw material source, at least in the LCI context (Saouter et al. 1998; Salimon et al. 2012).

Table 5.3 Green surfactants chemically synthesized from renewable raw materials summarized as per (Foley et al. 2012)

Surfactant	Raw material	Mode of synthesis	Use
Fatty acid methyl ester (FAME)	Triglycerides (TG) or fatty acids	TG hydrogenation and transesterification	Manufacture of biodiesel
Methyl ester sulfonate (MES)	Triglycerides	Sulfonation of FAME	Detergent industry, enhanced oil recovery
Fatty alcohols (FOH)	Triglycerides	FAME reduction	Intermediates in surfactant synthesis
Alkylpolyglucosides	Triglycerides and starch/glucose	FOH glycosidation	Laundry, dishwashing, agrochemicals, hard surface cleaning, cosmetics, micro-emulsions
Fatty acid sulfates	Triglycerides	FOH sulfonation	Personal care products, detergents
Fatty acids	Triglycerides	TG acid catalysis	Soaps
Mono and diglycerides	TG	Trans-esterification of TG or glycosylation of FAME	Emulsifiers in food, cosmetics, pharmaceuticals, textiles plastics
Sucrose esters	TG, sugar	Esterification of sucrose with either FAME/TG	Food, pharmaceuticals, detergents, personal care products
Methyl glucoside esters	TG, sugar	Esterification of methyl glucosides with FAME	Cosmetics, personal care products
Sorbitan esters	Glucose	Esterification of sorbitol (chemically derived or biotechnologically derived)	Food, pharmaceuticals, cosmetics, paints under trade name span and tween
Citrate esters	Sucrose, molasses, wood pulp	Esterification of biotechnologically derived citric acid	Agrochemicals, food industry, household products
Lactate esters	Carbohydrates or petrochemical	Esterification of lactic acid (chemically or biotechnologically produced)	Food
Gluconic acid, gluconolactones and gluconamides	Carbohydrate based	Biotechnologically derived by *Aspergillus niger*	–
Ascorbates	Carbohydrate based	Chemical modification or by fermentation	Antioxidants in drug formulations
Itaconic acid	Starch, molasses, lignocellulose, glycerol, citric acid, corn syrup	Biotechnologically produced by various fungi	Not used commercially
N-acetyl glutamate surfactants	Cane molasses, beet molasses, corn syrup, acetic acid	Thermal condensation of fatty amine or by fermentation	Detergents
Lysine based surfactants	Molasses, corn syrup	Chemical derivatisation of fermented lysine	Cosmetic care products, preservatives etc.
Betaines	Molasses, brown algae, tertiary amines	Chromatographic separation	Hair conditioners

Biosurfactants as Alternate to Synthetic Surfactants

Most bio-based surfactants are manufactured by chemical means; their preparation via bioprocessing is very attractive for future employment due to further enhancement of sustainability and potential savings in energy, downstream purification and disposal costs (Hayes 2012). Biosurfactants are biological compounds with high surface active properties (Georgia and Poe 1931), produced by microorganisms, plants, animals and humans (Christofi and Ivshina 2002). Biosurfactants are produced on microbial cells surfaces or excreted extracellularly and contain both hydrophilic and hydrophobic moieties. They have several advantages over the chemical surfactants, such as lower toxicity higher biodegradability (Zajic et al. 1977), better environmental compatibility (Georgiou et al. 1992), higher foaming ability (Razafindralambo et al. 1996), high selectivity and specific activity at extreme temperatures, pH and salinity (Velikonja and Kosaric 1993) and the ability to be synthesized from renewable feed stock (Desai and Banat 1997). In general, biosurfactants are more effective and efficient and their CMC is about 10–40 times lower than that of chemical surfactants, i.e., less surfactant is necessary to get a maximum decrease in surface tension (Desai and Banat 1997) and have higher EC_{50} than synthetic surfactants (Poremba et al. 1991). Biosurfactants constitutes an interesting alternative to the commercial chemical surfactants with potential use in several industries (Vaz et al. 2012).

Biosurfactants are of different types including glycolipids, lipopeptides, phospholipids, surface-active antibiotics, fatty acids/neutral lipids, polymeric surfactants and particulate (Muthusamy et al. 2008). These compounds are produced during the growth of microorganisms on water-soluble and water immiscible substrates. Diverse ranges of prokaryotic and eukaryotic microorganisms are capable of producing surfactants (Lang 2002). Bacterial surfactant – producing organisms include *Pseudomonas aeruginosa* (mono and di-rhamnolipid biosurfactants), *Corynebacterium*, *Nocardia* and *Rhodococcus* spp. (phospholipids, trehalose dimycolates/dicorynomycolates, glycolipids etc.), *Bacillus subtilis* (surfactin), *Bacillus licheniformis* (lipopeptide similar to surfactin), *Arthrobacter paraffineus* (trehalose and sucrose lipids) and others. Fungi involved in surfactant production include the yeasts *Torulopsis* spp. (sophorolipids) and *Candida* spp. (liposan, phospholipids) (Christofi and Ivshina 2002). The potential application of biosurfactants in Table 5.4 depicts the various fields in which the use of synthetic detergents can be replaced by biosurfactants.

Biosurfactant yield varies with organism, carbon source, nutrient level, aeration, agitation, dissolved oxygen, pH, temperature and fermentation protocol (batch/fed-batch, growing/resting cells). The large-scale production of these molecules was hindered because of low yields in production processes and high recovery and purification costs. Some practical approaches have been adopted to make biosurfactant production process economically attractive including use of cheaper raw materials, optimized and efficient bioprocesses and overproducing mutant and recombinant strains for obtaining maximum productivity (Muthusamy et al. 2008). Agro-industrial wastes such as olive oil mill effluent (Mercade et al. 1993), soap stock

Table 5.4 Potential applications of biosurfactants

Biosurfactant	Organism	Potential use	References
Rhamnolipid	*P. aeruginosa, P. chlororaphis strain NRRLB-30761, Burkholderia thailandensis, Burkholderia pseudomallei*	Wound healing	Stipcevic et al. (2006)
		Stabilization of nanoparticles	Kumar and Mamidyala (2011)
		Preparation of microemulsion	Nguyen and Sabatini (2009) and Xie et al. (2007)
		Antiagglomeration agent	York and Firoozabadi (2008)
		In cleaning soap mixtures	
		Cosmetic additives	Ishigami and Suzuki (1997)
		Biofungicide named Zonix	USEPA (2008)
		Pyrene degradation	Das and Mukherjee (2007)
		Degradation of hydrophobic compounds	Noordman and Janssen (2002)
Sophorolipid	*Candida bombicola, Candida apicola, Rhodotorula bogoriensis*	In cosmetics, food, cleaning and petroleum industry	Van Bogaert et al. (2007)
		Anticancer activity	Chen et al. (2006)
		Decrease pulmonary inflammation by decreasing IgE	Vakil et al. (2010)
		Antiviral activity against HIV, Herpes virus	Shah et al. (2005)
		Antibacterial activity of sophorolipid coated silver nanoparticles against both Gram-positive and G–negative bacteria.	Singh et al. (2009)
Surfactin	*Bacillus subtilis*	Inhibits fibrin clot formation that indicates its potential use in the pharmaceutical sector	Rodrigues et al. (2006)
		Antiadhesive in surgicals	Mireles et al. (2001)
Cyclic lipopeptides	*Bacillus subtilis*	In laundry detergent formulations	Mukherjee (2007)
Fengycin	*Bacillus subtilis*	Antifungal agents of crop pathogens	Ramarathnam et al. (2007)
Iturin	*Bacillus subtilis*	Antifungal agents of crop pathogens	Mizumoto et al. (2007)

(Shabtai 1990; Benincasa et al. 2002); molasses (Patel and Desai 1997), starch rich wastes (Nitschke and Pastore 2004) and vegetable oils (Makkar et al. 2011) are used for surfactant production. Optimized biosurfactant production by an integrated rational whole-cell biocatalyst and bioprocess design methodology, termed systems biotechnology has also been described (Muller and Hausmann 2011). A novel method of SDS based rhamnolipid synthesis is also been introduced yielding a high

Table 5.5 Commercially produced biosurfactants with their prospective uses

Company	Bio surfactant	Producer	Use
Ecover	Sophorolipids	*Candida bombicola*	Hard surface cleansers, laundry detergents
Jeneil	Rhamnolipids	*P. aeruginosa*	Zonix-Biofungicide Reco – cleaning oil storage tanks
Urumqi Unite Bio-Technology	Rhamnolipids	*P. aeruginosa*	Microbial enhanced oil recovery, environmental remediation etc.
MG Intobio	Sophorolipids	*Candida bombicola*	Soaps
Kemin products	Lecithins	Plants	Lysoforte – changes the environment in the digestive tract to enhance absorption of nutrients, including fat-soluble vitamins of animals aiming profitability of animal production
Athena Co. Ltd.	Saponins	Plant (*Saponaria officinalis*)	AT biosurfactant SP-
Etec environmental technologies	Nonionic biosurfactant	–	PetroSolv™ biosurfactant-remediate a wide range of contaminants like diesel, heating oil, and chlorinated solvents

substrate to product conversion ratio (Rebello et al. 2013). However, current biosurfactant research has advanced far ahead providing significant opportunities to replace chemical surfactants with sustainable biologically produced alternatives in bulk commercial products. Glycolipid biosurfactants such as sophorolipids are already available in sufficient yield to make their use feasible while rhamnolipids and mannosylerythritol lipids require further development (Marchant and Banat 2012). The growing demand for ecofriendly, truly bio-based surfactants, along with developments biosurfactant production has commercialized their production as shown in Table 5.5. The use of well aerated bioreactors has been shown to produce higher levels of the lactonic form of the sophorolipids (Casas and Garcia-Ochoa 1999; Guilmanov et al. 2002), a situation which is probably enhanced by the long fermentation time since the lactonisation step is the last in the synthetic pathway (Davila et al. 1997).

5.6 Conclusions

Tracing back from the ancient ashes, to modern petrochemical or nonrenewable raw material based surfactants, the surfactant industry is constantly evolving and expanding to a highly competitive sector yielding a myriad of brands to meet the various demands of mankind. Their consumption is increasing day by day with no limits and restrictions, equally contributed by domestic purposes and industry. Such accumulation of these silent toxicants to the ecosystem, could lead to drastic environmental problems including global warming, terrestrial and aquatic toxicity of the ecosystem and it inhabitants. Total banning of surfactants is impossible in such a modernised lifestyle needing surfactants in our food, cosmetics, cleansers etc.

Surfactants are quite often regarded harmless, on basis of its biodegradability and speculated low concentrations in the environment. But statistical analysis of surfactant concentrations worldwide reveals the fact that these pollutants are found in concentrations higher than their predicted no effect concentrations. Thus regarding surfactants as non-pollutants is a mistake. Visible manifestations of surfactant toxicity are available in the case of microbes, plants and animals.

The problem of surfactants toxicity should be thus addressed with cautiousness in every nation. Wise and limited usage of surfactants right from household level to large scale industries thus could reduce the intensity of surfactant pollution. The use of various physical, chemical and bioremediation strategies could help to reduce the toxicity of surfactants before their disposal into the environment. In such a scenario going Green by choosing the right surfactants, especially phosphate free and eco-friendly ones gains relevance. Ultimately the utilisation of biosurfactants could lower the extent of synthetic surfactants prevalence in environment and its associated toxicity.

References

Abboud MM, Khleifat KM, Batarseh M, Tarawneh KA, Al-Mustafa A, Al-Madadhah M (2007) Different optimization conditions required for enhancing the biodegradation of linear alkylbenzosulfonate and sodium dodecyl sulfate surfactants by novel consortium of *Acinetobacter calcoaceticus* and *Pantoea agglomerans*. Enzyme Microb Technol 41:432–439. doi:10.1016/j.enzmictec. 2007.03.011

Abd-Allah AMA (1995) Determination of long chain alkylbenzenes in sediment samples from Alexandria Coast, Egypt. Toxicol Environ Chem 47:83–88. doi:10.1080/02772249509358130

Abel PD (1974) Toxicity of synthetic detergents to fish and aquatic invertebrates. J Fish Biol 6:279–298. doi:10.1111/j.1095-8649.1974.tb04545.x

Abel PD (1976) Toxic action of several lethal concentrations of an anionic detergent on the gills of the brown trout (*Salmo trutta* L.). J Fish Biol 9:441–446. doi:10.1111/j.1095-8649.1976.tb04692.x

Abel PD, Skidmore JF (1975) Toxic effects of an anionic detergent on the gills of rainbow trout. Water Res 9:759–765. doi:10.1016/0043-1354(75)90068-8

Abu-Hassan MA, Kim JK, Metcalfe IS, Mantzavinos D (2006) Kinetics of low frequency sonodegradation of linear alkylbenzene sulfonate solutions. Chemosphere 62:749–755

Aizdaicher NA, Markina ZV (2006) Toxic effects of detergents on the alga *Plagioselmis prolonga* (Cryptophyta). Russ J Mar Biol 32:45–49. doi:10.1134/S1063074006010068

Aizdaicher NA, Reunova YA (2002) Effects of detergents on in vitro growth of diatom alga *Thalassiosira pseudonana*. Russ J Mar Biol 28:324–328. doi:10.1023/A:1020907501713

Allen SL, Allen JM, Licht BM (1965) Effects of Triton X-100 upon the activity of some electrophoretically separated acid phosphatases and esterases. J Histochem Cytochem 13:434–440. doi:10.1177/13.6.434

Amat AM, Arques A, Miranda MA, Segui S (2004) Photo-fenton reaction for the abatement of commercial surfactants in a solar pilot plant. Solar Energy 77:559–566

Ambily PS, Jisha MS (2011) Characterization of Alkyl sulphatase required for the biodegradation of sodium dodecyl sulphate (SDS). Eur J Exp Biol 1:41–49

Ambily PS, Jisha MS (2012) Biodegradation of anionic surfactant, sodium dodecyl sulfate by *Pseudomonas aeruginosa* MTCC 10311. J Environ Biol 33(4):717–720

Ankley GT, Burkhard LP (1992) Identification of surfactants as toxicants in a primary effluent. Environ Toxicol Chem 11:1235–1248

Ariyoshi T, Hasegawa H, Nanri Y, Arizono K (1990) Profile of hemoproteins and heme-metabolizing enzymes in rats treated with surfactants. Bull Environ Contam Toxicol 44:369–376. doi:10.1007/BF01701217

Ariyoshi T, Hasegawa H, Matsumoto H, Arizono K (1991) Effects of surfactants on the contents of metallothionein, heme and hemoproteins and on the activities of heme oxygenase and drug-metabolizing enzymes in rats pretreated with phenobarbital or β-naphthoflavone. Bull Environ Contam Toxicol 46:120–127. doi:10.1007/BF01688264

Asok AK (2011) Bioremediation of the anionic surfactant linear alkylbenzene sulphonate (las) by *Pseudomonas* sp. Isolated from soil. PhD thesis. Mahatma Gandhi University, Kottayam, India, p 192

Asok AK, Jisha MS (2012a) Biodegradation of the anionic surfactant linear alkylbenzene sulfonate (LAS) by autochthonous *Pseudomonas* sp. Water Air Soil Pollut 223(8):5039–5048. doi:10.1007/s11270-012-1256-8

Asok AK, Jisha MS (2012b) Assessment of soil microbial toxicity on acute exposure of the anionic surfactant linear alkylbenzene sulfonate. J Environ Sci Technol 5:354–363. doi:10.3923/jest.2012. 354.363

Azizullah A, Richter P, Jamil M, Hader DP (2012) Chronic toxicity of a laundry detergent to the freshwater flagellate *Euglena gracilis*. Ecotoxicology 21(7):1957–64. doi:10.1007/s10646-012-0930-3

Badot PM, Richard B, Lucot E, Badot MJ, Garrec JP (1995) Water disturbances and needle surface alterations in *Pinus halepensis* mill. Trees exposed to polluted sea-spray. In: Proceedings of the international symposium ecotoxicology of air compartment, Rouen, France, pp 179–189

Bandala ER, Pelaez MA, Salgado MJ, Torres L (2008a) Degradation of sodium dodecyl sulfate in water using solar driven fenton-like advanced oxidation processes. J Hazard Mater 151:578–584. doi:10.1016/j.jhazmat.2007.06.025

Bandala ER, Velasco Y, Torres LG (2008b) Decontamination of soil washing wastewater using solar driven advanced oxidation processes. J Hazard Mater 160:402–407. doi:10.1016/j.jhazmat.2008.03.011

Bantseev V, Mc Canna D, Banh A, Wong WW, Moran KL, Dixon DG, Trevithick JR, Sivak JG (2003) Mechanisms of ocular toxicity using the in vitro bovine lens and sodium dodecyl sulfate as a chemical model. Toxicol Sci 73:98–107. doi:10.1093/toxsci/kfg060

Barbieri E, Ngan PV, Gomes V (1998) The effect of SDS, sodium dodecyl sulfate, on the metabolism and swimming capacity of *Cyprinus carpio*. Revista brasileira de biologia 58:263–271

Bardach JE, Fujiya M, Holl A (1965) Detergents: effects on the chemical senses of the fish *Ictalurus natalis* (le Sueur). Science 148:1605–1607

Bateman TJ, Dodgson KS, White GF (1986) Primary alkyl sulfatase activities of the detergent-degrading bacterium *Pseudomonas* C12 B. Purification and properties of the P1 enzyme. Biochem J 236:401

Behler A, Biermann M, Hill K, Raths HC, Victor MES, Uphues G (2001) Industrial surfactant syntheses. Surfactant Sci Ser 100:1-44

Beigel C, Barriuso E, Calvet R (1998) Sorption of low levels of nonionic and anionic surfactants on soil: effects on sorption of triticonazole fungicide. Pestic Sci 54:52–60. doi:10.1002/(SICI)1096-9063(199809)54:1<52::AID-PS779>3.0.CO;2-Z

Beltran FJ, Garcia-Araya JF, Alvarez PM (2000a) Sodium dodecylbenzene sulfonate removal from water and wastewater. Kinetics of decomposition by ozonation. Ind Eng Chem Res 39:2214–2220. doi:10.1021/ie990721a

Beltran FJ, Garcia-Araya JF, Alvarez PM (2000b) Sodium dodecylbenzene sulfonate removal from water and wastewater. Kinetics of the integrated ozone-activated sludge system. Ind Eng Chem Res 39:2221–2227. doi:10.1021/ie9907223

Benincasa M, Contiero J, Manresa MA, Moraes IO (2002) Rhamnolipid production by *Pseudomonas aeruginosa* LBI growing on soapstock as the sole carbon source. J Food Eng 54:283–288. doi:10.1016/S0260-8774(01)00214-X

Benvegnu T, Sassi JF (2010) Oligomannuronates from seaweeds as renewable sources for the development of green surfactants. Topp Curr Chem 294:143–164)

Berna JL, Ferrer J, Moreno A, Prats D, Ruiz Bevia F (1989) The fate of LAS in the environment. Tenside Surfactants Deterg 26:101–107

Bhardwaj A, Hartland S (1993) Applications of surfactants in petroleum industry. J Dispers Sci Technol 14:87–116. doi:10.1080/01932699308943389

Bhatia M, Singh HD (1996) Biodegradation of commercial linear alkyl benzenes by *Nocardia amarae*. J Biosci 21:487–496

Bhatkhande DS, Pangarkar VG, Beenackers AACM (2002) Photocatalytic degradation for environmental applications – a review. J Chem Technol Biotechnol 77:102–116. doi:10.1002/jctb.532

Bhattacharya S, Mandal SS (1997) Interaction of surfactants with DNA. Role of hydrophobicity and surface charge on intercalation and DNA melting. BBA Biomembr 1323:29–44. doi:10.1016/S0005-2736(96)00171-X

Bird JA, Cain RB (1972) Metabolism of linear alkylbenzenesulfonates by a *Vibrio* sp. Biochem J 127(2):46

Boonyasuwat S, Chavadej S, Malakul P, Scamehorn JF (2005) Surfactant recovery from water using a multistage foam fractionator: Part I effects of air flow rate, foam height, feed flow rate and number of stages. Sep Sci Technol 40:1835–1853

Braaten B, Granmo A, Lange R (1972) Tissue swelling in *Mytilus edulis* L. induced by exposure to a nonionic surface active agent. Nor J Zool 20:137–140

Brandt KK, Hesselso M, Roslev P, Henriksen K, So J (2001) Toxic effects of linear alkylbenzene sulfonate on metabolic activity, growth rate, and microcolony formation of *Nitrosomonas* and *Nitrosospira* strains. Appl Environ Microbiol 67:2489–2498

Bravo RV, Jurado AE, Francisco MGJ, Reyes RA, Garcia LAI, Sampaio CJM, Fernandes P, Joaquim PFLJ (2006) Modification of the activity of an α-amylase from *Bacillus licheniformis* by several surfactants. Electron J Biotech 9(5):566–567

Bubenheim D, Wignarajah K, Berry W, Wydeven T (1997) Phytotoxic effects of gray water due to surfactants. J Am Soc Hortic Sci 122:792–796

Burnett CL, Bergfeld WF, Belsito DV, Klaassen CD, Marks JG, Shank RC, Slaga TJ, Snyder PW, Andersen FA (2011) Final report on the safety assessment of *Cocos nucifera* (coconut) oil and related ingredients. Int J Toxicol 30:5S–16S. doi:10.1177/1091581811400636

Carlsen L, Metzon MB, Kjelsmark J (2002) Linear alkylbenzene sulfonates (LAS) in the terrestrial environment. Sci Total Environ 290:225–230

Carswell ADW, Edgar AOR, Grady BP (2003) Adsorbed surfactants as templates for the synthesis of morphologically controlled polyaniline and polypyrrole nanostructures on flat surfaces: from spheres to wires to flat films. J Am Chem Soc 125:14793–14800. doi:10.1021/ja0365983

Casas JA, Garcia-Ochoa F (1999) Sophorolipid production by *Candida bombicola* : medium composition and culture methods. J Biosci Bioeng 88:488–494. doi:10.1016/S1389-1723(00)87664-1

Casellato S, Negrisolo P (1989) Acute and chronic effects of an anionic surfactant on some freshwater tubificid species. Hydrobiologia 180:243–252

Cater KC, Harbell JW (2006) Prediction of eye irritation potential of surfactant-based rinse-off personal care formulations by the bovine corneal opacity and permeability (BCOP) assay. Cutan Ocul Toxicol 25:217–233

Chahine L, Sempson N, Wagoner C (1997) The effect of sodium lauryl sulfate on recurrent aphthous ulcers: a clinical study. Compend Contin Educ Dent 18(12):1238–1240

Chaturvedi V, Kumar A (2010a) Bacterial utilization of sodium dodecyl sulfate. Int J Appl Biol Pharm Tech 3:1126–1131

Chaturvedi V, Kumar A (2010b) Isolation of sodium dodecyl sulfate degrading strains from a detergent polluted pond situated in Varanasi city, India. J Cell Mol Biol 2:103–111

Chaturvedi V, Kumar A (2010c) Toxicity of sodium dodecyl sulfate in fishes and animals a review. Int J Appl Biol Pharm Tech 2:630–633

Chaturvedi V, Kumar A (2011a) Diversity of culturable sodium dodecyl sulfate (SDS) degrading bacteria isolated from detergent contaminated ponds situated in Varanasi city, India. Int Biodeterior Biodegrad 65:961–971

Chaturvedi V, Kumar A (2011b) Isolation of a strain of *Pseudomonas putida* capable of metabolizing anionic detergent sodium dodecyl sulfate (SDS). Iran J Microbiol 3:47–53

Chaudhury BL, Meena L (2007) A environmental hazard – a case study of toxic bloom of *Microcystis (Anacystis)* spp. in Udaipur lakes, Rajasthan (India). J Herb Med Toxicol 1:55–59

Chen J, Song X, Zhang H, Qu Y, Miao J (2006) Sophorolipid produced from the new yeast strain *Wickerhamiella domercqiae* induces apoptosis in H7402 human liver cancer cells. Appl Microbiol Biotechnol 72:52–59. doi:10.1007/s00253-005-0243-z

Chen M, Huang S, Wang J (2008) Research of the present situation and development of nonionic surfactants used in pesticide in our country. Mod Agrochem 7:6

Chistyakov BE (2001) Theory and practical application aspects of surfactants. In: Fainerman VB, Mobius D, Miller R (eds) Surfactants chemistry, interfacial properties, applications. Studies in Interface Science 13:511–618

Christofi N, Ivshina IB (2002) Microbial surfactants and their use in field studies of soil remediation. J Appl Microbiol 93:915–929. doi:10.1046/j.1365-2672. 2002.01774

Chu W, Kwan CY (2003) Remediation of contaminated soil by a solvent/surfactant system. Chemosphere 53:9–15

Cirelli AF, Ojeda C (2008) Wastewater management in Greater Buenos Aires, Argentina. Desalination 218:52–61

CIRSL Sulfate (1983) Final report on the safety assessment of sodium lauryl sulfate and ammonium lauryl sulfate. J Am Coll Toxicol 2:127–181. doi:10.3109/10915818309142005

Clausen SK, Sobhani S, Poulsen OM, Poulsen LK, Nielsen GD (2000) Study of adjuvant effect of model surfactants from the groups of alkyl sulfates, alkylbenzene sulfonates, alcohol ethoxylates and soaps. Food Chem Toxicol 38:1065–1074. doi:10.1016/S0278-6915(00)00092-2

Cloves JM, Dodgson KS, White GF, Fitzgerald JW (1980) Specificity of P2 primary alkylsulfohydrolase induction in the detergent-degrading bacterium *Pseudomonas* C12B – effects of alkanesulfonates, alkyl sulfates and other related compounds. Biochem J 185:13

Conry T (1980) Consumer's guide to cosmetics. Ancor Press/Doubleday, Garden City, p 74

Cook TM, Goldman CK (1974) Degradation of anionic detergents in Chesapeake Bay. Chesap Sci 15:52–55

Cooper KJ, Earl LK, Harbell J, Raabe H (2001) Prediction of ocular irritancy of prototype shampoo formulations by the isolated rabbit eye (IRE) test and bovine corneal opacity and permeability (BCOP) assay. Toxicol In Vitro 15:95–103

Corma A, Iborra S, Velty A (2007) Chemical routes for the transformation of biomass into chemicals. Chem Rev 107:2411–2502

Corre G, Templier J, Largeau C, Rousseau B, Berkaloff C (1996) Influence of cell wall composition on the resistance of two *Chlorella* species (Chlorophyta) to detergents1. J Phycol 32:584–590

Cosovic B, Zutic V, Vojvodic V, Plese T (1985) Determination of surfactant activity and anionic detergents in seawater and sea surface microlayer in the Mediterranean. Mar Chem 17:127–139

Cserhati T, Forgacs E, Oros G (2002) Biological activity and environmental impact of anionic surfactants. Environ Int 28:337–348

Das K, Mukherjee AK (2007) Differential utilization of pyrene as the sole source of carbon by *Bacillus subtilis* and *Pseudomonas aeruginosa* strains: role of biosurfactants in enhancing bioavailability. J Appl Microbiol 102:195–203

Davila AM, Marchal R, Vandecasteele JP (1997) Sophorose lipid fermentation with differentiated substrate supply for growth and production phases. Appl Microbiol Biotechnol 47:496–501

Davison J, Brunel F, Phanopoulos A, Prozzi D, Terpstra P (1992) Cloning and sequencing of *Pseudomonas* genes determining sodium dodecyl sulfate biodegradation. Gene 114:19–24

De Neve G (2009) Power, inequality and corporate social responsibility: the politics of ethical compliance in the South Indian garment industry. Econ Polit Wkly 44:63–71

De Oliveira LL, Costa RB, Okada DY, Vich DV, Duarte IC, Silva EL, Varesche MB (2010) Anaerobic degradation of linear alkylbenzene sulfonate (LAS) in fluidized bed reactor by microbial consortia in different support materials. Bioresour Technol 101:5112–5122. doi:10.1016/j.biortech.2010. 01.141

Dehghani MH, Najafpoor AA, Azam K (2010) Using sonochemical reactor for degradation of LAS from effluent of wastewater treatment plant. Desalination 250:82–86

Desai JD, Banat IM (1997) Microbial production of surfactants and their commercial potential. Microbiol Mol Biol Rev 61:47–64

Deschenes L, Lafrance P, Villeneuve JP, Samson R (1996) Adding sodium dodecyl sulfate and *Pseudomonas aeruginosa* UG2 biosurfactants inhibits polycyclic aromatic hydrocarbon biodegradation in a weathered creosote-contaminated soil. Appl Microbiol Biotechnol 46:638–646

Dietrich MJ, Randall TL, Canney PJ (1985) Wet air oxidation of hazardous organics in wastewater. Environ Prog 4:171–177

Diriligen N, Ince N (1995) Inhibition effect of the anionic surfactant SDS on duckweed, LEMNA minor with considerations of growth and accumulation. Chemosphere 31:4185–4196

Doyle C (2010) Powerful choices podcast: dispelling cancer myths. www.cancer.org

Eagle SC, Barry BW, Scott RC (1992) Differential scanning calorimetry and permeation studies to examine surfactant damage to human skin. Cutan Ocul Toxicol 11:77–90

Eaton AD, Clesceri LS, Greenberg AE (1995) Standard methods for the examination of water and wastewater. American Public Health Association, Washington, DC

Eichhorn P, Rodrigues SV, Baumann W, Knepper TP (2002) Incomplete degradation of linear alkylbenzene sulfonate surfactants in Brazilian surface waters and pursuit of their polar metabolites in drinking waters. Sci Total Environ 284:123–134

Elsgaard L, Petersen SO, Debosz K (2001) Effects and risk assessment of linear alkylbenzene sulfonates in agricultural soil. 2. Effects on soil microbiology as influenced by sewage sludge and incubation time. Environ Toxicol Chem 20:1664–1672

Eniola KT (2012) Effect of nitrogen supplementation on aerobic degradation of LAS by consortia of bacteria. J Xenobiotics 2(1):24–27

Enomoto R, Suzuki C, Ohno M, Ohasi T, Futagami R, Ishikawa K, Komae M, Nishino T, Konishi Y, Lee E (2007) Cationic surfactants induce apoptosis in normal and cancer cells. Ann N Y Acad Sci 1095:1–6

Fainerman VB, Mobius D, Miller R (2001) Surfactants: chemistry, interfacial properties, applications. Studies in Interface Science 13:1–661

Farzaneh H, Fereidon M, Noor A, Naser G (2010) Biodegradation of dodecylbenzene sulfonate sodium by *Stenotrophomonas maltophilia* biofilm. Afr J Biotechnol 9:55–62

Feitkenhauer H, Meyer U (2002) Anaerobic digestion of alcohol sulfate (anionic surfactant) rich wastewater batch experiments. Part II: influence of the hydrophobic chain length. Bioresour Technol 82:123–129

Field JA, Leenheer JA, Thorn KA, Barber LLB, Rostad C, Macalady DL, Daniel SR (1992) Identification of persistent anionic surfactant-derived chemicals in sewage effluent and groundwater. J Contam Hydrol 9:55–78

Fitzgerald JW (1975) Secondary alkylsulfatases in a strain of *Comamonas terrigena*. Biochem J 149:477

Florence AT, Attwood D (2011) Physicochemical principles of pharmacy, 5th edn. Pharmaceutical Press, Royal Pharmaceutical Society, London

Foley PM, Beach ES, Zimmerman JB (2011) Algae as a source of renewable chemicals: opportunities and challenges. Green Chem 13:1399–1405

Foley P, Beach ES, Zimmerman JB (2012) Derivation and synthesis of renewable surfactants. Chem Soc Rev 41:1499–1518

Forsyth FR (1964) Surfactants as fungicides. Can J Bot 42:1335–1347

Fountain JC, Klimek A, Beikirch MG, Middleton TM (1991) The use of surfactants for in situ extraction of organic pollutants from a contaminated aquifer. J Hazard Mater 28:295–311

Fountain JC, Starr RC, Middleton T, Beikirch M, Taylor C, Hodge D (1996) A controlled field test of surfactant enhanced aquifer remediation. Ground Water 34:910–916

Fox K, Holt M, Daniel M, Buckland H, Guymer I (2000) Removal of linear alkylbenzene sulfonate from a small Yorkshire stream: contribution to great-er project 7. Sci Total Environ 251:265–275

Frazier Jr RH, Kuehne DL, Horn Jr W, Cantor J (1993) Method for enhancing the recovery of petroleum from an oil-bearing formation using a mixture including anionic and cationic surfactants. US patent 5,246,072

Furrer P, Plazonnet B, Mayer JM, Gurny R (2000) Application of in vivo confocal microscopy to the objective evaluation of ocular irritation induced by surfactants. Int J Pharm 207:89–98

Gadallah MAA (1996) Phytotoxic effects of industrial and sewage waste waters on growth, chlorophyll content, transpiration rate and relative water content of potted sunflower plants. Water Air Soil Pollut 89:33–47

Gadler P, Faber K (2007) New enzymes for biotransformations: microbial alkyl sulfatases displaying stereo-and enantioselectivity. Trends Biotechnol 25:83–88

Georgia FR, Poe CF (1931) Study of Bacterial fluorescence in various media: I. Inorganic substances necessary for bacterial fluorescence. J Bacteriol 22:349

Georgiou G, Lin SC, Sharma MM (1992) Surface-active compounds from microorganisms. Biotechnology 10:60–65

Ghai VU (2011) Say no to chemical detergents. http://EzineArticles.com/?expert=Vineet_U_Ghai

Ghazali R (2002) The effect of disalt on the biodegradability of methyl ester sulfonates (MES). J Oil Palm Res 14:45–50

Ghodrat M (2006) Lung surfactants. Am J Health Syst Pharm 63:1504–1521

Giannis A, Gidarakos E, Skouta A (2007) Application of sodium dodecyl sulfate and humic acid as surfactants on electrokinetic remediation of cadmium-contaminated soil. Desalination 211:249–260

Ginkel CG (1989) Complete degradation of xenobiotic surfactants by consortia of aerobic microorganisms. Biodegradation 7:151–164

Goel G, Kaur S (2012) A study on chemical contamination of water due to household laundry detergents. J Hum Ecol 38:65–69

Gonzalez S, Lopez-Roldan R, Cortina JL (2012) Presence and biological effects of emerging contaminants in Llobregat River basin: a review. Environ Pollut 161:83–92. doi:10.1016/j.envpol.2011.10.002

Gonzalez-Mazo E, Leon VM, Saez M, Gomez-Parra A (2002) Occurrence and distribution of linear alkylbenzene sulfonates and sulfophenylcarboxylic acids in several Iberian littoral ecosystems. Sci Total Environ 288:215–226

Goodwin RM, McBrydie HM (2000) Effect of surfactants on honey bee survival. N Z Plant Prot 53:230–234

Grant RL, Yao C, Gabaldon D, Acosta D (1992) Evaluation of surfactant cytotoxicity potential by primary cultures of ocular tissues: I. Characterization of rabbit corneal epithelial cells and initial injury and delayed toxicity studies. Toxicology 76:153–176

Guilmanov V, Ballistreri A, Impallomeni G, Gross RA (2002) Oxygen transfer rate and sophorose lipid production by *Candida bombicola*. Biotechnol Bioeng 77:489–494

Gunther P, Pestemer W (1992) Phytotoxicity of surfactants to higher plants. In: Hall JE, Sauerbeck DE, Hermit P (eds) Effects of organic contaminants in sewage sludge on soil fertility, plants and animals. 103–111

Guo A, Javni I, Petrovic Z (2000) Rigid polyurethane foams based on soybean oil. J Appl Polym Sci 77:467–473

Hall WS, Patoczka JB, Mirenda RJ, Porter BA, Miller E (1989) Acute toxicity of industrial surfactants to *Mysidopsis bahia*. Arch Environ Contam Toxicol 18:765–772

Hari AC, Paruchuri RA, Sabatini DA, Kibbey TCG (2005) Effects of pH and cationic and nonionic surfactants on the adsorption of pharmaceuticals to a natural aquifer material. Environ Sci Technol 39:2592–2598

Hashim MA, Kulandai J (1988) Biodegradation of a recalcitrant detergent Wastewater. In: Chemeca 88: Australia's bicentennial international conference for the process industries. Institution of Engineers, Barton; preprints of papers, p 107.

Hayes DG (2012) Bioprocessing Approaches to synthesize bio-based surfactants and detergents. In: Food and industrial bioproducts and bioprocessing. Wiley-Blackwell, Ames, pp 243–266
Hashim MA, Kulandai J (1988) Biodegradation of a recalcitrant detergent Wastewater. In: Chemeca 88: Australia's bicentennial international conference for the process industries. Institution of Engineers, Barton; preprints of papers, p 107
Hidaka H (1998) Photodegradation of surfactants with TiO_2 semiconductor for the environmental wastewater treatment. J Chem Sci 110:215–228
Hill K (2000) Fats and oils as oleochemical raw materials. Pure Appl Chem 72:1255–1264
Hill K, Rhode O (1999) Sugar-based surfactants for consumer products and technical applications. Fett-Lipid 101:25–33
Hoffman FA, Bishop JW (1994) Impacts of a phosphate detergent ban on concentrations of phosphorus in the James River, Virginia. Water Res 28:1239–1240
Holland PM, Rubingh DN (1992) Mixed surfactants systems. In: Mixed surfactant systems, vol 50, ACS symposium series. American Chemical Society, Washington, DC
Holt MS, Matthus E, Waters J (1989) The concentrations and fate of linear alkylbenzene sulfonate in sludge amended soils. Water Res 23:749–759
Hosseini F, Malekzadeh F, Amirmozafari N, Ghaemi N (2007) Biodegradation of anionic surfactants by isolated bacteria from activated sludge. Int J Environ Sci Tech 4:127–132
Hrabak A, Antoni F, Szabo MT (1982) Damaging effect of detergents on human lymphocytes. Bull Environ Contam Toxicol 28:504–511
Hsu YC (1963) Detergent (sodium lauryl sulfate)-splitting enzyme from bacteria. Nature 200:1091–1092. doi:10.1038/2001091b0
Huang L, Hao L, Yuan J, Liu Y, An Q (2010) Research progress on preparation and application of silicone surfactants for pesticide adjuvants. Silicone Material 1:014
Huddleston RL, Allred RC (1963) Microbial oxidation of sulfonated alkylbenzenes. Dev Ind Microbiol 4:24–38
Ikehata K, El-Din MG (2004) Degradation of recalcitrant surfactants in wastewater by ozonation and advancedoxidationprocesses:areview.OzoneSciEng26:327–343.doi:10.1080/01919510490482160
Ilardi JM, Schwartzkopf G, Dailey GG (1995) pH adjusted nonionic surfactant-containing alkaline cleaner composition for cleaning microelectronics substrates. United States J. T. Baker Inc., Phillipsburg, 5466389. http://www.freepatentsonline.com/5466389.html
Imai T, Tsuchiya S, Morita K, Fujimori T (1994) Surface tension-dependent surfactant toxicity on the green peach aphid, *Myzus persicae* (Sulzer) (Hemiptera: *Aphididae*). Appl Entomol Zool 29:389–393
Infante MR, Perez L, Pinazo A, Clapes P, Moran MC, Angelet M, Garcia MT, Vinardell MP (2004) Amino acid-based surfactants. C R Chimie 7:583–592. doi:10.1016/j. crci.2004.02.009
International Program on Chemical Safety – IPCS (1996) Linear alkylbenzene sulfonates and related compounds. http://www.inchem.org/documents/ehc/ehc/ehc169.htm
Ishigami Y, Suzuki S (1997) Development of biochemicals-functionalization of biosurfactants and natural dyes. Prog Org Coat 31:51–61
Issa AA, Ismail MA (1994) Effects of detergents on River Nile water microflora. J Islam Acad Sci 7(3):157–162
Ivankovic T, Hrenovic J, Gudelj I (2009) Toxicity of commercial surfactants to phosphate-accumulating bacterium. Acta Chim Slov 56:1003–1009
Jensen J (1999) Fate and effects of linear alkylbenzene sulfonates (LAS) in the terrestrial environment. Sci Total Environ 226:93–111. doi:10.1016/S0048-9697(98)00395-7
Jing-xin Y (2004) Application of phosphate-type surfactants in the textile industry. Textile auxiliaries 3:008
Jovanic BR, Bojovic S, Panic B, Radenkovic B, Despotovic M (2010) The effect of detergent as polluting agent on the photosynthetic activity and chlorophyll content in bean leaves. Health 2:395–399
Jovcic B, Venturi V, Davison J, Topisirovic L, Kojic M (2010) Regulation of the sdsA alkyl sulfatase of *Pseudomonas* sp. ATCC 19151 and its involvement in degradation of anionic surfactants. J Appl Microbiol 109:1076–1083

Kado CI, Liu ST (1981) Rapid procedure for detection and isolation of large and small plasmids. J Bacteriol 145:1365–1373

Kanchi S, Niranjan T, Babu Naidu K, Naidu Venkatasubba N (2012) Monitoring the Status of anionic surfactants in various water systems in urban and rural areas of Tirupati, Andhra Pradesh, South India. Int J Res Chem Environ 2:144–156

Kantin R, Baumgarten MGZ, Cabeda M, Beaumord AC, De Almeida TL (1981) Concentration of anionic detergents in Rio Grande water (south Brazil). Mar Pollut Bull 12:50–53

Kapoor Y, Howell BA, Chauhan A (2009) Liposome assay for evaluating ocular toxicity of surfactants. Invest Ophthalmol Vis Sci 50:2727–2735

Karsa DR (1990) Industrial applications of surfactants. Royal Society of Chemistry, Cambridge

Khalladi R, Benhabiles O, Bentahar F, Moulai-Mostefa N (2009) Surfactant remediation of diesel fuel polluted soil. J Hazard Mater 164:1179–1184

Khurana R (2002) Detergents: counting the cost of cleanliness. Toxic Link Fact Sheet 16:1–4

Klebensberger J, Rui O, Fritz E, Schink B, Philipp B (2006) Cell aggregation of *Pseudomonas aeruginosa* strain PAO1 as an energy-dependent stress response during growth with sodium dodecyl sulfate. Arch Microbiol 185:417–427

Knud-Hansen C (1994) Historical perspective of the phosphate detergent conflict. http://www.colorado.edu/conflict/full_text_search/AllCRCDocs/94-54.htm

Konnecker G, Regelmann J, Belanger S, Gamon K, Sedlak R (2011) Environmental properties and aquatic hazard assessment of anionic surfactants: physico-chemical, environmental fate and ecotoxicity properties. Ecotoxicol Environ Saf 74(6):1445–1460. doi:10.1016/j.ecoenv.2011.04.015

Korzenowski C, Martins MBO, Bernardes AM, Ferreira JZ, Duarte ECNF, De Pinho MN (2012) Removal of anionic surfactants by nanofiltration. Desalin Water Treat 44:269–275

Kostal J, Suchanek M, Klierova H, Demnerova K, Kralova B, McBeth DL (1998) *Pseudomonas* C12B, an SDS degrading strain, harbours a plasmid coding for degradation of medium chain length n-alkanes. Int Biodeter Biodegrad 42:221–228

Kotani M, Masamoto Y, Watanabe M (1994) An alternative study of the skin irritant effect of an homologous series of surfactants. Toxicol In Vitro 8:229–233

Kralova I, Sjoblom J (2009) Surfactants used in food industry: a review. J Dispers Sci Technol 30:1363–1383

Krishnakumar R (2010) A lake's last sigh?. Frontline 27 (11):04

Kristiansen IB, de Jonge H, Nornberg P, Mather-Christensen O, Elsgaard L (2003) Sorption of linear alkylbenzene sulfonate to soil components and effects on microbial iron reduction. Environ Toxicol Chem 22:1221–1228

Kuhm AE, Stolz A, Ngai KL, Knackmuss HJ (1991) Purification and characterization of a 1, 2-dihydroxynaphthalene dioxygenase from a bacterium that degrades naphthalenesulfonic acids. J Bacteriol 173:3795–3802

Kuhnt G (1993) Behavior and fate of surfactants in soil. Environ Toxicol Chem 12:1813–1820

Kumar CG, Mamidyala SK (2011) Extracellular synthesis of silver nanoparticles using culture supernatant of *Pseudomonas aeruginosa*. Colloids Surf B Biointerfaces 84(2):462–466. doi:10.1016/j.colsurfb.2011.01.042

Lang S (2002) Biological amphiphiles (microbial biosurfactants). Curr Opin Coll Interface Sci 7:12–20

Lara-Martin PA, Gomez-Parra A, Gonzalez-Mazo E (2008) Sources, transport and reactivity of anionic and non-ionic surfactants in several aquatic ecosystems in SW Spain: a comparative study. Environ Pollut 156:36–45

Lee BKH (1970) The effect of anionic and nonionic detergents on soil microfungi. Can J Bot 48:583–589

Lewis MA (1990) Chronic toxicities of surfactants and detergent builders to algae: a review and risk assessment. Ecotoxicol Environ Safe 20:123–140

Lewis MA (1991) Chronic and sublethal toxicities of surfactants to aquatic animals: a review and risk assessment. Water Res 25:101–113

Lewis MA (1992) The effects of mixtures and other environmental modifying factors on the toxicities of surfactants to freshwater and marine life. Water Res 26:1013–1023

Lewis MA, Hamm BG (1986) Environmental modification of the photosynthetic response of lake plankton to surfactants and significance to a laboratory-field comparison. Water Res 20:1575–1582

Li M-H (2008) Effects of nonionic and ionic surfactants on survival, oxidative stress, and cholinesterase activity of planarian. Chemosphere 70:1796–1803

Li Z, Zhang Z, Zhang X, Li H (2009) Application of surfactants in pesticide water dispersible granule. China Surfactant Deterg Cosmet 39(227):50–54

Lijun X, Bochu W, Zhimin L, Chuanren D, Qinghong W, Liu L (2005) Linear alkyl benzene sulfonate (LAS) degradation by immobilized *Pseudomonas aeruginosa* under low intensity ultrasound. Colloids Surf B Biointerfaces 40:25–29

Lillis V, Dodgson KS, White GF, Payne WJ (1983) Initiation of activation of a preemergent herbicide by a novel alkylsulfatase of *Pseudomonas putida* FLA. Appl Environ Microbiol 46:988–994

Lima TM, Procopio LC, Brandao FD, Leao BA, Totola MR, Borges AC (2011) Evaluation of bacterial surfactant toxicity towards petroleum degrading microorganisms. Bioresour Technol 102:2957–2964

Lindman B, Thalberg K (1992) Polymer-surfactant interactions-recent developments. In: Goddard ED, Ananthapadmanabhan KP (eds) Interactions of surfactants with polymers and proteins. CRC Press, Boca Raton, pp 203–276

Litz N, Doering HW, Thiele M, Blume HP (1987) The behavior of linear alkylbenzenesulfonate in different soils: a comparison between field and laboratory studies. Ecotoxicol Environ Saf 14:103–116

Liwarska-Bizukojc E, Miksch K, Malachowska-Jutsz A, Kalka J (2005) Acute toxicity and genotoxicity of five selected anionic and nonionic surfactants. Chemosphere 58:1249–1253

Long M, Ruan L, Li F, Yu Z, Xu X (2011) Heterologous expression and characterization of a recombinant thermostable alkylsulfatase (sdsAP). Extremophiles 15:293–301

Lopez J, Iturbe R, Torres LG (2005) Washing of soil contaminated with PAHs and heavy petroleum fractions using two anionic and one ionic surfactant: effect of salt addition. J Environ Sci Health A 39:2293–2306

Lopez-Vizcaino R, Saez C, Canzares P, Rodrigo MA (2012) Electrocoagulation of the effluents from surfactant-aided soil-remediation processes. Sep Purif Technol 98:88–93

Makkar RS, Cameotra SS, Banat IM (2011) Advances in utilization of renewable substrates for biosurfactant production. AMB Express 1:1–19

Maksimov VN, Parshikova TV (2006) Influence of surfactants on the photosynthetic activity of algae. Hydrobiol J 42:67–76

Mallatt J (1985) Fish gill structural changes induced by toxicants and other irritants: a statistical review. Can J Fish Aquat Sci 42:630–648

Manousaki E, Psillakis E, Kalogerakis N, Mantzavinos D (2004) Degradation of sodium dodecylbenzene sulfonate in water by ultrasonic irradiation. Water Res 38:3751–3759

Mantzavinos D, Burrows DM, Willey R, Lo Biundo G, Zhang SF, Livingston AG, Metcalfe IS (2001) Chemical treatment of an anionic surfactant wastewater: electrospray-MS studies of intermediates and effect on aerobic biodegradability. Water Res 35:3337–3344

Marchant R, Banat IM (2012) Biosurfactants: a sustainable replacement for chemical surfactants? Biotechnol Lett 34(9):1597–1605. doi:10.1007/s10529-012-0956-x

Marcomini A, Pojana G, Sfriso A, Alonso JMQ (2000) Behavior of anionic and nonionic surfactants and their persistent metabolites in the Venice Lagoon, Italy. Environ Toxicol Chem 19(8):2000–2007

Marin MG, Pivotti L, Campesan G, Turchetto M, Tallandini L (1994) Effects and fate of sediment-sorbed linear alkylbenzene sulfonate (LAS) on the bivalve mollusc *Mytilus galloprovincialis* Lmk. Water Res 28:85–90

Marrakchi S, Maibach HI (2006) Sodium lauryl sulfate-induced irritation in the human face: regional and age-related differences. Skin Pharmacol Physiol 19:177–180

Matsui S, Park H (2000) Morphological effects and ecotoxicity of nonionic and anionic surfactants to *Closterium ehrenbergii* using AGZI (algal growth and zygospore inhibition) test. Environ Eng Res 5(2):63–69

Matthijs E, Holt MS, Kiewiet A, Rijs GBJ (1999) Environmental monitoring for linear alkylbenzene sulfonate, alcohol ethoxylate, alcohol ethoxy sulfate, alcohol sulfate, and soap. Environ Toxicol Chem 18:2634–2644

McNamee P, Hibatallah J, Costabel-Farkas M, Goebel C, Araki D, Dufour E, Hewitt NJ, Jones P, Kirst A, Le Varlet B, Macfarlane M, Marrec-Fairley M, Rowland J, Schellauf F, Scheel J (2009) A tiered approach to the use of alternatives to animal testing for the safety assessment of cosmetics: eye irritation. Regul Toxicol Pharmacol 54:197–209

Medha MJ, Lee S (1996) Optimization of surfactant-aided remediation of industrially contaminated soils. Energy Sources 18:291–301

Mehrvar M, Tabrizi GB (2006) Combined photochemical and biological processes for the treatment of linear alkylbenzene sulfonate in water. J Environ Sci Health Part A 41:581–597

Mendez-Diaz JD, Sanchez-Polo M, Rivera-Utrilla J, Bautista-Toledo MI (2009) Effectiveness of different oxidizing agents for removing sodium dodecyl benzene sulfonate in aqueous systems. Water Res 43:1621–1629

Mercade ME, Manresa MA, Robert M, Espuny MJ, De Andres C, Guinea J (1993) Olive oil mill effluent (OOME). New substrate for biosurfactant production. Bioresour Technol 43:1–6

Merta J, Stenius P (1999) Interactions between cationic starch and mixed anionic surfactants. Colloids Surf A Physicochem Eng Asp 149:367–377

Minareci O, Ozturk M, Egemen O, Minareci E (2009) Detergent and phosphate pollution in Gediz River, Turkey. Afr J Biotechnol 8:3568–3575

Mireles JR, Toguchi A, Harshey RM (2001) *Salmonella enterica* serovar typhimurium swarming mutants with altered biofilm-forming abilities: surfactin inhibits biofilm formation. J Bacteriol 183:5848–5854

Mishra V, Lal R (2001) Enzymes and operons mediating xenobiotic degradation in bacteria. Crit Rev Microbiol 27:133–166

Mizumoto S, Hirai M, Shoda M (2007) Enhanced iturin A production by *Bacillus subtilis* and its effect on suppression of the plant pathogen *Rhizoctonia solani*. Appl Microbiol Biotechnol 75:1267–1274

Mortazavi SB, Khavanin A, Moussavi G, Azhdarpoor A (2008) Removal of sodium dodecyl sulfate in an intermittent cycle extended aeration system. Pak J Biol Sci 11:290–293

Mukherjee AK (2007) Potential application of cyclic lipopeptide biosurfactants produced by *Bacillus subtilis* strains in laundry detergent formulations. Lett Appl Microbiol 45:330–335

Muller MM, Hausmann R (2011) Regulatory and metabolic network of rhamnolipid biosynthesis: traditional and advanced engineering towards biotechnological production. Appl Microbiol Biotechnol 91(2):251–264. doi:10.1007/s00253-011-3368-2

Mungray AK, Kumar P (2008) Occurrence of anionic surfactants in treated sewage: risk assessment to aquatic environment. J Hazard Mater 160:362–370

Mungray AK, Kumar P (2009) Fate of linear alkylbenzene sulfonates in the environment: a review. Int Biodeterior Biodegrad 63:981–987

Muramoto S, Oki Y (1988) Effects of surface active agents on the salinity tolerance of water hyacinth (*Eichhornia crassipes*). J Environ Sci Health A 23:603–611

Muthusamy K, Gopalakrishnan S, Ravi TK, Sivachidambaram P (2008) Biosurfactants: properties, commercial production and application. Curr Sci 94:736–747

Myers D, Wiley J (2001) Surfactant science and technology, 3rd edn. Wiley Online Library, New York

Naldoni A, Schiboula A, Bianchi C, Bremner D (2011) Mineralisation of surfactants using ultrasound and the advanced Fenton process. Water Air Soil Pollut 215:487–495

Nguyen TT, Sabatini DA (2009) Formulating alcohol-free microemulsions using rhamnolipid biosurfactant and rhamnolipid mixtures. J Surfactants Deterg 12:109–115

Nielsen AD, Borch K, Westh P (2000) Thermochemistry of the specific binding of C12 surfactants to bovine serum albumin. BBA Protein Struct Mol Enzymol 1479:321–331

Nitschke M, Pastore GM (2004) Biosurfactant production by *Bacillus subtilis* using cassava-processing effluent. Appl Biochem Biotechnol 112:163–172
Noordman WH, Janssen DB (2002) Rhamnolipid stimulates uptake of hydrophobic compounds by *Pseudomonas aeruginosa*. Appl Environ Microbiol 68:4502–4508
Ostroumov SA (2003) Studying effects of some surfactants and detergents on filter-feeding bivalves. Hydrobiologia 500:341–344
Ostroumov SA, Widdows J (2006) Inhibition of mussel suspension feeding by surfactants of three classes. Hydrobiologia 556:381–386
Oya M, Hisano N (2010) Decreases in surface activities and aquatic toxicities of linear alkylbenzene sulfonate and alcohol ethoxylates during biodegradation. J Oleo Sci 59:31–39
Oya M, Takemoto Y, Ishikawa Y (2008) Large decrease in acute aquatic toxicity of linear alkylbenzene sulfonate in hard water and seawater by adding adsorbent. J Oleo Sci 57:15–21
Parr JF, Norman AG (1965) Considerations in the use of surfactants in plant systems: a review. Bot Gaz 126(2):86–96
Part P, Svanberg O, Bergstrom E (1985) The influence of surfactants on gill physiology and cadmium uptake in perfused rainbow trout gills. Ecotoxicol Environ Safe 9:135–144
Partearroyo MA, Ostolaza H, Goni FM, Barbera-Guillem E (1990) Surfactant-induced cell toxicity and cell lysis: a study using B16 melanoma cells. Biochem Pharmacol 40:1323–1328
Patel RM, Desai AJ (1997) Biosurfactant production by *Pseudomonas aeruginosa* GS3 from molasses. Lett Appl Microbiol 25:91–94
Patel MK, Theiss A, Worrell E (1999) Surfactant production and use in Germany: resource requirements and CO_2 emissions. Res Conserv Recycl 25:61–78
Payne WJ (1963) Pure culture studies of the degradation of detergent compounds. Biotechnol Bioeng 5:355–365
Payne WJ, Feisal VE (1963) Bacterial utilization of dodecyl sulfate and dodecyl benzene sulfonate. Appl Microbiol 11:339–344
Perez T, Sarrazin L, Rebouillon P, Vacelet J (2002) First evidences of surfactant biodegradation by marine sponges (Porifera): an experimental study with a linear alkylbenzene sulfonate. Hydrobiologia 489:225–233
Petrovic M, Fernandez-Alba AR, Borrull F, Marce RM, Mazo EG, Barcelo D (2002) Occurrence and distribution of nonionic surfactants, their degradation products, and linear alkylbenzene sulfonates in coastal waters and sediments in Spain. Environ Toxicol Chem 21:37–46
Petta TB, De Medeiros SRB, Do Egito EST, Agnez-Lima LF (2004) Genotoxicity induced by saponified coconut oil surfactant in prokaryote systems. Mutagenesis 19:441–444
Pittinger CA, Sellers JS, Janzen DC, Koch DG, Rothgeb TM, Hunnicutt ML (1993) Environmental life-cycle inventory of detergent-grade surfactant sourcing and production. J Am Oil Chem Soc 70:1–15
Pletnev MY (2001) Chemistry of surfactants. In: Studies in interface science. Elsevier 13:1–97
Poremba K, Gunkel W, Lang S, Wagner F (1991) Toxicity testing of synthetic and biogenic surfactants on marine microorganisms. Environ Toxicol Water Qual 6:157–163
Prakasam VR, Joseph ML (2000) Water quality of Sasthamcotta Lake, Kerala (India) in relation to primary productivity and pollution from anthropogenic sources. J Environ Biol 21:305–307
Prats D, Lopez C, Vallejo D, Varo P, Leon VM (2006) Effect of temperature on the biodegradation of linear alkylbenzene sulfonate and alcohol ethoxylate. J Surfactants Deterg 9:69–75
Ramarathnam R, Bo S, Chen Y, Fernando WGD, Xuewen G, De Kievit T (2007) Molecular and biochemical detection of fengycin-and bacillomycin D-producing *Bacillus* spp., antagonistic to fungal pathogens of canola and wheat. Can J Microbiol 53:901–911
Razafindralambo H, Paquot M, Baniel A, Popineau Y, Hbid C, Jacques P, Thonart P (1996) Foaming properties of surfactin, a lipopeptide biosurfactant from *Bacillus subtilis*. J Am Oil Chem Soc 73:149–151. doi:10.1007/BF02523463
Rebello S, Asok AK, Joseph SV, Joseph BV, Jose L, Mundayoor S (2013) Bioconversion of sodium dodecyl sulphate to Rhamnolipid by *Pseudomonas aeruginosa*: a novel and cost-effective production strategy. Appl Biochem Biotechnol 169(2):418–430. doi:10.1007/s12010-012-9988-x, Epub 2012 Dec 5

Ribelles A, Carrasco MC, Rosety M, Aldana M (1995) A histochemical study of the biological effects of sodium dodecyl sulfate on the intestine of the gilthead seabream, *Sparus aurata* L. Ecotoxicol Environ Saf 32:131–138

Rico-Rico A, Temara A, Behrends T, Hermens JLM (2009) Effect of sediment properties on the sorption of C12-2-LAS in marine and estuarine sediments. Environ Pollut 157:377–383

Rinallo C, Bennici A, Cenni E (1988) Effects of two surfactants on *Triticum durum desf.* plantlets. Environ Exp Bot 28:367–374

Rivera-Utrilla J, Mendez-Diaz J, Sanchez-Polo M, Ferro-Garcia MA, Bautista-Toledo I (2006) Removal of the surfactant sodium dodecylbenzene sulfonate from water by simultaneous use of ozone and powdered activated carbon: comparison with systems based on O_3 and O_3/H_2O_2. Water Res 40:1717–1725

Rodrigues L, Banat IM, Teixeira J, Oliveira R (2006) Biosurfactants: potential applications in medicine. J Antimicrob Chemother 57:609–618

Roig MG, Pedraz MA, Sanchez JM (1998) Sorption isotherms and kinetics in the primary biodegradation of anionic surfactants by immobilized bacteria: I. *Pseudomonas* C12B. J Mol Catal B Enzym 4:253–270

Romanelli MF, Moraes MCF, Villavicencio ALCH, Borrely SI (2004) Evaluation of toxicity reduction of sodium dodecyl sulfate submitted to electron beam radiation. Radiat Phys Chem 71:411–413

Rosen MJ (1989) Surfactants and interfacial phenomena, 3rd edn. Wiley, Hoboken, pp 2–5

Rosen MJ, Li F, Morrall SW, Versteeg DJ (2001) The relationship between the interfacial properties of surfactants and their toxicity to aquatic organisms. Environ Sci Technol 35:954–959

Rosety M, Ordonez FJ, Rosety-Rodriguez M, Rosety JM, Rosety I, Carrasco C, Ribelles A (2001) Acute toxicity of anionic surfactants sodium dodecyl sulfate (SDS) and linear alkylbenzene sulfonate (LAS) on the fertilizing capability of gilthead (*Sparus aurata* L.) sperm. Histol Histopathol 16(3):839–843

Rosety M, Ordonez FJ, Rosety-Rodriguez M, Rosety JM, Rosety I (2003) In vitro acute toxicity of anionic surfactant linear alkylbenzene sulfonate (LAS) on the motility of gilthead (*Sparus aurata* L.) sperm. Histol Histopathol 18:475–478

Rosety-Rodríguez M, Ordonez FJ, Roldan S, Rosety JM, Rosety M, Ribelles A, Carrasco C, Rosety I (2002) Acute effects of sodium dodecyl sulfate on the survival and on morpho-histochemical characteristics of the trunk kidney of juvenile turbot *Scophthalmus maximus* L. Eur J Histochem 46:179–184

Russell GL, Britton LN (2002) Use of certain alcohol ethoxylates to maintain protease stability in the presence of anionic surfactants. J Surfactants Deterg 5:5–10

Salager JL (2002) Surfactants types and uses. Fire p booket-E300-attaching aid in surfactant science and engineering in English. Laboratory of FIRP, Universidad dos las andes, Merida Venezuela 2:3

Salimon J, Salih N, Yousif E (2012) Industrial development and applications of plant oils and their biobased oleochemicals. Arab J Chem 5(2):135–145. doi:10.1016/j.arabjc.2010.08.007

Sanchez-Peinado MM, Rodelas B, Martinez-Toledo MV, Gonzalez-Lopez J, Pozo C (2009) Response of soil enzymes to linear alkylbenzene sulfonate (LAS) addition in soil microcosms. Soil Biol Biochem 41:69–76

Sanderson H, Dyer SD, Price BB, Nielsen AM, van Compernolle R, Selby M, Stanton K, Evans A, Ciarlo M, Sedlak R (2006) Occurrence and weight-of-evidence risk assessment of alkyl sulfates, alkyl ethoxysulfates, and linear alkylbenzene sulfonates (LAS) in river water and sediments. Sci Total Environ 368:695–712

Saouter G, Van Hoof T, Feijtel M, Stalmans JC, Uhl LHM, Vollebergt, Westra J (1998) Life cycle inventory on laundry detergents: analysis of the LCI profiles of liquid and powder detergents. In: 38th WFK international detergency conference Frefeld, Seidenweberhaus, Krefeld

Schleheck D, Cook AM (2005) ω-Oxygenation of the alkyl sidechain of linear alkylbenzenesulfonate (LAS) surfactant in *Parvibaculum lavamentivorans* T. Arch Microbiol 183:369–377

Schleheck D, Dong W, Denger K, Heinzle E, Cook AM (2000) An α-proteobacterium converts linear alkylbenzenesulfonate surfactants into sulfophenyl carboxylates and linear alkyldiphenyletherdisulfonate surfactants into sulfodiphenyl-ethercarboxylates. Appl Environ Microbiol 66:1911–1916

Schleheck D, Knepper TP, Fischer K, Cook AM (2004) Mineralization of individual congeners of linear alkylbenzene sulfonate by defined pairs of heterotrophic bacteria. Appl Environ Microbiol 70:4053–4063

Schleheck D, Knepper TP, Eichhorn P, Cook AM (2007) *Parvibaculum lavamentivorans* DS-1T degrades centrally substituted congeners of commercial linear alkylbenzene sulfonate to sulfophenyl carboxylates and sulfophenyl dicarboxylates. Appl Environ Microbiol 73:4725–4732

Schleheck D, von Netzer F, Fleischmann T, Rentsch D, Huhn T, Cook AM, Kohler H-PE (2010) The missing link in linear alkylbenzenesulfonate surfactant degradation: 4-sulfoacetophenone as a transient intermediate in the degradation of 3-(4-sulfophenyl) butyrate by *Comamonas testosteroni* KF-1. Appl Environ Microbiol 76:196–202

Schramm L (2000) Surfactants: fundamentals and applications in the petroleum industry. Cambridge University Press, Cambridge, pp 1–615

Schroeder D, Moeggenborg KJ, Chou H, Chamberlain JP, Hawkins JD, Carter P (2007) CMP method utilizing amphiphilic non-ionic surfactants. EP patent 1,560,890

Schweigert MK, Mackenzie DP, Sarlo K (2000) Occupational asthma and allergy associated with the use of enzymes in the detergent industry a review of the epidemiology, toxicology and methods of prevention. Clin Exp Allergy 30:1511–1518

Sebag H, Vanlerberghe G (1990) New polygylcerol ethers and their use in cosmetics and in pharmacy. US patent 4,946,670

Sequeira CAC (1994) Environmental oriented electrochemistry, vol 59, Studies in environmental science. Elsevier, Amsterdam

Setzkorn EA, Huddleston RL (1965) Ultraviolet spectroscopic analysis for following the biodegradation of hydrotropes. J Am Oil Chem Soc 42:1081–1084

Shabtai Y (1990) Production of exopolysaccharides by *Acinetobacter* strains in a controlled fed-batch fermentation process using soap stock oil (SSO) as carbon source. Int J Biol Macromol 12:145–152

Shah V, Doncel GF, Seyoum T, Eaton KM, Zalenskaya I, Hagver R, Azim A, Gross R (2005) Sophorolipids, microbial glycolipids with anti-human immunodeficiency virus and sperm-immobilizing activities. Antimicrob Agents Chemother 49:4093–4100. doi:10.1128/AAC.49.10.4093-4100.2005

Shivji GM, Segal L, McKenzie RC, Sauder DN (1994) Cutaneous toxicity of surfactants in normal human keratinocytes assessed by cytotoxicity, arachidonic acid release, and regulation of interleukin-1α mRNA. Toxicol Mech Method 4:193–203

Shu-de D (2009) Research progress on agricultural surfactants. J Anhui Agric Sci 37(7):2842–2843

Shukor MY, Husin WSW, Rahman MFA, Shamaan NA, Syed MA (2009) Isolation and characterization of an SDS-degrading *Klebsiella oxytoca*. J Environ Biol 30:129–134

Singh KL, Kumar A (1998) Short communication: *Bacillus cereus* capable of degrading SDS shows growth with a variety of detergents. World J Microbiol Biotechnol 14:777–779

Singh A, Van Hamme JD, Ward OP (2007) Surfactants in microbiology and biotechnology: Part 2. Application aspects. Biotech Adv 25:99–121

Singh S, Patel P, Jaiswal S, Prabhune AA, Ramana CV, Prasad BLV (2009) A direct method for the preparation of glycolipid -metal nanoparticle conjugates: sophorolipids as reducing and capping agents for the synthesis of water re-dispersible silver nanoparticles and their antibacterial activity. New J Chem 33:646–652

Sirisattha S, Momose Y, Kitagawa E, Iwahashi H (2004) Toxicity of anionic detergents determined by *Saccharomyces cerevisiae* microarray analysis. Water Res 38:61–70

Soberon-Chavez G, Campos J, Haidour A, Ramos JL, Ortigoza J (1996) Selection and preliminary characterization of a *Pseudomonas aeruginosa* strain mineralizing selected isomers in a branched chain dodecylbenzene sulfonate mixture. World J Microbiol Biotechnol 12:367–372

Sonc A, Grilc V (2004) Batch foam fractionation of surfactants from aqueous solutions. Acta Chim Slov 51:687–698

Stalmans M, Berenbold H, Berna JL, Cavalli L, Dillarstone A, Franke M, ... Van Sloten R (1995) European life-cycle inventory for detergent surfactants production. Tenside, Surfactants Detergents 32(2):84–109

Standard PG, Ahearn DG (1970) Effects of alkylbenzene sulfonates on yeasts. Appl Microbiol 20:646

Stipcevic T, Piljac A, Piljac G (2006) Enhanced healing of full-thickness burn wounds using di-rhamnolipid. Burns 32:24–34

Stow CA, Borsuk ME, Stanley DW (2001) Long-term changes in watershed nutrient inputs and riverine exports in the Neuse River, North Carolina. Water Res 35:1489–1499

Susmi TS, Rebello S, Jisha MS, Sherief PM (2010) Toxic effects of sodium dodecyl sulfate on grass carp *Ctenopharyngodon idella*. Fish Technol 47(2):157–162

Swisher RD (1963) Biodegradation of ABS in relation to chemical structure. J (Water Pollut Contr Fed) 35:877–892

Swisher RD (1967) Biodegradation of LAS benzene rings in activated sludge. J Am Oil Chem Soc 44:717–724

Takada H, Ishiwatari R, Ogura N (1992) Distribution of linear alkylbenzenes (LABs) and linear alkylbenzene sulfonates (LAS) in Tokyo Bay sediments. Estuar Coast Shelf Sci 35:141–156

Tharapiwattananon N, Scamehorn JF, Osuwan S, Harwell JH, Haller KJ (1996) Surfactant recovery from water using foam fractionation. Sep Sci Technol 31:1233–1258

Thomas OR, White GF (1989) Metabolic pathway for the biodegradation of sodium dodecyl sulfate by *Pseudomonas* sp. C12B. Biotechnol Appl Biochem 11:318–327

Thurnheer T, Zurrer D, Hoglinger O, Leisinger T, Cook AM (1990) Initial steps in the degradation of benzene sulfonic acid, 4-toluene sulfonic acids, and orthanilic acid in *Alcaligenes* sp. strain O-1. Biodegradation 1:55–64

Tiehm A (1994) Degradation of polycyclic aromatic hydrocarbons in the presence of synthetic surfactants. Appl Environ Microbiol 60:258–263

Tolls J, Kloepper-Sams P, Sijm DTHM (1994) Surfactant bioconcentration-a critical review. Chemosphere 29:693–717

Torres LG, Orantes JL, Iturbe R (2003) Critical micellar concentrations for three surfactants and their diesel-removal efficiencies in petroleum-contaminated soils. Environ Geosci 10:28–36

Tozum-Calgan SRD, Atay- Guneyman NZ (1994) The effects of an anionic and a non-ionic surfactant on growth and nitrogen fixing ability of a cyanobacterium, *Gloeocapsa*. J Environ Sci Health A Environ Sci Eng 29:355–369

Trehy ML, Gledhill WE, Mieure JP, Adamove JE, Nielsen AM, Perkins HO, Eckhoff WS (1996) Environmental monitoring for linear alkylbenzene sulfonates, dialkyltetralin sulfonates and their biodegradation intermediates. Environ Toxicol Chem 15(3):233–240

Tyman JHP (1979) Non-isoprenoid long chain phenols. Chem Soc Rev 8:499–537

USEPA (2008) Zonix. www.epa.gov/pesticides/biopesticides/ingredients/product/prod_110029.htm

Usman K, Adesina AA, Lucien FP, Waite TD (2001) Kinetics of the autoxidation of sodium dodecyl sulfate catalyzed by alumina-supported Co-Zn composite. Ind Eng Chem Res 40:5095–5101

Vakil H, Sethi S, Fu S, Stanek A, Wallner S, Gross R (2010) Sophorolipids decrease pulmonary inflammation in a mouse asthma model. Mod Pathol 23:392A

Van Bogaert INA, Saerens K, De Muynck C, Develter D, Soetaert W, Vandamme EJ (2007) Microbial production and application of sophorolipids. Appl Microbiol Biotechnol 76:23–34

Van de Plassche EJ, de Bruijn JHM, Stephenson RR, Marshall SJ, Feijtel TCJ, Belanger SE (1999) Predicted no effect concentrations and risk characterization of four surfactants: linear alkyl benzene sulfonate, alcohol ethoxylates, alcohol ethoxylated sulfates, and soap. Environ Toxicol Chem 18:2653–2663

Van Hylckama Vlieg JET, Janssen DB (2001) Formation and detoxification of reactive intermediates in the metabolism of chlorinated ethenes. J Biotechnol 85:81–102

Vaz DA, Gudina EJ, Alameda EJ, Teixeira JA, Rodrigues LR (2012) Performance of a biosurfactant produced by a *Bacillus subtilis* strain isolated from crude oil samples as compared to commercial chemical surfactants. Colloids Surf B 89:167–174

Velikonja J, Kosaric N (1993) Biosurfactants in food applications. In *Biosurfactants — production, properties and applications*, ed Kosaric, N. pp. 419–446. New York: Marcel Dekker.
Venhuis SH, Mehrvar M (2004) Health effects, environmental impacts, and photochemical degradation of selected surfactants in water. Int J Photoenergy 6:115–125
Verge C, Moreno A, Bravo J, Berna JL (2001) Influence of water hardness on the bioavailability and toxicity of linear alkylbenzene sulfonate (LAS). Chemosphere 44:1749–1757
Vian L, Vincent J, Maurin J, Fabre I, Giroux J, Cano JP (1995) Comparison of three in vitro cytotoxicity assays for estimating surfactant ocular irritation. Toxicol In Vitro 9:185–190
Vigon BW, Rubin AJ (1989) Practical considerations in the surfactant-aided mobilization of contaminants in aquifers. J (Water Pollut Contr Fed) 61(7):1233–1240
Vinther FP, Mortensen GK, Elsgaard L (2003) Effects of linear alkylbenzene sulfonates on functional diversity of microbial communities in soil. Environ Toxicol Chem 22:35–39
Volkering F, Breure AM, Rulkens WH (1997) Microbiological aspects of surfactant use for biological soil remediation. Biodegradation 8:401–417
Vreysen S, Maes A (2005) Remediation of a diesel contaminated, sandy-loam soil using low concentrated surfactant solutions. J Soil Sediments 5:240–244
Warson H, Finch CA (1998) Surfactants and polymers in aqueous solution 1edn Wiley Publishers.
Watanabe K (2001) Microorganisms relevant to bioremediation. Curr Opin Biotechnol 12:237–241
Wei G, Li Y, Du G, Chen J (2003) Effect of surfactants on extracellular accumulation of glutathione by *Saccharomyces cerevisiae*. Process Biochem 38:1133–1138
Weuthen M, Kawa R, Hill K, Ansmann A (1995) Long chain alkyl polyglycosides– a new generation of emulsifiers. Lipid/Fett 97:209–211
Wilhelm KP, Cua AB, Wolff HH, Maibach HI (1993) Surfactant-induced stratum corneum hydration in vivo: prediction of the irritation potential of anionic surfactants. J Invest Dermatol 101:310–315
Wilhelm KP, Freitag G, Wolff HH (1994) Surfactant-induced skin irritation and skin repair: evaluation of the acute human irritation model by noninvasive techniques. J Am Acad Dermatol 30:944–949
Williams J, Payne WJ (1964) Enzymes induced in a bacterium by growth on sodium dodecyl sulfate. Appl Microbiol 12:360–362
Wolfenbarger DANA, Lukefahr MJ, Lowry WL (1967) Toxicity of surfactants and surfactant-insecticide combinations to the bollworm, 1 tobacco budworm, 1 and pink bollworm 2, 3, 4. J Econ Entomol 60:902–904
Woltering DM (1984) The growth response in fish chronic and early life stage toxicity tests: a critical review. Aquat Toxicol 5:1–21
Xiaoli Y, Xuegang L, Changbin S, Shihong C, Qiang W (2000) The effects of surfactants on the activity of invertase and superoxide dismutase of soybean leaf in vitro and in vivo. Colloids Surf A 175:249–255
Xie Y, Ye R, Liu H (2007) Microstructure studies on biosurfactant-rhamnolipid/n-butanol/water/n-heptane microemulsion system. Colloids Surf A 292:189–195
Xu KP, Li XF, Fu-Shin XY (2000) Corneal organ culture model for assessing epithelial responses to surfactants. Toxicol Sci 58:306–314
Yadav JS, Lawrence DL, Nuck BA, Federle TW, Reddy CA (2001) Biotransformation of linear alkylbenzene sulfonate (LAS) by *Phanerochaete chrysosporium*: oxidation of alkyl side-chain. Biodegradation 12:443–453
Yamaguchi JY, Kanada A, Horimoto K, Oyama TM, Chikutei K, Nishimura Y, Yamamoto H, Ishida S, Okano Y, Oyama Y (2006) Modification of vulnerability to dodecylbenzenesulfonate, an anionic surfactant, by calcium in rat thymocytes. Environ Toxicol Pharmacol 22:234–239. doi:10.1016/j.etap. 2006. 03.011
Yeldho D, Rebello S, Jisha MS (2011) Plasmid-mediated biodegradation of the anionic surfactant sodium dodecyl sulfate, by *Pseudomonas aeruginosa* S7. Bull Environ Contam Toxicol 86:110–113

Ying GG (2006) Fate, behavior and effects of surfactants and their degradation products in the environment. Environ Int 32:417–431
York JD, Firoozabadi A (2008) Comparing effectiveness of rhamnolipid biosurfactant with a quaternary ammonium salt surfactant for hydrate anti-agglomeration. J Phys Chem B 112(3):845–851. doi:10.1021/jp077271h
Yuan S, Tian M, Lu X (2006) Electrokinetic movement of hexachlorobenzene in clayed soils enhanced by Tween 80 and β-cyclodextrin. J Hazard Mater 137:1218–1225
Yuksel E, Sengil IA, Ozacar M (2009) The removal of sodium dodecyl sulfate in synthetic wastewater by peroxi-electrocoagulation method. Chem Eng J 152:347–353
Zajic JE, Guignard H, Gerson DF (1977) Properties and biodegradation of a bioemulsifier from *Corynebacterium hydrocarboclastus*. Biotechnol Bioeng 19:1303–1320
Zhang W, Lo IMC (2006) EDTA-enhanced washing for remediation of Pb-and/or Zn-contaminated soils. J Environ Eng 132:1282
Zhong JF (1999) Application of surfactants in pharmacy. Peoples Medical Press, Beijing, pp 120–124
Zimmermann T, Kulla HG, Leisinger T (1982) Properties of purified Orange II azoreductase, the enzyme initiating azo dye degradation by *Pseudomonas* KF46. Eur J Biochem 129:197–203

Chapter 6
Cadmium, Lead, Thallium: Occurrence, Neurotoxicity and Histopathological Changes of the Nervous System

Concepción Nava-Ruíz and Marisela Méndez-Armenta

Contents

Abstract Metals, particularly heavy metals such as lead, cadmium and thallium between others, constitute significant potential threats to human health in both occupational and environmental settings. Metals toxicity most commonly involves the kidney, liver and mainly nervous system. Neurons in general have a high metabolic rate, which makes them more susceptible to different heavy metals producing changes in neuronal function may lead to secondary alterations in neuronal

C. Nava-Ruíz • M. Méndez-Armenta (✉)
Laboratory of Experimental Neuropathology, National Institute of Neurology and Neurosurgery, Manuel Velasco Suarez, Insurgentes Sur 3877, Tlalpan C.P. 14269, Mexico
e-mail: mmendezarmenta@hotmail.com

E. Lichtfouse et al. (eds.), *Pollutant Diseases, Remediation and Recycling*, Environmental Chemistry for a Sustainable World 4, DOI 10.1007/978-3-319-02387-8_6,

anatomy. Neuropathology is frequently used to evaluate the effects of toxic agents on nervous system organization and cellular components; thus, careful histologic evaluations increase our knowledge of the neurotoxicity of heavy metals. Structural changes can often be correlated with altered neurochemistry, behavior, and electrophysiologic function.

Keywords Heavy metals • Neurotoxicity • Central nervous system • Cadmium • Thallium • Lead • Histopathological

6.1 Introduction

Heavy metals are natural components of the Earth's crust; they cannot be easily degraded. Heavy metals, such as cadmium, lead, and thallium constitute a significant potential threat to human health because they are associated with adverse effects on health. Humans have used heavy metals in many different areas for thousands of years; this use influences their potential for health effects in at least two major ways: first, by environmental transport, that is, by human or anthropogenic contributions to air, water, soil, and food, and second by altering the speciation or biochemical form of the element (Beijer and Jernelov 1986). Metal-rich mine tailings, metal smelting, electroplating, gas exhausts, energy and fuel production, the downwash from power lines, intensive agriculture, and sludge dumping are the most important human activities that contaminate environment with large quantities of toxic metals (Raskin et al. 1994). The importance of these metals as environmental health hazards is readily evident from the fact that the three metals are ranked in the main on the current Agency for Toxic Substances and Disease Registry Priority List of Hazardous Substances (ATSDR 1999, 2007, 2008). The toxic nonessentials metals are characterized as having no demonstrated biological requirement in humans, and exposure is associated with recognizable toxicity, the exposition to these metals is associated with physiological, biochemical and several pathological problems and the severity of toxicity increase with increase in dosage.

Metal induced toxicity is amply reported in the literature; one of the major mechanisms behind heavy metal toxicity has been attributed to oxidative stress. Oxidative stress is caused by an imbalance between the production of reactive oxygen and a biological system's ability to readily detoxify the reactive intermediates or easily repair the resulting damage; toxic metals increase production of free radicals and decrease availability of antioxidant reserves to respond to the resultant damage (Valko et al. 2006). Redox active metals may undergo cycling reactions participating in the transfer of electrons between metals and substrates and therefore may play an important role in the maintenance of redox homeostasis (Lindeque et al. 2010). Metals such as cadmium, lead and thallium possess the ability to produce Reactive Oxygen Species (ROS); production of oxygen-, carbon-, sulfur-, nitrogen radicals, and lipid peroxide thus resulting in molecular damages including lipid, proteins and DNA producing alteration of cell homeostasis (Flora et al.2008; Hanzel and Verstraeten 2006).

Heavy metals generally have numerous effects on the both experimental and humans targeting organs and systems such as the haematopoietic system, liver, placenta, lung and kidney, mainly the central nervous system (Goyer and Clarkson 2001); they can directly influence behaviour by impairing mental and neurological function, influencing neurotransmitter production and utilization, and altering numerous metabolic body processes (Flora et al. 2008). Although the clinical toxicity of metals may be quite varied, most metals induce morphological changes through similar mechanisms; once the neuropathology is frequently used to evaluate the effects of a heavy metals on central nervous system organization and cellular components, identifies vulnerable sites or susceptible cell populations (Dorman 2000). The general aspects of morphological alterations of heavy metal intoxication are present in this chapter, however, more detailed reviews of the chemicals and their mechanisms of toxicity are available.

6.2 Cadmium

6.2.1 Sources and Exposure

Cadmium is considered one of the most toxic substances in the environmental, occurs in the earth's crust at a concentration of 0.1–0.5 ppm and is commonly associated with zinc, lead, and copper ores. There are estimates that 30,000 tons of cadmium are released into the environment each year, with an estimated 4,000–13,000 tons coming from human activities (ATSDR 2008). Natural as well as anthropogenic sources of cadmium, which include industrial emissions and the application of fertilizer and sewage sludge to farm land, increased cadmium environmental levels (ATSDR 2008). The first reports of severe health problems due to cadmium intoxication arose in the 1940s in Toyama Japan, where the Itai-itai disease was endemic and characterized by severe pain, bone fractures, proteinuria and severe osteomalacia, which appeared mainly among women as a consequence of eating cadmium-polluted rice (Nordberg 2004; Takebayashi et al. 2000).

Nowadays, cadmium is still an important industrial metal; it is extracted during the production of other metals, such as zinc, lead and copper. Cadmium components are used as stabilizers in PVC products, colour pigment, several alloys, anticorrosion agent and most commonly, in re-chargeable nickel-cadmium batteries which made up 18 % of the rechargeable battery market, down from 56 % in 1996 with global sales decreasing 16 % between 2005 and 2006, due to mainly that niquel-cadmium batteries have been replaced in some markets by lithium ion and nickel metal hydride batteries (USGS 2008). Other sources of concern are phosphate fertilizers, which may contain high concentrations of cadmium depending on the origin of the rock, and the application of contaminated sewage sludge as a soil amendment; likewise, this continue to be contaminated with cadmium as a result of the recycling of industrial waste sold as zinc sulphate or other materials for agricultural and home use fertilizers (Patrick 2003; Cupit et al. 2002).

Humans and other mammals are exposed to cadmium primarily through two sources. The first is the oral route through ingestion of contaminated water or food mainly fish, liver, grains, leafy vegetables, potatoes and other root vegetables (Castro-Gonzalez and Méndez-Armenta 2008; ATSDR 2008) the ingestion of contaminated food serves as the most non occupational source of exposure. The second source is through inhalation of cadmium particles during industrial activities or inhaled cigarette smoke, that is by far the largest source of cadmium exposure in the general human population; has been reported that one cigarette contains about 1–2 μg or more of cadmium, and 40–60 % of the cadmium in inhaled smoke generally pass through the pulmonary epithelium into systemic circulation (Saldivar et al. 1991; Stohs et al. 1997; ATSDR 2008). Epidemiological studies in different parts of the world have been concluded that the toxicity of cadmium increases with age, time of exposure, and gender (Jin et al. 2004; Gamo et al. 2006; Omarova and Phillips 2007); several studies have shown that children exposed to cadmium have alterations in their cognitive development as well as a decrease in their IQ due to elevated hair levels of cadmium (Thatcher et al. 1982; Ozden et al. 2007) while that in adult studies results are inconsistent, some showing neurological disorders (Viaene et al. 2000) cognitive alterations (Gao et al. 2008) and others not shown neurological alterations, a Swedish study on elderly persons did not find an association between blood cadmium level and cognitive function (Basun et al. 1994).

6.2.2 General Toxicology

In humans and experimental animals, have been reported that cadmium is able produce adverse effects a number of organ and tissues with distinct pathologies depending age, time of exposure, and gender. During the acute cadmium exposure, the general symptoms are relatively mild but, within a few days following exposure, severe pulmonary edema, hemorrhage, fulminate hepatitis, testicular injury, and lethality; whereas chronic exposure to cadmium often leads to renal dysfunction, immunotoxicity, anemia, osteoporosis, and bone fractures (Zalups and Ahmad 2003; Satarug et al. 2003; Liu et al. 2007). Few epidemiological studies are devoted to explore cadmium neurotoxicity in humans, the first report with neurobehavioral effects of cadmium exposure was published in 1989, Hart et al. (1989) shown that exist attention, psychomotor speed, and memory disorders whit neurological symptoms such as fatigue, mental irritability, headache, muscle weakness, syncope, insomnia, and anosmia in workers who are acutely or chronically exposed to cadmium. Viaene et al. (1999, 2000) was reported that workers exposed to cadmium for 15 years, shown diminished attention, psychomotor speed, memory, slowing of psychomotor functions, reduced visuomotor functioning, and increase in the risk of peripheral neuropathy. Although a few studies have specifically looked for neurological effects cadmium-induced neurotoxicity it has been observed in experimental animals; significantly decreased motor activity, aggressive, anxiety; weakness and muscle atrophy (ATSDR 2008) and decrease in bioelectrical activity in peripheral nerve (Papp et al. 2003). Prenatal cadmium exposure produces a delay in neuromotor and neurobehavioral development of neonates (neonatal reflexes) in terms of righting reflex, negative geotaxis, cliff avoidance, startle responses and vibrissa reflex (Kamel and El Razek 2011). Experimental evidence supporting that in adult rats only small amounts of cadmium reach the brain because of the selective permeability of the brain-blood barrier (BBB); however, studies conducted in developing rats suggest that cadmium is able to cross the BBB in newborn, inducing damage in the BBB and its primary effect on blood vessels might be due to differences in the BBB maturation (Wong and Klaassen 1982).

Blood cadmium is often used as a measure of recent exposure to this heavy metal, but blood cadmium not only reflects current exposure, but it has also been shown to be a good indicator of the lifetime body burden. In most studies, the half-life in humans is estimated to be between 15 and 30 years, this is particularly important in the kidney, where the half-life of cadmium exposure in the kidney has a considerable accumulation and damage occurs (Nordberg 2004; Suwazono et al. 2006). Cadmium concentration in urine has been proposed as an indirect biological indicator for cadmium accumulation in the kidney has been proposed that the population exposed showing evidence of renal damage associated with urinary cadmium concentrations not exceed 5 μg Cd/g creatinine; likewise β_2-microglobulin, retinol binding protein and albumin are considered good biomarkers of renal dysfunction by cadmium exposure (Kakkar and Jaffery 2005; Omarova and Phillips 2007).

The absorption of cadmium generally depends on the route of entry. Approximately 3–10 % of ingested cadmium is absorbed from the gastrointestinal system, whereas 50 % of inhaled cadmium is absorbed; therefore the pulmonary absorption of cadmium is relatively more efficient than the efficiency of gastrointestinal absorption (Zalups and Ahmad 2003). Although absorption via the gastrointestinal tract is less efficient, low dietary intakes of calcium, protein, zinc, iron and copper may increase cadmium absorption in the gut. Following exposure, cadmium is absorbed by the intestines and subsequently delivered to the liver by portal blood. In the liver, cadmium is taken up avidly from the sinusoidal blood by hepatocytes (ATSDR 2008; Liu et al. 2007; Zalups and Ahmad 2003). Albumin-cadmium complex taken up by hepatocytes are degraded in the lysosomes and the free cadmium release into the cytoplasm where it rapidly binds intracellular proteins via sulfhydryl groups including glutathione (Ikediobi et al. 2004; Leverrier et al. 2007) or the high affinity metal binding protein metallothionein to form cadmium-metallothionein complex; such as is release into hepatic circulation and this cadmium-metallothionein complex is delivered via systemic circulation to the kidneys where are affected adversely primarily renal tubular cells (Zalups and Ahmad 2003). The organs that store cadmium include the liver, kidney, testis, spleen, heart, lungs, thymus, salivary glands, epididymis, and prostate; this retention of cadmium is MT-dependent, cadmium is mainly accumulated in kidney and liver, due to their high MT concentrations and the low excretion rate of cadmium result long biological half life (Liu et al. 1998; Järup et al. 1998; Klaassen et al. 2009).

6.2.3 Mechanism of Neurotoxicity

It has been suggested that the mechanism of cadmium toxicity involves production of reactive oxygen species and free radicals. Mitochondrion is an important target of metal toxicity of cadmium, is associated with accumulation of this metal, inhibit the respiratory chain reaction, production of reactive oxygen species and the release of mitochondrial calcium (Dorta et al. 2003; Gobe and Crane 2010). It has been proposed that cadmium initially binds to protein thiols (natural anti-oxidant enzymes like superoxide dismutase, catalase and glutathione peroxidase) in mitochondrial membrane inhibiting electrons transference and oxidative phosphorilation with the decrease of ATP levels, affects mitochondrial permeability transition, leading to the breakdown of the mitochondrial potentials (Thévenod 2009; López et al. 2006).

Excessive reactive oxygen species production can generate to free radical attack of double bonds in membrane lipids and result in an increase in lipid peroxidation(LPO) that is a major consequence of cadmium induced oxidative stress (Valko et al. 2006; Manca et al. 1991; Stohs et al. 2001); itself is unable to generate free radicals directly, however, indirectly generates several reactive species of both oxygen as superoxide anion ($O^{2}\cdot-$), hydrogen peroxide (H_2O_2), hydroxy radical (·OH) and lipid radicals (·L), by replacing *Fe* and *Cu* in various cytoplasmic and membrane proteins and thus, increasing the amount of unbound free or chelated copper and iron ions which then participate in oxidative stress via Fenton reactions (Valko et al. 2006; Jomova and Valko 2011; Franco et al. 2009). Moreover, has been reported that the alterations in the activities (reduction or inhibition) of antioxidant enzymes such as Cu, Zn-SOD, catalase, glutathione peroxidase, glutathione reductase and glutathione-S-transferase should be considered as a significant event in the generation of free radicals (Antonio et al. 2003; El-Demerdash et al. 2004; El-Sokkary and Awadalla 2011; Shagirtha et al. 2011). Metallothionein (MT) is another protein involved in the protection against of oxidative injury, this protein is known to free-radical scavenging properties, maintaining zinc homeostasis and detoxification of heavy metals mainly cadmium; there are four isoforms of MTs (I,II,III,IV), MT I-II isoforms are expressed throughout the brain and spinal cord in protoplasmic and fibrous astrocytes whereas MT-III isoform is predominantly expressed in neurons (Sato and Kondoh 2002; Klaassen et al. 1999; Hidalgo et al. 2001).

On the other hand, experimental evidence has shown that cellular targets of cadmium including intracellular mobilization of second messengers and mainly calcium. Cadmium enters cells through membrane transporters involved in the uptake of metals such as Fe and Zn as well as Ca^{2+} channels (Zalpus and Ahmad 2003) acts as a competitive ion to Ca^{2+}, interact with calcium in ways as well: inhibition of plasma membrane calcium channels, inhibition of Ca2+-ATPases of the sarcoplasmic reticulum, and blocker of the Ca^{2+}-dependent neurotransmitter release (Méndez-Armenta and Rios 2007). Alterations in the metabolism of neurotransmitter have also been implicated in cadmium neurotoxicity, several authors has been reported that content of several neurotransmitters (glutamate, aspartate, glycine, gamma-aminobutyric acid, GABA and/or taurine) can be increased/decreased depending on brain region or developmental age in animals exposed to cadmium affecting the excitation/inhibition balance of the synaptic neurotransmission (Gutierrez-Reyes et al. 1998; Lafuente et al. 2001, 2003; Antonio et al. 2003; Esquifino et al. 2001; Minami et al. 2001).

Cadmium is able induces apoptosis/necrosis through the mitochondrial pathway and/or ROS formation (Lόpez et al. 2006); interfere with DNA repair processes (Waisberg et al. 2003) and enhance genotoxic damage mainly via an increase in oxidative stress by an attack of ROS that is considered to be one of the major sources of endogenous damage to DNA. Cadmium is both rodents and human carcinogen and has been classified as a human carcinogen by the International Agency for Research on Cancer (IARC) and the National Toxicology Program (IARC 1993). Recent controlled epidemiological case studies, which have demonstrated an association between occupational and/or environmental Cd exposure and lung, bladder and renal cancer (Schwartz and Reis 2000; Joseph 2009; Schwerdtle et al. 2010) (Fig. 6.1).

6.2.4 Morphological Changes on Central Nervous System

Although cadmium-induced neuropathological changes have not been clearly demonstrated in human, it has been observed in experimental studies. Some studies have shown morphological changes such as shrinkage and deeply stained neurons in the parietal cortex and hippocampus of adult rats exposed during 40 days to cadmium (El-Sokkary and Awadalla 2011); moreover marked gliosis, nuclear pyknosis, spongiform necrosis and lymphocytic inflammatory infiltrates and hemorrhage in Virchow space also were observed in rats exposed to cadmium (Shagirtha et al. 2011; Kaoud et al. 2010). In adult rats, high doses of cadmium produce damage in the brain cortex, cerebellum, caudate nuclei and putamen with extensive necrosis and hemorrhage; in the necrotic areas of the brain cortex and cerebellum, the endothelium of some blood vessels was slightly damaged (Gabbiani et al. 1967a; Wong and Klaassen 1982) and according to those authors, the CNS lesions produced by cadmium are located primarily in the white matter (Gabbiani et al. 1967b). The administration of cadmium initially affects the integrity and permeability of the vascular endothelium that is sensitive to cadmium-induced damage and the necrotic changes in nerve cells are only secondary to this effect, resulting in edema and interference with oxygen and nutrient uptake in the brain, mainly in developing rats (Wong and Klaassen 1982; Gwalteney-Brant 2002; Méndez-Armenta et al. 2001, 2003). Developing animals are most affected by cadmium exposition; in experimental studies developing rats have shown neuronal pyknosis, interstitial oedema, destroyed fibers and alteration of Purkinje cells mainly in parietal cortex, striatum, and cerebellum (Méndez-Armenta et al. 2001). The ultrastructural study shown on optic nerves of rats exposed to cadmium swollen mitochondria with disrupted cristae, dissolution of microtubule and axonal swelling (Fern et al. 1996). These studies have suggested that the age of the animals is very

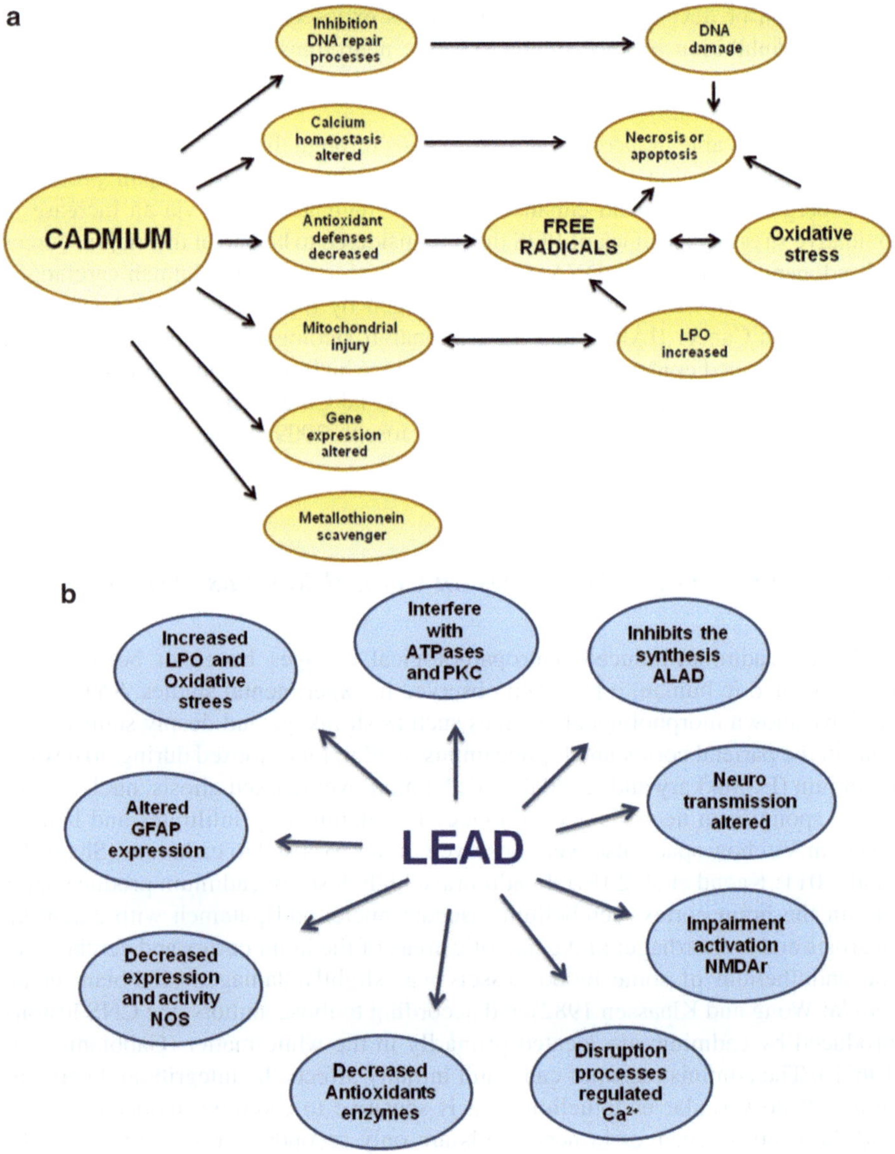

Fig. 6.1 A schematic representation summarizing the proposed pathways involved neurotoxicity of cadmium (**a**), lead (**b**) and thallium (**c**). *LPO* lipid peroxidation, *PKC* Protein Kinasa C, *NMDAr* N-Methyl-D-Aspartate receptor, *GFAP* Glial Fibrilar Acidic Protein, *ALAD* δ-aminolevulinic acid dehydratase, *NOS* Nitric Oxide Synthase, *ROS* Reactive Oxidative Species, *GPx* Glutathione Peroxidase, *SOD* Super Oxide Dismutase, Ca^{2+}; calcium

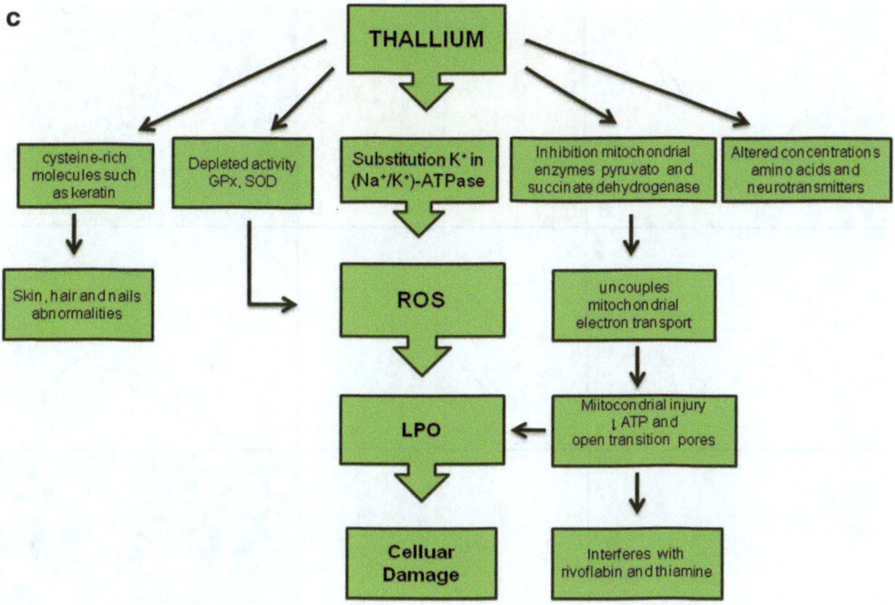

Fig. 6.1 (continued)

important to make them susceptible to the neurotoxic effects of cadmium due to the immaturity of the blood–brain barrier. Minor histopathological reports exist that document neurological effects of human exposure to cadmium. Autopsy findings of a 2-year-old boy, who died from sudden accidental showed marked cerebral swelling with herniation and histological evidence of marked cerebral edema with perivascular protein leakage, indicating blood–brain barrier disruption (Provias et al. 1994). Our results have shown minimal alteration of neuronal cells and neuropil in parietal cortex, hippocampus with pyramidal cells and Purkije cells of the cerebellum are observed (Fig. 6.2a–c respectively) in adult rats exposed to low cadmium doses. However, several changes are found in brain regions of rats following perinatal exposure to cadmium. Figure 6.2d, shown several eosinophilic neurons with pyknotic nuclei (ischemic cell damage), with slight interstitial oedema on the cerebral parenchyma in parietal cortex; similar findings were observed in cerebellum tissue, eosinophilic Purkinje cells with atrophy cellular (Fig. 6.2e) and interstitial oedema in the molecular layer (Fig. 6.2f). On the other hand, microscopic study of peripheral nerve of rats exposed subchronically to low doses of cadmium shown cells Schwann nuclei between nerve fibers (axon and myelin sheaths) and immunoreactivity myelin basic protein with the normal appearance of in peripheral nerve (Fig. 6.2g, h). However, observations of semithin sections of sciatic nerve showed myelinated and amyelinated axons with an incipient disorder from the interlamellar space (Fig. 6.2i).

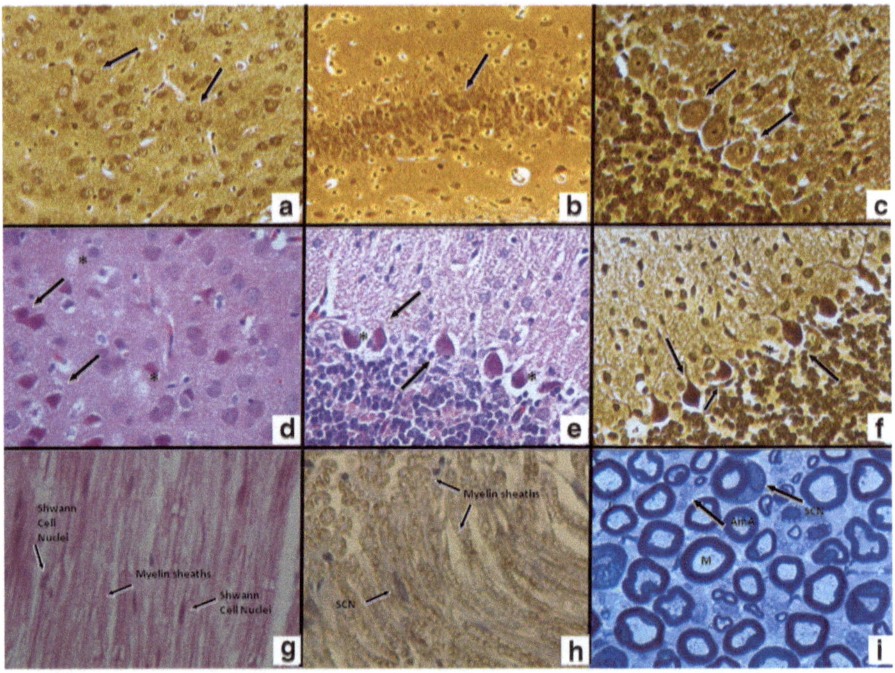

Fig. 6.2 Parietal Cortex (**a**) hippocampus (**b**) and Cerebellum (**c**) of adult rat exposed to cadmium. Parietal cortex (**d**) and cerebellum (**e** and **f**) of rat exposed to cadmium perinatally. Peripheral nerve of rats exposed chronically to cadmium (**g**, **h** and **i**). Neuronal cells and neuropil in parietal cortex, hippocampus and Purkije cells of the cerebellum are observed with minimal alterations (**a**, **b**, **c**). Several eosinophilic neurons with pyknotic nuclei (↑) and slight interstitial oedema (*) on the cerebral parenchyma in parietal cortex are observed (**d**). Cerebellum shows eosinophilic Purkinje cells with atrophy cellular and interstitial oedema in the molecular layer (**e** and **f**). Peripheral nerve show cells Schwann nuclei between axon and myelin sheaths (**g**) and immunoreactivity myelin basic protein with normal appearance (**h**). Semi-thin sections of peripheral nerve showed myelinated (**m**) and amyelinated axons (AmA) with an incipient disorder from the (↑) interlamellar space (**i**). H&E stain (**d**, **e** and **g**), Silver Stain, Rio–Hortega's method (**a**, **b**, **c** and **f**). Immunohistochemical stains for Myelin antibody (**h**), Toluidine blue stain (**i**)

6.3 Lead

6.3.1 Sources and Exposure

Lead is one of the most ubiquitous and it is detectable in practically all phases of the inert environment and in all biological systems. Human exposure to lead is common and results from the many uses of this metal due to its exceptional properties, such as ductility, high density, low chemical reactivity, easy extraction, and low cost, have made it broadly used since prehistory. Environmental levels of lead have increased more than 1,000-fold over the past three centuries as a result of human activity; the greatest increase occurred between the years 1950 and 1920. Lead is a naturally occurring element; it is a member of Group 14 (IVA) of the periodic table, is a bluish-gray heavy metal and it is usually found combined with two or more other elements to form lead compounds (ATSDR 2007).

The greatest potential for human exposure to lead arises from its previous use as an additive in gasoline, which resulted in its widespread dispersal throughout the environment, and its use as a pigment in both interior and exterior paints; in contrast, occupational exposure to lead occurs in workers employed in the lead smelting and refining industries, battery, manufacturing plants, plastics, printing industries and pigments to glaze ceramics. It is estimated that, between 0.5 and 1.5 million, workers are exposed to lead each year in his/her workplaces (ATSDR 2007). Pediatric lead poisoning is still an important public health problem for millions of children in the world. In South and Central America, 33–34 % of children have blood lead levels above 10 μg/dL (0.48 μmol/L) as compared with 7 % in North America. Another survey showed that one in every 20 children in the US has toxic blood levels. In China, because up to 23 % of the populations (377 million) are children, lead exposure is still a serious public concern (Wang and Zhang 2006; Jones et al. 2009). Eating, inhaling and coming into dermal contact with older paints containing lead are also common routes of exposure (Bridges and Zalpus 2005). Over the last few decades, however, lead emissions in developed countries have decreased markedly due to the introduction of unleaded gasoline (Järup 2003); despite efforts to reduce exposure through regulation, excessive lead exposure still persists.

Blood lead levels were revised in the general population over the past three decades and the results demonstrated the adverse effects of lead on child neurodevelopment; low level lead exposure can significantly induce motor dysfunctions and cognitive impairment in pediatric, especially if the exposures occur before the age of six. Lanphear et al. (2000) observed cognitive effects in children aged 6–16 years with blood lead levels <5 μg/dL in USA. Two other cross-sectional studies, in Detroit and Mexico City evaluated children at age 7 years and found an inverse relationship with BLLs <10 mg/dL and cognitive development (Chiodo et al. 2004; Kordas et al. 2006); consequently, the levels of lead considered hazardous were reduced to 40 μg/dL in 1971, 30 μg/dL in 1975, 10 μg/dL in 1991 (Bellinger and Bellinger 2006) and 1–2 μg/dL in 2004 (EPA 2006); however, lead is still a significant public health concern.

6.3.2 General Toxicology

Lead may enter the body through the intestines by ingestion; through the lungs by inhalation; through the skin by adsorption; or by direct swallowing and ingestion. The metal is absorbed into and transported by the bloodstream to other tissues. Once absorbed, is bound to red blood cells and transported by the bloodstream throughout tissues; lead is accumulated in high concentrations in bone, teeth, liver, lung, kidney, brain, and spleen, and it goes through the blood–brain barrier and the placenta (Goyer and Clarkson 2001; Gwalteney-Brant 2002). The biological half-life of lead may be considerably longer in children than in adults. Lead in blood has an estimated half-life of 35 days, in soft tissue 40 days and in bones 20–30 years (Papanikolaou et al. 2005). The blood lead level indicates a recent exposure, whereas bone lead level, which forms 90–95 % of lead burden in adults and 80–95 % of total lead in children indicate a chronic exposure (Kakkar and Jaffery 2005). The major route of excretion of absorbed lead is the urinary tract, usually with glomerular filtrate in the kidney; it can also be excreted with bile through the gastrointestinal tract (Goyer and Clarkson 2001). Lead has been shown to have serious effects in the skeletal, renal, hematopoietic, endocrine, male and female reproductive systems (Papanikolaou et al. 2005; ATSDR 2007). The nervous system is more susceptible to its effects particularly, the children and lead poisoning are still considered a primary environmental hazard to children (Goyer and Clarkson 2001; White et al. 2007). Epidemiological studies have shown that lead exposure during pregnancy is associated with reduced growth during infancy and early childhood as well as impair the cognitive development of the infant; the placenta cannot prevent the fetus from exposure to lead, as this metal is able to cross the placental barrier easily placenta-umbilical cord and can accumulate in fetal tissues including the brain (Osman et al. 2000; Villeda-Hernandez et al. 2001; Lafond et al. 2004; Benitez et al. 2009).

Acute lead exposure in adults can cause clinical early symptoms such as nausea, abdominal pain, irritability, headache, loss of memory, and hallucinations and may develop within weeks of initial exposure. Chronic lead poisoning in humans is most common that the acute form and affects gastrointestinal, neuromuscular, renal and haematological systems (ATSDR 2007). Neurological symptoms such as dullness, irritability, poor attention span, lethargy, fatigue, epigastric, constipation, vomiting, convulsions, paresthesia, and coma can be observed following prolonged exposure (Jarup 2003; ATSDR; 2007). Likewise, peripheral neuropathy is a classic manifestation reported by effect of chronic lead exposure, somatosensory evoked potentials as well as diminution of peripheral nerve function through conduction velocity of electrically stimulated nerves in the arm or leg of lead workers can be reported (Araki et al. 2000).

Although peripheral nerve damage and paralysis may still be reported in adult, the most common clinically toxicity is lead encephalopathy, coma, convulsions, and death occurring in children with very severe lead poisoning. Likewise, neurobehavioral and cognitive effects such as aggression, psychosis, confusion, memory and learning deficit, related to environmental long term lead exposure during neurodevelopment persist through childhood and adulthood (Bellinger and Needleman 2003; Bellinger and Bellinger 2006; Mazumdar et al. 2011). A number of studies provide convincing evidence that supporting the association between blood lead levels and intellectual impairment in children with deficit and cognitive development impairments, difficulties in learning and mental concentration, as well as reduced IQ scores, diminished school performance, deficits suggestive of Attention Deficit Disorder (ADD) at low levels of exposure (Canfield et al. 2003; Bellinger and Needleman

2003; Bellinger and Bellinger 2006; Mazumdar et al. 2011). Experimental studies have reported similar behavioral problems in lead-exposed primates and/or rats, demonstrating that deleterious effects such as distractibility, inability to inhibit inappropriate behavioral response and learning and/or memory impairment can be a consequence of developmental lead exposure (Rice 1993; Yang et al. 2003; Garcia-Arenas et al. 2004; EPA 2006).

6.3.3 Mechanism of Neurotoxicity

Lead can affect the nervous system through multiple pathways, several mechanisms have been proposed for lead toxicity: (1) the interaction of lead in the heme biosynthetic pathway, (2) potential mechanisms for lead induced oxidative stress; (3) the direct depletion of the level of available sulfhydryl antioxidant reserves and (4) interaction with calcium ion mimicking Ca^{2+} action and/or disrupting Ca^{2+} homeostasis.

Lead has high affinity for sulfhydryl groups and can inactivate enzymes, especially those involved in processes responsible for heme synthesis such as δ-aminolevulinic acid dehydratase (ALAD) that is the most sensitive enzyme to lead and ferrochelatase that is the last step of heme synthesis; the inhibition of ALAD lowers heme production and increases levels of the substrate delta-aminolevulinic acid (ALA), elevated levels of ALA induces ROS production (Flora et al. 2008; Flora 2009). ALA suppresses GABA-mediated neurotransmission by inhibiting its release and also possibly by competing with GABA at receptors (Anderson et al. 1996). Glutathione reductase (GR) is another specific sulfhydryl containing enzymes that are inhibited by lead further contributing to an impairment in pro-oxidant/antioxidant balance (Patrick 2006). Oxidative stress is considered an important molecular mechanism of lead neurotoxicity which occurs when generation of free radicals exceed the capacity of the antioxidant defense mechanism. Accumulation of ALA induced ROS generation in lead toxicity and leading a significant increase of lipoperoxidation (LPO) induced by H_2O_2 and iron and increases fluxes of O^{-2} and H_2O_2; it's due to direct interaction of lead with biological membranes; has been reported that neurons are particularly vulnerable to increases in ROS levels due to they have a reduced capacity to detoxify ROS (Antonio et al. 2003; Bennet et al. 2007) it is unclear whether oxidative stress is the cause or the consequence of the toxic effect of lead.

Lead exposure might induce changes in the activities of antioxidant enzymes, depressed levels of glutathione reductase, glutathione peroxidase, and glutathione-S-transferase correlated with depressed glutathione levels, this depletion of glutathione and the changes in the activities of various antioxidant enzymes indicative of lipid peroxidation have been implicated in lead-induced oxidative tissue damage (Ahamed and Siddiqui 2007). Several authors have reported that lead can induce significative decrease in the activity of superoxide dismutase (SOD) and catalase (CAT) in rat brains exposed to lead (Nehru and Kanwar 2004; Moreira et al. 2001).

Similar results were obtained by Wang et al. (2006), who found that the activity of SOD, Glutathione peroxidase (GPx) and reduced glutathione (GSH) were decreased significantly in 21-day-old rat brains exposed to lead during pregnancy.

On the other hand, one of the important mechanism of lead induced neurotoxicity is the interaction between lead and calcium, its due to that lead may gain entry into cells although one or more of the different types of calcium channels (Bridges and Zalups 2005; ATSDR 2007). Lead impairs normal calcium homeostasis and the two pathways that have received more attention are calmodulin and protein kinasa C (PKC), calcium induces a conformational change in calmodulin that converts the protein to its active form lead acts by displacing calcium ions bound to calmodulin; the activation of calmodulin by lead results altered production of Nitric Oxide (NO) through Nitric Oxide Synthase (NOS) pathway and protein phosphorylation altering cAMP messenger (Goyer and Clarkson 2001; Nava-Ruíz et al. 2012). Lead can alter the activity and expression of nNOS and eNOS in different brain regions primarily because lead can mimic calcium at binding sites, this binding may prevent the accessibility of calcium to NOS, thus leading to the decreased activity of nNOS/eNOS (Selvin-Testa et al. 1997; Nava-Ruíz et al. 2012); the possible deleterious effect of lead might be related to an interference that significantly affects the induction of Long Term Potentiation (LTP) in the hippocampus, mediated by NO regulation impairment the learning and memory processes (White et al. 2007; Toscano and Guilarte 2005).

Glutamate is the major excitatory neurotransmitter in the brain, can activate both ionotropic a-amino-3-hydroxy-5-methyl-4-isoxazolepropionic acid(AMPA) and metabotropic receptors, N-methyl-D-aspartate receptor (NMDAr) which upon activation open their associated ion channel to allow the influx of Ca^{2+} and Na^{+} ions. It is well known that activation the NMDAr plays an essential role in hippocampus mediated the learning and memory; the major cellular mechanism within the hippocampus believed to be responsible for acquisition of new memories is LTP and can be disrupted by inhibition of the NMDAr. Thus, impairment of the NMDAR has been shown to produce learning deficits on both the behavioral and cellular level (Toscano and Guilarte 2005). These functions are mediated by calcium entry through the NMDA; calcium Ca^{2+}, activates protein kinase A (PKA), mitogen-activated protein kinase (MAPK) and calcium/calmodulin-dependent protein kinase (CAMK) pathways, which converge at the cyclic-AMP-response element-binding protein (CREB) (Toscano and Guilarte 2005). The over-activation of NMDA receptors triggers an excessive entry of Ca^{2+}, initiating a series of cytoplasmic and nuclear processes that promote neuronal cell death. Lead is a potent, non-competitive antagonist of the NMDAr and has been shown to impair hippocampus-mediated learning in animal models of lead exposure the cognitive effects of Pb^{2+} exposure are believed to be mediated through its selective inhibition of the NMDAr, (Toscano and Guilarte 2005). Lead exposure during synaptogenesis in hippocampal neurons altered synaptic NMDAr composition and synapse development (Neala et al. 2011). Likewise, calcium disruption by lead affects release, synthesis, and storage of some neurotransmitters, impair the neurotransmission.

6.3.4 Morphologic Changes on Central Nervous System

Lead neurotoxicity occurs when the exposure to lead alters the normal activity of the CNS and causes morphological damage to the CNS. Neurotoxic actions of lead include apoptosis in hippocampal cells (Sharifi et al. 2002) and apoptotic neurodegeneration in the neonatal mouse brain on acute lead exposure (Dribben et al. 2011). Probability the primary lesion in acute lead encephalopathy is a breakdown of the blood-brain barrier (BBB) which does not fully develop until the middle of the first year of life (Costa et al. 2004). The endothelial cells are the first to be exposed to lead passage into the brain, are especially rich in mitochondria, making the BBB susceptible to lead poisoning. The capillaries may be dilated, narrowed necrotic or trombosed with consequent extravasation of fluid disruption develop in severe cerebral oedema and loss of neuronal cells (Gwalteney-Brant 2002; ATSDR 2007). Other neuropathological lesions by chronic lead intoxication include cytoplasmic vacuolization, hyperchromatic cells, chromatolysis, interstitial oedema, gliosis, in the prefrontal cortex, hippocampus, striatum and thalamus (Villeda-Hernandez et al. 2006) loss of pigmented neurons and the appearance of small shrunken cells (Moneim et al. 2011);degeneration and necrosis of cerebellar Purkije cells, hipommyelinitation and demyelination of nerve fibers (Soltaninejad et al. 2003). Ultrastructurally, Pb causes alterations in mitochondria, Golgi apparatus and increment of gliofilaments in astrocytes (Struzyñska et al. 2001). Exposition of lead during early life alters granule cell neurogenesis and morphology manifested by reductions in the length density of apical dendrites of granule cells in the dentate girus (Verina et al. 2007).

Animal studies reveal morphological changes in the oligodendroglia with a significant decrease in the formation of myelin, particularly during late gestational development and the postnatal period, in both the peripheral and the central nervous system in rats exposed to lead (Dabrowska-Bouta et al. 1999, 2008). The lead exposure causes commonly astrogliosis with an increases in the synthesis and concentration of glial protein fibrillary acidic protein (GFAP), that has been suggested as an early indicator of neurotoxic insult; animal studies have shown that the expression, synthesis, and concentration of GFAP, as well as the number of astrocytes, were increased in specific brain regions of rats exposed to lead and the most marked long-lasting effects occurred when the rats were exposed during the developmental period (Van Den Berg et al. 1996; Tiffany-Castiglioni and Qian 2001; Struzyñska et al. 2001). Peripheral neuropathy is a classic manifestation of lead toxicity; several humans and animal studies have been shown segmental demyelinitation, axonal degeneration, cell degeneration with swelling cytoplasm and hypertrophy Schwann cells in peripheral nerves (Gwalteney-Brant 2002; Goyer and Clarkson 2001; ATSDR 2007).

In our laboratory, we have observed in adult rats exposed acute to high doses of lead, eosinophilic neurons with slight interstitial oedema on the brain parenchyma (Fig. 6.3c, f), pyknotic nuclei (ischemic cell damage) of pyramidal neurons displayed highly condensed nuclei and eosinophilic cytoplasm located in the large

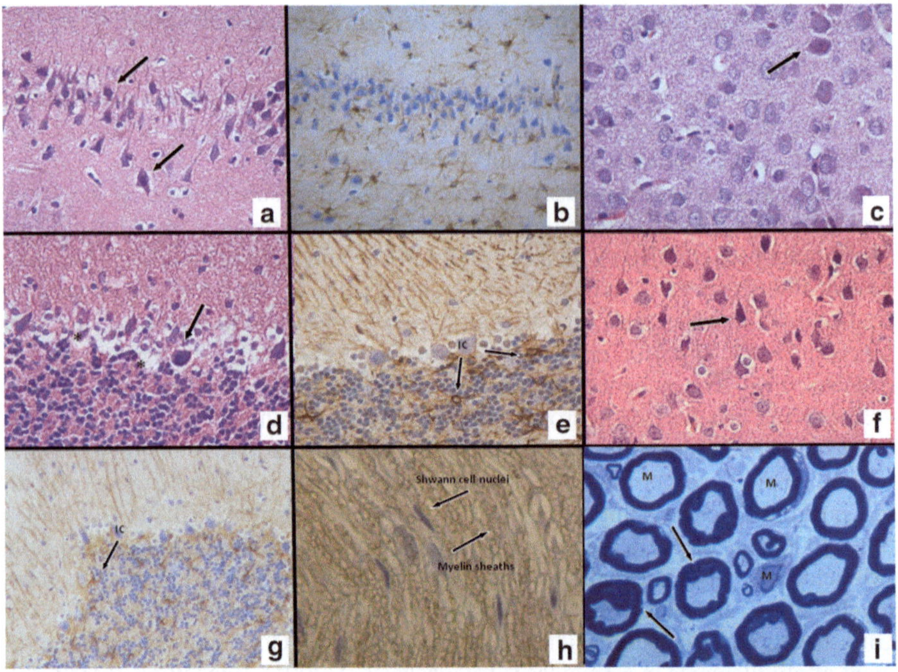

Fig. 6.3 Hippocampus (**a** and **b**), parietal cortex (**c** and **f**) and cerebellum (**d** and **e**) of adult rat exposed to acute to high doses of lead. Cerebellum of rat fetuses exposed to lead acetate during total gestation (**g**). Peripheral nerve of rats exposed chronically to lead (**h** and **i**). Pyknotic nuclei with eosinophilic cytoplasm of large pyramidal neurons (↑) (**a**); and eosinophilic neurons with slight interstitial oedema on the brain parenchyma (**c** and **f**) were observed. Cell death and dendritic breakdown are shown in Purkinje cells (**d**). An increase in characteristic staining pattern for GFAP in hippocampus (**b**) and cerebellum regions was observed (↑) (**e**). Similar patterns of immunoreactivity staining for GFAP was observed in the cerebellum of rat fetuses (**g**). Peripheral nerve show cells Schwann nuclei with vacuolization and edematous damage of myelin (**h**). In semi-thin sections myelinated axons (**m**) can be observed with separation of the myelin lamellae (↑) and irregular profile (protrusion) (↑). H&E stain (**a**, **c**, **d** and **f**), immunohistochemical stains for GFAP antibody (**b**, **e** and **g**) and Myelin antibody (**h**), Toluidine blue stain (**i**)

pyramidal neurons of the hippocampus with severe interstitial oedema (Fig. 6.3a), similar results include neuronal cell death and ultimately in axonal and dendritic breakdown were observed in Purkinje cells in cerebellum tissue (Fig. 6.3d). In lead encephalopathy gliosis may occur, the astrocytes can increase the number of cytoplasmic filaments (chiefly glial fibrillary acidic protein, GFAP) altered arrangement of cellular processes and it's has been suggested as an early indicator of neurotoxic insult. We found an increase in the characteristic staining pattern immunoreactivity for GFAP in the parietal cortex (data not shown), large pyramidal neurons of the hippocampus (Fig. 6.3b) and cerebellum of brain adult rats to high acute lead exposure (Fig. 6.3e); the similar immunoreactivity pattern was observed in the cerebellum of rat fetuses exposed prenatally to high doses of lead (Fig. 6.3g). Lead neuropathy in humans and experimental animals is manifested by Wallerian axonal degeneration and segmental demyelination, affecting primarily motor nerves. We observed immunohistochemical alteration with the myelin basic protein antibody (that may be used to assess myelin integrity) and edematous lesions characterized by vacuolization; while that cells Schwann nuclei between nerve fibers (axon and myelin sheaths) shown normal appearance (Fig. 6.3h). Semi-thin sections of sciatic nerve of rats exposed chronically to lead shown damage myelin with separation of the myelin lamellae (intramyelinic edema) with irregular profile (protrusion into the axoplasma) and loss of myelin sheaths (Fig. 6.3i).

6.4 Thallium

6.4.1 Sources and Exposure

Thallium is a naturally occurring trace element, widely distributed in the earth's crust. Pure thallium exists in nature but is usually found combined with other elements in inorganic compounds. Thallium forms compounds in both the monovalent and trivalent states; however, the monovalent state is the most stable; it's a highly toxic heavy, in the form of tasteless, odourless and colourless salts (Nriagu 1998). Thallium exists in the environment mainly combined with other elements (primarily oxygen, sulfur, and the halogens) in inorganic compounds. Release of thallium into the environment may occur as the result of human activities, with the emission of vapor and dust, liquid and solid waste; mainly from coal-burning power plants, cement factories, and ferrous and nonferrous smelting operations in which thallium is a trace contaminant of the raw materials, rather than from facilities producing or using thallium compounds; monitoring data indicate elevated thallium concentrations near industrial and commercial sources and hazardous waste sites (Nriagu 1998; John Peter and Viraraghavan 2005; Cvjetko et al 2010).

Humans may be exposed to thallium by ingestion, inhalation, or dermal absorption. However, the general population is exposed more frequently by ingestion of thallium-containing foods, especially homegrown fruits and green vegetables. Inhalation of contaminated air near emission sources or in the workplace may also contribute to thallium exposure of some individuals (ATSDR 1999). Occupational exposure to thallium may be significant for workers in smelters, power plants, cement factories, and other industries that produce or use thallium compounds or alloys (John Peter and Viraraghavan 2005; Cvjetko et al. 2010). Thallium has also been employed for clinical purposes as a depilatory as well as in the treatment of tuberculosis, syphilis and gonorrhea (Kazantzis 2000). The therapeutic use of these salts has been abandoned, but the hazards of an accidental ingestion of thallium-containing rodenticides, or its use in homicidal or suicidal ingestion, occur occasionally (Kazantzis 2000; Tsai et al. 2006). Likewise, thallium salts are still widely used as rodenticides and insecticides in some countries, despite the World Health Organization recommendation against its use in 1973 (Galvan-Arzate and Santamaria 1998).

6.4.2 General Toxicology

Thallium produces one of the most complex and severe toxicity, involving a wide range of the organs and tissues. Human exposure may take different routes; orally, by digestion of contaminated food, dermal or respiratory by inhalation of dust and fumes (Cvjetko et al. 2010). The high degree of absorption, high bioavailability, and slow excretion contribute to the high potential for toxicity of thallium such as is distributed throughout the body (Repetto et al. 1998). Thallium levels in urine, blood, and hair have been used as indications of exposure to thallium and the determination of thallium in urine has been the most widely used of biological indicators of thallium exposure (ATSDR 1999; Mulkey and Oehme 1993). Once it enters the body, thallium is absorbed through the skin and mucous membranes, is widely distributed throughout the body and accumulates in the bones, renal medulla, liver, heart, muscle, lung and, eventually in the central nervous system (Leonard and Gerberr 1997; John Peter and Viraraghavan 2005). Thallium can cross the placental barrier and is present in breast milk (Hoffman 2000) as well as can cross the blood-brain barrier and deposit in the brain (Galván-Arzate et al. 2000). Thallium is slowly excreted through the bile, urine and feces and may persist for many weeks or even months (Repetto et al. 1998). The biological half-life of thallium is estimated at approximately 10 days but up to 30 days in humans (Cvjetko et al 2010).

Thallium compounds can affect the respiratory, cardiovascular, and gastrointestinal systems, liver, kidneys, the male reproductive system, peripheral and central nervous systems; temporary hair loss has also been associated with ingestion of thallium in humans. Clinical symptoms of thallium poisoning can be classified as acute, subchronic and chronic depending on the time and level of exposure, the rate of absorption, age and individual susceptibility (Repetto et al. 1998). In humans, a triad of gastroenteritis, polyneuropathy, and alopecia is regarded as the classic syndrome of thallium intoxication (Tabandeh et al. 1994). Other symptoms include polyneuritis, encephalopathy, tachycardia, degenerative changes of heart, liver, kidney and, the characteristic dermatological such as alopecia, hyperkeratosis, and Mees' lines in nails (Mulkey and Oehme 1993; Tromme et al. 1998; Gwalteney-Brant 2002); while that major symptom of thallium poisoning includes anorexia, headache, pains in abdomen, paresthesia and hyperesthesia of extremity, ataxia, muscle weakness, cranial nerve palsies, convulsion, coma and even death may be caused (Van Kesteren 1994; ATSDR 1999; Tsai et al. 2006). Subacute and chronic intoxication is associated with cumulative thallium, in contrast no symptoms are observed in chronic exposure and the thallium intoxication may not be suspected until alopecia occurs (Repetto et al. 1998). After exposure for weeks or months, ataxia and paresthesia may be the outstanding symptoms; in time, the paresthesia may progress to evident peripheral neuropathy with weakness and atrophy of the associated musculature (Galván-Arzate and Santamaria 1998). Mental disturbances include poor attention, anxiety, depression, altered consciousness, hallucination, and paranoid syndromes (Tsai et al. 2006), as well as nerve conduction studies have shown abnormalities consistent with peripheral neuropathy with changes in somatosensory action potential, atrophy of the optic nerve with ophtalmoplegia and nystagmus (Limos et al. 1982; Tsai et al. 2006; Cvjetko et al. 2010).

6.4.3 Mechanism of Neurotoxicity

The exact mechanism that mediates thallium toxicity is unclear. However, the well-known cellular mechanisms of thallium toxicity include the substitution of potassium in many vital potassium-dependent processes due to the similarity of the ionics radius, as well as a high affinity for the sulfhydril or thiol group of mitochondrial membranes, inhibition of cellular respiration, interaction with B vitamins (especially rivoflabin and thiamine), alteration of intracellular calcium levels and increased ROS formation, and high levels of lipid peroxidation (Mulkey and Oehme 1993).

The substitution of potassium in the (Na^{+}/K^{+})-ATPase is due to that both univalent ions have similar ionic radii and the remarkable inability of cell membranes to distinguish between thallium and potassium (Mulkey and Oehme 1993; Galvan-Arzate and Santamaria 1998), thallium binds Na^{+}/k^{+}-ATAase with tenfold higher affinity than potassium, this leads to accumulation of thallium within the cell at the expense of potassium; substitution of thallium for potassium alters nerve cell excitability and muscle fiber contractibility (Repetto et al. 1998).

Another possible mechanism of thallium toxicity is its high capacity binds sulfhydril groups, from proteins and other enzymes, such as hydrolases, oxidoreductases, and transferases leading to a generalized poisoning (Ramsden 2002). Thallium further interferes with cellular respiration by binding sulfhydryl groups on mitochondrial enzymes such as pyruvato dehydrogenase and succinate dehydrogenase; and cysteine-rich molecules such as keratin, thallium increase the solubility of keratin and decrease its resistance to stretching-these alterations are manifested clinically as skin, hair and nail abnormalities (Gwalteney-Brant 2002). As mentioned above, it has been proposed that thallium uncouples mitochondrial electron transport, so inducing loss of mitochondrial transmembranal potential. This, in turn, increases the oxidant content in mitochondria with subsequent release of cytochrome C which impairs mitochondrial functioning, decreasing ATP production by reduction of riboflavin and a depletion of flavine and opening the transition pores leading uncoupling of the respiratory chain (Hanzel and Verstraeten 2006; Puga and Verstraeten 2008).

On the other hand, thallium affects metabolism of non-protein sulfhydryls such as glutathione which plays a crucial role in heavy metal toxicity; the reduction of glutathione levels as well as a depleted glutathione peroxidase (Hanzel et al. 2005) and super oxide dismutase (SOD) activity resulting in a increase in free radicals and lipid peroxidation induced damage to cellular membranes (Galvan-Arzate et al. 2005). Some groups have reported that thallium increase lipid peroxidation in various tissues and brain regions (Hasan and Ali 1981; Hanzel and Verstraeten 2006; Galvan-Arzate et al. 2000, 2005), as well as changes associated with the concentrations of amino acids(glutamic acid, aspartic acid) and other neurotransmitters (glutamic acid, dopamine, serotonin) in various regions of the rat brain (Galvan-Arzate and Santamaria 1998).

6.4.4 Morphologic Changes on Central Nervous System

Pathological studies in intoxicated humans and animals have revealed damage in brain and peripheral nerves. Neuropathological studies of the brain are limited and reveal edematous changes and vascular engorgement in the cerebral hemispheres, and chromatolytic changes in the motor cortex, globus pallidus, substantia nigra, and brainstem nuclei (Davies et al. 1981) as well as changes in the brain capillaries, cerebellar edema with pyknotic Purkinje cells and focal areas of necrosis also has been reported (Barroso-Moguel et al. 1994; Gwalteney-Brant 2002).

Pathological studies in intoxicated humans and animals have revealed focal axonal swelling and fragmentation; with swelling and degeneration of myelin sheaths, chromatolysis of the nerve cells, and secondary myelin loss of the peripheral nerve. Limos et al. (1982) used an electron microscope to study the effects of thallium intoxication in human muscle and peripheral nerve; they observed axons of myelinated fibers with markedly dilated myelin sheaths that showed the absence, severe alteration or disintegration of axoplasmic organelles; as well as swelling and vacuolation with complete loss of microtubules. Long peripheral nerve fibers, especially large sensory nerve tracts, appear to be affected to be affected preferentially.

Moreover, our results shown several neurons with hyperchromatic, shrunken perikaryon and corkscrew-shaped dendrite in cerebral parenchyma, as well as slight interstitial oedema in the neuropil (Fig. 6.4a, f); eosinophilic neurons with pyknotic nuclei (ischemic cell damage) in fascia dentate of hippocampal region (Fig. 6.4b, e) in adult rats exposed to acute thallium intoxication. In the cerebellar cortex, Purkinje cells with atrophic nuclei, loss of Purkije cells and interstitial oedema between molecular and granular layer has been observed (Fig. 6.4c, d). The expression of myelin in the peripheral nerve in thallium exposed rats shown Schwann nuclei between nerve fibers with myelin edematous lesions, characterized by marked extensive vacuolization (Fig. 6.4g, h). Likewise, observations of semi-thin sections of peripheral nerve showed myelinated and amyelinated axons with disorder from the interlamellar space, myelin condensate with protrusion into the axoplasma, segmental demyelinitation and several nerve fibers producing myelinopathy of rats treated with thallium (Fig. 6.4i).

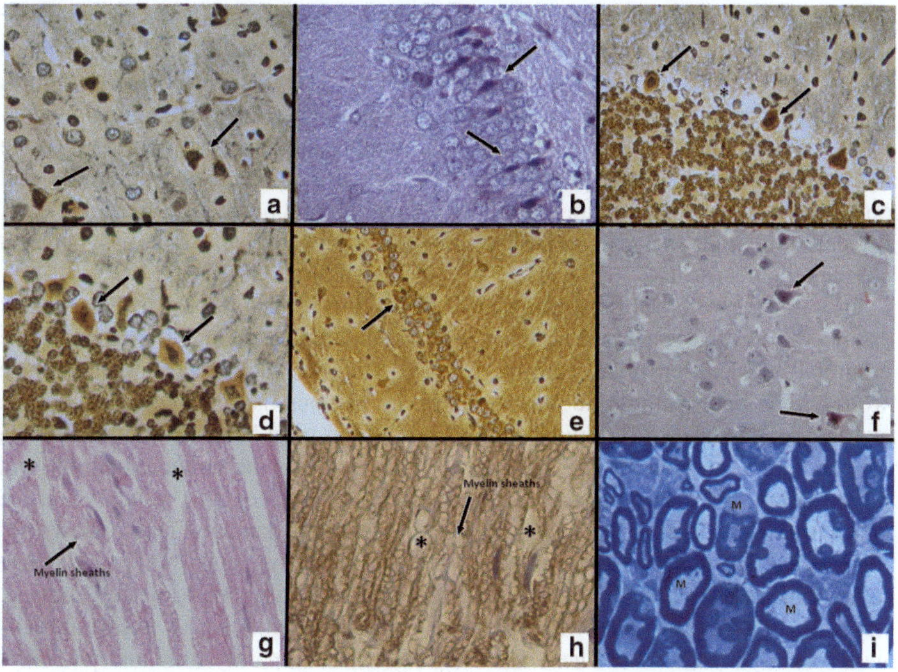

Fig. 6.4 Parietal cortex (**a** and **f**), hippocampus (**b** and **e**), cerebellum (**c** and **d**) of rats adult exposed to acute doses of thallium. Peripheral nerve of rat exposed chronically to low doses of thallium (**g**, **h** and **i**). Several neurons with hyperchromatic, shrunken perikaryon and corkscrew-shaped dendrite in the cerebral parenchyma with interstitial oedema were observed (**a** and **f**) as well as eosinophilic neurons with pyknotic nuclei in fascia dentate of hippocampal region (**b**). Purkinje cells with atrophic nuclei (↑), loss of Purkije cells and interstitial oedema (*) has been observed (**c** and **d**). Peripheral nerve show cells Schwann nuclei between axon and edematous myelin sheaths (**g**) and immunoreactivity stain myelin is altered by extensive vacuolization (↑) (**h**). Semi-thin sections showed myelinated axons (**m**) with alteration in the interlamellar structure, myelin condensate with protrusion (↑) and loss of myelin (**i**). H&E stain (**b**, **f** and **g**), Silver Stain, Rio–Hortega's method (**a**, **c**, **d** and **e**). Immunohistochemical stains for Myelin antibody (**h**), Toluidine blue stain (**i**)

6.5 Conclusion

Exposure of human populations to a wide variety of chemicals has generated concern about the potential neurotoxicity of new and existing chemicals or environmental agents. The complexity of the central nervous system is a key factor in its vulnerability to toxicant insult; the nerve cells or groups of cells greatly vary in morphology, function and characteristics from one region of the central nervous system to another. Neurotoxicology may be simply defined as the study of the adverse effects on the structural, biochemical and functional integrity of the nervous system; this effect that develops in the central nervous system or peripheral system or both following exposure to chemical agents during development or adulthood. Therefore, the objectives of neurotoxicology are the study of metabolism of neurotoxicants, identification their toxic effects, characterization of lesions and then correlate morphological changes with molecular, and biochemical mechanism. Experimental studies conducted in laboratory animals remain critical to the study of neurotoxicity. At present, morphological data are used commonly and correlate with neurochemical, electrophysiological, and behavioral findings; thus, the pathologist plays a critical role in the evaluation and determination whether a chemical is neurotoxic.

Presented here are three examples of heavy metals that access to the central/peripheral nervous systems: cadmium, lead and thallium. Although they share many common physical, chemical, and biochemical features, each heavy metal produce well-defined neuropathologic alterations in the nervous tissue. Studies in animals and human exposed to cadmium, lead and, thallium have been reported that these metals are able to increase the generation of free radical with an increment of lipid peroxidation, this increase may disrupt the balance between free radicals and the cell antioxidant defense leading induction of oxidative damage. This review provides some findings the neuropathology changes to heavy metals, such as: damage in neurons, interstitial oedema, pyknotic cells, chromatolysis, Purkinje cell damage, necrosis, cell degeneration with swelling cytoplasm, hypertrophy Schwann cells with loss myelin sheaths. These alterations are due to mainly time and pathways exposure as well as the differences toxicokinetics of several heavy metals. Finally, future research in the recognizing of the dynamics of pathological processes is needed, particularly in the central nervous system damage for complete understanding of the human health consequences to heavy metals exposure.

Despite the number of studies that have been carried out by different laboratories throughout the world in the last decades about the neurotoxicity of heavy metals is still necessary to explore further the molecular mechanisms of these metals. The use of new strategies such as confocal microscopy to determine the morphological alterations in several animal models both developing and adult animals should be correlated with these studies of molecular biology. All the efforts of researchers in different laboratories is important especially need to translate to the human condition in order for basic science to be most effective in benefiting human health.

References

Agency for Toxic Substance and Disease Registry (ATSDR) (1999) Toxicological profile for thallium. U.S. Department of Health and Humans Services, Public Health Service, Centers for Diseases Control, Atlanta

Agency for Toxic Substance and Disease Registry (ATSDR) (2007) Toxicological profile for lead. U.S. Department of Health and Humans Services, Public Health Service, Centers for Diseases Control, Atlanta

Agency for Toxic Substance and Disease Registry (ATSDR) (2008) Toxicological profile for cadmium. U.S. Department of Health and Humans Services, Public Health Service, Centers for Diseases Control, Atlanta

Ahamed M, Siddiqui MKJ (2007) Low level lead exposure and oxidative stress: current opinions. Clin Chim Acta 383(1–2):57–64

Anderson AC, Pueschel SM, Linakis JG (1996) Pathophysiology of lead poisoning. In: Pueschel SM, Linakis JG, Anderson AC (eds) Lead poisoning in children. P.H. Brookes, Baltimore, pp 75–96

Antonio MT, Corredor L, Leret ML (2003) Study of the activity of several brain enzymes like markers of the neurotoxicity induced by perinatal exposure to lead and/or cadmium. Toxicol Lett 143:331–340

Araki S, Sato H, Yokoyama K, Murata K (2000) Subclinical neurophysiological effects of lead: a review on peripheral, central, and autonomic nervous system effects in lead workers. Am J Ind Med 37(2):193–204

Barroso-Moguel R, Villeda-Hernández J, Méndez-Armenta M, Ríos C, Monroy-Noyola A (1994) Combined *D*-penicillamine and Prussian blue as antidotal treatment against thallotoxicosis in rats: evaluation of cerebellar lesions. Toxicology 89:15–24

Basun H, Lind B, Nordberg M, Nordström M, Björkstén KS, Winblad B (1994) Cadmium in blood in Alzheimer's disease and non-demented subjects: results from a population-based study. Biometals 7:130–134

Beijer K, Jernelov A (1986) Sources, transport and transformation of metals in the environment. In: Friberg L, Nordberg GF, Vouk VB (eds) Handbook on the toxicology of metals. General aspects, 2nd edn. Elsevier, Amsterdam

Bellinger CD, Bellinger MA (2006) Childhood lead poisoning: the torturous path from science to policy. J Clin Invest 116:853–957

Bellinger DC, Needleman HL (2003) Intelectual impairment and blood lead levels. N Engl J Med 349:500–502

Benitez MA, Mendez-Armenta M, Montes S, Rembao D, Sanin LH, Rios C (2009) Mother-fetus transference of lead and cadmium in rats: involvement of metallothionein. Histol Histopathol 24:1523–1530

Bennet C, Bettaiya R, Rajanna B, Baker L, Yallapragada PR, Brice JJ, White SL, Bokara KK (2007) Region specific increase in the antioxidant enzymes and lipid peroxidation products in the brain of rats exposed to lead. Free Radic Res 41(3):267–273

Bridges CC, Zalups RK (2005) Molecular and ionic mimicry and the transport to toxic metals. Toxicol Appl Pharmacol 204:274–308

Canfield RL, Kreher DA, Cornwell C, Henderson CR Jr (2003) Low level lead exposure executive functioning, and learning in early childhood. Child Neuropsychol 9(1):35–53

Castro-Gonzalez MI, Méndez-Armenta M (2008) Heavy metals: implications associated to fish consumption. Environ Toxicol Pharmacol 26:263–271

Chiodo LM, Jacobson SW, Jacobson JL (2004) Neurodevelopmental effects of postnatal lead exposure at very low levels. Neurotoxicol Teratol 26:359–371

Costa LG, Aschner M, Vitalone A, Syversen T, Soldin OP (2004) Developmental neuropathology of environmental agents. Annu Rev Pharmacol 44:87–110

Cupit M, Larsson O, de Meeûs C, Eduljee GH, Hutton M (2002) Assessment and management of risks arising from exposure to cadmium in fertilisers-II. Sci Tot Environ 291:189–206

Cvjetko P, Cvjetko I, Pavlica M (2010) Thallium toxicity in humans. Arh Hig Rada Toksikol 61:111–119

Dabrowska-Bouta B, Sulkowski G, Bartosz G, Walski M, Rafalowska U (1999) Chronic lead intoxication affects the myelin membrane status in the central nervous system (CNS) of adult rats. J Mol Neurosci 13:127–139

Dabrowska-Bouta B, Struzyñska L, Walski M, Rafalowska U (2008) Myelin glycoproteins targeted by lead in the rodent model of prolonged exposure. Food Chem Toxicol 46:961–966

Davis LE, Standefer JC, Kornfeld M, Abercrombie DM, Butler C (1981) Acute thallium poisoning: toxicological and morphological studies of the nervous system. Ann Neurol 10:38–44

Dorman DC (2000) An integrative approach to neurotoxicology. Toxicol Pathol 28(1):37–42

Dorta DJ, Leite S, DeMarco KC, Rodrigues T, Mingatto FE, Uyemura SA, Santos AC, Curti C (2003) A proposed sequence of events for cadmium-induced mitochondrial impairment. J Inorg Biochem 97:251–257

Dribben WH, Creeley CE, Farber N (2011) Low-level lead exposure triggers neuronal apoptosis in the developing mouse brain. Neurotoxicol Teratol 33:473–480
El-Demerdash FM, Yousef MI, Kedwany FS, Baghdadi HH (2004) Cadmium induced changes in lipid peroxidation, blood hematology, biochemical parameters and semen quality of male rats: protective role of vitamin E and beta-carotene. Food Chem Toxicol 42:1563–1571
El-Sokkary GH, Awadalla AE (2011) The protective role of vitamin c against cerebral and pulmonary damage induced by cadmium chloride in male adult albino rats. Open Neuroendocrinol J 4:1–8
Environmental Protection Agency (2006) U.S., Air quality criteria for lead Volume I and II of II. National Center for Environmental Assessment-RTO Office, Research Triangle Park
Esquifino AI, Seara R, Fernández-Rey E, Lafuente A (2001) Alternate cadmium exposure differentially affects the content of gamma-aminobutyric acid (GABA) and taurine within the hypothalamus, median eminence, striatum and prefrontal. Arch Toxicol 75:127–133
Fern R, Black JA, Ransom BR, Waxman SG (1996) Cd+–induced injury in CNS white matter. J Neurophysiol 76:3264–3273
Flora SJS (2009) Metal poisoning: threat and management. Al Ameen J Med Sci 2:4–26
Flora SJS, Mittal M, Mehta A (2008) Heavy metal induced oxidative stress and its possible reversal by chelation therapy. Indian J Med Res 128:501–523
Franco R, Sánchez-Olea R, Reyes-Reyes EM, Panayiotidis MI (2009) Environmental toxicity, oxidative stress and apoptosis: ménage à trios. Mutat Res 674:3–22
Gabbiani G, Baic D, Deziel C (1967a) Toxicity of cadmium for the central nervous system. Exp Neurol 18:154–160
Gabbiani G, Gregory A, Basic D (1967b) Cadmium-induced selective lesions of sensory ganglia. J Neuropath Exp Neurol 26:498–501
Galván-Arzate S, Ríos C (1994) Thallium distribution in organs and brain regions of developing rats. Toxicology 90:63–69
Galván-Arzate S, Santamaria A (1998) Thallium toxicity. Toxicol Lett 99:1–13
Galván-Arzate S, Martínez A, Medina E, Santamaría A, Rios C (2000) Subchronic administration of sublethal doses of thallium to rats: effects on distribution and lipid peroxidation in brain regions. Toxicol Lett 116:37–43
Galván-Arzate S, Pedraza-Chaverrí J, Medina-Campos ON, Maldonado PD, Vázquez-Román B, Ríos C, Santamaria A (2005) Delayed effects of thallium in the rat brain: regional changes in lipid peroxidation and behavioral markers, but moderate alterations in antioxidants, after a single administration. Food Chem Toxicol 43:1037–1045
Gamo M, Ono K, Nakanishi J (2006) Meta-analysis for deriving age- and gender-specific dose–response relationships between urinary cadmium concentration and β_2-microglabulinuria under environmental exposure. Environ Res 101:104–112
Gao S, Jin Y, Unverzagt FW, Ma F, Hall KS, Murrell JR, Cheng Y, JShen J, Ying B, Ji R, Matesan J, Liang C, Hendrie HC (2008) Trace element levels and cognitive function in rural elderly Chinese. J Gerontol A Biol Sci Med Sci 63:635–641
García-Arenas G, Ramírez-Amaya V, Balderas I, Sandoval J, Escobar ML, Ríos C, Bermúdez-Rattoni F (2004) Cognitive deficits in adult rats by lead intoxication are related with regional specific inhibition of cNOS. Behav Brain Res 149(1):49–59
Gobe G, Crane D (2010) Mitochondria, reactive oxygen species and cadmium toxicity in the kidney. Toxicol Lett 198:49–55
Goyer RA, Clarkson TW (2001) Toxic effects of metals. In: Klaassen CD (ed) Casarett and Doull's, toxicology: the basic science of poisons, 6th edn. McGraw-Hill, New York
Gutierrez-Reyes EY, Albores A, Ríos C (1998) Increase of striatal dopamine release by cadmium in nursing rats and its prevention by dexamethasone induced metallothionein. Toxicology 131:145–154
Gwalteney-Brant SM (2002) Heavy metals. In: Haschek WM, Rosseaux CG, Wallig AM (eds) Handbook of toxicologic pathology. Academic, New York
Hanzel CE, Verstraeten SV (2006) Thallium induces hydrogen peroxide generation by impairing mitochondrial function. Toxicol Appl Pharmacol 216(3):485–492

Hanzel CE, Villaverde SV, Verstraeten SV (2005) Glutathione metabolism is impaired in vitro by thallium (III) hydroxide. Toxicology 207:501–510
Hart RP, Rose CS, Hamer RM (1989) Neuropsycological effect of occupational exposure to cadmium. J Clin Exp Neuropsychol 11:933–943
Hasan M, Ali FS (1981) Effects of thallium, nickel and cobalt administration on the lipid peroxidation in different regions of the rat brain. Toxicol Appl Pharmacol 57:8–13
Hidalgo J, Aschner A, Zatta P, Vasák M (2001) Roles of the metallothionein family of proteins in the central nervous system. Brain Res Bull 55:133–145
Hoffman RS (2000) Thallium poisoning during pregnancy; a case report and comprehensive literature review. J Toxicol Clin Toxicol 38:767–775
Ikediobi CO, Badisa VL, Ayuk-Takem LT, Latinwo LM, West J (2004) Response of antioxidant enzymes and redox metabolites to cadmium-induced oxidative stress in CRL-1439 normal rat liver cells. Int J Mol Med 14:87–92
International Agency for Research on Cancer (IARC) (1993) Cancer monographs on the evaluation of the carcinogenic risks to humans, vol 58. IARC Scientific Pub, Lyon
Jarup L (2003) Hazards of heavy metal contamination. Brit Med Bull 68:167–182
Jarup L, Berglund M, Elinder CG, Nordberg G, Vahter M (1998) Health effects of cadmium exposure–a review of the literature and a risk estimate. Scand J Work Environ Health 24(S1):1–51
Jin T, Wu X, Tang Y, Nordberg M, Bernard A, Ye T, Kong Q, Lundström NG, Nordberg GF (2004) Environmental epidemiological study and estimation of benchmark dose for renal dysfunction in a cadmium-polluted area in China. BioMetals 17:525–530
John Peter AL, Viraraghavan T (2005) Thallium: a review of public health and environmental concerns. Environ Int 31:493–501
Jomova K, Valko M (2011) Advances in metal-induced oxidative stress and human disease. Toxicology 283:65–87
Jones RL, Homa DM, Meyer PA, Brody DJ, Caldwell KL, Pirkle JL, Brown MJ (2009) Trends in blood lead levels and blood lead testing among US children aged 1 to 5 years, 1988–2004. Pediatrics 123:e376–e385
Joseph P (2009) Mechanisms of cadmium carcinogenesis. Toxicol Appl Pharmacol 238:272–279
Kakkar P, Jaffery NF (2005) Biological markers for metal toxicity. Environ Toxicol Pharmacol 19:335–349
Kamel MM, El Razek AAH (2011) Perinatal exposure to cadmium affects neurobehavioural development and anxiety – like behaviour in rat offspring. Life Sci J 8:529–536
Kaoud HA, Kamel MM, Abdel-Razek AH, Kamel GM, Ahmed KA (2010) Neurobehavioural, neurochemical and neuromorphological effects of cadmium in male rats. J Am Sci 6:189–202
Kazantzis G (2000) Thallium in the environment and health effects. Environ Geochem Health 22:275–280
Klaassen CD, Liu J, Choudhuri S (1999) Metallothionein: an intracellular protein to protect against cadmium toxicity. Annu Rev Pharmacol Toxicol 39:267–294
Klaassen CD, Liu J, Diwan BA (2009) Metallothionein protection of cadmium toxicity. Toxicol Appl Pharmacol 238(3):215–220
Kordas K, Canfield RL, López P, Rosado JL, Vargas GG, Cebrián ME, Rico JA, Ronquillo D, Stoltzfus RJ (2006) Deficits in cognitive function and achievement in Mexican first-graders with low blood lead concentrations. Environ Res 100:371–386
Lafond J, Hamel A, Tasker L, Vaillancourt C, Mergler D (2004) Low environmental contamination by lead in pregnant women: effect on calcium transfer in human placental syncytiotrophoblasts. J Toxicol Environ Health A67(1069):1079
Lafuente A, Fernández-Rey E, Seara R, Pérez-Lorenzo M, Esquifino AI (2001) Alternate cadmium exposure differentially effects amino acid metabolism within the hypothalamus, median eminence, striatum and prefrontal cortex of male rats. Neurochem Int 39:187–192
Lafuente A, Gonzalez-Carracedo A, Romero A, Esquifito AI (2003) Effect of cadmium on 24-h variations in hypothalamic dopamine and serotonin metabolism in adult male rats. Exp Brain Res 149:200–206

Lanphear BP, Dietrich K, Auinger P, Cox C (2000) Cognitive deficits associated with blood lead concentrations <10 microg/dL in US children and adolescents. Public Health Rep 115:521–529

Leonard A, Gerberr GB (1997) Mutagenicity, carcinogenicity and teratogenicity of thallium compounds. Mutat Res 387:47–53

Leverrier P, Montigny C, Garrigos M, Champeil P (2007) Metal binding to ligands: cadmium complexes with glutathione revisited. Anal Biochem 371(215):228

Limos CL, Ohnishi A, Suzuki N, Kojima N, Yoshimura T, Goto I, Kuroiwa Y (1982) Axonal degeneration and focal muscle fiber necrosis in human thallotoxicosis: histopathological studies of nerve and muscle. Muscle Nerve 5:598–706

Lindeque JZ, Levanets O, Louw R, van der Westhuizen FH (2010) The involvement of metallothioneins in mitochondrial function and disease. Curr Protein Peptide Sci 11:292–309

Liu J, Liu Y, Habeebu SS, Klaassen CD (1998) Susceptibility of MT-null mice to chronic CdCl2-induced nephrotoxicity indicated that renal injury is not mediated by the CdMT complex. Toxicol Sci 46:197–203

Liu J, Goyer R, Waalkes MP (2007) Toxic effects of metals. In: Klaassen CD (ed) Casarett and Doull's toxicology the basic science of poisons, 7th edn. McGraw Hill, New York

López E, Arce C, Oset-Gasque MJ, Cañadas S, González MP (2006) Cadmium induces reactive oxygen species generation and lipid peroxidation in cortical neurons in culture. Free Radic Biol Med 40:940–951

Manca D, Ricard AC, Trotter B, Chevalier G (1991) Studies for lipid peroxidation in rat tissues following administration of low and moderate doses of cadmium chloride. Toxicology 67:303–323

Mazumdar M, Bellinger CD, Gregas M, Abanilla K, Bacic J, Needleman HL (2011) Low-level environmental lead exposure in childhood and adult intellectual function: a follow-up study. Environ Health 10:24–30

Méndez-Armenta M, Ríos C (2007) Cadmium neurotoxicity. Environ Toxicol Pharmacol 23:350–358

Méndez-Armenta M, Barroso-Moguel R, Villeda-Hernández J, Nava-Ruiz C, Ríos C (2001) Histopathological alterations in the brain regions of rats after perinatal combined treatment with cadmium and dexamethasone. Toxicology 161:189–199

Méndez-Armenta M, Villeda-Hernández J, Barroso-Moguel R, Nava-Ruiz C, Jiménez-Capdeville ME, Rios C (2003) Brain regional lipid peroxidation and metallothionein levels of developing rats exposed to cadmium and dexamethasone. Toxicol Lett 144:151–157

Minami A, Takeda A, Nishibaba D, Takefuta S, Oku N (2001) Cadmium toxicity in synaptic neurotransmission in the brain. Brain Res 894:336–339

Moneim AES, Dkhil MA, Al-Quraishy S (2011) Effects of flaxseed oil on lead acetate-induced neurotoxicity in rats. Biol Trace Elem Res 144:904–913

Moreira EG, Rosa GJ, Barros SB, Vassilieff VS, Vassilieff I (2001) Antioxidant defense in rat brain regions after developmental lead exposure. Toxicology 169(2):145–151

Mulkey JP, Oehme FW (1993) A review of thallium toxicity. Vet Hum Toxicol 35:445–453

Nava-Ruiz C, Méndez-Armenta M, Rios C (2012) Lead neurotoxicity: effects on brain nitric oxide synthase. J Mol Hist 43:553–563. doi:10.1007/s10735-012-9414-2

Neala PA, Worley PF, Guilarte TR (2011) Lead exposure during synaptogenesis alters NMDA receptor targeting via NMDA receptor inhibition. Neurotoxicology 32:281–289

Nehru B, Kanwar SS (2004) N-acetylcysteine exposure on lead induced lipid peroxidative damage and oxidative defense system in brain regions of rats. Biol Trace Elem Res 101:257–264

Nordberg FG (2004) Cadmium and health in the 21st century-historical remarks and trends for the future. BioMetals 17:485–489

Nriagu JO (1998) Thallium in the environment, vol 29, Wiley series in advances in environmental science and technology. Wiley, New York

Omarova A, Phillips CJC (2007) A meta-analysis of literature data relating to the relationships between cadmium intake and toxicity indicators in humans. Environ Res 103:432–440

Osman K, Akesson A, Berglund M, Bremme K, Schütz A, Ask K, Vahter M (2000) Toxic and essential elements in placentas of Swedish women. Clin Biochem 33:131–138

Ozden TA, Gökçay G, Ertem HV, Süoğlu OD, Kiliç A, Sökücü S, Saner G (2007) Elevated hair levels of cadmium and lead in school children exposed to smoking and in highways near schools. Clin Biochem 40:52–56

Papanikolaou NC, Hatzidaki EG, Belivanis S, Tzanakakis GN, Tsatsakis AM (2005) Lead toxicity update. A brief review. Med Sci Monit 11:329–336

Papp A, Nagymajtényi L, Dési I (2003) A study on electrophysiological effects of subchronic cadmium treatment in rats. Environ Toxicol Pharmacol 13:181–186

Patrick L (2003) Toxic metals and antioxidants: part II. The role of antioxidants in arsenic and cadmium toxicity. Altern Med Rev 8:106–128

Patrick L (2006) Lead toxicity part II. The role of free radical damage and the use of antioxidants in the pathology and treatment of lead. Altern Med Rev 11:114–127

Provias JP, Ackerley CA, Smith C, Becker LE (1994) Cadmium encephalopathy: a report with elemental analysis and pathological findings. Acta Neuropathol Berl 88:583–586

Puga MLC, Verstraeten VS (2008) Thallium (III)-mediated changes in membrane physical properties and lipid oxidation affect cardiolipin–cytochrome c interactions. Biochem Biophys Act 1778:2157–2164

Ramsden D (2002) Thallium. In: Waring RH, Steventon GB, Mitchell SC (eds) Molecules of death. Imperial College Press, London

Raskin I, Nanda-Kumar PBA, Dushenkov S, David E (1994) Bioconcentration of heavy metals by plants. Salt Curr Opin Biotechnol 5:285–290

Repetto G, Del Peso A, Repetto M (1998) Human thallium toxicity. In: Nriagu J (ed) Thallium in the environment, Advances in environmental science and technology. Wiley, New York, pp 167–199

Rice DC (1993) Lead-induced changes in learning: evidence for behavioral mechanisms from experimental animal studies. Neurotoxicol 14:167–178

Saldivar RL, Luna M, Reyes E, Soto R, Fortul T (1991) Cadmium determination in Mexican-produced tobacco. Environ Res 55:91–96

Satarug S, Baker JR, Urbenjapol S, Haswell-Elkins M, Reilly EBP, Williams JD, Moore MR (2003) A global perspective on cadmium pollution and toxicity in non occupationally exposed population. Toxicol Lett 137:65–83

Sato M, Kondoh M (2002) Recent studies on metallothionein: protection against of heavy metals and oxygen free radicals. Tohoku J Exp Med 196:9–22

Schwartz GG, Reis IM (2000) Is cadmium a cause of human pancreatic cancer? Cancer Epidemiol Biomark Prev 9:139–145

Schwerdtle T, Ebert F, Thuy C, Richter C, Mullenders LH, Hartwig A (2010) Genotoxicity of soluble and particulate cadmium compounds: impact on oxidative DNA damage and nucleotide excision repair. Chem Res Toxicol 23:432–442

Selvin-Testa A, Capani F, Loidl CF, Pecci-Saavedra J (1997) The nitric oxide synthase expression of rat cortical and hippocampal neurons changes after early lead exposure. Neurosci Lett 236:75–78

Shagirtha K, Muthumani M, Prabu M (2011) Melatonin abrogates cadmium induced oxidative stress related neurotoxicity in rats. Eur Rev Med Pharmacol Sci 15:1039–1050

Sharifi AM, Baniasadi S, Jorjani M, Rahimi F, Bakhshayesh M (2002) Investigation of acute lead poisoning on apoptosis in rat hippocampus in vivo. Neurosci Lett 329:45–48

Soltaninejad K, Kebriaeezadeh A, Minaiee B, Ostad SN, Hosseini R, Azizi E, Abdollahi M (2003) Biochemical and ultrastructural evidences for toxicity of lead through free radicals in rat brain. Hum Exp Toxicol 22:417–423

Stohs SJ, Bagchi D, Bagchi M (1997) Toxicity of trace elements in tobacco smoke. Inhal Toxicol 9:867–890

Stohs SJ, Bagchi D, Hassoun E, Bagchi M (2001) Oxidative mechanisms in the toxicity of chromium and cadmium ions. J Environ Pathol Toxicol Oncol 20:77–88

Struzyñska L, Bubko I, Walski M, Rafalowska U (2001) Astroglial reaction during the early phase of acute lead toxicity in the adult rat brain. Toxicology 165:121–131

Suwazono Y, Sand S, Vahter M, Filipsson AF, Skerfving S, Lidfeldt J, Akesson A (2006) Benchmark dose for cadmium-induced renal effects in humans. Environ Health Perspect 114:1072–1076

Tabandeh H, Crowston JG, Thompson GM (1994) Features of thallium poisoning. Am J Ophthalmol 117:243–245

Takebayashi S, Jimi S, Segawa M, Kiyoshi Y (2000) Cadmium induces osteomalacia mediated by proximal tubular atrophy and disturbances of phosphate reabsorption. A study of 11 autopsies. Pathol Res Pract 196:653–663

Thatcher RW, Lester ML, McAlaster R, Horst R (1982) Effects of low levels of cadmium and lead on cognitive functioning in children. Arch Environ Health 37:159–166

Thévenod F (2009) Cadmium and cellular signaling cascades: to be or not to be? Toxicol Appl Pharmacol 238:221–239

Tiffany-Castiglioni E, Qian Y (2001) Astroglia as metal depots: molecular mechanisms for metal accumulation, storage and release. Neurotoxicology 22:577–592

Toscano CD, Guilarte RT (2005) Lead neurotoxicity: from exposure to molecular effects. Brain Res Rev 49:529–554

Tromme I, van Neste D, Dobbelaere F, Bouffioux B, Courtin C, Dugernier T (1998) Skin signs in the diagnosis of thallium poisoning. J Dermatol 138:321–325

Tsai YT, Huang C, Kuo HC, Wang HM, Shen WS, Shih TS, Chu NS (2006) Central nervous system effects in acute thallium poisoning. NeuroToxicol 27:291–295

U.S. Geological Survey (USGS) (2008) Cadmium. Mineral commodity summaries. U.S. Geological Survey

Valko M, Rhodes CJ, Moncol J, Izakovic M, Mazur M (2006) Free radicals, metals and antioxidants in oxidative stress-induced cancer. Chem Biol Interact 160:1–40

Van den Berg KJ, Lammers JH, Hoogendijk EM, Kulig BM (1996) Changes in regional brain GFAP levels and behavioral functioning following subchronic lead acetate exposure in adult rats. Neurotoxicology 17:725–734

Van Kesteren RG (1994) Thallium. In: Vinken DJ, Bruyn GW, de Wolff FA (eds) Handbook of clinical neurology, vol 64, Intoxication of the nervous system. Elsevier North-Holland Biochemical Press, Amsterdam

Verina T, Rohde CA, Guilarte TR (2007) Environmental Pb2+ exposure during early life alters granule cell neurogenesis and morphology in the hippocampus of young adult rats. Neuroscience 145:1037–1047

Viaene MK, Roels HA, Leenders J, De Groof M, Swerts LJ, Lison D, Masschelein R (1999) Cadmium: a possible etiological factor in peripheral polyneuropathy. Neurotoxicology 20:7–16

Viaene MK, Masschelein R, Leenders J, De Groof M, Swerts LJ, Roels HA (2000) Neurobehavioural effects of occupational exposure to cadmium: a cross sectional epidemiological study. Occup Environ Med 57:19–27

Villeda-Hernández J, Barroso-Moguel R, Méndez-Armenta M, Nava-Ruíz C, Huerta-Romero R, Rios C (2001) Enhanced brain regional lipid peroxidation in developing rats exposed to low level lead acetate. Brain Res Bull 55:247–251

Villeda-Hernández J, Méndez-Armenta M, Barroso-Moguel R, Trejo-Solís MC, Guevara J, Ríos C (2006) Morphometric analysis of brain lesions in rat fetuses prenatally exposed to low level lead acetate: correlation with lipid peroxidation. Histol Histopathol 21:609–617

Waisberg M, Joseph P, Hale B, Beyersmann D (2003) Molecular and cellular mechanisms of cadmium carcinogenesis. Toxicology 192:95–117

Wang S, Zhang J (2006) Blood lead levels in children, China. Environ Res 101:412–418

Wang J, Wu J, Zhang Z (2006) Oxidative stress in mouse brain exposed to lead. Ann Occup Hyg 50:405–409

White LD, Cory-Slechta DA, Gilbert ME, Tiffany-Castiglioni E, Zawia NH, Virgolini M, Rossi-George A, Lasley SM, Qian YC, Basha MR (2007) New and evolving concepts in the neurotoxicology of lead. Toxicol Appl Pharmacol 225(1):1–27

Wong KL, Klaassen DC (1982) Neurotoxic effects of cadmium in young rats. Toxicol Appl Pharmacol 63:330–337

Yang Y, Ma Y, Ni L, Zhao S, Li L, Zhang J, Fan M, Liang C, Cao J, Xu L (2003) Lead exposure through gestation only caused long-term learning/memory deficits in young adult offspring. Exp Neurol 184(1):489–495

Zalups RK, Ahmad S (2003) Molecular handling of cadmium in transporting epithelia. Toxicol Appl Pharmacol 186:163–188

Chapter 7
Lead, Arsenic, Cadmium, Mercury: Occurrence, Toxicity and Diseases

Sunny O. Abarikwu

Contents

Abstract Toxic elements including lead (Pb), cadmium (Cd), mercury (Hg) and arsenic (As) are ubiquitous in the environment and exposure of humans through food and water as well as occupational sources are known to have adverse effect on male reproductive health, and may be involved in the aetiology of several of human

S.O. Abarikwu (✉)
Department of Chemical Sciences, College of Natural Sciences, Redeemer's University, KM 46 Lagos-Ibadan Expressway, Redemption City, Ogun State, Nigeria
e-mail: sabarikwu@yahoo.com; abarikwus@run.edu.ng

E. Lichtfouse et al. (eds.), *Pollutant Diseases, Remediation and Recycling*, Environmental Chemistry for a Sustainable World 4, DOI 10.1007/978-3-319-02387-8_7,

diseases including renal, neuronal, developmental retardation, cardiovascular and carcinogenesis. The results of the elemental analysis of these elements in biopsy materials (semen, urine, blood, nails, teeth and bone) indicate that their toxicity is associated to the metal overload in the body. However, inconsistencies exist in many cases and as such it is imperative to clarify the casuality of these associations before translating these data into public health impact. Moreover, the use of chelation therapy in medicine to reduce the concentrations of toxic metals in the body is a promising strategy in the treatment of metal toxicity.

Keywords Arsenic • Cadmium • Lead • Mercury • Chelation therapy • Toxicity • Human disease

7.1 Introduction

Heavy metals are metallic elements which have a high atomic weight and a density much greater (at least 5 times) than water. There are more than 20 heavy metals, and the majority of them are very toxic to living organisms. Cd, Pb, Hg and inorganic As are the most potentially toxic elements in the environment. They have been identified as the most probable causes of heavy metal-related diseases observed in primary care medicine (Hu 2000). Toxic metals have no known nutritional or beneficial effects on human health and, even worse, as in the case of Pb, As, Cd and Hg, may be toxic even at trace levels of exposure. Today mankind is exposed to the highest levels of these metals in recorded history. This is due to their industrial use, the unrestricted burning of coal, natural gas and petroleum, and incineration of waste materials worldwide. Consequently, they are ubiquitous in nature and present in air, water, and soil, so that some level of exposure is not readily preventable. Human exposure can occur through a variety of routes. Metals may be inhaled as dust or fume (tiny particulate matter, such as the lead oxide particles produced by the combustion of leaded gasoline). Some metals can be vaporized (e.g., mercury vapour in the manufacture of fluorescent lamps) and inhaled. Metals may also be ingested involuntarily through food and drink.

There is growing concern that exposure to heavy metals may adversely affect male reproductive outcomes, namely semen quality parameters (sperm concentration, motility, and morphology), and circulating reproductive hormone levels (Wirth and Mijal 2010). Metal can also disrupt signal transduction pathways that play important roles in cell growth and development (Evan and Vousden 2001; Valko et al. 2006) and are been implicated in a plethora of diseases including cancer, neurological diseases, cardiovascular and renal diseases (Fig. 7.1) (Jomova and Valko 2011). In terms of health assessment, the extent of exposure to a metal is best determined by measuring its internal concentration, and more preferably the biologically effective dose at the target organ (as opposed to environmental concentration). For a number of reasons, however, it is not always feasible to determine the internal or biologically effective dose of the metal at the target tissue. Therefore, the use of biological

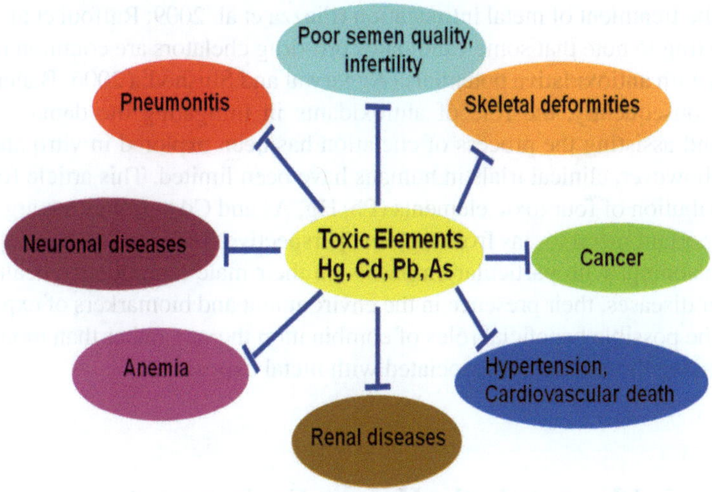

Fig. 7.1 Some disease conditions in humans cause by exposure to toxic elements

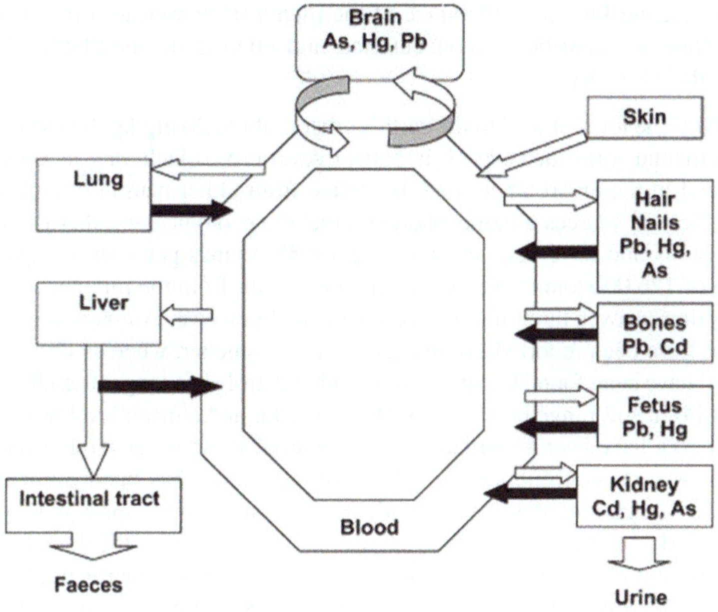

Fig. 7.2 Disposition of metals in humans (Source: Kakkar and Jaffery 2005)

indicators, also termed "biomarkers of exposure," is a way to link external exposure of a metal to internal dose (e.g., lead in blood and bone, arsenic and cadmium in urine, and mercury in maternal hair or umbilical cord blood). The disposition of metals in humans is shown in Fig. 7.2. Metal-chelation therapy in medicine is now being used in the treatment of metal intoxication (Piazza et al. 2009; Raffoul et al. 2007). It is interesting to note that some candidates pro-drug chelators are common food constituents with antioxidative potentials (Aggarwal and Shishodia 2006; Banerjee et al. 2006). Consequently, the role of antioxidants in mitigating the damage of toxic metals and assisting the process of chelation has been explored in vitro and animal studies; however, clinical trials in humans have been limited. This article focuses on the contribution of four toxic elements (Pb, Hg, As and Cd—as they are arguably the most important metal toxins from a global perspective) in the recent development on metal poisoning, with particular emphasis on their male reproductive health effects and other diseases, their presence in the environment and biomarkers of exposures as well as the possible beneficial roles of combination therapy rather than monotherapy in decreasing the health risk associated with metal exposure.

7.2 Toxic Elements in the Human Environment

Toxic elements occur naturally in our environment, especially in the Earth's crust, where they contribute to the balance of the planet. However, as a result of human activity they are distributed, concentrated and chemically modified, which may increase their toxicity.

Lead (Pb): The level of lead in the earth's crust is about 20 mg/kg. It is seldom found in its elemental form; however, it is part of several ores including its own (galena, PbS). Lead in the environment may be derive from either natural or anthropogenic sources. Natural sources of atmospheric Pb include geological weathering and volcanic emissions and have been estimated at 19,000 tonnes per year, compared to an estimate of 126,000 tonnes per year emitted to the air from the mining, smelting and consumption of over three million tonnes of lead per year. Ambient air levels over 10 μg/m^3 have been reported in urban areas near a smelter, whereas Pb levels below 0.2 μg/m^3 have been found in cities where leaded petrol is no longer used. Background levels of Pb in soil range between 10 and 70 mg/kg and a mean level near roadways of 138 mg/kg have been reported. Present levels of Pb in water rarely exceed a few μg/L; the natural concentration of Pb in surface water has been estimated to be 0.02 μg/L. The current annual worldwide production of Pb is approximately 5.4 million tons and continues to rise. Sixty percent of Pb is used for the manufacturing of batteries (automobile batteries, in particular), while the remainder is used in the production of pigments, glazes, solder, plastics, cable sheathing, ammunition, weights, gasoline additive, and a variety of other products. It is known to be the most common industrial metal that has become widespread in air, water, soil and food (Raikwar et al. 2008). Airborne Pb can be deposited on soil and water, thus reaching humans through the food chain and in drinking-water (Raikwar et al. 2008). A number of factors can modify the impact of Pb exposures. For example, water with a lower pH (such as drinking water stemming from the collection of untreated "acid rain") will leach more Pb out of plumbing connected by lead solder than more alkaline water. Lead from soil

Table 7.1 Toxic metal and their reactive forms

Metal	Toxicity
Arsenic	Inorganic arsenate [(As (+5)] or [As (+3)] are more toxic
Cadmium	All forms are toxic and need attention
Lead	Organic forms are more toxic and easily absorbed by the gastrointestinal tract
Mercury	Hg (II) Organomercurials mainly methylmercury; biologically magnified

Source: Michalke (2003)

tends to concentrate in root vegetables (e.g., onion) (Ward and Savage 1994) and leafy green vegetables (e.g., spinach). Individuals will absorb more Pb in their food if their diets are deficient in calcium, iron, or zinc (Mahaffey 1990).

Arsenic (As): In the environment As is usually found combined with other elements as inorganic and organic forms (Mishra et al. 2010). The most common oxidation numbers of arsenic are +5, +3 and −3 (Table 7.1), in which the element is able to form both inorganic and organic compounds both in water, soil or food and within the human body (Orloff et al. 2009). In combination with other elements such as oxygen, sulphur and chlorine, the element is referred to as inorganic arsenic and as combined with hydrogen and carbon as organic arsenic. Since most arsenic compounds lack colour or smell, the presence of arsenic is not immediately obvious in food, water or air, thus presenting a serious human health hazard given the toxic nature of the element. Arsenic is used in bronzing, hardening and improving the sphericity of shot, wood preservation, pyrotechnics, varieties of semiconductor devices solar cells, light-emitting diodes, lasers, and integrated circuits; and as pesticides. Arsenic is ubiquitous in nature and its abundance ranks twentieth in the Earth's crust, fourteenth in seawater and twelfth in the human body. Arsenicals usually occur in trace quantities in rock, soil, water, and air. Concentrations may be higher in certain areas as a result of weathering and anthropogenic activities including metal mining and smelting and during fossil fuel combustion (Das and Roy Chowdhury 2006). Inorganic arsenic is found in groundwater used as drinking-water in some areas of the world. Deep-water wells in Taiwan and Chile are now well-known to be contaminated with arsenic (Ferreccio et al. 2000). Water from relatively shallow tube wells in areas of Bangladesh, West Bengal, and other parts of the subcontinent has also been recently found to be heavily contaminated with As (Biswas et al. 1998). It was reported that almost 43 million people living in these areas have been drinking water that is laden with arsenic (Chowdhury et al. 2000). The WHO recommended limit for arsenic in water is 10 μg/L (Jomova et al. 2011). However, Arsenic concentrations in the range 50–3,200 μg/L have been measured (Bhattacharya et al. 2003). Background concentrations of As in soil range from 1 to 40 mg/kg, with a mean value of 5 mg/kg. Terrestrial plants may accumulate As by root uptake from the soil or by adsorption of airborne As deposited on the leaves. In fishes As ranged between 5–100 μg/g and reach to 100–250 μg/g in species at the top of the food chain (Jomova et al. 2011). The largest source of As and other metals is usually food, of which the main dietary forms are seafood, rice, mushrooms and poultry (Jomova et al. 2011). While there is more arsenic per se in seafood, this is mostly in an organic form called arsenobetaine which is much less harmful than others.

Cadmium (Cd): Cd occurs in nature at low concentrations, mainly associated with ores of zinc, Pb, and copper. Values for total Cd concentrations in unpolluted soils vary from <0.01 to 2.50 mg/kg. European studies reported that the Cd concentrations in European soils ranged from 0.06 mg/kg in Finland to 0.50 mg/kg in the UK, depending on the nature of the parent material (Jensen and Bro-Rasmussen 1992). EPA has established reference dose (RD) for Cd of 0.001 mg/kg/day for food and 0.0005 mg/kg/day for water. These amounts represent an estimated daily oral exposure that is likely not to cause adverse health effects (WHO 1992a; DEFRA 2002). Cd is employed in several industrial processes such as: protective coatings (electroplating) for metals like iron; preparation of Cd-Ni batteries, control rods and shields within nuclear reactors and television phosphors. Some Cd-compounds are used as stabilizers for PVC. For non-smoking population the major exposure pathway is through food. Cadmium is readily taken up by plants. Some root crops (carrots and parsnip) and some leafy crops (lettuce and spinach) are able to accumulate more Cd compared to other plant foods. Grain crops like rice and wheat can accumulate relatively high amounts of Cd (WHO 1992b). Furthermore, Cd is readily accumulated by many organisms, particularly by microorganisms and molluscs where the bio-concentration factors are in the order of thousands. Soil invertebrates also concentrate Cd markedly. The biological half life of Cd in humans is long ranging from 7 to 26 years in kidney and 3–4 months in blood. Due to its low rate of excretion, Cd accumulates over time in the kidneys, liver, semen, ovaries, and placenta (Akinloye et al. 2006; Wirth and Mijal 2010) with a preference for male reproductive organs (Wirth and Mijal (2010). Specific heavy-metal-binding proteins (metallothioneins) have been isolated from Cd-exposed organisms. The Cd level to which the general population is exposed has been estimated in a representative, population-based survey of the general USA population (Wirth and Mijal 2010). The estimated blood geometric mean for Cd, unadjusted for smoking, for individuals 20 years and older was 0.412 mg/L, and for all males 1 year and older was 0.403 mg/L.

Mercury (Hg): Mercury occurs in nature in mineral, cinnabar, metacinnabar and hypercinnabar in different chemical forms. Metallic Hg, mercurous and mercuric mercury (HgO, Hg^{+} and Hg^{2+}) respectively are widely used in some chemical, metal-processing, electrical-equipment, automotive, and building industries and in medical and dental services (Clarkson and Magos 2006). Mercurous and mercuric Hg form inorganic and organic compounds with other chemicals that can be readily absorbed through ingestion. All three forms of Hg are toxic to various degrees. The organic form, methyl mercury (MeHg), is the most toxic. When deposited in soil, organic mercury compounds are slowly broken down into inorganic compounds; conversely, inorganic mercury can be converted by microorganisms in soil and water into the organic compound methyl mercury, which is then bio-concentrated up the food chain; each level of the food chain has higher tissue concentrations than its prey. The reaction takes place at the water-sediment interface (Wright and Welbourn 2002) and is facilitated by low pH and high dissolved organic carbon.

Methyl mercury dissolves well in water, crosses biological membranes, and persists in fatty tissues of organisms. The major route for methyl mercury exposure for the general population is consumption of fish particularly, tuna, king mackerel, and swordfish. Thus, their consumption, though popular, should be limited to less than twice per week and entirely avoided by women during pregnancy (Mahaffey et al. 2004). Diet can be the main source of inorganic and organo-mercurials especially seafood while dental amalgams are the main exposure source to elemental Hg. Elemental (monatomic) or metallic mercury occurs in vapour and liquid forms and is very stable. Evaporation of Hg vapour from land and sea surfaces, from volcanic activity, or through burning fossil fuel has resulted in its worldwide distribution. In the upper atmosphere Hg vapour is oxidized to water-soluble ionic inorganic Hg and returns to earth in rain. Mercury levels at the top of the food chain are thousands or millions of times higher than in water or sediments (Wright and Welbourn 2002). Fishes containing more than 0.4 ppm Hg are unfit for human consumption and the critical urinary concentration of Hg has been suggested as 1–2 μg/mL. The intake of mercury as suggested by WHO is 43 μg (Krishnamurti 1987).

7.3 Biomarkers of Toxic Elements

7.3.1 Biomarkers: The Link Between Exposure Assessment and Disease Outcome

A biomarker can be broadly defined as any biological index capable of being measured, which is associated with or indicative of a defined biological endpoint such as a developmental or disease stage. In the field of environmental health, biomarkers are the measurements used in the diagnosis or risk assessment (Sanders et al. 2009). In many countries of the world, the overall objective of the biological monitoring of toxic metals in humans is to prevent adverse health effects (Sakai 2000).

Biomarkers are generally classified into three groups: biomarkers of exposure, effect, and susceptibility. The biomarkers of exposure and effect correspond to the monitoring of exposure and effect, respectively. Some of confounding factors in biological monitoring are related to the biomarkers of susceptibility. When chemicals enter the body, they produce internal doses as consequence of their disposition and metabolism. An exogenous substance or its metabolites or the product of an interaction between a chemical agent and some target tissues can be used for biomarkers of exposure. Biologically activated chemicals can bind functioning macromolecules such as enzymes, receptors, and DNAs, resulting in early effects of target tissues or organs. The early effect in biochemical process is usually reversible, and occurs within a short time after exposure (subclinical effect). It can be used for the biomarkers of effect, and may precede the altered structure or function of tissues, which may be irreversible and lead to clinical disease (Fig. 7.3). The toxicokinetics and toxicodynamics of chemicals entering

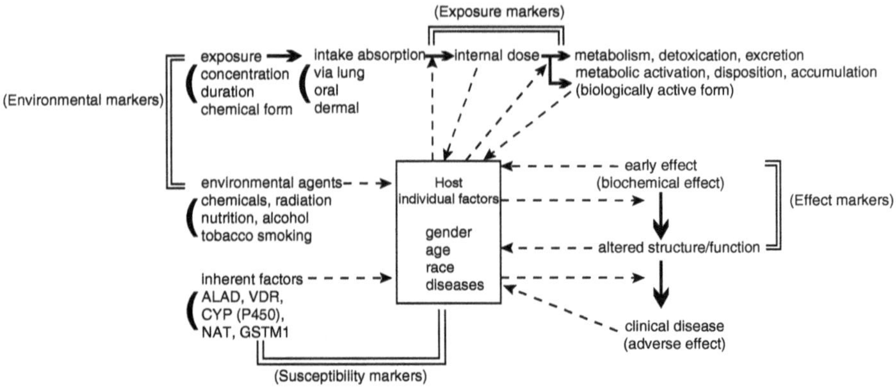

Fig. 7.3 Biomarkers of exposure, effects, and susceptibility in workers exposed to chemical substances (Source: Sakai 2000)

the body are affected by various individual factors, which are used as the biomarkers of susceptibility. They may also be defined as the indicators or measures of inherent or acquired abilities of individuals to respond to the specific xenobiotic exposure (Sakai 2000).

The "ideal" biomarker of exposure has several characteristics (Sanders et al. 2009) These include that the sample collection and analysis are simple, sensitive, and reliable; that the biomarker is specific for a particular type of exposure; that the exposure results in a reversible change; and that intervention or prevention of exposure is considered if exposure is confirmed by the biomarker. There should also be a well-established relationship between biomarker of exposure and outcome, in that the biomarker not only provides information about exposure levels but can also be predictive of an effect. For example, urinary Cd is directly correlated to the concentration of Cd in the renal cortex, which is one site for toxicant action of this metal. This is of immense value to an environmental carcinogen as means to identify risk factors for a disease outcome with much greater sensitivity. Biomarkers are therefore useful in the evaluation of progressive diseases that manifest their symptoms long after exposure to the initiating factor. In such cases, traditional early warning symptoms of developing disease may be lacking. When the disease becomes clinically apparent such as some cancers and neurological disorders, they may become irreversible. Therefore, detection of earlier events can provide a valuable timely warning of risk. Furthermore, a biomarker of exposure is also a measure of cumulative exposure to a metal—and also of metal actually existent in tissue, as occurs with chronic exposure for metals. However, such an approach may not be appropriate for metals that are not extensively accumulated in tissues, and it does not differentiate between metal present in a tissue in a sequestered or inactive form and metal engaged in toxic or pathological processes. There are environmental (water, air, soil, dust), occupational, medicinal, and dietary sources of metal exposure.

For this reason, use of biomarkers increases the need for comprehensive, multi-pathway assessments of exposure. Reference or background levels of biomarkers of exposure are essential for any assessment. Several metals, such as arsenic and selenium, are found naturally in the diet. Therefore, failure to consider dietary sources of metals may result in a misinterpretation of the exposure. For example, arsenobetaine is a non-toxic organic form of arsenic found naturally in shrimp and other seafood. The analysis of total unspeciated urinary arsenic of individuals who consume seafood, without recognition of their diet history, will lead to an overestimation of exposure to potentially toxic (inorganic) arsenic species. These markers can be obtained from different biological tissues and fluids (blood, urine, bone, tooth, hair, and nail e.t.c) by invasive and non-invasive methods. Non-invasive methods are preferred over invasive methods because of greater acceptability. They are expired air, saliva, semen, urine, sputum, hair, nails, faeces, and breast milk. Invasive methods may yield samples from blood, lung tissue, bone marrow, amniotic fluid, liver tissue, bone, follicular fluid or adipose tissue. The measurement of metals in these biological matrices is the primary means of quantifying biomarkers of exposure for toxic metals in humans and experimental animals.

7.3.2 Biomarkers of Arsenic Exposure

Measurements of the level of As in blood, urine, hair and nails have all been used as biological indicators of exposure to As (Jomova et al. 2011). However, As in the blood is metabolized within hours, therefore, the measurement of blood arsenic levels is not a good indicator of long-term exposure of individuals to arsenic (Jomova et al. 2011). There was also no correlation found between the level of As in the blood and the level of As in the drinking water of residents in several communities in the USA, where water levels ranged from about 6–125 μg/L (Levine 2012), supporting that measurement of blood As is not a reliable marker for As exposure. Since the majority of As is absorbed from the lungs or the gastrointestinal tract and excreted in the urine, normally within 1–2 days, measurement of urinary As levels is generally considered as the most reliable marker of acute arsenic exposure (Ng et al. 2005). The values obtained from such measurements are found to correlate well with exposures of populations living near industrial point sources of arsenic (Jones 2007). Furthermore, As is also known to influence the enzymes responsible for heme synthesis and degradation and activation of heme oxygenase (Jomova et al. 2011). For instance, arsenite was reported to induce de novo synthesis of heme oxygenase 1 (HO-1) in human lymphoblastoid cells, but it has not been determined whether the same response is induced in vivo. Animal tests have shown that arsenic poisoning increased urinary levels of uroporphyrin, coproporphyrin and bilirubin (Ng et al. 2005). These tests have also been shown to be applicable to human subjects (García-Vargas and Hernández-Zavala 1996). Hence, altered urinary levels of these heme-related compounds could serve as a sensitive biomarker of arsenic exposure.

7.3.3 Biomarkers of Mercury Exposure

Biomarkers of MeHg exposure, such as total mercury levels in hair or blood, are regarded as more accurate measures of human exposure than dietary assessment (i.e., of fish consumption) because MeHg concentrations vary both between and within fish species (Groth 2010). Although it is correlated with maternal hair, cord blood mercury may better reflect fetal exposure than maternal hair (Karagas et al. 2012). Mercury is excreted in breast milk, but it is not typically used as a matrix for assessing exposure, primarily because of low concentrations and variability in the proportion present as MeHg (García-Esquinas et al. 2011; Miklavcic et al. 2011). Tissues such as umbilical cord, meconium, placenta, and nail tissue are potentially useful matrixes for assessing mercury exposure, but have not been used widely in epidemiologic studies (Karagas et al. 2012). The largest survey of mercury levels in the general male USA population (1,127 men, mean age 52.8 years) occurred in the mid to late 1990s and reported mean total blood mercury of 2.55 mg/L (Kingman et al. 1998). Other more current surveys conducted in Europe that included men found lower levels (0.58 and 0.78 mg/L) (Wirth and Mijal 2010). Total blood Hg includes inorganic and organic forms, while urinary or plasma levels reflect inorganic mercury exposure (Wirth and Mijal 2010) and thus is not used as an indicator of MeHg exposure; however, in hair, nails, and blood, MeHg is the primary contributor to total mercury levels (Karagas et al. 2012). Even the best exposure biomarkers are imprecise measures of MeHg in target organs such as the fetal brain. Furthermore, the average coefficient of variation is about 25 % for cord blood mercury analysis and about twice that for maternal hair mercury (Grandjean et al. 2010). Typically, imprecision in an exposure measure will attenuate its calculated effect (Karagas et al. 2012); this highlights the potential for measurement of errors in MeHg exposure assessment to affect comparability of findings across studies.

7.3.4 Biomarkers of Lead Exposure

Lead toxicity is most commonly diagnosed through elevated blood levels. Blood levels of 10 μg/dL (equivalent to 0.48 μmol/L) or higher are considered toxic (Jomova and Valko 2011). Once absorbed, Pb accumulates in the blood, soft tissues, and bone with a half life of 35 days in blood to 20–30 years in bone (Sakai 2000). Blood Pb, mainly red cell Pb, is a representative of soft tissue Pb and most widely used as measures of body burden and absorbed (internal) doses of Pb. The relationship between Blood Pb and the concentration of Pb in exposure sources is curvilinear (Sakai 2000). The levels of plasma Pb are sharply elevated with a sudden intake or acute exposure to Pb and rapidly diminished by time elapse from it, indicating that plasma Pb is an index for very recent exposure (Barbosa et al. 2005). The half life is very short, maybe less than 1 h. The amount of urine Pb depends not only on the exposure conditions but also on the extent of body burden and kidney function. Urine Pb as well as plasma Pb increases exponentially with increasing blood Pb under a steady-state situation and are mainly a reflection of recent exposure (Sakai 2000). The amount of Pb excreted in urine or the plasma Pb level after administration of a chelating agent reflects the mobilizable pool of Pb which consists of mainly blood and soft tissue Pb with only a small fraction derived from bones (Sakai 2000). As the plasma fraction is rapidly exchangeable in the blood, the toxic effects of Pb are assumed to be primarily associated with plasma Pb (Barbosa et al. 2005). Although plasma Pb should be more germane than blood Pb to Pb exposure and distribution, little is known about the association between plasma Pb and clinical outcome. The determination of plasma Pb is problematic because erythrocyte haemolysis can shift the metal into the plasma and artificially increase plasma Pb levels. Thus, Blood Pb concentrations are currently regarded as the most reliable index of exposure to Pb.

Other currently available biomarkers of internal Pb dose have not yet been accepted by the scientific community as a reliable substitute for blood Pb measurement (Sanders et al. 2009). Nevertheless, in certain cases bone or teeth (for past exposures), faeces (for current gastrointestinal exposure), or urine (for organic Pb) are sometimes more useful than blood. Measurement of Pb in hair and teeth has been of practical use, particularly in epidemiological research. Hair and nails have many superficial advantages as a Pb exposure biomarker, especially as specimen collection is non-invasive and simple and specimens are very stable after collection, not requiring special storage conditions ((Schumacher et al. 1991). Nail Pb is considered to reflect long-term exposure because this compartment remains isolated from other metabolic activities in the body (Sanders et al. 2009). Because toenails are less affected than fingernails by exogenous environmental contamination, toenails have been preferred for Pb-exposure studies. The Pb concentration in nails depends on the age of the subject (Nowak and Chmielnicka 2000), and does not appear to depend on the subject's gender (Sanders et al. 2009). The teeth are superior to bone as an indicator of cumulative Pb exposure because the losses from teeth are much slower (Maneakrichten et al. 1991). Moreover, deciduous teeth are relatively easy to collect and analyze and are very stable for preservation purposes. In an early study, concentrations of blood Pb determined at regular 6- month intervals were related to the Pb concentrations in surface tooth enamel but correlated with saliva Pb only in the short term (Sanders et al. 2009). Saliva is a convenient source and therefore a potential substitute for blood as a biomarker for Pb exposure (Sanders et al. 2009). Nevertheless, saliva has not been generally accepted as a reliable biomarker of Pb exposure because of conflicting and unreliable saliva Pb measurements. For instance, previous data suggested an association between saliva Pb levels and blood Pb and plasma levels (Omokhodion and Crockford 1991) whereas recent data from a highly contaminated lead mining area in Thailand obtained a geometric mean for the blood Pb content as 24.03 μg/dL (range 11.80–46.60 μg/dL) whereas the saliva Pb content was 5.69 μg/dL (range 1.82–25.28 μg/dL). The data thus suggested that saliva is not a suitable material for biological monitoring with respect to Pb exposure. The collection of urine Pb is favoured for long-term biomonitoring, especially for occupational exposures. Urine Pb originates from plasma Pb that is filtered at the glomerular level and excreted through the kidneys. Consequently, it is thought that urine Pb levels adjusted for glomerular filtration rate can serve as a proxy for plasma Pb (Sanders et al. 2009). However, the correlation of urine Pb with blood Pb among workers occupationally exposed to Pb was close enough to suggest that urine Pb can be a good alternative to blood Pb on a group basis, but not close enough to allow urine Pb to predict blood Pb on an individual basis (Fukui et al. 1999). Micronuclei (MN), Sister chromatid exchanges (SCEs), high-SCE frequency cells (HFCs), and DNA-protein cross-links (DPCs) have also been shown to be reliable biomarkers for monitoring workers exposed to Pb and clearly indicate health effects from occupational exposure to Pb (Sanders et al. 2009). Also, the association between Pb exposure and DNA methylation may have implications for the mechanisms of action of Pb on health outcomes, and also suggests that changes in DNA methylation may represent a biomarker of past Pb exposure (Wright et al. 2010). Lead has also been reported to accumulate in male reproductive tissues and is considered a male reproductive toxicant (Wirth and Mijal 2010). In human study, the semen Pb was correspondingly increased with blood Pb in the subjects exposed to Pb over a period of 6 years (average) and 6 h daily. Moreover, workers occupationally exposed to Pb exhibited moderately high blood Pb levels associated with sexual disorders like decreased libido that was followed by an increased frequency of astheno, hypo- and teratospermia (Telisman et al. 2000; Benoff et al. 2000).

Several enzymatic processes responsible for heme synthesis can be used as biomarkers for the toxic effects of Pb, primarily delta-aminolaevulinic acid dehydratase (D-ALAD), which catalyzes the condensation of two molecules of 5-aminolevulinic acid (D-ALA) to form the heme precursor, porphobilinogen. As the activity of D-ALAD is inhibited by Pb binding, this enzyme is accepted as the most sensitive measurable biological index of Pb toxicity (Barbosa et al. 2005). An epidemiological survey of Pb exposure among children (Pb concentration >10 μg/dL) in India showed significantly suppressed levels of D-ALAD with respect to children with Pb concentration (<7 μg/dL) (Ahamed et al. 2005). A direct correlation between blood Pb levels, D-ALAD activity and erythrocyte levels of malondialdehyde (MDA) was observed among workers exposed to Pb. Inhibition of D-ALAD by Pb increases levels of the substrate D-ALA which is known to stimulate the formation of Reactive Oxygen Species (ROS) substantiated by the elevated levels of MDA. Thus, the, activity of D-ALAD in red blood cells is directly related to the concentration of Pb in blood and therefore may be used as a surrogate for the measurement of Pb in blood.

Highly accurate determination of low blood Pb level can be detected using an advanced analytical instrument such as inductively coupled plasma mass spectrometry, as well as graphite furnace atomic absorption spectrometry. Usual diagnostic determinations of clinical laboratory such as hematological test like Hb and Hct are not useful for the biological monitoring at low level exposure (Sakai 2000). The useful biological markers for evaluation of subclinical effects of Pb at low blood Pb levels such as 10–20 μg/dL include the inhibition of D-ALAD, pyrimidine nucleotidase, nicotinamide adenine dinucleotide synthetase (NADS), and the variation in some metabolite concentrations (e.g. D-ALA in urine, blood, or plasma, as well as coproporphyrin in urine, and zinc protoporphyrin in blood).

7.3.5 Biomarkers of Cadmium Exposure

Cadmium (Cd) can be measured in blood, liver, kidney, hair, faeces and urine. Blood levels of Cd correlate to recent exposures. Levels in whole blood of persons not occupationally exposed generally range from about 0.0035 to 0.0089 μmol/L for non-smokers and 0.012–0.035 μmol/L for smokers. Blood concentrations less than 0.089 μmol/L are considered acceptable in workers with occupational exposure to Cd. Workers with occupational exposure to Cd by inhalation may have blood levels up to 0.44 μmol/L. The levels of Cd found in liver and kidney cortex, as well as urinary levels, correlate to total body burden, which is the result of chronic exposure. In the general population, the average urinary Cd level is about 0.35 μg/g creatinine in non-smokers and values above 2 μg/g creatinine are rare. In populations with substantial environmental or occupational exposure, values can range up to 50 μg/g creatinine, even among individuals with no signs of renal dysfunction. Under conditions of chronic occupational or environmental exposure, the kidney is the critical target organ of Cd toxicity in the human organism (Bernard 2008). The Cd concentration in urine (CdU) has been proposed as an indirect biological indicator for Cd accumulation in the kidney. Elevated urinary albumin in the absence of an increase in low-molecular-weight protein excretion is an early indicator of glomerular damage and is observed at urine Cd concentrations from 3.6 to 4.2 μg urinary Cd/g creatinine (Mueller et al. 1998). The increase in albumin excretion and proteinuria is associated with cardiovascular disease (CVD), mortality, and renal disease (Gonick 2008), but these clinical implications have not been specifically associated with Cd toxicity.

The earliest manifestation of Cd-induced renal damage considered as critical consists in an increased urinary excretion of microproteins (molecular weight <40 kDa). Among these proteins, β2-microglobulin, retinol-binding protein and α1-microglobulin have been the most validated for the routine screening of tubular proteinuria. A modest increase in the urinary excretion of these proteins, as found at the early stage of Cd nephropathy (in the range of 300–1,000 μg/g creatinine for retinol-binding protein, is unlikely to compromise the renal function (Bernard 2008). Such a small increase might even be reversible after removal from Cd exposure. By contrast, when the urinary excretion of these proteins is increased by more than one order of magnitude, tubular dysfunction caused by Cd becomes irreversible and may be associated with a lower glomerular filtration rate (GFR) and an accelerated decline of the GFR with ageing (Bernard 2008). In occupationally exposed males, the CdU thresholds, for significant alterations of renal markers ranged, according to the marker, from 2.4 to 11.5 μg Cd/g creatinine. A threshold of I0 μg Cd/g creatinine (corresponding to 200 μg Cd/g renal cortex: the critical Cd concentration in the kidney) is confirmed for the occurrence of low-molecular-mass proteinuria (functional effect) and subsequent loss of renal filtration reserve capacity (Roels et al. 1999). The microproteinuria was found reversible, if the reduction or cessation of exposure to Cd occurred when tubular damage was still mild (β_2-microglobulinuria <1500 μg/g creatinine) and CdU had never exceeded 20 μg Cd/g creatinine. It is therefore required that to recommend that occupational exposure to Cd should not allow that CdU exceeds 5 μg Cd/g creatinine (Roels et al. 1999). In the general population with low-level Cd exposure, much lower no-adverse effect CdU levels were found, viz. 3 μg Cd/24 h for low-molecular-mass proteinuria and 2 μg Cd/24 h for hypercalciuria (Roels et al. 1999). As hypercalciuria may exacerbate the development of osteoporosis, especially in the elderly, a CdU of 2 μg Cd/g creatinine should be regarded as a measure of the maximum tolerable internal dose of Cd for individuals in the general population. Furthermore, renal effects may occur in people with urinary Cd levels <2 μg/g (Paschal et al. 2000) and recent findings indicate that changes in sensitive renal biomarkers may occur at lower urinary Cd levels than previously estimated among populations exposed to environmental Cd (Jarup et al. 2000).

Elevations in enzymes primarily of renal tubular origin, such as *N*-acetyl-β-D-glucosaminidase (NAG) and alanine aminopeptidase (AAP) have been observed at occupational Cd exposures of 3.7–6.3 μg urinary Cd/g creatinine (Kakkar and Jaffery 2005) However, some of the biomarkers of effect caused by Cd are not specific for Cd exposure. For instance, urinary level of β_2-microglubulin, a low molecular weight protein, widely used as a marker of renal tubular dysfunction rises naturally with age. Urinary N-acetyl-β-D-glucosamidase, a lysosomal enzyme present in high concentration in the proximal tubule, has a better correlation with urinary Cd levels than does β_2-microglobulin at low Cd exposure levels (urinary cadmium <10 μg/g creatinine). However, increased urinary NAG activity can result from effects other than nephrotoxicity. Urinary metallothionein correlates with Cd concentrations in liver, kidney and urine and relatively strong correlations have been found between metallothionein and Cd levels in urine in exposed humans. However, once renal damage becomes pronounced, urinary metallothionein levels increase sharply. Faecal Cd primarily reflects recently ingested Cd and is not a good indicator of past exposure. It can be used as a direct indicator of dietary intake because it is so poorly absorbed from the gastrointestinal tract. In addition, several novel biomarkers such as the kidney injury molecule-1, α-glutathione-S-transferase and insulin also offer the potential for improved biomonitoring of Cd-exposed populations. Other markers Elevations in the excretion of low-molecular-weight proteins, such as α1-microglobulin, or retinol-binding protein, have been used as indicators of damage to the tubular protein absorption capability.

7.4 Human Health Effects of Toxic Metals

7.4.1 Male Reproductive Health Effects of Arsenic

Studies investigating the effects of environmental levels arsenic on human male reproductive outcomes have been documented. A cross-sectional study of men attending infertility clinics in Michigan, USA found a significantly increased risk for low sperm motility, low semen volume with low levels of arsenic, after adjusting for smoking and age (Wirth and Mijal 2010). Similarly, experimental studies with animal models

Table 7.2 Health effects in environmentally exposed populations attributed to arsenic

Organ affected	Effects
Skin	Hyperpigmentation
	Hyperkeratosis
Lungs	Lung cancer[a]
Liver	Liver dysfunction
	Haemagioendothelioma
Cardiovascular system	Peripheral vascular disturbances leading to gangrene
Nervous system	Peripheral neuropathy
	Hearing defects
Haematopoietic system	Disturbed erythropoiesis with anaemia
Reproductive system	Increased frequency of spontaneous abortions

[a]The role of arsenic in these effects is equivocal
Source: Hutton (1987)

found harmful effects of arsenic on measures of male reproductive health. For example, arsenic were accumulated in the reproductive organs (testes, epididymis, seminal vesicles) of mice exposed to arsenic in drinking water leading to changes in sperm motility, epididymal sperm count and percent of sperm with morphological abnormalities with no sign of toxicity detected in the animals (Wirth and Mijal 2010; Pant et al. 2001). The activities of 3β-HSD and 17β-HSD (enzymes involved in testosterone metabolism), plasma and testicular testosterone and plasma luteinising hormone levels were decreased by low level arsenic (20–40 mg/L drinking water, 5 or 3 mg/k) in animal models (Wirth and Mijal 2010; Chang et al. 2007; Chiou et al. 2008). The suppressive effect of arsenic on spermatogenesis and gonadotrophin and testosterone release in rats has also been reported (Sarkar et al. 2003). Considering the substantial amount of arsenic detected in the reproductive organs, it is believed that arsenic has a direct effect on testicular tissue. However, because of the endocrine component in arsenic toxicity, several studies are of the opinion that the hypothalamus and brain may be the major targets of arsenic's effects, leading to hormone dysregulation and decreased sperm concentrations (Jana et al. 2006; Sarkar et al. 2003). Although the data is suggestive of possible effects of arsenic on male fertility, further research on the effects of exposure to environmental levels of arsenic on human semen quality is needed before any conclusions can be drawn.

7.4.2 *Other Health Effects of Arsenic*

In humans arsenic toxicity occurs due to ingestion of arsenic containing powders or solutions accidentally, suicide, homicide, or consumption of contaminated food or drinking water. Approximately 80–95 % of total arsenic is present as organic compounds thus do not cause any damage to health (arsenosugar, arsenolipids, etc.). The long term exposure to inorganic arsenic from contaminated water is responsible for various adverse health effects of arsenic in a large number of organs (Table 7.2) including tumours of the lung, skin, liver, bladder and kidney (Waalkes et al. 2004). Increased mortality due to lung cancer has been reported in humans living close to arsenic power plants (Smith et al. 2006). A follow up study on the increased prevalence of lung cancer and increased rate of spontaneous abortion reported in communities close to point sources of atmospheric arsenic concluded that it is not possible to say whether arsenic was the causative agent in those incidents (Kapaj et al. 2006). Similarly, affected populations, particularly those in Argentina and Chile, have revealed a close association between skin cancer and arsenic exposure (Hutton 1987). Skin lesions, peripheral neuropathy and anaemia are hallmarks of chronic arsenic exposure. Arsenic has been reported to be associated with Blackfoot disease in Taiwan, manifest by gangrene of the extremities, hypertension and serious impacts on the cardiovascular system, and even hepatic damage at high doses (Lee et al. 2003; Yoshida et al. 2004). The most characteristic effects of As in populations environmentally exposed to arsenic are periorbital swelling, hyperkeratosis of the palms and soles of the feet together with hyperpigmentation and hypopigmentation, particularly in areas not exposed to the sun. Skin tumours have also been commonly reported and these are often located on the hands and feet. Haemangioendothelioma of the liver, a very rare form of cancer, has been associated with long-term arsenic exposure in several instances. Most of the reported cases refer to individuals which had been prescribed fowler's solution, a trivalent arsenic medication.

Arsenic is also a potential risk factor for atherosclerosis, and there are epidemiological trials that inhaled inorganic arsenic can affect the cardiovascular system (Das et al. 2010). However, the interpretation of the data between high As exposure and cardiovascular outcomes was limited due to methodological constraints. Therefore, research on the relationship between As exposure and cardiovascular indices should be given priority in countries such as Taiwan (Jomova and Valko 2011). There is correlation between As exposure and type II diabetes mellitus (Walton et al. 2004). Some studies have reported association between intellectual deficiencies in children and exposure to arsenic has been found (Wang et al. 2007). For example, children exposed to arsenic contaminated drinking water (0.0034–0.0042 mg As/kg/day) showed decreased performance in the switching attention task and in tests of pattern memory. These neurological effects were manifested as changes in the levels of dopamine, norepinephrine, and 5-hydroxytryptamine in experimental rat models exposed to As in their drinking water for 16 weeks (Kannan et al. 2001). The molecular mechanism of arsenic toxicity and carcinogenesis is thought to involve genetic changes, the involvement of increased oxidative stress, enhanced cell proliferation and altered gene expression. For example elevated ROS formation was linked with the activation of the PI3K/Akt pathway and the subsequent induction of hypoxia-inducible factor-1 (HIF-1) alpha and vascular endothelial growth factor (VEGF) in prostate cancer cells exposed to As (Jomova and Valko 2011; Mishra et al. 2010). Therapeutically, Arsenic trioxide may be useful in the treatment of acute promyelocytic leukemia, a fast-growing cancer in which the bone marrow produces abnormal myeloblasts, which would normally develop into white blood cells that fight infection (Jomova and Valko 2011). Research effort is now made to decipher the underlying mechanisms of arsenic trioxide action and its

Table 7.3 Large scale cases of poisoning associated with exposure to inorganic arsenic

Type of incident	Episodes
Food inadvertently contaminated during preparation (acute and sub-acute effects)	Morinaga milk incident, Japan (1955). >12,000 cases, 130 deaths
	Soy sauce incident, Japan (1956). >400 cases
During water exposure (chronic effects)	Numerous incidents, with reports from Argentina, Chile, Taiwan, U.S.A., Canada and Japan. In excess of 0.25 million people exposed for several decades in Chile.
Communities close to point sources of airborne arsenic (chronic effects)	Evidence for effects on morbidity and mortality around copper smelters and pesticide plants. Single report of hearing effects in children living near a power plant in Czechoslovakia where high As (~ 1,000 μg/gm) lignite was burnt.
Exposure via medications (chronic effects)	Sodium arsenic (Fowler's solution) previously used in the treatment of psoriasis and leukaemia, resulting in severe effects in hundreds of patients.

Source: Hutton (1987)

interactions with different proteins to enhance its therapeutic potential ((Jomova and Valko 2011). In populations outside the occupational setting, the effects of arsenic may be divided into four categories (Table 7.3). The more serious of the two food poisoning incidents in Japan resulted from the accidental use of arsenic-contaminated sodium phosphate in the preparation of the dried milk (Morinag's Powdered Milk "MF") for infants. The incident resulted in the death of 600 new born babies, and out of the 624 survivors, 20 % were afflicted by severe mental retardation, severe hearing loss, developmental difficulties, and brain damage- related paralysis (Hutton 1987). Similarly, minor hearing loss was reported in children residing close to a power plant in Czechoslovakia burning lignite with an arsenic content of about 1,000 μg/g but this was not reported in children subjected to elevated arsenic exposure levels from a large copper smelter (Shargorodsky et al. 2011).

7.4.3 Male Reproductive Health Effects of Mercury

Mercury has been reported to cross the blood-testes barrier and accumulate in Sertoli and Leydig cells, in the testes of experimental animals (Wirth and Mijal 2010). This has been reported to negatively affect measures of male reproductive health (Tan et al. 2009). For instance in rats, the long term treatment with inorganic or organic mercury induced Leydig cell disintegration, inhibited the activity of 3β-HSD, decreased testosterone levels, decreased the number of epididymal sperm and motility (Rao and Gangadharan 2008). The decreases in the motile sperms were also reported in monkeys exposed to mercury. This was thought to be mediated via interference with the dynein/microtubule sliding assembly (Wirth and Mijal 2010). Similarly, in vitro treatment of human sperm with mercury (10.0–160.4 mg/L) induced membrane lipid peroxidation and DNA breaks, lowered sperm viability, and decreased the rate of the acrosome reaction leading to sperm dysfunction (Arabi and Heydarnejad 2007). The prominent feature of Mercury induced toxicity are: (1) depletion and clogging of different spermatogenic cells, (2) presence of pyknotic or karyotectic pachytene nuclei, (3) absence of nuclear chromatin at stage XII in dividing cells, (4) absence of noticeable lumen and (5) presence of vacuolated early elongated spermatid along with dispositioning of acrosome (Vachhrajani et al. 1990).

In humans methyl mercury has been detected in semen (Rignell-Hydbom et al. 2007) and these levels have correlated with poor male reproductive outcomes. For instance, male partners of infertile couples with abnormal sperm parameters (below World Health Organization cut-off levels for normal) had higher blood mercury levels than did men of proven fertility (Choy et al. 2002). The seminal fluid mercury level of male partners of infertile couples from Hong Kong was positively correlated with abnormal sperm morphology (Wirth and Mijal 2010). Subfertile men had approximately 40 % more methyl mercury in hair than did fertile age-matched men (4.5 parts per million (ppm) vs. 3.9 ppm) (Dickman et al. 1998). However, some studies because of the small number of participants, did not observe positive correlations between blood or semen mercury chloride levels and sperm motility, sperm concentration, and total sperm count (Wirth and Mijal 2010; Rignell-Hydbom et al. 2007). In one study conducted in an infertility clinic in Singapore, men with less than 40 % motile sperm and men with greater than 40 % motile sperm had no significant difference in their blood methyl mercury levels (Chia et al. 1992). Similarly, no adverse effects on sperm parameters or alterations in reproductive hormones were found in infertility patients in Michigan, USA with low blood mercury level (1.10 μg/L) (Wirth and Mijal 2010). It can be concluded that men with higher blood mercury concentrations (above 8 mg/L) were more likely to have lowered sperm parameters than men with a somewhat lower concentration of mercury. However, the influence of confounding factors in the small sample size study needs to be addressed especially where the infertile subjects are drawn from populations consuming higher levels of mercury contaminated fish that also contain other contaminants such as organochlorine compounds with known adverse reproductive effects.

7.4.4 Other Health Effects of Mercury

Mercury (Hg) and its compounds have caused a significant number of human fatalities in several accidents around the world. The majority of toxicity due to MeHg exposure involves the central nervous system and the areas mainly affected are those associated with the sensory, visual and auditory functions and those concerned with co-ordination. The effect is most commonly characterized by increased excitability and irritability, tremors, and gingivitis (Ozuah 2000). This initially appear as paraesthesia in the tongue, lips and distal extremities, while in more severe cases, blurred and constricted vision and ataxia may appear (Clarkson 2002). Methyl mercury can cause demyelination, autonomic dysfunction, sensory nerve conduction delay, abnormal neuronal migration, and abnormal central nervous system cell division (Patrick 2002). Rarely, significant exposure can cause acrodynia or "pink disease," involving a pink rash on the extremities, pruritis, paresthesias, and pain (Patrick 2002; Ozuah 2000). Elemental Hg can be oxidized to Hg^{2+}, which accumulates preferentially in the kidneys. Renal toxicity of inorganic mercury manifest as glomerulonephritis with proteinuria (glomerular and tubular) and nephritic syndrome or gastrointestinal problems (colitis, gingivitis, stomatitis, and excessive salivation) (Ozuah 2000). Ethyl mercury (fungicides, thimerosal in vaccines, and gamma-globulin) also causes renal and central nervous system toxicity and is deposited in the liver, kidneys, skin, brain, spleen, and plasma (Clarkson 2002). Similarly, dimethyl mercury can lead to central nervous system degeneration and death (Nierenberg et al. 1998) but the compound is luckily encountered only in specialized laboratories. Environmental levels of methyl mercury can cause harmful effects on the cardiovascular system. In a long-term prospective study, dietary Hg intake through fresh water fish and hair Hg content demonstrated significant correlation with increased risk for acute myocardial infarction (Salonen et al. 1995). Men with the highest hair mercury had a 2.9-fold increased risk for cardiovascular death. An examination of the same cohort found a correlation between hair mercury and increased risk for progression of carotid atherosclerosis (Patrick 2002). Prenatal exposure to methyl mercury has been correlated with significant blood pressure elevations in 7-year-old children as a result of maternal fish intake (Patrick 2002).

The developing nervous system of the fetus is more sensitive to low level MeHg than the adult and pre-natal exposure can result in neurotoxic effects in the infant in the absence of effects in the mother. Pregnant women are apparently more sensitive to methyl mercury than are other adults (Grandjean et al. 2010). Mothers exposed to mercury in the 1955 disaster in Minamata Bay, Japan, by long-term ingestion of contaminated fish, gave birth to infants with mental retardation, retention of primitive reflexes, cerebellar symptoms, and other abnormalities. These symptoms characterized the Hunter-Russell syndrome. The incident resulted in the death of over 100 infants (Jarup 2003). Similarly, pregnant women in Faroe Islands exposed to low level mercury through dietary intake of fish and whale meat was associated with decrements in motor function, language, memory, and neural transmission in their offspring (Grandjean et al. 1998; Murata et al. 1999). Following the several outbreaks of methylmercury poisoning, some follow up studies among the Canadian-Indians because of their elevated methylmercury intake from fish have reported an association between several neurological parameters and MeHg exposure (Jarup 2003). However, the conclusion of the study was questioned based on the influence of confounding factors on the putative association. Authors concluded that the evidence for a neurological effect of methylmercury in the study population was equivocal but that the exposure regimes in question may be at the threshold for effects (Jarup 2003). Similarly studies in other populations with heavy fish consumption in the Mediterranean, particularly in Italy, in Papua New Guinea, Peru, New Zealand and the Seychelles have failed to provide clinical evidence of methyl mercury intoxication (Mahaffey 2005). Mercury chloride intoxication was also characterized by the Iraqi epidemic in 1971–1972 where exposure resulted from consumption of bread prepared from grain dressed with alkyl mercury fungicides. The incident in question resulted in the poisoning of about 6,000 individuals and the deaths of over 500 in hospital (Maruyama et al. 2012). Previous outbreaks of this type together with the number of poisonings are as follows: Iraq (1956), about 100; Iraq (1960), about 1,000; Pakistan (1969), about 100; Guatemala (1963–1965), about 45; and Ghana (1967), about 150 (Paruchuri et al. 2010). Relatively modest levels of occupational mercury exposure, as experienced, for example, by dentists, have been associated with measurable declines in performance on neurobehavioral tests of motor speed, visual scanning, verbal and visual memory, and visuomotor coordination (Bittner et al. 1998). However, evidence from well-conducted studies is lacking that the small amount of Hg released from dental amalgams during chewing is capable of causing significant illnesses, such as multiple sclerosis, systemic lupus, or chronic fatigue syndrome (Grandjean et al. 1997).

The mechanism of mercury toxicity involves interrupting intracellular calcium homeostasis, disrupting membrane potential, altering protein synthesis, and interrupting the activity of enzymes that are needed for normal neurotransmission (Clarkson 2002). Mitochondrial damage from oxidative stress and lipid peroxidation may be the earliest sign of neurotoxicity with MeHg (Yee and Choi 1996). Microtubule destruction and the neurotoxic accumulation of serotonin, aspartate, and glutamate could all be involved in the mechanisms of MeHg neurotoxicity (Patrick 2002). In the brain both MeHg and inorganic Hg bind to various thiol-containing proteins (glutathione, cysteine, albumin, etc.). The binding and dissociation of these mercury-thiol complexes are thought to control the movement of mercury and its toxic effects in the body (Clarkson 2002).

7.4.5 Male Reproductive Health Effects of Lead

The reproductive health of animal models and humans are also affected by Pb. For instance, experimental animal studies have shown that Pb exposure can lead to testicular atrophy, changes in the weights of accessory glands, alterations in semen quality, spermatogenesis, and disruption of the hypothalamic-testicular pituitary axis (Thoreux-Manlay et al. 1995). Leydig cells appear to be a target as lead exposure results in suppression of testosterone synthesis (Thoreux-Manlay et al. 1995), while Sertoli cell function does not appear to be affected (Nathan et al. 1992). Pb accumulates preferentially in the epididymis, testis, and other accessory glands including the vas deferens, seminal vesicle and seminal ejaculate (Marchlewicz et al. 1993). Pb adversely affects the testicular tissues and steroidogenic processes either directly or through endocrinological system. However, in the majority of these studies, the doses of Pb used do not mimic the human occupational and environmental exposure conditions, thereby limiting the overall significance of the purported reproductive effects. The mechanism of toxicity could involve the ability of Pb to alter zinc bioavailability via disruption of methalothionein activity and its interference with calcium-mediated processes including the disruption of the blood-testis barrier in the zona adherens junctions (Wirth and Mijal 2010). Further, Pb is a redox metal that is well known to generate ROS and oxidative stress in a variety of systems. The generated ROS can inhibit the production of sulfhydryl antioxidants, damage nucleic acids and inhibit DNA repair, and initiate membrane lipid peroxidation and apoptosis (Wirth and Mijal 2010). In human study, the semen Pb was correspondingly increased with blood Pb in the occupationally exposed subjects (Wirth and Mijal 2010). Pb level is also a significant predictor for reproductive hormones; although this is less consistent than the association between blood Pb level and seminal parameters (Meeker et al. 2009). The high blood Pb level has been associated with decreased libido that was followed by reduced sperm count, poor sperm motility, and an increased frequency of atheno, hypo- and teratospermia (Wirth and Mijal 2010). The semen from these subjects showed decreased levels in certain seminal constituents like, fructose and succinic dehydrogenase, seminal zinc level, acid phosphatase activity, and citric acid level (Telisman et al. 2000). A decrease in D-ALAD, an indicator of long term Pb exposure, was associated with decreased seminal plasma zinc levels, indicating the adverse effects of Pb on prostate function (Wirth and Mijal 2010). Seminal plasma Pb levels were also correlated with fertilization rate and with two sperm function biomarkers (mannose receptor expression, mannose-induced acrosome reaction) in IVF patients (Benoff et al. 2003). It was reported that paternal Pb exposure was associated with an increased risk of infertility, defined as the non-occurrence of a marital pregnancy, and the risk of infertility increased with paternal Pb exposures (Sallmen et al. 2000). Many of these studies on the effects of environmental levels of Pb exposure on measures of male reproduction all found significant negative effects of low to moderate Pb exposure levels on human sperm parameters and reproductive hormone levels, an indication of Pb-induced male reproductive deficits in humans. Only in few studies were blood or seminal plasma Pb level and male reproductive outcomes not found to be correlated (Hovatta et al. 1998; Noack-Fuller et al. 1993), but considering the small number of participant recruited in most of these studies, large studies adjusted for appropriate confounders evaluating low Pb exposure levels are needed to confirm or refute these findings.

7.4.6 Other Health Effects of Lead

Lead has adverse behavioural, physiological, and biochemical effects on humans. The major health effects are manifest in three organ systems; the hematological system, the central nervous system (CNS) and the renal system (Table 7.4). Fetuses and children under the age of six are most vulnerable. The current recommended guideline for blood lead level is 50–100 ppb. But whether there is a threshold for effects is unknown (Landis and Yu 2003). Lead accumulates in bones and teeth, where it has a biological half-life of 20 years. As is the case with Cd, people who are exposed to Pb will never completely get rid of it. Although bones and teeth are

Table 7.4 Major health effects attributed to lead in environmentally exposed populations

Organ affected	Range of effects reported
Hematological system	Inhibition of D-ALAD and haem synthetase and corresponding accumulation of ALA and FEP (free erythrocyte protoporphyrin).
	At higher levels of exposure, reduced haem synthesis and anaemia
Nervous system	CNS impairment at moderate exposure in children reflected by inattention, cognitive difficulties, fine motor dysfunction and altered EEG patterns. Under heavy exposure, encephalopathy may arise.
	Effects on the peripheral nervous system indicated by reduced nerve conduction velocity.
Renal system	Functional impairment of the tubular region characterized by mild aminoaciduria, glycosuria and hyperphosphaturia. Morphological effects include mitochondria damage and intranuclear inclusion bodies. Long-term heavy exposure may result in irreversible nephropathy.

not harmed, they function as reservoirs for releasing Pb into the bloodstream where it then travels to target organs such as the brain. People with a calcium-deficient diet will accumulate more Pb in their bones; this results in greater lead toxicity (Landis and Yu 2003). The biochemical basis of the hematological effects of lead is thought to involve 2 key enzyme systems needed for the synthesis of haemoglobin, the iron-containing protein in red blood cells that binds with and transports oxygen to all cells in the human body. The enzyme catalysing the first step, D-ALAD and the enzyme, haem synthetase (ferrochelatase), that incorporate iron into protoporphyrin. When the exposure to Pb is sufficiently high, D-ALAD is inhibited in the erythrocytes and other haem forming tissues resulting to increased concentration of the substrate D-ALA in the body and urine. The accumulated D-ALA in the systemic circulation can pass the blood-brain barrier and is thought to be associated with neuromuscular and neurophysiological effects of lead (Sanders et al. 2009). When these enzymes are inhibited and haemoglobin synthesis is disrupted, there is insufficient transport of oxygen and anaemia is the result. Anaemic children fatigue easily and are less able to learn well (Landis and Yu 2003). Affected children may die from the neurological effects of Pb without even developing anaemia (Sanders et al. 2009). Lead can cross the placenta, resulting in miscarriages, stillbirths, and birth defects such as neurological damage. The neurological impacts are generally seen in children and include hyperactivity, poor attention span, low IQ, and severe encephalopathy, which can culminate in coma and death. Calcium transport across nerve cell membranes is necessary for the normal functioning of the nervous system. Lead mimics calcium, thereby inhibiting neurotransmission (Cunningham and Cunningham 2004; Landis and Yu 2003). Exposure of humans to Pb particularly at the critical developmental period is more likely to cause Alzheimer's disease later in life (Bashah and Reddy 2010). Lead exposure is also implicated in several disease burden including systemic effects such as hypertension, gastrointestinal effects (Ogunseitan and Smith 2007). The neuropsychological and neurobehavioural effects of Pb in children is of most concern in its environmental toxicity because these effects have been reported in children from the general population subjected to commonly occurring exposure regimes (Sanders et al. 2009). Study relevant to this issue has been carried out on children in the USA (Lanphear et al. 2000; Rohlman et al. 2005), the Federal Republic of Germany (Lidsky and Schneider 2003) and the United Kingdom (O'donohoe et al. 1998) and have all reported an association between increased Pb exposure and decreases in measurements of intelligence and behaviour. These data appear to provide sufficient evidence to infer a cause and effect relationship but the criticism is based on the fact that the study design of retrospective epidemiological investigations will never fully be able to control for confounding factors; apart the influence of possible bias in the recruitment of eligible subjects for the study. Occupational Pb exposure also causes renal effects including nephropathy. This occurs at exposure levels rarely encountered outside of the occupational setting. Renal dysfunction in children acutely intoxicated by Pb has also been reported but whether this effect is reversible or not remain to be determined.

7.4.7 Male Reproductive Health Effects of Cadmium

The reproductive health of animal models and humans are also affected by Cd. For example, environmental doses of Cd were found to stop spermiation, the final phase of sperm differentiation, reduced sperm concentration and motility in rats (Xu et al. 2001; Benoff et al. 2008). Cd also inhibits human chorionic gonadotrophin (hCG)-stimulated serum testosterone production in vivo and in vitro (Laskey and Phelps 1991). High dose of Cd exposure caused rapid testicular edema, haemorrhage and necrosis. Cd also exerted deleterious effect on the vascular structure of testis that may be the result of varying degrees of Cd induced ischemia. Even at doses that did not affect most organs, Cd was found to damage the testes within few hours (Wirth and Mijal 2010). Cd also has the ability to disrupt the blood-testis barrier thereby gaining entry into testicular cells (Wirth and Mijal 2010). Both Sertoli and Leydig have been identified as targets for Cd adverse effects (Wirth and Mijal 2010). Cd has also been shown to accumulate in the hypothalamus and pituitary and to decrease the level of prolactin (Lafuente et al. 2001). Thus, the mechanisms involved in the reproductive effects of Cd would involve both direct (via testicular and hypothalamus-pituitary toxicity) and indirect (via altered hormone secretion). Cd also interferes with zinc mediated metabolic processes (Bridges and Zalups 2005), an event that may lead to increase generation of ROS, induction of oxidative stress, DNA damage and cell death (Wirth and Mijal 2010). There is also evidence that environmental level of Cd could decrease semen quality and alter the level of reproductive hormones of humans. Studies of Asian men attending infertility clinics reported negative correlations between blood Cd level and sperm density, sperm motility and semen volume and positive correlations with abnormal sperm rates (Chia et al. 1992). However, potential confounding variables, including cigarette smoking, were not controlled in this study. In a Nigerian study, were these confounding factors were controlled, male partners of couples attending an infertility clinic had higher serum Cd level, seminal fluid Cd compared to men with a normal number of sperm. The serum levels of LH, FSH and prolactin were higher in the men with low or no sperm (Akinloye et al. 2006). The seminal fluid Cd level of male partners attending an Indian infertility clinic negatively correlated with sperm motility and sperm concentration. However, when confounding variables such as smoking was eliminated, seminal Cd level was no longer correlated with sperm motility (Pant et al. 2003). Similarly, adjustment for potential confounders also reported negative associations between blood Cd levels and testes size but not with sperm parameters among infertile Croatian men (Jurasovic et al. 2004). In this study, blood Cd level was positively associated with FSH, testosterone, and estradiol levels, and seminal fluid acid phosphatase, an indicator of prostate function. There was no adverse effect of blood Cd on sperm parameters in infertile patients in Michigan, USA (Wirth and Mijal 2010). However, Cd was positively associated with serum inhibin B but not FSH or Testosterone levels in a model adjusted for confounding factors (Wirth and Mijal 2010). Significant positive correlations between blood Cd levels and pathologic sperm, and LH, and testosterone levels and a negative correlation with prolactin level has been reported in a study population occupationally exposed to Cd (Telisman et al. 2000). Some studies have found no correlations between the semen Cd level and the sperm parameters; and that semen Cd level were similar in men with proven fertility, infertility patients with normal sperm parameters, and patients with abnormal semen parameters (Hovatta et al. 1998). There are also other studies were significant correlations were not found between blood and semen or sperm Cd levels (Kiziler et al. 2007). While it is difficult to interpret these data in the overall sense, it may appear that environmental levels of Cd adversely affect sperm parameters and alter hormone levels. The data from experimental animals are promising and would justify the need to clarify the relationship between adverse effects of low Cd exposure on semen parameters in human males

7.4.8 Other Health Effects of Cadmium

The public first became aware of the toxic effect of Cd on humans when people living close to a mine that had discharged untreated effluent containing Cd into the Jintsu River Basin (Fuchu, Japan) complained of joint, bone, and muscle pains, and kidney dysfunction. This combination of health problems was later called Itai-Itai (Ouch-Ouch) Disease and resulted in the deaths of some residents. The residents had been exposed to Cd-contaminated rice grown in these areas for up to 30 years (Landis and Yu 2003). Exposed humans will never get rid of all of the Cd in their bodies because it will take 20–30 years to get rid of 50 % of the Cd, 40–60 years to get rid of 75 % of the Cd, and 60–90 years to get rid of 87.5 % of the Cd (Landis and Yu 2003). As shown in Itai-Itai Disease, Cd effects on human health include skeletal deformities and bone loss, kidney damage, and generalized pain. These effects are still seen in elderly multiparous women (over 50 years of age) with poor nutritional status who had lived in the contaminated area for many years (over 50 years of age) (Kah et al. 2012).

Cadmium-induced nephropathy has been described also in industrial workers and in most population-based studies at Cd levels that is environmentally relevant. Most of the studies conducted on human populations and experimental animals have demonstrated that Cd exerts its renal toxicity at low levels that could exacerbate the age-related decline in renal function; in a dose-dependent manner, and the adverse effects occurring only when the renal Cd concentration reaches a critical threshold (Bernard 2008; Burbure et al. 2006; Lauwerys et al. 1991). One of the initial signs of renal dysfunction is an increased urinary excretion of proteins. Cd-induced proteinuria is generally considered to be characterized by the excretion of low molecular weight proteins, particularly α_2, β_2 and γ-globulins. This form of proteinuria is a reflection of impaired reabsorption function of the proximal tubules. Studies have shown that the development of Cd-induced proteinuria is predictive of an increased mortality by heart failure, cerebral infarction, nephritis and nephrosis (Bernard 2008) in humans subjected to long-term Cd-contaminated areas. There is some epidemiological evidence that diabetics are more susceptible to the nephrotoxic action of Cd, a finding consistent with animal studies (Buchet et al. 1990). The disturbances of calcium and phosphate metabolism accompanying Cd nephropathy may lead to bone demineralization, the formation of kidney stones and bone fractures. This was confirmed recently among Chinese farmers exposed to Cd from contaminated rice for more than 20 years (Nordberg et al. 2007). In this study, the bone mass density was decreased in postmenopausal women with elevated Cd in urine or blood as well as among men with elevated Cd in blood. By blocking Vitamin D synthesis, Cd prevents the bones from having a normal amount of calcium leading to a urinary waste of calcium and phosphate (Jarup et al. 1998). However, there has been conflicting results on the association between increased urinary Cd levels and increased risk of bone fractures in the general population with very low environmental exposure to Cd (Bernard 2008; Wallin et al. 2005).

Exposure to high level of Cd via inhalation can also results in acute pneumonitis with pulmonary oedema, which may be lethal. Long-term, high-level occupational exposure is associated with lung changes, primarily characterized by chronic obstructive airway disease. There is sufficient evidence that long-term occupational exposure to Cd (e.g. through cadmium oxide fume) contributes to the development of lung cancer in both human and animal studies (Bernard 2008; Waalkes 2003). Human studies on environmental-induced carcinogenesis in the occupationally exposed populations require a more careful analysis because of the possible effects of confounding variables. In a recent study, the relative risk of lung cancer was lower than in the past, when confounding variables (exposure to arsenic and nickel) were eliminated (Nordberg et al. 2007). There is limited evidence that Cd exposure in the general population can cause cancers of the prostate and kidney but this evidence is weaker compared to that on lung cancer (Zeng et al. 2004; Bernard 2008). Thus, the possibility that Cd can be involved in environmental carcinogenesis cannot be excluded (Straif et al. 2009). However, the mechanism of Cd carcinogenesis remains largely unknown, as the metal is not strongly genotoxic and does not cause direct genetic damage, epigenetic mechanisms and/or indirect genotoxic mechanisms such as a blockage of apoptosis, alterations in cell signaling or inhibition of DNA repair might be involved (Waalkes 2003).

7.5 Chelation Therapy of Toxic Elements in Medicine

Chelation therapy is a tool used in medicine for modifying metal concentrations in the body. It is therefore an attempt to prevent or reverse health problems of individuals exposed to high levels of metals. Chelation therapy of toxic metals involves the use of chelates injected into the blood, muscle or taken orally to bind metals that are present in toxic concentrations so they can be excreted from the body, most frequently in urine (Rogan et al. 2001). A chelating agent forms a stable complex with a toxic metal may shield the metal ion from biological targets, thereby reducing the toxicity, or it may expose the metal to the biological environment and prevent the metal from being scavenged by biological protective mechanisms and thereby increase the toxicity of the metal (Rogan et al. 2001). The characteristics of an ideal chelator include: (*i*) greater affinity for the toxic metal; (*ii*) low toxicity; (*iii*) ability to penetrate cell membrane; (*iv*) rapid elimination of metal; and (*v*) higher water solubility. One of the most frequently used chelators applied in the treatment of heavy metal toxicity is 2, 3-dimercaprol (BAL) (Blanusa et al. 2005). Dimercaprol has a dithiol group and has been used clinically as effective chelating agent against heavy metal toxicity since 1949. In humans and experimental models, the antidotal efficacy of dimercaprol is most effective when administered immediately after the exposure to toxic metals. Dimercaprol competes with the thiol groups of enzymes for binding the arsenic or other metals to form a stable metal-chelate which is then excreted from the body in the urine. Dimercaprol is lipophilic in nature and therefore distributes both extra-cellular and intra-cellular sites. Some of the major drawback in the use of dimercaprol against metal poisoning includes its tendency to accumulate As, Pb and Cd in some organs and exhibits side effects including nephrotoxicity and hypertension (Andersen 2004). Because of its lipophilic nature, dimercaprol requires deep intra-muscular injection that is extremely painful and allergic. Other side effects include vomiting, headache, lachrymation, rhinorrhea and salivation, profuse sweating, intense pain in the chest and abdomen and anxiety, and hemolysis (destruction of red blood cells) in patients with glucose-6-phophate deficiency (Flora et al. 2008).

The most effective chelators used in both paediatric and adult treatment of Pb toxicity are dimercaptosuccinic acid (DMSA) and calcium disodium ethylenediaminetetraacetic acid ($CaNa_2EDTA$). $CaNa_2EDTA$ chelates only extracellular Pb (not intracellular) and is frequently used in conjunction with BAL to increase its efficiency. In addition to urinary excretion of Pb, $CaNa_2EDTA$ is responsible for the excretion and depletion of essential metals like Zn, Cu, Fe, Co and Mn because of its relative lack of specificity. $CaNa_2EDTA$ has the tendency to accumulate Pb in the brain of experimental rats and exhibits neurotoxicity (Flora and Pachauri 2010).

$CaNa_2EDTA$ also causes adverse effect on the renal system (Aposhian et al. 1995). Thus $CaNa_2EDTA$ could not be regarded as a drug of choice against Pb poisoning. There is a lack of evidence of beneficial effects of chelating agents on Cd toxicity after prolonged exposure. Chelation therapy with $CaNa_2EDTA$ may be prescribed in the early period after acute Cd exposure. Progress in chelation therapy for Cd removal is rather difficult since this metal is tightly bound to metallothionein in the liver and kidneys (Jomova and Valko 2011). DMSA is a chemical derivative of dimercaprol and has been shown to be an effective chelator of toxic metal mainly Pb and As (Bradberry and Vale 2009; Jomova and Valko 2011). Few major advantages of DMSA include its low toxicity, oral administration and no redistribution of metal from one organ to another and do not increase the excretion of calcium, zinc or iron (Flora et al. 1997). However, DMSA has low efficiency to remove Pb from intracellular sites because of its lipophobic nature (Kalia and Flora 2005). DMSA can also cross the blood brain barrier, limiting thus its use in the treatment of the central nervous system. In cases of human As poisoning, DMSA was found ineffective in producing any clinical or biochemical benefits (Guha et al. 1998) and does not chelate Pb and As from hard tissues (Flora et al. 1997). DMSA administration is also associated with renal toxicity as a result of the excretion of large amount of chelated metals that pass through the renal tubules during therapy. Hemolytic anemia was observed in the course of DMSA chelation therapy in patients with chronic Pb intoxication (Jomova and Valko 2011). After termination of therapy, hematological values returned back to normal. New chelating agents that are analogues of DMSA such as monoisoamyl ester of DMSA (MiADMSA), monomethyl DMSA (MmDMSA) and monocyclohexyl DMSA (MchDMSA) are more effective than DMSA in reducing heavy metal burden from various organs in heavy metal exposed animals (Wang et al. 2006). They have better lipophilicity characteristic and might penetrate cells more readily than extra-cellularly acting chelating agent (Flora et al. 2004). Chelation therapy of Cd produced significant reductions in Cd levels with no redistribution of Cd in brain of experimental models (Jones et al. 1992). Thus, the oral administration of the analogues of DMSA improves their advantage in the clinical treatment of heavy metal toxicity. A new trend in the medical treatment of metal poisoning is the use of structurally different chelators in order to achieve a more effective removal of toxic metals. This is based on the assumption that various chelating agents are likely to mobilise toxic metals from different tissue compartments and therefore better results could be expected (Flora et al. 2002). For example, the combination of DMSA and $CaNa_2EDTA$ against Pb poisoning resulted in enhanced elimination of Pb without redistribution of Pb to any other organ in rats (Flora et al. 2002). Similarly, an enhanced elimination of As from blood and soft tissues was observed when DMSA and MiADMSA were administered together (Mishra et al. 2008). DMSA and MiADMSA in addition to acting as a metal chelator could also act as an antioxidant and protect against hepatic DNA damage (Flora et al. 2004). The combined administration of MiADMSA and $CaNa_2EDTA$ is also beneficial in reducing body Pb burden and neuronal cell death (Flora et al. 2007). It was also found that the combined administration of DMSA and MiADMSA was not only beneficial in decreasing Pb burden but also provide better clinical recoveries especially in the brain than monotherapy (Flora and Pachauri 2010).

In the event that oxidative stress can be partially implicated in toxicity of metals, a therapeutic strategy to increase the antioxidant capacity of cells may fortify the long term effective treatment. In this regard, antioxidant supplementation with chelating agents was found to show additive or synergistic effect that improved chelating ability. For example, the combination of taurine with DMSA or MiADMSA in arsenic-exposed animals was more effective in depleting blood, liver, kidney and brain Pb and As levels respectively (Flora and Pachauri 2010). Similarly, a combination of DMSA with alpha-lipoic acid in Pb-exposed animals was more effective in preventing oxidative damage as measured by alterations in erythrocyte membrane enzyme levels (Sivaprasad et al. 2004). A similar effect of improved chelating ability was observed for $CaNa_2EDTA$ administrated in conjunction with zinc (Batra et al. 1998). N-acetyl-cysteine (NAC) could be used both as preventive as well as therapeutic agent along with MiADMSA/DMSA in the prevention or treatment of Pb intoxication in rats (Pande et al. 2001). Vitamin C may facilitate the absorption and excretion of Pb in low-exposed subjects with higher vitamin C supplementation. Vitamin E alone or in combination with $CaNa_2EDTA$, was found to decrease the Pb-induced lipid peroxide levels of liver and brain in rats (Patra et al. 2001). Similarly, vitamin C in combination with MiADMSA is effective reducing liver and kidney As burden (Flora and Pachauri 2010). It was also found that combination treatment with the thiol chelator and natural antioxidant proved to be more beneficial than monotherapy in the recovery from Pb-induced oxidative stress and the body Pb level (Saxena and Flora 2006). Combination therapy with antioxidants or phytochemicals and chelating agent proved to be effective in mobilizing metal from soft as well as hard tissue (Flora and Pachauri 2010). Antioxidants compounds can therefore act as complimentary chelating agents (adjuvants) increasing the efficacy of a known chelator, or by acting independently (Fig. 7.4). Therefore, the co-administration of two chelating agents or when chelating agents are used in conjunction with antioxidants gives better efficacy in terms of recovery from metal toxicity and can be a standard strategy in treatment of heavy metal toxicity.

7.6 Conclusion

Multiple pathways of metals through air, food, water, e.t.c., account for its easy accessibility to man, and factors like bio-magnification along the food chain complicate the problem. Although, majority of the available information in the literature is on the adverse health effects of single exposures to metals, much is not known about the complex mechanisms of interaction between combined exposures of these metals in vivo. As the prevalence of metal exposure is increasingly recognized and identified in humans, the need for effective prevention and treatment will increase. Biological monitoring can be of great value in preventive programs especially when it is complemented with conventional monitoring of the different environmental media (e.g., air, food, water). Furthermore, the development of newer biomarkers with the help of

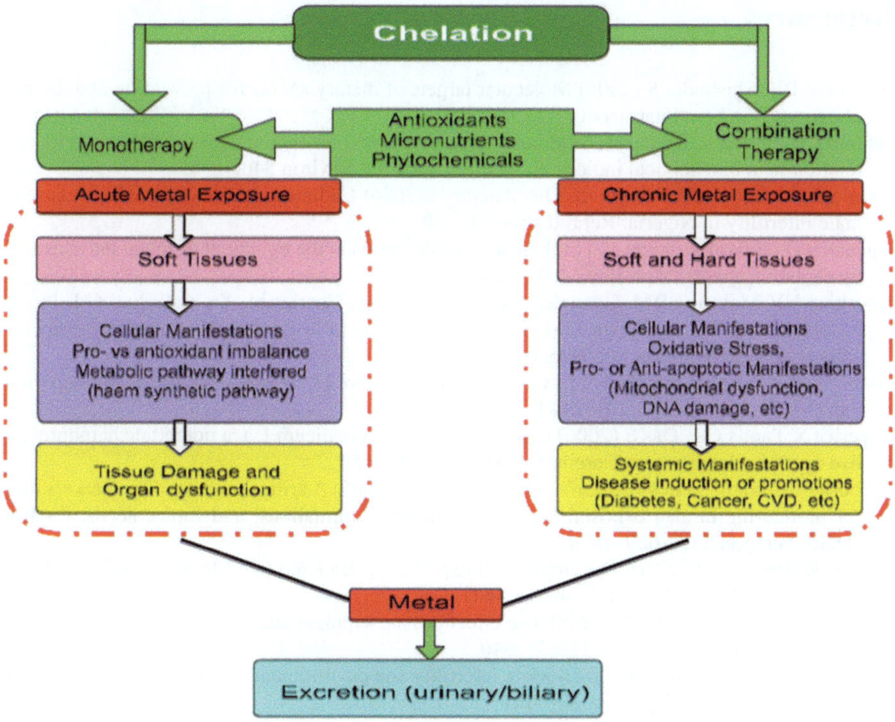

Fig. 7.4 Chelation therapy and the excretion of metals from soft and hard tissue of the body during acute and chronic exposure to toxic metals

advanced techniques and their use in environmental epidemiological studies can enable more informed risk assessment decision making. Preventive measures are now in place in some European countries, ranging from monitoring endogenous levels of metal ions in foods and drinks to detecting contamination during food preparation. The governments of respective developing countries should also ensure toxic metal free air, water and food by making strict laws regarding contaminating industrial units and ensuring proper disposal of toxic metal garbage. Media and non-governmental organizations should raise a voice against any negligence on part of government besides educating the masses about metal hygiene. Also, the advances of toxicology have improved our knowledge about human exposure to toxic metals and their health effects, and the detecting of the minutest deleterious effect of a chemical. This is very significant especially in today's industrialized world where every individual is carrying Pb burden in the bones, Hg in hair and Cd in semen. Although chelation therapy is an important tool in fighting metal storage disorders, the yet lack of larger clinical trials still offers controversy on its clinical therapeutic benefits. Further, newer therapeutic strategies should be investigated that may provide better therapeutic outcomes.

References

Aggarwal BB, Shishodia S (2006) Molecular targets of dietary agents for prevention and therapy of cancer. Biochem Pharmacol 71:1397–1421

Ahamed M, Verma S, Kumar A, Siddiqui MK (2005) Environmental exposure to lead and its correlation with biochemical indices in children. Sci Total Environ 346:48–55

Akinloye O, Arowojolu AO, Shittu OB, Anetor JI (2006) Cadmium toxicity: a possible cause of male infertility in Nigeria. Reprod Biol 6:17–30

Andersen O (2004) Chemical and biological considerations in the treatment of metal intoxications by chelating agents. Mini Rev Med Chem 4:11–21

Aposhian HV, Maiorino RM, Gonzales-Ramirez D, Zuniga-Charles M, Xu Z, Hrlbut KM, Junco-Munoz P, Dart RC, Aposhian MM (1995) Mobilization of heavy metals by newer, therapeutically useful chelating agents. Toxicology 97:23–38

Arabi M, Heydarnejad MS (2007) In vitro mercury exposure on spermatozoa from normospermic individuals. Pak J Biol Sci 10:2448–2453

Banerjee S, Panda CK, Das S (2006) Clove (Syzygium aromaticum L.), a potential chemopreventive agent for lung cancer. Carcinogenesis 27:1645–1654

Barbosa F Jr, Tanus-Santos JE, Gerlach RF, Parsons PJ (2005) A critical review of biomarkers used for monitoring human exposure to lead: advantages, limitations, and future needs. Environ Health Perspect 113:1669–1674

Bashah R, Reddy GR (2010) Developmental exposure to lead and late life abnormalities of nervous system. Indian J Exp Biol 48:636–641

Batra N, Nehru B, Bansal MP (1998) The effect of zinc supplementation on the effects of lead in the rat testis. Reprod Toxicol 12:535–540

Benoff S, Jacob A, Hurley IR (2000) Male infertility and environmental exposure to lead and cadmium. Hum Reprod Update 6:107–121

Benoff S, Centola GM, Millan C, Napolitano B, Marmar JL, Hurley IR (2003) Increased seminal plasma lead levels adversely affect the fertility potential of sperm in IVF. Hum Reprod 18:374–383

Benoff S, Auborn K, Marmar JL, Hurley IR (2008) Link between low-dose environmentally relevant cadmium exposures and asthenozoospermia in a rat model. Fertil Steril 89:e73–e79

Bernard A (2008) Cadmium and its adverse effects on human health. Indian J Med Res 128:557–564

Bhattacharya R, Chatterjee D, Nath B, Jana J, Jacks G, Vahter M (2003) High arsenic groundwater: mobilization, metabolism and mitigation – an overview in the Bengal delta plain. Mol Cell Biochem 253:347–355

Biswas BK, Dhar RK, Samanta G, Mandal BK, Chakraborti D, Faruk I, Islam KS, Chowdhury MM, Islam A, Roy S (1998) Detailed study report of Samta, one of the arsenic-affected villages of Jessore District, Bangladesh. Curr Sci 74:134–145

Bittner AC Jr, Echeverria D, Woods JS, Aposhian HV, Naleway C, Martin MD, Mahurin RK, Heyer NJ, Cianciola M (1998) Behavioral effects of low-level exposure to Hgo among dental professionals: a cross study evaluation of psychomotor effects. Neurotoxicol Teratol 20:429–439

Blanusa M, Varnai VM, Piasek M, Kostial K (2005) Chelators as antidotes of metal toxicity: therapeutic and experimental aspects. Curr Med Chem 12:2771–2794

Bradberry S, Vale A (2009) Dimercaptosuccinic acid (succimer; DMSA) in inorganic lead poisoning. Clin Toxicol 47:617–631

Bridges CC, Zalups RK (2005) Molecular and ionic mimicry and the transport of toxic metals. Toxicol Appl Pharmacol 204:274–308

Buchet JP, Lauwerys R, Roels H, Bernard A, Bruaux P, Claeys F, Ducoffre G, de Plaen P, Staessen J, Amery A, Lijnen P, Thijs L, Rondia D, Sartor F, Saint Remy A, Nick L (1990) Renal effects of cadmium body burden of the general population. Lancet 336:699–702

Burbure C, Buchet JP, Leroyer A, Catherine N, Haguenoer JM, Mutti A et al (2006) Renal and neurological effects of cadmium, lead, mercury and arsenic in children: evidence of early effects and multiple interactions at environmental exposure levels. Environ Health Perspect 114:584–590

Chang SI, Jin B, Youn P, Park C, Park JD, Ryu DY (2007) Arsenic-induced toxicity and the protective role of ascorbic acid in mouse testis. Toxicol Appl Pharmacol 218:196–203

Chia SE, Ong CN, Lee ST, Tsakok FH (1992) Blood concentrations of lead, cadmium, mercury, zinc, and copper and human semen parameters. Arch Androl 29:177–183

Chiou TJ, Chu ST, Tzeng WF, Huang YC, Liao CJ (2008) Arsenic trioxide impairs spermatogenesis via reducing gene expression levels in testosterone synthesis pathway. Chem Res Toxicol 21:1562–1569

Chowdhury UK, Biswas BK, Chowdhury TR, Samanta G, Mandal BK, Basu GC, Chanda CR, Lodh D, Saha KC, Mukherjee SK, Roy S, Kabir S, Quamruzzaman Q, Chakraborti D (2000) Groundwater arsenic contamination in Bangladesh and West Bengal, India. Environ Health Perspect 108:393–397

Choy CM, Lam CW, Cheung LT, Briton-Jones CM, Cheung LP, Haines CJ (2002) Infertility, blood mercury concentrations and dietary seafood consumption: a case-control study. BJOG 109:1121–1125

Clarkson TW (2002) The three modern faces of mercury. Environ Health Perspect 110:11–23

Clarkson TW, Magos L (2006) The toxicology of mercury and its chemical compounds. Crit Rev Toxicol 36:609–662

Cunningham WP, Cunningham MA (2004) Principles of environmental science: inquiry and applications. McGraw Hill Publishers, New York

Das S, Roy Chowdhury A (2006) Arsenic: a global monster. Sci Cult 27:230–237

Das AK, Sahu R, Dua TK, Bag S, Gangopadhyay M, Sinha MK, Dewanjee S (2010) Arsenic-induced myocardial injury: protective role of Corchorus olitorius leaves. Food Chem Toxicol 48:1210–1217

DEFRA (2002) Contaminants in soil: collation of toxicological data and intake values for humans. Cadmium. Environment Agency, Bristol

Dickman MD, Leung CK, Leong MK (1998) Hong Kong male sub-fertility links to mercury in human hair and fish. Sci Total Environ 214:165–174

Evan GI, Vousden KH (2001) Proliferation, cell cycle and apoptosis in cancer. Nature 411:342–348

Ferreccio C, Gonzalez C, Milosavjlevic V, Marshall G, Sancha AM, Smith AH (2000) Lung cancer and arsenic concentrations in drinking water in Chile. Epidemiology 11:673–679

Flora SJS, Pachauri V (2010) Chelation in metal intoxication. Int J Environ Res Public Health 7:2745–2788

Flora SJS, Pant BP, Tripathi N, Kannan GM, Jaiswal DK (1997) Distribution of arsenic by diesters of meso 2,3-dimercaptosuccinic acid during sub-chronic intoxication in rats. J Occup Health 39:119–123

Flora SJS, Dubey R, Kannan GM, Chauhan RS, Pant BP, Jaiswal DK (2002) meso 2, 3-dimercaptosuccinic acid (DMSA) and monoisoamyl DMSA effect on gallium arsenide induced pathological liver injury in rats. Toxicol Lett 132:9–17

Flora SJS, Mehta A, Rao PVL, Kannan GM, Bhaskar ASB, Dube SN, Pant BP (2004) Therapeutic potential of monoisoamyl and monomethyl esters of meso 2,3-dimercaptosuccinic acid in gallium arsenide intoxicated rat. Toxicology 195:127–146

Flora SJS, Saxena G, Mehta A (2007) Reversal of lead-induced neuronal apoptosis by chelation treatment in rats: role of ROS and intracellular Ca^{2+}. J Pharmacol Exp Ther 322:108–116

Flora SJS, Mittal M, Mehta A (2008) Heavy metal induced oxidative stress and its possible reversal by chelation therapy. Indian J Med Res 128:501–523

Fukui Y, Miki M, Ukai H, Okamoto S, Takada S, Higashikawa K, Ikeda M (1999) Urinary lead as a possible surrogate of blood lead among workers occupationally exposed to lead. Int Arch Occup Environ Health 72:516–520

García-Esquinas E, Pérez-Gómez B, Fernández MA, Pérez-Meixeira AM, Gil E, de Paz C, Iriso A, Sanz JC, Astray J, Cisneros M, de Santos A, Asensio A, García-Sagredo JM, García JF, Vioque J, Pollán M, López-Abente G, González MJ, Martínez M, Bohigas PA, Pastor R, Aragonés N (2011) Mercury, lead and cadmium in human milk in relation to diet, lifestyle habits and sociodemographic variables in Madrid (Spain). Chemosphere 85:268–276

García-Vargas GG, Hernández-Zavala A (1996) Urinary porphyrins and heme biosynthetic enzyme activities measured by HPLC in arsenic toxicity. Biomed Chromatogr 10:278–284

Gonick HC (2008) Nephrotoxicity of cadmium and lead. Indian J Med Res 128:335–352

Grandjean P, Guldager B, Larsen IB, Jorgensen PJ, Homstrup P (1997) Placebo response in environmental disease. Chelation therapy of patients with symptoms attributed to amalgam fillings. J Occup Environ Med 39:707–714

Grandjean P, Weihe P, White RF, Debes F (1998) Cognitive performance of children prenatally exposed to "safe" levels of methylmercury. Environ Res 77:165–172

Grandjean P, Satoh H, Murata K, Eto K (2010) Adverse effects of methylmercury: environmental health research implications. Environ Health Perspect 118:1137–1145

Groth E (2010) Ranking the contributions of commercial fish and shellfish varieties to mercury exposure in the United States: implications for risk communication. Environ Res 110:226–236

Guha Mazumder DN, Das Gupta J, Santra A (1998) Chronic arsenic toxicity in West Bengal-the worst calamity in the world. J Indian Med Assoc 96:4–7

Hovatta O, Venalainen ER, Kuusimaki L, Heikkila J, Hirvi T, Reima I (1998) Aluminium, lead and cadmium concentrations in seminal plasma and spermatozoa, and semen quality in Finnish men. Hum Reprod 13:115–119

Hu H (2000) Exposure to metals. Prim Care 27:983–996

Hutton M (1987) Human health concerns of lead, mercury, cadmium and arsenic. In: Hutchinson TC, Meema KM (eds) Lead, mercury, cadmium and arsenic in the environment. Wiley, New York, pp 53–68

Jana K, Jana S, Samanta PK (2006) Effects of chronic exposure to sodium arsenite on hypothalamo-pituitary-testicular activities in adult rats: possible an estrogenic mode of action. Reprod Biol Endocrinol 4:9

Jarup L (2003) Hazards of heavy metal contamination. Br Med Bull 68:167–182

Järup L, Berglund M, Elinder CG, Nordberg G, Vahter M (1998) Health effects of cadmium exposure – a review of the literature and a risk estimate. Scand J Work Environ Health 24(Suppl 1):1–51

Jarup L, Hellstrom L, Alfven T, Carlsson MD, Grubb A, Persson B, Pettersson C, Spang G, Schutz A, Elinder C-G (2000) Low level exposure to cadmium and early kidney damage: the OSCAR study. Occup Environ Med 57:668–672

Jensen A, Bro-Rasmussen R (1992) Environmental cadmium in Europe. Rev Environ Contam Toxicol 125:101–181

Jomova K, Valko M (2011) Advances in metal-induced oxidative stress and human disease. Toxicology 283:65–87

Jomova K, Jenisova Z, Feszterova M, Baros S, Liska J, Hudecova D, Rhodes CJ, Valko M (2011) Arsenic: toxicity, oxidative stress and human disease. J Appl Toxicol 31:95–107

Jones FT (2007) A broad view of arsenic. Poult Sci 86:2–14

Jones MM, Singh PK, Gale GR, Smith AB, Atkins LM (1992) Cadmium mobilization *in vivo* by intraperitoneal or oral administration of mono alkyl esters of meso 2, 3-dimercaptosuccinic acid. Pharmacol Toxicol 70:336–343

Jurasovic J, Cvitkovic P, Pizent A, Colak B, Telisman S (2004) Semen quality and reproductive endocrine function with regard to blood cadmium in Croatian male subjects. Biometals 17:735–743

Kah M, Levy L, Brown C (2012) Potential for effects of land contamination on human health. 1. The case of cadmium. J Toxicol Environ Health B Crit Rev 15:348–363

Kakkar P, Jaffery FN (2005) Biological markers for metal toxicity. Environ Toxicol Pharmacol 19:335–349

Kalia K, Flora SJS (2005) Strategies for safe and effective therapeutic measures for chronic arsenic and lead poisoning. J Occup Health 47:1–21

Kannan GM, Tripathi N, Dube SN, Gupta M, Flora SJ (2001) Toxic effects of arsenic (III) on some hematopoietic and central nervous system variables in rats and guinea pigs. J Toxicol Clin Toxicol 39:675–682

Kapaj S, Peterson H, Liber K, Bhattacharya P (2006) Human health effects from chronic arsenic poisoning– a review. J Environ Sci Health Part A 41:2399–2428

Karagas MR, Choi AL, Oken E, Horvat M, Schoeny R, Kamai E, Cowell W, Grandjean P, Korrick S (2012) Evidence on the human health effects of low-level methylmercury exposure. Environ Health Perspect 120:799–806

Kingman A, Albertini T, Brown LJ (1998) Mercury concentrations in urine and whole blood associated with amalgam exposure in a US military population. J Dent Res 77:461–471

Kiziler AR, Aydemir B, Onaran I, Alici B, Ozkara H, Gulyasar T, Akyolcu MC (2007) High levels of cadmium and lead in seminal fluid and blood of smoking men are associated with high oxidative stress and damage in infertile subjects. Biol Trace Elem Res 120:82–91

Krishnamurti CR (1987) The cycling of arsenic, cadmium, lead and mercury in India. In: Hutchinson C, Meema KM (eds) Lead, mercury, cadmium and arsenic in the environmental. Wiley, Chichester, pp 315–333

Lafuente A, Marquez N, Perez-Lorenzo M, Pazo D, Esquifino AI (2001) Cadmium effects on hypothalamic-pituitary-testicular axis in male rats. Exp Biol Med (Maywood) 226:605–611

Landis G, Yu M (2003) Introduction to environmental toxicology: impacts of chemicals upon ecological systems. CRC Press/Lewis Publishers, Boca Raton

Lanphear BP, Dietrich K, Auinger P, Cox C (2000) Cognitive deficits associated with blood lead concentrations <10μg/dl in US children and adolescents. Public Health Rep 115(521):529

Laskey JW, Phelps PV (1991) Effect of cadmium and other metal cations on in vitro Leydig cell testosterone production. Toxicol Appl Pharmacol 108:296–306

Lauwerys R, Bernard R, Buchet JP, Roels H, Bruaux P, Claeys F, Ducoffre G, De Plaen P, Staessen J, Amery A, Fagard R, Lijnen P, Thijs L, Rondia D, Sartor F, Saint Remy A, Nick L (1991) Does environmental exposure to cadmium represent a health risk? Conclusions from the Cadmibel study. Acta Clin Belg 46(4):219–225

Lee MY, Jung BI, Chung SM, Bae ON, Lee JY, Park JD, Yang JS, Lee H, Chung JH (2003) Arsenic-induced dysfunction in relaxation of blood vessels. Environ Health Perspect 111:513–517

Levine RL (2012) The need for congressional action to finance arsenic reductions in drinking water. J Environ Health 75:20–25

Lidsky TI, Schneider JS (2003) Lead neurotoxicity in children: basic mechanisms and clinical correlates. Brain 126:5–19

Mahaffey KR (1990) Environmental lead toxicity: nutrition as a component of intervention. Environ Health Perspect 89:75–78

Mahaffey KR (2005) Mercury exposure: medical and public health issues. Trans Am Clin Climatol Assoc 116:127–153

Mahaffey KR, Clickner RP, Bodurow CC (2004) Blood organic mercury and dietary mercury intake: National Health and Nutrition Examination Survey, 1999 and 2000. Environ Health Perspect 112:562–570

Maneakrichten M, Patterson C, Miller G, Settle D, Erel Y (1991) Comparative increases of lead and barium with age in human tooth enamel, rib, and ulna. Sci Total Environ 107(179):203

Marchlewicz M, Protasowicki M, Rozewicka L, Piasecka M, Laszczynska M (1993) Effect of long-term exposure to lead on testis and epididymis in rats. Folia Histochem Cytobiol 31:55–62

Maruyama K, Yorifuji T, Tsuda T, Sekikawa T, Nakadaira H, Saito H (2012) Methyl mercury exposure at Niigata, Japan: results of neurological examinations of 103 adults. J Biomed Biotechnol 2012:635075

Meeker JD, Rossano MG, Protas B, Diamond MP, Puscheck E, Daly D, Paneth N, Wirth JJ (2009) Multiple metals predict prolactin and thyrotropin (TSH) levels in men. Environ Res 109:869–873

Michalke B (2003) Element speciation definitions, analytical methodology, and some examples. Ecotoxicol Environ Saf 56:122–139

Miklavcic A, Cuderman P, Mazej D, Snoj Tratnik J, Krsnik M, Planinšek P, Osredkar J, Horvat M (2011) Biomarkers of low-level mercury exposure through fish consumption in pregnant and lactating Slovenian women. Environ Res 111:1201–1207

Mishra D, Mehta A, Flora SJS (2008) Reversal of hepatic apoptosis with combined administration of DMSA and its analogues in guinea pigs: role of glutathione and linked enzymes. Chem Res Toxicol 21:400–407

Mishra S, Dwivedi SP, Singh RB (2010) A review on epigenetic effect of heavy metal carcinogens on human health. Open Nutraceuticals J 3:188–193

Mueller PW, Price RG, Finn WF (1998) New approaches for detecting thresholds of human nephrotoxicity using cadmium as an example. Environ Health Perspect 106:227–230

Murata K, Weihe P, Araki S, Budtz-Jorgensen E, Grandjean P (1999) Evoked potentials in Faroese children prenatally exposed to methyl mercury. Neurotoxicol Teratol 21:471–472

Nathan E, Huang HF, Pogach L, Giglio W, Bogden JD, Seebode J (1992) Lead acetate does not impair secretion of Sertoli cell function marker proteins in the adult Sprague Dawley rat. Arch Environ Health 47:370–375

Ng JC, Wang JP, Zheng B, Zhai C, Maddalena R, Liu F, Moore MR (2005) Urinary porphyrins as biomarkers for arsenic exposure among susceptible populations in Guizhou province, China. Toxicol Appl Pharmacol 206:176–184

Nierenberg DW, Nordgren RE, Change MB, Siegler RW, Blayney MB, Hochberg F, Toribara TY, Cernichiari E, Clarkson T (1998) Delayed cerebellar disease and death after accidental exposure to dimethylmercury. New Engl J Med 338:1672–1676

Noack-Fuller G, De Beer C, Seibert H (1993) Cadmium, lead, selenium, and zinc in semen of occupationally unexposed men. Andrologia 25:7–12

Nordberg G, Nogawa K, Nordberg M, Friberg L (2007) Cadmium. In: Nordberg G, Fowler B, Nordberg M, Friberg L (eds) Handbook on toxicology of metals. Academic, New York, pp 65–78

Nowak B, Chmielnicka J (2000) Relationship of lead and cadmium to essential elements in hair, teeth, and nails and of environmentally exposed people. Ecotoxicol Environ Safe 46:265–274

O'donohoe J, Chalkley S, Richmond J, Barltrop D (1998) Blood lead in U.K. children ± time for a lower action level? Clin Sci 95:219–223

Ogunseitan OA, Smith TR (2007) The cost of environmental lead (Pb) poisoning in Nigeria. Afr J Environ Sci Tech 1:27–36

Omokhodion FO, Crockford GW (1991) Lead in sweat and its relationship to salivary and urinary levels in normal healthy subjects. Sci Total Environ 103:113–122

Orloff K, Mistry K, Metcalf S (2009) Biomonitoring for environmental exposures to arsenic. J Toxicol Environ Health B Crit Rev 12:509–524

Ozuah PO (2000) Mercury poisoning. Curr Probl Pediatr 30:91–99

Pande M, Mehta A, Pant BP, Flora SJS (2001) Combined administration of a chelating agent and an antioxidant in the prevention and treatment of acute lead intoxication in rats. Environ Toxicol Pharmacol 9:173–184

Pant N, Kumar R, Murthy RC, Srivastava SP (2001) Male reproductive effect of arsenic in mice. Biometals 14:113–117

Pant N, Banerjee AK, Pandey S, Mathur N, Saxena DK, Srivastava SP (2003) Correlation of lead and cadmium in human seminal plasma with seminal vesicle and prostatic markers. Hum Exp Toxicol 22:125–128

Paruchuri Y, Siuniak A, Johnson N, Levin E, Mitchell K, Goodrich JM, Renne EP, Basu N (2010) Occupational and environmental mercury exposure among small-scale gold miners in the Talensi-Nabdam District of Ghana's Upper East region. Sci Total Environ 408:6079–6085

Paschal DC, Burt V, Caudill SP, Gunter EW, Pirkle JL, Sampson EJ, Miller DT, Jackson RJ (2000) Exposure of the U.S. population aged 6 years and older to cadmium: 1988–1994. Arch Environ Contam Toxicol 38:377–383

Patra RC, Swarup D, Dwivedi SK (2001) Antioxidant effects of a-tocopherol, ascorbic acid and L-methionine on lead induced oxidative stress to the liver, kidney and brain in rats. Toxicology 162:81–88

Patrick L (2002) Mercury toxicity and antioxidants: Part I: Role of glutathione and alpha-lipoic acid in the treatment of mercury toxicity. Altern Med Rev 7:456–471

Piazza GA, Keeton AB, Tinsley HN, Gary BD, Whitt JD, Mathew B, Thaiparambil J, Coward L, Gorman G, Li Y, Sani B, Hobrath JV, Maxuitenko YY, Reynolds RC (2009) A novel sulindac derivative that does not inhibit cyclooxygenases but potently inhibits colon tumor cell growth and induces apoptosis with antitumor activity. Cancer Prev Res (Phila Pa) 2:572–580

Raffoul JJ, Sarkar FH, Hillman GG (2007) Radiosensitization of prostate cancer by soy isoflavones. Curr Cancer Drug Targets 7:759–765

Raikwar MK, Kumar P, Singh M, Singh A (2008) Toxic effect of heavy metals in livestock health. Vet World 1:28–30

Rao MV, Gangadharan B (2008) Antioxidative potential of melatonin against mercury induced intoxication in spermatozoa in vitro. Toxicol In Vitro 22:935–942

Rignell-Hydbom A, Axmon A, Lundh T, Jonsson BA, Tiido T, Spano M (2007) Dietary exposure to methyl mercury and PCB and the associations with semen parameters among Swedish fishermen. Environ Health 6:14

Roels HA, Hoet P, Lison D (1999) Usefulness of biomarkers of exposure to inorganic mercury, lead, or cadmium in controlling occupational and environmental risks of nephrotoxicity. Ren Fail 21:251–262

Rogan WJ, Dietrich KN, Ware JH, Dockery DW, Salganik M, Radcliffe J, Jones RL, Ragan NB, Chisolm JJ, Rhoads GG (2001) The effect of chelation therapy with succimer on neuropsychological development in children exposed to lead. New Engl J Med 344:1421–1426

Rohlman DS, Arcury TA, Quandt SA, Lasarev M, Rothlein J, Travers R, Tamulinas A, Scherer J, Early J, Marín A, Phillips J, McCauley L (2005) Neurobehavioral performance in preschool children from agricultural and non-agricultural communities in Oregon and North Carolina. Neurotoxicology 26:589–598

Sakai T (2000) Biomarkers of lead exposure. Ind Health 38:127–142

Sallmen M, Lindbohm ML, Nurminen M (2000) Paternal exposure to lead and infertility. Epidemiology 11:148–152

Salonen JT, Seppanen K, Nyyssonen K, Korpela H, Kauhanen J, Kantola M, Tuomilehto J, Esterbauer H, Tatzber F, Salonen R (1995) Intake of mercury from fish, lipid peroxidation, and the risk of myocardial infarction and coronary, cardiovascular, and any death in eastern Finnish men. Circulation 91:645–655

Sanders T, Liu Y, Buchner V, Tchounwou PB (2009) Neurotoxic effects and biomarkers of lead exposure: a review. Rev Environ Health 24:15–45

Sarkar M, Chaudhuri GR, Chattopadhyay A, Biswas NM (2003) Effect of sodium arsenite on spermatogenesis, plasma gonadotrophins and testosterone in rats. Asian J Androl 5:27–31

Saxena G, Flora SJS (2006) Changes in brain biogenic amines and heme- biosynthesis and their response to combined administration of succimers and *Centella asiatica* in lead poisoned rats. J Pharm Pharmacol 58:547–559

Schumacher M, Domingo JL, Llobet JM (1991) Corbella lead in children's hair, as related to exposure in Tarragona province, Spain. Sci Total Environ 104:167–177

Shargorodsky J, Curhan SG, Henderson E, Eavey R, Curhan GC (2011) Heavy metals exposure and hearing loss in US adolescents. Arch Otolaryngol Head Neck Surg 137:1183–1189

Sivaprasad R, Nagaraj M, Varalakshmi P (2004) Combined efficacies of lipoic acid and 2,3 dimercaptosuccinic acid against lead-induced lipid peroxidation in rat liver. J Nutr Biochem 15:18–23

Smith AH, Marshall G, Yuan Y, Ferreccio C, Liaw J, von Ehrenstein O, Steinmaus C, Bates MN, Selvin S (2006) Increased mortality from lung cancer and bronchiectasis in young adults after exposure to arsenic *in utero* and in early childhood. Environ Health Perspect 114:1293–1296

Straif K, Baan R, Grosse Y, Secretan B, El Ghissassi F, Bouvard V, Guha N, Freeman C, Galichet L, Cogliano V (2009) A review of human carcinogens—Part C: Metals, arsenic, dusts, and fibres. Lancet Oncol 10:453–454

Tan SW, Meiller JC, Mahaffey KR (2009) The endocrine effects of mercury in humans and wildlife. Crit Rev Toxicol 39:228–269

Telisman S, Cvitković P, Jurasović J, Pizent A, Gavella M, Rocić B (2000) Semen quality and reproductive endocrine function in relation to biomarkers of lead, cadmium, zinc, and copper in men. Environ Health Perspect 108:45–53
Thoreux-Manlay A, Velez de la Calle JF, Olivier MF, Soufir JC, Masse R, Pinon-Lataillade G (1995) Impairment of testicular endocrine function after lead intoxication in the adult rat. Toxicology 100:101–109
Vachhrajani KD, Roy Chowdhury A, Dutta KK (1990) Testicular toxicity of methylmercury. Reprod Toxicol 6:355–361
Valko M, Rhodes CJ, Moncol J, Izakovic M, Mazur M (2006) Free radicals, metals and antioxidants in oxidative stress-induced cancer. Chem Biol Interact 160:1–40
Waalkes MP (2003) Cadmium carcinogenesis. Mutat Res 533:107–120
Waalkes MP, Liu J, Ward JM, Diwan LA (2004) Mechanisms underlying arsenic carcinogenesis: hypersensitivity of mice exposed to inorganic arsenic during gestation. Toxicology 198:31–38
Wallin E, Rylander L, Jönssson BA, Lundh T, Isaksson A, Hagmar L (2005) Exposure to CB-153 and p, p′-DDE and bone mineral density and bone metabolism markers in middle aged and elderly men and women. Osteoporos Int 16:2085–2094
Walton FS, Harmon AW, Paul DS, Drobná Z, Patel YM, Styblo M (2004) Inhibition of insulin dependent glucose uptake by trivalent arsenicals: possible mechanism of arsenic-induced diabetes. Toxicol Appl Pharmacol 198:424–433
Wang J, Wu J, Zhang Z (2006) Oxidative stress in mouse brain exposed to lead. Ann Occup Hyg 50:405–409
Wang SH, Wang ZH, Cheng XT, Li J, Sang ZP, Zhang XD, Han LL, Qiao XY, Wu ZM, Wang ZQ (2007) Arsenic and fluoride exposure in drinking water: Children's IQ and growth in Shanyin County, Shanxi Province, China. Environ Health Perspect 115:643–647
Ward NI, Savage JM (1994) Metal dispersion and transportational activities using food crops as biomonitors. Sci Total Environ 146:309–319
WHO (1992a) Environmental Health Criteria 134. Cadmium. WHO, Geneva
WHO (1992b) Environmental Health Criteria 135. Cadmium. Environmental Aspects, WHO, Geneva
Wirth JJ, Mijal RS (2010) Adverse effects of low level heavy metal exposure on male reproductive function. Syst Biol Reprod Med 56:147–167
Wright DA, Welbourn P (2002) Environmental toxicology. Cambridge University Press, Cambridge
Wright RO, Schwartz J, Wright RJ, Bollati V, Tarantini L, Park SK, Hu H, Sparrow D, Vokonas P, Baccarelli A (2010) Biomarkers of lead exposure and DNA methylation within retrotransposons. Environ Health Perspect 118:790–795
Xu LC, Wang SY, Yang XF, Wang XR (2001) Effects of cadmium on rat sperm motility evaluated with computer assisted sperm analysis. Biomed Environ Sci 14:312–317
Yee S, Choi BH (1996) Oxidative stress in neurotoxic effects of methylmercury poisoning. Neurotoxicology 17:17–26
Yoshida T, Yamauchi H, Fan Sun G (2004) Chronic health effects in people exposed to arsenic *via* the drinking water: dose-response relationships in review. Toxicol Appl Pharmacol 198:243–252
Zeng X, Jin T, Jiang X, Kong Q, Ye T, Nordberg GF (2004) Effects on the prostate of environmental cadmium exposure-a cross sectional population study in China. Biometals 17:559–565

Chapter 8
Plants as Monitors of Lead Air Pollution

Mirjana Ristić, Aleksandra Perić-Grujić, Davor Antanasijević, Milica Ristić, Mira Aničić Urošević, and Milica Tomašević

Contents

Abstract The most important environmental problems are caused by intensive industrial activities, urbanization and population growth. The increase in air pollution resulting from the expanding use of fossil energy sources and the growth in the manufacture and use of chemicals has been accompanied by mounting public awareness of and concern about detrimental effects on health and the environment. Anthropogenic activities have greatly changed pollutant atmospheric concentrations and consequently, their availability and cycling. Airborne lead (Pb) is a constituent of atmospheric

M. Ristić • A. Perić-Grujić
Faculty of Technology and Metallurgy, University of Belgrade,
Karnegijeva 4, 11120 Belgrade, Serbia

D. Antanasijević
Innovation Centre of the Faculty of Technology and Metallurgy, University of Belgrade,
Karnegijeva 4, 11120 Belgrade, Serbia

M. Ristić
Johannes Kepler Universität, Altenberger Straße 69, 4040 Linz, Austria

M. Aničić Urošević • M. Tomašević (✉)
Institute of Physics, University of Belgrade, Pregrevica 118, 11080 Belgrade, Serbia
e-mail: milica.tomasevic@ipb.ac.rs

E. Lichtfouse et al. (eds.), *Pollutant Diseases, Remediation and Recycling*,
Environmental Chemistry for a Sustainable World 4, DOI 10.1007/978-3-319-02387-8_8,

particulate matter (PM), and as such it may be transported to great distances before being removed in deposition processes. Here we review the application of plants in (i) trace elements monitoring, (ii) biomonitoring of lead air pollution and (iii) identifying lead sources in the environment. The instrumental monitoring techniques lack information on impact of atmospheric pollutants on the living systems and hence, there has been an increasing interest in using indirect monitoring methods based on a response of living organisms that may act as trace element bioaccumulators. Since plants accumulate lead, as well as other trace elements, from the atmosphere, the indirect air pollution monitoring, using plants has gained importance in the last decades. Mosses, lichens, but also higher plants have been used for biomonitoring of various pollutants, including lead. Lead isotopic studies may provide a convenient approach for studying and tracing the sources of Pb pollution in different environmental compartments. The lead isotope approach for plant biomonitoring in source identification and biomonitoring species validity assessment has been discussed. It has been shown that airborne Pb is the most important source of accumulated Pb in plants. The overview of advantages using plants for lead air pollution monitoring is presented in this paper, as well as the determination of lead sources in the environment.

Keywords Air pollutants • Particulate matter • Deposition • Lead • Pollution sources • Pb isotope ratio • Leaded gasoline • Tracers • Plants • Bioindicators • Biomonitors • Tree leaves

List of Abbreviations

AEI	Average Exposure Indicator
Ag	Argentum
Al	Aluminium
APM	Airborne particulate matter
As	Arsenic
Ba	Barium
BB	Biesbosch National Park
Be	Beryllium
BD	Bulk Deposition
Cd	Cadmium
Co	Cobalt
CO	Carbon Monoxide
Cr	Chromium
Cu	Copper
DEFRA	Department for Environment, Food and Rural Affairs
EC	European Commission
EEA	European Environment Agency

ETC/CDS	European Topic Centre on Catalogue of Data Sources
FAAS	Flame Atomic Absorption Spectrometry
Fe	Iron
GEMET	General Multilingual Environmental Thesaurus
GFAAS	Graphite Furnace Atomic Absorption Spectrometry
GHGs	Greenhouse Gases
Hg	Mercury
ICP-MS	Inductively Coupled Plasma Mass Spectrometry
ICP-OES	Inductively Coupled Plasma Optical Emission Spectrometry
IUPAC	International Union of Pure and Applied Chemistry
JRBP	Jasper Ridge Biological Preserve
MC-ICP-MS	Magnetic Sector Inductively Coupled Plasma Mass Spectrometry
Mn	Manganese
NAA	Neutron Activation Analysis
Ni	Nickel
NH_3	Ammonia
NMVOC	Non-methane Volatile Organic Compound
NO_2	Nitrogen Dioxide
NO_x	Nitrogen Oxides (nitric oxide and nitrogen dioxide)
O_3	Ozone
Pb	Lead
PM	Particulate Matter
$PM_{2.5}$	Particulate Matter with aerodynamic diameter less than 2.5 μm
PM_{10}	Particulate Matter with aerodynamic diameter less than 10 μm
REF	Reference location
Q-ICP-MS	Quadrupole Inductively Coupled Plasma Mass Spectrometry
Se	Selenium
SEM–EDAX	Scanning Electron Microscopy Combined with Energy Dispersive X-ray Spectroscopy
SO_2	Sulfur Dioxide
SPM	Suspended particulate matter
Sr	Strontium
Th	Thorium
TIMS	Thermal Ionization Mass Spectrometry
U	Uranium
UN	United Nations
UNEP	United Nations Environment Programme
US EPA	United States Environmental Protection Agency
V	Vanadium
VOC	Volatile Organic Compound
WHO	World Health Organization
XRF	X-ray Fluorescence Analysis
Zn	Zinc

8.1 Introduction

Major environmental compartments or conceptual spheres are atmosphere, hydrosphere, lithosphere and biosphere. Flow of matter and energy in this dynamic system takes place within every sphere and also between each of them in both directions, as presented schematically in Fig. 8.1. Moreover, different processes, biotic and abiotic, taking place in these compartments are responsible for the transformations of matter and energy, making the system more complex and influencing the distribution and flow of matter and energy.

Intensive industrial activities, urbanization and population growth are recognized as the reasons for the most important environmental problems. The significant differences between environmental compartments influence the mobility, redistribution, transformations, storage, and monitoring of pollutants. Development of modern analytical techniques, as well as knowledge about the influence of different chemicals on the human health and the environment, have encouraged significant improvement in the field of pollutant detection and environmental protection.

Presence of chemicals, particulates or biological material in air can result in atmospheric pollution. Several effects of air pollution are even of global concern, with climate change and stratospheric ozone depletion, being the most serious environmental issues. One of the consequences of the current stage of industrialization and the demand for improved quality of life has been increased exposure to air pollution

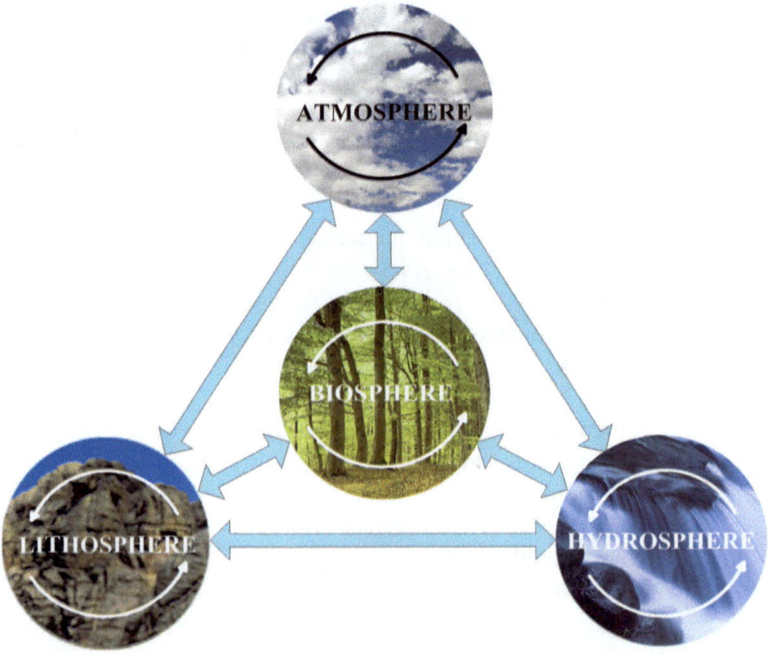

Fig. 8.1 The four environmental spheres and their interactions (Adapted from Husar 1994)

coming from industrial activities, traffic and energy production (WHO 2007). For example, the emission of greenhouse gases (GHGs) from various domestic and industrial sources is one of the major contributors towards global warming and climatic changes (Meinshausen et al. 2009; Radojević et al. 2013).

Since the major pollution smog events that occurred in the United Kingdom in the 1950s, the awareness of policy makers, economical stakeholders and the general public kept increasing at a steady pace over the last decades (Colette et al. 2011). The increase in air pollution resulting from the expanding use of fossil energy sources and growth in the manufacture and use of chemicals has been accompanied by mounting public awareness of and concern about its detrimental effects on health and the environment (WHO 2000). Namely, air pollution can cause acute and chronic adverse effects on human health via ingestion, dermal contact and inhalation exposure (Hu et al. 2012); the effects of air pollution on health are very complex (Pandey et al. 2005).

Undeniably, air pollution is a serious problem in many parts of the world, affecting people and the economy of countries, leading to forest decline, loss in agricultural production and diminished status of population health and the well-being of the whole environment.

Air pollutants originate from both natural and anthropogenic sources, and may be found outdoors and indoors in solid, liquid or gaseous form. Anthropogenic emissions of various pollutants (the only factors that can be controlled) interact between each other, as well as with biogenic (natural) emissions and local meteorology, that results in time and space varying pollutant levels, which also depend on the system boundary conditions (i.e. emission and meteorology of surrounding areas) (Carnevale et al. 2012).

Dealing with a certain topic usually requires proper definition; however in the field of air pollution many definitions originating from different sources are in use. Some of them are presented in the following text. Air pollution is defined as:

- The presence of contaminant or pollutant substances in the air that do not disperse properly and interfere with human health or welfare, or produce other harmful environmental effects (US EPA 1989).
- The presence of contaminant or pollutant substances in the air at a concentration that interferes with human health or welfare, or produces other harmful environmental effects (Gunn 1990). This definition is given in GEMET – General Multilingual Environmental Thesaurus, in 19 European languages, developed by the European Environment Agency (EEA) and the European Topic Centre on Catalogue of Data Sources (ETC/CDS), in collaboration with European Union member countries and the UNEP Infoterra.
- The presence of contaminant or pollutant substances in the air that do not disperse properly and that interferes with human health or welfare, or produces other harmful environmental effects (UN 1997).
- Presence of substances in the atmosphere resulting either from human activity or natural processes, in sufficient concentration, for a sufficient time and under circumstances such as to interfere with comfort, health, or welfare of persons or to harm the environment (Nordberg et al. 2007).

- One or more chemicals or substances in high enough concentrations in the air to harm humans, other animals, vegetation, or materials (Tyler Miller and Spoolman 2010).

Also, there are few definitions of air pollutant/pollutants:

- Any substance in air which could, if in high enough concentration, harm man, other animals, vegetation, or material. Pollutants may include almost any natural or artificial composition of matter capable of being airborne. They may be in the form of solid particles, liquid droplets, gases, or in combination of these forms (US EPA 1989).
- Substances in air that could, at high enough concentrations, harm human beings, animals, vegetation or material. Air pollutants may thus include forms of matter of almost any natural or artificial composition capable of being airborne. They may consist of solid particles, liquid droplets or gases, or combinations of these forms (UN 1997).
- Any air pollution agent or combination of such agents, including any physical, chemical, biological, radioactive (including source material, special nuclear material, and byproduct material) substance or matter which is emitted into or otherwise enters the ambient air. Such term includes any precursors to the formation of any air pollutant, to the extent the Administrator has identified such precursor or precursors for the particular purpose for which the term "air pollutant" is used (Clean Air Act, Subchapter 7602 Definitions Sec. 302).
- Pollutant shall mean any substance present in ambient air and likely to have harmful effects on human health and/or the environment as a whole (DIRECTIVE 2008/50/EC 2008).

Even the terms pollutant and contaminant are often used as synonyms; several definitions for the term contaminant are in use:

- Any particulate matter, gas, or combination thereof, other than water vapor (US EPA 1989).
- Any undesirable solid, liquid, or gaseous matter occurring, as a result of human activities, in a solid, liquid, or gaseous environmental medium, even without adverse effects being observed. In some contexts, as in relation to gas cleaning equipment, used as a synonym for "pollutant", especially on a small scale (Nordberg et al. 2009).

In order to facilitate their classification, air pollutants are often grouped in following categories: solids, sulfur compounds, volatile organic chemicals, particulate matter, nitrogen compounds, oxygen compounds, halogen compounds, radioactive compounds, and odors. They are also classified as primary, those emitted directly from a source, and secondary pollutants, which are formed in air by interaction between two or more primary pollutants, or by reaction with atmosphere constituents. Most primary pollutants, including particulate matter (PM) are emitted by anthropogenic activities (Pacyna 1986; Seinfeld and Pandis 1998; Pacyna and Pacyna 2001; Vukmirović et al. 2004). The urban atmosphere is subjected to large inputs of anthropogenic contaminants produced by both stationary sources and mobile sources related to traffic (Seinfeld and Pandis 1998; Tasić et al. 2006; Saeedi et al. 2012).

US EPA has set National Ambient Air Quality Standards for six principal pollutants, which are called "criteria" pollutants: carbon monoxide (CO), primary pollutant; lead (Pb), primary and secondary; nitrogen dioxide (NO_2), primary for averaging time of 1-h while primary and secondary on the annual basis; ozone (O_3), primary and secondary; particle pollution ($PM_{2.5}$ and PM_{10}), primary and secondary and sulfur dioxide (SO_2), primary and secondary. Since 1990, national annual air pollutant emissions in the United States have declined, with the greatest percentage drop in lead emissions (83 %). Direct $PM_{2.5}$ emissions have declined by more than half; PM_{10} and SO_2 emissions have declined by more than 60 %, and NO_x and VOC emissions have declined by more than 40 % (US EPA 2010).

The European Parliament has established the need to reduce atmospheric pollution in order to minimize harmful effects on human health, with particular attention paid to sensitive populations, and to reduce damage to the environment as a whole and this may be carried out by improving the monitoring and assessment of air quality and pollutant emissions, as well as providing information to the public (Vicente et al. 2012). Emissions of the main air pollutants in Europe declined significantly in the period 1990–2009, in particular sulphur dioxide (SO_2) and lead (Pb), resulting in improved air quality across the region. At present, PM and O_3 are Europe's most problematic pollutants in terms of harm to health (Office of the European Union 2011).

Also, it should be emphasized that with the increase of industrial activities in developing countries, the emission of air pollutants has substantially increased, often leading to a deterioration of the environmental quality in areas where no pollution measurements are available (Rodriguez et al. 2012).

8.2 Atmospheric Particulate Matter

Particulates in the air are a mixture of solid and liquid droplets that vary in size, morphology, and chemical composition. Unlike many other air pollutants, which are classified by their chemical composition, particles are classified according to their physical property. By size, they are often classified as "coarse" and "fine". In practical terms, a distinction is made between PM_{10} ("thoracic particles" that are particulates with aerodynamic diameters smaller than or equal to 10 μm and can penetrate into the lower respiratory system) and $PM_{2.5}$ ("respirable particles" that are particulates with aerodynamic diameters smaller than or equal to 2.5 μm and can penetrate into the deeper part of the lungs) (Brunekreef and Holgate 2002). A visual comparison of PM_{10} and $PM_{2.5}$ is shown in Fig. 8.2. The coarse fraction (particles >2.5 μm) is dominated by natural sources (geogenic material: fugitive and resuspended dust; biological material: pollen, bacteria), while the fine fraction (particles <2.5 μm) is dominated by anthropogenic emissions (Wall et al. 1988).

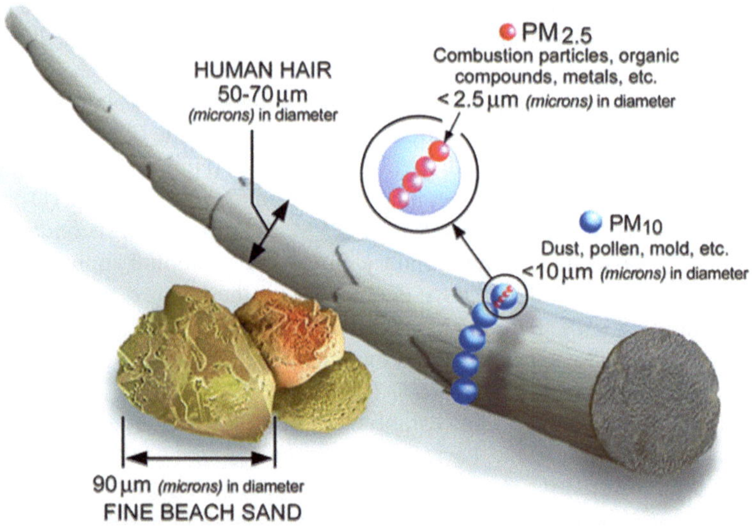

Fig. 8.2 Comparison of the dimensions of PM_{10}, $PM_{2.5}$ and a human hair (http://www.epa.gov/airquality/particlepollution/basic.html)

Directive 2008/50/EC of the European parliament and of the council of 21 May 2008 on ambient air quality and cleaner air for Europe, for the purpose of this Directive, defines 'PM_{10}': "shall mean particulate matter which passes through a size-selective inlet as defined in the reference method for the sampling and measurement of PM_{10}, EN 12341, with a 50 % efficiency cut-off at 10 µm aerodynamic diameter; and '$PM_{2.5}$' shall mean particulate matter which passes through a size-selective inlet as defined in the reference method for the sampling and measurement of $PM_{2.5}$, EN 14907, with a 50 % efficiency cut-off at 2.5 µm aerodynamic diameter" (DIRECTIVE 2008/50/EC 2008).

Particulate matter is emitted directly from a large variety of anthropogenic, biogenic and natural sources and formed in the atmosphere by chemical and physical processes from gas-phase precursors such as NMVOC, NO_x, SO_2 and NH_3 (Seinfeld and Pandis 1998). PM primarily generated by thermal and heating power plants, industry and transportation (Fig. 8.3), stands out as one of the most important pollutants.

Unlike other US criteria pollutants (O_3, CO, SO_2, NO_2 and Pb), PM is not a specific entity but is a mixture of particles from different sources and of different size, composition, and properties (Vallero 2008). Atmospheric PM, or aerosol, plays a central role in atmospheric processes (Fountoukis and Nenes 2007). After emission, PM is subjected to physical, chemical and photochemical transformations, which ultimately decide their fate and atmospheric concentrations.

Like other air pollutants, PM does not remain confined near the source of emission, but is spread over distances, depending upon topography and meteorological conditions, especially wind direction and speed, vertical and horizontal thermal

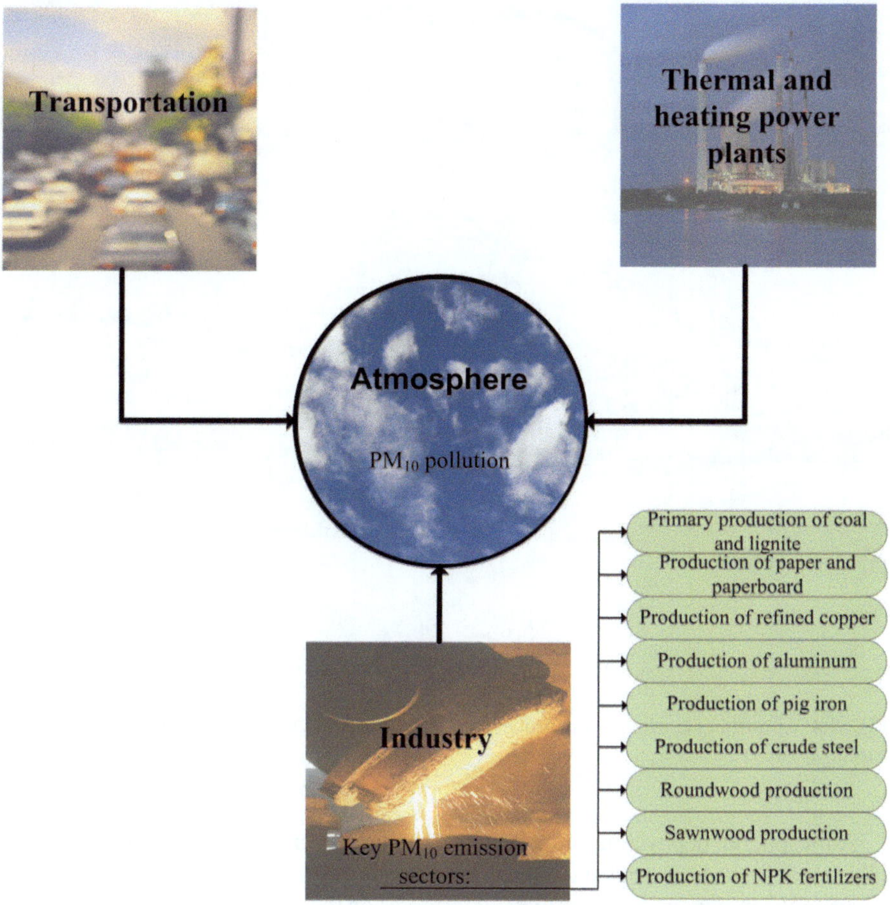

Fig. 8.3 The most important sources of particulate matter in atmosphere (Adapted from Antanasijević et al. 2013)

gradients. Large particles have a tendency to be removed quickly, near the emission sources, but particles of sizes less than 10 μm exhibit long term existence, from days to months, and may be transported to distant regions. Both transport and mobilization of PM have attracted much attention and differences in particulates regional and seasonal distribution are evident (Nriagu and Pacyna 1988; Alcamo 1992; Cong et al. 2010; Ewing et al. 2010). The elemental composition of airborne particulates at different locations may vary over almost three orders of magnitude indicating a strong effect of local sources. The highest PM concentrations are usually reported for urban and industrial areas. Particulate matter in urban areas is mainly made up of metals, organic compounds, materials of biological origin (Tasić et al. 2006) and elemental carbon. The particle core, which often forms bulk of the urban particulate matter, mainly comprises of elemental carbon (Pandey et al. 2005).

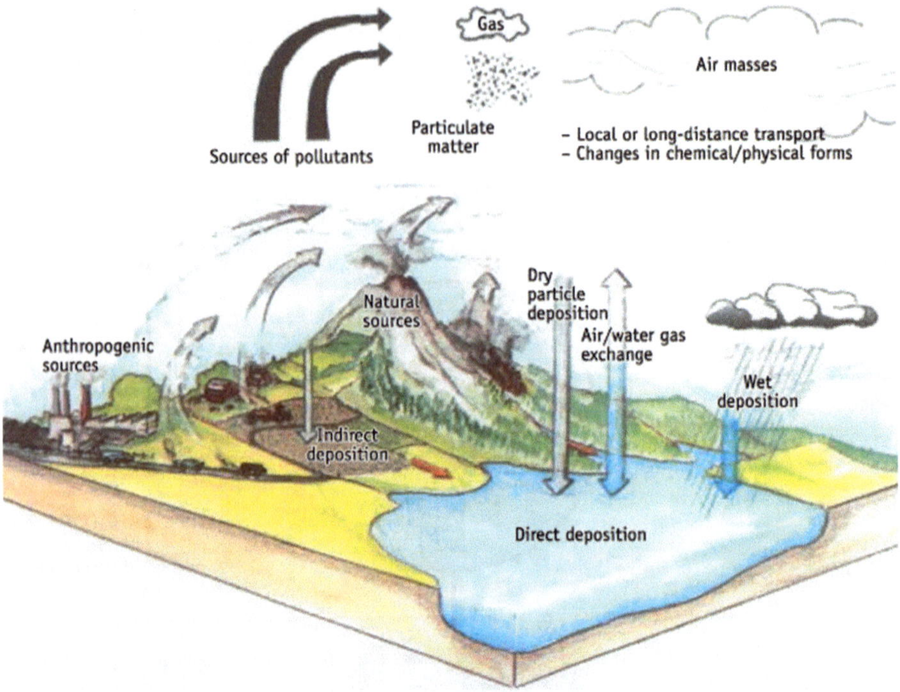

Fig. 8.4 Atmospheric deposition processes (http://www.ioe.ucla.edu/reportcard/article.asp?parentid=1497)

Atmospheric deposition of particulate matter is an important source of toxic substances in the terrestrial and aquatic environment (Fig. 8.4). Interaction with different natural components may result in toxic effects on the biosphere (Nriagu and Pacyna 1988; Bargagli 1998). Atmospheric deposition depends on the physical characteristics of the particles, predominantly their size and shape. While large particles precipitate quickly near the discharge sources, particles of sizes less than 10 μm and low density exist from days to months, and may be transported to distant regions. Among other factors, the magnitude of atmospheric deposition and the ratio of wet to dry deposition are controlled by emission sources, distance to emission sources, the sampling site, meteorological conditions, e.g. prevailing wind directions, type, frequency and amount of precipitation, and also on the morphological characteristics of depositing surfaces (Tasić et al. 2001; Avila and Rodrigo 2004; Motelay-Massei et al. 2005). Since either wet or dry modes may contribute equally to, or predominate in total deposition, it appears essential to include both, wet and dry deposition, in estimations of loading from atmospheric deposition. From a biogeochemical perspective, the characterization of bulk deposition (BD) is relevant for identifying the variability and sources of the atmospheric pollutants (Azimi et al. 2005).

Particulate matter induces serious health consequences for the exposed population. Numerous studies have shown that PM poses a great health threat because it can pass through the nose and throat and get deep into the lungs. Very small particles can penetrate deeper into the respiratory system and enter the blood circulation. Furthermore, these particles often contain toxic trace elements or organic compounds. Studies showed that chronic exposure to PM_{10} results in adverse health effects (Pope et al. 1995) in the form of increased risk of developing cardiovascular and respiratory diseases and lung cancer. Analysis of elemental content of PM_{10} allows an assessment of the health risks (both carcinogenic and non-carcinogenic) associated with exposure to PM_{10}-bound trace elements through inhalation. In addition, PM_{10} may be absorbed by aerial parts of vegetables and thus included into the food chain (Kabata-Pendias and Mukherjee 2007).

Even the range of the health effects is broad, affecting the respiratory and the cardiovascular system, especially for more sensitive groups as children and elderly (WHO 2006), for both, ultrafine and larger aerosol particles, it is unclear which physical and chemical properties actually determine their adverse health effects (particle size, structure, number, mass concentration, solubility, chemical composition, and individual components, etc.) (Pöschl 2005). It seems that the effect of ultrafine particles on health correlate better with particle surface area than with particle number or mass (Moshammer and Neuberger 2003; Sinclair and Tolsma 2004).

The harmful effect that airborne particulate matter may have on human health, leads to the need for strict governmental regulations on air particle concentrations. The US EPA included particulate matter among six common air pollutants in national ambient air quality standards within the clean air act legislations first established in 1963. In the European Union for $PM_{2.5}$, the 2008 Air Quality Directive has introduced a target value for human exposure based on the average exposure indicator (AEI), set at the national level (DIRECTIVE 2008/50/EC 2008).

8.3 Trace Element Air Pollution

Trace elements, especially heavy metals, are currently pollutants of great concern, due to the increased awareness of their potentially hazardous effects (Järup 2003; Wei and Yang 2010; Milosavljević et al. 2011). The expanding interest in the field of heavy metal research is associated with an increasing world production of metals as well as their common usage in the past century, and consequently, with their increasing emissions into the environment (Järup 2003; Marjanović et al. 2009; UNEP 2010). Major anthropogenic sources of trace metals include vehicular emissions (Cr, Mn, Fe, Cu, Zn, Ba, Pb), stationary fossil fuel combustion (V, Cr, Mn, Ni), non-ferrous metal production (Cu, Zn, Cd) and combustion of gasoline (Pb) (Schauer et al. 1996; Pacyna and Pacyna 2001; Bilos et al. 2001). Long-term excessive inputs of such elements may impose a burden on ecosystems and human health through various biogeochemical cycles (Cong et al. 2010).

Trace elements generally refer to elements that occur in natural and perturbed environments in small amounts and that, when present in excessive bioavailable concentrations, are toxic to living organisms (Blum et al. 2009). Trace metals are metals found in low concentrations, in mass fractions of ppm or less, in some specified source, e.g., soil, plants, tissue, ground water (Duffus 2002). Sometimes this term has confusing overtones of low nutritional requirement (by a specified organism).

During the past few decades, many studies on bulk (wet and dry) deposition of metals in urban areas have been widely used to estimate the influence of atmospheric inputs of metals on the surface environment (Golomb et al. 1997; Wong et al. 2003; Motelay-Massei et al. 2005; Azimi et al. 2005; Rossini et al. 2005; Sharma et al. 2008). Fluxes of metals (Cd, Cu, Ni, Pb, Hg) of anthropogenic origin were 3–4 times higher in the sites nearest to the urban and industrial area compared to the more distant sites.

The transport and mobilization of trace elements have also attracted much attention (Nriagu and Pacyna 1988; Alcamo 1992) and regional and seasonal differences in the element distribution are evident. Trace elements, such as Pb, Cd and Hg, can be transported over long distances by atmospheric flow and deposited far from emission sources (Nriagu and Pacyna 1988; Alcamo 1992). There is evidence that trace elements from anthropogenic sources are widely distributed in the atmosphere and occur even at high altitudes in mountain regions and in the Arctic (Kabata-Pendias and Mukherjee 2007).

Unlike most other toxic pollutants, metals can accumulate throughout the food chain due to their nonbiodegradability and thus have potentially detrimental effects on all living species (An et al. 2001; Milosavljević et al. 2010; Nica et al. 2012). Among the many inorganic pollutants originating from anthropogenic activities, heavy metals such as arsenic, cadmium, chromium, and lead, are of major concern due to their toxic and potentially carcinogenic characteristics (Melaku et al. 2008). Moreover, metal pollutants such as Cd, Cr, Ni and Pb have cumulative effects, causing growth retardation in children, kidney disease, cancer and many other adverse health effects (Jiries 2003). These, as well as Hg, Ni and Zn, are emitted into the air from both natural and anthropogenic sources (Fig. 8.5).

Some recent observations have shown decreasing trends for several metals, e.g. Fe, Pb, Cr, Cd, Hg, As, and Be in the environment. A significant reduction of the Pb content in airborne particulates has recently been reported for some urban sites due to the replacement of leaded with unleaded gasoline (Kabata-Pendias and Mukherjee 2007).

More than 40 trace elements are measured in atmospheric particulate matter samples, including lead (Bargagli 1998; Viana et al. 2008). Studies of spatial and temporal variation of atmospheric aerosol particles gained significance and resulted in an increased interest in the use of analytical techniques capable to measure the size, morphology and chemical composition of individual aerosol particles. Such data is essential for understanding the particle formation, transport, transformation and deposition mechanisms, as well as the impact of particles inhaled into the respiratory system on human health.

PM sampling is performed for a variety of purposes, including mass concentration determination for compliance with standards. Chemical analysis of filter deposits cannot be separated from the methods used to obtain the sample. Techniques commonly used for trace element analysis of environmental samples, like PM

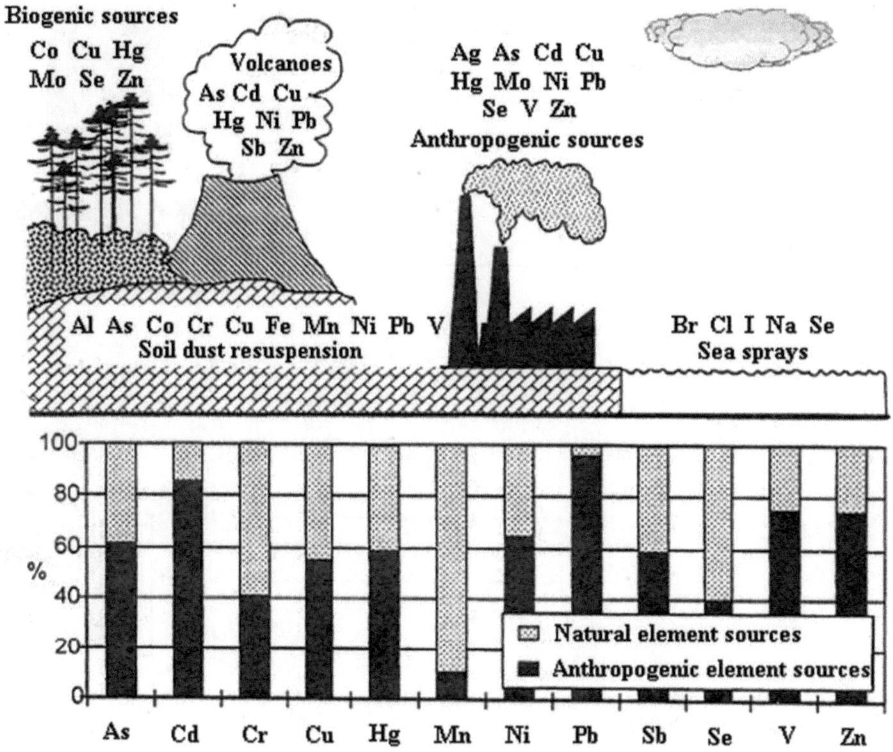

Fig. 8.5 The main natural and anthropogenic sources of atmospheric trace elements and the estimated relative proportions attributable to each of the two source types (Bargagli 1998)

include neutron activation analysis (NAA) and X-ray fluorescence analysis (XRF) as non-destructive techniques and flame atomic absorption spectrometry (FAAS), graphite furnace atomic absorption spectrometry (GFAAS), inductively coupled plasma optical emission spectrometry (ICP-OES) and inductively coupled plasma mass spectrometry (ICP-MS) as destructive ones. Analytical techniques XRF, FAAS and ICP-OES are used for medium to low level concentrations while GFAAS, NAA and ICP-MS are appropriate for ultra low level concentrations (Fig. 8.6).

8.3.1 Plants in Trace Elements Monitoring

The instrumental sampling methods are frequently limited by the high costs and the difficulties in carrying out extensive monitoring in time and space. It is well-known that plants accumulate trace elements from the atmosphere and have been used in a large number of studies, offering low-cost information about environmental quality with an easy sampling advantage (Markert 1993; Bargagli 1998; Weiss et al. 2003; Rucandio et al. 2011). Thus, there has been increasing interest in the use of indirect monitoring methods, such as the use of organisms that act as bioaccumulators.

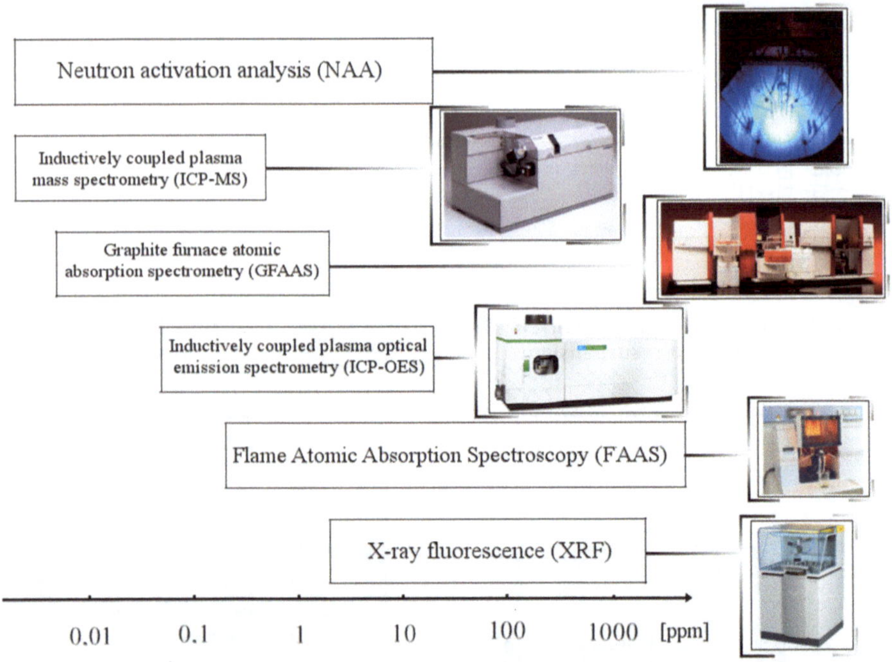

Fig. 8.6 Analytical techniques applied in trace element analysis of environmental samples (Adapted from Onjia 2007)

A bioindicator is an organism (or part of an organism or a community of organisms) that contains information on the quality of the environment (or a part of the environment). A biomonitor, on the other hand, is an organism (or part of an organism or a community of organisms) that contains information on the quantitative aspects of the quality of the environment (Markert 2007). Biomonitoring is a rapid and economical method that has commonly been used for assessing environmental quality and potentially detrimental effects of pollutants to the biosphere (Wolterbeek 2002; Szczepaniak and Biziuk 2003).

A perfect bioindicator is expected to have, among others, the following characteristics: (1) it has to accumulate high levels of pollutants without death; (2) it has to live a sessile style, thus definitely representing local pollution; (3) it has to have enough abundance and wide distribution for the repetitious sampling and comparison, (4) its life has to be long enough for the comparison between various ages; (5) it has to be easily sampled and easily raised in the laboratory (Zhou et al. 2008). Various effective bioindicators have been used so far for the assessment of the state of natural ecosystems (Fig. 8.7).

Air quality monitoring using biological material has been employed for more than a century (Serbula et al. 2012). The use of plants as a complementary tool to traditional (instrumental) methods of studying atmospheric pollution from

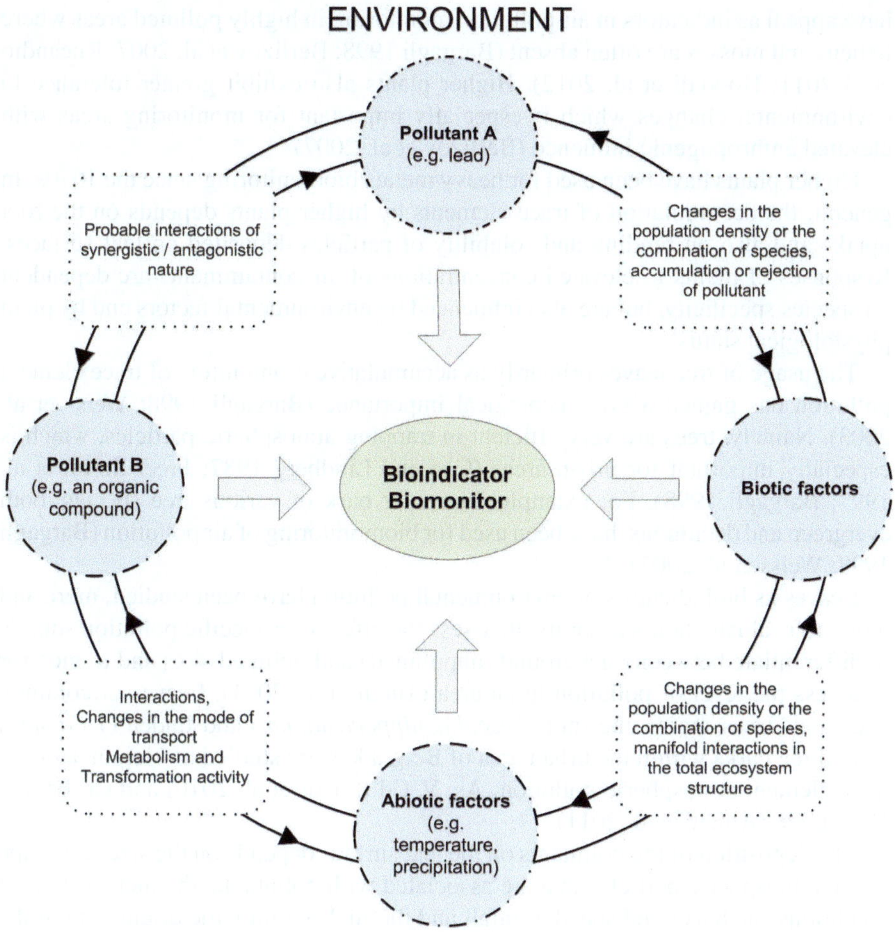

Fig. 8.7 Simplified representation of complex ecosystem interrelations with regard to a pollutant and resulting consequences for bioindication and biomonitoring (Adapted from Markert 1996)

anthropogenic and natural sources has become an established technique in the past 30–40 years because of the development of powerful analytical techniques (Markert 1995; Berlizov et al. 2007). Plants such as mosses and lichens are commonly regarded as the best bioindicators of air quality, as they can accumulate elements to a far greater level than it is necessary for their physiological needs (Rühling and Tyler 1968; Berg and Steinnes 1997; Harmens et al. 2008; Aničić et al. 2009a, b, c).

In industrial and urban areas higher plants can give even more complex information than non-biological samples. The main advantages of using higher plants for air monitoring are greater availability of the biological material, simplicity of species identification, sampling and treatment, harmless sampling and ubiquity of some genera, which makes it possible to cover large areas; therefore higher plants have appeal as indicators in air pollution monitoring in highly polluted areas where lichens and mosses are often absent (Bargagli 1998; Berlizov et al. 2007; Rucandio et al 2011; Hoodaji et al. 2012). Higher plants also exhibit greater tolerance to environmental changes which is especially important for monitoring areas with elevated anthropogenic influence (Berlizov et al. 2007).

Higher plants have been used for heavy metals biomonitoring since the 1970s. In general, the accumulation of trace elements by higher plants depends on the root uptake and also on binding and solubility of particles deposited on leaf surfaces. Responses of plants to elevated concentrations of air contaminants are dependent on species specificity, but are also influenced by environmental factors and by plant physiological status.

The usage of tree leaves primarily as accumulative biomonitors of trace element pollution has gained a great ecological importance (Bargagli 1998; Weiss et al. 2003). Namely, trees are very efficient in trapping atmospheric particles, which is especially important for urban areas (Coe and Lindberg 1987; Freer-Smith et al. 1997; Bargagli 1998). For example, leaves or bark of various tree species, both evergreen and deciduous, have been used for biomonitoring of air pollution (Bargagli 1998; Weiss et al. 2003).

Leaves as bioindicators of environmental pollution have been studied, more and more, to evaluate their suitability, to assess the effect of a specific pollution source, to differentiate between background (unpolluted) and polluted sites and to monitor or assess the level of pollution in an area (Turan et al. 2011). Leaves of common deciduous trees: horse chestnut (*Aesculus hippocastanum*) and linden (*Tilia* spp.) from three parks within the urban area of Belgrade were studied as bioindicators of trace element atmospheric pollution: As, V, Cd (Šućur et al. 2010) and Cr, Fe, Ni, Cu, Zn, Pb (Aničić et al. 2011).

The deposition of the pollutants on the leaf surface depends on the size and shape of the atmospheric particles that are associated with pollutants, the meteorological conditions, such as wind speed, rainfall and thermal stability, the orientation of the leaf surfaces and the structure of the canopy, while retention of the pollutants on the leaf depends on the adhesiveness of the particles, morphological characteristics of the leaf surface and the rate of removal from the leaf (Othman et al. 1997; Harrison and Yin 2000; De Nicola et al. 2008; Van Leeuwen and Vermiere 2007; Olowoyo et al. 2010).

Depending on tree species, the morphological characteristics of the leaves are greatly variable, influencing the amount of particles deposited on the leaf surface (Bargagli 1998; Weiss et al. 2003; Tomašević et al. 2004; Tomašević et al. 2005). Scanning electron microscopy combined with energy dispersive X-ray spectroscopy (SEM–EDAX) provides a powerful tool for physical and chemical characterization of individual particles deposited on different surfaces, including filters and leaves (Sharma and Srinivas 2009; US EPA 2002). As an example, scanning electron photomicrographs presented in Fig. 8.8 show a general appearance of adaxial and abaxial surfaces of hypostomatic leaves of *A. hippocastanum* and *C. colurna* sampled in a dry period, at the end of September 1997 in the Belgrade urban area (Tomašević et al. 2005).

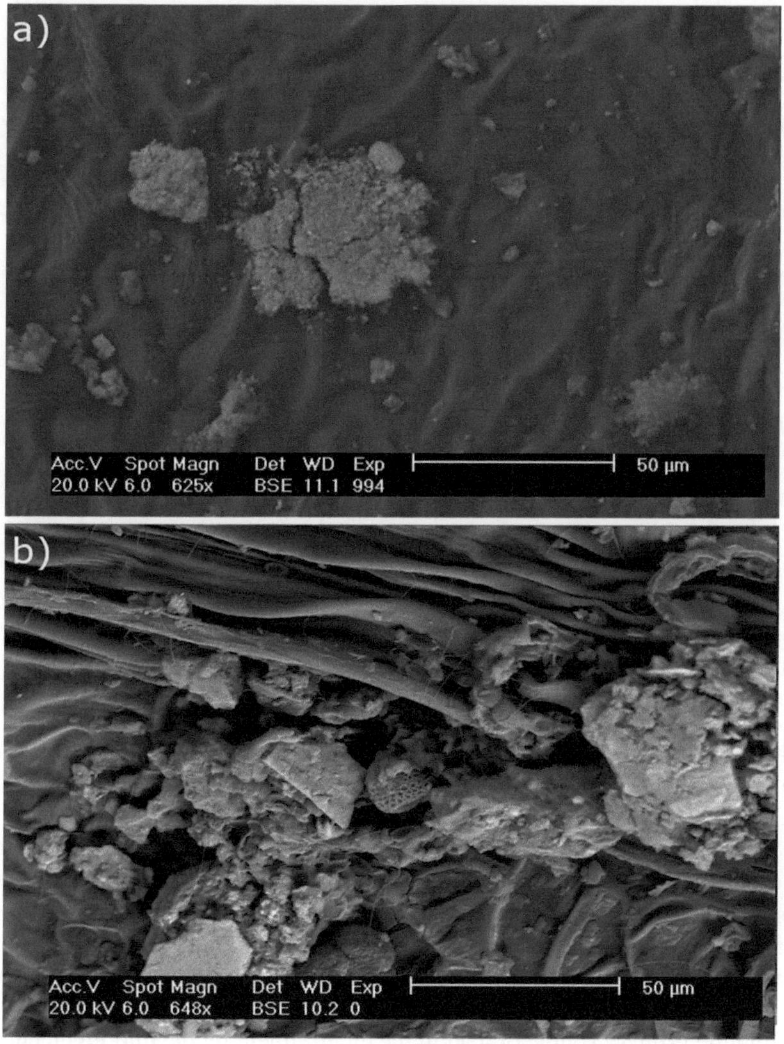

Fig. 8.8 Scanning electron micrographs of particles deposited on the adaxial (**a**) and abaxial (**b**) leaf surfaces of *A. hippocastanum* (Adapted from Tomašević et al. 2005)

8.4 Lead in the Environment

Although there was no Lead Age to rival the Copper, Bronze, and Iron ages, lead nevertheless has appeared in all phases of art, medicine, and technology and can claim first place among metals in tonnage produced or released to the environment (Nriagu 1998). Moreover, more than 95 % of present-day lead deposited in the environment is of anthropogenic origin (Marcantonio et al. 2002). However, our understanding of the fate of Pb air emissions is far from sufficient (Van der Gon and Appelman 2009).

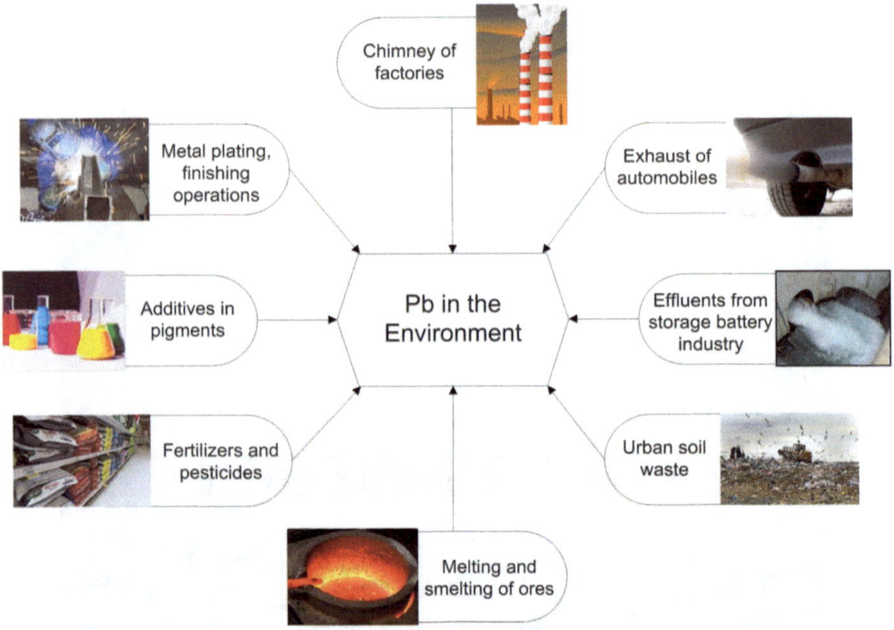

Fig. 8.9 Lead sources in the environment (Adapted from Sharma and Dubey 2005)

The main sources of lead emission and environmental pollution are presented schematically in Fig. 8.9 (adapted from Sharma and Dubey 2005). Many of them emit lead directly into the atmosphere, causing pollution and influencing the other three environmental spheres, lithosphere, hydrosphere and biosphere. Estimates of the emissions of individual sources of Pb indicate that the atmosphere is the major initial recipient and that the anthropogenic sources are at least 1–2 orders of magnitude greater than natural sources (Komárek et al. 2008). Atmospheric Pb is mostly present in fine, often sub-micrometric particles, and may be transported over long distances.

While the primary source of lead in the environment by the mid-eighteenth century was the combustion of Pb-containing coal, today the pollution caused by anthropogenic Pb originates mostly from mining, smelting, industrial uses, waste incineration, coal burning, and leaded gasoline. In the 1960s and 1970s, alkyl lead gasoline additives became the most important sources of atmospheric Pb pollution throughout the world, and potentially the most globally-widespread mechanism for the release of anthropogenic Pb into the environment has been the use of tetraethyl-Pb as an anti-knock additive in gasoline (Nriagu 1990; Hopper et al. 1991). Namely, from the 1930s onward, tetra-alkyl lead additives have been used as antiknock agents in petrol for motor vehicles and, by the 1970s, lead fallout from the combustion of petrol accounted for around 90 % of the atmospheric lead input into urban environments (Anagnostopoulou and Day 2006).

According to Facchetti (1989), and references therein, 50–60 % of the lead added to petrol is mobilized in the atmosphere, while most of the remainder adheres to the walls of the exhaust system from which it is expelled by mechanical and thermal shocks in the forms of easily sedimented particles. Namely, the fraction directly emitted by engine exhaust fumes is found in the form of fine particles, which can be transferred a long way from the emitting sources. Notwithstanding the decline in alkyl-lead emissions from petrol, this historic source of lead is still a threat to human health due to its small particle size, persistence in the environment, as well as resuspension and remobilization from contaminated soils and roadway grift (Duzgoren-Aydin and Weiss 2008).

It has been shown that lead is the most widely scattered toxic metal worldwide, persistent in the environment, with tendency to accumulate in living organisms, and as such is the topic in most current pollution studies (Hsu et al. 2006; Komárek et al. 2008; Cheng and Hu 2010; Sucharová et al. 2011). It enters the human body via inhalation of Pb-bearing particles and by ingestion of food and water contaminated with Pb (Wang and Zhang 2006).

Environmental lead toxicity is an old but persistent public health problem throughout the world, and children are more susceptible to lead than adults because of their hand to mouth activity, increased respiratory rates, and higher gastrointestinal absorption per unit body weight (Ahamed and Siddiqui 2007). As it is injurious to human health (US EPA 2006) the risk can be reduced only by lowering its emission levels. Even small amounts of Pb can be harmful, particularly to the nervous and hematopoietic systems, and can impair growth and mental function (Wang and Zhang 2006).

Due to increased knowledge of the significant negative impact of lead present in the environment, the emission of this element from different sources has been significantly reduced over the past 40 years. Despite the restrictions on leaded gasoline usage, aerosols in remote regions of the Canadian Arctic are still highly contaminated with anthropogenic Pb, suggesting an increase in Pb pollution from sources other than gasoline additives (Shotyk et al. 2005).

Despite regulatory measures adopted in many countries to limit Pb input into the environment, it continues to be one of the most serious global environmental and human hazards (Sharma and Dubey 2005). Even agricultural soil is contaminated because Pb was used as an agricultural chemical, such as Pb arsenate, which was used as a pesticide.

There are four naturally occurring, stable lead isotopes with mass numbers 204, 206, 207 and 208 (206, 207, 208 are radiogenic and 204 is primordial), which can be determined. ^{208}Pb derives from the decay of ^{232}Th, whereas ^{206}Pb and ^{207}Pb derive from decay of ^{238}U and ^{235}U, respectively (Eqs. 8.1, 8.2, and 8.3), so that variations in U/Th ratios of Pb sources, as well as their formation age, influence the ratio of ^{208}Pb to the other Pb isotopes (Bollhöfer and Rosman 2000; Chillrud et al. 2003). Due to the decay of parent isotopes the amount of stable lead isotopes is increased in comparison with the primordial level (Eqs. 8.4, 8.5, and 8.6). Despite the long half-lives of U and Th isotopes, high crustal U/Pb and Th/Pb ratios have resulted in increases of ~30 % in $^{206}Pb/^{207}Pb$ over 4.6 Ga of Earth history (Stacey and Kramers 1975). The increasing abundances of ^{208}Pb, ^{207}Pb and ^{206}Pb isotopes during the geological history of the Earth as a result of the ^{232}Th, ^{235}U and ^{238}U decay, respectively are observed.

$$^{238}_{92}U \rightarrow ^{206}_{82}Pb + 8\,^{4}_{2}He + 6\beta^{-} \tag{8.1}$$

$$^{235}_{92}U \rightarrow ^{207}_{82}Pb + 7\,^{4}_{2}He + 4\beta^{-} \tag{8.2}$$

$$^{232}_{90}Th \rightarrow ^{208}_{82}Pb + 6\,^{4}_{2}He + 4\beta^{-} \tag{8.3}$$

$$^{206}Pb_{today} = ^{206}Pb_{0} + \left(^{238}U \rightarrow ^{206}Pb\right) \tag{8.4}$$

$$^{207}Pb_{today} = ^{207}Pb_{0} + \left(^{235}U \rightarrow ^{207}Pb\right) \tag{8.5}$$

$$^{208}Pb_{today} = ^{208}Pb_{0} + \left(^{232}Th \rightarrow ^{208}Pb\right) \tag{8.6}$$

Since the development of high-resolution thermal ionization mass spectrometers (TIMS) by Nier in the late 1930s (Nier 1938), studies of lead isotopes have been common in many fields of geology (crust-mantle processes, geochronology, geochemistry, ore prospection and mining, and many others) (Kunert et al. 1999). Mixing of lead from different primary ores during industrial processing and recycling reduces the variations, but they are still large compared with the available precision of TIMS instruments (Bollhöfer and Rosman 2001). Nevertheless, exhaustive sample preparations, tedious sample handling protocols, long analysis periods (2–24 h), and inflexible analytical routines have made the use of TIMS unfeasible in some environmental studies where results are drawn from a large number of samples (Gwiazda et al. 1998; Marguí et al. 2006). It has been shown that inductively coupled plasma mass spectrometry (ICP-MS) is sufficiently accurate and precise to be used in environmental studies. ICP-MS with both a quadrupole mass analyzer-single collector (Q-ICP-MS) and a magnetic sector mass analyzer (MC-ICP-MS), have increasingly been used in high precision and accuracy isotope ratio measurement (Heumann et al. 1998; Becker and Dietze 2000). The reliable measurement of the isotope ^{204}Pb needs much care and specialized instrumentation and is time consuming while the ratios between the more abundant isotopes ^{206}Pb, ^{207}Pb and ^{208}Pb can be easily determined by ICP-MS (Reimann et al. 2012).

Sources of Pb are often difficult to identify in contamination studies. Lead isotopic analyses proved to be a very efficient tool for tracing the sources of local and global Pb pollution (Komárek et al. 2008). The utility of lead isotopes as an anthropogenic tracer was established by Chow and co-workers (Chow and Johnstone 1965; Chow and Earl 1970, 1972; Chow et al. 1973, 1975). Two common characteristics of anthropogenic lead isotope analysis are low concentrations within complex natural matrices and a large isotopic range (Reuer et al. 2003). Also, the abundances of Pb isotopes are not affected to a measurable extent by physical, chemical or biological processes. In source identification, the ratios of the isotopic concentrations of ^{208}Pb, ^{207}Pb, and ^{206}Pb detected in a sample, are matched to those of the potential sources, as the Pb ratios retain the isotopic composition of the ore from which it originates before being released into the environment (Ault et al. 1970).

The lead isotopic composition of tetraethyl- and tetramethyl-lead additives depends directly on the geological source of the Pb ores used in their manufacture (Erel et al. 1997). In this way, products that originate from specific Pb ore deposits have characteristic Pb isotope ratios (Faure 1997). Consequently, Pb oxides, Pb halides and organolead compounds emitted from burning of leaded petrol, reflect the composition of the Pb ore(s) used in its production (Chow et al. 1975). Thus, the ratios found in the environment may reflect the origins of the petrol supplies, but the positioning is often complicated by frequent changes in the origin of these supplies (Monna et al. 1997; Bollhöfer and Rosman 2001).

Very old lead ores, such as those from Broken Hill, Australia (ca. 2.5×10^9 years) contain small amounts of its radiogenic isotopes, while younger ores derived from high U/Pb sources, such as those mined in several regions of Missouri (ca. 2.5×10^8 years) have much higher proportions of ^{206}Pb, ^{207}Pb, and ^{208}Pb relative to ^{204}Pb (Chillrud et al. 2003). Australian type ores have radiogenic ratios of $^{206}Pb/^{207}Pb$ ($^{208}Pb/^{207}Pb$) ~1.05 (~2.33), while these ratios typical for Chinese Pb ores are in the range 1.13–1.16 and ~2.46, respectively (Bollhöfer and Rosman 2001). North American ores have lead isotope ratios which differ by as much as 6–25 %, depending on the isotope ratio (Chillrud et al. 2003). Namely, one major producer, Broken Hill in Australia, is characterized by a very low $^{206}Pb/^{207}Pb$ isotope ratio of 1.04, while the largest Pb deposits in the world, the Mississippi Valley type deposits from the United States, show a characteristically high $^{206}Pb/^{207}Pb$ isotope ratio of 1.4 (Reimann et al. 2012). Obviously, stable lead isotope ratios in natural lead deposits vary according to the radiogenic characteristics of their location (Chow et al. 1975). Because Pb deposits have characteristic Pb isotope compositions, the combination of Pb concentration and Pb-isotope ratios may in some cases be used as a "fingerprint" to trace Pb to its source (Reimann et al. 2012).

Isotopic measurement is superior to using concentrations alone in tracing pollutant sources because isotope ratios are significantly more sensitive tracers than elemental concentrations or ratios (Cheng and Hu 2010). Many authors have shown that lead isotopic studies may provide a convenient approach for studying and tracing the sources of Pb pollution in different environmental compartments (Monna et al. 1997; Simonetti et al. 2000; Doucet and Carignan 2001; Erel et al. 2001). Obviously, the use of variations in stable lead isotope ratios has been recognized as a useful technique for characterizing sources of lead contamination and transport pathways of Pb into the environment. Therefore, lead isotopes have been introduced as "fingerprints" of environmental pollution (Komárek et al. 2008; Cheng and Hu 2010).

Pb isotope ratios may vary according to their geological sources, and the isotopic compositions measured in the samples reflect those of the specific or mixed Pb sources. When lead pollution develops from multiple sources, it is also possible to apportion the contributions of the major sources based on their Pb isotopic compositions (Komárek et al. 2008). It is suggested that the Pb isotopic "fingerprinting" may be useful for efficient tracking of phase-out of leaded gasoline. The phasing out of leaded petrol, however, increases the relative contribution of multiple industrial Pb sources (mining and smelting, domestic and industrial fossil fuel combustion, waste incineration, etc.). Also, mixing of lead from different primary ores during industrial processing and recycling are adding further variety to the isotopic signatures that are appearing in the environment. Moreover, the isotopic composition of Pb quickly changes depending on the different sources of Pb, distances from industrial areas, traffic density, prevailing wind directions, rainfall intensity, etc. (Monna et al. 1997; Simonetti et al. 2000). Still, a multi-elemental approach using concentrations of other elements (e.g. Zn, Cd, Cu, As, Mn, Al, etc.) or other isotopic ratios (e.g. $^{87}Sr/^{86}Sr$) in aerosols can thus be helpful for a precise evaluation of the Pb origin (Doucet and Carignan 2001). However, in order to fully exploit the potential of Pb isotopes as a tracer of atmospheric pollution, the isotopic signatures must be well characterized in space and time.

8.5 Isotopic Composition in Plant Biomonitoring of Lead Air Pollution

Since the sources of lead observed in contamination studies are often difficult to identify and interpretation of isotopic data on aerosols is complicated, biomonitoring using plants in lead isotopic studies seems to be a powerful tool in source identification. Long term studies of Pb isotopic composition may be useful in assessment of dominant source contributions, and also to evaluate the validity of different plant species used for biomonitoring.

Pb uptake studies in plants have demonstrated that the roots have an ability to take up significant quantities of Pb whilst simultaneously greatly restricting its translocation to above ground parts (Lane and Martin 1977).

The root uptake of lead (Pb) by trees and the transfer of Pb by leaf litter deposition to the forest floor were investigated through a pot experiment with Norway spruce (Hovmand et al. 2009). Natural Pb and radio isotopic lead (^{210}Pb) were determined in needles and twigs and in the pot soil spiked with ^{210}Pb. The authors have shown that the internal circulation of lead (Pb) in the forest ecosystem, defined as Pb taken up by the tree roots translocated to the leaves and deposited on the forest floor with leaf litterfall is extremely small compared to the input from sources outside the forest, mainly atmospheric deposition.

The use of stable Pb isotopes for tracing Pb contamination within the environment has strongly increased the understanding of the fate of airborne Pb contaminants. For example, the study on the $^{206}Pb/^{207}Pb$ ratio in atmospheric deposition over

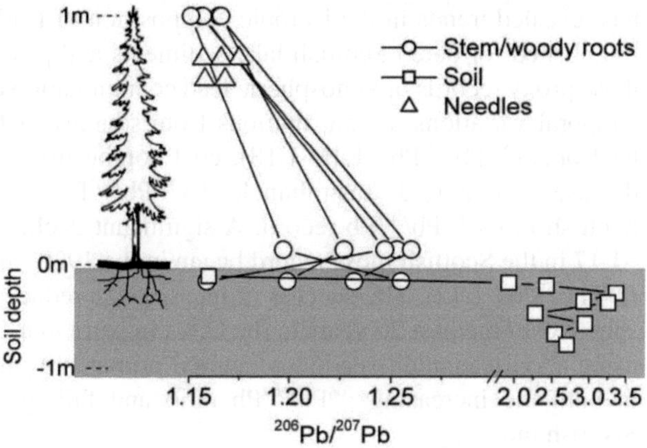

Fig. 8.10 The $^{206}Pb/^{207}Pb$ ratio of soil, fine roots and stemwood (including branches and shallow coarse roots) from five young spruce trees from the chronosequence (Adapted from Klaminder et al. 2008)

most parts of Norway showed that an overwhelming part of Pb in the humus layer of natural soils is derived from air pollution and anthropogenic contribution to surface soil Pb (Steinnes et al. 2005a). On the basis of stable Pb isotope ($^{206}Pb/^{207}Pb$ ratio) measurements of solid soil samples, stream water (from a mire outlet and a stream draining a forest dominated catchment) and components of *Picea abies* (roots, needles and stemwood) (Klaminder et al. 2008), and synthesis of some of their recent findings regarding the biogeochemistry of Pb within the boreal forest, Klaminder et al. (2008) concluded that the biogeochemical cycling of Pb in the present-day boreal forest ecosystem is dominated by Pb pollution from atmospheric deposition. The $^{206}Pb/^{207}Pb$ ratios of the mor layer (O-horizon), forest plants and stream water (mainly between 1.14 and 1.20) are similar to atmospheric Pb pollution (1.14–1.19), while the local geogenic Pb of the mineral soil (C-horizon) has high ratios (>1.30). Roots and basal stemwood of the analyzed forest trees have higher $^{206}Pb/^{207}Pb$ ratios (1.15–1.30) than needles and apical stemwood (1.14–1.18), which indicates that the latter components are more dominated by pollution derived Pb. Lower $^{206}Pb/^{207}Pb$ ratios for needles in comparison to roots and basal stemwood (Fig. 8.10) indicate that a significant part of the Pb is incorporated into the tree biomass directly from the atmosphere.

Spatial and temporal variation in the isotopic composition of atmospheric lead in Norwegian moss indicated various sources of lead (Rosman et al. 1998; Steinnes et al. 2005b). Determination of the $^{206}Pb/^{207}Pb$ isotope ratio in moss samples (*Polytrichum formosum*) from a 30–40 year old spruce forest (Hoerner Bruch) near Osnabrück (F.R.G.) has shown that this ratio has risen significantly from 1.131 in 1987 to 1.154 in 1996 due to a reduction of atmospheric inputs of lead from petrol (Kunert et al. 1999). The goal of this study was to distinguish different sources of atmospheric lead pollution by investigating lead isotope ratios. Based on the obtained results, it was possible to conclude that the sources of man-made atmospheric inputs of lead have changed over the 10-year investigated period.

Farmer et al. (2002) have investigated lead isotopic compositions in almost 200 Scottish *Sphagnum* moss samples collected over the past 170 years. The analysis has revealed trends in the isotopic composition of lead similar to those previously established for dated Scottish lake sediments and peat bogs, lending credibility to these proxy records of atmospheric lead contamination and deposition. The effect of temporal variations in contributions from sources such as smelting of indigenous lead ores ($^{206}Pb/^{207}Pb \sim 1.16$–$1.18$), coal combustion ($^{206}Pb/^{207}Pb \sim 1.17$–$1.19$), and the use of imported Australian lead ($^{206}Pb/^{207}Pb \sim 1.04$) was clearly seen in the Scottish moss $^{206}Pb/^{207}Pb$ record. A significant decline from a $^{206}Pb/^{207}Pb$ value of ~1.17 in the Scottish moss record began in the 1920s and continued until the 1980s ($^{206}Pb/^{207}Pb \sim 1.12$). The success of measures to reduce lead emissions to the atmosphere over the past 20 years in the U.K., in particular from petrol-engined vehicles using alkyl lead additives manufactured primarily from Australian lead, is evident in both the increasing $^{206}Pb/^{207}Pb$ ratio and falling lead concentration data for Scottish moss.

The $^{206}Pb/^{207}Pb$ isotope ratio was studied to investigate the viability of the sycamore (*Acer pseudoplatanus*) tree-ring analysis for the reconstruction of atmospheric lead pollution histories. Tree cores spanning 1892–2003 were collected from several sycamores from the eastern shore of Loch Lomond, Scotland, an area with no local point of lead emission (Patrick and Farmer 2006). The obtained results suggested that, in areas with no local point source of lead, accurate records of changes in atmospheric lead concentration and isotopic composition are not preserved in the annual rings of sycamore trees. Overall the data suggests that the sycamore tree-ring analysis is an unsuitable method for obtaining records of historical lead deposition in areas with no large local lead input, although it can reveal some information about the temporal and spatial influence of point source emitters. The unsuitability probably arises from the number of active annual rings in a single year, the post-uptake radial translocation of elements, the relative importance of the different routes of uptake, and the soil depth(s) from which trees draw nutrients.

The outermost bark layer of trees, predominantly Scots pine (*Pinus sylvestris*), sampled from non-urban locations throughout Scotland during 2002–2003 were analyzed for lead concentration and stable lead isotopes (Patrick and Farmer 2007). The bark $^{206}Pb/^{207}Pb$ values, 90 % of which lay between 1.10 and 1.14, were more akin to those recorded for the atmosphere (via rainwater, atmospheric particulate, moss, etc.) at various locations throughout Scotland during the 1990s, a decade over which the use of leaded petrol (mean $^{206}Pb/^{207}Pb = 1.076 \pm 0.011$) declined markedly before its complete withdrawal in 2000. This raises the question of the validity of tree bark as a monitor of current or very recent (e.g. <2 years) exposure. Because of its integrative capacity over a greater number of years, tree bark appears to be a less reliable medium than moss for measuring temporal trends in the isotopic composition of atmospheric lead.

In the review article, Komárek et al. (2008) also emphasized that it is possible to assume that bark pockets, compared to tree rings, are more suitable for Pb dendrochemical analysis used for studies dealing with the evolution of atmospheric pollution. Lead isotope ratio in tree bark pockets was determined as an indicator of past air pollution in four areas of the Czech Republic differing in Pb pollution levels (Conkova and Kubiznakova 2008). To apportion Pb emission sources, a three-isotope

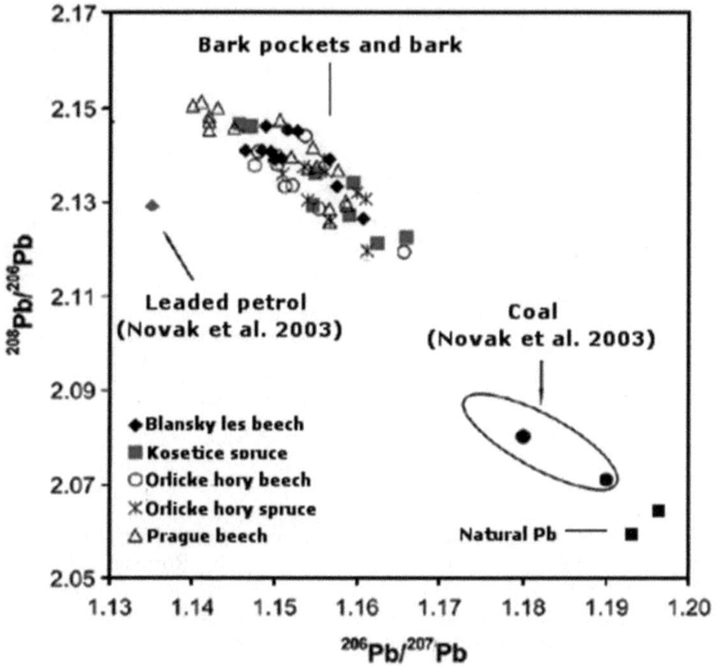

Fig. 8.11 Three-isotope plot for bark pockets and barks (Adapted from Conkova and Kubiznakova 2008)

plot that depicts the relationship between $^{206}Pb/^{207}Pb$ and $^{208}Pb/^{206}Pb$ in bark pockets is used (Fig. 8.11). The observed Pb isotopic composition of all bark pocket samples ranges between the isotopic signature of petrol (left upper part) and the values for natural Pb isotopes (right bottom part). Clearly, the isotope composition of beech bark pockets and bark in Prague Stromovka is strongly affected by automobile emissions. The isotopic ratios of spruce bark pockets, at Kosetice, Orlicke hory and Blansky les, approach the values of coal combustion and/or natural Pb. Five year intervals of bark pocket blocks provide records of the history of air pollution over the last century, which can be made even more precise with 1 year separation of bark pockets (Conkova and Kubiznakova 2008).

The contents of cobalt (Co), copper (Cu), manganese (Mn), lead (Pb) and the $^{206}Pb/^{207}Pb$ isotope ratios in the tree rings and soils collected at several locations affected by smelting and transportation in the vicinity of Kitwe (Copperbelt, Zambia) were determined; the results were interpreted in relation to potential sources of contamination such as smelter production, acidification of the environment, soil composition, raw material processing, and atmospheric suspended particulate matter (SPM) (Mihaljevič et al. 2011). The isotopic composition of the tree rings of the studied tree species varies in the interval 1.16–1.34 and the youngest parts of all the studied trees exhibit a low $^{206}Pb/^{207}Pb$ ratio (<1.17). The soil $^{206}Pb/^{207}Pb$ isotope ratio varies in the range 1.18–1.35. As the soils in the distant region have

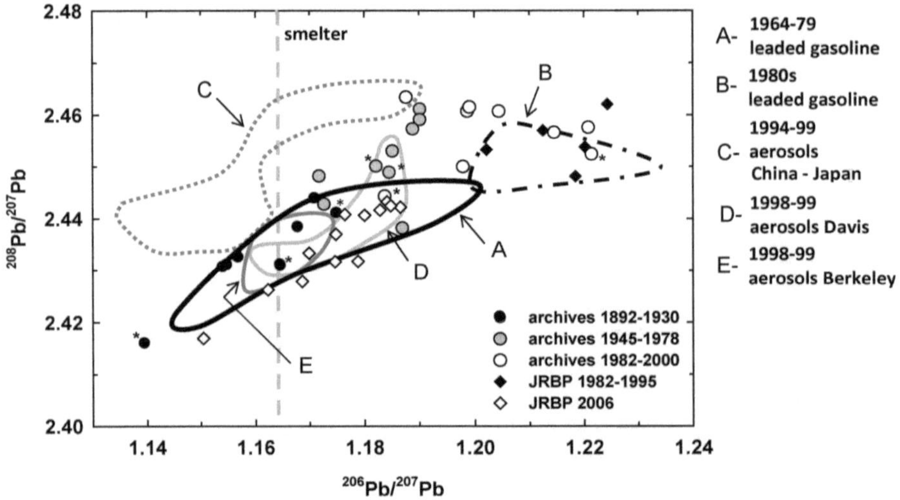

Fig. 8.12 Three-isotope plot for lichen and principal sources of industrial lead along Western North America from 1892 to 2006. Data are from Bollhöfer and Rosman (2001), Dunlap et al. (2008), Mukai et al. (2001), Rabinowitz and Wetherill (1972), Ritson et al. (1999), Steding et al. (2000), and references therein. Samples from the San Francisco Bay area are marked with an *asterisk* (archives) or represented by diamonds (JRBP) (Adapted from Flegal et al. 2010)

high $^{206}Pb/^{207}Pb$ ratios (>1.3) in the whole profile and simultaneously the youngest parts of the tree rings of tree species growing in this soil have a low $^{206}Pb/^{207}Pb$ ratio (<1.17), it can be assumed that the Pb in the youngest parts of the tree species is derived from absorption of SPM Pb through the bark rather than root uptake. It thus follows that the source of the Pb is probably interception (foliar, bark) uptake of this pollutant by trees.

Lichens and mosses have been widely used as biomonitors for assessing the atmospheric deposition of heavy metals and radionuclides (Sert et al. 2011). Flegal et al. (2010) have shown that historic and contemporary collections of epiphytic lichens may be employed to compare temporal variations in both atmospheric lead concentrations and isotopic compositions. The proposed industrial origins of lead in all of the lichen samples, dating from 1892 to 2006, are corroborated by their isotopic compositions ($^{208}Pb/^{207}Pb$ versus $^{206}Pb/^{207}Pb$) as presented in Fig. 8.12.

In the oldest sample from 1892 lead levels in lichen from the northern reach of the San Francisco Bay estuary were 9–12 μg/g and their isotopic composition corresponded to those of high lead emissions from the Selby smelter (e.g. $^{206}Pb/^{207}Pb$ 1.165) that were killing horses in adjacent fields at that time. By the mid-1950s lead isotopic compositions of lichens shifted to the more radiogenic leaded gasoline emissions (e.g., $^{206}Pb/^{207}Pb$ 1.18–1.22). Lead concentrations in the lichen peaked at 880 μg/g in 1976, corresponding with the maximum of leaded gasoline emissions in California in the 1970s. After that, lead concentrations in lichen declined to current levels, ranging from 0.2 to 4.7 μg/g. However, isotopic compositions of contemporary samples still correspond to those of previous leaded gasoline emissions in

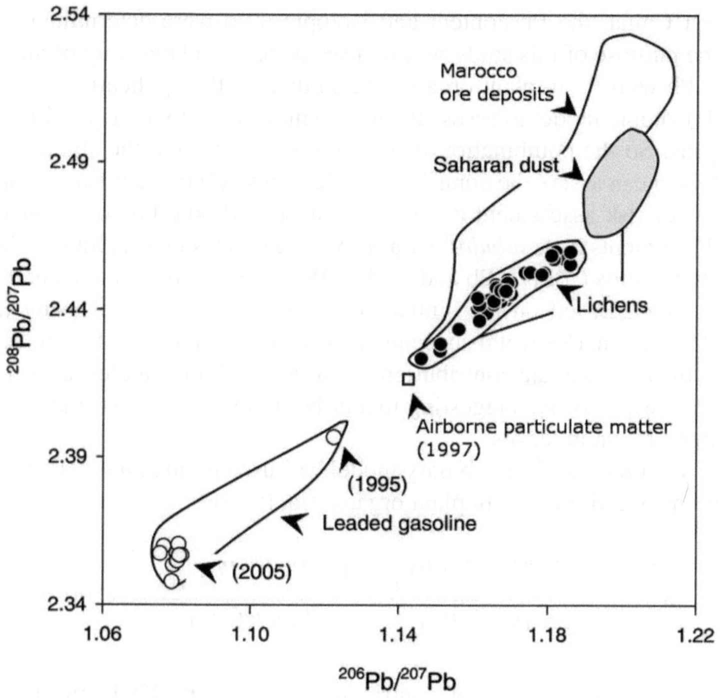

Fig. 8.13 $^{208}Pb/^{207}Pb$ vs $^{206}Pb/^{207}Pb$ in leaded gasoline (*open circles*), and in lichens (*filled circles*). Leaded gasoline sampled in 1995 and 2005 and airborne particulate matter sampled in 1997 (Bollhöfer and Rosman 2000) have also been presented as well as the isotopic compositions of Morocco ore deposits (Jébrak et al. 1998; Marcoux and Wadjinny 2005; Marcoux et al. 2008) and Saharan dust (Grousset et al. 1995; Bollhöfer and Rosman 2001; Abouchami and Zabel 2003) (Adapted from Monna et al. 2012)

California. This correspondence is consistent with other observations that attest to the persistence of environmental lead contamination from historic industrial emissions in central California.

The possibility of using lichens as monitors of atmospheric pollution around Agadir (Southwestern Morocco) as a case study predating leaded-free gasoline was examined (Monna et al. 2012). These new, homogeneous values for gasoline derived lead improve and update the scarce isotopic database of potential lead sources in Morocco, and may be of great value to future environmental surveys on the presence of lead in natural reservoirs, where it persists over time (e.g., soils and sediments). None of the collected lichens in 2005 isotopically matches modern leaded gasoline or the gasoline sold 10 years before sampling. Their position on the plot (Fig. 8.13) is rather the result of a complex mixture of several components, including anthropogenic and natural sources. The present study provides reference levels of local air pollution just before lead was finally phased out from gasoline.

In samples of aerosols, wild plants (*Aster subulatus*) and the corresponding soils, collected during the summer of 2009 in urban Nanjing, a typical mega-city located in the East China, the Pb content and isotopic ratio were determined (Hu et al. 2011). The purpose of this study was to investigate the relationship of airborne and soilborne Pb with Pb in plant organs, and to discuss the application of the simple binary Pb isotopic model to assess the approximate contributions of airborne Pb in plant organs. So the combination of Pb isotope tracing and the simple binary Pb isotope model can assess the contribution of airborne Pb in plant leaves and may be of interest for risk assessment of the exposure to airborne Pb contamination. The order of Pb contents in *A. subulatus* was roots > leaves > stems. A linear relationship of Pb isotope ratios ($^{206}Pb/^{207}Pb$ and $^{208}Pb/^{206}Pb$) among soil, plant and aerosol samples was found. Based on the simple binary Pb isotopic model using the mean $^{206}Pb/^{207}Pb$ ratios in TSP (total suspended particulates) and in SSA (size-segregated aerosols), the approximate contributions of airborne Pb in plant leaves were 72.2 % and 65.1 %, respectively, suggesting that airborne Pb is the most important source of Pb accumulation in leaves.

In this research, the simple binary model was used to quantitatively estimate the contributions of airborne Pb in plant organs (Eq. (8.7)).

$$X_{airborne} = \frac{\left({}^{206}Pb/{}^{207}Pb\right)_{plant} - \left({}^{206}Pb/{}^{207}Pb\right)_{soilborne}}{\left({}^{206}Pb/{}^{207}Pb\right)_{airborne} - \left({}^{206}Pb/{}^{207}Pb\right)_{soilborne}} \times 100\ \% \qquad (8.7)$$

where $X_{airborne}$ is the procentual contribution of airborne Pb in the plant organ, $(^{206}Pb/^{207}Pb)_{plant}$ is the isotopic composition of Pb in the plant organs, $(^{206}Pb/^{207}Pb)_{airborne}$ is the isotopic composition of airborne Pb, and $(^{206}Pb/^{207}Pb)_{soilborne}$ is the isotopic composition of soilborne Pb.

Pb isotope ratios ($^{206}Pb/^{207}Pb$ and $^{208}Pb/^{206}Pb$) in the studied aerosol samples, plant organs and soils are shown in Fig. 8.14.

Lead isotope ratios of all soil, plant and aerosol samples demonstrated an approximately linear relationship, with the soil samples at one end, the atmospheric particle samples at the other end and the plant samples between the two. The ranges of soilborne and airborne Pb isotope ratios did not overlap with each other (Fig. 8.14). The ranges of Pb isotope ratios in the roots overlapped partially with those in the soil, also leaf ratios overlapped partially with airborne Pb, while stem samples intercrossed the Pb isotope ratio in airborne source and roots. The linear relationship of Pb isotope ratios among the various samples indicates that there were two primary sources of plant Pb: soilborne and airborne Pb in the studied case. Therefore, approximate contributions of airborne Pb in plant organs can be calculated using the simple binary Pb isotope ratio model.

The extent to which Pb enters into plants via leaves depends on the ability of leaves to absorb Pb from aerial sources and resuspended soil particles, which in turn depends on the specific leaf morphology (Sharma and Dubey 2005). Generally the order of Pb contents accumulated in plant organs is roots > stems > leaves. During plant growth (about several months), Pb concentrations and Pb isotope ratios of the atmospheric particle samples show a wide range of variation, which might cause some uncertainty in the quantitative estimates of airborne Pb for plants (Hu et al. 2011).

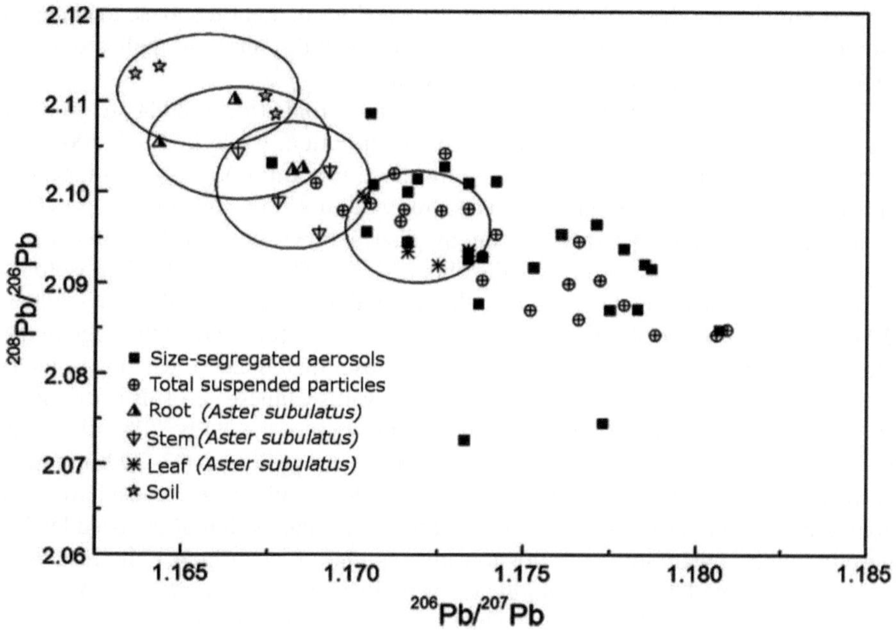

Fig. 8.14 Relationship between $^{206}Pb/^{207}Pb$ and $^{208}Pb/^{206}Pb$ in aerosols, plant organs and soils (Adapted from Hu et al. 2011)

Long term studies of Pb isotopic composition in leaves may be useful in assessment of dominant source contributions, and also to evaluate the validity of different plant species used for biomonitoring purposes (Tomašević et al. 2013).

Previous investigations of trace element concentrations in tree leaves sampled in the Belgrade urban area, in a multi-year period (2002–2006), showed that *Aesculus hippocastanum* and *Tilia* spp. could be used as appropriate biomonitors of trace elements atmospheric deposition (Aničić et al. 2011). For some elements (Pb, Cu, etc.), this biomonitoring observation correlated with the atmospheric trace element concentrations, as assayed by the bulk atmospheric deposition measurements. The Pb atmospheric pollution in Belgrade was mostly associated with traffic emissions (Rajšić et al. 2008). The focus of the work performed by Tomašević et al. (2013) was the determination of a relationship between the Pb concentration and the stable isotopic ratios ($^{206}Pb/^{207}Pb$ and $^{208}Pb/^{207}Pb$) in leaves of common urban deciduous trees *Aesculus hippocastanum* and *Tilia* spp. in a biomonitoring approach for source identification within the Pb atmospheric contamination studies. The study took place over a period of several years (2002–2006, 2009) when leaded gasoline was still in use in the Belgrade urban area, but an increasing number of vehicles using non-leaded gasoline were evident during the investigated time span. The obtained Pb isotopic ratios in leaves indicated the major contribution of leaded gasoline to the leaf Pb content; even more, the leaf Pb isotope ratios followed the switch (from Australian – 2002/2003/2004 to Chinese origin – 2005/2006) of the Pb ore used as the gasoline lead additive through the investigated years. Leaf response to changed isotopic ratios in the environment indicates the possibility of their application as a biomonitor for tracing Pb sources in the environment.

For the samples of *A. hippocastanum*, the values from 2002 to 2004 are grouped close to the values of the Australian Pb ore distribution, while in 2005–2006 and later (2009) isotopic ratios correlated with the Chinese Pb ore. Concerning *Tilia* spp. there was no similar regularity, except for the years 2006 and 2009; in previous years the values were relatively far from the distribution shown for Australian aerosols. In leaves of *A. hippocastanum* the obtained $^{206}Pb/^{207}Pb$ and $^{208}Pb/^{207}Pb$ isotopic ratios in 2002–2004 were grouped near the Australian aerosols distribution, while in 2004, when this type of leaded fuel was last in use, the ratios were at the edge of the cluster – the most distant from Australian, but closer than in 2005, which was intermediary, and then in 2006, were shifted more towards Chinese aerosols distribution (Fig. 8.15). Thus, it may be suggested that the increase of the $^{206}Pb/^{207}Pb$ ratio in the investigated tree leaves during the studied multi-year period (2002–2006) is only partly a consequence of the reduced use of the leaded gasoline, i.e. atmospheric lead contamination, but also due to the changed isotopic composition of the used leaded gasoline with higher isotopic ratios towards the end of experiment. The results of this study indicated *A. hippocastanum* as the more appropriate biomonitor than *Tilia* spp. in the Pb source identification, showing more consistent data obtained on leaf Pb isotopic composition and the isotopic ratios of gasoline used throughout the studied years.

In order to determine soil plant interactions, lead isotopes ratio in soil and different plant species was determined.

Lead isotope ratios were used to trace the origin of Pb in a soil–plant (*Urtica dioica*)–snail (*Cepaea nemoralis*) food chain in two polluted locations in the floodplains of the rivers Meuse and Rhine (Biesbosch National Park) and one reference location in the Netherlands (Notten et al. 2008). Lead in the atmosphere contributed substantially to Pb pollution and Pb transfer in plant leaves and snails in all locations despite the current low concentrations of Pb in the atmosphere (Fig. 8.16).

Sucharová et al. (2011) investigated lead concentrations and lead isotope ratios in samples of soil B-horizon, forest- floor humus (O-horizon), grass (*Avenella flexuosa*) and spruce (*Picea abies)* needles (2nd year) collected at the same locations evenly spread over the territory of the Czech Republic at an average density of 1 site/300 km^2. Median Pb concentrations differ widely in the four materials: soil B-horizon: 27 mg/kg (3.3–220 mg/kg), humus: 78 mg/kg (19–1,863 mg/kg), grass: 0.37 mg/kg (0.08–8 mg/kg) and spruce needles: 0.23 mg/kg (0.07–3 mg/kg). The median value of the $^{206}Pb/^{207}Pb$ isotope ratio in the soil B-horizon is 1.184 (variation: 1.145–1.337). In both humus and grass the median value for the $^{206}Pb/^{207}Pb$ isotope ratio is 1.162 (variation: 1.130–1.182), in spruce needles the median ratio is 1.159 (variation: 1.116–1.186). The results of this study demonstrate that humus and B-horizon are de-coupled, the humus layer is part of the biosphere, the B horizon, however, carries the geochemical signature of the lithosphere. Local variation in Pb concentration and in the Pb-isotope ratios is higher in the B-horizon than in the O-horizon. None of the

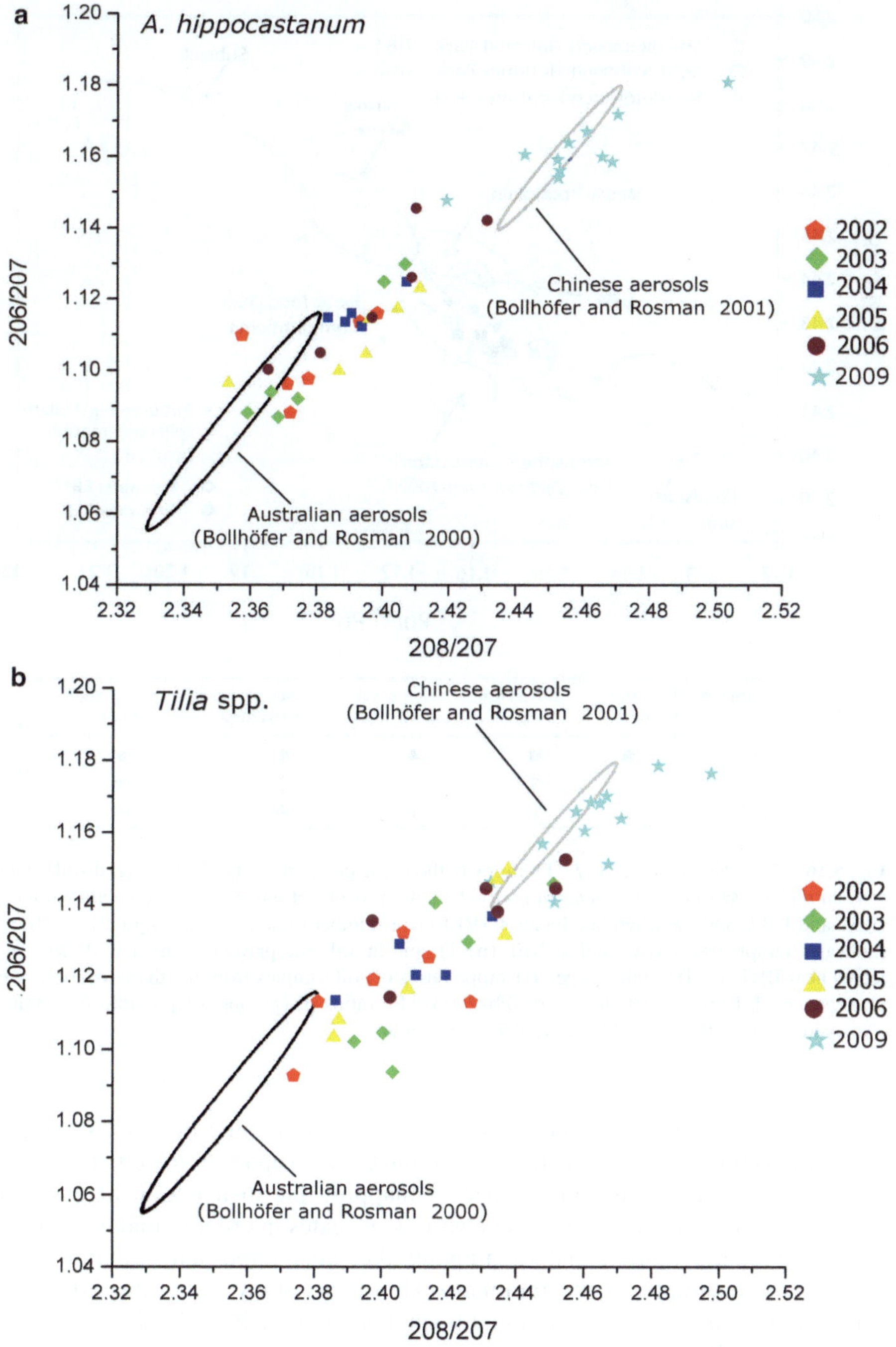

Fig. 8.15 Scatter plots of $^{206}Pb/^{207}Pb$ versus $^{208}Pb/^{207}Pb$ obtained for the *A. hippocastanum* (**a**) and *Tilia* spp. (**b**) samples (Adapted from Tomašević et al. 2013)

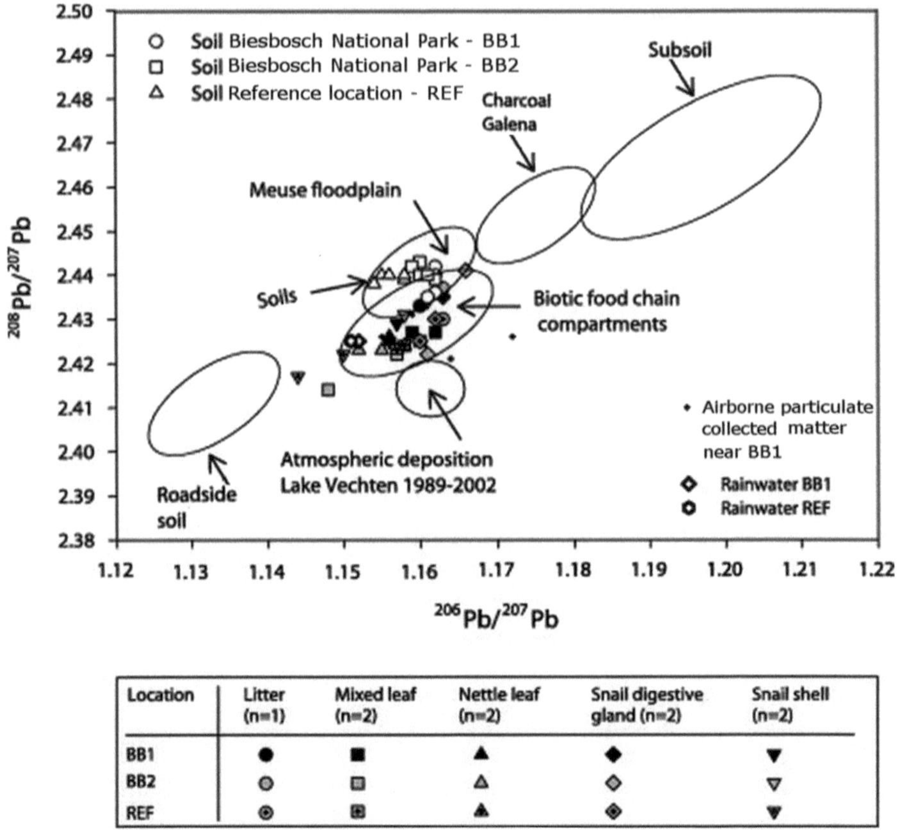

Fig. 8.16 $^{208}Pb/^{207}Pb$ versus $^{206}Pb/^{207}Pb$ ratios in the biotic compartments of a terrestrial soil–plant (*Urtica dioica*)–snail (*Cepaea nemoralis*) food chain in two locations in Biesbosch National Park (BB1 and BB2) and the reference location (REF) (for number of samples: see figure legend); in rainwater sampled near BB1 and in REF (n = 1); and in airborne particulate matter (APM) collected near BB1 (n = 4). Anthropogenic isotope ratios of soil samples from the three locations are also displayed. Ellipses include $^{208}Pb/^{207}Pb$–$^{206}Pb/^{207}Pb$ ratios of (groups) of possible Pb origins (Adapted from Notten et al. 2008 and references therein)

maps provide evidence of the importance of traffic-related emissions for the observed isotope ratios on the territory of the Czech Republic. Results from this project demonstrated the importance of mapping the distribution of element concentrations and isotope ratios at a variety of scales in order to understand the impact of contamination and to pinpoint the most important contamination sources. For example, at the regional scale of this study (c. 1 site/ 300 km^2) urban areas, traffic and even a large Pb-smelter can remain invisible.

Soil, vegetables and rainwater from three vegetable production bases in the Guiyang area, southwest China, were analyzed for Pb concentrations and isotope compositions to trace its sources (Li et al. 2012). The objective of the study was to

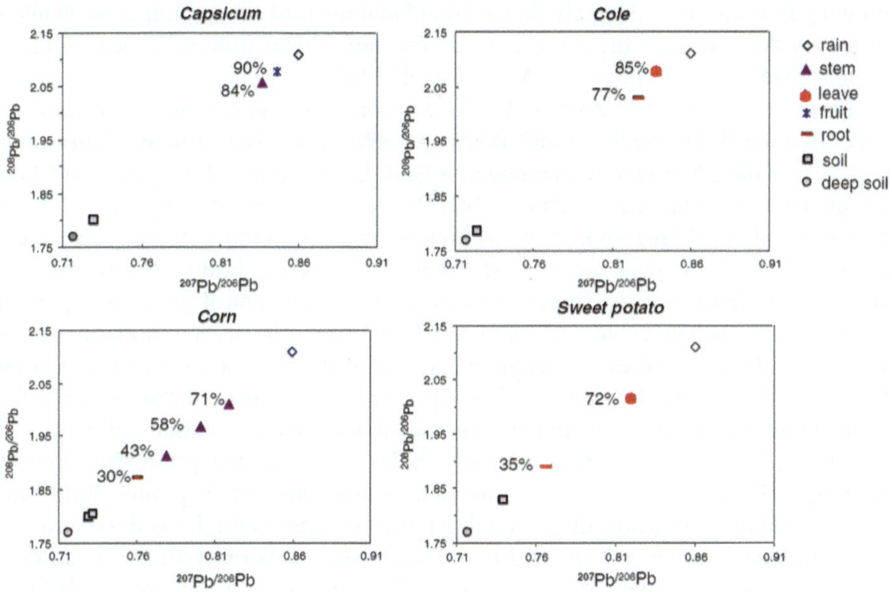

Fig. 8.17 $^{207}Pb/^{206}Pb$ versus $^{208}Pb/^{206}Pb$ diagrams of different vegetable parts from Ma-ling village around Guiyang, southwest China. Percentages denote the relative contributions of atmospheric (or anthropogenic) Pb to total Pb in different vegetable parts (Adapted from Li et al. 2012)

evaluate the biogeochemical behavior of Pb in soil-vegetable systems. The relative contribution of natural and anthropogenic sources of Pb accumulated in vegetables was determined (Fig. 8.17).

Distinctive differences in Pb isotopic composition existed between vegetables and soils, involving another Pb source, most likely to be from atmospheric Pb deposition. The isotope ratios of vegetable samples, including roots, stems, leaves, and fruits, spanned a broader range than those of soil samples, and were intermediate between those of soils and rainwater. This suggests that Pb in vegetables originates from both atmospheric/anthropogenic and background/geogenic sources. China vehicle exhausts, with the Pb isotope ratios of 0.8634–0.8788 for $^{207}Pb/^{206}Pb$ and 2.116–2.136 for $^{208}Pb/^{206}Pb$ after leaded gasoline was phased out in Shanghai, were almost in the straight line formed by the obtained data. Thus, the end member of anthropogenic Pb in vegetables of Guiyang, southwest China likely originated from Chinese ores, coals, and vehicle exhaust.

Since mining and smelting of Pb ores are significant sources of this element in the environment, the topic of researches dealing with lead isotopes were, among others, investigation of the impact of this polluted areas to Pb content in plants originating from the mining/smelting area.

On the basis of the lead content and the isotope ratio in *B. davidii* specimens collected from two abandoned Pb/Zn mining areas in Spain, it was concluded that the areas studied can become an important source of contamination of the vegetation growing in these areas. Namely, it was found that the lead accumulation in plants is primarily derived from the mining operations rather than from other lead sources such as leaded-petrol emissions (Marguí et al. 2006).

Trace element (Pb, Cd, Zn, Cu, Ag, As, Se) contents were determined in caps and stipes of three different edible mushroom species (*Boletus edulis* Bull. Fr., *Xerocomus badius* Fr. Gilb., *Xerocomus chrysenteron* Bull. Quél.), originating from a smelter-polluted area (Komárek et al. 2007). Total Pb concentration in the organic soil horizons reached 36,234 mg Pb kg^{-1}, while high contents of Pb (up to 165 mg kg^{-1}) were determined in the studied species, exceeding the regulatory limits. Pb isotopic data showed that Pb originating from the recent air pollution control residues is present mainly in the exchangeable/acid-extractable fraction of the organic horizons and is taken up by fruiting bodies; especially in the case of *B. edulis*, where fast Pb accumulation occurs. Due to the high species-dependent variations of metal contents, the studied mushrooms are not suitable as bioindicators of environmental pollution.

Lead/cadmium contamination in vegetables grown in the peri-urban area of Nanjing, China was assessed and the route for metals entering into plants was investigated through lead isotopic tracing (Hu and Ding 2009). Pb isotope ratios in the examined vegetables indicated that besides soil, airborne Pb may be another source of lead in the peri-urban and mining/smelting sites. It was concluded that the Pb isotopic content in plants is possibly a combination of soilborne Pb uptaken through the soil-to-root pathway, and airborne Pb assimilated via the atmosphere-to-leaf pathway.

Bi et al. (2009) investigated the distribution and Pb and Cd sources in maize plants (*Zea mays* L.) grown in a typical zinc smelting impacted area of southwestern China. Representative maize plant samples grown on soils with different Pb concentrations were selected for Pb isotope analyses. Pb isotope ratios were very similar in the three leaf ($^{206}Pb/^{207}Pb$, 1.185–1.186; $^{208}Pb/^{206}Pb$, 2.090–2.094) and grain ($^{206}Pb/^{207}Pb$, 1.180–1.181; $^{208}Pb/^{206}Pb$, 2.091–2.104) tissues, but differed greatly in the roots ($^{206}Pb/^{207}Pb$, 1.182–1.209; $^{208}Pb/^{206}Pb$, 2.050–2.098) and stalks ($^{206}Pb/^{207}Pb$, 1.172–1.193; $^{208}Pb/^{206}Pb$, 2.080–2.114). In Fig. 8.18 (Bi et al. 2009 and references therein), the maize samples were distributed along a line, which is similar to that defined by the soils and ores, indicating the mixed origins of Pb in the maize samples from zinc smelting emissions and local natural background.

Based on the Pb isotopic compositions of the ores, Bi et al. (2009) estimated that the $^{206}Pb/^{207}Pb$ and $^{208}Pb/^{206}Pb$ ratios of the atmospheric particles emitted from the zinc smelting operations were about 1.187 and 2.104, respectively. These ratios were very similar to those of the maize leaf samples, and thus demonstrated that Pb in maize leaves may have mainly originated from the atmospheric deposition of the smelting flue gas dust. This is consistent with the reported results that plant leaves can uptake substantial amount of Pb directly from the atmosphere (Haar 1970; Buchauer 1973; Tjell et al. 1979; Dollard 1986; Harrison and Chirgawi 1989; Klaminder et al. 2005). Pb isotope data revealed that the foliar uptake of atmospheric Pb was the dominant pathway for Pb to the leaf and grain tissues of maize, while Pb in the stalk and root tissues was mainly derived from root uptake.

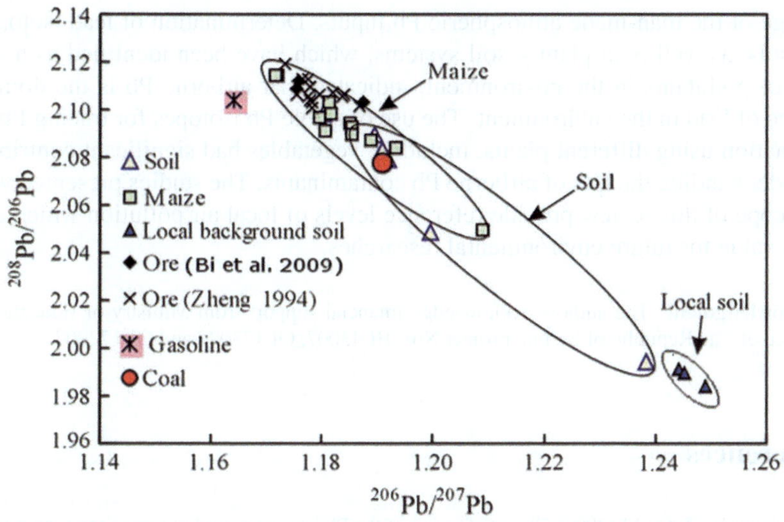

Fig. 8.18 A plot of $^{206}Pb/^{207}Pb$ versus $^{208}Pb/^{206}Pb$ for the analyzed samples. Ore data (denoted with x) is from Zheng (1994); gasoline data is from Zhu et al. (2001) and Gao et al. (2004); coal data is from Gao et al. (2004) (Adapted from Bi et al. 2009)

A combined vegetation and soil survey was undertaken within the Rookhope Catchment in the North Pennines (an area of historical Pb and Zn mining and smelting) to assess the extent of Pb contamination in soils, grass and heather as an aid to understanding potential bioaccumulation in grouse, sheep and cattle (Chenery et al. 2012). The isotopic analysis has demonstrated that there is an almost complete mixing of Pb between the labile and non-labile pools of Pb and generally high lability, allowing the grass access to the reservoirs of 'Broken Hill type' Pb and increasing its content in the vegetation. These results demonstrate that although petrol Pb was phased out in 2000, it resides in soil reservoirs and is highly persistent in the environment.

8.6 Conclusion

Airborne particles as significant sources of heavy metals have been implicated in adverse human health outcomes and environmental deterioration. Application of plants as biomonitors and bioindicators is of significant scientific interest. Even more, for certain pollutants they can be used in the field of source determination. The use of different species for lead air pollution monitoring is discussed, for those without root system as mosses and lichens, as well as for higher plants. Plant leaf characteristics like size, surface properties and plant species influence the accumulation of trace elements from the environment. On the basis of spatial and temporal variation in isotopic composition, the sources of lead were indicated, as well as the change of the man-made atmospheric Pb inputs. Determination of lead isotopes in aerosols, as well is in plant – soil systems, which have been identified as a major sink of pollutants in the environment, indicated that airborne Pb is the dominant source of lead in the environment. The use of stable Pb isotopes for tracing Pb contamination using different plants, including vegetables had significant contribution in understanding the fate of airborne Pb contaminants. The studies presented within the scope of this review provide reference levels of local air pollution which are of great value for future environmental researches.

Acknowledgment The authors acknowledge financial support from Ministry of Education and Science of The Republic of Serbia, project Nos. III 43007, OI 173028 and OI 172007.

References

Abouchami W, Zabel M (2003) Climate forcing of the Pb isotope record of terrigenous input into the Equatorial Atlantic. Earth Planet Sci Lett 213:221–234. doi:10.1016/S0012-821X(03)00304-2

Ahamed M, Siddiqui M (2007) Environmental lead toxicity and nutritional factors. Clin Nutr 26:400–408. doi:10.1016/j.clnu.2007.03.010

Alcamo J (1992) Transboundary air pollution in Central Europe and Eastern Europe. In: Alcamo J (ed) Coping with crisis in Eastern Europe's environment. IIASA Book, Parthenon Publishing Group, Carnforth

An HK, Park BY, Kim DS (2001) Crab shell for the removal of heavy metals from aqueous solution. Water Res 35:3551–3556. doi:10.1016/S0043-1354(01)00099-9 doi:10.1016/S0043-1354(01)00099-9#doilink

Anagnostopoulou MA, Day JP (2006) Lead concentrations and isotope ratios in street dust in major cities in Greece in relation to the use of lead in petrol. Sci Total Environ 367:791–799. doi:10.1016/j.scitotenv.2006.01.040

Aničić M, Tasić M, Frontasyeva MV, Tomašević M, Rajšić S, Strelkova LP, Popović A, Steinnes E (2009a) Active biomonitoring with wet and dry moss: a case study in an urban area. Environ Chem Lett 7:55–60. doi:10.1007/s10311-008-0135-4

Aničić M, Tasić M, Frontasyeva MV, Tomašević M, Rajšić S, Mijić Z, Popović A (2009b) Active moss biomonitoring of trace elements with *Sphagnum girgensohnii* moss bags in relation to atmospheric bulk deposition in Belgrade, Serbia. Environ Pollut 157:673–679. doi:10.1016/j.envpol.2008.08.003

Aničić M, Tomašević M, Tasić M, Rajšić S, Popović A, Frontasyeva MV, Lierhagen S, Steinnes E (2009c) Monitoring of trace element atmospheric deposition using dry and wet moss bags: accumulation capacity versus exposure time. J Hazard Mater 171:182–188. doi:10.1016/j.jhazmat.2009.05.112

Aničić M, Spasić T, Tomašević M, Rajšić S, Tasić M (2011) Trace elements accumulation and temporal trends in leaves of urban deciduous trees (*Aesculus hippocastanum* and *Tilia* spp.). Ecol Ind 11:824–830. doi:10.1016/j.ecolind.2010.10.009

Antanasijević D, Pocajt V, Povrenović D, Perić-Grujić A, Ristić M (2013) PM10 emission forecasting using artificial neural networks and genetic algorithm input variables optimization. Sci Total Environ 443:551–559. doi:1016/j.scitotenv.2012.10.110

Ault WU, Senechal RG, Erlebach WE (1970) Isotopic composition as a natural tracer of lead in the environment. Environ Sci Technol 4:305–313. doi:10.1021/es60039a001

Avila A, Rodrigo A (2004) Trace metal fluxes in bulk deposition, throughfall and stemflow at two evergreen oak stands in NE Spain subject to different exposure to the industrial environment. Atmos Environ 38:171–180. doi:10.1016/j.atmosenv.2003.09.067

Azimi S, Rocher V, Muller M, Moilleron R, Thevenot S (2005) Sources, distribution and variability of hydrocarbons and metals in atmospheric deposition in an urban area (Paris, France). Sci Total Environ 337:223–239. doi:10.1016/j.scitotenv.2004.06.020

Bargagli R (1998) Trace elements in terrestrial plants: an ecophysiological approach to biomonitoring and biorecovery. Springer, Heidelberg

Becker JS, Dietze HJ (2000) Precise and accurate isotope ratio measurements by ICP-MS. Fresenius J Anal Chem 368:23–30. doi:10.1007/s002160000481

Berg T, Steinnes E (1997) Use of mosses (*Hylocomium splendens* and *Pleurozium schrebery*) as biomonitors of heavy metal deposition: from relative to absolute deposition values. Environ Pollut 98:61–71. doi:10.1016/S0269-7491(97)00103-6

Berlizov AN, Blum OB, Filby RH, Malyuk IA, Tryshyn VV (2007) Testing applicability of black poplar (*Populus nigra* L.) bark to heavy metal air pollution monitoring in urban and industrial regions. Sci Total Environ 372:693–706. doi:10.1016/j.scitotenv.2006.10.029

Bi X, Feng X, Yang Y, Li X, Shin GPY, Li F, Qiu G, Li G, Liu T, Fu Z (2009) Allocation and source attribution of lead and cadmium in maize (*Zea mays* L.) impacted by smelting emissions. Environ Pollut 157:834–839. doi:10.1016/j.envpol.2008.11.013

Bilos C, Colombo JC, Skorupka CN, Rodriguez Presa MJ (2001) Sources, distribution and variability of airborne trace metals in La Plata City area, Argentina. Environ Pollut 111:149–158. doi:10.1016/S0269-7491(99)00328-0

Blum WEH, Horak O, Mentler A, Purshenreiter M (2009) Environmental and ecological chemistry, vol II, Trace elements, Editor Sabljic A, Encyclopedia of Life Support Systems (EPLSS) ISBN: 978-1-84829-693-3

Bollhöfer A, Rosman KJR (2000) Isotopic source signatures for atmospheric lead: the Southern Hemisphere. Geochim Cosmochim Acta 64:3251–3262. doi:10.1016/S0016-7037(00)00436-1

Bollhöfer A, Rosman KJR (2001) Isotopic source signatures for atmospheric lead: the Northern Hemisphere. Geochim Cosmochim Acta 65:1727–1740. doi:10.1016/S0016-7037(00)00630-X

Brunekreef B, Holgate ST (2002) Air pollution and health. Lancet 360:1233–1242. doi:10.1016/S0140-6736(02)11274-8

Buchauer MJ (1973) Contamination of soil and vegetation near a zinc smelter by zinc, cadmium, copper, and lead. Environ Sci Technol 7:131–135. doi:10.1021/es60074a004

Carnevale C, Finzi G, Guariso G, Pisoni E, Volta M (2012) Surrogate models to compute optimal air quality planning policies at a regional scale. Environ Model Software 34:44–50. doi:10.1016/j.envsoft.2011.04.007

Chenery SR, Izquierdo M, Marzouk E, Klinck B, Palumbo-Roe B, Tye AM (2012) Soil-plant interactions and the uptake of Pb at abandoned mining sites in the Rookhope catchment of the N. Pennines, UK – A Pb isotope study. Sci Total Environ 433:547–560. doi:10.1016/j.scitotenv.2012.03.004

Cheng H, Hu Y (2010) Lead (Pb) isotopic fingerprinting and its applications in lead pollution studies, a review. Environ Pollut 158:1134–1146. doi:10.1016/j.envpol.2009.12.028

Chillrud SN, Hemming S, Shuster EL, Simpson HJ, Bopp RF, Ross JM, Pederson DC, Chaky DA, Tolley LR, Estabrooks F (2003) Stable lead isotopes, contaminant metals and radionuclides in upper Hudson River sediment cores: implications for improved time stratigraphy and transport processes. Chem Geol 199:53–70. doi:10.1016/S0009-2541(03)00055-X

Chow TJ, Earl JL (1970) Lead aerosols in the atmosphere: increasing concentrations. Science 169:577–580

Chow TJ, Earl JL (1972) Lead isotopes in North American coals. Science 176:510–511

Chow TJ, Johnstone MS (1965) Lead isotopes in gasoline and aerosols of Los Angeles Basin, California. Science 147:502–503

Chow TJ, Bruland KW, Bertine K, Soutar A, Koide M, Goldberg ED (1973) Lead pollution: records in southern California coastal sediments. Science 181:551–552

Chow TJ, Snyder CB, Earl JL (1975) Isotope ratios as pollutant source and behaviour indicators. In: Proceedings of a symposium on Isotope ratios as pollutant source and behaviour indicators. Proceedings series, international atomic energy agency, Vienna, 18–22 November 1974, pp 95–108

Coe JM, Lindberg SE (1987) The morphology and size distribution of atmospheric particles deposited on foliage and inert surfaces. J Air Pollut Control Assoc 37:237–243

Colette A, Granier C, Hodnebrog Ø, Jakobs H, Maurizi A, Nyiri A, Bessagnet B, D'Angiola A, D'Isidoro M, Gauss M, Meleux F, Memmesheimer M, Mieville A, Rouïl L, Russo F, Solberg S, Stordal F, Tampieri F (2011) Air quality trends in Europe over the past decade: a first multi-model assessment. Atmos Chem Phys 11:11657–11678. doi:10.5194/acpd-11-11657-2011

Cong Z, Kang S, Zhang Y, Li X (2010) Atmospheric wet deposition of trace elements to central Tibetan Plateau. Appl Geochem 25:1415–1421. doi:10.1016/j.apgeochem.2010.06.011

Conkova M, Kubiznakova J (2008) Lead isotope ratios in tree bark pockets: an indicator of past air pollution in the Czech Republic. Sci Total Environ 404:440–445. doi:10.1016/j.scitotenv.2008.04.025

De Nicola F, Maisto G, Prati MV, Alfani A (2008) Leaf accumulation of trace elements and polycyclic aromatic hydrocarbons (PAHs) in *Quercus ilex* L. Environ Pollut 153:376–383. doi:10.1016/j.envpol.2007.08.008

Directive 2008/50/EC of the European Parliament and of the Council of 21 May 2008 on ambient air quality and cleaner air for Europe. Off J L152/1, available at: http://eur-lex.europa.eu/LexUriServ/LexUriServ.do?uri=OJ:L:2008:152:0001:0044:EN:PDF

Dollard GJ (1986) Glasshouse experiments on the uptake of foliar applied lead. Environ Pollut Ecol Biol 40:109–119. doi:10.1016/0143-1471(86)90078-4

Doucet F, Carignan J (2001) Atmospheric Pb isotopic composition and trace metal concentration as revealed by epiphytic lichens: an investigation related to two altitudinal sections in eastern France. Atmos Environ 35:3681–3690. doi:10.1016/S1352-2310(00)00510-0

Duffus JH (2002) "Heavy metals" – a meaningless term? (IUPAC technical report). Pure Appl Chem 74:793–807. doi:10.1351/pac200274050793

Dunlap C, Steding DD, Bouse R, Unruh J, Alpers C, Flegal AR (2008) Anthropogenic sources of lead in the Sacramento and San Joaquin drainage basins. Geochim Cosmochim Acta 72:5935–5948. doi:10.1016/j.gca.2008.10.006

Duzgoren-Aydin NS, Weiss AL (2008) Use and abuse of Pb-isotope fingerprinting technique and GIS mapping data to assess lead in environmental studies. Environ Geochem Health 30:577–588. doi:10.1007/s10653-008-9179-4

Erel Y, Veron A, Halicz L (1997) Tracing the transport of anthropogenic lead in the atmosphere and in soils using isotopic ratios. Geochim Cosmochim Acta 61:4495–4505. doi:10.1016/S0016-7037(97)00353-0

Erel Y, Dubowski Y, Halicz L, Erez J, Kaufman A (2001) Lead concentrations and isotopic ratios in the sediments of the sea of Galilee. Environ Sci Technol 35:292–299. doi:10.1021/es001399u

Ewing SA, Christensen JN, Vancuren RA, Cliff SS, Depaolo DJ (2010) Pb isotopes as an indicator of the Asian contribution to particulate air pollution in Urban California. Environ Sci Technol 44:8911–8916. doi:10.1021/es101450t

Facchetti S (1989) Lead in petrol, the isotopic lead experiment. Acc Chem Res 22:370–374. doi:10.1021/ar00166a005

Farmer JG, Eades LJ, Atkins H, Chamnerland DF (2002) Historical trends in the lead isotopic composition of archival *Sphagnum* mosses from Scotland (1838–2000). Environ Sci Technol 36:152–157. doi:10.1021/es010156e

Faure G (1997) Principles and applications of geochemistry: a comprehensive textbook for geology students. Prentice-Hall Inc., Upper Saddle River

Flegal AR, Gallon C, Hibdon S, Kuspa ZE, Laporte LF (2010) Declining – but persistent – atmospheric contamination in Central California from the resuspension of historic leaded gasoline emissions as recorded in the Lace Lichen (*Ramalina menziesii* Taylor) from 1892 to 2006. Environ Sci Technol 44:5613–5618. doi:10.1021/es100246e

Fountoukis C, Nenes A (2007) ISORROPIA II: a computationally efficient thermodynamic equilibrium model for K^+–Ca^{2+}–Mg^{2+}–NH_4^+–Na^+–SO_4^{2-}–NO_3^-– Cl^-–H_2O aerosols. Atmos Chem Phys 7:4639–4659. doi:10.5194/acp-7-4639-2007

Freer-Smith PH, Holloway S, Goodman A (1997) The uptake of particulates by an urban woodland: Site description and particulate composition. Environ Pollut 95:27–35. doi:10.1016/S0269-7491(96)00119-4

Gao ZY, Yin G, Ni SJ, Zhang CJ (2004) Geochemical feature of the urban environmental lead isotopeinChenducity.CarsologicaSinica23:267–272.doi:cnki:ISSN:1001–4810.0.2004-04-002 002, in Chinese with English abstract

Golomb D, Ryan D, Eby N, Underhill J, Zemba S (1997) Atmospheric deposition of toxics onto Massachusetts Bay – I. Metals. Atmos Environ 31:1349–1359. doi:10.1016/S1352-2310(96)00276-2

Grousset FE, Quétel CR, Thomas B, Donard OFX, Lambert CE, Guillard F, Monaco A (1995) Anthropogenic vs. Lithogenic origins of trace elements (As, Cd, Pb, Rb, Sb, Sc, Sn, Zn) in water column particles: Northwestern Mediterranean Sea. Mar Chem 48:291–310. doi:10.1016/0304-4203(94)00056-J

Gunn SWA (1990) Multilingual dictionary of disaster medicine and international relief. Kluwer Academic Publishers, Dordrecht

Gwiazda R, Woolard D, Smith D (1998) Improved lead isotope ratio measurements in environmental and biological samples with a double focusing magnetic sector inductively coupled plasma mass spectrometer (ICP-MS). J Anal Atomic Spectrom 13:1233–1238. doi:10.1039/A804062A, Paper

Haar GT (1970) Air as a source of lead in edible crops. Environ Sci Technol 4:226–229. doi:10.1021/es60038a008

Harmens H, Norris AD, Koerber RG, Buse A, Steinnes E, Ruhling A (2008) Temporal trends (1990–2000) in the concentration of cadmium, lead and mercury in mosses across Europe. Environ Pollut 151:368–376. doi:10.1016/j.envpol.2007.06.043

Harrison RM, Chirgawi MB (1989) The assessment of air and soil as contributors of some trace metals to vegetable plants. I. Use of a filtered air growth cabinet. Sci Total Environ 83:13–34. doi:10.1016/0048-9697(89)90003-X

Harrison RM, Yin J (2000) Particulate matter in the atmosphere: which particle properties are important for its effects on health? Sci Total Environ 249:85–101

Heumann KG, Gallus SM, Radlinger G, Vogl J (1998) Precision and accuracy in isotope ratio measurements by plasma source mass spectrometry. J Anal Atomic Spectrom 13:1001–1008. doi:10.1039/A801965G, Paper

Hoodaji M, Ataabadi M, Payam Najafi P (2012) Biomonitoring of airborne heavy metal contamination. In: Mukesh K (ed) Air pollution – monitoring, modelling, health and control. InTech. doi:10.5772/32963. ISBN 978-953-51-0381-3

Hopper JF, Ross HB, Sturges WT, Barrie LA (1991) Regional source discrimination of atmospheric aerosols in Europe using the isotopic composition of lead. Tellus B 43:45–60. doi:10.1034/j.1600-0889.1991.00004.x

Hovmand MF, Nielsen SP, Johnsen I (2009) Root uptake of lead by Norway spruce grown on ^{210}Pb spiked soils. Environ Pollut 157:404–409. doi:10.1016/j.envpol.2008.09.038

Hsu S-C, Liu SC, Jeng W-L, Chou CCK, Hsu R-T (2006) Lead isotope ratios in ambient aerosols from Taipei, Taiwan: identifying long-range transport of airborne Pb from the Yangtze Delta. Atmos Environ 40:5393–5404. doi:10.1016/j.atmosenv.2006.05.003

Hu X, Ding Z (2009) Lead/cadmium contamination and lead isotopic ratios in vegetables grown in peri-urban and mining/smelting contaminated sites in Nanjing, China. Bull Environ Contam Toxicol 82:80–84. doi:10.1007/s00128- 008-9562-y

Hu X, Zhang Y, Luo J, Xie M, Wang T, Lian H (2011) Accumulation and quantitative estimates of airborne lead for a wild plant (*Aster subulatus*). Chemosphere 82:1351–1357. doi:10.1016/j.chemosphere.2010.11.079

Hu X, Zhang Y, Ding Z, Tijian W, Hongzhen L, Yuanyuan S, Wu J (2012) Bioaccessibility and health risk of arsenic and heavy metals (Cd, Co, Cr, Cu, Ni, Pb, Zn and Mn) in TSP and PM2.5 in Nanjing, China. Atmos Environ 57:146–152. doi:10.1016/j.atmosenv.2012.04.056

Husar RB (1994) Ecosystem and the biosphere: metaphors for human-induced material flows. In: Ayres RU, Simonis UE (eds) Industrial metabolism: restructuring for sustainable development. United Nations University Press, Tokyo/New York

Järup L (2003) Hazards of heavy metal contamination. Br Med Bull 68(1):167–182. doi:10.1093/bmb/ldg032

Jébrak M, Marcoux E, Nasloubi M, Zaharaoui M (1998) From sandstone- to carbonate- hosted stratabound deposits: an isotope study of galena in the Upper-Moulouya District (Morocco). Miner Deposita 33:406–415. doi:10.1007/s001260050158

Jiries A (2003) Vehicular contamination of dust in Amman, Jordan. Environmentalist 23:205–210. doi:10.1023/B:ENVR.0000017390.93161.99

Kabata-Pendias A, Mukherjee AB (2007) Trace elements from soil to human. Springer, New York

Klaminder J, Bindler R, Emteryd O, Renberg I (2005) Uptake and recycling of lead by boreal forest plants: quantitative estimates from a site in northern Sweden. Geochim Cosmochim Acta 69:2485–2496. doi:10.1016/j.gca.2004.11.013

Klaminder J, Bindler R, Renberg I (2008) The biogeochemistry of atmospherically derived Pb in the boreal forest of Sweden. Appl Geochem 23:2922–2931. doi:10.1016/j.apgeochem.2008.04.007

Komárek M, Chrastný V, Štíchová J (2007) Metal/metalloid contamination and isotopic composition of lead in edible mushrooms and forest soils originating from a smelting area. Environ Int 33:677–684. doi:10.1016/j.envint.2007.02.001

Komárek M, Ettler V, Chrastný V, Mihaljevič M (2008) Lead isotopes in environmental sciences: a review. Environ Int 34:562–577. doi:10.1016/j.envint.2007.10.005

Kunert M, Friese K, Weckert V, Markert B (1999) Lead Isotope systematics in *Polytrichum formosum*: an example from a biomonitoring field study with mosses. Environ Sci Technol 33:3502–3505. doi:10.1021/es981352x

Lane SD, Martin ES (1977) A histochemical investigation of lead uptake in *Raphanus sativus*. New Phytol 79:281–286

Li F-L, Liu C-Q, Yang Y-G, Bi X-Y, Liu Z-Y, Zhao Z-Q (2012) Natural and anthropogenic lead in soils and vegetables around Guiyang city, southwest China: a Pb isotopic approach. Sci Total Environ 431:339–347. doi:10.1016/j.scitotenv.2012.05.040

Marcantonio F, Zimmerman A, Xu Y, Canuel E (2002) A Pb isotope record of mid- Atlantic US atmospheric Pb emissions in Chesapeake Bay sediments. Mar Chem 77:123–132. doi:10.1016/S0304-4203(01)00081-0

Marcoux E, Wadjinny A (2005) The Ag–Hg Zgounder ore deposit (Jebel Siroua, Anti- Atlas, Morocco): a Neoproterozoic epithermal mineralization of the Imiter type. C R Geosci 337:1439–1446. doi:10.1016/j.crte.2005.09.005

Marcoux E, Belkabir A, Gibson HL, Lentz D, Ruffet G (2008) Draa Sfar, Morocco: a Visean (331 Ma) pyrrhotite-rich, polymetallic volcanogenic massive sulphide deposit in a Hercynian sediment-dominant terrane. Ore Geol Rev 33:307–328. doi:10.1016/j.oregeorev.2007.03.004

Marguí E, Iglesias M, Queralt I, Hidalgo M (2006) Lead isotope ratio measurements by ICP-QMS to identify metal accumulation in vegetation specimens growing in mining environments. Sci Total Environ 367:988–998. doi:10.1016/j.scitotenv.2006.03.036

Marjanović DM, Vukčević MM, Antonović GD, Dimitrijević IS, Jovanović MĐ, Matavulj MM, Ristić ĐM (2009) Heavy metals concentration in soils from parks and green areas in Belgrade. J Serb Chem Soc 74:697–706. doi:10.2298/JSC0906697M

Markert B (1993) Instrumental analysis of plants. In: Markert B (ed) Plants as biomonitors. Indicators for heavy metals in terrestrial environment. VCH, Weinheim, pp 65–103

Markert B (1995) Sampling and sample pretreatment. In: Quevauviller PH (ed) Quality assurance in environmental monitoring. VCH, Weinheim/New York, pp 215–254

Markert B (1996) Instrumental element and multi-element analysis of plant samples, methods and applications. Wiley, New York

Markert B (2007) Definitions and principles for bioindication and biomonitoring of trace metals in the environment. J Trace Elem Med Biol 21(S1):77–82. doi:10.1016/j.jtemb.2007.09.015

Meinshausen M, Meinshausen N, Hare W, Raper SCB, Frieler K, Knutti R, Frame DJ, Allen MR (2009) Greenhouse-gas emission targets for limiting global warming to 2 °C. Nature 458:1158–1162. doi:10.1038

Melaku S, Morris V, Raghavan D, Hosten C (2008) Seasonal variation of heavy metals in ambient air and precipitation at a single site in Washington, DC. Environ Pollut 155:88–98. doi:10.1016/j.envpol.2007.10.038

Mihaljevič M, Ettler V, Šebek O, Sracek O, Kříbek B, Kyncl T, Majer V, Veselovský F (2011) Lead isotopic and metallic pollution record in tree rings from the Copperbelt mining–smelting area, Zambia. Water Air Soil Pollut 216(657):668. doi:10.1007/s11270-010-0560-4

Milosavljević BN, Ristić ĐM, Perić-Grujić AA, Filipović MJ, Štrbac BS, LjZ R, Kalagasidis Krušić TM (2010) Hydrogel based on chitosan, itaconic acid and methacrylic acid as adsorbent of Cd^{2+} ions from aqueous solution. Chem Eng J 165:554–562. doi:10.1016/j.cej.2010.09.072

Milosavljević BN, Ristić ĐM, Perić-Grujić AA, Filipović MJ, Štrbac BS, LjZ R, Kalagasidis Krušić TM (2011) Sorption of zinc by novel pH-sensitive hydrogels based on chitosan, itaconic acid and methacrylic acid. J Hazard Mater 192:846–854. doi:10.1016/j.jhazmat.2011.05.093

Monna F, Lancelot JR, Croudace I, Cundy A, Lewis JT (1997) Pb isotopic signature of urban air in France and United Kingdom: implications on Pb pollution sources. Environ Sci Technol 31:2277–2286. doi:10.1021/es960870+

Monna F, Bouchaou L, Rambeau C, Losno R, Bruguier O, Dongarrà G, Black S, Chateau C (2012) Lichens used as monitors of atmospheric pollution around agadir (Southwestern Morocco) – a case study predating lead-free gasoline. Water Air Soil Pollut 223:1263–1274. doi:10.1007/s11270-011-0942-2

Moshammer H, Neuberger N (2003) The active surface of suspended particles as a predictor of lung function and pulmonary symptoms in Austrian school children. Atmos Environ 37:1737–1744. doi:10.1016/S1352-2310(03)00073-6

Motelay-Massei A, Ollivon D, Tiphagne K, Garban B (2005) Atmospheric bulk deposition of trace metals to the Seine river Basin, France: concentrations, sources and evolution from 1988 to 2001 in Paris. Water Air Soil Pollut 164:119–135. doi:10.1007/s11270-005-1659-x

Mukai H, Tanaka TA, Fujii Y, Zeng Y, Hong J, Tang S, Guo H, Xue Z, Sun J, Zhou D, Xue J, Zhao G, Zhai J, Gu ZP (2001) Regional characteristics of sulfur and lead isotope ratios in the atmosphere at several Chinese urban sites. Environ Sci Technol 35:1064–1071. doi:10.1021/es001399u

Nica VD, Bura M, Gergen I, Harmanescu M, Bordean D-M (2012) Bioaccumulative and conchological assessment of heavy metal transfer in a soil-plant-snail food chain. Chem Cent J 6:55. doi:10.1186/1752-153X-6-55

Nier AO (1938) Variations in the relative abundances of the isotopes of common lead from various sources. J Am Chem Soc 60:1571–1576. doi:10.1021/ja01274a016

Nordberg M, Templeton MD, Andersen O, Duffus HJ (2007) Glossary of terms used in toxicology. Pure Appl Chem 79:1153–1344. doi:10.1351/pac200779071153

Nordberg M, Templeton MD, Andersen O, Duffus HJ (2009) Glossary of terms used in ecotoxicology. Pure Appl Chem 81:829–970. doi:10.1351/PAC-REC-08-07-09

Notten MJM, Walraven N, Beets CJ, Vroon P, Rozema J, Aerts R (2008) Investigating the origin of Pb pollution in a terrestrial soil–plant–snail food chain by means of Pb isotope ratios. Appl Geochem 23:1581–1593. doi:10.1016/j.apgeochem.2008.01.010

Nriagu JO (1990) The rise and fall of leaded gasoline. Sci Total Environ 92:13–28. doi:10.1016/0048-9697(90)90318-O

Nriagu JO (1998) Tales told in lead. Science 281:1622–1623

Nriagu JO, Pacyna JM (1988) Quantitative assessment of worldwide contamination of air, water and soils by trace metals. Nature 333:134–139. doi:10.1038/333134a0

Office of the European Union (2011) Air quality in Europe – 2011 report. Publications Office of the European Union, Luxembourg

Olowoyo JO, van Heerden E, Fisher JL, Baker C (2010) Trace metals in soil and leaves of *Jacaranda mimosifolia* in Tshwane area, South Africa. Atmos Environ 44:1826–1830. doi:10.1016/j.atmosenv.2010.01.048

Onjia A (2007) Analytical techniques for determination and monitoring of chemical substances affecting the corrosion. Struct Integr Life 7:79–82

Othman I, Al-Oudat M, Al-Masri MS (1997) Lead levels in roadside soils and vegetation of Damascus city. Sci Total Environ 207:43–48

Pacyna JM (1986) Atmospheric trace elements from natural and anthropogenic sources. In: Nriagu JO, Davidson CI (eds) Toxic metals in the atmosphere. Wiley, New York

Pacyna JM, Pacyna EG (2001) An assessment of global and regional emissions of trace metals to the atmosphere from anthropogenic sources worldwide. Environ Rev 9:269–298. doi:10.1139/a01-012

Pandey JS, Kumar R, Devotta S (2005) Health risks of NO_2, SPM and SO_2 in Delhi (India). Atmos Environ 39:6868–6874. doi:10.1016/j.atmosenv.2005.08.004

Patrick GJ, Farmer JG (2006) A stable lead isotopic investigation of the use of sycamore tree rings as a historical biomonitor of environmental lead contamination. Sci Total Environ 362: 278–291. doi:10.1016/j.scitotenv.2005.12.004

Patrick GJ, Farmer JG (2007) A lead isotopic assessment of tree bark as a biomonitor of contemporary atmospheric lead. Sci Total Environ 388:343–356. doi:10.1016/j.scitotenv.2007.07.047

Pope CA III, Dockery DW, Schwartz J (1995) Review of epidemiological evidence of health effects of particulate air pollution. Inhal Toxicol 7:1–18

Pöschl U (2005) Atmospheric aerosols: composition, transformation, climate and health effects. Angew Chem Int Ed 4:7520–7540. doi:10.1002/anie.200501122

Rabinowitz MB, Wetherill GW (1972) Identifying sources of lead contamination by stable isotope techniques. Environ Sci Technol 1972:705–709. doi:10.1021/es60067a003

Radojević DM, Pocajt VV, Popović IG, Perić-Grujić AA, Ristić MD (2013) Forecasting of greenhouse gas emissions in Serbia using artificial neural networks. Energ Source Part A 35:733–740. doi:10.1080/15567036.2010.514597

Rajšić S, Mijić Z, Tasić M, Radenković M, Joksić J (2008) Evaluation of the levels and sources of trace elements in urban particulate matter. Environ Chem Lett 6:95–100. doi:10.1007/s10311-007-0115-0

Reimann C, Flem B, Fabian K, Birke M, Ladenberger A, Negrel P, Demetriades A, Hoogewerff J, The GEMAS Project Team (2012) Lead and lead isotopes in agricultural soils of Europe – the continental perspective. Appl Geochem 27:532–542. doi:10.1016/j.apgeochem.2011.12.012

Reuer KM, Boyle AE, Grant CB (2003) Lead isotope analysis of marine carbonates and seawater by multiple collector ICP-MS. Chem Geol 200:137–153. doi:10.1016/S0009-2541(03)00186-4

Ritson P, Bouse R, Flegal AR, Luoma SN (1999) Stable lead isotopic analyses of historic and contemporary lead contamination of San Francisco Bay estuary. Mar Chem 64:71–83. doi:10.1016/S0304-4203(98)00085-1

Rodriguez JH, Wannaz ED, Salazar MJ, Pignata ML, Fangmeier A, Franzaring J (2012) Accumulation of polycyclic aromatic hydrocarbons and heavy metals in the tree foliage of *Eucalyptus rostrata*, *Pinus radiata* and *Populus hybridus* in the vicinity of a large aluminium smelter in Argentina. Atmos Environ 55:35–42. doi:10.1016/j.atmosenv.2012.03.026

Rosman KR, Ly C, Steinnes E (1998) Spatial and temporal variation in isotopic composition of atmospheric lead in Norwegian Moss. Environ Sci Technol 32:2542–2546. doi:10.1021/es9710215

Rossini P, Guerzoni S, Molinaroli E, Rampazzo G, De Lazzari A, Zancanaro A (2005) Atmospheric bulk deposition to the lagoon of Venice Part I. Fluxes of metals, nutrients and organic contaminants. Environ Int 31:959–974. doi:10.1016/j.envint.2005.05.006

Rucandio IM, Petit-Domínguez DM, Fidalgo-Hijano C, García-Gimenéz R (2011) Biomonitoring of chemical elements in an urban environment using arboreal and bush plant species. Environ Sci Pollut Res 18:51–63. doi:10.1007/s11356-010- 0350-y

Rühling Å, Tyler G (1968) An ecological approach to lead problem. Bot Notiser 121:321–342

Saeedi M, Li YL, Salmanzadeh M (2012) Heavy metals and polycyclic aromatic hydrocarbons: pollution and ecological risk assessment in street dust of Tehran. J Hazard Mater 227–228: 9–17. doi:10.1016/j.jhazmat.2012.04.047

Schauer JJ, RoggeWF HLM, Mazurek MA, Cass GR, Simoneit BRT (1996) Source apportionment of airborne particulate matter using organic compounds as tracers. Atmos Environ 30: 3837–3855. doi:10.1016/1352-2310(96)00085-4

Seinfeld JH, Pandis SN (1998) Atmospheric chemistry and physics from air pollution to climate change. Wiley, New York

Serbula MS, DjD M, Kovacevic MR, Ilic AA (2012) Assessment of airborne heavy metal pollution using plant parts and topsoil. Ecotox Environ Safe 76:209–214. doi:10.1016/j.ecoenv.2011.10.009

Sert E, Uğur A, Özden B, Murat Saç M, Camgöz B (2011) Biomonitoring of ^{210}Po and ^{210}Pb using lichens and mosses around coal-fired power plants in Western Turkey. J Environ Radioact 102:535–542. doi:10.1016/j.jenvrad.2011.02.005

Sharma P, Dubey RS (2005) Lead toxicity in plants. Braz J Plant Physiol 17:35–52. doi:10.1590/S1677-04202005000100004

Sharma SG, Srinivas MS (2009) Study of chemical composition and morphology of airborne particles in Chandigarh, India using EDXRF and SEM techniques. Environ Monit Assess 150:417–425. doi:10.1007/s10661-008-0240-7

Sharma RK, Agrawal M, Marshall MF (2008) Atmospheric deposition of heavy metals (Cu, Zn, Cd and Pb) in Varanasi City, India. Environ Monit Assess 142:269–278. doi:10.1007/s10661-007-9924-7

Shotyk W, Zheng J, Krachler M, Zdanowicz C, Koerner R, Fisher D (2005) Predominance of industrial Pb in recent snow (1994–2004) and ice (1842–1996) from Devon Island, Arctic Canada. Geophys Res Lett 32, L21814. doi:10.1029/2005GL023860

Simonetti A, Gariépy C, Carignan J (2000) Pb and Sr isotopic composition of snowpack from Québec, Canada: inferences on the sources and deposition budgets of atmospheric heavy metals. Geochim Cosmochim Acta 64:5–20. doi:10.1016/S0016-7037(99)00207-0

Sinclair AH, Tolsma D (2004) Associations and lags between air pollution and acute respiratory visits in an ambulatory care setting: 25-month results from the aerosol research and inhalation epidemiological study. J Air Waste Manage Assoc 54:1212–1218

Stacey JS, Kramers JD (1975) Approximation of terrestrial lead isotope evolution by a two-stage model. Earth Planet Sci Lett 26:207–221. doi:10.1016/0012- 821X(75)90088-6

Steding DJ, Dunlap CE, Flegal AR (2000) New isotopic evidence for chronic lead contamination in the San Francisco Bay estuary system: implications for the persistence of past industrial lead emissions in the biosphere. Proc Natl Acad Sci U S A 97:11181–11186

Steinnes E, Sjøbakk TE, Donisa C, Brännvall M-L (2005a) Quantification of pollutant lead in forest soils. Soil Sci Soc Am J 69:1399–1404. doi:10.2136/sssaj2004.0095

Steinnes E, Berg GA, Hjelmseth H (2005b) Atmospheric deposition of lead in Norway: spatial and temporal variation in isotopic composition. Sci Total Environ 336:105–117. doi:10.1016/j.scitotenv.2004.04.056

Sucharová J, Suchara I, Reimann C, Boyd R, Filzmoser P, Englmaier P (2011) Spatial distribution of lead and lead isotopes in soil B-horizon, forest-floor humus, grass (*Avenella flexuosa*) and spruce (*Picea abies*) needles across the Czech Republic. Appl Geochem 26:1205–1214. doi:10.1016/j.apgeochem.2011.04.009

Šućur K, Aničić M, Tomašević M, Antanasijević D, Perić-Grujić A, Ristić M (2010) Urban deciduous tree leaves as biomonitors of trace element (As, V, and Cd) atmospheric pollution in Belgrade, Serbia. J Serb Chem Soc 75:1453–1461. doi:10.2298/JSC100319079S

Szczepaniak K, Biziuk M (2003) Aspects of the biomonitoring studies using mosses and lichens as indicators of metal pollution. Environ Res 93:221–230. doi:10.1016/S0013-9351(03)00141-5

Tasić DM, Rajčić FS, Novaković TV, Tomašević NM (2001) Contribution to the methodology of dry deposition measurements. Fresen Environ Bull 10:305–309

Tasić M, Đuric-Stanojevic B, Rajšic S, Mijic Z, Novakovic V (2006) Physico-chemical characterization of PM10 and PM2.5 in the Belgrade Urban Area. Acta Chim Slov 53(3):401–405

Tjell JC, Hovmand MF, Mosbaek H (1979) Atmospheric lead pollution of grass grown in a background area in Denmark. Nature 280:425–426

Tomašević M, Rajšić S, Đorđević D, Tasić M, Krstić J, Novaković V (2004) Heavy metals accumulation in tree leaves from urban areas. Environ Chem Lett 2:151–154. doi:10.1007/s10311-004-0081-8

Tomašević M, Vukmirović Z, Rajšić S, Tasić M, Stevanović B (2005) Characterization of trace metal particles deposited on some deciduous tree leaves in an urban area. Chemosphere 61:753–760. doi:10.1016/j.chemosphere.2005.03.077

Tomašević M, Antanasijević D, Aničić M, Deljanin I, Perić-Grujić A, Ristić M (2013) Lead concentrations and isotope ratios in urban tree leaves. Ecol Ind 24:504–509. doi:10.1016/j.ecolind.2012.08.007

Turan D, Kocahakimoglu C, Kavcar P, Gaygısız H, Atatanir L, Turgut C, Sofuoglu CS (2011) The use of olive tree (*Olea europaea* L.) leaves as a bioindicator for environmental pollution in the Province of Aydın, Turkey. Environ Sci Pollut Res Int 18:355–364. doi:10.1007/s11356-010-0378-z

Tyler Miller G, Spoolman ES (2010) Environmental science. Brooks/Cole, Belmont

UNEP (2010) Assessing the environmental impacts of consumption and production: priority products and materials, a report of the working group on the environmental impacts of products and materials to the international panel for Sustainable Resource Management. Hertwich E, van der Voet E, Suh S, Tukker A, Huijbregts M, Kazmierczyk P, Lenzen M, McNeely J, Moriguchi Y. ISBN: 978-92-807-3084-5. Available at: http://www.unep.org/resourcepanel/Portals/24102/PDFs/PriorityProductsAndMaterials_Report.pdf

United Nations (1997) Glossary of environment statistics, studies in methods, Series F, No. 67. Department for Economic and Social Information and Policy Analysis, New York. ISBN 92-1-161386-8

US EPA (1989) Environmental acronyms, abbreviations and glossary of terms. Executive Enterprises, Inc., New York

US EPA (2002) Guidelines for the application of SEM/EDX analytical techniques to particulate matter samples. US Environmental Protection Agency. EPA-600/R- 02–070, Research Triangle Park

US EPA (2006) Lead human exposure and health risk assessments and ecological risk assessment for selected areas, Pilot Phase, External Review Draft Technical Report, ICF International Research Triangle Park, NC. Available at: http://www.epa.gov/ttn/naaqs/standards/pb/data/PbNAAQS_pilot_riskassessment_technical-report.pdf

US EPA (2010) Our nation's air, status and trends through 2010, U.S. Environmental Protection Agency, Office of Air Quality Planning and Standards Research, 2012 –EPA-454/R-12-001, Research Triangle Park

Vallero DA (2008) Fundamentals of air pollution. Elsevier Inc., London, p 60

Van der Gon HD, Appelman W (2009) Lead emissions from road transport in Europe. Sci Total Environ 407:5367–5372. doi:10.1016/j.scitotenv.2009.06.027

Van Leeuwen CJ, Vermiere TG (2007) Risk assessment of chemicals: an introduction, 2nd edn. Springer, Dordrecht, pp 106–111

Viana M, Kuhlbusch TAJ, Querol X, Alastuey A, Harrison RM, Hopke PK, Winiwarter W, Vallius M, Szidat S, Prévôt ASH, Hueglin C, Bloemen H, Wåhlin P, Vecchi R, Miranda AI, Kasper-Giebl A, Maenhaut W, Hitzenberger R (2008) Source apportionment of particulate matter in Europe: a review of methods and results. J Aerosol Sci 39:827–849. doi:10.1016/j.jaerosci.2008.05.007

Vicente AB, Sanfeliu T, Jordan MM (2012) Assesment of PM10 pollution episodes in a ceramic cluster (NE Spain): proposal of a new quality index for PM10, As, Cd, Ni and Pb. J Environ Manage 108:92–101. doi:10.1016/j.jenvman.2012.04.032

Vukmirović Z, Unkašević M, Lazić L, Tošić I, Rajšić S, Tasić M (2004) Analysis of the Saharan dust regional transport. Meteorol Atmos Phys 85(4):265–273. doi:10.1007/s00703-003-0010-6

Wall SM, John W, Ondo JL (1988) Measurement of aerosol size distributions for nitrate and major ionic species. Atmos Environ 22:1649–1656. doi:10.1016/0004- 6981(88)90392-7
Wang S, Zhang J (2006) Blood lead levels in children, China. Environ Res 101:412–418. doi:10.1016/j.envres.2005.11.007
Wei B, Yang L (2010) A review of heavy metal contaminations in urban soils, urban road dusts and agricultural soils from china. Microchem J 94:99–107. doi:10.1016/j.microc.2009.09.014
Weiss P, Offenthaler I, Öhlinger R, Wimmer J (2003) Higher plants as accumulative bioindicators. In: Markert B, Breure AM, Zechmeister HG (eds) Bioindicators & biomonitors, principles, concepts and applications, vol 6. Elsevier, Amsterdam, pp 465–500
WHO (2000) Air quality guidelines for Europe, WHO regional publications, European series, No. 91. World Health Organization Regional Office for Europe, Copenhagen
WHO (2006) Air quality guidelines – global update 2005. World Health Organization Regional Office for Europe, Copenhagen
WHO (2007) Health risks of heavy metals from long-range transboundary air pollution. Joint task force on the health aspects of long-range transboundary air pollution. International Government Publication, Geneva
Wolterbeek B (2002) Biomonitoring of trace element air pollution: principles, possibilities and perspectives. Environ Pollut 120:11–21. doi:10.1016/S0269- 7491(02)00124-0
Wong CSC, Li XD, Zhang G, Qi SH, Peng XZ (2003) Atmospheric deposition of heavy metals in the Pearl River Delta, China. Atmos Environ 37:767–776. doi:10.1016/S1352-2310(02)00929-9
Zheng C (1994) An approach on the source of ore-forming metals of lead–zinc deposits in north-western part, Guizhou. J Guilin Coll Geol 14:113–124. doi:cnki:SCN:451214PT.0.1994-02-000. (in Chinese with English abstract)
Zhou Q, Zhang J, Fu J, Shi J, Jiang G (2008) Biomonitoring: an appealing tool for assessment of metal pollution in the aquatic ecosystem. Anal Chim Acta 606:135–150. doi:10.1016/j.aca.2007.11.018
Zhu B, Chen Y, Peng J (2001) Lead isotope geochemistry of the urban environment in the Pearl River Delta. Appl Geochem 16:409–417. doi:10.1016/S0883 2927(00)00047-0

Chapter 9
Carcinogenic Nitrosamines: Remediation by Zeolites

Yan Yan Li, Mi Mi Wan, and Jian Hua Zhu

Contents

Abstract Cancer is a terrible killer that destroys the health and career of humans. Among various factors that induce cancer and tumor, environment pollution is the chief criminal because it affects the human life timelessly and universally. Human cancer is influenced by life style, smoking and taking processed meats as well as overdrinking beer are harmful for health because of the existence of *N*-nitrosamines. *N*-Nitrosamines are compounds with an *N*-nitroso group N-NO. They have been

---

Y.Y. Li • M.M. Wan • J.H. Zhu (✉)
Key Laboratory of Mesoscopic Chemistry of MOE, College of Chemistry and Chemical Engineering, Nanjing University, Nanjing 210093, China
e-mail: Liyanyan151@yeah.net; Mimiwanellen@gmail.com; jhzhu@netra.nju.edu.cn

E. Lichtfouse et al. (eds.), *Pollutant Diseases, Remediation and Recycling*, Environmental Chemistry for a Sustainable World 4, DOI 10.1007/978-3-319-02387-8_9,

recognized as strong carcinogens or potent carcinogens, and the nitrosamines in environment such as tobacco smoke cause serious health hazard to people. Tobacco smoking is associated with the cancers of lungs, larynx, oral cavity and pharynx, pancreas, kidney, and bladder. Consequently, the health consequences of exposure to environmental tobacco smoke has been a cause of scientific and public health concern because of increasing evidence showing that acute respiratory illness, chronic respiratory symptoms are associated with exposure to tobacco smoke. Nevertheless, no matter how often antismoking campaigns are shown, some people will never give up cigarette so that smoking is still a serious environmental problem in many countries.

Tobacco-specific *N*-nitrosamines (TSNA) are the most abundant carcinogens identified in tobacco and tobacco smoke, and the health hazard caused by the TSNA in cigarette smoke has been well established. Besides, *N*-nitrosamines were known many years ago to be present in food treated with sodium nitrite, which made fish meal hepatotoxic to animals through formation of *N*-nitrosodimethylamine (NDMA). It is necessary to remove *N*-nitrosamines in environment in order to protect the health of people. However, it is difficult to selectively capture carcinogens in the environment where numerous constituents are mixed. For instance capturing and degrading *N*-nitrosamines in complex media such as tobacco smoke is a challenge for chemists. Here zeolite is a solution because it is a molecular sieve that can selectively eliminate carcinogens. Zeolite is a class of alumino-silicates, characterized by ordered microporous structures with acidic-basic sites on the surface and energetic field inside the pores. Zeolites have many applications, mostly as adsorbents and catalysts because of their unique shape-selectivity. This review includes some latest research progresses in adsorption and catalytic degradation of *N*-nitrosamines by zeolite materials, either fundamental or pilot researches, emphatically the special selectivity of zeolite toward *N*-nitrosamines along with the specific degradation manners of the carcinogens. Many efforts on elevating the efficiency of zeolite with metal oxides modifiers are described and the research progress on the removal of *N*-nitrosamines in smoke is summarized. Finally the development of new functional materials to control the pollution of *N*-nitrosamines is prospected.

Keywords Zeolite • *N*-nitrosamines (nitrosamines) • Selective adsorption • Cigarette smoke • Environmental protection • Tobacco-specific nitrosamines • Catalytic degradation • Health care • Human cancer

List of Abbreviations

CHO	Chinese hamster ovary
CPTES	chloropropyl triethoxy silane
FTIR	Fourier transform infrared spectroscopy
GC	gas chromatograph
MPTES	3-mercaptopropyltriethoxysilane

NAB	*N*′-nitrosoanabasine
NAT	*N*′-nitrosoanatabine
NDMA	*N*-nitrosodimethylamine
NDPA	*N*-Nitrosodiphenylamine
NHMI	*N*-nitrosohexamethyleneimine
NNK	4-(methylnitrosamino)-1- (3-pyridyl)-1-butanone
NNN	*N*′-nitrosonornicotine
NPYR	*N*-nitrosopyrrolidine
SEM	Scanning Electron Microscopy
TG-MS	thermogravimetric-mass spectrometry
TPSR	temperature programmed surface reaction
TSNA	tobacco specific nitrosamines
VNA	volatile nitrosamines

9.1 Introduction

Development of modern industry ineluctably causes an increasingly serious pollution in the environment where human live in, constituting a catastrophic health risk including cancer and tumor. Anti-cancer is thus one of the challenges faced scientists in the realm of life science, and removal of carcinogen from environment is an important step for this struggle (Koide et al. 1999; Terry and Rohan 2002; Zhou and Zhu 2005). Nitrosamines are probably the most widespread carcinogens or potential carcinogens, they proverbially exist in workplace, processed meats, cigarette smoke and beer (Izquierdo-Pulido et al. 1996; Levallois et al. 2000; Hiramoto et al. 2001; Altkofer et al. 2005), and they are the chemical compounds characterized with *N*-nitroso group (N-NO). In other words, these compounds are amines with two organic groups (R) and one NO group bonded to central nitrogen (Filho et al. 2003), and they can be divided into three types: volatile nitrosamines (VNA), non-volatile nitrosamines and tobacco specific nitrosamines (TSNA) that are found only in tobacco products. Most of nitrosamines have been well recognized teratogens and carcinogens in animals for about 60 years and they are considered potentially carcinogens in humans (Mervish 1995; Hecht 1999), and actually many nitrosamines, even in trace amounts, can induce tumors in a variety of organs, including liver, lung, kidney, bladder, pancreas, esophagus and tongue depending the species (Lijinsky 1999). Human cancer relates to life style; apart from the health hazard caused by salted fish and preserved food, chemical exposure such as tobacco smoke and diet contributed to more than 65 % of human cancers (Doll and Peto 1981a). Especially, tobacco smoking was associated with about one third of U.S. cancer deaths (Doll and Peto 1981b), or of all cancers of males in the world (Parkin et al. 1994), not only cancers of the lungs, but also of the larynx, oral cavity and pharynx, oesophagus, pancreases, kidneys, and bladder. Cigarette smoke is a complex mixture of over 5,200 identified chemicals (Hoffmann et al. 1997; Branton et al. 2009), of which at least 150 are known to have specific toxicological properties and can be termed as “tobacco smoke toxicants” (Meier and

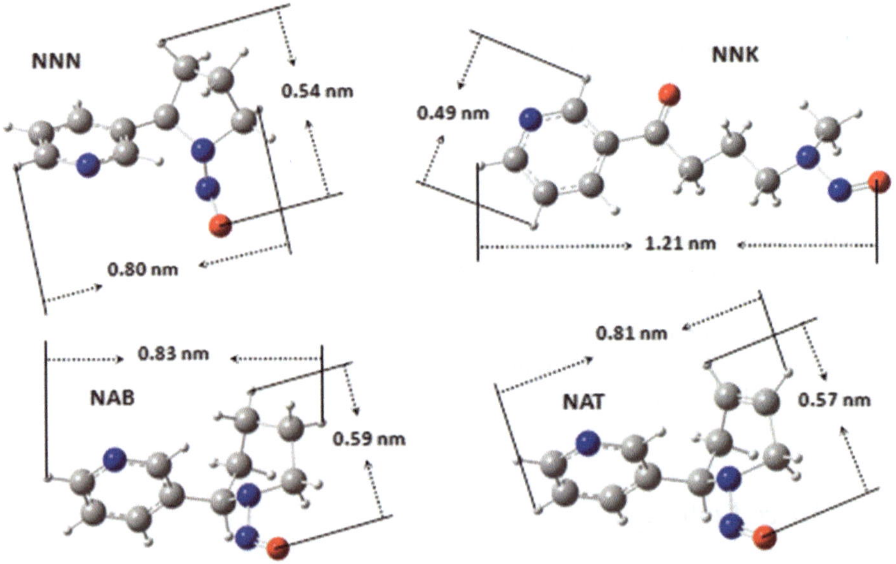

Fig. 9.1 The molecular structure of four tobacco specific nitrosamines (TSNA) (Lin et al. 2012b)

Siegmann 1999). Among these toxicants (Fig. 9.1) TSNA are well known to be the strong carcinogens, including *N'*-nitrosonornicotine (NNN), 4-(methylnitrosamino)-1-(3-pyridyl)-1-butanone (NNK), *N'*-nitrosoanatabine (NAT) and *N'*-nitrosoanabasine (NAB) (Lee et al. 1996; Baker 1999; Hecht 1999; Koide et al. 1999).

Volatile nitrosamines also exist in the vapor or semivolatile phase of mainstream smoke inhaled by smokers (Meier and Siegmann 1999; Baker 1999; Zhuang et al. 2006). And most of volatile nitrosamines such as *N*-nitrosodimethylamine (NDMA) and *N*-nitrosopyrrolidine (NPYR) are also carcinogenic, and they constitute a serious health risk for either smoker or the people around (Xu et al. 2003a; Sharp et al. 2005). Therefore, it is crucial to develop new functional materials and technologies to control the nitrosamines pollutants in environment, reducing the nitrosamines content of cigarette smoke and removing nitrosamines in food (Meier and Siegmann 1999; Ahn et al. 2003; Jo et al. 2003). For this purpose, lots of materials and techniques, including porphyrin, denitrifying bacteria, and other bacteria, are tried to remove nitrosamines in tobacco smoke. What's more, modified filter tips have been used in cigarette to reduce the harm of smoking for long time (Hoffmann et al. 1997; Baker 1999; Branton et al. 2009). However, it is undoubtedly a severe challenge to capture nitrosamines in complex system such as tobacco smoke that contains thousands compounds, and thus the specific functional materials with selectivity are urgently required.

A zeolite is crystalline aluminosilicates with a three-dimensional framework structure that forms uniformly sized pores of molecular dimensions (Maesen 2007), and its special structure is characterized by ordered microporous structures with acidic-basic sites on the surface and energetic field inside the pores (Fig. 9.2). With

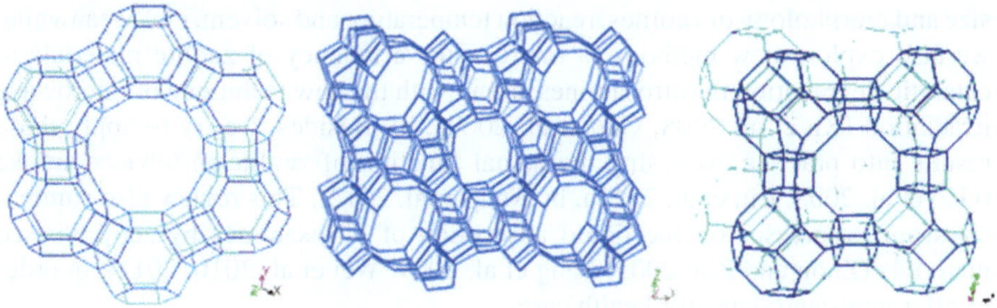

Fig. 9.2 The pore structure of zeolite FAU (*left*), MFI (*middle*) and LTA (*right*) (Yang et al. 2007)

the unique shape selectivity, zeolites have been successfully and extensively utilized in petrochemical and chemical industry as adsorbents, catalysts, and catalyst supports (Weitkamp and Hunger 2007; van Bakkum and Kouwenhoven 2007). And recently the application of zeolites has been extended to life science and environment protection (Weiner 1997; Colella 2007). Actually, zeolites are generally infiltrated our daily live, from purification of indoor air, water softening and decontamination to controllable releases of drugs; and capture of nitrosamines in complex system like tobacco smoke is not the exception of the potential application of zeolites. In practice zeolites can efficiently trap the volatile nitrosamines in gas stream at ambient temperature even when the contact time between the adsorbent and the carcinogen is shorter than 0.1 s (Zhou et al. 2007; Gu et al. 2010; Yang et al. 2010a), in which the pore size of zeolite plays a crucial role to govern the instantaneous adsorption. Zeolites also adsorb nitrosamines in aqueous solution and the adsorption isotherms can be fitted with Freundlich equation (Zhu et al. 2001; Zhou and Zhu 2005). Moreover, several new functional materials can be fabricated by the modification of zeolite with various physical or chemical methods to elevate the performance of nitrosamines trap (Xu et al. 2003b; Cao et al. 2007b; Yang et al. 2010b). Zeolites and other aluminosilicates are also used as tobacco additives (Meier and Siegmann 1999; Gao et al. 2008; Lin et al. 2012a, b). For instance Cu-ZSM-5 zeolite was used as the catalyst to reduce the amount of NO and NOx in mainstream smoke (Cvetkovic et al. 2002), either added into the cigarette filter or directly mixed with tobacco fibers in cigarette. Apart from these significant practical progresses, capture of nitrosamines also spurs the fundamental research on the fabrication of special porous functional materials as the candidate of powerful adsorbent and catalyst (Zhou et al. 2011a, b, c, 2012), along with the development of new method to detect the nitrosamines in environment (Xu et al. 2003a; Lin et al. 2012a). The emphasis of research is focused on the new material with designed pore architectures and customized surface functionalities.

This short review will cover the issues and developments on the removal of nitrosamines in environment, especially in tobacco products by zeolites. Our group has dedicated lots of effort to the fundamental study on the adsorption and catalytic degradation of nitrosamines on zeolites in gas and liquid phases, carrying out a systematic analysis on the influence of various factors such as the pore size and morphology of zeolites, reaction temperature and solvent, etc. Meanwhile, we also explore new methods to improve the efficiency of zeolite materials in catalytic degradation of nitrosamines, along with the new technique of microwave irradiation (Xu et al. 2008; Gao et al. 2009a, b). Besides, we try to apply these results into practice, assessing the actual function of zeolite in tobacco smoke (Gao et al. 2008; Lin et al. 2012a, b; Wang et al. 2012). This review also summarizes our progresses on the liquid adsorption of nitrosamines by zeolite-based materials (Zhou and Zhu 2005; Yang et al. 2009; Wei et al. 2010, 2011), in order to offer new candidates for health care.

9.2 Adsorption of Nitrosamines on Zeolites

9.2.1 Gaseous Adsorption of Volatile Nitrosamines

It is important to develop a reliable method for quickly assessing the efficiency of zeolites in adsorption of volatile nitrosamines in environmental. There are two common adsorptive techniques, gravimetric and volumetric adsorption methods often used to evaluate the capability of zeolites in gaseous adsorption. However, the results obtained in these conventional adsorption processes cannot depict the real adsorptive behavior of a zeolite in the application of trapping nitrosamines in environment, especially in the capture of nitrosamines in tobacco smoke for three reasons (Zhou et al. 2007). The first, the concentrations of adsorbate in the conventional adsorptive process are kept relatively steady, while the concentration of nitrosamines in the side stream and mainstream smoke of cigarette or an aeration system of room or building is ceaselessly varied. The second, the adsorbent in the conventional adsorptive process can contact with the adsorbate for quite a long time, several minutes or hours. Nevertheless, for those zeolites put in a cigarette filter to capture nitrosamines in smoke, they only have a rather short contact time. In the filter of cigarette with diameter of around 8 mm, the flow rate of the gas flow that passes through the filter can reach 36 cm sec^{-1} under the standard cigarette-smoking test conditions (ISO 1991). Thus, zeolites placed in a 29-mm-long filter will have less than 0.1 s to contact the nitrosamines in the gas stream. For sampling environmental tobacco smoke, the rate of flow varies from ten's of ml min^{-1} to less than 28 l min^{-1} (Guerin et al. 1992), causing a fluky contact time for the zeolite adsorbents. The third, in many applications of zeolites such as the additives in cigarette filters or the aeration system, intermittent adsorption instead of continues adsorption is predominant, which is much different from that in conventional adsorptive process. For these reasons, an instantaneous rather than steady adsorption method should be adopted to critically examine the adsorption of nitrosamines by zeolite in the gas stream. Therefore, a new test method, similar to concentration pulse chromatography (Harlick and Tezel 2000), was proposed to study the instantaneous adsorption of volatile nitrosamines on zeolites (Zhou et al. 2007). This instrument can be hermetically operated (Fig. 9.3), suitable for the test of volatile nitrosamines with

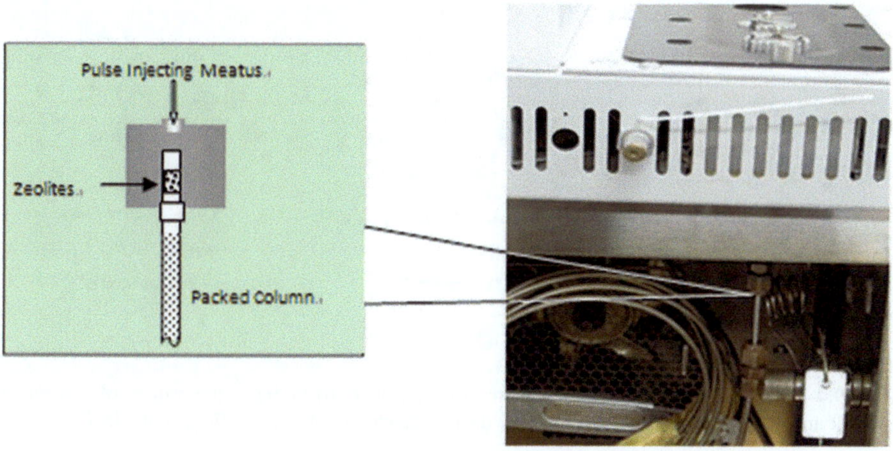

Fig. 9.3 Schematic representation of adsorption in gas chromatography (*left*) and diagram of experimental setup (*right*) (Zhou et al. 2007)

toxicity, mutagenic and carcinogenic properties. Likewise, it is able to be used for the investigation involving adsorption of other carcinogenic compounds such as polycyclic aromatic hydrocarbon. Instantaneous adsorption of volatile nitrosamines was performed in a stainless steel micro-reactor, and one of its ends was inserted deeply into the injector port of gas chromatograph (GC), while another end connects with the separation column in the GC. Samples were filled in the reactor where the temperature could be accurately controlled. The nitrosamine solution was pulse injected, and passed through the zeolite bed and then to the packed column, finally detected with thermal conductivity detector at the column outlet. To avoid environment contamination, exhaust gas was treated by an oil of vitriol to destroy nitrosamines. The decrement in the ratio of solute to solvent represents the amount of nitrosamines adsorbed by zeolite (Zhou et al. 2007). With the assistance of this special instrument, the influence of adsorption temperature and flow rate of carrier gas on the capability of zeolite in adsorbing volatile nitrosamines was examined. And the adsorption of mixed nitrosamines on zeolites was also investigated to be compared with that of individual component.

N-nitrosopyrrolidine (NPYR) was chosen as adsorbate to study the gaseous adsorption of zeolites, because it is a typical volatile nitrosamine in food and tobacco smoke (Baker 1999; Ahn et al. 2002; Byun et al. 2004). NPYR has a structure of five-member ring with a molecular diameter of 0.42×0.54 nm^2 (shown in Fig. 9.4). Several common zeolites, such as NaZSM-5 (the ratio of silicon to aluminum is 13 and 26), NaA, KA and NaY zeolites, show the different characteristics in instantaneous adsorption experiments: NaY, which had the largest pore size (0.74 nm) and the pore volume (0.31 cm^3 g^{-1}) among the zeolites used (Xu et al. 2003a; Zhou et al. 2007) exhibited the highest capability to capture volatile nitrosamines in gas stream (Fig. 9.5a). And the weakly adsorbed moisture did not hinder the adsorption of

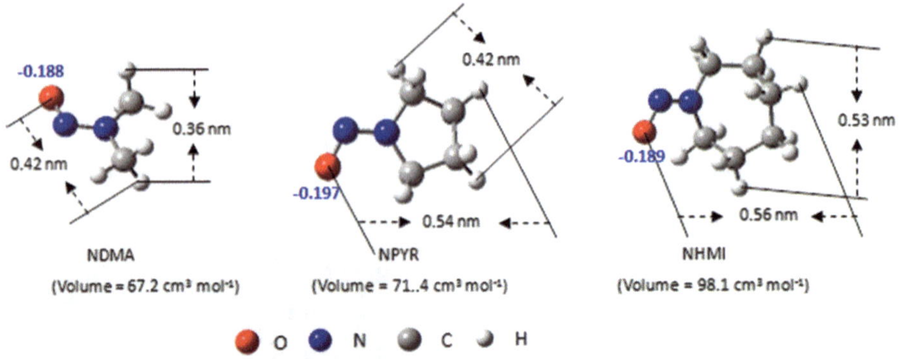

Fig. 9.4 Theoretically optimized geometry of volatile nitrosamines (the values of molecular weight and density in bracket, and the *blue* data is the charge of atom) (Zhou et al. 2007)

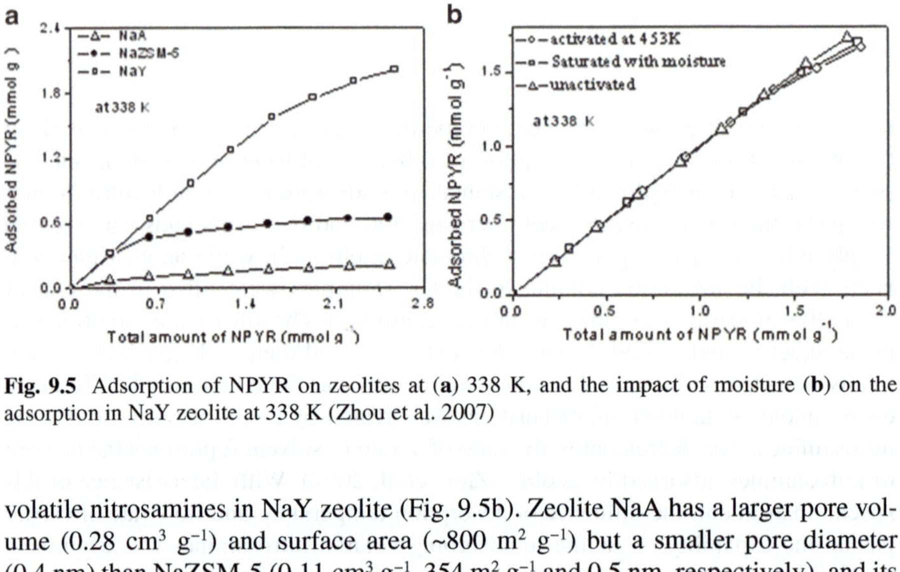

Fig. 9.5 Adsorption of NPYR on zeolites at (**a**) 338 K, and the impact of moisture (**b**) on the adsorption in NaY zeolite at 338 K (Zhou et al. 2007)

volatile nitrosamines in NaY zeolite (Fig. 9.5b). Zeolite NaA has a larger pore volume (0.28 $cm^3\ g^{-1}$) and surface area (~800 $m^2\ g^{-1}$) but a smaller pore diameter (0.4 nm) than NaZSM-5 (0.11 $cm^3\ g^{-1}$, 354 $m^2\ g^{-1}$ and 0.5 nm, respectively), and its instantaneous adsorption of volatile nitrosamines was restricted by the narrow pore size (Fig. 9.6) (Zhu et al. 2001; Yun et al. 2004). As the accumulated amount of NPYR that passed through the zeolite rose to 2.5 mmol g^{-1}, 78.3 % was adsorbed by NaY, NaZSM-5 could trap 25.6 % but NaA only adsorbed 8.3 %. Since the molecular size of NPYR is close to the pore diameter of NaZSM-5, this zeolite exhibits some unusual properties in the adsorption. When the adsorption temperature was raised from 338 to 453 K, the proportion of NPYR trapped by NaY decreased from 97.5 to 83.0 %, the data of NaA lowered from 11.4 to 8.2 % when 1.5 mmol g^{-1} NPYR passed over it. However, NaZSM-5 adsorbed 43.9 % at 453 K, more than that at 338 K (35.4 %), which was a strange phenomenon. When the flow rate of the carrier gas was doubled from 15 to 30 mL min^{-1}, NaZSM-5 also trapped more

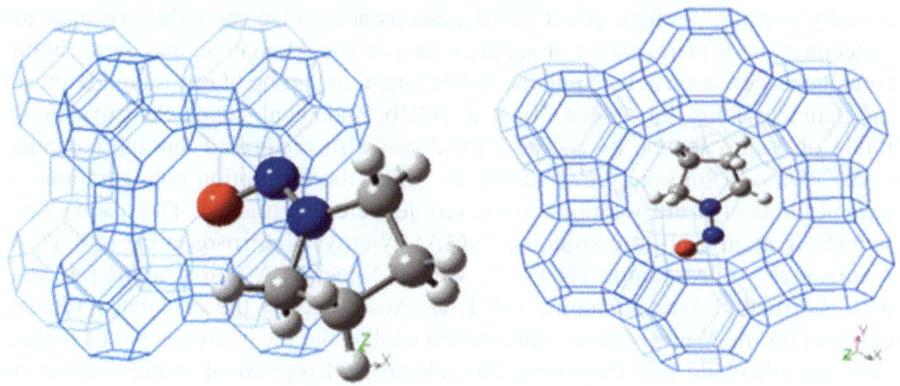

Fig. 9.6 The adsorption manner of NPYR in (*left*) A and (*right*) Y zeolites (Zhou et al. 2007)

NPYR (Zhou et al. 2007). Freundlich equation can sensitively depict the difference in the adsorption of volatile nitrosamines with different structures in various zeolites for an overall analysis of the adsorption isotherms, but most of the isotherms deviate Dubinin-Radushkevich or Langmuir equations (Zhou et al. 2007; Zhu et al. 2001). Also, Freundlich equation can sensitively reveal the impact of temperature, flow rate of carrier gas, and the co-existence of other nitrosamines on the adsorption of volatile nitrosamines in air stream.

Another two volatile nitrosamines were used to depict how the molecular structure of volatile nitrosamines affected the gaseous adsorption of zeolite. One is *N*-nitrosodimethylamine (NDMA), a strong carcinogen with a molecular size of 0.42×0.36 nm^2, and another is *N*-nitrosohexamethyleneimine (NHMI), the strong carcinogen with a seven-member ring and size of 0.56×0.54 nm^2 (Yun et al. 2004). It was found that the adsorption would be dramatically declined once the molecular size of nitrosamine exceeded the pore of zeolite, and the bulky nitrosamines might be chemisorbed on the external surface of zeolite with the nitroso functional penetrating into the pore mouth (Yun et al. 2004; Zhou et al. 2004). For the mixture of NDMA and NPYR on different zeolites, NaY preferentially adsorbed the more negative charged molecule NPYR instead of the smaller one NDMA, while NaZSM-5 and NaA zeolites willingly captured NDMA because of its smallness. Two major factors govern the adsorption, the geometric matching content and the electrostatic interaction between the adsorbate and the adsorbent, the former determines whether the molecule can be adsorbed by the zeolite or not, while the latter spurs and accelerates the adsorption (Zhou et al. 2007). The structure and conformation of nitrosamine molecule, together with the pore structure of zeolite, determine the adsorbate-adsorbent geometric matching content. For the given nitrosamine, the zeolite with a pore size slightly larger than its molecular diameter can exhibit the highest adsorption capability (Zhu et al. 2001), since the effective attracting range of the cation in zeolite to pull the nitrosamine molecule is limited within about 0.05 nm (Yang et al. 2007). For a given zeolite, however, its Al content actually exerts a stronger effect on its gaseous adsorptive capability, because the concentration of metal cation of zeolite relates to the Al content, and these cations form the electrostatic field to pull the N-NO functional group of nitrosamines toward the channels as Fig. 9.6 shown (Xu et al. 2003b; Zhou et al. 2004). Selective recognition of *N*-nitrosamine by zeolite ZSM-5 was also confirmed through computational method (Pinisakul et al. 2008), in which the adsorption configurations of several kinds of *N*-nitrosamines on the zeolite were investigated. These adsorbates include *N*-methyl-N-nitrososamine (NMA), *N*-ethyl-*N*-nitrososamine (NEA), *N,* *N-dimethyl-N*-nitrosoamine (NDMA), *N*-ethyl-*N*-methyl-*N*-nitrosoamine (NEMA) and *N,N*-diethyl-*N*-nitrosoamine (NDEA). According to the calculated binding energies for the *N*-nitrosamines adsorbed in zeolite and the hydrogen bonds formed between adsorbate and adsorbent, the selective adsorption of *N*-nitrosamine on ZSM-5 is successfully proved.

9.2.2 Impact of Cation in Zeolite on the Gaseous Adsorption of Volatile Nitrosamines

As aforementioned, the cation of zeolite plays an important role in gaseous adsorption of volatile nitrosamines, because it provides an essential electrostatic affinity on the N-NO group of nitrosamine. However, cation also occupies the space within the narrow channel of zeolite to hinder the entering and diffusion of adsorbate more or less, thus it is necessary to understand the factual impact of the cation on the adsorption of nitrosamines by zeolite. For this purpose, four samples of zeolite A with different cations, CsA, KA, NaA, and CaA are chosen to study the special selectivity of zeolite towards nitrosamines (Yang et al. 2008). These samples have similar particle diameter and structure, and the only difference is the alkali metal ion so that their pore size varies from 0.2 to 0.5 nm (Fig. 9.7). In the adsorption of NPYR at 338 K, NaA zeolite showed the highest adsorption capacity, and the adsorption difference between NaA and others would be enlarged when the adsorption temperature rose to 453 K (Yang et al. 2008), mirroring the strong influence of cation type on the adsorption of zeolite A. Although CaA zeolite has the largest pore size (0.5 nm), it fails to selectively adsorb NPYR because of the competition of dichloromethane solvent (Cao et al. 2007a). It is known that the contact time between zeolite and NPYR is less than 0.1 s in the instantaneous adsorption (Zhou et al. 2007; Gu et al. 2010; Yang et al. 2010a), so the competition adsorption of solvent will severely limit the NPYR adsorption on zeolite. Instead, zeolite NaA has a pore size of 0.4 nm that matches with the smaller molecular diameter of NPYR, which is advantageous for the adsorption (Fig. 9.7). Distribution of Ca^{2+} cation in zeolite A badly affects the adsorption. There is no Ca^{2+} cation locates in the window of 8-member ring in the zeolite, which makes the electrostatic force of CaA weaker than other analogues hence it cannot capture nitrosamine efficiently in gas stream. Another sample of zeolite A, CsA has a density larger than NaA about 74.5 % in the dehydrated form, and its number of crystal cells is about 58 % of that of NaA in 1 g

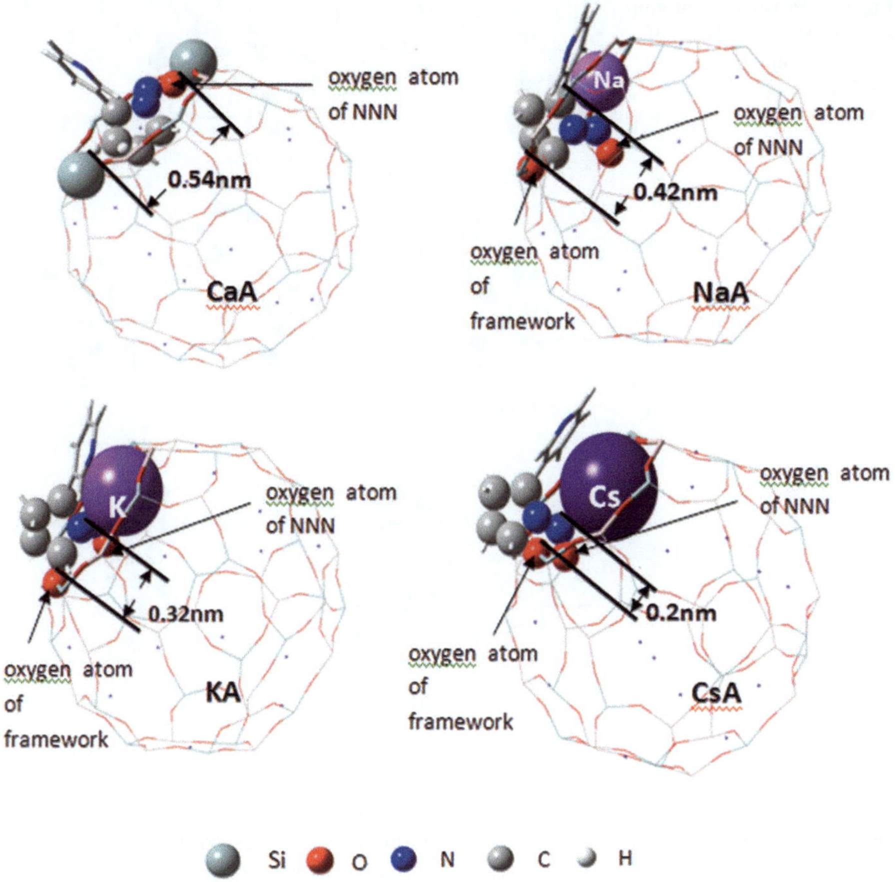

Fig. 9.7 The possible adsorption manner of NNN in zeolite A (Yang et al. 2008)

hydrated sample (Yang et al. 2008). As the result, the number of pore opening in CsA sample is about 40 % less than that of NaA zeolite. Likewise, 1 g sample of KA has the number of pores about 9.6 % less than NaA zeolite. These differences in the pore number of per gram sample among the CsA, NaA, KA and CaA should be considered for evaluating their adsorption performance.

N′-nitrosonornicotine (NNN) is one of tobacco-special nitrosamines (TSNA), with 0.80 nm of length and 0.54 nm of width (Fig. 9.1). Very strangely, this bulky nitrosamine could be adsorbed by zeolite CsA, KA and NaA as proven in FTIR experiment but CaA failed due to the competitive adsorption of dichloromethane solvent. In principle, NNN may be adsorbed in zeolites by three possible ways depending on the pore diameter of the adsorbent. For the zeolite such as Y, ZSM-5 and CaA with relative large pore size, NNN can insert the 5-member ring into the channel, which requires the aperture of zeolite to be 0.541 nm at least. If

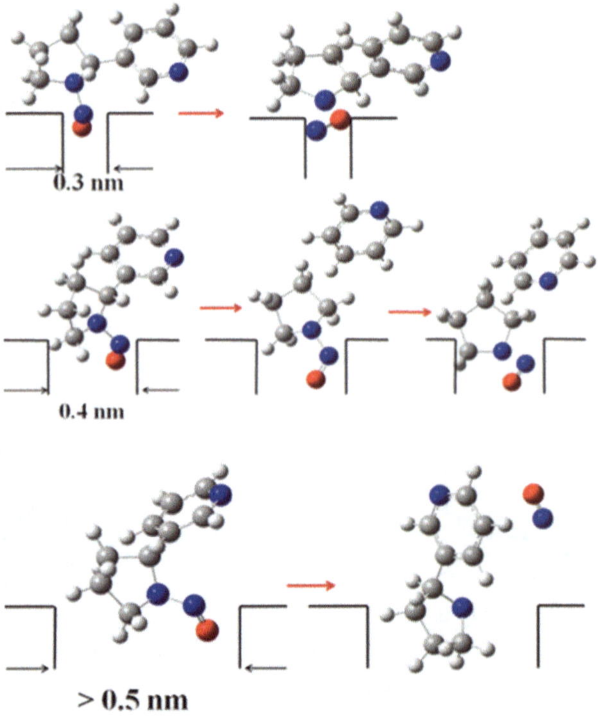

Fig. 9.8 Possible adsorption and degradation manner of NNN in zeolite (Cao et al. 2007a)

the 5-member ring and pyridine ring of the NNN molecule are on the same plane, the whole molecule can totally enter the channel of zeolites whose pore diameter is larger than 0.54 nm. Unfortunately, these two rings are not on the same plane of the NNN molecule. Consequently, if the aperture of zeolite is similar to 0.541 nm, NNN molecule can only insert the 5-member ring into the channel due to the steric constraints, leaving the residual parts clipped in the pore mouth of zeolite (Fig. 9.8). Alternatively, in the case of zeolite Y and ZSM-5, it is possible for NNN to enter the channel of zeolite by way of inserting the pyridine ring into the pore first, because the pore diameter of about 0.56 nm seems enough for housing the whole adsorbate. Zeolite NaA and KA only have a small pore thereby the most possible adsorptive manner of NNN is to penetrate its N-NO group in the channel of zeolite, which only needs a diameter of 0.123 nm. On the basis of these facts, it is clear that the adsorption manner of NNN with the pyridine ring is forbidden in zeolite NaA or KA, since the aperture of adsorbent is narrower than the diameter of pyridine ring (0.43 nm). Among these zeolites, NaA showed a special adsorption manner: its pore size lets the NNN to be clipped in the pore mouth, making the C-C bond linking the five-member ring and pyridine ring breakable (Cao et al. 2007a; Yang et al. 2008). Adsorption of NPYR and NNN on zeolite A reveals the unusual adsorptive property of zeolite NaA and its alkali metal cation exchanged analogues towards nitrosamines. Due to the limitation of small pore,

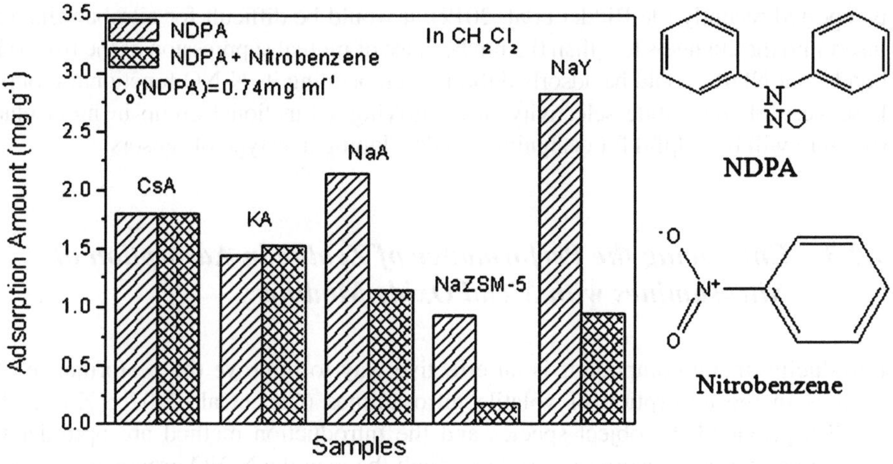

Fig. 9.9 Adsorption of *N*-Nitrosodiphenylamine (NDPA) by zeolites in dichloromethane with or without nitrobenzene at 277 K (Yang et al. 2009)

NNN or NPYR cannot enter the channel of zeolite A (Wu et al. 2008), subsequently the nitrosamines will be adsorbed into the zeolite in the way through inserting the N-NO group inside the pore while the rest of the molecule retains on the external surface of the zeolite. Compared with other zeolites such as ZSM-5 or Y, zeolite A has the smallest pore size and the largest density of cation to provide the confinement and attraction to volatile nitrosamines, avoiding desorption of the adsorbate since the high cation density in the zeolite helps to anchor the nitrosamines through strong interaction (Yang et al. 2008).

An unusual subtle selectivity is found on zeolite KA and CsA whose pore sizes are smaller than 0.4 nm. These two zeolites can accurately identify *N*-nirosodiphenylamine (NDPA), the bulky nitrosamine with two phenyl groups, and nitrobenzene (Yang et al. 2009) no matter in dichloromethane or benzene solution (Fig. 9.9). They are able to selectively adsorb the larger molecule NDPA, similar to the "molecular trapdoor" mechanism reported recently (Shang et al. 2012), and its adsorption capacity is not influenced by nitrobenzene. No such geometric limitation is observed on NaA and NaZSM-5 due to their increased pore size, and their adsorption of NDPA will be retarded by nitrobenzene. Clearly the pore structure of CsA and KA zeolites can identify two kinds of functional groups in molecules, that is to say, their adsorption capacity depended on the geometrical configuration difference between -N-NO and -NO_2 (Yang et al. 2009) since the largest diameter of nitrobenzene molecule is about 0.44 nm, and the distance between two oxygen atoms in -NO_2 functional group is 0.36 nm, which exceeds the pore size of KA and CsA. Instead, the N-NO group of NDPA molecule has a diameter of only 0.2 nm so it can be inserted into zeolite pore mouth. If NDPA and nitrobenzene were adsorbed into zeolite channels through "plug-in mode" (Yun et al. 2004; Zhou et al. 2004), similar to the "close fit" mechanism that is reported recently (de Ridder et al. 2012), it would be difficult for nitrobenzene to insert into the channels less than 0.3 nm because of mutual suppression of the two NO bonds, but NDPA could be adsorbed through embedding its N-NO functional group. Discovery of this subtle selectivity of identifying a functional group in molecular structure will be helpful for designing and developing new type of sensors.

9.2.3 Enhancing the Performance of Zeolite in Adsorption of Nitrosamines with Metal Oxide Modifiers

Introducing metal compounds is an effective way to improve the performance of zeolites in the adsorption of volatile nitrosamines (Cao et al. 2007b; Xu et al. 2003b), provided the object species and the introduction method are optimized. Many nitrosamines adsorb on zeolite through the way the N-NO groups insert into the zeolite channels (Zhu et al. 2003), whose manner is similar to the adsorption of NOx in zeolite. Therefore, the methods help to adsorb nitrous oxide on zeolite is also beneficial for the adsorption of nitrosamines. Copper is the best element to promote adsorption of nitrogen oxides on zeolite, so it is utilized to modify zeolite through impregnation to form highly dispersed copper oxide particles in the porous host (Xu et al. 2004). These copper oxide nanoparticles in zeolite will attract the N-NO groups of nitrosamine and thus promote the adsorption. As is evident from FTIR spectra, the N-NO functional group of nitrosamines has a strong interaction with the copper species dispersed in zeolite NaY so that the carcinogenic compounds can be much easily selectively adsorbed (Xu et al. 2003b). Loading 3 % (weight percent) of copper oxide can increase the ratio of NPYR adsorbed on NaY from 57 to 75 % at 453 K. For the zeolite NaZSM-5 with smaller pore size, loading 3 % of copper oxide enables it to adsorb the same amount of NDMA as that by NaY zeolite. Impregnation of zeolite with zirconia or cobalt oxide also exhibits a similar positive effect in nitrosamines adsorption (Zhuang et al. 2006; Cao et al. 2007b). However, these metal oxides inevitably occupy the space inside the channel of zeolite, which may cause a negative effect on the entering and diffusion of adsorbate, so the further study is performed on the NaY zeolite ion exchanged and impregnated with cooper oxide to assess their actual structure-property variation caused by different modifications (Gu et al. 2008a). Ion exchange with copper cation adjusts the distribution of cation and the pore size of zeolite delicately: replacing sodium cation of NaY zeolite with copper cation through ion exchange can keep its pore structure unchanged, but the absolute number of total cation is reduced because one copper ion exchanges two sodium ions. Again, sodium ion has a radius of 0.099 nm, larger than that of Cu^{2+} (0.073 nm) hence substituting two Na^{+} ions with one Cu^{2+} enables the zeolite to have a slightly larger pore size. In succeeding adsorption the nitrosamine molecule will be attracted by the fewer bivalent cation instead of the relative more univalent cations while the pore opening became a little bit wider, reducing the collision frequency with cation. As the result, there is no promoted adsorption ability on the copper exchanged zeolites (Gu et al. 2008a), which differs

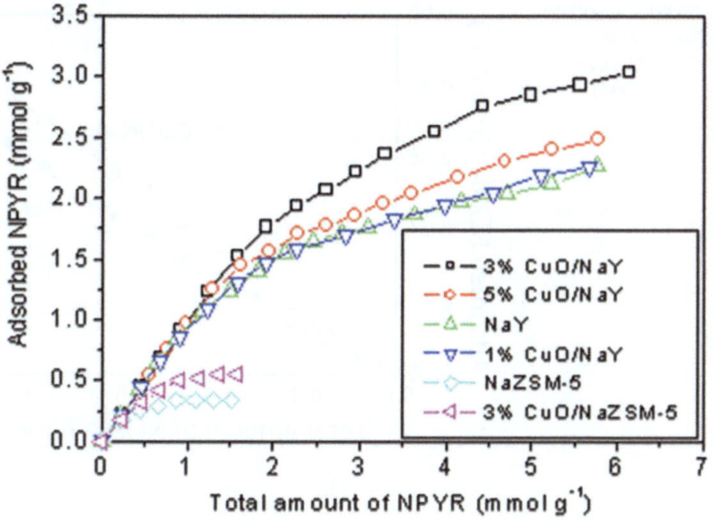

Fig. 9.10 Influence of loading copper on the selective adsorption of NPYR over zeolite NaY and NaZSM-5 (Xu et al. 2004)

from CoY zeolite that exhibits a little bit enhanced ability in adsorbing NPYR at 338 K (Cao et al. 2007b).

Impregnation method not only remains most of the Na^+ ions in zeolite NaY but also introduces additional copper species in the resulting composite (Xu et al. 2003b, 2004). These modifiers have an electrostatic affinity towards the N-NO group of nitrosamines, but cover the inherent active sites of zeolite and block the channel more or less. Also, incorporation of the guest reduces the proportion of zeolite in adsorbent. For instance 1 g of NaY sample consists of 1 g zeolite, but 1 g of 5%CuO/NaY sample contains 0.95 g zeolite. Since zeolite is crucial to capture the nitrosamines in gas stream, there is no promotion but decline in adsorptive ability observed on CuO/NaY samples at 338 K (Gu et al. 2008a). Again, the more guest loaded, the larger the capacity decreased. Different situation appears once the adsorption temperature rose to 453 K, CuO/NaY is superior to NaY for trapping volatile nitrosamines (Fig. 9.10) because the copper modifier interacts stronger with nitrosamines to suppress desorption of NPYR from NaY zeolite (Xu et al. 2003b, 2004; Gu et al. 2008a). Similar positive effect is also observed in NaA zeolite, as shown in Fig. 9.11. Moreover, the 3 wt.-% of copper oxide is proven to be the optimal value (Xu et al. 2004). For the given porous support, the optimal amount of modifier is determined by the host structure, it should offer the strong electrostatic interaction to promote the adsorption of nitrosamines while kept the zeolite channel unblocked. This value (3 wt.-%) is below the monolayer dispersion threshold of copper oxide in zeolite NaY therefore the modifier is well dispersed and accessible for nitrosamines adsorbate.

Fig. 9.11 NDMA adsorbed on zeolite NaA and the copper modified NaA at 453 K (Xu et al. 2004)

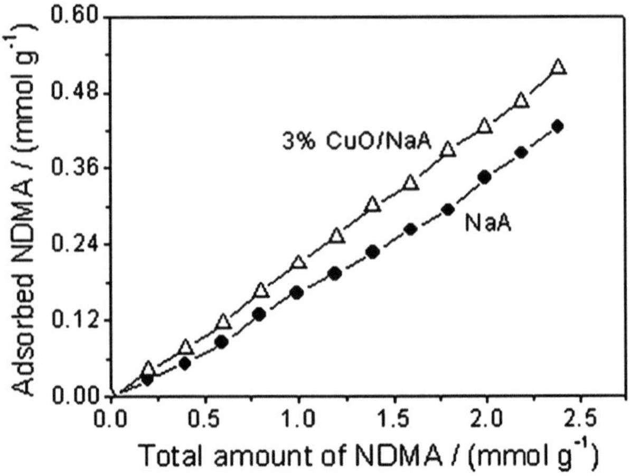

Ferric oxide can be loaded on zeolites NaY through solid state grinding, microwave irradiation and impregnation methods. Since the exchange of Na^+ of zeolite NaY by Fe^{2+} or Fe^{3+} cation during impregnation reduces the absolute number of ionic site, the modified zeolites show a declined capability in instantaneous adsorption of NPYR at 453 K (Cao et al. 2008). In contrary, the 3%Fe_2O_3/NaY prepared with microwave irradiation exhibits a higher adsorption capacity than NaY, it traps 86 % of NPYR when the accumulated amount of NPYR achieves 1.82 mmol g^{-1} but NaY adsorbs about 77 %.

Cobalt modified NaY zeolites can trap more volatile nitrosamines in instantaneous adsorption as total amount of NPYR exceeds 1.0 mmol g^{-1} (Cao et al. 2007b). Zeolite NaY can capture about two thirds of the NPYR in gaseous phase when the amount of NPYR passed through the adsorbent accumulates to 2.66 mmol g^{-1}, while loading 1 wt.-% of cobalt oxide on NaY enhances this proportion to 74 %. 3CoY sample, the NaY zeolite modified with cobalt oxide of 3 wt.-%, shows the highest adsorptive capability and 92 % of NPYR is trapped at 338 K, one third more than that of the parent zeolite. Loading more cobalt oxide on NaY cannot elevate the adsorptive capability further. 5CoY sample only adsorbs 72 % of NPYR whereas the adsorption ability of 10CoY is lower than that of NaY. The radii of cobalt and oxygen ions are 0.04 and 0.14 nm, respectively (Dean 1999). If the cobalt oxide is equally dispersed in zeolite to form the monolayer, the actual pore size of the NaY will be reduced from 0.74 to 0.46 nm, which is harmful for fast adsorption of NPYR since the molecule size of NPYR is 0.56 nm (Xu et al. 2004). To keep the modified channel accessible for NPYR, the residual pore size should be about 0.6 nm, that is, only half of the pore wall could be coated by cobalt oxide. Otherwise, the entry and diffusion of NPYR will be obstructed. The threshold for monolayer of cobalt oxide on NaY is about 6 wt %; consequently loading 3 wt % becomes the best amount. Also, calcination conditions affect the distribution of cobalt cations in zeolite NaY. Calcination at 673 K in nitrogen enables cobalt cations to exist in both octahedral and tetrahedral

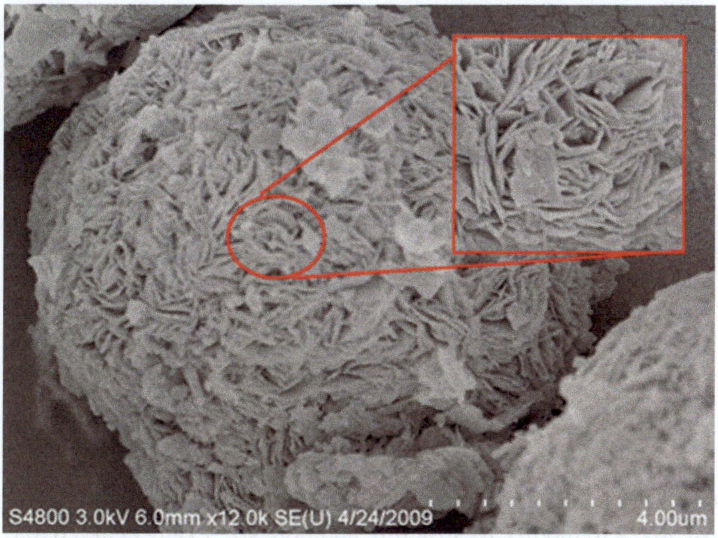

Fig. 9.12 Scanning Electron Microscopy (SEM) image of MCM-22 zeolite (Yang et al. 2010a)

symmetries and a lot of them will be located in the sodalite cage of zeolite NaY, but the calcination at 773 K accompanied with plenty of oxygen simplifies the state of the cobalt cations to be octahedral symmetry, migrating to the supercage of NaY. The location of cobalt cations in zeolite NaY badly influences the adsorption of NPYR in airflow hereby the sample loaded with 3 wt% of cobalt oxide and calcined at 673 K exhibits the highest adsorption capability of NPYR at 338 K.

9.2.4 The Influence of Zeolite Morphology on the Gaseous Adsorption of Nitrosamines

The influence of morphology of mesoporous materials on their adsorption of nitrosamines has been studied (Zhou et al. 2011a, b, c, 2012), but few is understood on the effect of zeolites morphology. Actually, nitrosamine molecule cannot be captured in gas stream until it flies very close to the pore mouth of zeolite because the electrostatic interaction of cation in zeolite is limited within 0.05 nm (Yang et al. 2007). Is it possible to adjust the movement of nitrosamine molecule and increase its collision frequency with zeolite? MCM-22 is chosen for the study because of its specific rose-like morphology (Fig. 9.12, Yang et al. 2010a; Wu et al. 2009). Zeolite MCM-22 has two independent pore systems, one involves two-dimensional (Fig. 9.13), sinusoidal channels (0.4×0.5 nm^2) with the circular 10-membered rings like ZSM-5 (Ravishankar et al. 1995), and another consists of

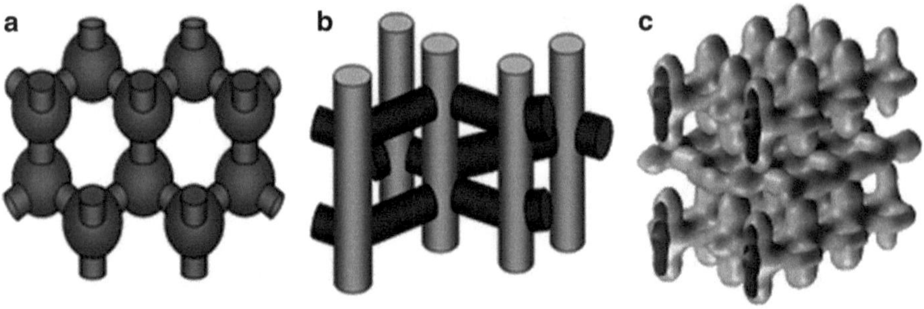

Fig. 9.13 Structures of zeolites (**a**) NaY, (**b**) ZSM-5 and (**c**) MCM-22 (Yang et al. 2010a)

12-membered large cylindrical supercages ($0.7 \times 0.7 \times 1.8\ nm^3$) similar to that of NaY (Daems et al. 2005). And desilication technique is employed to create mesopore in MCM-22 to modify its structure and pore size (Yang et al. 2010a). Surface morphology of zeolite assuredly affects its performance in the adsorption of nitrosamines. MCM-22 traps all NPYR in gas stream at 338 K until the accumulated amount of NPYR reaches 1.5 mmol g^{-1}, but zeolites NaZSM-5 losses its ability at the amount of 0.58 mmol g^{-1} while NaY is inferior once the amount of NPYR exceeds 1.5 mmol g^{-1}. In the case that of 2.28 mmol g^{-1} NPYR, NaZSM-5 adsorbs about 28 % while NaY traps 80 % but MCM-22 captures 88 %. Desilicated MCM-22 shows the highest adsorptive ability of NPYR at 338 K, and the formation of mesopores is preponderant for adsorbing NPYR at low temperature. When the flow rate of carrier gas increases from 10 to 30 mL min^{-1}, MCM-22 samples traps more NPYR, similar to NaZSM-5 (Zhou et al. 2007), and the desilicated sample adsorbed more NPYR (0.57 mmol g^{-1}) than it parent (0.29 mmol g^{-1}). As the adsorption temperature rises to 453 K and the amount of NPYR accumulates to 1.0 mmol g^{-1}, MCM-22 captures 0.92 mmol g^{-1}, and NaY and NaZSM-5 adsorb 0.86 and 0.52 mmol g^{-1} respectively.

For the instantaneous adsorption, any change in surface roughness or morphology of zeolite will cause significant differences on its adsorption behavior because rough surfaces can provide more conducive sites for adsorption and collision (Reitmeier et al. 2008; Zhou et al. 2012). For NPYR gaseous adsorption, interception of NPYR and holding in the carcinogen inside channels are two key links. At low temperature such as 338 K, the former is the dominant, special morphology such as the rich surface curvature has obvious advantage in the process of nitrosamines adsorption. And the formation of mesopores in zeolite can also promote the absorption (Yang et al. 2010a). At a high temperature such as 453 K, it is quite important to suppress desorption of nitrosamines which are already adsorbed in zeolites (Xu et al. 2003a) hence the constraints from channels will play a major role. The electrostatic attraction from cations of zeolites as well as the confinement of channels determine the adsorption performances of the samples: the ideal hierarchical adsorbent should have not only the wide channels to lead the streams pass quickly (Zhou et al. 2012), but also the zeolites structures in their pore walls, as a result, electrostatic field formed by cations can strongly attract nitrosamines, making them constrained by fine geometric function in the sub-nanoscale channels, and thus the selective adsorption of nitrosamines can be realized through the combination of diffusion effect, cage effect, molecular electric moment and coulomb field. MCM-22 exhibits a high ability in the instantaneous adsorption of volatile nitrosamines at 453 K due to its inherent Al content and the specific porous structure containing bell-mouthed pore mouth connected with the slightly narrow channel that are beneficial to intercept nitrosamines in air flow (Yang et al. 2010a).

9.2.5 *Liquid Adsorption of Nitrosamines by Zeolites*

Apart from gaseous adsorption of volatile nitrosamines, zeolites can capture the carcinogens in organic or aqueous solutions. NaY has the largest surface area (766 $m^2\ g^{-1}$) and the largest pore size (0.74 nm) among commercial zeolites thereby it exhibits the largest adsorption capacity of nitrosamines in dichloromethane solution and the amount of adsorption (Qe) increases as the residual concentration (Ce) is raised (Fig. 9.14, Zhu et al. 2001). In the case of that the Ce of NDMA is around 2,000 mg L^{-1}, the Qe value on NaY zeolite (175 mg g^{-1}) is higher than that on SiO_2 (25 mg g^{-1}) or Al_2O_3 (10 mg g^{-1}). And this difference of their Qe value (7–17 times) is higher than that of their surface area (2–4 times). The difference between the adsorptive capacity of NaY and other zeolites becomes more and more obvious as the adsorbate is changed from NDMA to NYPR or NMHI. Nitrosamines have a weak basicity and adsorb favorably on acidic materials (Ma et al. 2000), therefore acid zeolite HZSM-5 adsorbs more nitrosamines than NaZSM-5 though their surface area and pore size are almost same. On the other hand, competition of solvent weakens the adsorption of nitrosamines. NaY can adsorb 59 mg g^{-1} of NDMA at Ce of 320 mg L^{-1} in organic solution, but only 1.05 mg g^{-1} at Ce of 478 mg L^{-1} in aqueous solution (Zhu et al. 2001). NaA zeolite is another hydrophilic adsorbent, but the difference in two solvents is unobvious. For instance 3.39 mg g^{-1} of NDMA is adsorbed by NaA at Ce of 432 mg L^{-1} in dichloromethane solution, but about 0.84 mg g^{-1} adsorbed at Ce of 478 mg L^{-1} in aqueous solution. Furthermore, more NPYR can be trapped on NaA in aqueous solution (3.80 mg g^{-1}) than that in dichloromethane solution (2.80 mg g^{-1}) in the case of low Ce (423 mg L^{-1}). Reverse difference is observed when NDMA and NPYR were adsorbed on low Al content zeolite NaZSM-5 hence it is favorable to use hydrophobic zeolite to remove nitrosamines from aqueous solution while using hydrophilic zeolite for separate nitrosamines from organic solutions. In general the degree of geometrical matching between the channel opening of a zeolite and the molecular diameter of nitrosamine determines the adsorption manner by the zeolite (Cao et al. 2007a, b; Zhou et al. 2007). The molecular dimensions of NDMA is 0.42 nm, NPYR is 0.54 nm and NHMI is 0.56 nm, and these critical dimensions are comparable with the pore size of zeolite (Shen et al. 2000). Therefore the adsorption capacity of these zeolites decreases in following: NDMA>NPYR>NHMI, revealing the "sieve effect" for

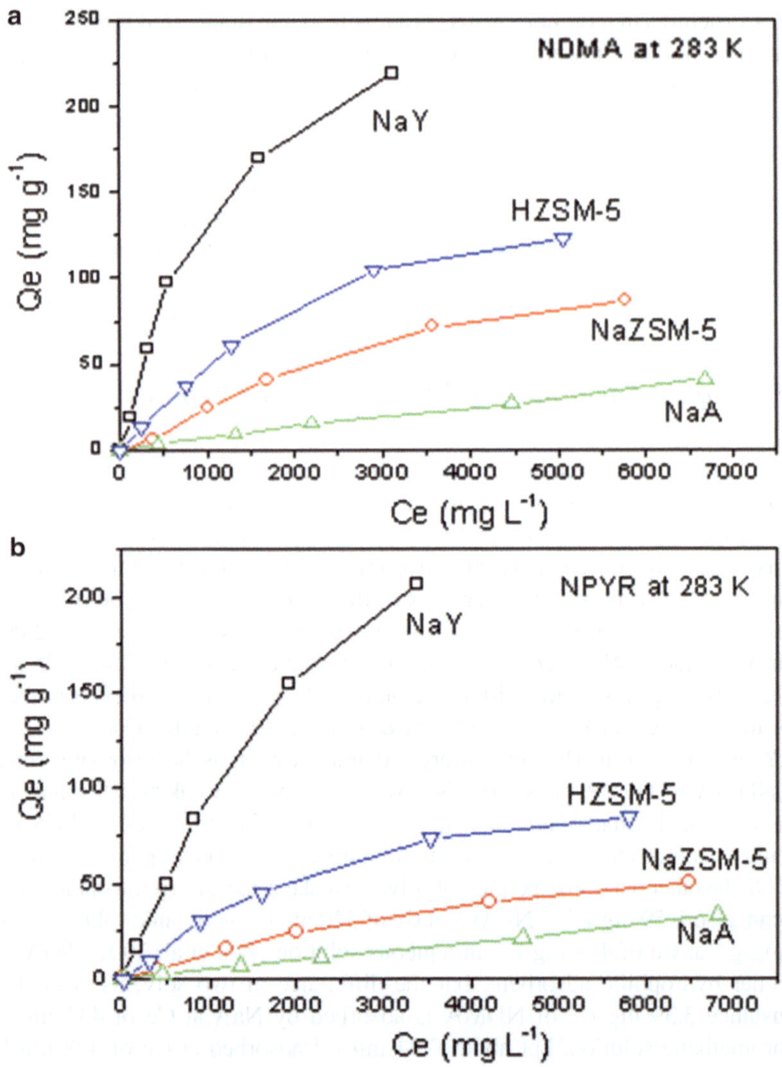

Fig. 9.14 Adsorption of nitrosamines of (**a**) NDMA, and (**b**) NPYR at 283K on zeolite adsorbents in dichloromethane (Zhu et al. 2001)

trap of nitrosamines in solution. Some strange phenomena are observed in aqueous solution where NaY adsorbs the NPYR (1.05 mg g^{-1}) similar to NDMA (1.08 mg g^{-1}), while NaA zeolites adsorb NPYR (3.80 mg g^{-1}) three time more than NDMA (0.80 mg g^{-1}). These facts imply the existence of some unknown factors on the adsorption of nitrosamines in solution (Zhu et al. 2001), and further investigation is required. In the liquid adsorption of NPYR at 277 K, MCM-22 zeolite adsorbed two times more than NaY and NaZSM-5 (Yang et al. 2010a). In addition to the special channels and

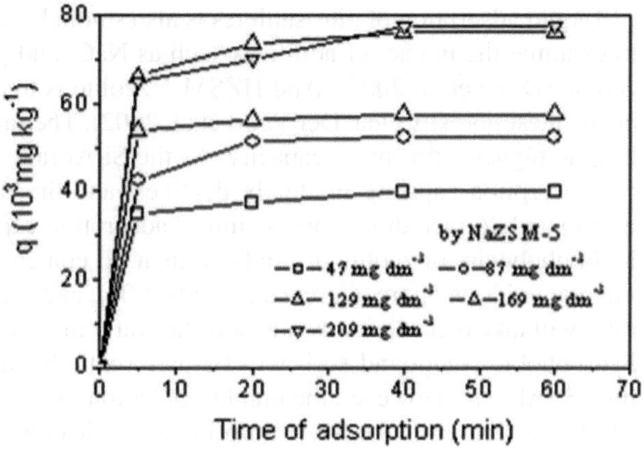

Fig. 9.15 Adsorption of NPYR by NaZSM-5 zeolite in acidic solution with different initial concentration (Zhou and Zhu 2005)

proper hydrophilic/hydrophobic nature of MCM-22, its essential grain shape owns compact chip morphology and rich surface curvature to increase the collision probability between the nitrosamine and active sites, significantly improving the adsorption. For removal of neutral nitrosamines in demineralised water, mordenite MOR200 and zeolite ZSM-5 show a higher efficacy than dealuminated NaY (DAY) and activated carbon. DAY is hydrophilic zeolite so that it does not show appreciable removal of any of the nitrosamines. When nitrosamines are adsorbed from surface water, there is no influence competition with, or pore blockage by natural organic matter components on nitrosamine removal for ZSM-5 zeolite, in contrast to activated carbon. A relationship is found between the Stokes diameter of nitrosamines, and their removal by ZSM-5 and MOR200, indicating that a "close fit" adsorption mechanism more likely than hydrophobic interaction in these zeolites (de Ridder et al. 2012).

Many carcinogenic agents like nitrosamines or their precursors enter human stomach through diet and drinking, so it is necessary to trap the nitrosamines in gastric juice. However, the liquid adsorption results obtained in neutral solution cannot be simply applied to the gastric juice that is acidic. And many zeolites cannot be used in acidic solution because of their inherent feature. Consequently, some liquid adsorption experiments of nitrosamines are performed in hydrochloric acidic solution at the $pH = 1$ (Fig. 9.15), similar to that of gastric juice, and NPYR is selected as the adsorbate (Zhou and Zhu 2005). As expected, NaY and NaA zeolite are dissolved in the solution within 30 min while NaZSM-5, ZSM-11 and Hβ are stable. And NPYR is found to be also stable in the acidic solution for 3 h. The NPYR uptake process on NaZSM-5 zeolite obeys the first-rate Lagegren equation, and the adsorption curve of ZSM-5 can be fitted with Langmuir formula. Zeolites are superior to silica and mesoporous silica SBA-15 for the adsorption of nitrosamines in acidic solution, and among them Hβ zeolite has the largest adsorption capacity owing to its big pore size and surface area (Zhou and Zhu 2005). NPYR is hydrophobic, and the NaZSM-5 zeolite with a high Si/Al ratio is also hydrophobic so that the sample with the Si/Al of 12.5 adsorbs less NPYR (70.0 mg g^{-1}) than that with Si/Al of 26 (77.5 mg g^{-1}).

Liquid adsorption of nitrosamines is also studied in fasting artificial gastric juice to examine the impact of additives such as NaCl and glycine on the adsorption of zeolite (Dong et al. 2007). And HZSM-5 zeolite is treated with alkali corrosion to create mesoporosity (Van Der Voort et al. 2002). The sample with a Si/Al ratio of 38 has the highest adsorption capacity. As the Si/Al ratio of zeolite further increases, the adsorption capacity gradually declines, and the samples with the Si/Al ratio of 500 and 26 incredibly have a similar adsorption capacity. The hydrophilicity or hydrophobocity of zeolite depends on their Al content that governs the polarity of internal voids and pore (Garcia and Roth 2002), and thus the zeolites with low Si/Al ratio will take precedence over adsorbing water in artificial gastric juice. NPYR is a hydrophobic compound so it will be preferentially adsorbed on the zeolites with high Si/Al ratio. However, the number of cation to generate static electronic field in ZSM-5 decreases as the Si/Al ratio increases, leading to the decline of adsorption ability. Two impacts compromise to give an optimal promotion for the adsorption on the sample with Si/Al ratio of 38 where the cations provide enough electrostatic induction to pull NPYR but the competitive adsorption of water is not too serious (Dong et al. 2007). Since the hydrated radii of the cation bigger than 0.3 nm in the channels of zeolite will hinder the transport of adsorbate to decrease the adsorption capability (Rabo et al. 1966), acidic HZSM-5 zeolite is more favorable for adsorption of NPYR in artificial gastric juice. Nonetheless, the adsorption of zeolite in artificial gastric juice is different from in acid solution, and its adsorption isotherm is accord with Freundlich equation instead of Langmuir equation (Dong et al. 2007). Creation of mesopores in ZSM-5 zeolite is beneficial for mass transport inside the channel to promote the adsorption of nitrosamines in artificial gastric juice, but zeolite surfaces become more uneven. At the same time, the amino acetic acid and sodium chloride in solution interferes with the interaction between nitrosamines and zeolites, resulting in the deviation of adsorption isotherms of NPYR from Langmuir equation. On the other hand, in the case that both NPYR and lead ions exist in artificial gastric juice, different zeolites have different adsorption performances (Tao et al. 2010): zeolite NaZSM-5 (Si/Al = 26) captures more nitrosamines (4.36 mg g^{-1}) than lead ion (0.42 mg g^{-1}), whilst natural Na-clinoptilolite shows a selectivity toward lead ion in the solution, trapping more lead ions (6.88 mg g^{-1}) than nitrosamine (1.04 mg g^{-1}). Based on these results, a versatile zeolite material is reported to release NO at first and then capture the nitrosamines in mimic gastric juice (Wei et al. 2010).

Zeolite is able to selectively adsorb tobacco-special nitrosamines (TSNA) in organic solvents such as dichloromethane. For promoting the adsorption of bulky nitrosamines, MCM-22 zeolite is treated with alkali solution to create mesopores for accelerating the mass transfer of adsorbate, then sulfonic acid is used to establish acid sites in these mesopores (Yang et al. 2010b). MCM-22 sample has a high adsorption capacity towards NNN in dichloromethane solution, and it can trap 55.3 % of *N′*-nitrosonornicotine (NNN) at the initial concentration of 1.02 mmol L^{-1}. Hierarchical structure increases the adsorptive capability of MCM-22, and the corresponding value is 51.7 %; again, incorporating $-SO_3H$ group in the sample significantly elevates its capacity to 74.9 %. Nonetheless, not all of the organic

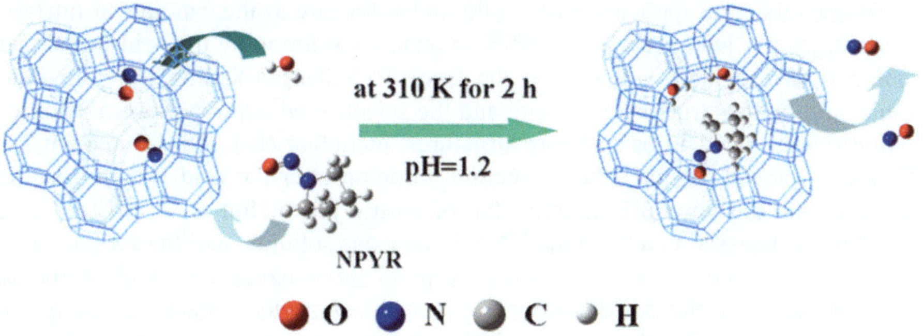

Fig. 9.16 Releasing NO and capturing nitrosamines by zeolite in mimic gastric juice (Wei et al. 2009b)

functional groups can promote the adsorption ability of MCM-22 toward the NNN in dichloromethane solution. The sample modified with chloropropyl triethoxy silane (CPTES) or 3-mercaptopropyltriethoxysilane (MPTES) does not show such high adsorption ability (Yang et al. 2010b). Contrarily, grafting small amount -SO_3H group, less than 4 wt.-%, in the sample of alkaline treated MCM-22 significantly promotes the adsorption towards TSNA even in the aqueous tobacco extract solution, and the resulting sample can trap up to 54 % of TSNA in the solution. The composition of extractable liquid of cut tobacco is more complex (Wei et al. 2009a), but many zeolites can still selectively adsorb TSNA in this solution. The bulky TSNA molecules only insert their N-NO groups into the channels of NaA zeolite due to the small pore size of the zeolite (Zhu et al. 2003; Yun et al. 2004; Zhou et al. 2004), NaA only adsorbed 0.58 μmol g^{-1} of TSNA, which is much lower than that of NaY zeolite (1.37 μmol g^{-1}). The adsorption capacity of NaZSM-5 is about 1.11 μmol g^{-1}, same as that of MCM-22 whose pores opening is similar to ZSM-5 but pore structure different. These results indicate the crucial role played by the pore size of zeolite in the liquid adsorption, since the limitation of narrow channel on the mass transfer of adsorbate is unavoidable. Acidic zeolite such as HZSM-5 and Hβ exert a higher capability than their basic analogues NaZSM-5 and Naβ for the liquid adsorption, because the proton in acidic zeolite can promote the adsorption of nitrosamines through forming hydrogen bond with the N atom in the nitroso group of nitrosamines (Meier and Siegmann 1999; Xu et al. 2003a; Wu et al. 2008). Calcium ion has a usual promotion on this liquid adsorption. Actually CaZSM-5 shows the enhanced adsorption capability of 1.41 μmol g^{-1} nitrosamines to exceed either NaZSM-5 or HZSM-5, and higher adsorption ability is observed on CaA zeolite, because it can trap 1.68 μmol g^{-1} of nitrosamines in the solution, even exceeded activated carbon (Wei et al. 2009a). This may result from the specific high cation concentration of zeolite A that provides a powerful electrostatic interaction toward nitrosamines (Gao et al. 2008; Yang et al. 2008). On the basis of these studies, new efficient materials are developed to remove 70 % TSNA in the extractable liquid of cut tobacco (Fig. 9.16. Wei et al. 2009b).

To open the new application of zeolite in health care as the capturer of nitrosamines in gastric juice, capture of NNN in aqueous solution by the acid-immerged zeolite is investigated, which is in the face of challenges arising from the acid immersion on the structure of zeolite and the selective adsorption. Several zeolites with different Si/Al ratios and pore structures, including NaY, ZSM-5 and MCM-22, are employed to study their structure variation after the acid leaching in the solution with the same pH value as that of gastric juice (Hou et al. 2013). These samples are assessed to adsorb the NNN in aqueous solution, and the zeolites with Si/Al ratio of about 13 can keep their adsorption performance after acid treatment, rather, those with the Si/Al ratio of around 10 show the enhanced adsorption capability in deed. Besides, zeolite NaY and its dealuminated analogue NTY are proven to be safe for the cell of 293-T, and NTY can trap the NNN in the aqueous solution to reduce toxicity so that it is the valuable candidate to trap nitrosamines in stomach juice (Hou et al. 2013).

9.2.6 Reducing Nitrosamine Level of Smoke by Zeolites

Although many zeolites show the capacity of selectively adsorbing nitrosamines from gas stream in laboratory study, it is unclear whether they can capture nitrosamines in the mainstream smoke of tobacco, since tobacco smoke is an aerosol composed of volatile agents in the vapor phase along with lots of semi- and non-volatile compounds in the particulate matter. In fact, zeolites had the ability to adsorb volatile nitrosamines in mainstream smoke of cigarettes (Xu et al. 2003a), and their adsorption capacity depends on the pore structure and surface acidity since protons form hydrogen bonds with the N atoms of nitroso-groups in nitrosamine. NaY and HZSM-5 are the best adsorbents while NaZSM-5 is better than NaA zeolite, and the residual nitrosamines in the smoke of cigarettes after adsorption on zeolite is very low (Xu et al. 2003a). HZSM-5 zeolite not only adsorbs more nitrosamines than NaZSM-5 but also keeps a high activity for the degradation of nitrosamines without interference of competitive adsorption of other components in smoke. The nitrosamines adsorbed in NaY zeolite start to desorb at 425 K, so that NaY seems suitable to be used in the filter of cigarette as an adsorbent to remove nitrosamines, because the temperature of filter usually is below 373 K at which the desorption of nitrosamines cannot occur on the zeolite (Meier and Siegmann 1999; Xu et al. 2003a). To prevent desorption of nitrosamines from zeolite at high temperatures, copper oxide is used as the modifier for NaY, NaZSM-5, and NaA to strengthen their ability to adsorb volatile nitrosamines (Xu et al. 2003b, 2004). Due to the special interaction from the copper species in channel, nitrosamines are easily trapped by the modified zeolite so that about 70 % of the nitrosamines in smoke are eliminated. 1%CuO/NaY and 3%CuO/NaY composites filter more volatile nitrosamines than zeolite NaY in mainstream smoke, and 3%CuO/NaY is proven to be the best one. On the basis of gaseous adsorption of nitrosamines, a new bifunctional zeolite composite is fabricated as the cigarette additive to release menthol in the burning tobacco while adsorb the nitrosamines in the smoke (Gu et al. 2008b).

Fig. 9.17 SEM images of (**A**) zeolite NaA and (**B**) CAS-1 samples (Gao et al. 2009b)

Trapping tobacco specific nitrosamines in smoke is a hard task for zeolite, because most TSNA exist in the particles in tobacco smoke (Baker 1999) and the sizes of their particles exceed the micropores of zeolites. However, zeolite-like calcosilicate CAS-1 can effectively intercept the particles and thus reduces the TSNA level of mainstream smoke in the range of 30–60 % when it is added into the cigarette filter (Gao et al. 2009b). Addition of zeolite NaA of 30 mg cig^{-1} in the filter of cigarette decreases about 11 % of TSNA in smoke, but the reduction can achieve about 30 % once 10 mg CAS-1 is added into the filter of cigarette, superior to the function of 30 mg zeolite NaA. As the amount increases to 20 and 30 mg per cigarette, the decrease of TSNA achieves 55 % and 58 % respectively. Due to its special morphology (Fig. 9.17), CAS-1 can intercept the bulky particle similar to the cobweb capturing the flying inspects. Besides, CAS-1 is not the simply physical filter, since the Ca^{2+} in framework provides a strong affinity to attract the N-NO group of nitrosamines. The fiber-like morphology of CAS-1 realizes the effective interception of particles in smoke, better than common zeolite, while the plenty of cations plus zeolite-like microporous structure in CAS-1 enables the TSNA adhered on particles to be captured.

9.3 Researches of Nitrosamines Degradation on Zeolite

9.3.1 Catalytically Decomposition of Nitrosamines on Zeolite

Zeolite has been tried as the cigarette additive to destroy the harmful components in smoke including TSNA and volatile nitrosamines such as NDMA (Meier and Siegmann 1999). NaY zeolite showed a high capability in the gaseous adsorption of volatile nitrosamines (Xu et al. 2003a; Zhou et al. 2007), but it is not usually considered as a catalyst material. In the potential application of zeolite as the catalysts for cigarette or the purifier for vent-pipe of building, there are numerous compounds that will competitively adsorb in zeolite to occupy active sites (Meier and Siegmann 1999; Xu et al. 2003a). What impact will be caused by the pre-adsorbed impurities on the

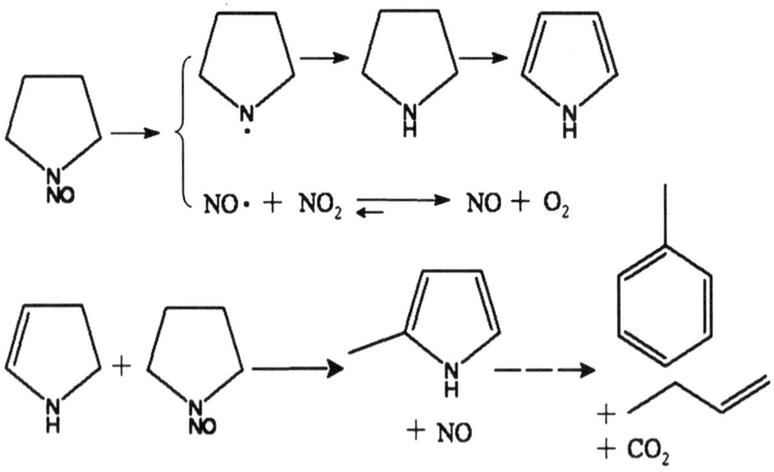

Fig. 9.18 The possible degradation of NPYR on basic zeolite similar to that of pyrolysis (Wu et al. 2008)

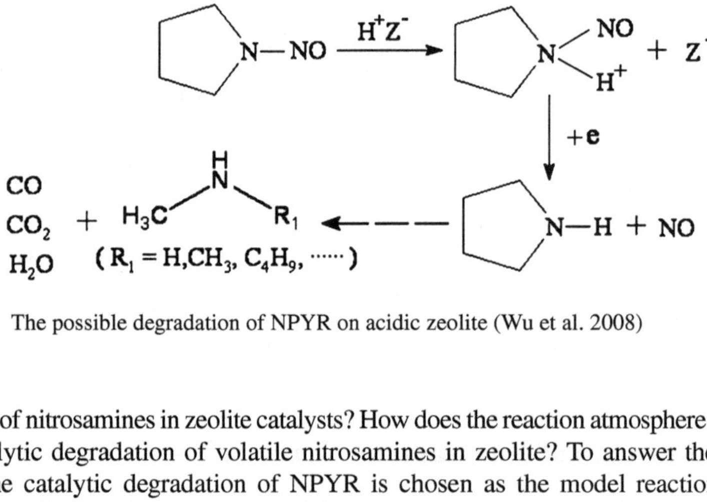

Fig. 9.19 The possible degradation of NPYR on acidic zeolite (Wu et al. 2008)

reaction of nitrosamines in zeolite catalysts? How does the reaction atmosphere influence the catalytic degradation of volatile nitrosamines in zeolite? To answer these questions, the catalytic degradation of NPYR is chosen as the model reaction, Y and ZSM-5 zeolites as the catalysts, and the temperature programmed surface reaction (TPSR) is employed to examine the variation of degradation products of nitrosamines in zeolite via the different temperatures (Wu et al. 2008). NPYR can be decomposed to form NO, methylamine, dimethylamine and N-methl-1-butanamine in HY zeolite at 773 K, but produces more complex products such as pentanedinitrile, 1-methyl-1-H-pyrrole, toluene, 1H-pyrrole and 1-H-2-methyl-pyrrol in NaY zeolite at 773 K (Fig. 9.18). Nitrogen and CO, water and propylene are the main species formed on NaY while the portion of NO, N_2O and CO_2 in the products is small. Contrarily, majority of the products on zeolite HY are nitrogen-containing species and CO_2 or CO, only few hydrocarbons are detected due to the coking on acidic zeolite (Fig. 9.19).

Decomposition of NPYR on HZSM-5 forms more products than that on zeolite HY, including toluene and 2-butanamine. On zeolite NaZSM-5 the nitrosamine is degraded to N_2O, 1-H-1-methyl-pyrrole and 1-H-pyrrole (Wu et al. 2008). Zeolite NaY can adsorb and degrade more NPYR than NaZSM-5 and in general the acidic zeolite shows a higher activity than its basic analogue to decompose nitrosamines. Moreover, all of the adsorbed NPYR could be degraded in the TPSR tests of NPYR on zeolite NaZSM-5 or NaY in nitrogen. NO and N_2O as well as pyrrolidine fragments appear in the range of 450–650 K, 2-methyl-1H-pyrrole and toluene emerge at 591 K and 667 K respectively, but their amounts are very small. Desorption of NO emerges on NaZSM-5 zeolite around 587 and 734 K, accompanied by propylene and 1H-pyrrole. The moisture pre-adsorbed in zeolite NaY does not hinder the adsorption of volatile nitrosamines in gas stream at 338 K (Zhou et al. 2007), but suppresses the adsorption at 298 K, and the capacity is lost about 18 % on NaY, 23 % on HY, 21 % on NaZSM-5 and 33 % on HZSM-5. Besides, moisture has minor impact on the degradation of nitrosamines on NaY, but more serious on ZSM-5 zeolites (Wu et al. 2008).

Different carrier gasses used in TPSR experiment affect the products distribution of nitrosamine degradation. In oxygen atmosphere, most NO_2 is produced from the oxidation of NO gasses (Wu et al. 2008); nonetheless it comes from disproportionation reactions of N-O group in nitrosamines in inert and reductive atmospheres. NO_2 gas can form varieties of adsorption species on zeolite, for instance NOx starts to strip at about 673 K, and decomposition of Na^+-NOy^-, Al^{3+}-NOy^- nitrate and nitrite begins at above 793 K. For the species pre-adsorbed, *n*-Dipropylamine occupies the active sites of NaZSM-5 zeolite, interrupting the degradation of nitrosamines; pyridine blocks the channels to hinder the diffusion of degradation products of nitrosamines. However, water does not affect the catalytic degradation of NPYR on NaY and HY zeolite. Consequently, NaY zeolite does not need to be protonated before performing the catalysis, and it is still active to catalyze nitrosamine degradation even in inert atmosphere (Wu et al. 2008). Again, the similar catalytic behavior of NaZSM-5 in N_2 and in air confirms the catalytic ability of zeolites with sodium ion; therefore protonation of them prior to catalytic reaction (Meier and Siegmann 1999) is not necessary for degradation of volatile nitrosamines. In fact, NPYR can be adsorbed by ZSM-5 or Y zeolites (Zhou et al. 2007). The N–N bond in the N–NO group is the weakest bond in the structure of nitrosamine, hereby the degradation of NPYR starts from the cleavage of N–NO bond to release NO or NO^+ (Cheng et al. 1998; Miura et al. 2000), and the NOx and other fragments formed in the pyrolysis procedure can further react with each other at high temperature like 773 K to give complex products through different reaction paths. Either reaction temperature or products distribution of NPYR has been changed on zeolite (Zhu et al. 2001). NO emerges near 450 K on NaY zeolite, much earlier than that in pyrolysis (593 K) meanwhile several other nitrogen oxides are formed resulting from the further reaction of NO. Since NaY exhibits a higher activity than NaZSM-5 in the TPSR process of NPYR to degrade more nitrosamines, it may be utilized as the candidate for the catalyst used in cigarette. Also, acidic zeolites such as HY can be used to eliminate the nitrosamines in smoke (Meier and Siegmann 1999; Wu et al. 2008).

To study how the bulky TSNA with a molecular size exceeded the pore mouth of zeolite is catalytically decomposed in the zeolite, *in situ* Fourier transform infrared spectroscopy (FTIR) and thermogravimetric-mass spectrometry (TG-MS) are chosen for the research and NNN is selected as the reactant (Cao et al. 2007a). Degradation of NNN takes place on NaY zeolite at 423 K, causing new IR bands such as bridged nitrates at 1,590 cm^{-1} and giving the band of isolated nitrosonium ion or nitrite NNN (2,243 cm^{-1}) at 473 K. The bands of $-CH_2$, C-H and NNN itself fade at 523 K because of both degradation of NNN and desorption of the products. Degradation of NNN occurs in HZSM-5 zeolite at 423 K due to the promotion of protons in zeolite, and the degrade manner of NNN adsorbed in zeolite depends on the pore structure of the adsorbent. Neither pyridine (m/e = 79) nor NPYR (m/e = 100) is detected during the TG-MS process of the NNN adsorbed in NaY, and they are absent in that of copper modified NaY, HY, NaZSM-5 and HZSM-5 zeolites. Clearly the NNN in these zeolites does not divide into two parts that contain pyridine and 5-member-ring respectively. Some aromatic compounds appear in the degradation of NNN on acidic zeolites, showing the reformation of the pyridine rings of NNN molecules plus the fractures of pyrrole ring and five-member ring. Also, a lot of the fragments of NNN deposit in zeolite, forming the coke of 0.4 % and 1.2 % on HZSM-5 and HY, respectively. NaA zeolite is an exception on which the fragments of NPYR and pyridine as well as pyrrolidine appear. This strange phenomenon originates from the special pore diameter of NaA zeolite that enables part of the 5 member ring to enter the channel with the *N*-nitroso group while the rest of the molecule is clipped on the pore mouth due to the C-C bond between the 5 member ring and the pyridine ring (Fig. 9.8, Cao et al. 2007a). In case that the sample is heated, the clipped NNN molecule will be divided into two parts in the joint bond. For zeolite KA with the smaller pore size than NaA, the degradation of NNN still begins from cleavage of N-NO band. In principle, degradation of NNN in zeolite sensitively depends on the pore diameter of catalyst. If the pore size of zeolite is smaller than 0.4 nm, NNN inserts its *N*-nitroso group inside the narrow channel while for the zeolite with a pore diameter equal to or larger than 0.5 nm, NNN molecule enters in the channel, whole or partially. In both cases the adsorbed NNN molecule will be degraded from the rupture of N-N bond in *N*-nitroso group (Fig. 9.8), forming nitrogen oxides and the residual carbon-containing parts. If the pore size of zeolite is between 0.32 and 0.54 nm, NNN will be clipped in the pore mouth, decomposing through the rupture of the C–C bond linking five-member ring and pyridine ring to form pyridine fragments (Cao et al. 2007a).

Apart from the catalytically degradation at high temperature, nitrosamines can be decomposed in zeolite with the assistance of microwave irradiation (Fig. 9.20, Xu et al. 2008). *N*-Nitrosodiphenylamine (NDPA) is a possible cancer-causing substance in humans, its N–NO bond in molecule is relatively a weak one with the bond energy of 11 kcal mol^{-1}. With two rigid phenyls connected to the N-NO functional group, NDPA molecule has a molecular size larger than NNN and it usually adsorbs on zeolite through the way of inserting its N-NO group into zeolite channel (Zhou et al. 2004). Without addition of zeolite, no NDPA is degraded in dichloromethane solution under the irradiation of 45 s no matter how the power of microwave changed

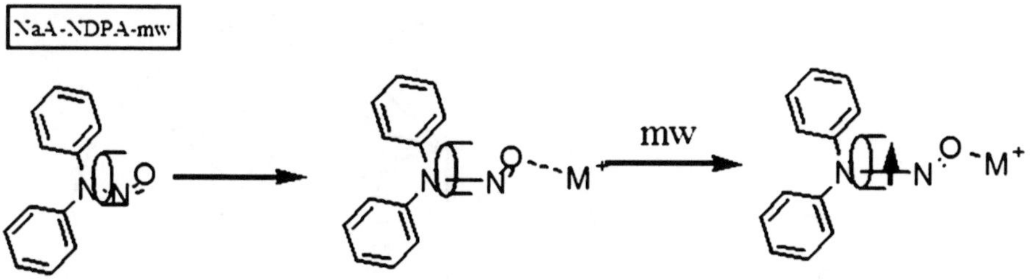

Fig. 9.20 The possible manners of NDPA adsorbed and degraded in zeolites NaA by microwave (Xu et al. 2008)

from 0.1 to 1 kW. Addition of NaY zeolite into the solution starts the decomposition of NDPA at the low microwave power of 0.3 kW, and the percentage of degraded nitrosamines increases quickly as the power achieves to 0.8 kW. Large pore size of zeolite is beneficial for the adsorption and decomposition of NDPA. Under the condition of irradiating 45 sec with the power of 1 kW, the percentage of NDPA adsorbed and decomposed in zeolite NaA, NaZSM-5 and NaY reaches 4.1 %, 16.6 % and 22.3 %, respectively (Xu et al. 2008). Ion exchange of zeolite NaA with Ca^{2+} can increase the conversion of NDPA from 4.1 to 15.4 %. Also, the moisture in zeolite will promote degradation of NDPA under the microwave irradiation. Neither NPYR nor NNN in solution could be degraded when the microwave irradiated them with 1 kW for 45 s; and less NNN was degraded in NaY zeolite than NDPA because of its larger N-NO bond energy than that of NDPA (Miura et al. 2000). The volatility of NPYR suppressed its microwave-induced degradation in NaY zeolite: although NaY adsorbs all of the NPYR in solution, about 45 % desorbs while only 1.4 % is degraded. The narrow channel and the plenty cation of NaA tightly anchor the nitrosamine molecules. It also adsorbs all nitrosamines in solution, but no desorption of NPYR occurs while 11.2 % is degraded under microwave irradiation.

There are two reasons to explain the promotion of microwave on the NDPA degradation in zeolites. The first is heating effect, since zeolite cations can transfer heat energy to the adsorbed nitrosamines to degrade them. The second is the limited-domain effect, because nitrosamines are constrained in the channels after being adsorbed, while microwave irradiation makes the surrounding water molecules moving with a high speed to bump the carcinogen, promoting the degradation of nitrosamines. Thus, two factors are crucial for the microwave-assisted degradation of nitrosamines. Firstly, the geometric matching degree between the configurations of zeolite and the nitrosamines should be high, in order to constrains the movement of nitrosamines in zeolite. Secondly, zeolite should have necessary cations inside channels to form strong electrostatic field forces attracting nitrosamines and weakening their N-N bonds. N-NO bond is known to be the fragile bond in NDPA molecule, and it becomes rather weaker due to the electrostatic attraction from zeolite cations in channels after NDPA is adsorbed on zeolite, which makes N-N bond much brittle. After adding zeolite in NDPA solution and irradiating with microwave, blue free radicals are observed on zeolite, indicating the broken of N-N bond in

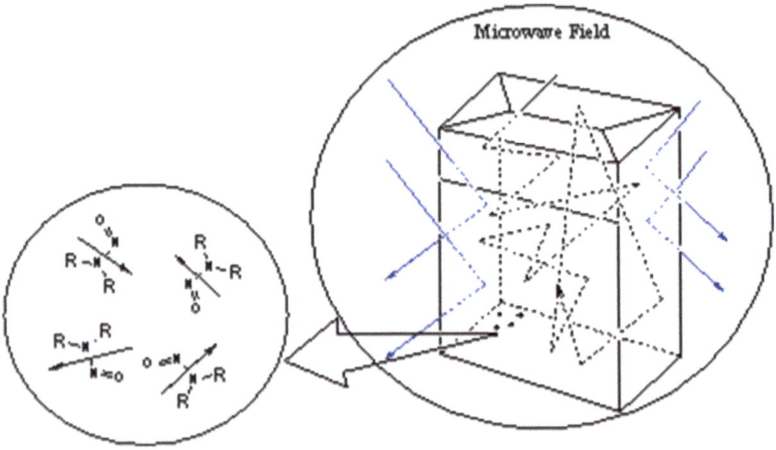

Fig. 9.21 Regulation of package case for the microwave-induced degradation of nitrosamines in cigarette (Gao et al. 2009a)

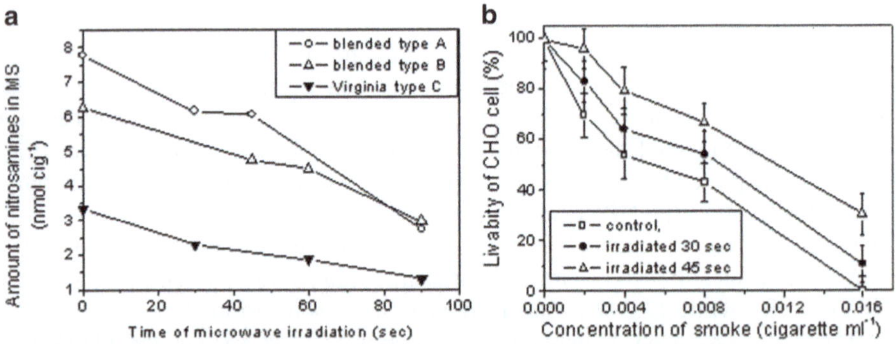

Fig. 9.22 Impact of microwave irradiation on (**a**) nitrosamines and (**b**) CHO cell test in the main stream smoke of cigarette (Gao et al. 2009a)

NDPA molecule (Zhou et al. 2004). And it should be pointed out that the fracture of N-N bond removes the carcinogenicity of nitrosamines. Based on the results of fundamental study, microwave-degradation of nitrosamines is developed to be a new technology for reducing the harm compounds in tobacco smoke (Gao et al. 2009a, b). The finished cigarettes are directly irradiated with microwave in the optimized conditions (Fig. 9.21), and the N-N bond of nitrosamines in tobacco fibers is fractured. After the irradiation of 90 s with the microwave of 1 kW, 25–30 % of nitrosamines in tobacco and 50–60 % in mainstream smoke of the cigarette are reduced (Fig. 9.22). Moreover, less mutagenic activity and higher CHO cell livability are found for the irradiated cigarette according to the results of in-vitro tests, mirroring the delinced biological toxicity of smoke (Gao et al. 2008, 2009a, b).

9.3.2 Researches of Metal Oxide Modified Zeolite Catalysts Applied to the Removal of Nitrosamines

Zeolites can be modified with impregnation of copper nitrate solution followed by calcination, in order to highly disperse CuO nanoparticles in the porous host (Xu et al. 2003b, 2004). These copper oxide particles strengthen the attraction towards N-NO group, not only promoting 20 % or 60 % more NPYR to be adsorbed on NaY or NaZSM-5 at 453 K, but also elevating its catalytic degradation in TPSR test. The total amount of NOx product formed in TPSR experiment reflects the number of active centers in zeolite while the T_{max}, a temperature at which the maximum concentration of NOx appears, indicates the activity of the active centers (Shen et al. 2000). NaY zeolite can degrade 0.31 mmol g^{-1} of NPYR with a T_{max} around 573 K. Introducing CuO of 1 wt.-% enhances the catalytic activity about 8 % while loading 3 wt.-% makes the activity further increased about 12 %. In the latter case the T_{max} is lowered from 573 to 553 K, implying more NPYR are decomposed at relatively lower temperature. Rather, 5%CuO/NaY exhibits the best performance since it degrades the NPYR 33 % more than NaY. Further loading more copper does not create a higher activity on zeolite (Xu et al. 2004). Loading CuO on NaZSM-5 and NaA zeolites also dramatically enhance their catalytic activity. The amount of NOx detected on 3%CuO/NaZSM-5 is thrice of that on NaZSM-5 while fourfold NOx forms on 3%CuO/NaA compared with NaA zeolite alone. However, their T_{max} is unchanged that differs from CuO/NaY catalysts. The incorporated copper species cannot have their optimal catalysis until they located in a suitable microenvironment provided by the zeolite host, especially the right geometric space to form adsorbate-adsorbent optimal interaction. In TG-MS experiment the desorbed NPYR from NaY zeolite (m/z = 100) in TPSR process from 430 to 580 K is 1.67 % in comparison with those degraded to NO (Xu et al. 2004). This proportion is reduced to one third on the sample of 3%CuO/NaY (Fig. 9.23). Clearly copper modified zeolite catalytically decomposes instead of desorbs nitrosamines at high temperature, which is crucial to reduce environmental pollution.

Introducing ferric modifier also elevates the activity of NaY zeolite in nitrosamines decomposition. Modification of NaY zeolite with ferric oxide of 1 % (w/w) increases the nitrogen oxides products desorbed in TPSR test from 0.308 to 0.474 mmol g^{-1} (Cao et al. 2008), but the climax of desorption shifts to 593 K. Nonetheless, 0.433 mmol g^{-1} of nitrogen oxides products are detected on 3%Fe_2O_3/NaY while the climax of desorption decreases to near 533 K. Further addition of ferric oxide on zeolite NaY makes less NPYR to be decomposed. According to TG-MS analysis, the desorbed NPYR (m/z = 100) from zeolite NaY in the process of TPSR from 450 to 580 K is 1.67 % in comparison with that degraded to NO, and this proportion is lowered to 0.76 % on 3%Fe_2O_3/NaY due to its stronger catalytic ability, almost all of desorbed NPYR molecules are degraded before they escape from the porous adsorbent. Modification with ferric oxide also affects degradation products of NPYR on NaY zeolite. Strong MS signals of N_2O and 1-H-pyrrole are observed in the spectrum of NaY, resulting from the rupture of N-NO bond (Cao

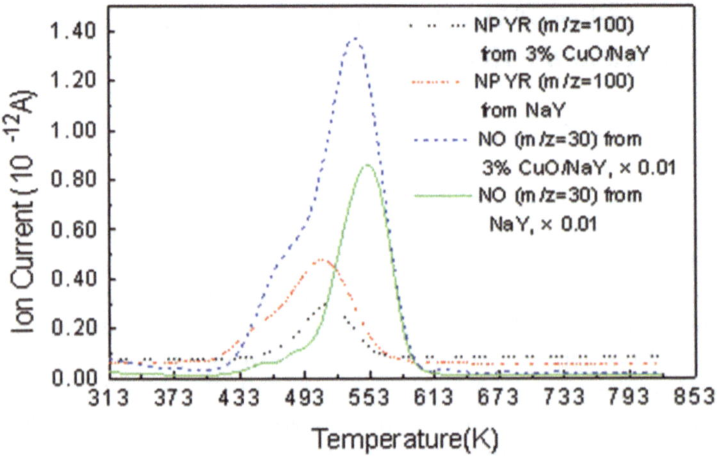

Fig. 9.23 MS spectrum of NPYR and NO desorbed from zeolite NaY and 3%CuO/NaY sample that adsorbed NPYR and was heated in TPSR process (Xu et al. 2004)

et al. 2007b) whereas N_2O absents but propionitrile becomes the predominant product in the ferric oxide modified zeolite. Besides, the Si/Fe ratio of 3%Fe_2O_3/NaY fresh sample is 7.61, much larger than the calculated value (25.9), which means the enrichment of ferric species in the external surface and pore opening of zeolite. This delicate heterogonous distribution of guest in zeolite is invisible for common XRD technique but can be detected by XPS method. After the TPSR process, the surface Si/Fe ratio of the sample changes from 7.6 to 5.2, implying the migration of ferric species from channel towards external surface of zeolite. At the same time, the Si/Al ratio of the sample is slightly changed from 2.45 to 2.80, indicating the overcast of some aluminium by the ferric species. Surfaces enrich and migration of ferric oxide will change the surface properties of zeolite, because the ferric guest can improve the capture of nitrosamines through the electrostatic interaction with the N-NO group of nitrosamines (Xu et al. 2004). Ferric modification has complex effect on zeolite NaZSM-5. N_2O and 1H-pyrrol are detected at lower temperature on 3%Fe_2O_3/NaZSM-5 than NaZSM-5 zeolite in TG-MS test, but the formation of 2-mehtyl-1-H-pyrrole and propionitrile is suppressed. No desorption of NPYR is detected in the range of 400–823 K.

In the TPSR experiment on the cobalt modified NaY above 500 K, NPYR is first divided into a nitric oxide radical and pyrrolidine radical, which will quickly change into NO and pyrrolidine (Fig. 9.24, Cao et al. 2007b). Through the disproportionation of nitric species, NO is converted to N_2O and NO_2, while the Co^{2+} cations in zeolites catalyze the decomposition of NOx to form N_2. At the same time, dehydrogenation of pyrrolidine occurs to form 1H-pyrrole and 2-methyl-1H-pyrrole, which involves the creaking of the five ring of NPYR, because this is the only source of methyl. As the temperature raises over 600 K, new MS signals of aromatic cyclic compound such as aniline emerge.

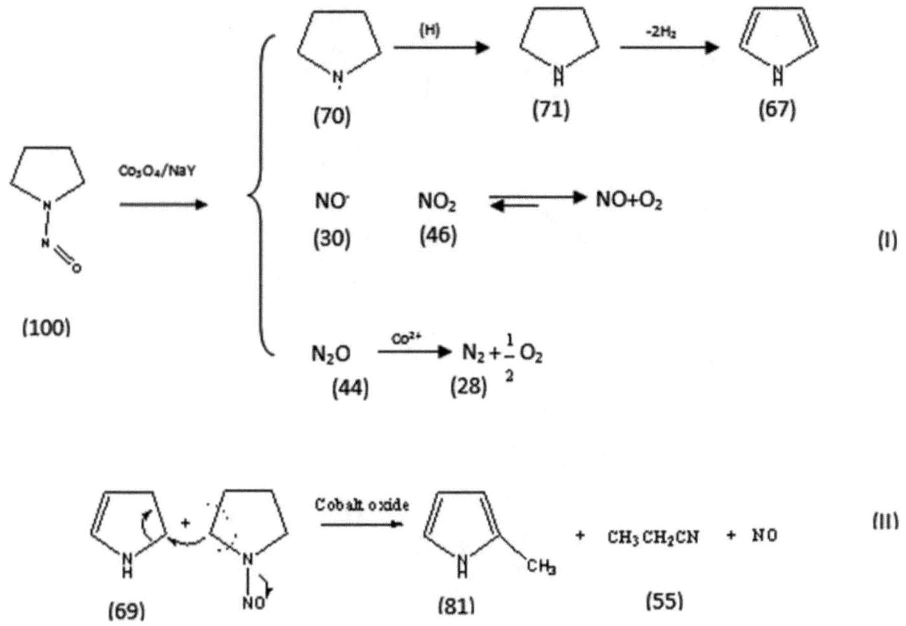

Fig. 9.24 The hypothesis on the decomposition routine of NPYR in the cobalt modified zeolite (The number in the bracket is the m/e value of the compound, Cao et al. 2007b)

Degradation of nitrosamines usually releases nitrogen oxides NOx due to the rupture of N–NO band. Zirconia is thus chosen to replace copper oxide for zeolites to obviously reduce the release of NOx during the decomposition of nitrosamines (Zhuang et al. 2006). Zirconia is the amphoteric material to have both weak acidity and weak basicity, and nitrosamines also possess both weak acidity and weak basicity. Consequently, establishing a suitable chemical environment in zeolite by use of zirconia modifier may promote the decomposition of nitrosamines. Dispersing zirconia of 3 wt% on NaY through microwave-irradiation enhances the activity of the zeolite in decomposition of nitrosamine. The NPYR adsorbed in ZrO_2/NaY begins to degrade and liberate NOx at 473 K when nitrogen is used as the carrier gas, giving rise to a maximum concentration of NOx near 593 K on the sample of 5%ZrO_2/NaY and 10%ZrO_2/NaY with the amount of 78 and 82 $\mu mol \cdot g^{-1}$, respectively. As the reaction is finally hold at 773 K for 0.5 h, another unconspicuous desorption of nitrogen oxides appears on 5%ZrO_2/NaY with a relative low concentration of about 6.6 $\mu mol \cdot g^{-1}$. This phenomenon becomes obvious in 10%ZrO_2/NaY and the corresponding value increases to 11.8 $\mu mol \cdot g^{-1}$. As a comparison, NOx of 53 $\mu mol \cdot g^{-1}$ emerges on NaY at 613 K, but further desorption is absent when the sample is hold at 773 K (Zhuang et al. 2006). The NOx desorption from ZrO_2/NaY at 773 K in the TPSR process of NPYR originates from the inherent feature of zirconia, since zirconia itself can adsorb nitrosamines and decomposes them at around 733 K, giving rise to a maximum concentration of 29 $\mu mol \cdot g^{-1}$, and these nitrogen oxides

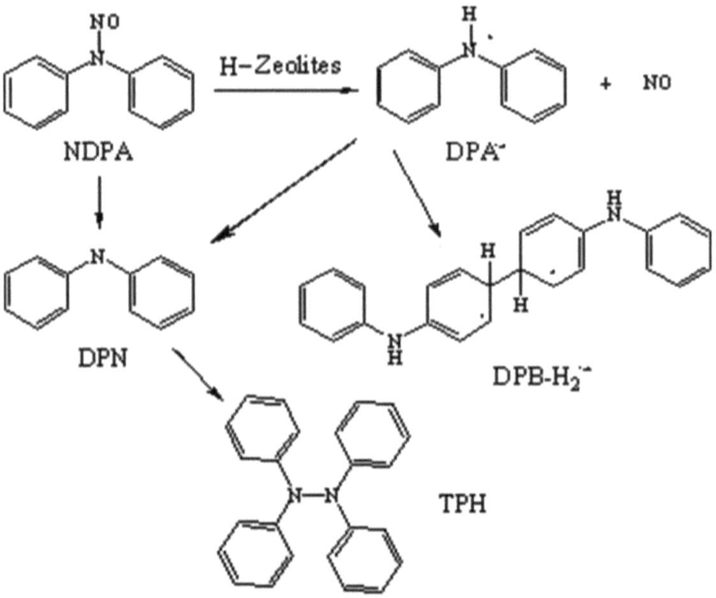

Fig. 9.25 The possible degradation pathway of NDPA in acidic zeolites at room temperature (Zhou et al. 2004)

continuously desorb when the sample is hold at 773 K. When the carrier gas of TPSR is changed to air, however, the maximum concentration of NOx released from 10%ZrO_2/NaY is abruptly decreased to below 10 μmol g^{-1} and similar phenomenon is also observed on 5%ZrO_2/NaY sample. Contrarily, no such decline is found on NOx desorbed from NaY zeolite where the concentration of NOx at 633 K increases from 49.4 to 54.5 μmol·g^{-1} instead. Although the reason why the zirconia-modified zeolite NaY can strongly suppress the release of NOx during the decomposition of nitrosamines is still enigmatic, discovery of this unwonted feature gives a clue for preparation of new functional materials.

Apart from the degradation process of nitrosamines through thermal catalysis, a strange degradation of NDPA on zeolite at room temperature is reported for the first time (Zhou et al. 2004), in which they proved that metal ions in zeolite had effects on the catalytic activity of the sample. In addition, the degradation manner of nitrosamines on acidic zeolite is started from the cleavage of N-NO band without interference of side-reactions like pyrolysis (Fig. 9.25). Acidity of zeolite is the key factor determining its ability for degradation of NDPA, and among the zeolites evaluated Hβ is the most effective catalyst with an activity much higher than that of others at ambient temperature. Bronsted acid sites are the main active centres while existence of metal ion such as Ca^{2+} or Ba^{2+} in zeolite is helpful to trap the NOx products and to prevent from secondary environment pollution.

9.3.3 *Degradation of Nitrosamines in Tobacco Smoke by Zeolites*

Zeolite was tried to be added into blend cigarette as catalyst to reduce carcinogenic compounds in smoke in 1999 (Meier and Siegmann 1999). NaY zeolite catalysts were activated when they were approached by the hot zone in the burning cigarette where temperatures could reach around 1,073 K. Of the nitrosamines some 40–50 % could be eliminated from the mainstream (inhaled by smokers), and around 50–70 % from the side stream (Meier and Siegmann 1999). In a pilot experiment performed lately, the additives consisting of zeolite NaA were added into alginate-based slurry and sprayed onto the tobacco before the Chinese Virginia cigarette manufacturing, in order to prepare the test sample cigarettes containing 3 % (w/w) of zeolite (Gao et al. 2008). NaA shows a typical cubic crystal with an average particle size of ca. 4 μm. After the zeolite was added onto the tobacco fibers, it did not change the color of the tobacco. In fact, the zeolite added into the tobacco rod is more effective in reducing the volatile nitrosamines than the TSNAs in the smoke of cigarette: the mean reduction for nitrosamines in smoke is around 61 % and the majority of the reduction is for volatile nitrosamines. Nonetheless, the reduction of TSNA levels is variable at different periods post manufacturing: it is about one third in the mainstream smoke after 6 months and the value declines to about one fifth after 14 months. After 18 months or longer, the decrease of TSNAs becomes insignificant but the mean reduction of total nitrosamines is still above 30 % (Table 9.1, Gao et al. 2008). The total toxicity of smoke is evaluated in some in-vitro and in-vivo investigations. Less mutagenic activity and higher CHO cell livability are found for the test cigarette containing zeolite in comparison with the control, and no significant damage is detected in 30-day animal experiments when exposed to the smoke of the cigarettes. Zeolite additive does not form particles in the burned tobacco to escape into smoke so that the suspicion whether the zeolites in cigarette cause risks for the health in the inhalation of these particles by the smoker, can be excluded (Wang et al. 2012). Actually, either the TSNA or the total nitrosamines level

Table 9.1 The analysis data of TSNA in the mainstream smoke of Chinese Virginia cigarettes (Gao et al. 2008)

Detected after manufactured	Sample	NNN (ng cig^{-1})	NAT (ng cig^{-1})	NAB (ng cig^{-1})	NNK (ng cig^{-1})	Total (ng cig^{-1})	Decrease (%)
6 month	Control	16.3	53.4	34.5	26.7	130.9	–
	Test[a]	19.5	32.3	11.5	20.7	84.0	35.8
14 month	Control	13.1	35.5[b]	–	15.8	64.4	–
	Test[a]	8.3	29.2[b]	–	11.2	48.7	24.4
18 month	Control	12.2	41.4	10.4	17.0	81.0	–
	Test[a]	14.1	37.7	5.5	17.9	75.2	7.2
22 month	Control	5	13	4	5	27	–
	Test[a]	5	11	3	5	24	11.1

[a]The test sample cigarette contains 3 wt.-% of zeolite additive in its tobacco rod
[b]This is the amount of NAB+NAT

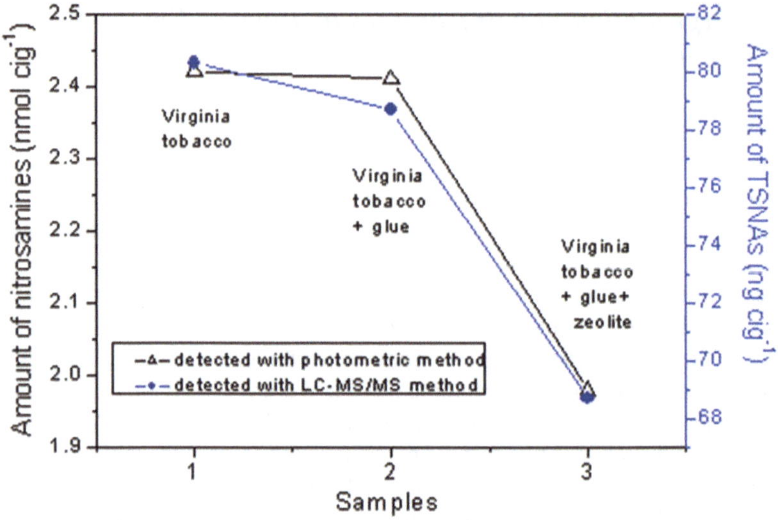

Fig. 9.26 Variation of nitrosamines content in the mainstream smoke of cigarettes detected by different methods (Wang et al. 2012)

of the mainstream smoke is lowered by adding zeolites into the tobacco rod of cigarette and this phenomenon is proven by two nitrosamines analysis methods (Fig. 9.26, Wang et al. 2012). LC-MS/MS is sensitive to monitor the variation of TSNA level in smoke, while the photometric method can give the same trend through detecting the mean amount of nitrosamines (Xu et al. 2003a). Zeolite HZSM-5 is the most competitive candidate to reduce the nitrosamines level of cigarette consisting of Virginia type tobacco, but the nitrosamines reduction by zeolite additive is more obvious in mainstream smoke than in side-stream smoke, no matter the additive is added through mechanical mixing or suspension spraying methods (Wang et al. 2012).

Several zeolites are used to assess the impact of pore size, as well as acid-base property of additives on the reduction of TSNA content in smoke (Lin et al. 2012a), and they have different distribution manners on the cut tobacco leafs and their efficiency is different. Zeolite NaA, NaZSM-5 and HZSM-5 eliminates about 18–19 % of total TSNA content while NaY removes 31 %. Pyrolysis of TSNA such as NNN starts from the dissociation of nitroso group, and forms less or non-carcinogenic fragments (Fig. 9.27). Less NNN (16 %) is removed than NAT (23 %) by NaA zeolite, whilst the reduction of four components of TSNAs is similar on ZSM-5. Contrarily, more NNN (35 %) in smoke is eliminated by NaY zeolite than NAT (28 %). On the other hand, most of zeolite additives can keep their structure after tobacco combustion. Structure collapse may occur in puff that causes the high temperature up to 1,432 K while the smoldering below 873 K is safe for these catalysts (Lin et al. 2012a).

NaZSM-5 and NaY zeolites are modified by ferric species through aqueous ion-exchange and impregnation methods. Since the Fe cations in zeolite framework are

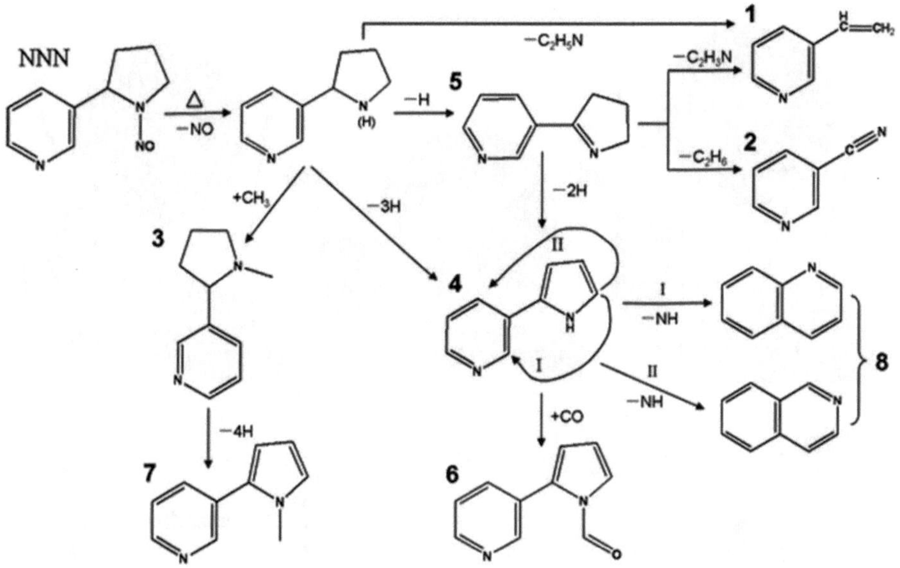

Fig. 9.27 The possible pyrolysis process of NNN (Lin et al. 2012a). (*1*) 3-vinyl pyridine; (*2*) 3-cyano pyridine; (*3*) 3-(N-methylpyrrolidine-2-yl) pyridine; (*4*) 3-pyrrol-2-yl pyridine; (*5*) 3-(4, 5-dihydro-3H-pyrrol-2-yl) pyridine; (*6*) 3-(N-carbonylpyrrol-2-yl) pyridine; (*7*) 3-(N-methylpyrrol-2-yl) pyridine; (*8*) quinoline and isoquinoline

more active than Fe_2O_3 nanoparticles in catalyzing the decomposition of nitrosamines, ion exchanged ferric zeolite shows a higher activity of degrading nitrosamines than its impregnated analogue in both TPSR test and smoke. Moreover, the sample with a higher activity to degrade nitrosamine in TPSR experiment will show the better performance in selectively removing TSNA in tobacco smoke. Ferric Y zeolite can selectively reduce 26 % of total TSNA in mainstream smoke of Burley type tobacco, and it still has a considerable activity to selectively remove TSNA in smoke (18.0 %) after stored for 2 months (Lin et al. 2012b). On the reduction of four TSNA components, NAB is quite less than others so that other three components, NNN, NNK and NAT, are focused. No preference is observed on NaZSM-5 and NaY, but ion-exchanged ferric zeolites have the orderliness of NNN>NNK>NAT. It should be pointed out that the function of zeolite is badly limited in tobacco smoke with complex chemical composition. In laboratory adsorption of volatile nitrosamine such as NPYR, both NaY and NaZSM-5 zeolites can capture almost all of the carcinogens when 0.2 mmol g^{-1} of NPYR passes at 593 K (Zhou et al. 2007), but they trap less than 2 μmol g^{-1} of TSNA in tobacco-extract aqueous solution (Wei et al. 2009a). Practically, tobacco smoke is the nightmare of catalyst because the smoke contains thousands of compounds (Meier and Siegmann 1999; Baker 1999) and the contact time of catalyst to targets is less than 0.1 s (Gao et al. 2009b). Moreover, most TSNA exist in the particulate matter of smoke and these particles

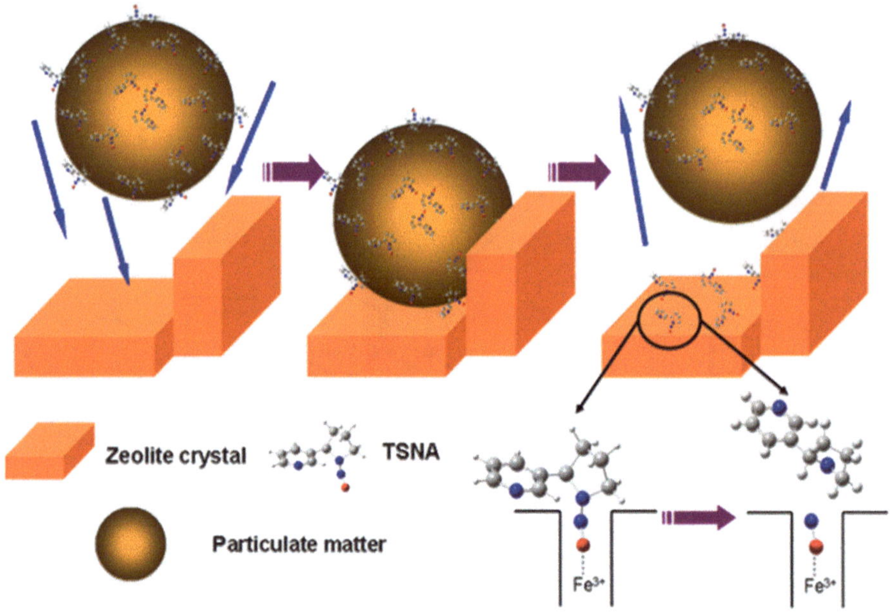

Fig. 9.28 The mechanism of action on removing TSNA from particulate matter in smoke by ferric zeolite (Lin et al. 2012b)

have average size of micrometer (Zhou et al. 2009) hence only the TSNA on the surface of these particles can be captured by zeolite (Fig. 9.28). These difficulties make the *in situ* removal of TSNA in tobacco smoke much more difficult than that in ordinary laboratory test. Ferric zeolites give the promising data that about quarter of TSNA can be *in situ* eliminated selectively in smoke. Further, the selective reduction of TSNA in tobacco smoke by zeolite follows the trend of catalytic ability in decomposing nitrosamine in TPSR test (Lin et al. 2012b). Thus, TPSR test can be used for preliminary screening the numerous cig-cat candidates in order to seek new efficient catalyst for eliminating TSNA pollutants.

9.4 Conclusion

Adsorption and catalytic degradation of nitrosamines by zeolites are two effective strategies for removal of the carcinogens in environment, which derives from the unique selectivity of zeolites. The electrostatic field inside channel and the pore size of zeolite are two determinative factors governing the selective adsorption of nitrosamines, the former drives the nitrosamine molecule inserting closely into zeolite channels, wholly or partly, while the latter provides the fine geometric constraints to imprison the carcinogen. Owing to these special factors, zeolite is able to not only selectively adsorb/catalyze the volatile nitrosamines whose molecular diameter is less than or equal to the pore mouth of zeolite, but also capture/degrade the bulky nitrosamines like TSNA whose cross section exceeds the channel size through special group-embedded mode; Besides, some zeolites can even distinguish the tiny differences between the function group of nitrosamines and nitrobenzene. Understanding of these new characteristics of zeolite will promote the design and development of new versatile materials, while introducing modifiers such as metal oxides can greatly improve the performance of zeolites, supplying many powerful candidates for the potential application of zeolite in environment protection and life science. Furthermore, the principle of selective adsorption is utilized to fabricate efficiency nitrosamines-trap on other porous materials such as mesoporous silica and activated carbon. Apart from the foundational study to analyze the mechanism of selective adsorption of nitrosamines from the molecule level, some pilot researches are performed to apply zeolite controlling the pollution cause by smoking.

Researches of selective adsorbing and degrading nitrosamines also promote the development of new technology to deal with environmental pollutants. The fact that N-NO bond of nitrosamines is easily breakable once they are adsorbed in zeolite not only reveals the new strategy of degrading nitrosamines through microwave irradiation in limited domain, but also implies an new concept of eliminating environment pollution: it is possible to transform or exploit the environment pollutants by "lop off" their harmful groups. Through the strategy of transforming and utilizing the large numbers of pollutants, people can protect their ecological environment more effectively.

The study on the selective elimination of nitrosamines by zeolites is in progress, in order to fabricate more effective materials for the environment protection such as controlling the smoking-caused pollution. Meanwhile, the continued efforts are focused on the development of new materials with hierarchical structures included but not limited to regular porous structures, in order to reduce the pressure drops in high-speed airflow and capture the pollutants with different molecular sizes. In addition, it will be necessary and attractive to establish the selective adsorption function on those non-molecular sieve materials such as polymers, fabricating specific active sites in them with suitable micro-environment through nanotechnology, and enduing them with new capability of selectively trapping target toxicant such as nitrosamines in complex system. These studies will deepen the comprehension and utilization of the superfine spaces inside zeolites, exploring the potential application of selective adsorption and catalysis.

Acknowledgments Financial support from NSF of China (20273031, 20373024, 20673053, 20873059, 21173117, and 21273106), 863 projects of Chinese Science Committee (2008AA06Z327), the Scientific Research Foundation of Graduate School and the Analysis Centre of Nanjing University is gratefully acknowledged.

Miss Yan Yan Li is a PhD student in the College of Chemistry and Chemical Engineering of Nanjing University (China). Her study is focused on the new porous adsorbents to trap CO_2 at high temperature. Tel: 86–18260086595; e-mail: Liyanyan151@yeah.net

Miss Mi Mi Wan is a PhD candidate in the College of Chemistry and Chemical Engineering of Nanjing University (China). Her research is focused on zeolites and their application in enzyme immobilization and drug delivery systems. Tel: 86–15850573185; e-mail: Mimiwanellen@gmail.com

Dr. Jian Hua Zhu is a professor of Nanjing University, member of international zeolite society, and the director of zeolite committee in Jiangsu Province (China). He studied in University of Auckland (New Zealand) from 1987 to 1988, and got his Ph.D. degree from Nanjing University in 1989. He studies the selective adsorption and catalysis of zeolite and mesoporous materials, and their application in environment protection and life science. Tel: 86-25-83595848; Fax: 86-25-83317761; e-mail: jhzhu@netra.nju.edu.cn

References

Ahn HJ, Yook HS, Rhee MS, Lee HC, Cho YJ, Byun MW (2002) Application of gamma irradiation on breakdown of hazardous volatile *N*-nitrosamines. J Food Sci 67(2):596–599. doi:10.1111/j.1365-2621.2002.tb10644.x

Ahn HJ, Kim JH, Jo C, Yook HS, Lee HJ, Byun MW (2003) *N*-nitrosamine reduction in salted and fermented anchovy sauce by ionizing irradiation. Food Control 14(8):553–557. doi:10.1016/S0956-7135(02)00132-9

Altkofer W, Braune S, Ellendt K, Kettl-Grömminger Steiner G (2005) Migration of nitrosamines from rubber products – are balloons and condoms harmful to the human health? Mol Nutr Food Res 49(3):235–238. doi:10.1002/mnfr.200400050

Baker RR (1999) Smoke chemistry. In: Layten D, Nielsen MT (eds) Tobacco: production, chemistry and technology. Blackwell Science, Oxford, pp 398–439

Branton P, Lu A, Schüth F (2009) The effect of carbon pore structure on the adsorption of cigarette smoke vapour phase compounds. Carbon 47(4):1005–1011. doi:10.1016/j.carbon.2008.12.003

Byun MW, Ahn H-J, Kim J-H, Lee J-W, Yook H-S, Han S-B (2004) Determination of volatile N-nitrosamines in irradiated fermented sausage by gas chromatography coupled to a thermal energy analyzer. J Chromatogr A 1054(1–2):403–407. doi:10.1016/j.chromato.2004.07.096

Cao Y, Yun ZY, Yang J, Dong X, Zhou CF, Zhuang TT, Yu Q, Liu HD, Zhu JH (2007a) Removal of carcinogens in environment: adsorption and degradation of *N'*-nitrosonornicotine (NNN) in zeolites. Microporous Mesoporous Mater 103(1–3):352–362. doi:10.1016/j.micromeso.2007.01.058

Cao Y, Zhuang TT, Yang J, Liu HD, Huang W, Zhu JH (2007b) Promoting zeolite NaY as efficient nitrosamines trap by cobalt oxide modification. J Phys Chem C 111(2):538–548. doi:10.1021/jp064978f

Cao Y, Wu ZY, Zhuang TT, Wang HJ, Liu HD, Zhou SL, Wang Y, Xue J, Zhu JH (2008) Eliminating carcinogenic pollutants in environment: improving zeolite NaY as capturer of nitrosamines through ferric oxide modification. Int J Chem Eng 1(1):1–12

Cheng JP, Xian M, Wang K, Zhu X, Yin Z, Wang PG (1998) Heterolytic and homolytic Y-NO bond energy scales of nitroso-containing compounds: chemical origin of NO release and NO capture. J Am Chem Soc 120(39):10266–10267. doi:10.1021/ja982086q

Colella C (2007) Natural zeolites and environment. In: Čejka J, van Bekkum H, Corma A, Schüth F (ed) Introduction to zeolite science and practice, 3rd rev edn, Studies in Surface Science and Catalysis 168. Elsevier BV, Amsterdam, pp 999–1036

Cvetkovic N, Adnadjevich B, Nikolic M (2002) Catalytic reduction of NO and NOx content in tobacco smoke. Beitr zur Tabakforsch Int 20(1):43–48

Daems I, Methivier A, Leflaive P, Fuchs AH, Baron GV, Denayer JFM (2005) Unexpected Si:Al effect on the binary mixtures liquid phase adsorption selectivities in faujasite zeolites. J Am Chem Soc 127(33):11600–11601. doi:10.1021/ja053375z

de Ridder DJ, Verberk JQJC, Heijman SGJ, Amy GL, Van Dijk JC (2012) Zeolites for nitrosamine and pharmaceutical removal from demineralised and surface water: mechanisms and efficacy. Sep Purif Technol 89:71–77. doi:10.1016/j.seppur.2012.01.025

Dean JA (1999) Lange's handbook of chemistry, 15th edn. McGraw-Hill Book Company, New York, p 4.43

Doll R, Peto R (1981a) The cause of cancer. Oxford Medical Publications/Oxford University Press, Oxford, pp 1226–1235

Doll R, Peto R (1981b) The cause of cancer: quantitative estimates of avoidable risks of cancer in the United States today. J Natl Cancer Inst 66(6):1191–1308

Dong X, Zhou CF, Yue MB, Zhang CZ, Huang W, Zhu JH (2007) New application of hierarchical zeolite in life science: fast trapping nitrosamines in artificial gastric juice by alkaline-tailored HZSM-5. Mater Lett 61(14–15):3154–3158. doi:10.1016/j.matlet. 2006.11.024

Filho PJS, Rios A, Valcárcel M, Zanin KD, Caramão EB (2003) Determination of nitrosamines in preserved sausages by solid-phase extraction-micellar electrokinetic chromatography. J Chromatogr A 985:503–512. PII: S0021-9673(02)01825-3

Gao L, Wang Y, Xu Y, Zhou SL, Zhuang TT, Wu ZY, Zhu JH (2008) New strategy to reduce the harm of smoking: reducing the *N*-nitrosamines level of mainstream smoke by zeolite NaA and in-vitro and in-vivo investigations. Clean Soil Air Water 36(3):270–278. doi:10.1002/clen.200700134

Gao L, Wang Y, Xu JH, Xu Y, Cao Y, Yun ZY, Hong C, Zhu JH (2009a) New attempt to reduce the harm of smoking: reducing the nitrosamines level in tobacco smoke by microwave irradiation. Clean Soil Air Water 37(1):31–38. doi:10.1002/clen.200800104

Gao L, Cao Y, Zhou SL, Zhuang TT, Wang Y, Zhu JH (2009b) Eliminating carcinogenic pollutants in environment: reducing the tobacco specific nitrosamines level of smoke by zeolite-like calcosilicate. J Hazard Mater 169:1034–1039. doi:10.1016/j.hazmat.2009.04.042

Garcia H, Roth HD (2002) Generation and reactions of organic radical cations in zeolites. Chem Rev 102(11):3947–4007. doi:10.1021/cr980026x

Gu FN, Zhuang TT, Cao Y, Zhou CF, Zhu JH (2008a) Effect of copper cation on the adsorption of nitrosamines in zeolite. Solid State Sci 10:1658–1665. doi:10.1016/j.solidstatesciences.2008.03.002

Gu FN, Cao Y, Wang Y, Zhu JH (2008b) Zeolite multifunctional materials as the menthol carrier and nitrosamines trapper. Stud Surf Sci Catal 174:565–568

Gu FN, Wei F, Yang JY, Wang Y, Zhu JH (2010) Fabrication of hierarchical channel wall in Al-MCM-41 mesoporous materials to enhance their adsorptive capability: why and how? J Phys Chem C 114:8431–8439. doi:10.1021/jp1009143

Guerin MR, Jenkins RA, Tomkins BA (1992) Chemistry of environmental tobacco smoke: composition and measurement. Lewis Publishers, London, pp 91–239

Harlick PJE, Tezel FH (2000) A novel solution method for interpreting binary adsorption isotherms from concentration pulse chromatography data. Adsorption 6(4):293–309. doi:10.1023/A:1026556831351

Hecht SS (1999) DNA adduct formation from tobacco-specific N-nitrosamines. Mutat Res 424:127–142. PII: S0027-5107(99)00014-7

Hiramoto K, Ohkawa T, Kikugawa K (2001) Release of nitric oxide together with carbon-centered radicals from N-nitrosamines by ultraviolet light irradiation. Free Radicle Res 35:803–813

Hoffmann D, Djordjevic MV, Hpoffmann I (1997) The changing cigarette. Preventive Medicine, Article No. PM970183, 26:427–434. 0091-7435/97

Hou Q, Lin WG, Wei F, Yang JY, Zhu JH (2013) Utilizing the acid immersion to elevate the performance of zeolite in liquid adsorption of *N*'-nitrosonornicotine (NNN). Solid State Sci 15(2):143–151. doi:10.1016/j.solidstatesciences.2011.11.001

International Organisation for Standardisation (1991) ISO 3308: routine analytical cigarette-smoking machine – definition and standard condition

Izquierdo-Pulido M, Barbour JF, Scanlan RA (1996) *N*-nitrosodimethylamine in Spanish beers. Food Chem Toxicol 34(3):297–299. 0278-6915(95)00116-6

Jo C, Ahn HJ, Son JH, Lee JW, Byun MW (2003) Packaging and irradiation effect on lipid oxidation, color, residual nitrite content, and nitrosamine formation in cooked pork sausage. Food Control 14:7–12. PII: S0956-7135(02)00045-2

Koide A, Fuwa K, Fuukawa F, Hirose M, Nishikawa A, Mori Y (1999) Effect of cigarette smoke on the mutagenic activation of environmental carcinogens by rodent liver. Mutagen Res 428:165–176. PII: S1383-5742(99)00044-7

Lee CK, Fulp C, Bombick BR, Doolittle DJ (1996) Inhibition of mutagenicity of *N*-nitrosamines by tobacco smoke and its constituents. Mutat Res 367:83–92. SSDI: 0165-1218(95)00078-X

Levallois P, Ayotte P, Van Maanen JMS, Desrosiers T, Gingras S, Dallinga JW, Vermeer ITM, Zee J, Poirier G (2000) Excretion of volatile nitrosamines in a rural population in relation to food and drinking water consumption. Food Chem Toxicol 38:1013–1019. PII: S0278-6915(00)00089-2

Lijinsky W (1999) *N*-nitroso compounds in the diet. Mutat Res 443:129–138. PII: S1383-5472(99)00015–0

Lin WG, Zhou Y, Cao Y, Zhou SL, Wan MM, Wang Y, Zhu JH (2012a) Applying heterogeneous catalysis to health care: *In situ* elimination of tobacco-specific nitrosamines (TSNA) in smoke by molecular sieves. Catal Today. doi: 10.1016/j.cattod.2012.07.045

Lin WG, Zhou Y, Gu FN, Zhou SL, Zhu JH (2012b) Catalytic degradation of tobacco-specific nitrosamines by ferric zeolite. Appl Catal B 129:301–308. doi:10.1016/j.apcatb.2012.09.039

Ma LL, Shen B, Zhu JH, Xia JR, Xu QH (2000) Modifying NaY zeolite with metal oxide by microwave irradiation: influence on the adsorption and decomposition of *N*-nitrosamines. Chin Chem Lett 11(7):649–652

Maesen T (2007) The zeolite scene-an overview. In: Čejka J, van Bekkum H, Corma A, Schüth F (ed) Introduction to zeolite science and practice, 3rd rev edn, Studies in Surface Science and Catalysis 168. Elsevier, Amsterdam, pp 1–12

Meier MW, Siegmann K (1999) Significant reduction of carcinogenic compounds in tobacco smoke by the use of zeolite catalysts. Microporous Mesoporous Mater 33(1–3):307–310. PII: S1387-1811(99)00162-6

Mervish SS (1995) Role of *N*-nitroso compounds (NOC) and N-nitrosation in etiology of gastric, oesophageal, nasopharyngeal and bladder cancer and contribution to cancer of known exposure to NOC. Cancer Lett 93(1):17–48. doi:10.1016/0304-3835(95)03786-V

Miura M, Sakamoto S, Yamaguichi K, Ohwada T (2000) Influence of structure on N-NO bond cleavage of aliphatic *N*-nitrosamines. Tetrahedron Lett 41:3637–3641. PII: S0040-4039(00)00431-7

Parkin DM, Pisani P, Lopez AD, Masuyer E (1994) At least one in seven cases of cancer is caused by smoking. Global estimate for 1985. Int J Cancer 59(4):494–504. doi:10.1002/ijc.2910590411

Pinisakul A, Kritayakornupong C, Ruangpornvisuti V (2008) Molecular modeling of nitrosamines adsorbed on H-ZSM-5 zeolite: an ONIOM study. J Mol Model 14:1035–1041. doi:10.1007/s00894-008-0346-4

Rabo JA, Angell CL, Kasai PH, Schomaker V (1966) Studies of cations in zeolites: adsorption of carbon monoxide; formation of Ni ions and Na_3-$^{4+}$ centers Na_4^{3+}. Discuss Faraday Soc 41:328–349

Ravishankar R, Bhattacharya D, Jacob NE, Sivasanker S (1995) Characterization and catalytic properties of zeolite MCM-22. Microporous Mater 4(1):83–93. doi:10.1016/0927-6513(94)00086-B

Reitmeier SJ, Gobin OC, Jentys A, Lercher JA (2008) Enhancement of sorption process in the zeolite HZSM-5 by postsynthetic surface modification. Angew Chem Int Ed 47:1–7. doi:10.1002/anie.200803869

Shang J, Li G, Singh R, Gu Q, Nairn KM, Bastow TJ, Medhekar N, Doherty CM, Hill AJ, Liu JZ, Webley PA (2012) Discriminative separation of gases by a "molecular trapdoor" mechanism in chabazite zeolites. J Am Chem Soc 134:19246–19253. doi:10.1021/ja309274y

Sharp JO, Wood TK, Alvarez-Cohen L (2005) Aerobic biodegradation of *N*-nitrosodimethylamine (NDMA) by axenic bacterial strains. Biotechnol Bioeng 89(5):608–618. doi:10.1002/bit.20405

Shen B, Ma LL, Zhu JH, Xu QH (2000) Decomposition of *N*-nitrosamines over zeolites. Chem Lett 4:380–381. doi:10.1246/cl.2000.380

Tao YF, Qiu Y, Fang SY, Liu ZY, Wang Y, Zhu JH (2010) Trapping the lead ion in multi-components aqueous solution by natural clinoptilolite. J Hazard Mater 180:282–288. doi:10.1016/j.hazmat.2010.04.028

Terry PD, Rohan TE (2002) Cigarette smoking and the risk of breast cancer in women: a review of the literature. Cancer Epidemiol Biomark Prev 11(10):953–971

van Bakkum H, Kouwenhoven HW (2007) Progress in the use of zeolites in organic synthesis. In: Čejka J, van Bekkum H, Corma A, Schüth F (ed) Introduction to zeolite science and practice, 3rd rev edn, Studies in Surface Science and Catalysis 168. Elsevier, Amsterdam, pp 947–998

Van Der Voort P, Ravikovitch PI, De Jong KP, Benjelloun M, Van Bavel E, Janssen AH, Neimark V, Weckhuysen BM, Vansant EF (2002) A new templated ordered structure with combined micro- and mesopores and internal silica nanocapsules. J Phys Chem B 106(23):5873–5877. doi:10.1021/jp025642i

Wang Y, Cao Y, Gao L, Zhuang TT, Zhu JH (2012) Evaluation of zeolite additive to reduce the nitrosamines level of cigarette smoke. Int J Chem Eng 5(1):9–22

Wei F, Gu FN, Zhou Y, Gao L, Yang J, Zhu JH (2009a) Modifying MCM-41 as an efficient nitrosamine trap in aqueous solution. Solid State Sci 11:402–410. doi:10.1016/j.solidstatesciences.2008.07.001

Wei F, Yang JY, Gao L, Gu FN, Zhu JH (2009b) Capturing nitrosamines in tobacco-extract solution by hydrophobic mesoporous silica. J Hazard Mater 172:1482–1490. doi:10.1016/j.hazmat.2009.08.016

Wei F, Yang JY, Hou Q, Zhu JH (2010) A new strategy of releasing nitric oxide from the moisture-saturated zeolite. New J Chem 34(12):2897–2905. doi:10.1039/c0nj00445f

Wei F, Hou Q, Yang JY, Zhu JH (2011) Multifunctional NO-delivery vessel derived from aminopropyl-modified mesoporous zeolites. J Colloid Interface Sci 356:526–535. doi:10.1016/j.jcis.2011.01.019

Weiner HC (1997) Oral tolerance: immune mechanisms and treatment of autoimmune diseases. Immunol Today 18(7):335–343. doi:10.1016/S0167-5699(97)01053-0

Weitkamp J, Hunger M (2007) Acid and base catalysis on zeolites. In: Čejka J, van Bekkum H, Corma A, Schüth F (ed) Introduction to zeolite science and practice, 3rd rev edn, Studies in Surface Science and Catalysis 168. Elsevier, Amsterdam, pp 787–836

Wu ZY, Wang HJ, Ma LL, Xu J, Zhu JH (2008) Eliminating carcinogens in environment: degradation of volatile nitrosamines by zeolites Y and ZSM-5. Microporous Mesoporous Mater 109:436–444. doi:10.1016/j.micromeso.2007.05.066

Wu YJ, Ren XQ, Wang J (2009) Facile synthesis and morphology control of zeolite MCM-22 via a two-step sol-gel route with tetraethyl orthosilicate as silica source. Mater Chem Phys 113 (2–3):773–779. doi:10.1016/j.matchemphys.2008.08.008

Xu Y, Zhu JH, Ma LL, Ji A, Wei YL, Shang XY (2003a) Removing nitrosamines from main-stream smoke of cigarette by zeolite. Microporous Mesoporous Mater 60:125–137. doi:10.1016/s1387-1811(03)00334-2

Xu Y, Yun ZY, Zhu JH, Xu JH, Liu HD, Wei YL, Hui KJ (2003b) Trapping volatile nitrosamines with the copper incorporated zeolites. Chem Commun 2003(15):1894–1895. doi:10.1039/b304322c

Xu Y, Liu HD, Zhu JH, Yun ZY, Xu JH, Wei YL (2004) Removal of volatile nitrosamines with copper modified zeolites. New J Chem 28(2):244–252. doi:10.1039/b307557e

Xu JH, Wen JJ, Wu ZY, Zhou CF, Zhu JH (2008) Microwave induced degradation of nitrosamines in zeolites. Asian Pac J Chem Eng 3:481–488. doi:10.1002/apj.176

Yang J, Ma LL, Shen B, Zhu JH (2007) Capturing nitrosamines in environment by zeolite. Mater Manuf Process 22(5–6):750–757. doi:10.1080/10426910701385382

Yang J, Zhou Y, Wang HJ, Zhuang TT, Cao Y, Yun ZY, Zhu JH (2008) Capturing nitrosamines by zeolite A: molecular recognizing in sub-nanometer space. J Phys Chem C 112(17):6740–6748. doi:10.1021/jp7119729

Yang J, Dong X, Zhou Y, Yue MB, Zhou CF, Wei F, Zhu JH (2009) Selective adsorption of zeolite towards nitrosamines in organic solution. Microporous Mesoporous Mater 120:381–388. doi:10.1016/j.micromeso.2008.12.006

Yang J, Zhou Y, Yang JY, Lin WG, Wu YJ, Lin N, Wang J, Zhu JH (2010a) Capturing nitrosamines by zeolite MCM-22: effect of zeolite structure and morphology on adsorption. J Phys Chem C 114:9588–9595. doi:10.1021/jp9116497

Yang J, Yang JY, Zhou Y, Wei F, Lin WG, Zhu JH (2010b) Hierarchical functionalized MCM-22 zeolite for trapping tobacco specific nitrosamines (TSNAs) in solution. J Hazard Mater 179:1031–1036. doi:10.1016/j.hazmat.2010.03.108

Yun ZY, Xu Y, Xu JH, Wu ZY, Wei YL, Zhou ZP, Zhu JH (2004) An *in situ* FTIR investigation on the adsorption of nitrosamines in zeolites. Microporous Mesoporous Mater 72(1–3):127–135. doi:10.1016/j.micromeso.2004.04.012

Zhou CF, Zhu JH (2005) Adsorption of nitrosamines in acidic solution by zeolites. Chemsophere 58:109–114. doi:10.1016/j.chemosphere.2004.08.056

Zhou CF, Yun ZY, Xu Y, Wang YM, Chen J, Zhu JH (2004) Adsorption and room temperature degradation of *N*-nitrosodiphenylamine on zeolites. New J Chem 28(7):807–814. doi:10.1039/b401004c

Zhou CF, Cao Y, Zhuang TT, Huang W, Zhu JH (2007) Capturing volatile nitrosamines in gas stream by zeolites: why and how? J Phys Chem C 111(11):4347–4357. doi:10.1021/jp067474q

Zhou Y, Gao L, Gu FN, Yang JY, Yang J, Wei F, Wang Y, Zhu JH (2009) Novel functional mesoporous material derived from 3D net-linked SBA-15. Chem Euro J 15:6748–6757. doi:10.1002/chem.200801606

Zhou Y, Gu FN, Gao L, Yang JY, Lin WG, Yang J, Wang Y, Zhu JH (2011a) 3D net-linked mesoporous silica monolith: new environmental adsorbent and catalyst. Catal Today 166:39–46. doi:10.1016/j.cattod.2010.04.152

Zhou Y, Yang J, Yang JY, Gu FN, Wang Y, Zhu JH (2011b) One-pot synthesis of novel ferric cubic mesoporous silica (*Im3m* Symmetry) and its high efficient performance of adsorption. J Mater Chem 21(36):13895–13901. doi:10.1039/c1jm11877c

Zhou Y, Lin WG, Yang J, Gao L, Lin N, Yang JY, Hou Q, Wang Y, Zhu JH (2011c) Controlling the primary particle evolution process towards silica monoliths with tunable hierarchical structure. J Colloid Interface Sci 364:594–604. doi:10.1016/j.jcis.2011.08.061

Zhou Y, Lin WG, Wan MM, Yang J, Zhu JH (2012) Novel selective adsorbent derived from hierarchical rockery-like MCM-41 monolith. J Mater Chem 22(44):23633–23641. doi:10.1039/c2jm35226e

Zhu JH, Yan D, Xia JR, Ma LL, Shen B (2001) Attempt to adsorb *N*-nitrosamines in solution by use of zeolites. Chemosphere 44(5):949–956. PII: S0045-6535(00)00560-9

Zhu JH, Zhou SL, Xu Y, Cao Y, Wei YL (2003) Ordered mesoporous materials. Novel catalyst for degradation of *N*'-nitrosonornicotine. Chem Lett 32(4):338–339. CL-021049

Zhuang TT, Xu JH, Xia JR, Cao Y, Zhou SL, Wang Y, Liu HD, Chun Y, Zhu JH (2006) New development in nanoporous composites: novel functional materials for capturing nitrosamines in airstreams. J Nanomater, Article ID 54909. doi:10.1155/JNM/2006/54909

Chapter 10
Dioxins and Furans: Sources, Impacts and Remediation

Ackmez Mudhoo, Ganeshan Thayalan, Nella Jane Muthoora, Manisha Nishi Muthoora, Bibi Zainab Oozeer, Yogeshwari Pooja Rago, Mukshada Priyadarshini Ramphul, Andini Keshni Valaydon, and Sunil Kumar

Contents

A. Mudhoo (✉) • G. Thayalan • N.J. Muthoora • M.N. Muthoora
B.Z. Oozeer • Y.P. Rago • M.P. Ramphul • A.K. Valaydon
Faculty of Engineering, Department of Chemical and Environmental Engineering,
University of Mauritius, Réduit, Mauritius
e-mail: ackmezchem@yahoo.co.uk

S. Kumar
Kolkata Zonal Laboratory, National Environmental Engineering Research Institute
(NEERI), Council of Scientific and Industrial Research (CSIR),
I-8, Sector "C", East Kolkata, New Township, EM Bypass, Kolkata,
West Bengal 700107, India

E. Lichtfouse et al. (eds.), *Pollutant Diseases, Remediation and Recycling*,
Environmental Chemistry for a Sustainable World 4, DOI 10.1007/978-3-319-02387-8_10,

Abstract Dioxins and furans are persistent, highly hazardous and ubiquitous compounds that have become the focus of many investigations due to the contamination of food and the ecosystem. Dioxins and furans refer to a class of polyhalogenated aromatic hydrocarbons generated by various chemical processes occurring often at high temperature. The main sources of dioxins and furans have been classified into four categories, namely: incineration, combustion, industrial processes and natural processes such as volcanic eruptions. Dioxins and furans are present in flue gas and fly ash from combustion, in contaminated soils, sediments and water bodies. The expression toxic equivalency (TEQ) factor has been defined to assess their toxicity. Dioxins and furans have adverse effects on human health because they are bioaccumulated in tissues and cells. They are carcinogenic and cause reproductive alterations. They are a threat to the ecosystem because they persist in nature, affecting marine and terrestrial living organisms, and subsequently ending up in the food chain. We review remediation techniques of dioxins and furans. The most common techniques are landfill cap system and deep well injections, high temperature incineration and cement kilns. This chapter focuses on the principal destruction, removal and remediation techniques based on flue gas, fly ash and soil treatment. Basic techniques are selective catalyst reduction technologies, adsorption by means of activated carbon, base catalysed reactions, photolytic degradation processes, thermal desorption processes, alkaline polyethylene glycolate (APEG) process, bioslurry treatment and vitrification.

Keywords Dioxins • Furans • Toxic equivalency factor • Removal techniques

10.1 Introduction

10.1.1 Dioxin and Furans

Dioxins and furans are perceived as a pressing issue of great concern that is wreaking havoc worldwide due to their ability to travel throughout the globe and gnaw at the environment as well as human health. Dioxins and furans refer to a class of structurally and chemically related polyhalogenated aromatic hydrocarbons that include polychlorinated dibenzo-*p*-dioxins (PCDDs), polychlorinated dibenzo-*p*-furans (PCDFs) and polychlorinated biphenyls (PCBs) (Kulkarni et al. 2008). They are viewed as persistent environmental compounds and their presence has been detected in all compartments of the ecosystem, including water (Rodriguez et al. 2008).

Kulkarni et al. (2008) stated that the presence of dioxins and furans was first reported in fly ash samples by Dutch and Swiss scientists in the year 1977 and 1978 respectively. Nevertheless, these persistent pollutant gained public attention 4 years later during the explosion at the ICMESA factory in Seveso, Italy, when these toxic compounds were spread and deposited over an area of 2.8 km^2 (Wilson 1982). As per Hirashi (2003), these chloroaromatic compounds have been identified to be toxic, carcinogenic and also consist of long half-lives. Furthermore, they are categorized among the most problematic environmental pollutants. Dioxins and furans are unwanted by-products of many chemical industrial processes and of all combustion processes. They occur as a mixture of congeners and so far, 210 possible congeners have been identified. Amongst these compounds, 17 congeners have chlorine atoms at least in the position 2, 3, 7 and 8 of the parent molecule which are persistent and lipophilic in nature. Consequently, dioxins and furans accumulate in fatty tissues of humans and animals as well as in carbon-rich matrices such as soils and sediments.

10.1.2 Chemical Structure and Types

Dioxins or "dioxin-like" compounds include the chemical class of polychlorinated dibenzo-*p*-dioxins (PCDDs), polychlorinated dibenzo-*p*-furans (PCDFs), polybrominated dibenzo-*p*-dioxins (PBDDs) or brominated dibenzo-*p*-dioxins(BDD), polybrominated dibenzofurans (PBDFs) or brominated dibenzofurans (BDFs) and PCBs (Patel et al. 2003). In fact, there are 75 different PCDDs, 135 different PCDFs and 209 PCBs (Patel et al. 2003). Only 7 of the 75 PCDD congeners, 10 of the 135 PCDF congeners and 12 of the 209 PCB congeners have dioxin-like toxicity (Kulkarni et al. 2008). These 29 different PCDDs, PCDFs and PCBs exhibit similar toxic effects as they are all bonded to a particular common molecule known as the aryl hydrocarbon or the "Ah" receptor (Patel et al. 2003). The 2, 3, 7, 8-tetrachlorodibenzo-*p*-dioxin (TCDD) is the most widely studied of the dioxin compounds as it has the greatest affinity for the Ah receptor. The two key factors in toxicity are namely: the number of chlorine in the molecule and the positions where the chlorines are attached; those species which have three or fewer chlorines lack dioxin-like toxicity whilst chlorine molecules at positions 2,3,7, and 8 are critical for toxicity (Moriguchi et al. 2003).

10.1.3 Chemical and Physical Properties

Dioxins have an affinity for particulates and readily partition to particles in air, water and soil. The more toxic compounds are the 2,3,7,8- substituted tetra-, penta- and hexachloro compounds.

The main physical properties of dioxins are summarised in Table 10.1.

Generally, the dioxin-like compounds have low solubility in water, low vapour pressure and high octanol-water partition coefficients. They are also resistant to chemical degradation under normal environmental conditions. The most toxic congener in mammals, 2,3,7,8-TCDD, has a photodegradation half-life of 1 year on surface soil and a half-life of 10 years in the body. The atmospheric lifetime of 2,3,7,8-TCDD using gas phase reaction with the hydroxyl radical (OH•) is 0.8–2 days (Bhawan 2007).

Table 10.1 Physical properties of dioxins (Kao 1999)

Physical property	Typical value
Molecular weight	322 g/mol for a tetra substituted dioxin
	460 g/mol for an octa substituted dioxin
Solubility in water	19.3 ng/L
	Dioxins have a very high solubility in organic solvents
Melting point	590 K
Vapour pressure at 298 K	7.4×10^{-10} mmHg
Partition coefficients for octanol/water:	1.4×10^{6}

10.2 Sources of Dioxins and Furans

The past studies have revealed that the dioxins and furans emissions in the atmosphere have mainly been of anthropogenic nature. The various sources of dioxins and furans emissions can broadly be classified as follows.

10.2.1 Incineration Processes

The incineration process can be considered as being one of the highest anthropogenic sources of dioxins and furans emission. It is a thermal waste reduction method which is currently being used in various countries. The incineration process is in fact closely related to the combustion reaction which occurs within the process. In an attempt to further investigate the sources of dioxins and furans, Kulkarni et al. (2008) categorised the various type of incineration processes based on the origin of materials used. Poor operating conditions were first found to be the major cause of dioxins and furans emissions from the incineration industries. However, McKay (2002) related the presence of dioxins and furans to the incomplete combustion of organic waste, the thermal degradation of the chlorinated-compounds and lastly, the formation of dioxins and furans precursors in the post incineration zone.

Abad et al. (2002) studied dioxin abatement strategies and mass balance at a municipal waste management plant. The authors reported that stack gas emission levels of PCDD/Fs around 3.26 ng I-TEQ/Nm^3 were determined with the use of only an electrostatic precipitator while the levels decreased below 0.1 ng I-TEQ/Nm^3 with the use of a new air pollution control system. Abad et al. (2002) also discussed that the presence of significant quantities of dioxins and dioxin precursors existed in municipal solid waste around 50 ng I-TEQ/kg (Abad et al. 2002). Mininni et al. (2007) investigated the emissions of dioxins and furans from a hospital and cemetery waste incinerator. Raw gases were sampled in the afterburning chamber and samples of slags from the combustion chamber and fly ashes from the bag filter were also collected and analysed. The authors reported PCDD/Fs emission factors for air spanned from 2.3 up to 44 μg I-TEQ/t of burned waste, whereas those through solid residues (mainly fly ashes) were in the range 41–3,700 μg I-TEQ/t.

10.2.2 Industrial Sources

Pulp and Paper Mills

Dioxins in paper mills are mainly formed during the production of bleached paper. The latter contains some naturally occurring phenolic compounds which undergo chlorination during the paper manufacturing process to form dioxins and furans.

Metallurgy Industry

The metallurgy industry is a large contributor to the formation of dioxins and furans in the atmosphere, through the production of metals. These pollutants are generally released in the different metals manufacturing processes, namely: the production of non-ferrous metals, alloys, metal casting and the iron and steel metals. Holtzer et al. (2007) investigated the possibilities of formation of dioxins and furans in metallurgical processes and reported PCDD/F emissions were around 0.5–6.5 mg I-TEQ/t of liquid steel in ore sintering processes, 0.001–0.004 mg I-TEQ/t of liquid steel in pig iron production in blast-furnaces and 0.001–0.006 mg I-TEQ/t of liquid steel in converter steel plant. Buekens et al. (2001) reviewed dioxins formation from the thermal and metallurgical processes and reported that the off-gases from iron ore sintering and metal smelting operations contain dioxins in the range 0.3–30 ng I-TEQ/Nm^3. Anderson and Fisher (2002) studied the sources of dioxins in the steel industry in the United Kingdom and stated that the annual release of dioxins is estimated at around 500–4,000 g I-TEQ.

Chemical Industry

The chemical synthesis of chloro-organic chemicals has for a long time been considered as being the major source of dioxins and furans production. It has been observed that dioxins and furans are released as by-products during the synthesis of different chloro-containing compounds, such as the manufacturing of phenoxy-herbicides, chlorinated aliphatic compounds, chlorinated phenols and chlorine bleached for wood pulp (McKay 2002). The use of the synthesised chloro-chemical containing compounds in different manufacturing industries is known to release a high level of dioxins and furans. It would appear that there is a direct correlation between the use of chloro-organic chemicals and the emissions of dioxins and furans. In an attempt to investigate the level of dioxins and furans emissions, Weber et al. (2008) revealed striking figures about the unprecedented release of dioxins and furans that resulted from the use of chloro-organics pesticides as well as the

Table 10.2 Sources of dioxins and furans incineration, metallurgical and chemical processes

Industrial process(es)	Region	Levels detected	References
Incomplete combustion of medical waste	Taiwan	In fly ash: 6.2 I-TEQ	Wang et al. (2012)
Incomplete combustion of medical waste	Italy	In exhaust gas: 0.29 ng I-TEQ/Nm^3 and 31.5 mg I-TEQ/kg in fly ash	Mininni et al. (2007)
Incomplete combustion process cemetery waste		In exhaust gas 0.54 ng I-TEQ/Nm^3 and 5.7 mg I-TEQ/kg in fly ash	
Incomplete combustion of municipal solid waste	USA	Less than 12 g TEQ per year since 2002	Psomopoulos et al. (2009)
Waste incineration	Portugal	14.6 g TEQ per year	Quina et al. (2011)
Incomplete coal combustion processes at high temperature in the steel and iron industry	Northern China	PCDD/Fs in summer 0.74 pg/m^3 and 1.97 pg/m^3 during winter	Li et al. (2011)
Iron sintering process	China	For the year 2007–2009 the level recorded are: 2,070 g, 2,212 g, and 2,307 g WHO-TEQ, respectively.	Tian et al. (2012)
Iron ore sintering plant	Poland	1.10–1.32 ng	Grochowalski et al. (2007)
Metal production ferrous and non ferrous	Portugal	15.5 g TEQ per year	Quina et al. (2011)
Pesticide manufacturing plant	China	964 pg I-TEQ/g	Zhu et al. (2008)
Wood pulp chemical manufacturing	China	39.5 pg I-TEQ/g	Zhu et al. (2008)
Manufacturing of pesticides	Australia	Concentration range detected 0.020–2,100 ng	Holt et al. (2010)
Heat and power generation	Portugal	20.1 g TEQ per year	Quina et al. (2011)
E-waste recycling	China	6,523 fg/m^3	Wong et al. (2007)

chlorine bleaches for wood treatment. Table 10.2 equally provides a summary of processes and quantities of dioxins and furans detected in a several regions.

10.2.3 Other Sources

There exists other sources of emissions of dioxins and furans in the atmosphere; these are at times referred as reservoirs. The reservoir source is often described as being the accumulation of dioxins and furans in soil, landfills and even vegetation and can eventually be redistributed in the atmosphere (Kulkarni et al. 2008). Basically, some of the reservoir sources are as follows.

Vegetation Sources

There have been several cases of incidental release of dioxins and furans, which has led to contamination of the food chain over the past decades. In 1998, Germany revealed relatively high levels of dioxins present in milk whilst Belgium, in 1999, was held responsible for the release high levels of dioxins and furans in poultry, bovine and pigs meat. The major source of contaminant in both cases was identified as the livestock fed to animals (Hays and Aylward 2003). This had led to massive destruction of poultry and other meat products. These historical evidences revealed the various sources of unknown exposure to high levels of dioxins in the food and hence threatened the food security. Reiner et al. (2006) reported that the highest level of dioxins and furans was recorded around the 1970s and could only be reduced in the 1990s with implementation of air pollution controls.

Accidental Sources

Some natural sources of emissions dioxins and furans are namely, forest fires and fumes emanating from volcanoes eruptions as well as natural combustion processes such as open burning of materials (Kulkarni et al. 2008; Reiner et al. 2006). However, it has to be pointed that the contribution to dioxins and furans emissions from these sources are considered as being relatively insignificant when compared to the other the anthropogenic sources (Reiner et al. 2006).

Biological and Photochemical Processes

McKay (2002) reported that dioxins and furans are also released in the atmosphere by the actions of microorganisms on chlorinated compounds under certain specified environmental conditions. Typical examples of these processes cited were mainly composting, sewage decomposition and photolysis of chlorinated compounds.

10.3 Health Effects and Environmental Impacts of Dioxins and Furans

10.3.1 Toxic Equivalency Factor (TEF)

As stated earlier, dioxins and furans, PCBs as well as other related halogenated aromatic compounds constitute a group of lipophilic, chemically stable environmental contaminants with low volatility which have been identified in fatty tissues of animals and humans. In fact, it has been demonstrated that dioxins and furans species exhibit several toxic responses on human health and are detrimental to the environment (Ahlborg et al. 1994). The number of chlorine molecules in the dioxins and furans species determines the level of toxicity; those with fewer chlorines lack dioxin-like toxicity as compared to species with higher chlorine content. The position where the chlorine molecules are attached is also an instrumental factor that affects toxicity of dioxins and furans. Chlorine at positions 2,3,7 and 8 are critical for toxicity (Patel et al. 2003).

As dioxins and furans normally prevail in the environment in the form of complex mixtures of congeners, the concept of toxic equivalents (TEQs) has been employed to simplify risk assessment and regulatory control in order to gauge the toxicity of such hazardous compounds. TCDD, being the most toxic congener, is rated as 1.0 whilst other less toxic congeners are expressed as a fraction of this reference value (Kulkarni et al. 2008). TCDD is classified as being a probable human carcinogen and reproductive toxin yet evaluated by the United States Environmental Protection Agency. The toxicity of dioxins and furans is mediated through the Ah receptor. The Toxic Equivalency Factor (TEF) which represents the ratio of the toxicity of dioxins and furans congeners to that of TCDD is used, assuming that the effects are additive and act via a common mechanism to exhibit toxicity (Boening 1998; Kerkvliet 2002; Kulkarni et al. 2008). In the 1990s, NATO/CCMS came up with the TEF system for dioxins and furans that was adopted internationally and termed as International – TEFs (I-TEFs), with the support of the United States Environmental Protection Agency. Various dioxins and furans species that are less potent than TCDD may be as toxic as TCDD depending on their respective concentrations. TEQ is generally obtained when the TEF is multiplied by the congener concentration level. The dioxins and furans are thus, converted to an equipotent concentration of TCDD; the product is commonly referred to as a "TCDD equivalent." Although coplanar – PCBs have less potency than dioxins and furans, their concentrations can often exceed those of TCDD such that their contribution to total TEQ is significant (Kang et al. 1997; Patterson et al. 1994; Kulkarni et al. 2008). So far, only 7 of the 75 different dioxins, 10 of the 135 furans and 12 of the 209 PCBs have been reported to exhibit dioxin-like toxicity and were rated TEFs. Recently, World Health Organization (WHO) experts have been consensus TEFs to facilitate human and wildlife risk assessments (Kulkarni et al. 2008).

10.3.2 Health Effects

Human health effects from exposure to dioxins and furans have been studied by scientists for about two decades. Since the rate of degradation of dioxins and furans is very slow, they persist in the environment for a long period of time. PCDDs, PCDFs, and PCBs are ubiquitous environmental contaminants, which persist and bio-accumulate through the food chain in high fat foods such as dairy products, eggs, animal fats, and some fish. Table 10.3 reports the levels of dioxins and furans species in human body fluids and tissues. In 1990, a WHO working group concluded that 90 % of the daily dioxin intake (from background contamination) results

Table 10.3 Levels of dioxins and furans species in human body fluids and tissues

Species	Living organism type	TEQ (pg/g)	References
2,3,7,8 TCDD	In the blood Vietnamese Agent Orange exposed	101	Schecter et al. (2001), Schecter et al. (2006), and Srogi (2008)
	In the fat of PCP-exposed person in USA	33	Ryan and Lizotte (1987), Schecter et al. (2006), and Srogi (2008)
	In the blood of incinerator workers in Japan	6.4	Schecter et al. (1999), Schecter et al. (2006), and Srogi (2008)
PCDD	In the blood Vietnamese Agent Orange exposed	7	Schecter et al. (2001), Schecter et al. (2006), and Srogi (2008)
	In the fat of PCP-exposed person in USA	374	Ryan and Lizotte (1987), Schecter et al. (2006), and Srogi (2008)
	In the blood of incinerator workers in Japan	96	Schecter et al. (1999), Schecter et al. (2006), and Srogi (2008)
PCDF	In the blood Vietnamese Agent Orange Exposed	3	Schecter et al. (2001), Schecter et al. (2006) and Srogi (2008)
	In the fat of PCP-exposed person in USA	202	Ryan and Lizotte (1987), Schecter et al. (2006), and Srogi (2008)
PCDD/PCDF	In human adipose tissue from Turkish men	9.2	Çok et al. (2007)
	In human adipose tissue from Finnish men	29	Kiviranta et al. (2005)
PCB	In human adipose tissue from Turkish men	6.6	Çok et al. (2007)
	In human adipose tissue from Finnish men	20.7	Kiviranta et al. (2005)

from ingestion. Other exposures also include industrial accidents (Baccarelli et al. 2002) and several miscellaneous exposures. Dr Linda S. Birnbaum (Director of the National Institute of Environmental Health Sciences of the National Institutes of Health (NIH), and National Toxicology Program; http://www.niehs.nih.gov/about/od/director/index.cfm), a leading science expert on dioxins, led the United States Environmental Protection Agency's dioxin assessment process, which concluded that no safe level of exposure for humans were detected for certain dioxins. The study on one particular dioxin, TCDD, concludes that it is a developmental toxicant that causes skeletal deformities, kidney defects and weakened immune responses in the offspring of animals exposed to it during pregnancy. Moirangthem et al. (2013) have found that exposure to environmental contaminants such as TCDD is proinflammatory in wounded tissues, disrupts normal healing and ultimately produces in a poorly healed wound. Chronic exposure to dioxin may result in various types of cancer and porphyria – a disorder in enzyme synthesis. People exposed to PCBs in the air for a long time have experienced irritation of the nose and lungs, and skin irritations, such as skin lesions. Other adverse effects associated with dioxins include bio chemical liver test abnormalities, soft tissue, sarcomas or elevated blood lipids (Mitrou et al. 2001). Industrial exposure to dioxin may lead to neuro-toxic effects involving polyneuropathy of lower extremities.

Many scientists believe that dioxin exposure is responsible for developing breast cancer and reduced sperm count in men. Organic chemicals like PCBs, PCDDs and PCDFs, are known as "Gender bender". Genetic, reproductive, and developmental effects have been observed in many animal studies although species differ dramatically in sensitivity to these chemicals (Cole et al. 2006; Huff et al. 1994). Dioxins and furans also concentrate in breast milk so that human infants receive a dose of greater magnitude that can be endured by an average adult. Such exposure to new-borns is of great concern because it occurs at their most vulnerable stage of development. In Guiyu, China, an area famous for its informal electronics recycling activities, the WHO estimates that the daily intake of dioxins and furans by breast fed infants exceeds guidelines by 11–25 times (Chan et al. 2007).

Kogevinas (2001) has made an excellent review of the human health effects of dioxins. Kogevinas (2001) has concluded that dioxins are proven to be carcinogenic in animals and humans. In humans, excess risks were observed for all cancers, without any specific cancer predominating. In specific cohorts, excess risks have been observed for reproductive cancers like breast female, endometrium, breast male, testes. However, but, overall, the pattern is inconsistent. In animals, endocrine, reproductive and developmental effects are among the most sensitive to dioxin exposure (Bruns-Weller et al. 2010). Decreased sperm counts in rats and endometriosis in rhesus monkeys occur at concentrations 10 times higher than current human exposure (Kogevinas 2001). Shen et al. (2009) conducted a study to analyze the speciation of polychlorinated dibenzo-dioxins/furans (PCDD/Fs) and polychlorinated biphenyl contamination of human adipose tissue from China. Shen et al. (2009) found that the total PCDD/Fs concentrations in human adipose tissue ranged from 33.9 to 504 pg/g lipid. Corresponding values for dioxin-like PCBs ranged from 4.1 to 125 ng/g. The mean total WHO toxicity equivalent (TEQ) values for PCDD/Fs and PCBs in human adipose tissue were 9.22 and 16.2 pg/g lipid, respectively. Everett and Thompson (2012) have found that pre-diabetes with glycohemoglobin (A1c) was associated with PCB 126, PCB 118 and having one or more compounds elevated. Total diabetes was associated with six of the eight individual dioxin based compounds tested, and was associated with having four or more compounds elevated. Everett and Thompson (2012) also reported toxic equivalency ≥81.58 TEQ fg/g was associated with total diabetes, but was not associated with A1c 5.9–6.4 %. Having multiple compounds elevated appeared to be important for total diabetes, whereas for pre-diabetes with A1c 5.9–6.4 %, having a single compound elevated appeared most important. Diabetes plus A1c ≥5.9 % was associated with 34.16–81.57 TEQ fg/g and with ≥81.58 TEQ fg/g, indicating that half the population has elevated risk for this combination of conditions. The existence of a peripheral neuropathy after exposure to PCDDs is still discussed, as studies concerning dioxin effects on the peripheral nervous system are rare and contradictory. Thömke et al. (1999) conducted clinical and neurophysiological examinations in 156 dioxin exposed workers from one pesticide producing plant. Thömke et al. (1999) found that workers having chloracne had a significantly higher exposure against PCDDs as documented by back calculated lipid levels. These workers complained significantly more often of sexual impotence (28.6 % compared to 5.8 % of workers without chloracne, $P < 0.001$), had significantly more frequent clinical signs of a sensory neuropathy restricted to the legs, had significantly more frequent neurophysiologic abnormalities, and had significantly lower mean amplitudes of the motor compound muscle potential of the peroneal nerve. Organochlorine chemicals may also contribute to an increased risk of non-Hodgkin lymphoma (NHL) within non-occupationally exposed populations. Among these chemicals, dioxins and furans were mainly released by municipal solid waste incinerators (MSWIs) until a recent past in France, a source of exposure that is of public concern. Viel et al. (2011) investigated organochlorines and the risk of NHL among neighbors of a French MSWI with high levels of dioxin emissions (Besançon, France), using serum concentrations to assess exposure. Evidence from Viel et al. (2011) indicated an increased NHL risk associated with cumulative WHO_{1998}-toxic equivalency factor concentrations (dioxins, 95 % CI = 1.03–1.26; furans, 95 % CI = 1.03–1.35; dioxin-like PCBs, 95 % CI = 1.00–1.07; and total TEQ, 95 % CI = 1.01–1.05), as well as with non dioxin-like PCBs (95 % CI = 1.01–1.05, per 10 ng/g lipid). The study of Viel et al. (2011) provided strong and consistent support for an association between serum cumulative WHO_{1998}-TEQ concentrations, at levels experienced by people residing in the vicinity of a polluting MSWI, and risk of NHL.

Food is contaminated by polychlorinated dibenzo-p-dioxins and furans, polychlorinated biphenyls and polybrominated diphenylethers worldwide. Previous data show elevated intakes in children (Karjalainen et al. 2012). Karjalainen et al. (2012) have recently determined intakes of persistent organic pollutants (POPs) in Finnish children. Karjalainen et al. (2012) calculated that the long-term upper-bound dioxin intakes ranged between 0.1 and 12.8 pg $WHO_{PCDD/F\text{-}PCB}$-TEQ/kg bw/d (min and max). An immediate tolerable daily intake (TDI) for $WHO_{PCDD/F\text{-}PCB}$-TEQs of 4.0 pg/kg bw/d was exceeded by 2.5–7.5 % of the children. Congener-specific analyses indicated a typical Finnish adult exposure pattern of the children to polychlorinated biphenyls and polybrominated diphenylethers (PBDEs). Long-term daily PCDD/Fs, PCBs and PBDEs intakes among Finnish children varied greatly between individuals and ages. Karjalainen et al. (2012) also found that in each age group of the study population, there was a proportion of children with their $WHO_{PCDD/F\text{-}PCB}$-TEQ intake exceeding considered safe limits set for adults. Based on the exposure profile reported herein, Karjalainen et al. (2012) advised that children should be clearly considered as a specific sub-population in food-mediated contaminant risk assessment. Fish consumption has been classified as one of the primary pathways of exposure to polychlorinated dibenzo-*p*-dioxins, dibenzofurans and biphenyls. Scott et al. (2009) have evaluated tissue levels of the 17 laterally substituted PCDD/Fs, 12 dioxin-like PCBs, and 97 non-dioxin-like PCBs in a number of wild-caught and farm-raised catfish collected throughout southern Mississippi. Scott et al. (2009) found that the total lipid-adjusted TEQ and non-dioxin-like PCB concentrations in wild-caught catfish fillets were significantly higher than concentrations in farm-raised fillet samples. The percent contribution of PCDDs, PCDFs, and dioxin-like PCBs to mean total TEQ varied between wild-caught and farm-raised samples as well as by collection site for wild-caught catfish. Regardless of whether samples were farm-raised or wild-caught, estimated cancer risks associated with consumption of these catfish were less than 27.0×10^{-6}. Overall, results of Scott et al. (2009) indicated that levels of dioxin-like compounds and PCBs in Mississippi catfish were similar to those measured in more recent studies in the United States and that levels of these compounds appeared to be decreasing in this food source.

10.3.3 Environmental Impacts

In the post Second World War period, persistent organic pollutants (POPs) were recognised as chemical pollutants, capable of persisting in the environment for a long time, migrating in air, water, soil and sediments. These when accumulated to high levels may harm human health and wildlife (Manahan 2000; El-Shahawi et al. 2010). These chemical pollutants possess a combination of chemical and physical properties such that, once released into the environment, they remain stable for remarkably long periods of time due to their ability to resist photolytic, chemical and biological degradation (El-Shahawi et al. 2010). Organic pollutants such as dioxin are a good illustration of a persistent chemical in soil and capable of strongly sorbed onto mineral or organic solid phases due to their high hydrophobicity property. The widespread distribution of POPs across the globe results from the transportation at low concentrations by movement of fresh and marine waters and due to their semi-volatility property, they are transported over long distances, even to regions where they have never been used (El-Shahawi et al. 2010). POPs travel towards colder areas and they tend to settle due to colder temperatures. Since they cannot be easily broken down by the temperature effect, they are apt to remain in the area and move from the air and water into the soil and plants, then to animals and humans (El-Shahawi et al. 2010). In a study conducted in Korea, Ok et al. (2002) found that the dominant homologues of PCDD/Fs in pine needles were the lower chlorine-substituted compounds such as tetra chlorinated dibenzo-p-dioxins and chlorinated dibenzofurans, and the distribution ratios of PCDD/Fs decreased with an increase of the number of chlorine substituents. Ok et al. (2002) equally observed that the homologue profiles of pine needle samples were similar to PCDD/Fs profiles of the vapor phase in the ambient air, and thus the pine needles absorbed the vapor phase of PCDD/Fs from air. The results from the study of Ok et al. (2002) suggested that pine needles could be used as an indicator of the atmospheric contamination for PCDD/Fs in the regions studied. Fries et al. (2002) detected concentrations of PCDD/Fs in wood as toxic equivalents ranged from 10 to 320,000 pg/g. These concentrations were closely related to the concentrations of pentachlorophenol (PCP), indicating that analysis for PCP could provide an economical method to identify wood with high concentrations of PCDD/Fs. PCDD/Fs are also present in abundance in the environment, found throughout all primary media, namely, air, soil, waters and sediments, and secondary media such as fish, and sea food, meat, milk, and dairy products (Mukerjee 1998; Schuhmacher et al. 2000). However, the bulk of the environmental burden of PCDD/Fs exists in soils, the primary supply of which is by atmospheric deposition. Studies have revealed that the PCDD/Fs in soils increase as a function of atmospheric deposition (Schuhmacher et al. 2000).

Pollution problems are often only belatedly recognised because their damage takes years to be noted. However, due to their widespread distribution in the environment and their carcinogenicity, various remediation techniques have been developed and employed in the removal of environmental pollutants. Environmental pollution arising from electronic waste (e-waste) disposal and recycling has received considerable attention in recent years. Treatment, at low temperatures, of e-wastes that contain polyvinylchloride and related polymers can release PCDD/Fs. Although several studies have reported trace metals and PBDEs released from e-waste recycling operations, environmental contamination and human exposure to PCDD/Fs from e-waste recycling operations are less well understood. In a study conducted in the region of Taizhou, Eastern China, Ma et al. (2008) found that concentrations of total PCDD/Fs were high in all of the matrices analyzed and ranged from 30.9 to 11,400 pg/g for shredder waste, 3,460–9,820 pg/g dry weight for leaves, 2,560–148,000 pg/g dry weight for workshop-floor dust, and 854–10,200 pg/g dry weight for soils. After analysis of surface soils from a chemical industrial complex (a coke-oven plant, a coal-fired power plant, and a chlor-alkali plant) in Shanghai, Ma et al. (2008) reported concentrations of total PCDD/Fs in surface soil (44.5–531 pg/g dry weight) from the chemical industrial complex were lower than the concentrations found in soils from e-waste recycling plants, but higher than the concentrations found in agricultural soils. Agricultural soils from six cities in China contained low levels of total PCDD/Fs and ranged 3.44–33.8 pg/g dry weight. Leung et al. (2007) have analyzed surface soils and combusted residue from a village located in southeast China, which has been intensely involved in the dismantling and "recycling" of computer parts (e-waste) for the past decade for PBDEs and PCDD/Fs. Total PBDE concentrations were highest in combusted residue of plastic chips and cables collected from a residential area (33,000–97,400 ng/g, dry weight), in soils from an acid leaching site (2,720–4,250 ng/g, dry weight), and a printer roller dump site (593–2,890 ng/g, dry weight). Leung et al. (2007) identified e-waste activities, acid leaching and open burning as emitters of the highest concentrations of PBDEs and PCDD/Fs. The study of Leung et al. (2007) was a first in its kind to show conclusively that the crude processing of e-waste has become one of the major contributors of PBDEs and PCDD/Fs to terrestrial environmental pollution.

Im et al. (2002) have reported total concentrations of PCDD/Fs and of 2,3,7,8-tetrachlorinated dibenzo-p-dioxin (TeCDD) equivalents (TEQs) in soils from Changwon and Masan Cities, Korea ranging from 35 to 121 400 pg/g (dry weight) and from 0.2 to 3,720 pg of I-TEQ/g, respectively. 61 % of soils needed measures such as investigations of source identification, soil decontamination, and/or soil removal. Im et al. (2002) equally reported that total concentrations of PCDD/Fs were greater at or near four industrial sites, which were concerned with the steel industry, petrochemical-related industry, and industrial waste incineration, than other areas. Soil collected from a site 50 m from an open-burning industrial waste incinerator in an industrial complex was heavily contaminated, containing a total concentration of PCDD/Fs of 121,400 pg/g (dry weight). PCDD/Fs were also detected in soils from the top of a 200 m mountain indicating a wide dispersal of PCDD/Fs by atmospheric transport from point source areas. Furthermore, Im et al. (2002) equally inferred from a wide range of PCDD/Fs isomers detected in soils from many other locations that multitude of sources such as incineration of industrial wastes (car tyres, scrap wires, plastics, papers, and emission of automobile exhaust) accounted for these species. Andersson and Ottesen (2008) have identified a number of sources for PCDD/Fs in the soil samples through congener profiles. Their estimations showed that the municipal solid waste incinerator (MSWI) and domestic wood burning are the largest PCDD/Fs pollution sources within their study area (Trondheim, Norway). Bakoglu et al. (2005) made PCDD/Fs analyses in soil and ambient air samples taken from Kocaeli, an industrialized area of Turkey. Their results showed that the levels of PCDD/Fs pollution were comparable to the levels observed in the various urban areas in the world. PCDD/Fs concentrations in surface soils ranged between 0.4 and 4.27 pg I-TEQ/kg while those in ambient air were between 23 and 563 fg/m^3. Bakoglu et al. (2005) later assessed combustion activities present in the area to be the most significant source of the PCDD/Fs pollution based on the congener and homologue profiles and statistical analysis of the data.

10.4 Removal of Dioxins and Furans

Due to the toxicity and chemical inertness of dioxins and furans and their subsequent impacts on health as well as the eco-system, their removal from polluted environments is considered to be one of the most challenging problems in environmental technology. In the past, several attempts have been made to contain and destroy these hazardous persistent compounds by storage, burial and/or burning in combustion systems. Nevertheless, further and intensive research and development have led to the advent of new and more effective technologies with higher destruction and removal efficiencies of dioxins and furans.

10.4.1 Conventional Removal Technologies

Some of the main conventional techniques that have been addressed in this chapter are: landfill cap system and deep well injections, high temperature incineration and cement kilns, respectively.

Landfill Cap System and Deep Well Injections

The most common alternative for dioxins and furans disposal is dredging a hazardous waste landfill with capping. Landfill capping is simply a multi complex barrier layer system made up of soils and geosynthetic materials which is used to contain the hazardous dioxins and furans compounds. This technique thus reduces the potential for redistribution and circulation of dioxins and furans compounds in the environment by minimizing the infiltration and generation of leachate. However, landfill capping is only considered as a containment method rather than a destruction technology. It is not totally effective as it allows certain constituents to escape either by leaching in ground water or by vaporising in the air. Compounds such as PCBs are known to escape from landfills by volatilizing into the surrounding air and to evaporate more rapidly with increased moisture in soils, sediments and even with an increase relative humidity of air (Chiarenzelli et al. 1998). Hence, this containment method is considered as being unsuitable when dealing with for high persistent compounds such as dioxins and furans.

Deep well injection is a liquid waste disposal technique that is used for dioxins and furans removal. The hazardous materials are simply forced deep into the ground between two impermeable layers far below fresh water aquifer. However, once these hazardous chemicals are injected into these layers, it is practically impossible to track their movement. Their traces are detectable only when they have already contaminated ground water. The long term chemical impacts of these injected substances are still unknown. They may react with underground rocks; clay, sand water brines, oil or gas and the subsequent effect of these reactions on the natural biota are still unknown. In fact, United Nation Food and Agriculture Organization (1996) designated deep well injection as "unsuitable because of the environmental risk and lack of control". It should also be noted that contaminants that are injected in the environment tend to return back to affect human beings. In addition, a study undergone by United States Environmental Protection Agency reveals there is certain earthquake hazard associated with such technology. For these reasons, the use of deep well injection technique is less appealing and is nowadays rarely used.

High Temperature Incineration

High temperature incineration is the most applied remedial technology for the treatment of dioxins and furans. It is a destructive ex situ method of treatment. The contaminated soil or dioxins and furans contaminated waste is fed into an incinerator, under controlled condition, at a temperature between 870 and 1,200 °C and in the presence of oxygen. The high temperature medium will enhance the breaking down of dioxins and furans which will thus combust into innocuous substances. According to research done by the United States Environmental Protection Agency, the destruction and removal efficiency of dioxin in treated waste, using such technology, may exceed 99.9999 % (United States Congress 1991). The United States Environmental Protection Agency has also approved high efficiency incinerators to destroy PCBs with concentrations above 50 ppm (Rahuman et al. 2000). Technical requirements like 2-s residence time at 1,200 °C and 3 % of excess oxygen, and 1.5-s residence time at 1,600 °C and 2 % of excess oxygen in the stack gases (Rahuman et al. 2000), should be met for the treatment of liquid dioxin. The most noteworthy incineration technologies, in relation to dioxin treatment are rotary kilns, liquid injections, fluidized bed/circulating fluidized bed and high-temperature fluid wall destruction (advanced electric reactor). Only incineration has been fully tested and approved for use at specific sites by the regulators. Extensive research is being made in the development of incineration technologies and there are nowadays new promising technologies such as infrared thermal destruction, plasma arc pyrolysis, supercritical water oxidation, and in situ vitrification.

Cement Kilns

Cement kilns have been used for destruction of hazardous wastes for many years. Such practice is most common in countries that lack space for landfill site, such as Japan, Norway and Switzerland. More recently, modern kilns have also been used in some developing countries that lack of waste disposal and incineration infrastructure for economic reasons. Cement kiln provides high temperature (about 1,450 °C), long residence time and a carefully facility capable of high destruction efficiency. The highly alkaline conditions, provided mainly by the limestone, are ideal for the decomposition of chlorinated organic waste. Thus, chlorinated liquids, chlorine and sulphur are neutralized in the form of chlorides and sulphates. There are no liquid or solid residuals formed since all the residuals are bound within the product. The contaminated waste can be fed into the cement kiln either in dry form or wet slurry form. In general, additional modification should be made to the design of cement kiln. The contaminated waste should be fed at the middle of the kiln by a properly designed hopper. It cannot be fed into the cold end of the kiln as contaminant will vaporise, resulting to low destruction efficiency. In case the waste is fed into the firing end of the kiln, an appropriate pre-treatment process, such as thermal desorption, is required, thus increasing the operating cost. The use of cement kilns for waste destruction may be less desirable than other approaches, such as recycling or reprocessing, but is a useful alternative to landfill capping. However, waste destruction in cement kilns must meet strict environmental, health and safety standards and must not impair the quality of final product. The process must be precisely controlled when destroying such wastes and emissions regularly measured. According to a study done at Norwegian cement kiln (a dry cement process), the destruction level of PCBs was found to be at least 99.999 % with no adverse effect on the quality of the exhaust gas (Benestad 1989). However, the contribution of these PCBs contaminated wastes to the exhaust gases can be said to be relatively minor only if the portion of the wastes used to the main energy or raw material stream is in minor quantity.

Karstensen (2008) studied the formation, release and control of dioxins in cement kilns and reported that PCDD/Fs emissions vary considerably among different kilns, ranging from 0.004 to 50 ng TEQ/m^3. However, the author also found out that among more than 2,000 PCDD/F cement kiln measurements evaluated in the study, the data indicated that modern cement kilns can meet an emission level of 0.1 ng

I-TEQ/m^3. Ames et al. (2012) investigated the profiles of PCDD/Fs congener in cement kiln emissions and reported that the emissions vary considerably depending on the design and operating conditions of the kiln and the fuels and raw materials fed into the kiln. Among a total of 152 stack-tests performed, the authors reported that the use of coal to fuel the kilns was found to generate significantly different emission-profiles relative to the use of petroleum coke but the addition of hazardous wastes, as a supplemental fuel, did not significantly alter profiles. Loo (2008) evaluated around 2,200 dioxin/furan stack emission measurements collected from various sources and reported that most cement kilns can meet an emission level of 0.1 ng TEQ/Nm^3 if primary measures such as process-integrated measures are applied. The author also reported that the reduction of the temperature at the inlet of the air pollution control system below 200 °C was found to limit dioxin formation and emission in all types of cement kilns.

10.5 Emerging Technologies

Research have revealed that the main modes of entry of dioxins and furans into the eco-system are from flue gas and fly ash originating from combustion and incineration processes and from contaminated soils, sediments or water bodies due to industrial and other sources. In this context, various state-of-the-art technologies have been studied for the destruction, removal, remediation and prevention of toxic dioxins and furans release based on the media through which they prevail in the environment. Table 10.4 provides a useful summary of some of the technologies and/or techniques which have been employed to destroy a variety of dioxins and furans species. More details of these destruction methods are provided downstream in this chapter.

10.5.1 Treatment of Flue Gas

Flue gas generated from incineration and combustion processes are major and potential sources of bulk dioxins and furans emissions in the environment. Some of the main techniques that have studied for the treatment of flue gas are, namely: the electron beam technology, selective catalytic destruction and carbon adsorption removal technologies.

Electron Beam Technology

Since the late 1980s, the electron beam technology for volatile organic compounds removal from flue gases has been investigated for industrial applications (Kim et al. 2003a). The technology of electron beam has the following features namely:

Table 10.4 Destruction of dioxins and furans using thermal desorption, plasma technology, supercritical water oxidation, and vitrification

Species	Destruction technique	Key features	References
PCBs and PCDD/Fs contaminated sediments	Thermal desorption	PCB congeners were removed by more than 99.9 and 32 % from sediments and effluent gas respectively. 98.4 and 90.3 % of PCDD/Fs removal efficiencies were attained mainly via vaporization and degradation from sediments and effluent gas respectively, while the high concentrations of PCDFs resulting from the process triggered an increase in the TEQ values	Zhao et al. (2012)
PCDD/Fs in soil bed (laboratory scale)	Thermal desorption based on a zone combustion process	98.6 % decomposition of OCDD, used as model homologous model was reached using stable combustion of coke in the soil bed. No conversion to other homologues of PCDDs was noted. The decomposition process was found to increase by pre-treating the soil (drying, pre-granulation and addition of limestone)	Harjanto et al. (2002)
PCBs contaminated sediments	Thermal desorption	48–70 % PCBs were degraded from the sediments, however, the formation of PCDFs from thermal desorption led to an increase in TEQ values. Volatilization of the PCDFs was found to decrease with increasing number of chlorine atoms	Sato et al. (2010)
Dioxins and furans contaminated feedstock	Plasma arc pyrolysis	Very high destructive efficiencies were achieved (six to eight times). Resulting emissions of dioxins and furans after plasma arc pyrolytic treatment were found to comply with emissions norms of 0.1 $ngNm^3$	Rae (2003)
PCDD/Fs in simulated flue gas (bench scale)	Non-thermal plasma (Gliding arc plasma)	Results obtained are as follows: 25–79 % destruction of PCDD/Fs; 64 and 62 % reduction of total concentration; TEQ values were decreased by 62 and 61 %	Yan et al. (2007)

(continued)

Table 10.4 (continued)

Species	Destruction technique	Key features	References
PCBs waste mixture from conventional incinerators	Plasma pyrolysis (Non-transferred DC steam (H_2O) plasma)	Conversion of a mixture of 27 % PCB and 73 % CCl_4 led to a total toxic concentration of around 0.056 ng TEQ/Nm^3 of PCDD/PCDF, indicating the efficiency of the steam plasma torch process as compared to conventional incineration processes in minimizing production of dioxins and furans as toxic by-products	Kim et al. (2003a, b)
Dioxins contained fly ash	Non-thermal nanosecond plasma	The extent of detoxification was found to increase as a function of increasing toxicity of the dioxin congeners. Around 20–80 % destruction efficiencies was achieved on an overall. The highly toxic 2,3,7,8-TCDD showed 81 % destruction	Zhou et al. (2003)
Gaseous dioxin-like compounds	Dielectric barrier discharge (DBD) plasma	20 % water vapour in the simulated gas stream resulted in around 74 % of PCDDs and 89 % of PCDFs conversion. The destruction process proceeded via UV, electrons and OH radicals generated by the DBC plasma process. All these destruction pathways significantly promoted destruction of PCDD/Fs	Hung et al. (2010)
PCBs	Supercritical water oxidation using hydrogen peroxide	With hydrogen peroxide as oxidant, the final concentration of 3-PCB in collected effluent was 0.3 ppb, attaining 99.9 % conversion of H_2O_2 while conversion of oxygen was found to be 14.3 %. Over 99 % 3-PCB was destroyed by SCWO in a flow-reactor system by more than stoichiometric demand of hydrogen peroxide	Hatakeda et al. (1999)
PCBs	Supercritical water oxidation in the presence of methanol	The extent of biphenyl disappearance ranged from 3 to 97 % under different SCWO conditions. The degradation process was found to follow second-order kinetics	Anitescu and Tavlarides(2005)

(continued)

Table 10.4 (continued)

Species	Destruction technique	Key features	References
Decachlorobiphenyl (10-CB)	Supercritical water oxidation with and without Na_2CO_3	Greater than 93 % dissolution of 10-CB was obtained in micro reactors in excess oxygen. Higher degradation was obtained in SCW- Na_2CO_3 treatment strategy (100 %) as compared to system without Na_2CO_3 where 99.2 % 10-CB was destroyed	Fang et al. (2004)
PCB-contaminated mineral transformer oil	Supercritical oxidation in excess oxygen	The organic content of the complex mixture of hydrocarbons and PCBs was decreased up to 99.6 %. PCB concentration after SCWO treatment was below the detection limit	Marulanda and Bolaños (2010)
PCBs, PCDD/Fs from solid-waste incinerator.	Plasma vitrification	High removal efficiency, greater than 99.9 % of dioxins and furans was noted. High immobilization of heavy metals was also achieved in the silicate matrix	Tu et al. (2010)
PCDD/Fs contaminated fly ash from municipal solid waste incinerators	Plasma vitrification	Complete destruction of PCDD/Fs was observed. Vitrification by thermal plasma furnace resulted in 100 % destruction efficiency in terms of toxic equivalent unlike vitrification carried out by conventional resistance furnace	Wang et al. (2010)

secondary pollution is completely eliminated as there is direct decomposition of dioxins and the temperature does not need to be controlled. Additionally, the process is simple and can be easily installed in existent incinerators (Kulkarni et al. 2008).

A preliminary investigation on the influence of electron beam irradiation on polycyclic aromatic hydrocarbons (PAH) as organic pollutants has been carried out in Poland. The results indicated a remarkable reduction of PAH concentration under electron beam influence (Chmielewski et al. 2002). The decrease of PAH concentration in the flue gas can be explained by PAH destruction in the oxidation processes, adsorption on the surface of by-products and new compound formation in radical process (Ostapczuk et al. 2008).

During the electron beam treatment, fast electrons penetrate through a material while continuously transferring their energy to it. Upon gaining energy the molecules in the material become excited and ionized. Free radicals like the OH• are formed (Kikuchi and Pelovski 2008). However, in the electron beam process, other reactions are also possible, like the charge transfer reactions where M can be N^+, N_2^+, O^+, O_2^+ or NO^+ formed in primary ionization of the flue gas molecules by the electron beam:

$$\mathrm{PAH} + \mathrm{M}^+ \rightarrow \mathrm{PAH}^+ + \mathrm{M}$$

The PAH molecules show a strong tendency to adsorb on the dust surface and with the addition of ammonia in the electron beam flue gas treatment, powdered fertilizer is formed. In the irradiation chamber, the precipitation process is initiated. The PAH concentration in the gas phase is decreased by being adsorbed on the surface of aerosols there (Ostapczuk et al. 2008). Consequently, the reduction of PAH concentration is caused mainly by the oxidation process initialised by irradiation and the decomposition process becomes much more rapid and effective due to the higher concentration of the active species (Ostapczuk et al. 2008).

Selective Catalytic Reduction

Liljelind et al. (2001) showed that selective catalytic reduction catalysts are effective in the decomposition of PCDD/Fs. The catalysts used are vanadium and tungsten oxides on titanium or platinum based substrate to reduce NO_x emissions via selective catalytic reduction (Yang et al. 2008). The V_2O_5/TiO_2 catalysts offer high total oxidation activity and excellent resistance against chlorine poisoning (Debecker et al. 2011). When the selective catalytic reduction catalysts are applied, NO_x and PCDD/Fs are removed simultaneously. This occurs by increasing the oxidation potential of the catalysts by increasing the vanadium content (Liljelind et al. 2001). Ide et al. (1996) showed that the TiO_2-based V_2O_5-WO_3 catalysts are applied to control NO_x, via the selective catalytic reduction of NO with NH_3 injection and they are also effective in the decomposition of PCDD/Fs (Chang et al. 2007). Goemans et al. (2004) proposed the following reactions for the oxidation of PCDD/Fs over a $TiO_2/V_2O_5/WO_3$ catalyst:

$$\begin{aligned} \mathrm{Cl_2H}_n\mathrm{Cl}_{8-n}\mathrm{O}_2 + (9+0.5_n)\mathrm{O}_2 &\rightarrow (\mathrm{TiO_2/V_2O_5/WO_3}) \\ &\rightarrow (n-4)\mathrm{H_2O} + 12\mathrm{CO_2} + (8-n)\mathrm{HCl} \\ \mathrm{Cl_2H}_n\mathrm{Cl}_{8-n}\mathrm{O}_2 + (9.5+0.5_n)\mathrm{O}_2 &\rightarrow (\mathrm{TiO_2/V_2O_5/WO_3}) \\ &\rightarrow (n-4)\mathrm{H_2O} + 12\mathrm{CO_2} + (8-n)\mathrm{HCl} \end{aligned}$$

Kim et al. (2008) found that dechlorination of PCDD/Fs by zerovalent iron (ZVI) is thermodynamically feasible, but useful rates of reaction have not been previously reported. In their work, Kim et al. (2008) have showed that ZVI (both micro- and nanosized ZVI, without palladization) could dechlorinate PCDD congeners with four or more chlorines in aqueous systems, but the reaction was too slow to achieve complete dechlorination within a practical period of time. In contrast, palladized nanosized ZVI (Pd/nFe) rapidly dechlorinated PCDDs, including the mono- to tetrachlorinated congeners. The rate of 1,2,3,4-tetrachloro dibenzo-p-dioxin (1,2,3,4-TeCDD) degradation using Pd/nFe was about three orders of magnitude faster than 1,2,3,4-TeCDD degradation using unpalladized ZVI. Cobo et al. (2009) examined the reducing power of 2-Propanol and molecular H_2 (in methanol (MeOH) and MeOH-water) for the liquid phase hydrodechlorination of dioxins over 2 wt% Pd/γ-Al_2O_3. Different amounts of NaOH were added to the reaction mixtures. The 2-propanol and H_2(g)/MeOH systems presented similar hydrodechlorination activity. Notwithstanding, Pd sintering and graphitic carbon directly bonded to Pd on catalyst surface was observed on samples used with H_2(g)/MeOH. Cobo et al. (2009) found that the addition of water to H_2(g)/MeOH decreased Pd sintering and favored dissolution of sodium compounds. However, dioxin degradation efficiency diminished. By contrast, 2-propanol acting both as reducing agent and solvent provided hydrogen to the hydrodechlorination reaction, avoided metal sintering and Pd-C formation. Besides, almost complete dioxin degradation under mild reaction conditions was obtained.

In pilot plants, a temperature lower than 210 °C might be sufficient for the effective catalytic destruction of the PCDD/Fs but in field tests, higher temperatures of 240–260 °C are generally required (Ide et al. 1996; Liljelind et al. 2001). It is indicated by the study of Weber et al. (2001) that the catalytic decomposition and removal efficiencies of semi-volatile compounds like the PCDD/Fs depend on the volatility and oxidative behaviour of the compounds. These two properties are closely related to the chlorination level of the compound. As the chlorination level increases, the volatility of the PCDD/Fs decreases (Yang et al. 2008). It is indicated by Finocchio et al. (2006) that the adsorption and absorption techniques of activated carbon applied to industrial flue gas streams only transfer the PCDD/Fs from the vapour phases to solid or liquid phases. However, the application of catalytic systems leads to the real destruction of the PCDD/Fs with the formation of carbon oxides and HCl. Hence, the technology of selective catalytic reduction is a promising alternative for the end-of-pipe treatment (Chang et al. 2007).

Carbon Adsorption Removal Technology

The carbon adsorption technology is one of the most common dioxins and furans removal technology current used in the treatment of flue gases emanating from incineration processes. Basically, the adsorption process can be described as being the accumulation of molecules on free surface area of provided by adsorbents of carbon sources (Everaert and Baeyens 2004). In an attempt to improve the performance of this particular treatment, several studies ranging from the variation of operating parameters to be modifications of the adsorption process on a pilot plant scale. This experiment has led to three mains types of carbon adsorption technologies over the years. These are classified as follows: the fixed bed adsorption, the moving bed adsorption and the entrained bed adsorption (Everaert and Baeyens 2004). All these three techniques have the same adsorption mechanism yet the difference between them lies in the type of adsorbent bed (Table 10.5). Different types of adsorbents are readily available on the market; typical cited examples are lignite, activated carbon and coke (Everaert et al. 2002). The most recurrent absorbent which is used in flue gas treatment is

Table 10.5 Removal of dioxins and furans species using fixed bed, moving bed and entrained bed activated carbon adsorption

Species	Adsorption method	Main findings	Merits and limitations	References
PCDD/F in flue gas	Fixed bed activated carbon filter	The total PCDD/F concentrations in the flue gas were reduced to 258–934 pg/Nm^3 (6–37 pg TEQ/Nm^3) after the AC unit, corresponding to dioxin removal rates of 40–69 % (57 % on average) on total PCDD/F and of 42–84 % (58 % on average) on the TEQ basis	Sufficient adsorbent is provided for bed operations as long as several years before the bed become saturated. The fixed bed process causes more pressure drop than the carbon injection process	Karademir et al. (2004)
PCDD/F in untreated and purified flue gas respectively	Entrained-flow carbon adsorption: Wet scrubber, bag filter with Activated Carbon injection	A removal efficiency of 96.8 % at AC dosage of 50 mg/Nm^3 was achieved	Easy handling, effectiveness and low capital cost Can result in high carbon consumption because of the low (less than 3 %) utilization efficiency	Yan et al. (2006) and Kim and Kim (2004)
PCDD/F in raw gas and clean gas respectively	Fixed-bed adsorption using MEDISORBON/ KOMBISORBON adsorber: a mixture of zeolite, carbon and inert material	PCDD/Fs in (a) raw gas: 0.3 ng TEQ/Nm^3; (b) clean gas: 0.05 ng TEQ/Nm^3. Removal efficiency of 83 % was achieved	The whole bed is replaced after a few years and can hence be used effectively	Yan et al. (2006)

(continued)

Table 10.5 (continued)

Species	Adsorption method	Main findings	Merits and limitations	References
PCDD/F	Moving bed counter-current activated coke process	AC consumption: 500 t per year per plant, PCDD/F stack emission: 0.015 ng TEQ/Nm^3; AC consumption: 5 mm of bed height per day; Removal efficiency as high as 98.8 %	Very high dioxin removal efficiencies exceeding 99 % can be achieved with the moving-bed process They do not require a large pressure drop and can be retrofitted into many sites The used carbon can be injected back into the furnace and burned, eliminating a disposal problem and costs The massive volume of carbon present in the filter is a potential fire hazard	Yan et al. (2006) and Karademir et al. (2004)
PCDD/F in flue gas	Entrained-flow carbon adsorption: baghouse with activated injection	PAC injection provides PCDD/PCDF removal efficiencies of 77–80 % at ESP equipped facilities	High removal efficiency in the order of 80–90 % is achieved High operating costs, high consumption of activated carbon as well as the inevitable disposal problem of the residues	Chandler AJ and Associates LTD (2007)
PCDD/F in flue gas	Fixed bed carbon adsorption	Removal efficiency of PCDD/F is over 99 %	Fixed bed processes are blocked due to moisture absorption and corrosion	Kulkarni et al. (2008)
PCDD in untreated flue gas	Activated carbon in fixed bed	The dioxin level of the flue gases was reduced to between 0.098 and 0.054 ng TEQ Nm^{-3} which was below the emission limits of 0.1 ng TEQ Nm^{-3}	Increased reduction of dioxins and heavy metals using a homogenous powder produced by mixing lime with activated carbon or lignite coke as an adsorbent	Parizek et al. (2008)

2,378-substituted PCDD/F	Removal of PCDD/F by activated carbons derived from the pyrolysis of refuse derived fuel, textile waste and scrap tyre using the fixed-bed reactor under a simulated flue gas at 275 °C with a reaction period of 4 days	The total PCDD/F toxic equivalent removal efficiencies in the exhaust gas stream were 58 %, 57 %, 64 % and 52 %, respectively by each type of activated carbon The removal of the PCDDs was much higher with an average of 85 % compared to PCDFs at 41 %	The activated carbons derived from the pyrolysis of waste, represented a good potential as a control material for PCDD/F emissions in waste incinerator flue gases However, the formation of substantial amounts of PCDD/F caused by the activated carbons investigated, increase their toxicity Further advance treatments are required prior to the disposal of the used activated carbons	Hajizadeh et al. (2011)
PCDD/F in flue gas	Multi-layer adsorption technology	The PCDD/F removal efficiencies based on International Toxic Equivalence (I-TEQ) basis were all higher than 99 % as the gas stream passed through the first layer of adsorption bed	Can attain very high removal efficiencies Removal efficiency is reduced when the moisture content in the gas stream is high. Moisture affects the adsorption efficiency The presence of impurities in the gas stream is also likely to lower the removal efficiency	Hung et al. (2011)

activated carbon (Table 10.5). Research (as summarized for some representative studies in Table 10.5) demonstrates that the use of activated carbon as adsorbents offered a better adsorption ability when compared to the other types of adsorbents (Inoue and Kawamoto 2005). Likewise, the activated carbon's popularity is mainly due to its high removal efficiency reaching almost 99 % and its relatively low capital cost. Everaert et al. (2002) claimed that the size of the adsorbents is one of the factors which greatly affect the adsorption ability. Porous adsorbent are generally used for this process and the adsorption ability is generally governed by the available surface area of the pores.

Fixed Bed Adsorption

The fixed bed adsorption, often characterised as a complex technology, consists of passing the flue gas through a fixed layer of adsorbent for the adsorption of the dioxins and furans to occur. In this particular process, the absorbents are used on a continuous basis and are only disposed of when the saturation point of the absorbents is reached (Everaert and Baeyens 2004). This method of adsorption was previously used in the past for the removal of dioxins and furans in stack gas; it is however, currently being used to a lesser extent. Its unpopularity was mainly due to its adsorption problems encountered. The removal efficiency of the fixed bed adsorption process is known to reach about 83 % (McKay 2002), However past studies revealed that the actual removal efficiency of dioxins and furans using this technology could only reached on average 60 %. Kulkarni et al. (2008) and McKay (2002) accounted for this drop in removal efficiency by the moisture content present. These gases react with the activated carbon and hence absorb the moisture present in the flue gas. Moisture is also known to cause some blocking problems over the fixed adsorption bed. It was also reported that, pressure drop is likely to be encountered in across fixed bed adsorption resulting from the use of deep adsorbents beds. However, this pressure drop effect can be reduced by using granular or pelletized absorbents having particle sizes ranging from 1 to 4 mm (Everaert and Baeyens 2004).

A multilayer bean shaped activated carbon adsorption process was recently investigated by Pao et al. (2010a). This innovative technique was proposed in the attempt to reduce the dioxins and furans emissions while monitoring the amount of contaminants present in the ash. In this particular process, the solid phase present in the flue gas was first removed using fibre glass and then passed through multilayers of bean shaped activated carbon. The overall efficiency of the adsorption process was found to be greater than 99 %.

Moving Bed Adsorption

This type of adsorption process became a popular means of air pollution control device on the market once the limitations of the fixed bed adsorption process were identified (McKay 2002). The moving bed adsorption consists ensuring turbulent contact either in the cross flow or in the counter current direction of the system between the adsorbents and the flue gas. This method is generally considered as being a batch wise process, in which the absorbents are replaced after a given adsorption time. An effective adsorption of contaminants has been observed using this method (Everaert and Baeyens 2004). The average removal efficiency that could be achieved using a moving bed adsorption process was about 98.8 %. However, the use of a moving bed adsorption process could lead to the formation of fine sorbent particles and eventually increase the amount of dioxins in the stacks. McKay (2002) mentioned that particular emphasis should be laid on this technique to avoid the formation of such fine sorbent particles.

Entrained Bed Adsorption

The entrained bed adsorption technology is the most common type of treatment that is currently being used in municipal waste incineration for gas polishing and is often described as being a relatively simple treatment process when compared to others (Everaert and Baeyens 2004). This technology consists of uniformly injecting activated carbon into the flue gas prior to a filtration step. The solid phase impurities present in the flue gas are retained in the filter bed, while the gas phase contaminants are then adsorbed on the surface of the activated carbon and hence changed to a solid phase and are recovered in the bag filter in the form of cake. The partitioning of the gas and solid phase in a flue gas stream is of prime importance when dealing with the entrained flow adsorption process. A change in the gas to solid phase ratio is likely to alter the removal efficiency of the adsorption process.

Various studies revealed that a relatively high removal efficiency of dioxins and furans of about 96–98 % can generally be achieved with the entrained flow activated carbon adsorption (McKay 2002; Pao et al. 2010a). Authors also claimed that improvements in the Entrained bed adsorption process could be made to achieve much higher removal efficiency. The injection of a mixture of AC and mixed lime under controlled amounts could in fact lead to removal efficiency greater than 99 % (Aristizábal et al. 2008). Kim et al. (2006) proposed an innovative use of activated carbon adsorption for dioxins and furans removal using dual filter bags on a pilot scale. This technique consist of passing the flue gas through a first filter bag to remove the fly ash present and injecting the activated carbon at the entrance of the second filter bag. This type of filter bag is known to have a much high efficiency of dioxins and furans removal. Although, the activated carbon injection is considered as a proven technology and can achieve high relatively high removal efficiency, its use has recently been questioned. Concerns were raised about the possibility of having high level of dioxins and furans in the fly ash as well as flue gas. Pao et al. (2006) demonstrated that the entrained flow activated carbon only adsorbs the dioxins and furans present in the vapour phase which are later collected in solid phase. This only shifts the dioxins and furans emission problems from the vapour phase to the solid phase.

It can be noted that the entrained flow adsorption remains one of the most preferred method on the market. This can be explained by the fact that the entrained flow has a process time which is 10 times smaller than that of the other two carbon adsorption techniques (Kulkarni et al. 2008). It also has to be pointed out that the use of the moving and fixed bed adsorption have been restricted for the small and medium gas treatment scale mainly because of recurring clogging problems encountered (Everaert and Baeyens 2004). It remains a fact that, the entrained flow adsorption technique not only owes its popularity to its rather simplistic design but also to its highest adsorption ability when compared to the two other techniques. This high adsorption capacity can be explained by the size of pulverised particles used. The smaller the particle size the higher is the internal porosity and hence the higher is the adsorption of dioxins and furans. The entrained flow uses particle size of 0.1 mm whilst the two other of 4 mm (Everaert and Baeyens 2004).

10.5.2 Treatment of Fly Ash

Fly ash refers to the thick solid residues that are produced as a result of incineration processes of hazardous wastes, municipal solids, medical wastes and sewage sludge amongst others. They contain high levels of dioxins and furans that are detrimental to the eco-system as they bio-accumulate in the food web. Kulkarni et al. (2008) reported that fly ash are classified as a potential hazardous material as per the environment protection legislation of many countries. The main methods for the destruction of dioxins and furans in fly ash are as follows: plasma technology, supercritical water oxidation, mechanochemical destruction technologies and UV irradiation (photolytic destruction).

Plasma Technology

Plasma is considered to be the fourth state of matter, consisting of a mixture of electrons, ions and neutral particles, although overall it is electrically neutral. The degree of ionisation of a plasma is the proportion of atoms that have lost (or gained) electrons and, in the case of thermal plasmas of interest for this review, this is controlled mostly by temperature (Gomez et al. 2009). Plasma technology involves the creation of a sustained electrical arc by the passage of electric current through a gas in a process referred to as electrical breakdown. Because of the electrical resistivity across the system, significant heat is generated, which strips away electrons from the gas molecules resulting in an ionised gas stream, or plasma (Gomez et al. 2009). At 2,000 °C gas molecules dissociate into the atomic state and when the temperature is raised to 3,000 °C, gas molecules lose electrons and become ionised. In this state, gas has a liquid-like viscosity at atmospheric pressure and the free electric charges confer relatively high electrical conductivities that can approach those of metals (Gomez et al. 2009). Plasma technology is considered as being a promising emerging technology in the treatment of dioxins and furans. Yan et al. (2007) described the plasma technology as being promising technique for the destruction of the dioxins and furans. Plasma technology with regards to waste and dioxins and furans reduction can be characterised into two main categories which are thermal plasma and non-thermal plasma.

Thermal Plasma

Thermal plasma can be described as the use of heat to melt fly ash produced from waste reduction into usable products such as slag (Yan et al. 2007). However this type of plasma technology is known to be energy intensive and is often characterised as a complex one. Typical cited examples of thermal plasma technology are the plasma pyrolysis, plasma gasification and plasma vitrification.

Plasma Pyrolysis/Plasma Gasification

Vyas et al. (2011) described both plasma gasification and plasma pyrolysis as processes that involving the thermal reduction of carbonaceous solids to simpler molecules such as methane, carbon dioxins and hydrogen. The only difference arising between these two methods is that plasma pyrolysis takes place in an oxygen starved environment whilst, the gasification occurs in a very limited amount of air to prevent combustion to occur. These methods are often described as a state of art technology used for the safe disposal of medical waste as well as plastic waste (Vyas et al. 2011). A simple comparison between the various dioxins and furans treatments methods revealed that plasma pyrolysis has higher removal efficiency. It also offers a higher solid reduction together with the production of syn gases which can be subsequently reused (Vyas et al. 2011).

Plasma Vitrification

Plasma vitrification is a good illustration of a thermal treatment method. In thermal plasma vitrification, the heat generated from plasma is used to process hazardous wastes containing metals, inorganic oxides and/or organics at temperatures above 1,500 °C (Cheng et al. 2002). Metal bearing wastes are melted and organic contaminants are thermally destroyed. The plasma vitrification yields a glass-like, leach-resistant monolith slag, which is environmentally safe for landfill disposal and/or can be reused as glass-ceramic for construction materials, such as interior and exterior wall cladding or ordinary floor tile applications (Cheng et al. 2002). The volume of the vitrified product is typically 20–45 % less than the original volume of the waste (Bonizzoni 1999).

PCBs are destroyed by the high temperatures used during vitrification.. The electron flame of a plasma vitrification breaks down the electron bonds of dioxins and furans and separates them into their basic chemistry (inorganic simple compounds and elements i.e. chlorine, sulphur, metal vapors, metal liquids, glass and carbon etc.). This is called disassociation, it is like electrolysis, and it is a reduction oxidation reaction where oxygen leaves one molecule and attaches to another. Thus, the dioxin molecules cease to exist.

The high temperature conditions that are reached in plasma gasification result in the decomposition of organic compounds into their elemental constituents, forming a high-energy synthesis gas, consisting mainly of hydrogen and carbon monoxide. On the one hand, tar, char and dioxins are broken down, resulting in a synthesis gas that is cleaner compared to conventional gasification processes. The inorganic fraction (glass, metals, silicates, heavy metals) on the other hand, is melted and converted into a dense, inert, non-leaching vitrified slag. The synthesis gas can be used for efficient production of electricity and/or heat, or second generation liquid (bio)fuels (e.g. Fischer Tropsch diesel) (Malkow 2004). The vitrified slag should be inert for leaching processes and consequently applicable as, for example, a building material additive (Lapa et al. 2002).

Non-thermal Plasma

The non-thermal plasma is described as the pollution control method which uses electron beams and corona discharges. This method when compared to the other types of treatment has been found to be much cheaper with higher removal efficiency. Pao et al. (2010b) revealed that the use of a non-thermal dielectric barrier discharge could lead to the reduction of dioxins and furans reduction. Likewise, it also has to be noted that non thermal plasma can have a destruction efficiency ranging from 80 to 20 % depending on the different dioxins and furans isomers (Zhou et al. 2003).

Gliding Arc Plasma Technology

This technology is described as innovative non thermal plasma method which offers various advantages with regards the control of dioxins and furans, volatile organic compounds (VOCs, SO_x and NO_x) emissions (Yan et al. 2007). Plasma gliding arc is used for the formation of energetic radicals are then made to collide with dioxins and furans molecules. These collisions hence, help to break down the dioxins and furans molecules. The potential reduction of the dioxins and furans using the gliding arc technology was found to vary from 25 to 79 % on a pilot scale study (Yan et al. 2007). In an attempt to further improve the efficiency of removal, a multistage gliding are reactor is now under studies.

Supercritical Water Oxidation

Supercritical water oxidation is the oxidation process in which the water occurs above its critical point. It uses the supercritical water as a reaction medium to exploit the unique solvating properties to provide enhanced solubility of reactants and permanent gases a single phase entrainment free of interphase, mass transfer limitations, and faster reaction kinetics and increased selectivity to complete oxidation products. (Super critical water oxidation is a promising technology for the clean and efficient decontamination of many organic wastes. It can rapidly and efficiently destroy organic substances into H_2O and CO_2. Furthermore a typical supercritical water oxidation process does not produce any NO_x or SO_x since the prevailing temperatures of the reactor is too low and does not favour the production of these components. Even if the incoming pollutant contains any ammonia or any other nitrogen compound the latter will be converted to N_2 and N_2O). The main disadvantages of the supercritical water oxidation process are namely: corrosion and plugging problems that occur in the reactor vessels and the equipment. The corrosion is mainly due to formation of acid by product in the reactor. Plugging occurs due to the formation of salt precipitates in the reactor vessels.

The existing supercritical water oxidation systems comprise:

1. Batch system: The reactor in this system has two temperature zones namely an upper part and a lower part to keep the reactor above critical temperature and the lower part to dissolve the salt precipitant at subcritical temperature
2. Continuous flow: It is the most common for the treatment of organic waste matter. It has been developed from tubular designs.
3. Transpiring wall reactor: It consists of a dual shell having an outer pressure resistant vessel and an inner porous vessel. The porous wall will allow the passage of supercritical water to form a thin protecting water film in the inner walls of the reactor. The system liner is used to protect the reactor against corrosion and salt deposition.
4. Floating type reactor: It consists of a pressure resistant vessel and an inner porous vessel. The supercritical water oxidation degradation occurs in the inner non-porous vessel.

Effectiveness of the treatment of dioxins and furans through the supercritical water oxidation process in oils and wastewater sludge has been demonstrated (Donghai et al. 2012). Supercritical water oxidation treatment can also be applied to the treatment of transformer oils heavily laden with PCBs. PCBs are persistent organic pollutants that degrade into the atmosphere to form dioxins and furans. Supercritical water oxidation processes have been successfully implemented in South-Korea for the treatment of such waste.

Mechanochemical Destruction

Mechanochemical destruction of dioxins and furans describes the physical and chemical changes that occur during the aggregation caused by mechanical energy

(Xiuying et al. 2010) Mechanical energy is transferred from milling bodies to the solid bodies through shear stresses, compression and milling. A significant part of the milling energy is converted to the heat and a minor part is used to induce breaks, stretches and compression at microscopic and macroscopic levels. Mechanochemical degradation can be easily performed using ball mills that are available in different sizes (Zheng et al. 2010). Pollutants are eliminated directly inside the contaminated material regardless of the complex nature and structure of the material.

Ultraviolet Irradiation

Ultraviolet light is a powerful and clean oxidising agent that can be used for the treatment of dioxins and furans. The process is also known as photolytic oxidation. It makes use free radicals from ultraviolet irradiation which brings the pollutants molecules to an excited state. A chain reaction then occurs with the participation of an oxygen (O_2) molecule to degrade the dioxins and furans.

It works properly to remediate contaminated groundwater but not soils directly because the photodegradation of dioxins, for example, is negligible. In order to be suitable for soil treatment, it needs to be performed in conjunction with a solubility-enhancement method such as extraction using surfactants or solvent-washing. The photodegradation is fast when used in combination with ethanol or catalyst such as Titanium oxide film. Photolytic oxidation can also be performed in conjunction with other oxidative techniques, using Fenton's reagent for example (Pérez et al. 2002; Lee et al. 2005).

Ultraviolet irradiation has been successfully implemented in the removal of dioxins and furans from liquid lab wastes. Typical destruction efficiencies of 89 % can be typically achieved with these treatment processes. This technique is also used in fly ash treatment as well. According to a study done on polychlorinated dibenzothiophenes (PCDTs) from sediment, soil, and fly ash samples collected in Japan (Satoshi et al. 2006), PCDTs can be photodegraded by ultraviolet radiation. The photodegradation process was accompanied by dechlorination. The same study was also done on polychlorinated dibenzofurans and it has been found that the photodegradation rates of PCDTs isomers such as 2,3,7,8-TeCDT and OCDT were lower than the rates for the corresponding PCDF isomers (2,3,7,8-TeCDF and OCDF) (Satoshi et al. 2006). Hence, it can be concluded that PCDTs are more stable than PCDFs to photodegradation.

10.5.3 Remediation of Dioxins- and Furans-Contaminated Soil and Sediments

Remediation techniques aim at lessening the degree of contamination at a hazardous water site or land and most importantly to prevent further deterioration of the environment and limit exposure of hazardous pollutants to human and other forms of life. Remediation technologies depend on the type of soils, the type and extent of contamination and also the natural processes that may occur at the site. Due to the various sources of pollution faced, numerous soils and sediments are contaminated with dioxins and furans and various researches have been made to address to these environmental problems. Some of innovative remediation technologies that have been addressed in this chapter include solvated electron technology, bioslurry, enhanced bioremediation technologies, subcritical water treatment, base catalysed dechlorination and thermal desorption amongst others.

Bioslurry

According to Lodolo et al. (2001), bioslurry consists of a proper technique for sites that necessitates greater process control, more complete and faster degradation rates. A slurry is formed by mixing the contaminated soils with water, hence, allowing the microorganisms to come into contact with the contaminants. The slurry is then fed into a bioreactor whereby a controlled amount of air is provided for mixing and aerating; inoculation may be carried out in order to improve the treatment.

As per Rahuman et al. (2000), slurry processes can be conducted faster than biological processes with the optimization of conditions such as temperature, nutrient concentration and aeration. It has been reported that the treated slurry is convenient for direct land application, similar to composted soils. In addition, the time taken for clean-up has turned out to be less than 12 months. Slurry-phase bioreactors find their applications in the remediation of soils and sludge contaminated with explosives, petroleum hydrocarbons, petrochemicals, solvents, pesticides among others. Studies conducted by Lodolo et al. (2001) and Rahuman et al. (2000) have demonstrated that bioslurry is favoured over *in-situ* biological techniques for heterogeneous soils, low permeability soils and areas where underlying groundwater is hard to capture.

Bioremediation

Bioremediation consist of the transformation of contaminants into fewer complex and possibly less toxic molecules by naturally occurring microbes, by enzyme systems, or by genetically engineered microorganisms. This process can be conducted in situ or in a reaction vessel, under anaerobic or aerobic conditions, and alone or together with other treatment methods; however, it may take several months or years to achieve complete contaminant removal. A suitable microbial population and appropriate characteristics such as moisture and oxygen levels, organic content, temperature, pH, food source availability and possible degradation pathways should be taken into consideration to ensure the survival of the microorganisms and consequently, the success of a bioremediation project. Nevertheless, there are few studies which demonstrate that contaminants have been destroyed or removed at levels higher than 90 % although considerable laboratory and field work has been reported in each of these major areas (United States Congress 1991).

Enhanced bioremediation, also known as bio-augmentation or bio-stimulation, comprises of a process during which the biodegradation rate of contaminated soil is increased by the addition of nutrients and oxygen. Stimulation of the activity of microflora and fauna may be caused by circulating water-based solutions through the contaminated soils and/or addition of indigenous/inoculated microorganisms, engineered microbial species or seeding with pollutant degrading bacteria in order to improve biodegradation of contaminants or immobilization of inorganic contaminants. Despite the fact that this process could be carried out in anaerobic conditions, it is more beneficial when oxygen is not limiting and therefore preventing the formation of persistent by-products such as vinyl chloride resulting from the anaerobic degradation of trichloroethylene. This technique can be used *in situ* to treat soils contaminated with different pollutants such as petroleum hydrocarbons, solvents, pesticides, wood preservatives and/or nitrotoluenes. (Lodolo et al. 2001; Rahuman et al. 2000).

Biodegradation of dioxins has been extensively studied in several microorganisms, and details concerning biodiversity, biodegradation, biochemistry and molecular biology of this process have accumulated during the last three decades (Chang 2008). There are several microbial mechanisms responsible for biodegradation of dioxins, including oxidative degradation by dioxygenase-containing aerobic bacteria, bacterial and fungal cytochrome P-450, fungal lignolytic enzymes, reductive dechlorination by anaerobic bacteria, and direct ether ring cleavage by fungi containing etherase-like enzymes (Chang 2008). Many attempts have been made to bioremediate PCDD/Fs using this basic knowledge of microbial dioxin degradation. Nam et al. (2008) have examined the ability of a microbial biocatalyst which is a mixture of four bacterial and five fungal dioxin-degrading strains to bioremediate PCDD/Fs from contaminated municipal solid waste incinerator (MSWI) fly ash under laboratory solid-state fermentation conditions. Nam et al. (2008) deduced that treatment of MSWI fly ash with the microbial biocatalyst for 21 days resulted in a 68.7 % reduction in total toxic PCDD/Fs. Further analyses revealed that the microbial biocatalyst also removed 66.8 % of the 2,3,7,8-substituted congeners from the fly ash. Denaturing gradient gel electrophoresis also showed that all of the bacterial and fungal strains composing this dioxin-degrading microbial mixture were maintained under the dioxin treatment conditions. Angular dioxygenase, cytochrome P450, lignin peroxidase, and dehalogenase are known as dioxin-metabolizing enzymes (Sakaki and Munetsuna 2010). All of these enzymes have metal ions in their active centers, and the enzyme systems except for peroxidase have each distinct electron transport chain. Dehalogenase system appeared to be quite promising because dehalogenases are capable of degradation of PCDDs with more than four chlorine substituents, whereas the other three enzyme systems prefer low-chlorinated PCDDs. On the other hand, protein engineering of angular dioxygenase, lignin peroxidase, and cytochrome P450 based on their tertiary structures has great potential to generate highly efficient dioxin-metabolizing enzymes (Sakaki and Munetsuna 2010). Sakaki and Munetsuna (2010) have recently successfully generated 2,3,7,8-tetrachlorodibenzo-*p*-dioxin (TCDD)-metabolizing enzyme by site-directed mutagenesis of cytochrome P450. Jaiswal et al. (2011) have enriched and isolated an alkalotolerant Pseudomonas strain from effluent of the pulp and paper industry. Jaiswal et al. (2011) found this strain was able to degrade dibenzofuran and utilize it as a sole source of energy and carbon. The gas chromatography-mass spectrometry based detection of various intermediary metabolites of biodegradation suggested the involvement of angular as well as lateral pathway of dibenzofuran biodegradation. This strain was found to have an optimum utilization (~85 %) of dibenzofuran (200 mg/L) within 36 h at pH 10 at 40 °C. The growth of the strain was supported by a wide range of environmental conditions supporting that it could be used for in situ bioremediation of dioxin-like compound. Nakamiya et al. (2002) investigated the potential of denitrifying activated sludge to degrade highly chlorinated dioxins from a landfill leachate treatment plant in Japan. Denitrifying activated sludge was cultivated in a 700-mL bioreactor under denitrifying conditions by adding 2.0 ng of a mixture of 4- to 8-chlorinated dioxins from fly ash. After 7 days cultivation, the authors reported about 90 % of dioxins were degraded. Nam et al. (2005) studied the ability of *Sphingomonas wittichii* strain RW1 to remove PCDDs from fly ash and reported that incubation of real fly ash collected from municipal waste incinerators with strain RW1 for 15 days resulted in a 75.5 % reduction in toxic PCDDs while dead strain RW1 cells caused a 20.2 % reduction in toxic PCDDs, indicating that adsorption onto biomass was an important factor in dioxin elimination. Cathum et al. (2006) investigated the usefulness of cyclodextrins (CDs) for the removal of PCDD/Fs in soil and water. Five CDs were selected and evaluated for their ability to trap PCDD/Fs in soil and water over a 28-day period. Among the 5 CDs used, hydroxypropyl-γ-cyclodextrin removed maximum of PCDD/Fs (80 % of the total) while α-cyclodextrin and β-cyclodextrin removed only 45 and 50 % of the total PCDD/Fs after 28 days. Hidayat and Tachibana (2013) have recently studied the degradation of 2,4,8-trichlorodibenzofuran (2,4,8-TCDF) by the isolate F0607, identified as belonging to the genus *Cerrena* in liquid culture medium. Strain F0607 showed good decolorization and production of ligninolytic enzymes. The degradation of 2,4,8-TCDF per-unit biomass was observed to increase during the incubation period in presence of 1 or 10 mg/L 2,4,8-TCDF and the mycelial biomass increased with the increase in 2,4,8-TCDF concentration.

Solvated Electron Technology

This technology makes use of solvated electron solutions which are produced by dissolving alkali or alkaline earth metals such as sodium, lithium or calcium in liquid anhydrous ammonia at room temperature but in a pressurized system. (Lodolo et al. 2001) An alternative way of preparing the reagent is at sub-ambient temperatures with lower pressure control requirements (McDowall et al. 2004). As the electrons are freed, a dark blue solution of solvated electrons is produced; these electrons act as dehalogenating agents and are also known to be the most powerful reducing agents. Halogenated compounds, which have a powerful affinity for free electrons, are mixed with the solvated solution and are instantaneously neutralized. Furthermore, this method has been claimed to be applicable to treat halogenated hydrocarbons, pesticides, dioxins, PCBs, herbicides, chlorofluorocarbons, and chemical warfare agents. Lodolo et al. (2001) have also reported that waste have been successfully treated in bulk pure material, soils, sludge, sediments, porous and non-porous surfaces, oils, contaminated vessels, hardware, and contaminated clothing. The process design makes use of a cement mixer like reactor in which contaminated material and liquid ammonia are mixed. The soil and clays are completely dispersed by the ammonia and hence, the contaminants are washed from the soil. A reactive metal charge, usually calcium is added after a brief mixing. Subsequently, the contaminants are rapidly dehalogenated by the electrons which are released from the calcium (Mitoma et al. 2009). Ammonia is recovered for further use, and the soil is deodorized. The decontaminated soil is appropriate for return to the site. Interestingly, treated soil (assuming that there is no other contaminant such as heavy metals) is not only cleaned but also nitrogen-enriched from the ammonia bath. Thus, the soil is suitable for agricultural use since is enriched in nitrogen from trace amounts of residual ammonia (Lodolo et al. 2001; Rahuman et al. 2000; McDowall et al. 2004).

Subcritical Water Treatment

Processes occurring in aqueous phase under subcritical conditions have gained much research interest because of the relatively lower costs and lower consumption of energy for extraction of pollutants when compared to other methods (Nose et al. 2007; Hashimoto et al. 2004; Teo et al. 2010; Chang et al. 2011). During subcritical water treatment, water is held in liquid state above a temperature of 100 °C by applying a pressure known as subcritical water. It has properties which are similar to the organic solvents and can also act as a benign medium. Applications of subcritical water treatment include the extraction of dioxins and furans and other organic pollutants from soil and sediment (Kulkarni et al. 2008). According to Kulkarni et al. (2008), 99.4 % extraction of dioxins was observed at a temperature of 350 °C within 30 min. Nevertheless, the treatment was more time consuming at lower temperatures.

It has been reported that solubility dioxins increases by several orders of magnitude when temperature is raised from 20 to 250 °C (Shelepchikov et al. 2005; Yang et al. 1998; Soyfer et al. 1999). In such cases water has a high tendency in becoming a good solvent for extraction of these chemicals from soil. The photolytic or another dechlorination reaction may be then used for detoxication (Choudhry and Webster 1985). Shelepchikov et al. 2005 proposed that a combined extraction/dechlorination process in subcritical conditions could be more beneficial for the detoxification of dioxins. Yak et al. (1999) have reported positive and conclusive results of dechlorination of PCBs supported on iron powder in an aqueous phase at 250–300 °C and Yang et al. (1998) equally showed that water in subcritical conditions could be used for the extraction of PCDD/Fs from soils through desorbing dioxins from soil particles and transferring them to the surface of the reducing agent.

Zero valent iron has been accepted as one of the most efficient means of environmental remediation. In addition, it is inexpensive, easy to handle and effective in treating a wide range of chlorinated compounds or heavy metals. It has been broadly applied in-situ, ex-situ or as part of a controlled treatment process in wastewater, drinking water soil amendment stabilization and mine tailing applications (Kulkarni et al. 2008). Kluyev et al. (2002) have developed an efficient method for reductive dechlorination of PCDDs and remediation of contaminated soils using zero valent iron as the dechlorination agent and subcritical water as reaction medium and extractive solvent. It is found that the zerovalent iron could be applied for stepwise dechlorination of octachlorinated dibenzo-*p*-dioxin (OCDD) on various matrices in subcritical water. By using iron powder as matrix higher chlorinated congeners were practically completely reduced to less than tetra-substituted homologues. A significant part of residual OCDD, when spiked in soils, formed less chlorinated congeners. The solubility of OCDD was increased by a 4–6 orders over its solubility at ambient conditions.

Base-Catalysed De-chlorination

The base-catalysed de-chlorination is a chemical de-halogenation process which involves the addition of an alkali or alkaline earth metal carbonate, bicarbonate or hydroxyl to the contaminated medium (Oku et al. 1995; Taniguchi et al. 1997; Chen et al. 1997; Kulkarni et al. 2008). The base-catalysed decomposition is normally carried out as a batch process where destruction of POPs are effected under reductive conditions with no production of dioxins and furans, or existing quantities of these are destroyed (Rae 2003). Base-catalysed de-chlorination is initiated in a medium temperature thermal desorber at temperature ranging from 315 to 426 °C and the alkali is added in proportions ranging from 1 to about 20 % by weight (Kulkarni et al. 2008). The process involves the addition of a hydrogen donor compound in order to provide hydrogen ions for the reaction in case these are not previously present in the contaminated medium. The purpose of adding a metal carbonate or bicarbonate to the reaction is to promote lower temperature desorption and to partially destroy the chlorinated organics (Rahuman et al. 2000). The base-catalysed de-chlorination process chemically detoxifies the chlorinated organic contaminants by transferring the hydride ions from the rich hydrogen donor substance to the chlorinated molecules. Thus, the chloride ions are expulsed from the contaminated medium and replaced by hydrogen (Kulkarni et al. 2008).

Base-catalysed de-chlorination can be effectively used to treat contaminated soils containing organic pollutants which are hazardous such as dioxins and furans (Rahuman et al. 2000). Compounds such as PCBs which at elevated temperatures, may react with oxygen resulting in formation of even more hazardous materials such as dioxins are especially suitably treated by base-catalysed de-chlorination. However, high clay or water content, acidity or high natural organic content of the soil may limit or interfere with the treatment effectiveness (Rahuman et al. 2000). Studies have shown that Na/NH_3 treatment can be used to enhance base-catalysed de-chlorination and hence successfully remediate PCBs and dioxin- contaminated soils which depend on the amount of water present and the concentration of the hazardous substrate. Ca/NH_3 treatment can also be used for the same effect. However, laboratory studies revealed that Ca is less efficient than Na in the case where the contaminated soils contain considerable amount of water (Pittman and He 2002; Kulkarni et al. 2008). Studies conducted on the treatment of PBCs by base-catalysed de-chlorination have reported that concentration of these pollutants in contaminated soil were reduced from as high as 4,000 ppm to less than 2 ppm.

Base-catalysed de-chlorination process can normally involve direct de-halogenation or can be linked to a pre-treatment such as thermal desorption. Combined thermal desorption and base-catalysed de-chlorination processes yield a relatively small quantity of a condensed volatile phase for separate treatment by the base-catalysed de-chlorination process (Lodolo et al. 2001). Ukisu and Miyadera (2004) reported that the dechlorination of 1,2,3,4-tetrachlorodibenzo-*p*-dioxin proceeded completely with Pd/Al_2O_3 at 35 °C to give a chlorine-free product, dibenzo-p-dioxin. However, the same reaction catalyzed by Pd/C was not complete since un-reacted substrate and partially dechlorinated products remained on the catalyst. Mitoma et al. (2004) found that the concentration for each isomer of PCDD/Fs and co-PCBs was reduced by 98.32–100 % conversions.

Mechano-chemical Dehalogenation Degradation

Mechano-chemical dehalogenation degradation can efficiently be used to treat high strength wastes containing POPs. In this technology, mechanical energy is used to initiate reaction for removal of pollutants from the contaminated medium. Mechanical energy is produced by milling bodies and is transferred to the solid system by compression or shear stress, depending on the device used. Most of the energy generated is converted into heat, with a minor amount being used for performing a reaction or to induce breaks, stretches and compression. Usually, mechano-chemical dehalogenation degradation can be easily performed using ball mills. Depending on the amount of materials to be treated, these are available in different sizes and designs (Kulkarni et al. 2008). Intimate contact among particles is brought about by the intense mixing occurring in the ball mills. Energy for local chemical reactions is provided by a highly localized tribo-plasma which is formed at the point of contact between two grinding balls during collision. This grinding has the effect of reducing the particle sizes and increases surface areas for reaction to occur. Also, new reactive surfaces are exposed during particle fraction and by the introduction of dislocations, thus, increasing surface reactivity. Free radicals are also produced during milling and these can react with neighbouring compounds. Removal of pollutants is done directly inside the contaminated matrix, regardless of the complex structure and the pollutant's nature (Kulkarni et al. 2008).

Mechano-chemical dehalogenation is one of the degradation techniques, carried out ex-situ in an enclosed ball bill, for destruction of poly-halogenated compounds.

In this technology, mechanical energy is used to promote reductive halogenation of contaminants by utilizing a base metal (typically aluminium, zinc or iron) and a hydrogen donor (alcohols, ethers, hydroxide and hydrides) to generate non-hazardous organics and metal salts as by-products. Mechano-chemical dehalogenation degradation can successfully be used for treating contaminated compounds as well as pure or extremely concentrated contaminants and their mixtures at moderately low temperatures, irrespective of their state. Experiments conducted on a laboratory scale showed that poly-halogenated pollutants such as PCBs, PCP were degraded to their parent hydrocarbons, that is, biphenyl and phenols respectively, in high yield by using magnesium, aluminium or sodium metals with a low acidic hydrogen source. For instance, the use of magnesium as a base metal contributes in dechlorinating PCBs in soils to harmless chloride and their parent hydrocarbon biphenyls with removal efficiency above 90 % (Kulkarni et al. 2008). Co-grinding POPs with calcium oxide or other metal oxides/hydrides is another means of Mechano-chemical dehalogenation degradation referred to as mechano-chemicaldechlorination. Electron transfer from O_2^- sites on the CaO particles' surfaces to the organic compounds is induced by the mechanical stresses caused by the ball mills. This then allows the organic anion particles to undergo self-dissociation of the chlorine-carbon bond effectively. There are several factors upon which the dechlorination process is dependent of, for example, moisture and quartz addition. This method has proven to be a successful one in destroying halogenated organic compounds, namely, POPs including chlorinated benzene, PCDD/Fs, PCBs and pentachlorophenol, as compared to dehalogenation by mechanochemical reaction, which aims only at achieving detoxification by stripping off the organic halogens without destroying the molecule.

Mechano-chemical dehalogenation degradation techniques offer several economic and ecological benefits. No harmful emissions are released into the atmosphere since the toxic compounds can be converted into usable products by the strikingly benign reaction conditions and the ball mills do not require much of energy input. It is considered as a very flexible method since it comprises of a portable mill and a washing tank with a filter which can be utilize at any location for disposal of organic pollutants (Kulkarni et al. 2008).

Soil Washing

Soil washing is a remediation process in which contaminants are separated from the contaminated matrix by transferring them into another medium, for example, from contaminated soil to a liquid, or from a contaminated liquid to an adsorbent by making use of suitable solvents (Jonsson 2009). It is basically an ex-situ technology wherein fine soil (clay and silt) is separated from coarser soil (sand and gravel) by this water-based process based on the fact that most contaminants tend to bind, either physically or chemically and, sorb to clay, silt and organic particles, that is, to the fine soil. These particles are in turn attached to the coarser particles by compaction and adhesion (Pavel and Gavrilescu 2008). In so doing, the contaminants are concentrated only into a smaller volume of soil, thus significantly reducing the volume of most contaminated materials. Soil washing is most commonly used in combination with other technologies such as incineration or bio-remediation as means of further treatment for the smaller volume of contaminated soil. Else, the latter can be disposed of according to federal environmental regulations. The larger volume on the other hand, is considered to be free from toxic components and can be returned clean to the site (Pavel and Gavrilescu 2008). This separation is accomplished by mechanically mixing, washing and rinsing the soil so as to remove the contaminants. The solvents used (usually water, occasionally combined with other solvents) are based on their ability to solubilize specific group of contaminants, taking into consideration their environmental and health impacts (Asante-Duah 1996; Feng et al. 2001; Chu 2003; Urum et al. 2003). The use of low molecular weight alcohol solvents has been proven to be a feasible extraction method for removal of dioxins from contaminated soils (Nam et al. 2001; Lee and Hosomi 2001; Jonsson 2009). The combined use of water and ethanol at room temperature extracting about 85 % of PCPs has been shown by Amid et al. (1994). The use of liquid propane as solvent can extract 98 % of PCBs from contaminated soil. Another study showed that liquefied propane and dimethyl other has decreased the dioxins level in soil by 98.8 % (Sahle-Demessie et al. 2000; Kunichika et al. 2003). Temperature effects also play an important role in soil washing and this has been shown in a study for treating PCDD/Fs- contaminated soil, in which ethanol at its boiling point, that is, 78.3 °C, extracted about 76 % of dioxins from the soil (Kunichika et al. 2003). Recently, the use of surfactants and other additives, such as cyclodextrins, as alternative soil washing processes have been developed and found to be feasible for extraction of more hydrophobic compounds from the soil, such as organic contaminants. These additives act as flushing solutions for dioxins and furans removal from the soil by substantially increasing desorption and solubilization through micellization and surface tension reduction (Saichek and Reddy 2004), with cyclodextrins being non-toxic, biodegradable and has low sorption to the solid phase at a wide range of pH range, as compared to regular surfactants (Maturi and Reddy 2006).

Fenton's Reagent Oxidation

Chemical oxidation has proved to be a useful destruction technique for a wide range of organic pollutants including octa-chlorodibenzo-p-dioxins, nitrophenols, petroleum hydrocarbons, chlorinated ethylene, chlorophenols and chlorinated biphenyls (Aunola 2004; Mariñosa 2007). Advanced oxidation processes (AOPs) have been developed which make use of ozone, Fenton's reagent, permanganate and persulphate. It should be noted that all AOPs are characterised by the same chemical feature which is the production of hydroxide radicals, however the only difference lies in chemistry in producing these. Fenton process is based on the use of oxidizing agents such as hydrogen peroxide and/or oxygen and a catalyst which is commonly iron salt or oxide. Oxidation of Fe^{2+} to Fe^{3+} and highly reactive hydroxyl radicals (OH•)

are formed by the oxidation of iron and H_2O_2 in acidic solution (Mariñosa 2007). These free radicals are able to attract dioxins- contaminated environmental media by heterogeneous catalytic oxidation. In soil remediation, despite the fact that iron oxide is present naturally in the ground, addition of iron slats or solid iron may be required. However, this process is highly affected by the type of soil (Mariñosa 2007). Studies have revealed the efficiency of Fenton reactions when used for organic pollutants removal and especially dioxins and furans species as illustrated by data in Table 10.6. More than 80 % of reduction when treating PCDD/Fs has been showed by Fenton's oxidation with a ratio of 8:1 for Fe^{2+}/H_2O_2 (Mariñosa 2007). Fenton has also been proven effective in PCBs removal, with an 87 % removal in treatment of contaminated lake sediments, with its pH adjusted with SO_4^{2-} instead of Cl^- (Manzano et al. 2001; Aunola 2005; Fronduti 2005; Mariñosa 2007). The treatment, however, has a high dependency on the ratios and nature of PCBs mixtures (Dercovà et al. 1999; Tuhkanen 2001; Arienzo et al. 2000; Mariñosa 2007). Other advances of Fenton treatment have also been developed including photo-Fenton process, the most commonly used, in which a reaction is initiated by irradiation of light of wavelength around 18–400 nm; electrochemical Fenton process which is combined with dechlorination by zero-valent iron with Fenton's reagent to supply soluble Fe^{2+} ions by using electric current and sacrificial electrodes; ethanol-Fenton whereby hydrogen peroxide and iron are added to a liquid phase of ethanol which improves the reaction for hydrophobic contaminants (Mariñosa 2007).

In-situ Photolytic Destruction Technologies

Photolysis is potentially an essential process for the removal and destruction of dioxins and furans from soils and sediments. In-situ photolysis refers to the photodegradation of dioxins and furans species in soils and sediments upon exposure to a light source – normally sunlight under specific conditions of time- and through the use of appropriate solvents. As per Choi et al. (2000), dioxins and furans have a strong UV absorption band in the range of 230–290 nm, making it possible to use sunlight as radiation source to favour the excitation of dioxins and furans molecules. The excited state may also be reached by accepting energy from an excited donor molecule by sensitized photolysis. Reactions may also occur with reactive oxygen species such as singlet oxygen or hydroxyl radicals (OH•) that are formed by the photochemical interactions with other organic molecules such as humic acid (Kim and O'Keefe 2000). The primary degradation pathway for photolysis of dioxins and furans involves the absorption of UV radiation and followed by the subsequent cleavage of C–Cl and C–O bonds, resulting in the formation of aryl radical species. Dechlorination reactions take place leading to the production of innocuous non-chlorinated dioxins and furans products (Wu and Ng 2008). The toxicity of dioxins and furans is thus reduced. The influence of the degree of chlorination on dioxins and furans molecules during photodegradation suggests that more rapid photodegradation occurs for lower

Table 10.6 Oxidative degradation of dioxins and furans species using Fenton's reagent oxidation

Species	Experimental conditions	Degradation performance	References
Polychlorinated Biphenyls (PCBs): PCB_{28}, PCB_{52}, PCB_{101}, PCB_{138}	Fenton's reagent (Fe^{2+}: H_2O_2 molar ratio 1: 1000)	Complete disappearance was observed for PCB_{28} and PCB_{52}; high removal > 99.5 % were achieved after 96 h of process. Complete mineralisation was not reached signifying moderate effectiveness of Fenton process. Penta-chlorinated biphenyl PCB_{101} was reported to degrade faster than hexa-chlorinated PCB_{138}; a residual of 23 % of the latter compound indicated more rapid degradation of low-chlorinated compounds	Prządo et al. (2007)
PCBs contaminated loamy soil	Modified Fenton's reagent	Catalysed H_2O_2 propagation used for ex situ treatment was found to be more effective in treating soil with higher iron oxide and manganese oxide	Ahmad et al. (2011)
2,3,7,8-tetrachlorodibenzo-p-dioxin (TCDD) contaminated soils	Combined Fenton and bioremediation	Up to 99 % TCDD was oxidised into chloro phenols and chloro benzenes after pre-treatment with Fenton's reagent. Under anaerobic conditions alone, 53 % TCDD degradation was achieved while no degradation occurred in the aerobic conditions	Kao and Wu (2000) and Venny et al. (2012)
Artificial mixture of DELOR 103 mixture of polychlorinated biphenyls (Mono, di-, tri- and tetra-chloro-biphenyls)	Fenton's reagent followed by bioremediation	The best result was obtained in a solution of DELOR 103, containing Fe^{2+}/H_2O_2 with a molar ratio of 1:1,000. The oxidation effect of Fenton's reagent on PCB congeners was noted to decrease with increasing number of chlorine substituents in the biphenyl molecule and molecular weight but decreased with lower octanol/water partition coefficient	

(continued)

Table 10.6 (continued)

Species	Experimental conditions	Degradation performance	References
PCDDs and PCDFs contaminated landfill leachates arising from a municipal solid waste	Electrokinetic-Fenton (EK) technology	Chemical oxidation demand reduction was approximately 90 % in both methods individual after 180 min. However, the behaviour of PCDD/Fs varied in terms of oxidation efficiency.73 and 71 % destruction of 1, 2, 3, 4, 6, 7, 8-HpCDD and OCDD were achieved by EK respectively and total toxic equivalent value dropped by 58 %. However, an increase in the concentration of several PCDD/Fs was detected	Vallejo et al. (2013) and Venny et al. (2012)

chlorinated dioxins and furans. Antonaraki et al. (2002) explains that the increase in chlorine atoms in the dioxins and furans species blocks favourable positions susceptible for OH attacks as well as decreases the electron density on benzene rings, making the species less likely for electrophilic OH attack. The rate of photodegradation hence decreases with an increase in number of chlorine atoms.

Photodegradation rates are slower in the soil as compared to photolytic reactions in the atmosphere and water bodies (Isosaari 2004; Kieatiwong et al. 1990; McPeters and Overcash 1993). As reported by Miller et al. (1989), photodegradation of dioxins and furans in soils is restricted to the surface layers less than 0.2–1 mm deep (Isosaari 2004). Most photolysis has been recorded in the depth of 0.06–0.13 mm; direct photolysis is limited to a photic depth of soil of 0.2–0.4 mm whilst indirect photolysis occurs a little deeper (Sinkkonen and Paasivirta 2000; Hebert and Miller 1990; Miller et al. 1989). Such behaviours can be accounted by the fact that the dioxins and furans molecules are strongly bonded to the solid surfaces and by the shielding effect of particles (Isosaari 2004; Kieatiwong et al. 1990; McPeters and Overcash 1993). Consequently, solvent addition becomes necessary to promote solubilisation and transport of the dioxins and furans species into the solvent prior to the photodegradation process. A low-toxicity organic solvent is generally sprayed on the soil surface for this purpose (Kulkarni et al. 2008).

The selection of an appropriate solvent depends on the solubility and interaction of the dioxins and furans, its diffusive properties, its environmental acceptability and its UV absorption. Strongly absorbing solvents tend to absorb most of the UV light within a few mm or even less (Isosaari 2004). Several investigations have revealed that direct photolysis of dioxins and furans takes place in hydrogen donor organic solvents (Kim and O'Keefe 2000; Crosby et al. 1971; Dobbs and Grant 1979). The hydrogen atoms supplied by the hydrogen donors has the capability of stabilizing photolytically generated aryl radicals (Isosaari 2004; Hung and Ingram 1990). It has been found that dioxins and furans on soil surfaces decompose rapidly after being sprayed with various organic solvents such as hexane, cyclohexane and isooctane amongst others (Kulkarni et al. 2008; Balmer et al. 2000; Goncalves et al. 2006). The major transport mechanism of the dioxins and furans in in-situ photolysis occurs by the diffusive and convective upward movement of dioxins and furans as the volatile solvents are evaporated. The effectiveness of this process is dependent on the sunlight available for photodegradation and the convective transport to the surface. These two factors are rate determining (Kulkarni et al. 2008). The transport mechanism determines whether any reduction of dioxins and furans take place in deeper solvent layers or in solvent-amended soils. Great importance has also been attributed to correct timing for solvent additions to avoid downward migration of dioxins and furans and optimize photodegradation on the soil surface (Isosaari 2004; Dougherty et al. 1993). Furthermore, the irradiation time plays a crucial role in the photodegradation of dioxins and furans and affects the efficiency of the process. Experimentally determined results have shown that the longer the irradiation time, the higher the degradation of dioxins and furans (Isosaari 2004). Nevertheless, the efficiency of the photolysis process is lower when direct sunlight is used compared to laboratory induced UV radiations. Additionally, the degradation period of dioxins and furans is longer (measured in days) with the application of direct sunlight than induced black light in laboratories.

Other factors that are bound to affect the photolysis process are the soil properties and the presence of photochemical sensitizers and semiconductors in the soil amongst others. Photochemical sensitizers such as humic acid, enhance the photodegradation process whilst the presence of semiconductors such as titanium dioxide (TiO_2), catalyses the photochemical reaction. It has been reported that furans degrade more rapidly than dioxins. Furans have different photodegradation rates compared to dioxins in particular solvents by virtue of the difference in their molecular structures as well as the hydrogen donor potential of the solvents (Isosaari 2004; Niu et al. 2003). Kim and O'Keefe (2000) have demonstrated that polychlorinated dibenzofurans have lower photodegradation rates in pure water rather than in an acetonitrile/water solvent. The reverse applies for PCDDs which have higher photodegradation rates in an acetonitrile/water solvent compared to pure water. A probable explanation to such a phenomenon would be due to the structure of PCDFs which have only one oxygen atom and are more polarized than PCDDs; the former are therefore more reactive in pure water than in polar solutions such as acetonitrile/water solvent. PCDDs consist of more delocalized electrons and are less polarized resulting in lower reactions with pure water.

Isosaari (2004) investigated the possibility of using vegetable oil as solvent for the photolysis of dioxins and furans. Indeed, it was proven that vegetable oils are efficient at dissolving soil-bound dioxins and furans, transporting them towards the surface and acting as hydrogen sources for reductive dechlorination. Dissolution of dioxins and furans occurs relatively fast. It was noted that the hydrogen donor properties of vegetable oils made them better solvents for dioxins and furans photodegradation in comparison with toluene and hexane. A rapid photodegradation rate of dioxins and furans was demonstrated by vegetable oils through longer contact time and periodical mixing of the soil-vegetable oil mixture. The most effective photodegradation occurred at an efficiency of 99 % over 22 h. Nevertheless, the type and cost of vegetable oil could eventually influence the photodegradation process. In-situ photolysis offers a good substitute for energy intensive conventional treatment methods as it entails the use of free renewable and pollution-free solar energy. It is a rapid, efficient and economically feasible technique for rendering toxic dioxins and furans innocuous thus minimizing their health and environmental impacts.

Thermal Desorption

Thermal desorption is the separation process whereby the soil is heated for either in-situ or ex-situ remediation. The process involves heating the soil at high temperatures to allow constituents of organic pollutants – PCBs, chlorinated solvents and organochlorine pesticides – to volatize and desorb from the soil. This technique is normally coupled with a post treatment-destruction technology such as an afterburner, a catalytic oxidation chamber, a carbon adsorption unit or a condenser to subsequently treat any volatile component or toxic emission emanating from the soil, prior to discharge into the atmosphere. Afterburners and oxidisers ensure complete destruction of the organic compounds while the carbon adsorption units and condensers trap the organic constituents for subsequent treatment or disposal.

Ex-situ treatment processes have been most commonly practised, but have often resulted in the generation of potentially toxic products of incomplete combustion which eventually led to the formation of dioxins and furans, either as air emission or particulates collected in an off-gas treatment system or even as residuals remaining in the soil. On studying the thermal desorption of soil contaminated with PCBs, Sato et al. (2010) demonstrated that despite a decrease of 48–70 % was recorded in the PCBs levels, thermal desorption triggered the formation of high levels of dioxins and furans, causing the TEF of the products to increase. It was also noted that furans with high number of chlorine atoms volatized less readily as compared to those with lower chlorine atoms. A possible explanation to this phenomenon was based on the positive correlation that was developed between the vapour pressure and ratio of volatilized furans from soil sediments. Consequently, it was deduced that vapour pressure correlated positively with the number of chlorine atoms, thus confirming that furans with lower chlorine atoms are more likely to be volatilized first.

In-situ thermal desorption is an innovative technology, aimed at minimizing the emission of toxic dioxins and furans and ensuring their complete destruction. It involves treating contaminated soil through a combination of heat and vacuum; an array of vertical or horizontal thermal wells is used to heat the soil up to 900 °C resulting in heat transfer by thermal conduction, causing VOCs and semi-volatile species (SVOCs) in the soil to vaporize due to evaporation, steam distillation and boiling in the soil layers. As vaporized water and contaminants are drawn counter-currently to the heat flow, the soil acts as a packed-bed reactor that ensures complete destruction of VOCs, SVOCs and dioxins and furans in-situ by oxidation and pyrolysis. According to Baker and Lachance (2003), typical destruction efficiencies of contaminants in the soil reach up to 99 %. Most of the contaminants are destroyed in the soil before the vapour reaches the extraction wells and is conveyed to the surface. In-situ thermal desorption are generally connected to an air quality control system (AQC) whereby any untreated contaminants in the extracted vapour stream undergo complete destruction. AQC systems usually consist of a thermal oxidizer, a heat exchanger, an acid-gas scrubber, granulated activated carbon adsorbers (GAC) and vacuum blowers.

The composition and chemical structure of the organic compounds, the temperature of the desorber systems and the residence for vapours travelling through the hot soil layers to the extractor wells influence the decomposition of the organic pollutants, causing some of the constituents to decompose completely or partially (Kulkarni et al. 2008). The temperature of the system influences the volatility of the organic constituents. The chemical structure of the contaminants affects their volatility as explained by Sato et al. (2010). Whilst ex-situ treatment are performed in reactors within few seconds or minutes, its in-situ counterpart necessitate longer residence times ranging from hours to days as large volumes of soil can be effectively treated. Longer residence time ensures complete destruction of the dioxins and furans in the system by pyrolysis and oxidation (Stegemeier and Vinegar 2001).

When comparing the in-situ and ex-situ technologies, it has been observed that the ex-situ process involves exposing the soil to high temperatures for a short time period, resulting in the production of off gases at high rates and temperatures whereby technologies such as GAC become expensive to be employed. Additionally, toxic dioxins and furans are generated and are not fully treated. The in-situ process eliminates the above mentioned lacunae as they have the ability to completely destroy (in-situ) and not generate dioxins and furans. Any dioxins and furans persisting in the extracted vapours are efficiently eliminated in the AQC system. In-situ thermal desorption have proved to be more robust and effective relative to dioxins and furans treatments in comparison to ex-situ thermal desorption.

Catalytic Hydrogenation

Catalytic hydrogenation represents a convenient method for the destruction of chlorinated wastes or the dehalogenation of aryl halides through the use of hydrogen gas over noble metal catalysts under mild conditions. The experimental simplicity, good yields and high purity of products of this technique have made it one of the most promising and innovative technologies for the disposal of toxic waste streams containing chlorinated organic compounds (Xia et al. 2009).

The process involves rapid and high conversion of dioxins and furans or toxic organic chloride in solutions consisting of an organic solvent and a base (usually NaOH) in the presence of carbon-supported noble metal catalysts under ambient conditions (Ukisu and Miyadera 2002). The chemistry that underlies the process is based on the sorption of the dioxins and furans species in the solvent and the hydrogen particles onto the active site of the catalyst followed by the gradual substitution of the chlorine atoms in the dioxins and furans species by hydrogen. This process is also commonly known as detoxification. An almost chlorine-free product is normally yielded and hydrogen chloride (HCl) is generated as a by-product. The organic solvents act as a reaction medium for the dissolution of dioxins and furans or chlorine compounds and represent a hydrogen donor which generates hydride species (H^-) that displaces the chloride anions (Cl^-). HCl produced during the reaction is adsorbed and accumulated onto the surface of the catalyst, eventually blocking its active site. The base is therefore added to the organic solvent to neutralize and eliminate HCl to minimize its inhibitory effects (Xia et al. 2003).

A wide variety of organic solvents may be employed, including alcohols (ethanol and isopropanol), alkanes (cyclohexane and n-heptane) and arenes (benzene, toluene, and pyridine) amongst others. A study on the catalytic hydrogenation of triclosan which has similar structures as dioxins and furans, by Xia et al. (2003), demonstrated that the rate of dechlorination in the catalytic hydrogenation process generally decreases in order of alcohols > alkanes > arenes when used as solvents. This is explained by the electronic effect of the solvents whereby the donation of a lone pair of electron by the organic solvent triggers a strong chemical action on the catalyst surface. Alcohols such as isopropanol can provide a lone pair as well as act as a hydrogen donor, at high dechlorination rates.

Ukisu and Miyadera (2002) investigated the impact of several catalysts on the rate of dechlorination and compared the efficiency of palladium-carbon (Pd/C) and rhodium-carbon (Rh/C) catalysts. It was observed that milder conditions were required for Pd/C to achieve a dechlorination efficiency of greater than 75 % at 23 °C as compared to Rh/C which yielded higher chlorine free species at a higher temperature of 35 °C. A change in temperature of the reaction affected the dechlorination yield substantially (Ghaffar and Tabata 2009). Normally, an increase in temperature increases the yield of the reaction while the reaction time is decreased. Typically, the use of Pd/C catalysts at a temperature of 50 °C under normal pressures may achieve a complete destruction of over 99 % of dioxins and furans (Xia et al. 2003). Nevertheless, noble metal catalysts are particularly susceptible to poisoning by a range of elements found in the contaminated species. Nitrogen atoms, for instance, are notorious poisons for many catalytic processes. Eventually, hydrogenation catalysts based on metal sulphides have been developed for PCBs destruction as they have proven to be robust and tolerant against catalyst poisoning (Rahuman et al. 2000). Aresta et al. (2008) proved that highly chlorinated dioxins and furans or chlorine compounds have a slower rate of dehalogenation as compared to those of low chlorinated compounds. Catalytic hydrogenation can be aptly used to detoxify soils, wastewater streams as well as fly ash or incinerator wastes that have been contaminated with dioxins and furans or toxic chlorine compounds such as PCBs. Furthermore, the separation and recovery of the catalyst is easily accomplished in such systems making them more effective.

Vitrification

A newly developing technology that could be potentially used for the destruction dioxins and furans in contaminated soils is the in-situ vitrification method also known as Geomelt (Rae 2003). It has been widely used for contaminated site and waste treatment, including mixed low-level radioactive wastes, PCBs, pesticides and a wide range of heavy metals. Thompson (2002) stated that this technology can be applied in several configurations ranging from deep subsurface in-situ treatment to above ground batch plants. Basically, vitrification involves the electric melting of contaminated soils and wastes that result in the destruction, removal or permanent immobilization of the hazardous species. The melting process is initiated by the passage of an electric current in the contaminated soil by means of graphite electrodes at relatively low voltage. As dry soil is a poor electrical conductor, the process also involves introducing a conductive path through a mixture of graphite and glass frite. Normally, square arrays of graphite electrodes are placed at regular distances in the soil. The voltage is regulated to achieve the desired melt and the melting temperature typically ranges from 1,400 to 12,000 °C depending on the materials being treated. The melt grows outwards and downwards until the power is switched off once the waste volume or contaminated zone has been treated. The melt is then allowed to cool.

Thompson (2002) explained that organic contaminants like PCBs and pesticides are destroyed by pyrolysis and dechlorination reactions at elevated temperatures in reducing conditions around the melt and that no organic residuals are left behind at the high temperatures involved. Indeed as verified by the Environmental Protection Agency, no PCB migration takes place from the zone of contamination into surrounding soils. It has been demonstrated that vitrification of organic species in the soil has attained typical destruction and removal efficiencies of 99.99 %. Pollutants, such as heavy metals and radionuclide wastes that cannot be destroyed by heat are encapsulated within the glass and are permanently immobilized in the resulting vitrified product, so that they cannot leach into the surrounding soil or groundwater (Kulkarni et al. 2008). Any volatile components such as mercury or VOCs are captured by an off-gas treatment system. The vitrified product consists of a mixture of glass and crystalline materials and resembles volcanic obsidian. It is typically five to ten times stronger than concrete and is extremely leach resistant.

Two other configurations have been developed based on the conventional in-situ vitrification process, namely: the subsurface planar configuration and in-container vitrification, respectively. The subsurface planar configuration is used to treat subsurface contamination in-situ, including buried wastes or structures such as buried tanks. Two planar melts are normally formed and both grow horizontally attacking the wastes and structures from the sides (Thompson 2002). Contaminants located above the planar melts down into the melts during the treatment providing a bottoms-up type of treatment approach. In-container vitrification consists of treating the contaminated zone in batches or stages in refractory-lined steel containers. After each batch treatment, the melted waste is allowed to cool and solidify in the container prior to being discharged to a disposal site. Both extensions of the vitrification technique are to be far more effective for waste treatment as compared to the conventional method. The in-container vitrification technique was effectively applied in Japan, to treat an abandoned waste incinerator that was heavily contaminated with dioxins and furans and PCBs. High destruction and removal efficiencies were achieved and the technology has proven to be relatively simple and robust. In-situ vitrification therefore represents an essential alternative for the treatment of dioxins and furans.

APEG PLUS Technology

APEG PLUS technology is a successful technology used to detoxify soil, contaminated with dioxins and PCBs. This technology has been approved by Environmental Protection Agency's Office of Toxic Substances. APEG is an alkoxide formed during the reaction between alkali metal hydroxide ('A') and polyethylene glycol (PEG). The overall process reaction is known as glycolate dehalogenation. The glycolate dehalogenation process is achieved through five steps namely: preparation, reaction, separation, washing and dewatering. APEG hence stands for alkali polyethylene glycolate.

The preparation stage consists of sieving the excavated contaminated soil to remove any debris or large extraneous material. The sieved soil is then transferred to a reactor vessel where it is blended and heated up to 300 °F (United States Congress 1991) for several hours with APEG. During the reaction phase, chlorine atoms from halogenated aromatic compounds are removed to form products that are less harmful and safer in the environment. In particular, APEG reacts with the PCBs in a dimethyl sulphoxide carrier to form glycol ether and an alkali metal salt, which are both water soluble and relatively non-toxic. Vapour produced during the reaction, is condensed to water and gaseous contaminants. The water is reused in later process whereas the gaseous contaminants are processed through activated carbon filter for safe emission. The contaminated carbon from filter is normally incinerated.

The product slurry leaving the reactor is further processed though a separator to remove any excess APEG reagent remaining which can be recycled back into the system. It then washed with recycling water from condenser, to remove the water soluble products (ether and metal salts). Finally, the soil is sent to a dewatering unit for oil and water separation. The resulting supernatant effluent from dewatering unit and washed water are treated prior to discharge. The clean soil obtained is tested to evaluate the contaminant concentration and will be transferred to initial location only if the concentration is within the required norms, or else it is recycled back to the system again.

APEG-PLUS process is a proven successful technology for treating PCBs to concentrations less than 2 milligram per kilogram (mg/kg) and dioxins to concentrations less than 1 microgram per kilogram (14 μg/kg) (United States Congress 1991). Other advantages of this technology include the low emission from treatment process and the mobility and self-containment of process equipment. In addition, emissions from the treatment process are minimal. Since, the main products are in nontoxic water-soluble forms, there is no need for incineration with subsequent air emissions.

In the past, critics claimed that such APEG technology was too costly for soil bioremediation According to hypothetical scenarios developed by Gilson Remediation Corp. and Environmental Protection Agency's RREL, the costs incurred from APEG treatment is nearly \$300 per ton of soil treated, with about two-thirds of this resulting from the purchase of reagents (United States Congress 1991). However, there is significant advances to improve the process economy in terms reagent by employing water rather than costly reagent to wet the soil. In general, APEG-PLUS processing costs ranged between \$100 and \$800 per ton for PCB-contaminated soil. However, this process costs may vary depending on factors such as the nature and volume of soil treated, characteristics of the site, and clean-up levels required (United States Congress 1991). Base-catalysed de-chlorination process is cheaper and faster than APEG. In addition, APEG is not an appropriate method to remediate soil contaminated with dioxins, while base-catalysed de-chlorination can be used to treat aqueous dioxin-contaminated soils (Haglund 2007). Although, APEG technology is converting dioxins and furans to less harmful and safer form, the proven effectiveness of the technology for a particular site or waste does not ensure that it will be effective at all sites.

10.6 Conclusion

The occurrence, sources and transport of dioxins and furans into the environment with their associated adverse effects are major issues that give rise to concerns. Being highly stable compounds that persist in the environment, dioxins and furans are bound to lead to chronic widespread human exposure and have severe adverse impacts on the ecosystem. The most common and severe health problems include birth defects and impaired child development, cancer, endocrine disruption, endometriosis, immune system damage, neurological damage and reproductive system damage amongst others. Among the various routes of occurrence of dioxins, incineration, combustion processes and reservoirs are the major and important ones. Contamination of soils, sediments or groundwater with PCBs mostly originates from hazardous waste disposal sites, municipal landfills, accidental spills and inappropriate underground storage. Thus, due to high environmental concerns of POPs, attempts to avoid generation of dioxins have been made by the implementation of stringent regulations. For instance, high levels of contaminants such as dioxins emanating from the production of substances such as pesticides were banned in an effort to lessen the adverse effects associated with the toxic emissions. Initiatives have also been taken by chemical plants wherein processes have been reviewed and altered thereby revealing significant reduction in the release of toxic pollutants. Among the various corrective techniques adopted, state-of art remediation technologies have also been developed for reducing dioxins and furans formation and emissions. The main technologies that have been studied are summarised below.

The selective catalytic reduction system can effectively reduce the PCDD/Fs emissions. The process, being a simple one, can be designed to meet almost any specific requirement like the installed catalyst volume. No residues, apart from a very small amount of spent catalyst, after several years, are formed and this can be also recycled. The electron beam flue gas treatment technology has been applied in the full industrial scale. This system contributes to the environmentally friendly use of lignite; thus ensuring global stability, as well as agricultural productivity via the fertilizer supply. The carbon adsorption removal technique remains one of the most commonly used technique. Researchers have been able to demonstrate that a high removal efficiency can be achieved using this removal techniques while at the same time monitoring the operating parameters such as temperature, particle size. Emphasis is also being laid on the research sector for innovative and efficient removal adsorption process. The most widely used the carbon adsorption technique is the entrained flow bed adsorption when compared to the other types of adsorption. As far as the innovative sector is concerned, the use of the dual filter bags and the multilayer been shaped offers a promising for the removal of dioxins and furans.

However, the use of this technology has been openly criticised in the past. This particular technology has proved to be very effective in terms to purifying the gas streams containing dioxins and furans. A relatively high removal efficiency can be easily achieved at low cost with activated carbon, yet it also has to be pointed out that, this technique can only can be considered as temporary solution to the dioxins and furans treatment, as the impurities present in the flue gas is only transferred from the gas phase to the solid phase. The disposal of residual solids in the long run can be a problem. Therefore, further research field could thus be explored to provide permanent treatment possibilities and limiting the effects of dioxins and furans on the environment. The use of plasma technology on the market for treatment of dioxins and furans is still limited when compared to other technology such as adsorption or remediation. It would appear that its relative high capital investment, complexity and energy usage contributes to its unpopularity. However, it has to be noted that this treatment offers the complete destruction of dioxins and furans and aims at recycling optimum products.

Supercritical water oxidation is a promising technology which allows for the efficient decontamination of organic wastes containing dioxins and furans. The supercritical water oxidation provides an added advantage in the sense that it does not produce any toxic byproduct while treating toxic wastes. The problems associated with the supercritical water oxidation process are mainly corrosion that might arise in the reactor vessels and in the case oxidants are used in the process the possible high cost of these oxidants. This process is highly efficient in the destruction of dioxins and furans. However little research exists on the applicability of mechanochemistry as a treatment option, more emphasis has been laid on the possible aggregates that could be used in the enhancement of the mechanochemical process efficiency. The disadvantage of this process lies in the prohibitive maintenance and purchase cost of the equipment that are used.

Little research exists in the potential of the ultraviolet irradiation as a possible treatment option for dioxins and furans removal. However, ultraviolet irradiation treatment technology remains a powerful and clean option. Studies have shown that Bioslurry can effectively be used in the remediation of soils and sludge which are contaminated with explosives, petroleum hydrocarbons, petrochemicals and pesticides. Furthermore, bioslurry has proved to be favorable over in-situ biological techniques for soils which are heterogeneous and have low permeability. However, it should be noted that dewatering soil fines after treatment can be expensive. Enhanced bioremediation can successfully be used in-situ to treat soils which are contaminated with pollutants such as petroleum hydrocarbons, pesticides and nitrotoluenes. Nevertheless, this technique is not suitable for low permeability soil and this technology is not effective at low temperatures. The environmental risk associated with this technology is that some POPs may be formed under anaerobic conditions. In addition, the mobilization of contaminants may affect surrounding environment (air and groundwater).

At present, the Solvated electron technology is a well-developed technology working at commercial scale. Furthermore, it has proved to be successful in reducing the level of dioxins and furans in soil. Nevertheless, the cost of this technology is still high compared with other technologies to treat POPs and the technology is available only through one provider, hence, limiting access to the market place. An extraction of 99.4 % of dioxins has been achieved with the subcritical water treatment at a temperature of 350 °C within 30 min. However, it should be noted that this technique is less efficient at lower temperatures. Since, subcritical water treatment is still considered to be an innovative technology. Further studies need to be carried out in order to conclude that it can successfully be used for the reduction or removal of dioxins and furans.

Base-catalysed de-chlorination processes have demonstrated to be effective in the remediation of liquids, soils, sludge and sediments contaminated with chlorinated organic compounds most particularly dioxins and furans and PCBs, taking into account also the factors which may influence or interfere with the treatment efficiency. It is an easily portable process and safely operated which employs off the shelf equipment and thus utilizing less space and less time consuming. Base-catalysed de-chlorination processes coupled with thermal desorption has also been proved to work effectively. Mechano-chemical dehalogenation could be a novel innovative ex-situ dioxins remediation and decontamination process. The application of this particular technique can effectively and readily treat wastes containing wide range of organic pollutants in one step only, reducing the need of waste handling with its risks associated with it. Operating at low temperatures has the advantage of decreasing the potential for dioxins and other toxic organic formation as wells as increasing safety and reducing energy consumption. Soil washing processes possess potential as remediation method for a wide range of soils. It is an effective, efficient and fast treatment technology. Nevertheless, although it can be applied to a limited quantity of soils, it also offers the potential for application to soils contaminated with organic and inorganic contaminants, radionuclides and heavy metals.

The Fenton reaction has demonstrated to rapidly degrade many organics via hydroxyl radicals. However, optimal conditions for Fenton oxidation should clearly be determined. Such type of chemical treatment is typically cost effective, not energy intensive and can be easily completed in short periods of time with relatively small amount of energy utilization. The application of photolytic techniques for the removal of dioxins and furans from soil and sediments is economical but its effectiveness relies on the availability of the sun for photodegradation. The selection of appropriate organic solvents and the irradiation time are equally essential parameters that affect the efficiency of the process. In-situ photolysis, nevertheless, represents an important process to render toxic dioxins and furans innocuous. Thermal desorption has proved to be an effective in-situ treatment technology that ensures complete destruction of dioxins and furans and prevent their formation during the destruction process. Although the ex-situ process is common, the latter causes more pollution through formation of dioxins and furans of high TEF. Research has demonstrated that the in-situ thermal desorption technology is gradually gaining impetus as compared to the conventional process due to its high treatment efficiency.

The potential of destroying toxic chlorinated compounds in fly ash, organic wastes, contaminated soil sediments and wastewater streams by catalytic hydrogenation has been recognized for many years. Further research and development have shown that this technology can be effectively used to treat dioxins and furans using Pd/C catalysts under mild conditions or it may be coupled with other removal techniques such as thermal desorption. The yield and rate of the reaction are dependent on the selection of an appropriate organic solvent, catalyst, the reaction temperature and the number of chlorine atoms in the contaminant species.

In-situ vitrification is an interesting developing technology that could treat soils contaminated with dioxins and furans. Its ability to heat and melt contaminants at very high temperatures and allowing them to cool to form glass is an asset to trap any toxic pollutant within the glass to prevent them from leaching the soil. The high temperatures involved are excellent means to ensure complete destruction of dioxins and furans. The two emerging techniques subsurface planar configuration and in-container vitrification have proven to be more effective than the conventional method and could further be investigated for the removal of dioxins and furans. APEG-PLUS is able to detoxify materials containing dioxin and PCBs. It has been tested in a number of sites, but it is still viewed as a highly experimental technology. APEG was highly efficient under laboratory conditions and more work is still required in the field.

Based on critical analysis made for the different remedial techniques discussed, it can be inferred that the technology to be used is governed by various essential factors. dioxins and furans sources, control measures and the potential impact of the remedial technique are determining factors for the choice of remediation strategy which should be economical, yet effective. Techniques of dioxin reduction such as selective catalytic removal of dioxins and furans, supercritical water oxidation, photolytic degradation, vitrification, mechanochemical processes and bioremediation seem to have high potential and more emphasis should be laid for more research for new approaches and development of sustainable remedial technologies.

Acknowledgements We wish to express our deepest gratitude to all the researchers whose valuable data have been adapted and have added considerable significance and substance to this chapter. We are also grateful Dr Eric Lichtfouse, Dr Jan Schwarzbauer, Dr Didier Robert and the anonymous reviewers for their suggestions and criticisms which have benefited the manuscript, and brought it to its present form.

References

Abad E, Adrados MA, Caixach J, Rivera J (2002) Dioxin abatement strategies and mass balance at a municipal waste management plant. Environ Sci Technol 36:92–99

Ahlborg UG, Becking GC, Birnbaum LS, Brouwer A, Derks HJGM, Feeley M, Color G, Hanberg A, Larsen JC, Liem AKD, Safe SH, Schlatter C, Wvern F, Younes M, Yrjäinheikki E (1994) Toxic equivalency factors for dioxin-like PCBs. Chemosphere 28:1049–1067

Ahmad M, Simon MA, Sherrin A, Tuccillo ME, Ullman JL, Teel AL, Watts RJ (2011) Treatment of polychlorinated biphenyls in two surface soils using catalyzed H_2O_2 propagations. Chemosphere 84:855–862

Ames M, Zemba S, Green L, Botelho MJ, Gossman D, Linkov I, Oliveira JP (2012) Polychlorinated dibenzo-p-dioxin and furan (PCDD/F) congener profiles in cement kiln emissions and impacts. Sci Total Environ 419:37–43

Amid PK, Julie A, Makram TS, Steven IS (1994) Solvent washing of PCP-contaminated soils with anaerobic treatment of wash fluids. Water Environ Res 66:692–697

Anderson DR, Fisher R (2002) Sources of dioxins in the United Kingdom: the steel industry and other sources. Chemosphere 46:371–381

Andersson M, Ottesen RT (2008) Levels of dioxins and furans in urban surface soil in Trondheim, Norway. Environ Pollut 152:553–558

Anitescu G, Tavlarides LL (2005) Oxidation of biphenyl in supercritical water: reaction kinetics, key pathways, and main products. Ind Eng Chem Res 44:1226–1232

Antonaraki S, Androulaki E, Dimotikali D, Hiskia A, Papaconstantinou E (2002) Photolytic degradation of all chlorophenols with polyoxometallates and H_2O_2. J Photochem Photobiol A Chem 148:191–197

Aresta M, Dibenedetto A, Fragale C, Giannoccaro P, Pastore C, Zammiello D, Ferragina C (2008) Thermal desorption of polychlorobiphenyls from contaminated soils and their hydrodechlorination using Pd- and Rh-supported catalysts. Chemosphere 70:1052–1058

Arienzo M, Chiarenzelli J, Scrudato R, Pagano J, Falaga L, Connor B (2000) Iron–mediated reactions of polychlorinated biphenyls in electrochemical peroxidation process (ECP). Chemosphere 44:1339–1346

Aristizábal B, Cobo M, Hoyos A, Montes de Correa C, Abalos M, Martínez K, Abad E, Rivera J (2008) Baseline levels of dioxin and furan emissions from waste thermal treatment in Colombia. Chemosphere 73:S171–S175

Asante-Duah DK (1996) Managing contaminated sites: problem diagnosis and development of site restoration. Wiley, New York

Aunola T (2004) In-situ remediation of contaminated soil by Fenton's reaction. Licentiate seminar. Environmental Engineering and Biotechnology, Tampere University of Technology

Aunola T (2005) Chemical oxidation of polychlorinated biphenyls in contaminated lake sediment by modified Fenton's reaction. Environmental Engineering and Biotechnology, Tampere University of Technology

Baccarelli A, Mocarelli P, Jr. Patterson DG, Bonzini M, Pesatori AC, Caporaso N, Landi MT (2002) Immunologic effects of dioxin: new results from Seveso and comparison with other studies. Environ Health Perspect 110:1169–1173

Baker RS, Lachance JC (2003) Performance relative to dioxins of the in-situ thermal destruction (ISTD) soil remediation technology. In: Proceedings of dioxin, Boston

Bakoglu M, Karademir A, Durmusoglu E (2005) Evaluation of PCDD/F levels in ambient air and soils and estimation of deposition rates in Kocaeli, Turkey. Chemosphere 59:1373–1385

Balmer ME, Goss KU, Schwarzenbach RP (2000) Photolytic transformation of organic pollutants on oil surfaces – an experimental approach. Environ Sci Technol 34:1240–1245

Benestad C (1989) Incineration of hazardous waste in cement kilns. Waste Manag Res 7:351–361

Bhawan P (2007) Study of environmental contamination with polychlorinated dioxins. Central Pollution Control Board, New Delhi

Boening DW (1998) Toxicity of 2,3,7,8-tetrachlorodibenzo-p-dioxin to several ecological receptor groups: a short review. Ecotoxicol Environ Saf 39:155–163

Bonizzoni G (1999) Destruction of organo-halogenated compounds by DC non-transferred arc plasma. Pure Appl Chem 71:1879–1887

Bruns-Weller E, Knoll A, Heberer T (2010) High levels of polychlorinated dibenzo-dioxins/furans and dioxin like-PCBs found in monitoring investigations of sheep liver samples from Lower Saxony, Germany. Chemosphere 78:653–658

Buekens A, Stieglitz L, Hell K, Huang H, Segers P (2001) Dioxins from thermal and metallurgical processes: recent studies for the iron and steel industry. Chemosphere 42:729–735

Cathum SJ, Dumouchel A, Brown CE, Punt M (2006) Proceedings of the annual international conference on soils, sediments, water and energy 11, Amherst, pp 243–257

Chan JKY, Guan HX, Ying X, Ying L, Chen LX, Sheng CW, Wong C, Leung C, Wong MH (2007) Body loadings and health risk assessment of polychlorinated dibenzo-p-dioxins and dibenzofurans at an intensive electronic waste recycling site in China. Environ Sci Technol 41:7668–7674

Chandler AJ & Associates Ltd (2007) Review of dioxins and furans from incineration in support of a Canada-wide standard review. Canadian Council of Ministers of the Environment Inc., Winnipeg

Chang Y-S (2008) Recent developments in microbial biotransformation and biodegradation of dioxins. J Mol Microbiol Biotechnol 215:152–171

Chang MB, Chi KH, Chang SH, Yeh JW (2007) Destruction of PCDD/Fs by SCR from flue gases of municipal waste incinerator and metal smelting plant. Chemosphere 66:1114–1122

Chang MS, Shen JY, Yang SH, Wu GJ (2011) Subcritical water extraction for the remediation of phthalate ester-contaminated soil. J Hazard Mater 192:1203–1209

Chen ASC, Gavaskar AR, Alleman BC, Massa A, Timberlake D, Drescher EH (1997) Treating contaminated sediment with a two-stage base-catalyzed decomposition (BCD) process: bench-scale evaluation. J Hazard Mater 56:287–306

Cheng TW, Chu JP, Tzeng CC, Chen YS (2002) Treatment and recycling of incinerated ash using thermal plasma technology. Waste Manage 22:485–490

Chiarenzelli J, Scrudato R, Bush B, Carpenter D, Bushart S (1998) Do large-scale remedial and dredging events have the potential to release significant amounts of semivolatile compounds to the atmosphere. Environ Health Perspect 106:47–49

Chmielewski AG, Ostapczuk A, Zimek Z, Licki J, Kubica K (2002) Electron beam treatment in flue gas from coal combustion. Radiat Phys Chem 71:441–444

Choi W, Hong SJ, Chang YS, Cho Y (2000) Photocatalytic degradation of polychlorinated dibenzo-p-dioxins on TiO_2 film under UV or solar light irradiation. Environ Sci Technol 22:4810–4815

Choudhry GG, Webster GRB (1985) Environmental photochemistry of PCDDs Part 1. Kinetics and quantum yields of the photodegradation of 1,2,3,4,7 Penta and 1,2,3,4,7,8 Hexachlorodibenzo-p-dioxin in aqueous acetonitrile. Chemosphere 14:9–26

Chu W (2003) Remediation of contaminated soils by surfactant-aided soil washing. Pract Periodical Hazard Toxic Radioact Waste Manage 7(1):19–24

Cobo M, Conesa JA, Montes de Correa C (2009) Effect of the reducing agent on the hydrodechlorination of dioxins over 2 wt% Pd/γ-Al_2O_3. Appl Catal B Environ 92:367–376

Çok I, Donmez MK, Satiroğlu MH, Aydinuraz B, Henkelmann B, Kolatik J, Schramm KW (2007) Concentration of polychlorinated dibenzo-p-dioxins, polychlorinated dibenzofurans and dioxin-like PCBs in human adipose tissue from Turkish men. Chemosphere 66:1955–1961

Cole P, Trichopoulos D, Pastides H, Starr T, Mandel JS (2006) Dioxin and cancer: a critical review. Regul Toxicol Pharmacol 38:378–388

Crosby DG, Wong AS, Plimmer JR, Woolson EA (1971) Photodecomposition of chlorinated dibenzo-p-dioxins. Science 173:748–749

Debecker DP, Delaigle R, Hung PC, Buekens A, Gaigneaux EM, Chang MB (2011) Evaluation of PCDD/F oxidation catalysts: confronting studies on model molecules with tests on PCDD/F--containing gas stream. Chemosphere 82:1337–1342

Dercovà K, Vrana B, Tandlich R, Subova L (1999) Fenton's type reaction and chemical pre-treatment of PCBs. Chemosphere 39:2621–2628

Dobbs AJ, Grant C (1979) Photolysis of highly chlorinated dibenzo-p-dioxin by sunlight. Nature 278:163–165

Donghai X, Shuzhong W, Xingying T, Yanmeng G, Yang G, Yuzhen W, Jie Z (2012) Design of the first pilot scale plant of China for supercritical water oxidation of sewage sludge. Chem Eng Res Des 90:288–297

Dougherty EJ, McPeters AL, Overcash MR, Carbonell RG (1993) Theoretical analysis of a method for in situ decontamination of soil containing 2,3,7,8-tetrachlorodibenzo-*p*-dioxin. Environ Sci Technol 27:505–515

El-Shahawi MS, Hamza A, Bashammakh AS, Al-Saggaf WT (2010) An overview on the accumulation, distribution, transformations, toxicity and analytical methods for the monitoring of persistent organic pollutants. Talanta 80:1587–1597

Everaert K, Baeyens J (2004) Removal of PCDD/F from flue gases in fixed or moving bed adsorbers. Waste Manage 24:37–42

Everaert K, Baeyens J, Degrève J (2002) Removal of PCDD/F from incinerator flue gases by entrained-phase adsorption. J Air Waste Manag Assoc 52:1378–1388

Everett CJ, Thompson OM (2012) Associations of dioxins, furans and dioxin-like PCBs with diabetes and pre-diabetes: is the toxic equivalency approach useful? Environ Res 118:107–111

Fang Z, Xu SK, Butler IS (2004) Destruction of decachlorobiphenl using supercritical water oxidation. Energy Fuels 18:1257–1265

Feng D, Lorenzen L, Aldrich C, Mare PW (2001) Ex-situ diesel contaminated soil washing with mechanical methods. Miner Eng 14(9):1093–1100

Finocchio E, Busca G, Notaro M (2006) A review of catalytic processes for the destruction of PCDD and PCDF from waste gases. Chemosphere 62:12–20

Fries GF, Feil VJ, Zaylskie RG, Bialek KM, Rice CP (2002) Treated wood in livestock facilities: relationships among residues of pentachlorophenol, dioxins, and furans in wood and beef. Environ Pollut 116:301–307

Fronduti C (2005) Chemical oxidation of PCBs contaminated sediments by Fenton treatment. Seminar paper, Institute of Environmental Engineering and Biotechnology, Department of Environmental Engineering, Tampere University of Technology

Ghaffar A, Tabata M (2009) Dechlorination/detoxification of aromatic chlorides using fly ash under mild conditions. Waste Manage 29:3004–3008

Goemans M, Clarysse P, Joannès J, De Clercq P, Lenaerts S, Matthys K, Boels K (2004) Catalytic NOx reduction with simultaneous dioxin and furan oxidation. Chemosphere 54:1357–1365

Gomez E, Amutha Rani D, Cheeseman CR, Deegan D, Wise M, Boccaccini AR (2009) Thermal plasma technology for the treatment of wastes: a critical review. J Hazard Mater 161:614–626

Goncalves C, Dimou A, Sakkas V, Alpendurada MF, Albanis TA (2006) Photolytic degradation of quinalphos in natural waters and on soil matrices under simulated solar irradiation. Chemosphere 64:1375–1382

Grochowalski A, Lassen C, Holtzer M, Sadowski M, Hudyma T (2007) Determination of PCDDs, PCDFs, PCBs and HCB emissions from the metallurgical sector in Poland. Environ Sci Pollut Res 14(5):326–332

Haglund P (2007) Methods for treating soils contaminated with polychlorinated-p-dioxins, dibenzofurans, and other polychlorinated aromatic compounds. AMBIO 36:467–474

Hajizadeh Y, Onwudili JA, Williams PT (2011) Removal potential of toxic 2378-substituted PCDD/F from incinerator flue gases by waste-derived activated carbons. Waste Manag 31:1194–1201

Harjanto S, Kasai E, Terui T, Nakamura T (2002) Behavior of dioxin during thermal remediation in the zone combustion process. Chemosphere 47:687–693

Hashimoto S, Watanabe K, Nose K, Morita M (2004) Remediation of soil contaminated with dioxins by subcritical water extraction. Chemosphere 54:89–96

Hatakeda K, Ikushima Y, Sato O, Aizawa T, Saito N (1999) Supercritical water oxidation of polychlorinated biphenyls using hydrogen peroxide. Chem Eng Sci 54:3079–3084

Hays S, Aylward L (2003) Dioxin risks in perspective: past, present, and future. Regul Toxicol Pharmacol 37:202–217

Hebert VR, Miller GC (1990) Depth dependence of direct and indirect photolysis on soil surfaces. J Agric Food Chem 38:913–918

Hidayat A, Tachibana S (2013) Degradation of 2,4,8-trichlorodibenzofuran by a new isolate of Cerrena sp. F0607. Int Biodeterior Biodegrad 77:51–55

Hirashi A (2003) Biodiversity of dioxin-degrading microorganisms and potential utilization in bioremediation. Microbes Environ 18(3):105–125

Holt E, Weber R, Stevenson G, Gaus C (2010) Polychlorinated dibenzo-p-dioxins and dibenzofurans (PCDD/Fs) impurities in pesticides: a neglected source of contemporary relevance. Environ Sci Technol 44(14):5409–5415

Holtzer M, Dańko J, Dańko R (2007) Possibilities of formation of formation of dioxins and furans in metallurgical processes as well as methods of their reduction. METABK 46:285–290

Huff J, Lucier G, Tritscher A (1994) Carcinogenicity of TCDD experimental, mechanistic, and epidemiologic evidence. Annu Revue Pharmacol Toxicol 343–72

Hung LS, Ingram JLL (1990) Effect of solvents on the photodegradation rates of octachlorodibenzo-p-dioxin. Bull Environ Contam Toxicol 44:380–386

Hung PC, Chang SH, Chi KH, Chang MB (2010) Degradation of gaseous dioxin-like compounds with dielectric barrier discharges. J Hazard Mater 182:246–251

Hung PC, Lo WC, Chi KH, Chang SH, Change MB (2011) Reduction of dioxin emission by a multi-layer reactor with bead-shaped activated carbon in simulated gas stream and real flue gas of a sinter plant. Chemosphere 82:72–77

Ide Y, Kashiwabara K, Okada S, Mori T, Hara M (1996) Catalytic decomposition of dioxin from MSW incinerator flue gas. Chemosphere 32:189–198

Im SH, Kannan K, Giesy JP, Matsuda M, Wakimoto T (2002) Concentrations and profiles of polychlorinated dibenzo-p-dioxins and dibenzofurans in soils from Korea. Environ Sci Technol 36:3700–3705

Inoue K, Kawamoto K (2005) Fundamental adsorption characteristics of carbonaceous adsorbents for 1,2,3,4-tetrachlorobenzene in a model gas of an incineration plant. Environ Sci Technol 39(15):5844–5850

Isosaari P (2004) Polychlorinated dibenzo-p-dioxin and dibenzofuran contamination of sediments and photochemical decontamination of soils. Academic dissertation, University of Kuopio, Finland

Jaiswal PK, Kohli S, Gopal M, Thakur IS (2011) Isolation and characterization of alkalotolerant Pseudomonas sp. strain ISTDF1 for degradation of dibenzofuran. J Ind Microbiol Biotechnol 38:503–511

Jonsson S (2009) The influence of soil and contaminant properties on the efficiency of physical and chemical soil remediation methods. Department of Chemistry, Umeå University, Finland

Kang DH, Tepper A, Patterson DG (1997) Coplanar PCBs and the relative contribution of coplanar PCBs, PCDDs, and PCDFs to the total 2,3,7,8-TCDD toxicity equivalents in human serum. Chemosphere 35:503–511

Kao J (1999) Treatment of dioxin contaminated soils by oxidation and bioremediation technologies. National Sun Yat-Sen University, Institute of Environmental Engineering, Kaohsiung

Kao CM, Wu MJ (2000) Enhanced TCDD degradation by Fenton's reagent peroxidation. J Hazard Mater 74:197–211

Karademir A, Bakoglu M, Taspinar F, Ayberk S (2004) Removal of PCDD/Fs from flue gas by a fixed-bed activated carbon filter in a hazardous waste incinerator. Environ Sci Technol 38:1201–1207

Karjalainen AK, Hirvonen T, Kiviranta H, Sinkko H, Kronberg-Kippilä C, Virtanene SM, Hallikainen A, Leino O, Kniph M, Veijola R, Simell O, Tuomisto JT (2012) Long-term daily intake estimates of polychlorinated dibenzo-p-dioxins and furans, polychlorinated biphenyls and polybrominated diphenylethers from food in Finnish children: risk assessment implications. Food Addit Contam A 29:1475–1488

Karstensen KH (2008) Formation, release and control of dioxins in cement kilns. Chemosphere 70:543–560

Kerkvliet NI (2002) Recent advances in understanding the mechanisms of TCDD immunotoxicity. Int J Immunopharmacol 2:277–291

Kieatiwong S, Nguyen LW, Hebert VR, Hackett M, Miller GC, Miille MJ, Mitzel R (1990) Photolysis of chlorinated dioxins in organic solvents and on soils. Environ Sci Technol 24:1575–1580

Kikuchi R, Pelovski Y (2008) Low-dose irradiation by electron beam for the treatment of high-SO_x flue gas on a semi-pilot scale-consideration of by-product quality and approach to clean technology. Process Saf Environ Protect 87:135–143

Kim BH, Kim JH (2004) Removal of PCDDs/Fs from a solid waste incinerator using a dual bag filter system with recycled activated carbon. In: Proceedings of the 3rd international conference on combustion, incineration/pyrolysis and emission control, Kuala Lumpur, pp 313–318

Kim MK, O'Keefe PW (2000) Photodegradation of polychlorinated dibenzo-p-dioxins and dibenzofurans in aqueous solutions and in organic solvents. Chemosphere 41:793–800

Kim SW, Park HS, Kim HJ (2003a) 100 kW steam plasma process for treatment of PCBs (polychlorinated biphenyls) waste. Vacuum 70:59–66

Kim H, Hakoda T, Kojima T (2003b) Decomposition of gas–phase diphenyl ethyl at 473 K by electron beam generated plasma. Appl Phys 36:1–9

Kim B, Lee S, Maken S, Song H, Park J, Min B (2006) Removal characteristics of PCDDs/Fs from municipal solid waste incinerator by dual bag filter (DBF) system. Fuel 86:813–819

Kim J-H, Tratnyek PG, Chang Y-S (2008) Rapid dechlorination of polychlorinated dibenzo-p-dioxins by bimetallic and nanosized zerovalent iron. Environ Sci Technol 42:4106–4112

Kiviranta H, Tuomisto JT, Tuomisto J, Tukiainen E, Vartiainen T (2005) Polychlorinated dibenzo-p-dioxins, dibenzofurans, and biphenyls in the general population in Finland. Chemosphere 60:854–869

Kluyev N, Cheleptchikov A, Brodsky E, Soyfer V, Zhilnikov V (2002) Reductive dechlorination of polychlorinated dibenzo-p-dioxins by zerovalent iron in subcritical water. Chemosphere 46:1293–1296

Kogevinas M (2001) Human health effects of dioxins: cancer, reproductive and endocrine system effects. Hum Reprod Update 7:331–339

Kulkarni PS, Crespo JG, Afonso CAM (2008) Dioxin sources and current remediation technologies – a review. Environ Int 34:139–153

Kunichika N, Tohru F, Kazuei I (2003) Evaluation of the optimal washing conditions for dioxin–contaminated soils from the circumference of an incinerator. J Mater Cycles Waste Manage 5:63–68

Lapa N, Santos Oliveira JF, Camacho SL, Circeo LJ (2002) An ecotoxic risk assessment of residue materials produced by the plasma pyrolysis/vitrification (PP/V) process. Waste Manag 22:335–342

Lee B-D, Hosomi M (2001) Ethanol washing with distillation to remediate benzo (a) pyrene– contaminated field soil. J Chem Eng Japan 34:1061–1064

Lee B-D, Okutsu S, Nakai S, Hosomi M (2005) Remediation of dioxins–contaminated soil by successive ethanol washing photodegradation. Chem Eng Japan 38:360–365

Leung AOW, Luksemburg WJ, Wong AS, Wong MH (2007) Spatial distribution of polybrominated diphenyl ethers and polychlorinated dibenzo–p–dioxins and dibenzofurans in soil and combusted residue at Guiyu, an electronic waste recycling site in Southeast China. Environ Sci Technol 41:2730–2737

Li X, Li Y, Zhang Q, Wang P, Yang H, Jiang G, Wei F (2011) Evaluation of atmospheric sources of PCDD/Fs, PCBs and PBDEs around a steel industrial complex in northeast China using passive air samplers. Chemosphere 84:957–963

Liljelind P, Unsworth J, Maaskant O, Marklund S (2001) Removal of dioxins and related aromatic hydrocarbons from flue gas streams by adsorption and catalytic destruction. Chemosphere 42:615–623

Lodolo A, Gonzalezvalencia E, Miertus S (2001) Overview of remediation technologies for persistent toxic substances. Arh Hig Rada Toksikol 52:253–280

Loo WV (2008) Dioxin/furan formation and release in the cement industry. Environ Toxicol Pharmacol 25:128–130

Ma J, Kannan K, Cheng J, Horii Y, Wu Q, Wang Y (2008) Concentrations, profiles, and estimated human exposures for polychlorinated dibenzo–p–dioxins and dibenzofurans from electronic waste recycling facilities and a chemical industrial complex in eastern China. Environ Sci Technol 42:8252–8259

Malkow T (2004) Novel and innovative pyrolysis and gasification technologies for energy efficient and environmentally sound MSW disposal. Waste Manage 24:53–79

Manahan S (2000) Environmental chemistry, 7th edn. Lewis Publishers/CRC Press LLC, London

Manzano MA, Perales J, Sales D, Quiroga JM (2001) Catalyzed hydrogen peroxide treatment of polychlorinated biphenyl contaminated sandy soils. Water Air Soil Pollut 154:57–69

Mariñosa PC (2007) Fenton treatment of PCB-contaminated surfaces, Tampere University of Technology, Department of Environmental Technology

Marulanda V, Bolaños G (2010) Supercritical water oxidation of heavily PCB–contaminated mineral transformer oil: laboratory–scale data and economic assessment. J Supercrit Fluids 54:258–265

Maturi K, Reddy KR (2006) Simultaneous removal of organic compounds and heavy metals from soils by electrokinetic remediation with a modified cyclodextrin. Chemosphere 63:1022–1031

McDowall R, Boyle C, Graham B (2004) Review of emerging, innovative technologies for the destruction and decontamination of POPs and the identification of promising technologies for use in developing countries, GF/8000-02-02-2205

McKay G (2002) Dioxin characterisation, formation and minimisation during municipal solid waste (MSW) incineration: review. Chem Eng J 86:343–368

McPeters AL, Overcash MR (1993) Demonstration of photodegradation by sunlight of 2,3,7,8-tetrachlorodibenzo-p-dioxin in 6 cm soil columns. Chemosphere 27:1221–1234

Miller GC, Hebert VR, Miille MJ, Mitzel R, Zepp RG (1989) Photolysis of octachlorodibenzo-*p*-dioxinon soils: production of 2,3,7,8-TCDD. Chemosphere 18:1265–1274

Mininni G, Sbrilli A, Braguglia CM, Guerriero E, Marani D, Rotaroti M (2007) Dioxins, furans and polycyclic aromatic hydrocarbons emissions from a hospital and cemetery waste incinerator. Atmos Environ 41:8527–8536

Mitoma Y, Uda T, Egashira N, Simion C, Tashiro H, Tashiro M, Fan X (2004) Approach to highly efficient dechlorination of PCDDs, PCDFs, and coplanar PCBs using metallic calcium in ethanol under atmospheric pressure at room temperature. Environ Sci Technol 38:1216–1220

Mitoma Y, Egashira N, Simion C (2009) Highly effective degradation of polychlorinated biphenyls in soil mediated by a Ca/Rh bicatalytic system. Chemosphere 74:968–973

Mitrou PI, Dimitriadis G, Raptis SA (2001) Review article–toxic effects of 2, 3, 7, 8-tetrachlorodibenzo-p-dioxin and related compounds. Eur J Intern Med 12:406–411

Moirangthem V, Katz WS, Su W, Choi E-Y, Cameron Dingle RW, Zeigler GM, Everson WV, Jennings CD, Gong M, Swanson HI (2013) Impact of 2,3,7,8-tetrachlorodibenzo-p-dioxin on cutaneous wound healing. Exp Toxicol Pathol 65:61–67

Moriguchi T, Motohashi H, Hosoya T, Nakajima O, Takahashi S, Ohsako S, Aoki Y, Nishimura N, Tohyama C, Fujii-Kuriyama Y, Yamamoto M (2003) Distinct response to dioxin in an arylhydrocarbon receptor (AhR) – humanized mouse. Proc Natl Acad Sci U S A 10:5652

Mukerjee D (1998) Health impact of polychlorinated dibenzo p-dioxins: a critical review. J Air Waste Manage Assoc 48:157–165

Nakamiya K, Furuichi T, Ishii K (2002) Isolation of a fungus from denitrifrying activated sludge that degrades highly chlorinated dioxins. J Mater Cycles Waste Manage 4:127–134

Nam P, Kapila S, Liu O, Tumiatti W, Porciani A, Flanigan V (2001) Solvent extraction and tandem dechlorination for decontamination of soil. Chemosphere 43:485–491

Nam IH, Hong HB, Kim YM, Kim BH, Murugesan K, Chang YS (2005) Biological removal of polychlorinated dibenzo-p-dioxins from incinerator fly ash by *Sphingomonas wittichii* RW1. Water Res 39:4651–4660

Nam I-H, Kim Y-M, Murugesan K, Jeon J-R, Chang Y-Y, Chang Y-S (2008) Bioremediation of PCDD/Fs-contaminated municipal solid waste incinerator fly ash by a potent microbial biocatalyst. J Hazard Mater 157:114–121

Niu J, Chen J, Henkelmann B, Quan X, Yang F, Kettrup A, Schramm KW (2003) Photodegradation of PCDD/Fs adsorbed on spruce (Piceaabies (L.) Karst.) needles under sunlight irradiation. Chemosphere 50:1217–1225

Nose K, Hashimoto S, Takahashi S, Noma Y, Sakai SI (2007) Degradation pathways of decabromodiphenyl ether during hydrothermal treatment. Chemosphere 68:120–125

Ok G, Ji S-H, Kim S-J, Kim Y-K, Park J-H, Kim Y-S, Han Y-H (2002) Monitoring of air pollution by polychlorinated dibenzo-p-dioxins and polychlorinated dibenzofurans of pine needles in Korea. Chemosphere 46:1351–1357

Oku A, Tomari K, Kamada T, Yamada E, Miyata H, Aozasa O (1995) Destruction of PCDDs and PCDFs. A convenient method using alkali-metal hydroxide in 1,3-dimethyl-2-imidazolidinone (DMI). Chemosphere 31:3873–3878

Ostapczuk A, Janusz L, Chmielewski AG (2008) Polycyclic aromatic hydrocarbons in coal combustion flue gas under electron beam irradiation. Radiat Phys Chem 77:490–496

Pao CH, Shu HC, Moo BC (2006) Characteristics of PCDD/F distributions in vapor and solid phases and emissions from the Waelz process. Environ Sci Technol 40:1770–1775

Pao CH, Wei CL, Kai HC, Shu HC, Moo BC (2010a) Reduction of dioxin emission by a multi-layer reactor with bead-shaped activated carbon in simulated gas stream and real flue gas of a sinter plant. Chemosphere 82:72–77

Pao CH, Wei CL, Kai HC, Shu HC, Moo BC (2010b) Degradation of gaseous dioxin-like compounds with dielectric barrier discharges. J Hazard Mater 182:246–251

Parizek T, Bebar L, Stehlik P (2008) Persistent pollutants emission abatement in waste-to-energy systems. Clean Technol Environ Policy 10:147–153

Patel S, Kaminski MD, Nuñez L (2003) Polychlorodibenzo-p-dioxin and polychlorodibenzo-furan removal and destruction. University of Chicago, Chicago

Patterson DG, Todd GD, Turner WE, Maggio V, Alexander LR, Needham LL (1994) Levels of non-ortho-substituted (coplanar), mono-ortho-substituted and diortho- substituted polychlorinated-biphenyls, dibenzo-p-dioxins, and dibenzofurans in human serum and adipose-tissue. Environ Health Perspect 102:195–204

Pavel LV, Gavrilescu M (2008) Overview of ex-situ decontamination techniques for soil. Environ Eng Manage J7:815–834

Pérez M, Torrades F, García-Hortal J, Domènech X, Peral J (2002) Removal of organic contaminants in paper pulp treatment effluents under Fenton and photo-Fenton conditions. Appl Catal 36:63–74

Pittman CU, He JB (2002) Dechlorination of PCBs, CAHs, herbicides and pesticides neat and in soils at 25 °C using Na/NH_3. J Hazard Mater 92:51–62

Prządo D, Kafarski P, Steininger M (2007) Studies on degradation of polychlorinated biphenyls by means of Fenton's reagent. Pol J Environ Stud 16:881–887

Psomopoulos CS, Bourka A, Themelis NJ (2009) Waste-to-energy: a review of the status and benefits in USA. Waste Manage 29:1718–1724

Quina MJ, Pedro RS, Gando-Ferreira LM, Quinta-Ferreira RM (2011) A national inventory to estimate release of polychlorinated dibenzo-p-dioxins and dibenzofurans in Portugal. Chemosphere 85:1749–1758

Rae ID (2003) Alternative technologies for destruction of PCB and other POPs. Handb Environ Chem 3:426–438

Rahuman MSM, Pistone L, Trifirò F, Miertus S (2000) Destruction technologies for polychlorinated biphenyls. (PCBS). ICS–UNIDO publications. Proceedings of expert group meetings on POPs and pesticides contamination: Remediation technologies (April 2000) and on clean technologies for the reduction and elimination of POPs (May 2000)

Reiner E, Clement R, Okey A, Marvin C (2006) Advances in analytical techniques for polychlorinateddibenzo-p-dioxins, polychlorinated dibenzofurans and dioxin-like PCBs. Anal Bioanal Chem 386:791–806

Rodriguez C, Cook A, Devine B, Buynder PV, Lugg R, Linge K, Weinstein P (2008) Dioxins, furans and PCBs in recycled water for indirect potable use. Int J Environ Res Public Health 5(5):356–367

Ryan JJ, Lizotte R (1987) Human tissue levels of PCDDs and PCDFs from a fatal pentachlorophenol poisoning. Chemosphere 16:1989–1996

Sahle-Demessie E, Grosse DW, Bates ER (2000) Solvent extraction and soil washing treatment of contaminated soils from wood- preserving site. Remediation 10:85–110

Saichek RE, Reddy KR (2004) Evaluation of surfactants/cosolvents for desorption/solubilization of phenanthrene in clayey soils. Int J Environ Stud 61:587–604

Sakaki T, Munetsuna E (2010) Enzyme systems for biodegradation of polychlorinated dibenzo-p-dioxins. Appl Microbiol Biotechnol 88:23–30

Sato T, Todoroki T, Shimoda K, Terada A, Hosomi M (2010) Behaviour of PCDDs/PCDFs in remediation of PCBs–contaminated sediments by thermal desorption. Chemosphere 80:184–189

Satoshi N, Saeko K, Yugo N, Masaaki H (2006) Polychlorinated dibenzothiophenes in Japanese environmental samples and their photo degradability and dioxin-like endocrine–disruption potential. Chemosphere 67:1852–1857

Schecter A, Miyata H et al (1999) Chloracne and elevated dioxin and dibenzofurans levels in the blood of two Japanese municipal incinerator workers and of the wife of one worker. Organohalog Comp 44:247–250

Schecter A, Cramer P, Boggess K, Stanley J, Päpke O, Olson J (2001) Intake of dioxins and related compounds from food in the U.S. population. J Toxicol Environ Health 63:1–18

Schecter A, Birnbaum L, Ryan JJ, Constable JD (2006) Dioxins: an overview. Environ Res 101:419–428

Schuhmacher M, Granero S, Rivera J, Muller L, Llobet JM, Domingo JL (2000) Atmospheric deposition of PCDD/Fs near an old municipal solid waste incinerator: levels in soil and vegetation. Chemosphere 40:593–600

Scott LLF, Staskal DF, Williams ES, Luksemburg WJ, Urban JD, Nguyen LM, Haws LC, Birnbaum LS, Paustenbach DJ, Harris MA (2009) Levels of polychlorinated dibenzo-p-dioxins, dibenzofurans, and biphenyls in southern Mississippi catfish and estimation of potential health risks. Chemosphere 74:1002–1010

Shelepchikov AA, Kluyev NA, Brodsky ES, Zhilnikov VG (2005) Catalytic dechlorination of polychlorinated dibenzo-p-dioxins and dibenzofurans in water under subcritical conditions. In: Lunin V, Tundo P, Tundo P, Lokteva E (eds) Green chemistry in Russia. Poligrafica, Venezia, pp 123–131

Shen H, Han J, Tie X, Xu W, Ren Y, Ye C (2009) Polychlorinated dibenzo-p-dioxins/furans and polychlorinated biphenyls in human adipose tissue from Zhejiang Province, China. Chemosphere 74:384–388

Sinkkonen S, Paasivirta J (2000) Degradation half-life times of PCDDs, PCDFs and PCBs for environmental fate modelling. Chemosphere 40:943–949

Soyfer VS, Shelepchikov AA, Kluyev NA, Rudenko BA (1999) Water in subcritical condition – an effective solvent for the extraction of dioxins from soils. Organohalog Comp 41:425–429

Srogi K (2008) Levels and congener distributions of PCDDs, PCDFs and dioxin-like PCBs in environmental and human samples: a review. Environ Chem Lett 6:1–28

Stegemeier LG, Vinegar HJ (2001) Thermal conduction heating for in-situ thermal desorption of soils. CRC Press, Boca Raton

Taniguchi S, Murakami A, Hosomi M, Miyamura A, Uchida R (1997) Chemical decontamination of PCB-contaminated soil. Chemosphere 34:1631–1637

Teo CC, Tana SN, Yong JWH, Hew CS, Ong ES (2010) Pressurized hot water extraction (PHWE). J Chromatogr A 1217:2484–2494

Thömke F, Jung D, Besser R, Röder R, Konietzko J, Hopf HC (1999) Increased risk of sensory neuropathy in workers with chloracne after exposure to 2,3,7,8-polychlorinated dioxins and furans. Acta Neurol Scand 100:1–5

Thompson LE (2002) Mixed waste treatment cost analyses for a range of Geomelt vitrification process configurations. AMEC Earth and Environmental, Inc., Richland

Tian B, Huang J, Wang B, Deng S, Yu G (2012) Emission characterization of unintentionally produced persistent organic pollutants from iron ore sintering process in China. Chemosphere 89:409–415

Tu X, Yu L, Yan J, Cen K, Chéron BG (2010) Plasma vitrification of air pollution control residues from municipal solid–waste incineration. Plasma Sci 38:3319–3325

Tuhkanen T (2001) Advanced oxidation of PCDD/Fs, PCBs and PAHs in contaminated soil/ash matrix. In: International conference on practical applications in environmental geotechnology EcoGeo 2000. Geological Survey of Finland, Special paper 32, pp 129–133

Ukisu Y, Miyadera T (2002) Dechlorination of polychlorinated dibenzo-p-dioxins catalysed by noble metal catalysts under mild conditions. Chemosphere 46:507–510

Ukisu Y, Miyadera T (2004) Dechlorination of dioxins with supported palladium catalysts in 2-propanol solution. Appl Catal A Gen 271:165–170

Urum K, Pekdemir T, Gopur M (2003) Optimum conditions for washing of crude oil-contaminated soil with biosurfactant solutions. Process Saf Environ Protect Tran Inst Chem Eng B 81(3):203–209

US Congress, Office of Technology Assessment (1991) Dioxin treatment technologies and background paper, OTA-BP-O-93

Vallejo M, San Román MF, Irabien A, Ortiz I (2013) Comparative study of the destruction of polychlorinated dibenzo-p-dioxins and dibenzofurans during Fenton and electrochemical oxidation of landfill leachates. Chemosphere 90:132–138

Venny, Gan S, Ng HK (2012) Current status and prospects of Fenton oxidation for the decontamination of persistent organic pollutants (POPs) in soils. Chem Eng J 213:295–317

Viel J-F, Floret N, Deconinck E, Focant J-F, De Pauw E, Cahn J-Y (2011) Increased risk of non-Hodgkin lymphoma and serum organochlorine concentrations among neighbors of a municipal solid waste incinerator. Environ Int 37:449–453

Vyas DS, Dave UB, Parekh HB (2011) Plasma pyrolysis: an innovative treatment to solid waste of plastic material. In: National conference on recent trends in engineering & technology, NCRTET, Gujarat

Wang Q, Yan JH, Chi Y, Li XD, Lu SY (2010) Application of thermal plasma to vitrify fly ash from municipal solid waste incinerators. Chemosphere 78:626–630

Wang YF, Wang LC, Hsieh LT, Li HW, Jiang HC, Lin YS, Tsai CH (2012) Effect of temperature and CaO addition on the removal of polychlorinated dibenzo-p-dioxins and dibenzofurans in fly ash from a medical waste incinerator. Aerosol Air Qual Res 12:191–199

Weber R, Plinke M, Xu Z, Wilken M (2001) Destruction efficiency of catalytic filters for polychlorinated dibenzo-p-dioxin and dibenzofurans in laboratory test and field operation – insight into destruction and adsorption behavior of semivolatile compound. Appl Catal 31:195–207

Weber R, Gaus C, Tysklind M, Johnston P, Forter M, Hollert H, Heinisch E, Holoubek I, Smith ML, Masunaga S, Moccarelli P, Santillo D, Seike N, Symons R, Torres JPM, Verta M, Varbelow G, Vijgen J, Watson A, Costner P, Woelz J, Wycisk P, Zennegg M (2008) Dioxin- and POP-contaminated sites—contemporary and future relevance and challenges. Environ Sci Pollut Res 15:363–393

Wilson DC (1982) Lessons from Seveso. Chem Br 18:499–504

Wong MH, Wu SC, Deng WJ, Yu XZ, Luo Q, Leung AOW, Wong CSC, Luksemburg WJ, Wong AS (2007) Mini-review export of toxic chemicals a review of the case of uncontrolled electronic-waste recycling. Environ Pollut 149:131–140

Wu CH, Ng HY (2008) Photodegradation of polychlorinated dibenzo-p-dioxins and polychlorinated dibenzofurans: direct photolysis and photocatalysis process. J Hazard Mater 151:507–514

Xia C, Xu J, Wu W, Luo Q, Chen J, Zhang Q, Liang X (2003) Catalytic hydrodechlorination of 2,4,4-trichloro-2-hydroxydiphenylether under mild conditions. App Catal B Environ 45:281–292

Xia C, Liu Y, Xu J, Yu J, Qin W, Liang X (2009) Catalytic hydrodechlorination reactivity of monochlorophenols in aqueous solutions over palladium/carbon catalyst. Catal Commun 10:456–458

Xiuying G, Dong X, Guanghong D, Peng M (2010) A review of mechanochemistry applications in waste management. Waste Manage 30:4–10

Yak HK, Wenclawiak BW, Cheng IF, Doyle JG, Wai CM (1999) Reductive dechlorination of polychlorinated biphenyls by zerovalent iron in subcritical water. Environ Sci Technol 33:1307–1310
Yan JH, Peng Z, Sy L, Li XD, Cen KF (2006) Removal of PCDDs/Fs from municipal solid waste incineration by entrained-flow adsorption technology. J Zhejiang Univ Sci 7:1862–1903
Yan JH, Peng Z, Lu SY, Li XD, Cen KF (2007) Destruction of PCDD/Fs by gliding arc discharges. J Environ Sci 19:1404–1408
Yang Y, Berlghazi M, Lagadec A, Miller DJ, Hawthorne SB (1998) Elution of organic solutes from different polarity sorbents using subcritical water. J Chromatogr A 810:149–159
Yang CC, Chang SH, Hong BZ, Chi KH, Chang MB (2008) Innovative PCDD/F-containing gas stream generating system applied in catalytic decomposition of gaseous dioxins over V_2O_5-WO_3/TiO_2-based catalysts. Chemosphere 73:890–895
Zhao L, Hou H, Shimoda K, Terada A, Hosomi M (2012) Formation pathways of polychlorinated dibenzofurans (PCDFs) in sediments contaminated with PCBs during the thermal desorption process. Chemosphere 88:1368–1374
Zheng P, Qiong D, Yang Zhao S, Chen J, Xinhua G, Jianhua Y (2010) Characterization of mechanochemical treated fly ash from a medical waste incinerator. J Environ Sci 22(10):1643–1648
Zhou YX, Yan P, Cheng ZX, Nifuku M, Liang XD, Guan ZC (2003) Application of non-thermal plasmas on toxic removal of dioxin-contained fly ash. Powder Technol 135:345–353
Zhu J, Hirai Y, Sakai S, Zheng M (2008) Potential source and emission analysis of polychlorinated dibenzo-p-dioxins and polychlorinated dibenzofurans in China. Chemosphere 73:S72–S77

Index

E. Lichtfouse et al. (eds.), *Pollutant Diseases, Remediation and Recycling*, Environmental Chemistry for a Sustainable World 4, DOI 10.1007/978-3-319-02387-8,

MIX
Papier aus verantwortungsvollen Quellen
Paper from responsible sources
FSC® C105338

If you have any concerns about our products,
you can contact us on
ProductSafety@springernature.com

In case Publisher is established outside the EU,
the EU authorized representative is:
Springer Nature Customer Service Center GmbH
Europaplatz 3, 69115 Heidelberg, Germany

Printed by Libri Plureos GmbH
in Hamburg, Germany